WITHDRAWN BY THE
UNIVERSITY OF MICHIGAN

Quantum Mechanics

Ernest S. Abers
University of California, Los Angeles

PEARSON

Prentice Hall

PEARSON EDUCATION, INC., *Upper Saddle River, New Jersey 07458*

Library of Congress Cataloging-in-Publication Data

Abers, Ernest S. (Ernest Stephen)
Quantum Mechanics / Ernest S. Abers
  p. cm.

Includes bibliographical references and index.
ISBN 0-13-146100-1
1. Quantum Theory   I. Title.

QC174.12.A27 2004
530.12–dc21   2003051437

Acquisitions Editor: *Erik Fahlgren*
Editor-in-Chief: *John Challice*
Vice President of Production and Manufacturing: *David W. Riccardi*
Executive Managing Editor: *Kathleen Schiaparelli*
Assistant Managing Editor: *Beth Sweeten*
Associate Editor: *Christian Botting*
Production Editor: *Debra Wechsler*
Manufacturing Buyer: *Alan Fischer*
Manufacturing Manager: *Trudy Pisciotti*
Executive Marketing Manager: *Mark Pfaltzgraff*
Marketing Assistant: *Melissa Berringer*
Director of Marketing: *John Tweeddale*
Editorial Assistant: *Andrew Sobel*
Copy Editor: *Daphne Hougham*
Art Director: *Jayne Conte*
AV Editor: *Jessica Einsig*
Cover Designer: *Geoffrey Cassar*

© 2004 by Pearson Education, Inc.
Pearson Education, Inc.
Upper Saddle River, New Jersey 07458

All rights reserved. No part of this book may be reproduced, in any form or by any means, without permission in writing from the publisher.

Printed in the United States of America
10 9 8 7 6 5 4 3 2

ISBN 0-13-146100-1

Pearson Education LTD., *London*
Pearson Education Australia PTY, Limited, *Sydney*
Pearson Education Singapore, Pte. Ltd
Pearson Education North Asia Ltd, *Hong Kong*
Pearson Education Canada, Ltd., *Toronto*
Pearson Educación de Mexico, S.A. de C.V.
Pearson Education—Japan, *Tokyo*
Pearson Education Malaysia, Pte. Ltd

To Sylvia

Contents

Preface — xiii

1 Classical Mechanics — 1
- 1.1 Newton's Laws, the Action, and the Hamiltonian — 1
  - 1.1.1 Newton's Law and Lagrange's Equations — 1
  - 1.1.2 Hamilton's Principle — 2
  - 1.1.3 Canonical Momenta and the Hamiltonian Formulation — 4
- 1.2 Classical Space-Time Symmetries — 6
  - 1.2.1 The Space-Time Transformations — 6
  - 1.2.2 Translations — 8
  - 1.2.3 Rotations — 9
  - 1.2.4 Rotation Matrices — 11
  - 1.2.5 Symmetries and Conservation Laws — 12
- Problems — 14

2 Fundamentals of Quantum Mechanics — 19
- 2.1 The Superposition Principle — 19
  - 2.1.1 The Double-Slit Experiment — 19
  - 2.1.2 The Stern-Gerlach Experiment — 22
- 2.2 The Mathematical Language of Quantum Mechanics — 24
  - 2.2.1 Vector Spaces — 24
  - 2.2.2 The Probability Interpretation — 27
  - 2.2.3 Linear Operators — 27
  - 2.2.4 Observables — 30
  - 2.2.5 Examples — 31
- 2.3 Continuous Eigenvalues — 35
  - 2.3.1 The Dirac Delta Function — 35
  - 2.3.2 Continuous Observables — 36
  - 2.3.3 Fourier's Theorem and Representations of $\delta(x)$ — 37
- 2.4 Canonical Commutators and the Schrödinger Equation — 38
  - 2.4.1 The Correspondence Principle — 38
  - 2.4.2 The Canonical Commutation Relations — 42
  - 2.4.3 Planck's Constant — 43
- 2.5 Quantum Dynamics — 44
  - 2.5.1 The Time-Translation Operator — 44
  - 2.5.2 The Heisenberg Picture — 45
- 2.6 The Uncertainty Principle — 47
- 2.7 Wave Functions — 49
  - 2.7.1 Wave Functions in Coordinate Space — 49
  - 2.7.2 Momentum and Translations — 49
  - 2.7.3 Schrödinger's Wave Equation — 52
  - 2.7.4 Time-Dependent Free Particle Wave Functions — 53
- Problems — 55

3 Stationary States — 62
- 3.1 Elementary Examples — 62
  - 3.1.1 States with Definite Energy — 62
  - 3.1.2 A Two-State System — 63
  - 3.1.3 One-Dimensional Potential Problems — 66
- 3.2 The Harmonic Oscillator — 68
  - 3.2.1 The Spectrum — 69
  - 3.2.2 Matrix Elements — 71
  - 3.2.3 The Ground-State Energy — 72
  - 3.2.4 Wave Functions — 73
- 3.3 Spherically Symmetric Potentials and Angular Momentum — 74
  - 3.3.1 Spherical Symmetry — 74
  - 3.3.2 Orbital Angular Momentum as a Differential Operator — 75
  - 3.3.3 The Angular Momentum Commutator Algebra — 76
  - 3.3.4 Classification of the States — 80
- 3.4 Spherically Symmetric Potentials: Wave Functions — 80
  - 3.4.1 Spherical Coordinates and Spherical Harmonics — 80
  - 3.4.2 The Radial Wave Equation — 82
- 3.5 Hydrogenlike Atoms — 84
  - 3.5.1 The Symmetries — 84
  - 3.5.2 The Energy Spectrum — 86
  - 3.5.3 The Radial Wave Functions — 88
- Problems — 91

4 Symmetry Transformations on States — 102
- 4.1 Introduction — 102
  - 4.1.1 Symmetries and Transformations — 102
  - 4.1.2 Groups of Transformations — 103
  - 4.1.3 Classical and Quantum Symmetries — 105
- 4.2 The Rotation Group and Algebra — 105
  - 4.2.1 Representations of Groups — 105
  - 4.2.2 Representations of the Generators of Rotations — 106
  - 4.2.3 Generators in an Arbitrary Direction — 107
  - 4.2.4 Commutators of the Generators — 108
  - 4.2.5 Explicit Form of the Finite Dimensional Representations — 111
  - 4.2.6 Summary — 112
- 4.3 Spin and Rotations in Quantum Mechanics — 113
  - 4.3.1 Rotations and Spinless Particles — 113
  - 4.3.2 Spin — 114
  - 4.3.3 The Spin-Zero Representation — 116
  - 4.3.4 The Spin-Half Representation — 116
  - 4.3.5 Euler Angles — 117
  - 4.3.6 The Spin-One Representation — 119
  - 4.3.7 Arbitrary j — 119
- 4.4 Addition of Angular Momenta — 120
  - 4.4.1 Spin and Orbital Angular Momentum — 120
  - 4.4.2 Two Simple Examples — 121

- 4.5 Clebsch-Gordan Coefficients 122
  - 4.5.1 Definition of the Coefficients 122
  - 4.5.2 Spin Half + Spin Half 123
  - 4.5.3 Spin Half + Angular Momentum One 125
  - 4.5.4 Spin Half + Angular Momentum l 126
  - 4.5.5 The General Rule 127
  - 4.5.6 Recursion Relation for the Coefficients 129
  - 4.5.7 The Clebsch-Gordan Series 129
- Problems 130

5 Symmetry Transformations on Operators — 138
- 5.1 Vector Observables 138
  - 5.1.1 Symmetries, Lifetimes, and Selection Rules 138
  - 5.1.2 Vector Operators Under Rotations 140
  - 5.1.3 Spherical Components of Vector Observables 141
  - 5.1.4 Selection Rules for Matrix Elements of Vectors 142
- 5.2 Tensor Observables 144
  - 5.2.1 Cartesian Tensor Operators 144
  - 5.2.2 Spherical Tensor Components 146
  - 5.2.3 Higher Rank Spherical Tensors 147
  - 5.2.4 Selection Rules and the Wigner-Eckart Theorem 148
- 5.3 Discrete Symmetries 151
  - 5.3.1 Reflections and Parity 151
  - 5.3.2 Reversal of the Direction of Motion 153
  - 5.3.3 Identical Particles 157
- 5.4 Internal Symmetries: Isospin 159
- Problems 164

6 Interlude — 171
- 6.1 External Magnetic Fields 171
  - 6.1.1 Natural Units 171
  - 6.1.2 Gauge Invariance 172
  - 6.1.3 Constant Magnetic Fields and Landau Levels 173
  - 6.1.4 Magnetic Moment 176
  - 6.1.5 The Hydrogen Atom in a Magnetic Field 177
- 6.2 The Density Matrix 178
  - 6.2.1 Definition 178
  - 6.2.2 Example: Thermodynamic Equilibrium 180
  - 6.2.3 Example: Spin-Half Systems 181
  - 6.2.4 Spin Magnetic Resonance 182
- 6.3 Neutrino Interference 185
  - 6.3.1 Neutrinos 185
  - 6.3.2 Neutrino Mixing 186
  - 6.3.3 Neutrino Oscillations and the Mass Splitting 187
  - 6.3.4 Solar Neutrinos 189
  - 6.3.5 Neutrino Oscillations in Matter 189
- 6.4 Measurements in Quantum Mechanics 191

viii   Contents

         6.4.1   Wave-Function Collapse 191
         6.4.2   The EPR Paradox 191
         6.4.3   Bell's Inequality 193
   Problems 195

7   Approximation Methods for Bound States   202

   7.1   Bound-State Perturbation Theory 202
         7.1.1   The Perturbation Expansion 202
         7.1.2   Example: Harmonic Oscillator 205
   7.2   Static External Electric Fields 206
         7.2.1   Perturbation of the First Excited Level 207
         7.2.2   Polarizability of the Ground State 208
   7.3   Fine Structure of the Hydrogen Atom 212
         7.3.1   The Spin-Orbit Coupling 212
         7.3.2   Correction to Energy Levels 214
         7.3.3   The Relativistic Kinetic Energy Correction 215
         7.3.4   The Fine Structure of the Hydrogen Atom 216
         7.3.5   External Magnetic Field Again 217
         7.3.6   The Hyperfine Structure of the Hydrogen Atom . . 218
   7.4   Other Atoms 220
         7.4.1   The Ground State of Helium 220
         7.4.2   The Perturbation Method for the Helium Atom . . 222
   7.5   The Variational Method 223
         7.5.1   The General Method 223
         7.5.2   The Helium Atom 225
         7.5.3   The Eigenvalue-Variational Scheme 228
   7.6   Molecules 229
         7.6.1   The Born-Oppenheimer Approximation 229
         7.6.2   The Hydrogen Molecular Ion 233
   7.7   The WKB Method 237
         7.7.1   Turning Points and Connection Formulas 238
         7.7.2   The Linear Approximation 240
         7.7.3   Bound States 243
         7.7.4   Tunneling through a Barrier 247
   Problems 248

8   Potential Scattering   264

   8.1   Introduction 264
         8.1.1   Kinematics of Scattering 264
         8.1.2   Scattering and Wave Functions 265
   8.2   The Scattering Amplitude 269
         8.2.1   Equation for the Scattering Amplitude 269
         8.2.2   The Born Series 270
         8.2.3   Spherically Symmetric Potentials 270
         8.2.4   The Optical Theorem 272
         8.2.5   The Refractive Index 273
   8.3   Partial Waves 275

		8.3.1 Expansion of a Plane Wave in a Legendre Series 276
		8.3.2 Partial Wave Expansion of $\psi(\mathbf{r})$ 278
		8.3.3 Calculation of the Phase Shift 280
	8.4	The Radial Wave Function 281
		8.4.1 The Integral Equation 281
		8.4.2 Partial Wave Green's Functions 281
		8.4.3 Scattering by an Impenetrable Sphere 283
	Problems 284

9 Transitions   288

9.1 Transitions in an External Field. 288
    9.1.1 Time-Dependent Perturbations 288
    9.1.2 The Semiclassical Method 288
9.2 The Transition Matrix 291
    9.2.1 The Transition Matrix 292
    9.2.2 The Lippmann-Schwinger Equation 295
    9.2.3 Relation to the Scattering Amplitude 297
9.3 Scattering and Cross Sections 298
    9.3.1 The Scattering Matrix 298
    9.3.2 The Transition Probability 299
    9.3.3 Cross Sections 300
    9.3.4 Scattering of Electrons by Atoms 302
    9.3.5 Scattering with Recoil 303
    9.3.6 Identical Particle Scattering 305
9.4 Decays of Excited States 307
    9.4.1 Lowest-Order Transition Rates 308
    9.4.2 Time Dependence of the Initial State 310
    9.4.3 Distribution of the Final States 314
Problems 316

10 Further Topics in Quantum Dynamics   324

10.1 Path Integration 324
    10.1.1 The Propagator as an Integral over Paths 324
    10.1.2 The Free Particle Propagator 327
    10.1.3 The Harmonic Oscillator. 328
    10.1.4 The Euclidean Formalism 331
    10.1.5 The Ground-State Energy 332
10.2 Path Integration: Some Applications 334
    10.2.1 The Born Series 334
    10.2.2 External Fields and Gauge Invariance 337
    10.2.3 The Aharonov-Bohm Effect 338
10.3 Berry's Phase 341
    10.3.1 Origin of the Phase. 341
    10.3.2 Example: Electron in a Precessing Magnetic Field 343
    10.3.3 The General Formula 344
    10.3.4 Two States near a Degeneracy 347
    10.3.5 Fast and Slow Coordinates 348

x   Contents

        10.3.6 The Aharonov-Bohm Effect Again 351
   Problems 352

11 The Quantized Electromagnetic Field   356
   11.1 The Classical Electromagnetic Field Hamiltonian 356
        11.1.1 Maxwell's Equations and the Transverse Gauge Condition . . 356
        11.1.2 The Independent Modes 358
        11.1.3 The Classical Hamiltonian 360
        11.1.4 The Canonical Coordinates 361
   11.2 The Quantized Radiation Field 362
        11.2.1 The Heisenberg Picture 362
        11.2.2 Canonical Quantization 362
        11.2.3 Photons 364
   11.3 Properties of the Quantum Electromagnetic Field 365
        11.3.1 The Momentum of the Field 365
        11.3.2 The Angular Momentum of the Field 366
        11.3.3 The Photon Spin 367
   11.4 Electromagnetic Decays of Excited States 369
        11.4.1 The Unperturbed Hamiltonian 369
        11.4.2 The Vector Potential Interaction 369
        11.4.3 The Spin Interaction 370
        11.4.4 The Rate for Photon Emission 370
        11.4.5 Multipole Matrix Elements 372
   11.5 Examples 373
        11.5.1 Decay of the 2P State of Atomic Hydrogen 373
        11.5.2 Hyperfine Emission 374
   11.6 Absorption and Stimulated Emission of Radiation 378
        11.6.1 Periodic Boundary Conditions 378
        11.6.2 Absorption 379
        11.6.3 Stimulated Emission 381
        11.6.4 The Blackbody Formula 382
   11.7 Scattering of Photons by Atoms 383
        11.7.1 The Photoelectric Effect 383
        11.7.2 Elastic Scattering of Photons 385
        11.7.3 Scattering by a Free Electron 391
   11.8 The Casimir Effect 394
        11.8.1 The Ground-State Energy of the Electromagnetic Field . . . 394
        11.8.2 The Casimir Force with an Elementary Cutoff 397
        11.8.3 The General Calculation 400
   Problems 402

12 Relativistic Wave Equations   407
   12.1 Lorentz Transformations 407
        12.1.1 Four-Vectors and Tensors 407
        12.1.2 Lorentz Transformations 408
        12.1.3 Spin 411
   12.2 Vector and Scalar Fields 413

		12.2.1 The Electromagnetic Field 413
		12.2.2 The Klein-Gordon Equation 414
	12.3	Relativistic Spin-Half Equations 417
		12.3.1 Two Component Spin-Half Equations 417
		12.3.2 The Dirac Equation 420
		12.3.3 Free Particle Solutions 421
		12.3.4 Probability Current and Hole Theory 423
	12.4	Dirac Electron in an Electromagnetic Field 423
		12.4.1 Second-Order Form of the Dirac Equation 423
		12.4.2 The Gyromagnetic Ratio 424
		12.4.3 The Nonrelativistic Limit and the Fine Structure 425
	12.5	The Dirac Hydrogen Atom 428
		12.5.1 Second-Order Equation 428
		12.5.2 Spherically Symmetric Potentials 431
		12.5.3 Radial Equations 432
		12.5.4 The Hydrogen Atom 433
	Problems 435

13 Identical Particles 438

	13.1	Nonrelativistic Identical-Particle Systems 438
		13.1.1 Creation and Annihilation Operators for Bosons 438
		13.1.2 Creation and Annihilation Operators for Fermions 443
	13.2	Elementary Applications 444
		13.2.1 Ideal Gas Distributions 444
		13.2.2 Ideal Electron Gas 446
		13.2.3 Collapsed Stars 448
	13.3	Relativistic Spinless Particles 453
		13.3.1 The Neutral Scalar Field 453
		13.3.2 The Classical Theory 453
		13.3.3 The Quantum Theory 455
		13.3.4 Charged Particles 456
	13.4	The Quantized Dirac Field 458
		13.4.1 The Dirac Action 458
		13.4.2 The Plane Wave Expansion 459
	13.5	Interacting Relativistic Fields 462
		13.5.1 Normal Ordering 462
		13.5.2 Example: The ϕ^4 interaction 463
	Problems 465

APPENDICES

A Mathematical Tools 470

	A.1	Miscellaneous Tools 470
		A.1.1 The Dirac Delta Function 470
		A.1.2 The Levi-Civita Symbol 472
		A.1.3 Some Integrals 472

xii Contents

        A.1.4 The Trapezoidal Approximation Series 474
    A.2 Special Functions 476
        A.2.1 Gamma Function 476
        A.2.2 Legendre Polynomials 479
        A.2.3 Solutions to the Free Radial Equation 484
        A.2.4 Hermite Polynomials 487
        A.2.5 Bessel Functions 489
    A.3 Orthogonal Curvilinear Coordinates 491
        A.3.1 Vector Calculus in Orthogonal Curvilinear Coordinates . . . 491
        A.3.2 Hydrogen Atom in Parabolic Coordinates 495
        A.3.3 Elliptic Coordinates 498

B Rotation Matrices   **500**
    B.1 Rotation Matrices—I 500
        B.1.1 Rotation Matrices and Spherical Harmonics 500
        B.1.2 The Explicit Form of the Rotation Matrices 502
        B.1.3 The Projection Theorem 507
    B.2 Rotation Matrices—II 508
        B.2.1 Averages over Products of Rotation Matrices 508
        B.2.2 The Wigner-Eckart Theorem Again 510

C SU(3)   **512**
    C.1 The Group and Algebra 512
    C.2 Some Representations 513

D References   **516**

    Index   **519**

Preface

The principles of quantum mechanics were formulated by many people during a short period of time at the beginning of the twentieth century. Max Planck wrote down his formula for the spectrum of blackbody radiation and introduced the constant that now bears his name in 1900. By 1924, through the work of Einstein, Rutherford and Bohr, Schrödinger and Heisenberg, Born, Dirac, and many others, the principles of quantum mechanics were discovered much as we know them today. They have become the framework for thinking about most of the phenomena that physicists study, from simple systems like atoms, molecules, and nuclei to more exotic ones like neutron stars, superfluids, and elementary particles.

This book is a text for an advanced course in quantum mechanics and, indeed, started out as notes for a graduate course at UCLA. Usually students in any field of physics must study quantum mechanics at this level before undertaking more specialized subjects.

The first part covers some of the formalism of quantum mechanics, especially the mathematics of rotations and other symmetries. It begins with a brief review of the Hamiltonian formulation of classical mechanics, which has become a trustworthy guide to finding the form of the quantum rules. The second chapter explains how the canonical quantum rules follow from the superposition principle and some form of the correspondence principle. It ends with the Schrödinger equation and the uncertainty principle.

The third chapter is about stationary states and the energy eigenvalue problem, with particular emphasis on spherical symmetry. It includes the theory of orbital angular momentum and the famous hydrogen atom problem. The latter will serve as a wonderful example over and over again.

The next two chapters are about the role of symmetry transformations in quantum mechanics, and how they restrict the possible values of some observables. There is a detailed discussion of three-dimensional rotations, the general theory of angular momentum, addition of angular momentum and selection rules. A good understanding of rotations in quantum mechanical systems is important for what follows. Rotations are an example for all sorts of other symmetries we have discovered or invented. The techniques learned in this context can be recycled many times.

These first five chapters contain the mathematical foundation of our subject. I have tried to be fairly rigorous, understanding that this is the students' second course in quantum mechanics.

There follows a brief interlude containing a miscellany of short subjects: magnetic field interactions, measurement and probability, the density matrix, and a recently discovered example of a simple quantum system, neutrino oscillations.

The rest is application. There is a section on bound-state perturbation theory, with the hydrogen atom as an example. There is a brief discussion of the variational principle, important in the theory of atomic and molecular structure, and of the WKB method. Transitions are introduced next in the context of potential scattering, with some applications to atoms and nuclei.

Next I have chosen a topic that students seem to enjoy learning about but which is hard to find in much detail in most textbooks at this level. This is the theory of transitions in general and, in particular, decay rates for excited states. There is an introduction to path integration and a section on geometric phases.

Then comes the theory of photons, the quantized electromagnetic field. Historically, this subject came first. The blackbody spectrum and the photoelectric effect were explained in terms of photons, quanta of the electromagnetic field, more than two decades before a real theory was available. Now, with the full power of the machinery of quantum mechanics in hand, we can understand completely those observations that puzzled Planck and his contemporaries. The quantum theory of the electromagnetic field is a useful subject to learn in its own right, and it is a good introduction to the methods used in both many body physics and elementary particle physics.

Next there is a chapter on relativistic wave equations, developed in the spirit of the earlier discussion of rotational symmetry, but here the symmetry is Lorentz invariance. I conclude with the occupation number space description of systems of identical particles, with a few applications.

I have tried to show the details of most mathematical calculations, and tried not to claim that one line follows easily from another unless my experience is that an average student will actually find this to be true. For the same reason I have included in the appendix derivations of many mathematical formulas even though most of them can be found in standard works.

If you want to learn quantum mechanics from this book, you need some preparation. Only the most extraordinary student could be expected to get through this material without the benefit of an introductory course, though in principle it is possible. You should also have studied classical mechanics and some mathematical methods at an introductory level.

The quantum mechanics "prerequisite" is to know what Schrödinger's wave equation is and how to use it. That means knowing how to find the bound states of a given potential in one and three dimensions, about tunneling problems, transmission and reflection coefficients, momentum and energy eigenfunctions, the elementary theory of the harmonic oscillator and the hydrogen atom. You can learn about them in more detail in some of the books listed among the references at the back.

The mathematical prerequisite is minimal. The quantum chapter of the book of nature is written in the language of linear algebra, which is the mathematical formulation of the superposition principle. I do not expect you to have studied Hilbert spaces or group theory previously. Pieces of the mathematics of linear vector spaces are presented as the need arises. But you should already know a little about vector calculus in curvilinear coordinates, and elementary concepts of vector space methods such as eigenvalues, Hermitean and unitary matrices, changes of basis, eigenfunction expansions, and so forth. I shall repeat the definitions of these tools, but this is not the place to learn them for the first time. A nodding acquaintance with complex numbers is also useful. More advanced parts of complex analysis, including the residue theorem, will be touched on only in the later parts.

You should also know some undergraduate-level classical mechanics, in particular the central force problem and the Lagrangian and Hamiltonian formalisms.

I avoid mentioning Poisson brackets in the body of the text, but as they provide an important insight into the structure of quantum mechanics, several problems are devoted to them.

Over the years I have assembled a collection of problems for the graduate quantum mechanics course. Some of the problems fill in gaps in the exposition. Most are the way to learn the tools of our trade. Occasionally the problems develop some themes not explained thoroughly in the body of the text. For a few of the problems you need to have access to a computer and know how to use it. Most are to be done analytically. You must work out many of the problems if you want to understand what is going on.

I have enjoyed collaborating with the many people at Prentice Hall and their associates who worked to turn the manuscript into this book. My thanks to Erik Fahlgren, my acquisitions editor, to Debra Wechsler, the production editor, to Daphne Hougham, who copyedited the manuscript, to Andrew Sobel, Bayani DeLeon, Adam Lewenberg, and many others whose names I do not know.

I am indebted to those who read earlier drafts with care and made suggestions for improvement, almost all of which I have included in the final version. Many thanks especially to Mike Berger (Indiana University), John Donoghue (University of Massachusetts), Colin Gay (Yale University), Maarten Golterman (San Francisco State University), Herbert Hamber (University of California, Irvine), Thomas Mehen (Duke University), Chandra Raman (Georgia Institute of Technology), Serge Rudaz (University of Minnesota), and several anonymous reviewers.

I thank the Department of Physics and Astronomy at UCLA for granting me the time to complete this manuscript, and the very many students over the years who suggested improvements or corrections in earlier versions. Finally, I thank my colleagues at UCLA and elsewhere for their criticism, advice, encouragement, and conversations about quantum mechanics. I am particular grateful for the discussions I have had with Sudip Chakravarty, John M. Cornwall, Robert Cousins, Carlos A. A. de Carvalho, Eric d'Hoker, Robert Finkelstein, Graciela Gelmini, Noah Graham, Alex Kusenko, Richard Norton, Shmuel Nussinov, Silvia Pascoli, Hidenori Sonoda, and E. Terry Tomboulis.

Ernest S. Abers
Los Angeles

CHAPTER 1

Classical Mechanics

1.1 NEWTON'S LAWS, THE ACTION, AND THE HAMILTONIAN

The superposition principle and the probability interpretation determine the mathematical framework of quantum mechanics. But, like Newton's three laws of classical mechanics, these two ideas do not tell us what the observables or how they change with time. For this physics still leans on classical mechanics in the form of the correspondence principle; not just the obvious idea that classical mechanics comes out of quantum mechanics in the appropriate limit, but the deeper idea that classical Hamiltonian mechanics leads to the correct quantum rules.

The equations of motion in quantum mechanics look like the equations of motion in classical mechanics. This resemblance has been a powerful tool for guessing the rules in quantum mechanics. The method, called **canonical quantization**, is based on Hamilton's "canonical" formulation of classical mechanics in terms of p's and q's. There is also a Lagrangian approach, the powerful path integral formulation of quantum mechanics, but even there the underlying postulates are stated in the Hamiltonian formalism.

Canonical quantization is not guaranteed to work—its predictions must be tested experimentally. And there are quantum mechanical ideas like spin and parity that have no close classical analog. Nevertheless it has proved a powerful idea, predicting correctly the behavior of quantum systems and, as a bonus, giving us hope that if the classical theory is consistent, then so is the corresponding quantum theory.

In this chapter I will review very briefly Hamiltonian classical mechanics with a particular view to those features we will use in quantum mechanics.

1.1.1 Newton's Law and Lagrange's Equations

Let us start where one always starts, from Newton's laws. The simplest mechanical systems have vector coordinates $\mathbf{r}_n$, one vector for each particle. The kinetic energy is

$$T = \frac{1}{2}\sum_n m_n \dot{\mathbf{r}}_n^2 = \frac{1}{2}\sum_{n,i} m_n \dot{r}_{n,i}^2 \tag{1.1}$$

where $r_{n,i}$ are the three Cartesian coordinates of the n-th particle. The potential energy $V(\mathbf{r}_n, \dot{\mathbf{r}}_n, t)$ is a function of the coordinates and perhaps the velocities. In the simplest cases V is independent of $\dot{\mathbf{r}}_n$ and also of the time. For these systems the **Lagrangian** is defined as

$$L = T - V \tag{1.2}$$

L is a function of the coordinates and velocities. It depends on the time, since the coordinates and velocities depend on the time. The i-th component of the force

1

Chapter 1 Classical Mechanics

on the n-th particle is

$$F_{n,i} = -\frac{\partial V}{\partial r_{n,i}} = \frac{\partial L}{\partial r_{n,i}} \tag{1.3}$$

while its momentum is

$$p_{n,i} = m_n \dot{r}_{n,i} = \frac{\partial L}{\partial \dot{r}_{n,i}} \tag{1.4}$$

In computing these partial derivatives we mean to vary L with respect to $r_{n,i}$, $\dot{r}_{n,i}$, and its explicit dependence on t, not its numerical value as a function of t. The latter is the variation dL/dt.

With these definitions Newton's third law

$$\mathbf{F}_n = \frac{d\mathbf{p}_n}{dt} = m_n \ddot{\mathbf{r}}_n \tag{1.5}$$

can be written

$$\frac{d}{dt}\frac{\partial L}{\partial \dot{r}_{n,i}} = \frac{\partial L}{\partial r_{n,i}} \tag{1.6}$$

or more concisely

$$\frac{d}{dt}\frac{\partial L}{\partial \dot{r}_k} = \frac{\partial L}{\partial r_k} \tag{1.7}$$

The index k now runs over all n and i. Equations (1.7) are Lagrange's equations.

1.1.2 Hamilton's Principle

Lagrange's equations make it easy to write the laws of motions in coordinates other than Cartesian coordinates, like spherical coordinates or confocal hyperbolic coordinates. One way to do that is to start in Cartesian coordinates and use the chain rule. The **action** provides a more elegant solution.

In general the Lagrangian can be a function of the time t as well as the coordinates and the velocities. Define the action as a functional of the coordinates and velocities:[1]

$$S = \int_{t_1}^{t_2} L(r_1 \ldots r_k, \dot{r}_1 \ldots \dot{r}_k, t)\, dt = \int_{t_1}^{t_2} L(r, \dot{r}, t)\, dt \tag{1.8}$$

In the second form the symbols r and $\dot{r}$ stand for $3n$ functions each. The endpoints of the integral are arbitrary fixed times. The value of the action depends on how the system gets from the configuration at t_1 to the configuration at t_2, subject only to the condition that it has the same coordinates at the start and at the finish. For any such path described by $r(t)$ the action has some value. Only one of these paths is a solution to Newton's law, and that one is an extremum of the action. To prove this, expand S about the correct path, keeping the values at the endpoints fixed. To first order,

$$S_o + \delta S = \int_{t_1}^{t_2} L(r + \delta r, \dot{r} + \delta \dot{r}, t)\, dt$$

$$= S_o + \sum_k \int_{t_1}^{t_2} \frac{\partial L}{\partial r_k} \delta r_k\, dt + \sum_k \int_{t_1}^{t_2} \frac{\partial L}{\partial \dot{r}_k} \delta \dot{r}_k\, dt + \cdots \tag{1.9}$$

---

[1] A functional is a special kind of function, one whose argument is a function but whose value is a number.

Section 1.1 Newton's Laws, the Action, and the Hamiltonian

But $\delta \dot{r}_k(t) = d\delta r_k/dt$, so

$$\delta S = \sum_k \int_{t_1}^{t_2} \left(\frac{\partial L}{\partial r_k} \delta r_k + \frac{\partial L}{\partial \dot{r}_k} \frac{d}{dt} \delta r_k \right) dt$$

$$= \sum_k \int_{t_1}^{t_2} \left(\frac{\partial L}{\partial r_k} \delta r_k + \frac{d}{dt}\left(\frac{\partial L}{\partial \dot{r}_k} \delta r_k\right) - \left(\frac{d}{dt}\frac{\partial L}{\partial \dot{r}_k}\right) \delta r_k \right) dt \quad (1.10)$$

The middle term is zero because $\delta r_k = 0$ at the endpoints. So

$$\delta S = \sum_k \int_{t_1}^{t_2} \left(\frac{\partial L}{\partial r_k} \delta r_k - \left(\frac{d}{dt}\frac{\partial L}{\partial \dot{r}_k}\right) \delta r_k \right) dt = 0 \quad (1.11)$$

Newtonian mechanics is equivalent to the statement that the classical physical path is the one that minimizes (more precisely extremizes) the action.

Equation (1.11) remains true under a reparametrization of the coordinates and velocities. Let q_k be any $3n$ functions of the r_k, and write L as a function of q_k and $\dot{q}_k$:

$$0 = \delta S = \delta \int_{t_1}^{t_2} L(q, \dot{q}, t)\, dt \quad (1.12)$$

where q and $\dot{q}$ stand for the new collection of $3n$ coordinates and velocities. Equation (1.12) is **Hamilton's principle**. It has the form of a standard problem in the calculus of variations, like the brachistochrone problem first solved by Newton. The same steps as above, in the reverse order, lead to the **Euler-Lagrange equations** in any coordinates:

$$\boxed{\frac{d}{dt}\frac{\partial L}{\partial \dot{q}_k} = \frac{\partial L}{\partial q_k}} \quad (1.13)$$

Of course equation (1.13) can also be derived from Newton's law by manipulating the chain rule for partial derivatives, or from d'Alembert's principle.

Hamilton's principle makes it easy to describe systems where the force depends on the velocity, not just the position, of the particles. For velocity-independent potentials as above, equation (1.11) restates Newton's second law. While not all dynamical systems are so simple that they can be written like this in terms of a potential function, *all physical systems we know of can be written in terms of an action functional and a Lagrangian.*

Example: Charged particle in an electromagnetic field

The most important example that is not trivial is a particle moving in a prescribed electromagnetic field. In terms of the electrostatic potential $\phi(\mathbf{r}, t)$ and the magnetic vector potential $\mathbf{A}(\mathbf{r}, t)$, the fields are[2]

$$\mathbf{B} = \boldsymbol{\nabla} \times \mathbf{A} \quad \text{and} \quad \mathbf{E} = -\boldsymbol{\nabla}\phi - \frac{1}{c}\frac{\partial \mathbf{A}}{\partial t} \quad (1.14)$$

The correct Lagrangian for a single charged particle is

$$L = \frac{1}{2}m\dot{\mathbf{r}}^2 - q\phi(\mathbf{r}, t) + \frac{q}{c}\mathbf{A}(\mathbf{r}, t) \cdot \dot{\mathbf{r}} \quad (1.15)$$

---

[2] I use Gaussian (cgs) units throughout.

4 Chapter 1 Classical Mechanics

where q is the particle's electric charge. Here is the demonstration:

$$\frac{\partial L}{\partial r_i} = -q\frac{\partial \phi}{\partial r_i} + \frac{q}{c}\sum_j \dot{r}_j \frac{\partial A_j}{\partial r_i}$$

$$\frac{\partial L}{\partial \dot{r}_i} = m\dot{r}_i + \frac{q}{c}A_i \qquad (1.16)$$

$$\frac{d}{dt}\frac{\partial L}{\partial \dot{r}_i} = m\ddot{r}_i + \frac{q}{c}\sum_j \frac{\partial A_i}{\partial r_j}\dot{r}_j + \frac{q}{c}\frac{\partial A_i}{\partial t}$$

The Euler-Lagrange equations are

$$m\ddot{r}_i = -q\frac{\partial \phi}{\partial r_i} + \frac{q}{c}\sum_j \left(\frac{\partial A_j}{\partial r_i} - \frac{\partial A_i}{\partial r_j}\right)\dot{r}_j - \frac{q}{c}\frac{\partial A_i}{\partial t} \qquad (1.17)$$

The middle two terms on the right can be written[3]

$$\frac{q}{c}\sum_j \left(\frac{\partial A_j}{\partial r_i}\dot{r}_j - \frac{\partial A_i}{\partial r_j}\dot{r}_j\right) = \frac{q}{c}\sum_{j,k}\epsilon_{kij}\dot{r}_j(\boldsymbol{\nabla}\times\mathbf{A})_k = \frac{q}{c}(\mathbf{v}\times\mathbf{B})_i \qquad (1.18)$$

while the remaining two terms are

$$-q\left(\frac{\partial \phi}{\partial r_i} + \frac{1}{c}\frac{\partial A_i}{\partial t}\right) = qE_i \qquad (1.19)$$

Hamilton's principle for this Lagrangian is solved by

$$m\ddot{\mathbf{r}} = q\left(\mathbf{E} + \frac{1}{c}(\mathbf{v}\times\mathbf{B})\right) \qquad (1.20)$$

Equation (1.20) is the Lorentz force law.

1.1.3 Canonical Momenta and the Hamiltonian Formulation

The momentum canonically conjugate to q_k is defined by

$$p_k = \frac{\partial L}{\partial \dot{q}_k} \qquad (1.21)$$

For a Lagrangian of the form $L = T - V(r)$, the canonical momentum is the same as the mechanical momentum, but not in general. For the electromagnetic example above, the canonical momenta are

$$p_k = \frac{\partial T}{\partial \dot{r}_k} + \frac{q}{c}A_k(\mathbf{r},t) \qquad \text{or} \qquad \mathbf{p} = m\mathbf{v} + \frac{q}{c}\mathbf{A} \qquad (1.22)$$

It is useful to express Hamilton's principle in terms of the q's and the p's. To do that one makes a Legendre transformation analogous to those used in elementary

---
[3] ϵ_{ijk} is the totally antisymmetric Levi-Civita symbol. See Appendix A.1.

Section 1.1 Newton's Laws, the Action, and the Hamiltonian

thermodynamics. Think of L as a functional with p instead of $\dot{q}$ as one of the independent variables. Then define the Hamiltonian as

$$H(p, q, t) = \sum_k p_k \dot{q}_k - L(p, q, t) \tag{1.23}$$

Example: Particle in a potential

In the simple potential case,

$$H = m \sum_k p_k \dot{q}_k - L = m \sum_k \left(\frac{p_k}{m}\right)^2 - L = 2T - T + V = T + V \tag{1.24}$$

The value of H will almost always be the energy, as it is here, but it is the functional dependence of H on the p's and the q's that is important.

Example: Charged particle in an electromagnetic field

For a particle in an electromagnetic field, the *value* of the Hamiltonian is indeed the energy:

$$\begin{aligned} H = \mathbf{p} \cdot \dot{\mathbf{r}} - L &= \left(m\mathbf{v} + \frac{q}{c}\mathbf{A}\right) \cdot \dot{\mathbf{r}} - \frac{1}{2}mv^2 + q\phi - \frac{q}{c}\mathbf{A} \cdot \mathbf{v} \\ &= \frac{1}{2}mv^2 + q\phi \end{aligned} \tag{1.25}$$

The magnetic field $\mathbf{B}$ does not contribute to the value of the energy, but it does enter the equations of motion. In terms of $\mathbf{p}$ and $\mathbf{r}$ the Hamiltonian is

$$H = \frac{1}{2m}\left(\mathbf{p} - \frac{q}{c}\mathbf{A}\right)^2 + q\phi \tag{1.26}$$

Hamilton's Equations

Expand the Hamiltonian for arbitrary q_k and p_k about the physical values:

$$\begin{aligned} \delta H &= \sum_k \left(\dot{q}_k \delta p_k + p_k \delta \dot{q}_k - \frac{\partial L}{\partial q_k}\delta q_k - \frac{\partial L}{\partial \dot{q}_k}\delta \dot{q}_k\right) \\ &= \sum_k \left(\dot{q}_k \delta p_k - \frac{\partial L}{\partial q_k}\delta q_k\right) \end{aligned} \tag{1.27}$$

If the p_k and q_k are solutions to the Euler-Lagrange equations (1.13), then

$$\delta H = \sum_k (\dot{q}_k \delta p_k - \dot{p}_k \delta q_k) \tag{1.28}$$

Since the p_k and q_k are varied independently,

$$\boxed{\dot{q}_k = \frac{\partial H}{\partial p_k} \quad \text{and} \quad \dot{p}_k = -\frac{\partial H}{\partial q_k}} \tag{1.29}$$

These are Hamilton's equations of motion. They have exactly the same content as equations (1.13).

Example: Particle in a potential
For a mechanical particle in Cartesian coordinates, $H = T + V$, and equations (1.29) are

$$\dot{r}_k = \frac{1}{m} p_k \quad \text{and} \quad \dot{p}_k = -\frac{\partial V}{\partial r_k} \tag{1.30}$$

Differentiate one more

$$\ddot{r}_k = \frac{1}{m}\dot{p}_k = -\frac{1}{m}\frac{\partial V}{\partial r_k} \tag{1.31}$$

or $m\ddot{\mathbf{r}} = \mathbf{F}$.

Example: Charged particle in an electromagnetic field
For a particle in an electromagnetic field, the equations of motion are

$$\begin{aligned}\dot{p}_k &= -\frac{\partial H}{\partial r_k} = \frac{1}{m}\sum_i \left(p_i - \frac{q}{c} A_i\right) \frac{q}{c}\frac{\partial A_i}{\partial r_k} - q\frac{\partial \phi}{\partial r_k} \\ \dot{r}_k &= \frac{\partial H}{\partial p_k} = \frac{1}{m}\left(p_k - \frac{q}{c}A_k\right)\end{aligned} \tag{1.32}$$

Differentiate:

$$\begin{aligned}m\ddot{r}_k = \left(\dot{p}_k - \frac{q}{c}\dot{A}_k\right) &= q\sum_i \frac{v_i}{c}\left(\frac{\partial A_i}{\partial r_k} - \frac{\partial A_k}{\partial r_i}\right) - q\left(\frac{\partial \phi}{\partial r_k} + \frac{1}{c}\frac{\partial A_k}{\partial t}\right) \\ &= q\left(E_k + (\frac{\mathbf{v}}{c}\times \mathbf{B})_k\right)\end{aligned} \tag{1.33}$$

Again, one gets the Lorentz force law in the nonrelativistic limit.

1.2 CLASSICAL SPACE-TIME SYMMETRIES

Ever since Galileo discovered the principle of relativity, the invariance of the laws of nature under certain changes of the coordinate system has been a fundamental tenet of physics. That the laws are the same even if you rotate or translate the coordinates is the mathematical statement of the principle. No experiment can discover the origin of coordinates or the orientation of the coordinate axes; they are arbitrary.

In both classical and quantum mechanics, **symmetries** are closely connected to conservation laws. Why is that? One way to solve a problem in classical mechanics is to find coordinates and momenta that make the Hamiltonian H independent of some of the coordinates. Then the conjugate momenta will be constants of the motion. For instance if H is independent of the differences of the positions of the particles, the total momentum is conserved. If H is independent of the angle of a rotation in some plane, that component of angular momentum is conserved.

1.2.1 The Space-Time Transformations

For isolated physical systems, these transformations are all symmetries in the sense that when the coordinates undergo these transformations, for isolated systems, the form of the physical laws is unchanged.

Translations shift the coordinates by a vector **a**. There is no absolute origin to the coordinate system, and no experiment can find out where it is. Time translations change the value of a coordinate to its value at a different time. There is no absolute meaning to the setting of the clock or the calendar, and so no experiment can find out when $t = 0$ was or will be.

Rotations by an angle θ about the direction of a unit vector $\hat{n}$ are described by a 3×3 real orthogonal matrix $\bar{R}$, one whose transpose is its inverse. Usually the word "rotation" excludes reflections, and so is reserved for real orthogonal 3×3 matrices that are *unimodular*, that is, have determinant $+1$. There is no absolute preferred orientation of the coordinate system. Boosts change the coordinates to ones moving with a uniform velocity relative to the first, and neither of the two frames of reference is preferred by nature. This is the principle of relativity that Galileo discovered.

> Note: The nonrelativistic form of the boosts given here is not really a symmetry of nature. In the real world Galilean boosts are replaced by the relativistic (Lorentz) form. We will study those in detail in Chapter 12.

Reflections change the sign of all three coordinates, and time inversion reverses the direction of the clocks. Reflection invariance means there is no absolute meaning to the distinction between left-handed coordinate systems and right-handed ones. Time-inversion invariance means that if you recorded the time of all events by a clock that is running backward, the laws of nature would be satisfied as well.

All the fundamental physical laws are invariant under translations, time translations, rotations, and Lorentz boosts. The most common laws are invariant under all the transformations listed in Table 1.1, but the weak interactions are not invariant under $\mathcal{R}$, and (probably) not even $\mathcal{T}$. The discrete symmetries are valid for electromagnetic, nuclear, and even gravitational forces.

Quantum systems have these symmetries and more: The additional ones are often called *internal* symmetries because they have nothing obvious to do with space and time. Examples of internal symmetries are charge conjugation or matter-antimatter symmetry, isotopic spin, and electromagnetic gauge invariance.

Table 1.1 shows what some symmetry transformations do to the Cartesian coordinates x, y, and z of a particle.[4]

Observables are usually functions of the momenta as well as the coordinates, so you have to know how the momenta transform too. It is not enough to know how the coordinates transform under a symmetry. In Cartesian coordinates the momenta transform as in Table 1.2.

The form of the equations of motion is unchanged under the transformations indicated, provided they are indeed symmetry transformations. Except for those that involve the time explicitly—time translation, time inversion, and the boosts—this means that the Hamiltonian must be invariant under the transformation.

What properties of a transformation are independent of the quantity being transformed? These are the universal structure of the transformations themselves,

---

[4] Rather than change coordinate systems as one does in general relativity, I will always define symmetries of a system by what a transformation does to observables. This is the "active" point of view. The opposite convention just replaces all the transformations with their inverses.

8 Chapter 1 Classical Mechanics

Translations:	$T(\mathbf{a})$	$\mathbf{r}_i \to \mathbf{r}_i + \mathbf{a}$
Time translations:	$U(t_o)$	$\mathbf{r}_i(t) \to \mathbf{r}_i(t + t_o)$
Rotations:	$R(\hat{\mathbf{n}}, \theta)$	$r_{i,m} \to \sum_{n=1}^{3} \bar{R}_{mn}(\hat{\mathbf{n}}, \theta) r_{i,n}$
Galilean boosts:	$G(\mathbf{v})$	$\mathbf{r}_i \to \mathbf{r}_i + \mathbf{v}t$
Reflections:	$\mathcal{R}$	$\mathbf{r}_i \to -\mathbf{r}_i$
Time inversion:	$\mathcal{T}$	$\mathbf{r}_i(t) \to \mathbf{r}_i(-t)$

TABLE 1.1: The space-time symmetries of classical physics: Coordinate transformations

Translations:	$T(\mathbf{a})$	$\mathbf{p}_i \to \mathbf{p}_i$
Time translations:	$U(t_o)$	$\mathbf{p}_i(t) \to \mathbf{p}_i(t + t_o)$
Rotations:	$R(\hat{\mathbf{n}}, \theta)$	$p_{i,m} \to \sum_{n=1}^{3} \bar{R}_{mn}(\hat{\mathbf{n}}, \theta) p_{i,n}$
Galilean boosts:	$G(\mathbf{v})$	$\mathbf{p}_i \to \mathbf{p}_i + m\mathbf{v}$
Reflections:	$\mathcal{R}$	$\mathbf{p}_i \to -\mathbf{p}_i$
Time inversion:	$\mathcal{T}$	$\mathbf{p}_i \to -\mathbf{p}_i(-t)$

TABLE 1.2: The space-time symmetries of classical physics: Momentum transformations

the rules for composing them. Under composition, the transformations form a **group**.[5] The composition rules, or group multiplication, are geometrical and can be taken over directly into quantum mechanics.

1.2.2 Translations

Now I want to examine translations and rotations in more detail. Translations are the simplest, since they all commute with one another:

$$T(\mathbf{a}_1)T(\mathbf{a}_2) = T(\mathbf{a}_1 + \mathbf{a}_2) = T(\mathbf{a}_2 + \mathbf{a}_1) = T(\mathbf{a}_2)T(\mathbf{a}_1) \qquad (1.34)$$

You can verify that this rule holds for translations on $\mathbf{r}$ and $\mathbf{p}$ given above.

The properties of any continuous group of transformations are almost completely determined by the properties of very small transformations. For translations, this means that if you expand $T(\mathbf{a})$ in a power series in $\mathbf{a}$, you only have to learn about the first-order term in the expansion. Geometrically, the large translations are determined by the small ones because any $\mathbf{a}$ can in principle be found by composing, or integrating, a large number of small translations.

Let A be any observable. Under a small transformation $T(\boldsymbol{\epsilon})$, $\delta\mathbf{r}_k = \boldsymbol{\epsilon}$ and

---

[5]The definition of a group can be found in Section 4.1.

$\delta \mathbf{p}_k = 0$. Then $A \to A + \delta A$ (plus higher order terms) where

$$\delta A = \sum_n \frac{\partial A}{\partial r_n} \delta r_n + \sum_n \frac{\partial A}{\partial p_n} \delta p_n = \sum_{k,m} \frac{\partial A}{\partial r_{k,m}} \epsilon_m = \boldsymbol{\epsilon} \cdot \sum_k \boldsymbol{\nabla}_k A \tag{1.35}$$

1.2.3 Rotations

Rotations do not all commute with one another, so they are both more complicated and more interesting. A theorem (due to Euler) says that any rotation leaves some axis unchanged; so any rotation can be described by a unit vector $\hat{\boldsymbol{n}}$ and an angle θ of rotation about $\hat{\boldsymbol{n}}$. That is, any of the 3×3 matrices $\bar{R}$ has an eigenvector $\hat{\boldsymbol{n}}$ with eigenvalue unity: $\bar{R}\hat{\boldsymbol{n}} = \hat{\boldsymbol{n}}$. I have been using this result in the notation for the rotation matrices $\bar{R}(\hat{\boldsymbol{n}}, \theta)$.

The rotation matrices are worth learning about in some detail: Rotations about the z-axis have the form[6].

$$\bar{R}(\hat{\boldsymbol{n}}_z, \theta) = \begin{pmatrix} \cos\theta & -\sin\theta & 0 \\ \sin\theta & \cos\theta & 0 \\ 0 & 0 & 1 \end{pmatrix} \tag{1.36}$$

Approximate the matrix in equation (1.36) to first order in the angle:

$$\bar{R}(\hat{\boldsymbol{n}}_z, \epsilon) = \mathbf{1} + \begin{pmatrix} 0 & -\epsilon & 0 \\ \epsilon & 0 & 0 \\ 0 & 0 & 0 \end{pmatrix} \tag{1.37}$$

The transformed coordinate is

$$x' = x - \epsilon y \qquad y' = y + \epsilon x \qquad z' = z \tag{1.38}$$

or

$$\mathbf{r}' = \mathbf{r} + \delta \mathbf{r} \tag{1.39}$$

where

$$\delta \mathbf{r} = \epsilon \hat{\boldsymbol{n}}_z \times \mathbf{r} + \cdots \tag{1.40}$$

There is nothing special about the z-axis. The result is general:

$$\mathbf{r}' = \mathbf{r} + \epsilon \hat{\boldsymbol{n}} \times \mathbf{r} + \cdots \tag{1.41}$$

or in components

$$\delta r_i = \epsilon \sum_{j,k} \epsilon_{ijk} n_j r_k + \cdots \tag{1.42}$$

Similarly, the first order change in the momentum is

$$\delta p_i = \epsilon \sum_{j,k} \epsilon_{ijk} n_j p_k + \cdots \tag{1.43}$$

---

[6] I reserve the notation $\bar{R}(\theta, \hat{\boldsymbol{n}})$ for these real 3×3 matrices, to distinguish them from the abstract operation $R(\theta, \hat{\boldsymbol{n}})$ that appears in Tables 1.1 and 1.2. This is not a standard notation, but I find it very convenient.

10 Chapter 1 Classical Mechanics

First Order Transformations of Vectors
Examples of vectors observables are $\mathbf{r}$, $\mathbf{p}$, and $\mathbf{L}$. Any three observables V_i components of a **vector** provided that under a rotation,

$$V_i \to \sum_j \bar{R}_{ij} V_j \qquad (1.44)$$

Then under a small rotation,

$$\delta V_i = \epsilon \sum_{j,k} \epsilon_{ijk} n_j V_k \qquad (1.45)$$

or $\delta \mathbf{V} = \epsilon \hat{\mathbf{n}} \times \mathbf{V}$. Observables that like the energy are invariant under rotations, are called **scalars**.

Vector and Scalar Observables
In classical mechanics all observables are functions of the r_k and the p_k, so the transformation properties of other observables can be deduced from the rules for $\mathbf{r}$ and $\mathbf{p}$. For example, the angular momentum of a particle is $\mathbf{L} = \mathbf{r} \times \mathbf{p}$. Under rotations (exercise!)

$$L_m \to \sum_{n=1}^{3} \bar{R}(\hat{\mathbf{n}}, \theta)_{mn} L_n \qquad (1.46)$$

Fields
Fields—like the electric or the gravitational field—are observables that are functions of space and time. They also transform in definite ways. The electrostatic potential $\phi(\mathbf{r}, t)$ introduced in equation (1.14) is an example of a scalar field. If you rotate a field, the value at some point $\mathbf{r}$ after rotation is the same value it had before rotation at the point that rotates into $\mathbf{r}$. That is,

$$\phi(\mathbf{r}, t) \to \phi(\mathbf{r}'', t) \qquad \text{where} \qquad r_i'' = \sum_j \bar{R}_{ij}^{-1} r_j \qquad (1.47)$$

Another way to write this rule is $\phi(\mathbf{r}, t) \to \phi'(\mathbf{r}, t)$, where

$$r_i' = \sum_j \bar{R}_{ij} r_j \qquad \text{and} \qquad \phi'(\mathbf{r}', t) = \phi(\mathbf{r}, t) \qquad (1.48)$$

The transformation rule for fields like the magnetic vector potential is $A_i(\mathbf{r}, t) \to A_i'(\mathbf{r}, t)$, where

$$r_i' = \sum_j \bar{R}_{ij} r_j \qquad \text{and} \qquad A_i'(\mathbf{r}', t) = \sum_j \bar{R}_{ij} A_j(\mathbf{r}, t) \qquad (1.49)$$

Reflections and Rotations
Ordinary vectors, like $\mathbf{r}$ and $\mathbf{p}$, are odd under reflections. Ordinary scalars, like $\mathbf{p}^2/2m$, are even under reflections. Vectors that have the opposite sign from $\mathbf{r}$ and $\mathbf{p}$ under reflections, or scalars that have the opposite sign from the energy under reflections, are called **pseudovectors** and **pseudoscalars**. Under a reflection $\mathbf{L} \to$

$(-\mathbf{r}) \times (-\mathbf{p}) = \mathbf{L}$. The angular momentum is a pseudovector. But the magnetic vector potential is odd under a reflection.

$$\mathbf{A}(\mathbf{r},t) \to -\mathbf{A}(-\mathbf{r},t) \tag{1.50}$$

$\mathbf{A}(\mathbf{r},t)$ and any fields that transform like it are ordinary vector fields. Without the minus sign in the reflection rule, they are pseudovector fields. $\mathbf{E}$ is a vector field, while $\mathbf{B}$ is a pseudovector field.[7]

1.2.4 Rotation Matrices

The structure of the rotation matrices will be important. Two rotations do not in general commute:

$$\bar{R}(\hat{\mathbf{n}}_1, \theta_1)\bar{R}(\hat{\mathbf{n}}_2, \theta_2) \neq \bar{R}(\hat{\mathbf{n}}_2, \theta_2)\bar{R}(\hat{\mathbf{n}}_1, \theta_1) \tag{1.51}$$

But two rotations about the same axis do commute:

$$\bar{R}(\hat{\mathbf{n}}, \theta_1)\bar{R}(\hat{\mathbf{n}}, \theta_2) = \bar{R}(\hat{\mathbf{n}}, \theta_1 + \theta_2) = \bar{R}(\hat{\mathbf{n}}, \theta_2)\bar{R}(\hat{\mathbf{n}}, \theta_1) \tag{1.52}$$

All the rotations about a given axis $\hat{\mathbf{n}}$ are therefore powers of some matrix. It is useful to write that matrix in an exponential form:

$$\bar{R}(\hat{\mathbf{n}}, \theta) = \left(e^{-i\bar{J}(\hat{\mathbf{n}})}\right)^\theta = e^{-i\theta \bar{J}(\hat{\mathbf{n}})} \tag{1.53}$$

for some matrix $\bar{J}(\hat{\mathbf{n}})$. Don't be alarmed by the notation. The exponential of a matrix has a precise meaning as a power series

$$e^M = I + M + \frac{1}{2}M^2 + \frac{1}{6}M^3 + \cdots \tag{1.54}$$

and the series converges for all M. The matrix $\bar{J}(\hat{\mathbf{n}})$ is said to be the **generator** of the rotations around that axis.

Since the rotation matrices are real, the $\bar{J}$ must be imaginary. And since they are orthogonal,

$$\left(e^{-i\bar{J}}\right)^T = e^{-i\bar{J}^T} = e^{i\bar{J}} \tag{1.55}$$

so that $\bar{J}(\hat{\mathbf{n}})^T = -\bar{J}(\hat{\mathbf{n}})$. The $\bar{J}$ matrices are all Hermitean (that is why physicists put the factor i in the definition). Each of them is an antisymmetric, imaginary 3×3 matrix. Since such matrices have zero along the main diagonal, they are all some linear combinations of only three independent matrices. These basis matrices can be taken to be the generators of rotations around the three orthogonal Cartesian axes. They can be found explicitly by expanding r'_i as a function of r_i to first order in ϵ:

$$r'_i = \sum_j \left(e^{-i\epsilon \bar{J}(\hat{\mathbf{n}})}\right)_{ij} r_j = r_i - \sum_j i\epsilon \bar{J}(\hat{\mathbf{n}})_{ij} r_j + \cdots \tag{1.56}$$

Compare the coefficient of ϵ in equation (1.56) with equation (1.41):

$$\sum_{ij} \epsilon_{ijk} n_i r_j = -i \sum_j \bar{J}(\hat{\mathbf{n}})_{kj} r_j \tag{1.57}$$

---

[7] Vectors and pseudovectors are sometimes called polar vectors and axial vectors, respectively.

Since **r** and $\hat{n}$ are arbitrary, the generator $\bar{J}_i$ of rotations about one of the coordinate axes is

$$(\bar{J}_i)_{jk} = \bar{J}(\hat{n}_i)_{jk} = i\epsilon_{ikj} = -i\epsilon_{ijk} \tag{1.58}$$

and in a general direction,

$$\bar{J}(\hat{n}) = \sum_i n_i \bar{J}_i = \hat{n} \cdot \bar{J} \tag{1.59}$$

Equation (1.59) is an important result. For most purposes, it reduces the study of rotations to the study of these three matrices:

$$\bar{J}_x = \begin{pmatrix} 0 & 0 & 0 \\ 0 & 0 & -i \\ 0 & i & 0 \end{pmatrix} \quad \bar{J}_y = \begin{pmatrix} 0 & 0 & i \\ 0 & 0 & 0 \\ -i & 0 & 0 \end{pmatrix} \quad \bar{J}_z = \begin{pmatrix} 0 & -i & 0 \\ i & 0 & 0 \\ 0 & 0 & 0 \end{pmatrix} \tag{1.60}$$

The commutator algebra of these three matrices is fundamental. It is a simple exercise to work it out from the explicit forms (1.60). The result is

$$[\bar{J}_x, \bar{J}_y] = i\bar{J}_z \qquad [\bar{J}_y, \bar{J}_z] = i\bar{J}_x \qquad [\bar{J}_z, \bar{J}_x] = i\bar{J}_y \tag{1.61}$$

or succinctly

$$\boxed{[\bar{J}_i, \bar{J}_j] = i \sum_k \epsilon_{ijk} \bar{J}_k} \tag{1.62}$$

1.2.5 Symmetries and Conservation Laws

Consider a physical system described by a Lagrangian $L(q, \dot{q}, t)$, where as usual q stands for a collection of coordinates q_k. If you make a transformation on the coordinates, like one of the space-time transformations listed above, then to first order there is some change δq_k in the coordinates, and consequently a change in the Lagrangian

$$\delta L = \sum_k \left[\frac{\partial L}{\partial q_k} \delta q_k + \frac{\partial L}{\partial \dot{q}_k} \delta \dot{q}_k \right] \tag{1.63}$$

Suppose δL is the total time derivative of some function of the coordinates, velocities, and the time:

$$\delta L = \frac{d}{dt} \delta \Omega(q, \dot{q}, t) \tag{1.64}$$

Then the action, the integral of the Lagrangian, will be invariant. This rule must hold as a consequence of the functional form of L, for any $q_k(t)$ whether or not they satisfy the equations of motion. Let

$$G = \sum_k p_k \delta q_k - \delta \Omega. \tag{1.65}$$

where $p_k = \partial L / \partial \dot{q}_k$ are the conjugate momenta. Then

$$\frac{d}{dt} G = \sum_k \left[\frac{d}{dt}\frac{\partial L}{\partial \dot{q}_k} \delta q_k + \frac{\partial L}{\partial \dot{q}_k} \delta \dot{q}_k \right] - \sum_k \left[\frac{\partial L}{\partial q_k} \delta q_k + \frac{\partial L}{\partial \dot{q}_k} \delta \dot{q}_k \right] = 0 \tag{1.66}$$

where now I have let the q_k and the $\dot{q}_k$ satisfy the equations of motion (1.13). G is a constant, and there is a conservation law associated with every continuous transformation that leaves the action invariant.

Example: Time translations

Under a time translation $U(\epsilon)$ as defined in Table (1.1), the coordinates transform as $q_k(t) \to q_k(t + \epsilon)$, or

$$\delta q_k = \epsilon \dot{q}_k \quad \text{and} \quad \delta \dot{q}_k = \epsilon \ddot{q}_k \tag{1.67}$$

and

$$\delta L(q, \dot{q}, t) = \sum_k \left[\frac{\partial L}{\partial q_k} \dot{q}_k + \frac{\partial L}{\partial \dot{q}_k} \ddot{q}_k \right] \epsilon \tag{1.68}$$

For a general Lagrangian

$$\frac{d}{dt} L(q, \dot{q}, t) = \sum_k \left[\frac{\partial L}{\partial q_k} \dot{q}_k + \frac{\partial L}{\partial \dot{q}_k} \ddot{q}_k \right] + \frac{\partial L}{\partial t} \tag{1.69}$$

Suppose L has no explicit time dependence: $\partial L / \partial t = 0$. Then

$$\epsilon \frac{d}{dt} L = \delta L \tag{1.70}$$

For these transformations the function $\delta \Omega$ defined in equation (1.64) above is

$$\delta \Omega = \epsilon L \tag{1.71}$$

and the conserved combination is

$$G = \sum_k p_k \delta q_k - \epsilon L = \sum_k p_k \dot{q}_k \epsilon - \epsilon L = \epsilon H \tag{1.72}$$

Since ϵ is a constant, time translation invariance implies the conservation of the Hamiltonian. (Of course you knew that all the time. If L, hence H, depends explicitly on the time, energy is not usually conserved.)

Example: Translations

Write the Lagrangian in Cartesian coordinates. Under an infinitesimal translation $T(\epsilon)$ the coordinates change as

$$\delta r_{m,i} = \epsilon_i \quad \text{and} \quad \delta \dot{r}_{m,i} = 0 \tag{1.73}$$

so that

$$\delta L = \sum_{m,i} \frac{\partial L}{\partial r_{m,i}} \epsilon_i \tag{1.74}$$

Suppose L is invariant under translations: $\delta L = 0$. Then the function $\delta \Omega$ in (1.64) is any constant, which I take to be zero. The conserved quantity is

$$G = \left[\sum_{m,i} p_{m,i} \right] \epsilon_m = \mathbf{P} \cdot \boldsymbol{\epsilon} \tag{1.75}$$

The total momentum is conserved as a consequence of translation invariance.

Example: Rotations

Under a rotation about $\hat{n}$ by the small angle ϵ the coordinates and momenta change as

$$\delta r_{m,i} = \epsilon \sum_{jk} \epsilon_{ijk} n_j r_{m,k} \quad \text{and} \quad \delta \dot{r}_{m,i} = \epsilon \sum_{jk} \epsilon_{ijk} n_j \dot{r}_{m,k} \tag{1.76}$$

so that now

$$\delta L = \epsilon \sum_{ijkm} \epsilon_{ijk} n_j \left[\frac{\partial L}{\partial r_{m,i}} r_{m,k} + \frac{\partial L}{\partial \dot{r}_{m,i}} \dot{r}_{m,k} \right] \tag{1.77}$$

If L is invariant under rotations, then $\delta L = 0$. Again $\delta \Omega = 0$, and the conserved quantity is

$$G = \sum_{m,i} p_{m,i} \delta r_{m,i} = \epsilon \sum_{ijk} \epsilon_{ijk} p_{m,i} n_j r_{m,k} = \epsilon \sum_m (\mathbf{r}_m \times \mathbf{p}_m) \cdot \hat{n} = \epsilon \mathbf{L} \cdot \hat{n} \tag{1.78}$$

Rotational invariance implies that the total angular momentum $\mathbf{L}$ is conserved.

Noether's theorem

These results are special cases of **Noether's theorem**, which says there is a conservation law for every continuous symmetry. Noether's theorem is very powerful and general, and has many applications in quantum field theory, statistical mechanics, and general relativity.

PROBLEMS

1.1. Spherical Coordinates

Consider a particle in three dimensions described by the usual spherical coordinates r, θ, ϕ:

$$x = r \sin\theta \cos\phi \quad y = r \sin\theta \sin\phi \quad z = r \cos\theta$$

Let the particle move in a central potential $V(r)$, that is, V depends on the radial coordinate r only, and is independent of θ and ϕ. Treat r, θ, and ϕ as canonical coordinates.

(a) Write the Lagrangian in these coordinates.
(b) What are Lagrange's equations?
(c) Find the canonically conjugate momenta p_r, p_θ, and p_ϕ, in terms of the coordinates and their time derivatives.
(d) Write the Hamiltonian and the three sets of Hamilton's equations.
(e) Express the square of the physical angular momentum, $\mathbf{L}^2$, in terms of the spherical coordinates and conjugate momenta, and then use Hamilton's equations to show that $\mathbf{L}^2$ is conserved.

1.2. Classical Theory of the Kepler Problem

The results of this problem will be important for the quantum mechanical Kepler problem in Section 3.5. Let the Hamiltonian for a single three-dimensional particle (e.g., a planet in the sun's gravitational field) be

$$H = \frac{\mathbf{p}^2}{2m} - \frac{K}{r} \quad (K > 0)$$

Define the **Runge-Lenz** vector $\mathbf{A}$ by

$$\mathbf{A} = \mathbf{L} \times \mathbf{p} + \frac{Km\mathbf{r}}{r}$$

where $\mathbf{L} = \mathbf{r} \times \mathbf{p}$ is the angular momentum.

(a) Using the equation of motion for $\mathbf{p}$ and the fact that $d\mathbf{L}/dt = 0$ in any central force problem, show that $\mathbf{A}$ is also a constant of the motion, that is, that

$$\frac{d\mathbf{A}}{dt} = 0$$

(b) Show that $\mathbf{A} \cdot \mathbf{L} = 0$.

(c) Since $\mathbf{L}$ is a constant, the motion is in a plane. Show that $\mathbf{A}$ lies in the plane of the orbit, and for closed orbits points in the direction of the maximum value of r (aphelion for the earth), as in Figure 1.1.

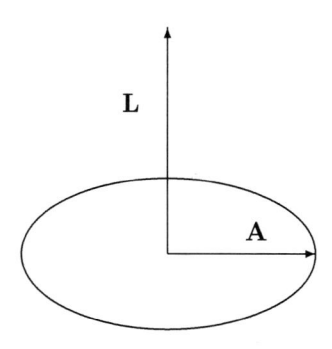

FIGURE 1.1: The Runge-Lenz vector and the angular momentum for a classical elliptical orbit

(d) Compute $\mathbf{A}^2$ in terms of $\mathbf{L}^2$, K, and the energy E.

(e) An ellipse has a semimajor axis a and a semiminor axis b. Let c be the distance from the center to the focus. The **eccentricity** ϵ is defined as c/a and is between 0 and 1 for a closed orbit. In plane polar coordinates, the equation of an ellipse with the origin at a focus is

$$r = \frac{l}{1 - \epsilon \cos \phi}$$

where $l = b^2/a$. r has its maximum value when $\phi = 0$, that is, when $\mathbf{r}$ is parallel to $\mathbf{A}$, so

$$\mathbf{A} \cdot \mathbf{r} = Ar \cos \phi$$

Show that the orbit is an ellipse, with eccentricity

$$\epsilon = \sqrt{1 + \frac{2EL^2}{mK^2}}$$

where E is the total energy, and for a closed orbit, $E < 0$, $\epsilon \leq 1$.

(f) What happens if $\epsilon > 1$?

Discussion: You have solved the Kepler problem by finding another constant of the motion besides the energy and angular momentum. The geometrical meaning of the conservation of **A** is that the aphelion (to which **A** points) never moves, that is, the orbit is closed. Closed orbits are not a general property of central forces. They are a special symmetry of the inverse-square force law (and a few others) that leads to new conserved observables.

This is the shortest way to solve the inverse-square force problem, as well as leading directly to the discussion of the degeneracy of the Kepler problem in quantum mechanics.

1.3. Poisson Brackets: I

Discussion: Poisson brackets are a variation on the Hamiltonian formulation of classical mechanics. They provide a formal way to get the quantum mechanical commutation relations for systems that can be described classically. You may have learned about Poisson brackets in an undergraduate classical mechanics course, but probably did not.

Any classical observable $A(p,q,t)$ is some function of the coordinates, the momenta, and perhaps the time.[8] The **Poisson bracket** between two observables is defined as

$$[A,B]_{PB} = \sum_k \left(\frac{\partial A}{\partial q_k} \frac{\partial B}{\partial p_k} - \frac{\partial B}{\partial q_k} \frac{\partial A}{\partial p_k} \right)$$

The Poisson bracket product is neither commutative nor associative, but like vector cross products and matrix commutators, satisfies the **anticommutativity** rule

$$[A,B]_{PB} = -[B,A]_{PB}$$

and the **Jacobi identity**:

$$[[A,B]_{PB}, C]_{PB} + [[C,A]_{PB}, B]_{PB} + [[B,C]_{PB}, A]_{PB} = 0$$

Cross products, matrix commutators, and Poisson brackets are all examples of **Lie algebras**.

Problem: Prove the anticommutativity rule and the Jacobi identity for Poisson brackets.

1.4. Poisson Brackets: II

(a) If q_i and p_i are a canonical collection of coordinates and momenta, show that

$$[q_i, q_j]_{PB} = [p_i, p_j]_{PB} = 0 \quad \text{and} \quad [q_i, p_j]_{PB} = \delta_{ij}$$

(b) Prove that the time dependence of any classical observable is

$$\frac{dA(q,p,t)}{dt} = [A,H]_{PB} + \frac{\partial A}{\partial t}$$

where $H(q,p,t)$ is the Hamiltonian. (This is why Poisson brackets are useful in classical mechanics.)

1.5. Classical Translations and Poisson Brackets

Suppose a classical system is subject to a small translation, that is, $\delta \mathbf{r} = \boldsymbol{\epsilon}$ and $\delta \mathbf{p} = 0$. Show that for any observable A

$$\delta A = [A, \boldsymbol{\epsilon} \cdot \mathbf{P}]_{PB}$$

---

[8] Here p and q are shorthand for $p_1, p_2 \ldots p_n$ and $q_1, q_2 \ldots q_n$.

for some function **P** of the coordinates and momenta. **P** is the **generator** of translations. What is it explicitly?

1.6. Classical Rotations and Poisson Brackets

(a) Suppose a classical system is subject to a rotation about a direction $\hat{n}$ by a small angle ϵ. Show that the change (to first order in ϵ) of an observable $A(q,p,t)$ is

$$\delta A = \epsilon [A, \hat{n} \cdot \mathbf{L}]_{PB}$$

where **L** is the total angular momentum. Assume there are no velocity-dependent potentials (i.e., that $\mathbf{p} = m\mathbf{v}$).

(b) Let A_i be the three components of some classical vector observable. From part (a) compute the Poisson bracket

$$[L_i, A_j]_{PB}$$

in terms of the components or **A** (**L** is still the total angular momentum).

(c) For a vector observable **A** as in part (b), what is $\left[L_i, \mathbf{A}^2\right]_{PB}$?

1.7. Poisson Bracket Algebra of the Runge-Lenz Vector

The components of the Runge-Lenz vector **A**, the angular momentum vector $\mathbf{L} = \mathbf{r} \times \mathbf{p}$, and the Hamiltonian for the Kepler problem (see Problem 1.2) form a closed seven-dimensional Poisson bracket algebra. From Problem 1.4

$$[L_i, H]_{PB} = [A_i, H]_{PB} = 0$$

(a) Using the result of Problem 1.6 compute

$$[L_i, L_j]_{PB} \quad \text{and} \quad [L_i, A_j]_{PB}$$

(b) Complete the calculation of the Poisson bracket algebra by computing

$$[A_i, A_j]_{PB} = ?$$

Note: This last can be done in a page and a half, but it is easy to get lost; don't spend more than two hours on it.

1.8. Noether's Theorem and Galilean Boosts

Consider a three-dimensional system with N particles described by the Lagrangian

$$L = \sum_{i=1}^{N} \frac{m_i \dot{\mathbf{r}}_i^2}{2} - V(\mathbf{r})$$

where **r** stands for $\mathbf{r}_1, \mathbf{r}_2, \ldots$. Suppose furthermore that there are no *external* forces, that is, in V the coordinates occur only in the combinations $\mathbf{r}_i - \mathbf{r}_j$. Then L itself is invariant under translations and the total momentum is conserved.

(a) What is the change of L under an infinitesimal Galilean transformation (by a velocity $\mathbf{V} = \boldsymbol{\epsilon} = \epsilon\hat{n}$) of the coordinates $\mathbf{r}_i$ and the velocities $\dot{\mathbf{r}}_i$?

(b) Find the function $\delta\Omega$ defined in equation (1.64) for this Lagrangian.

(c) What is the conserved quantity G defined in equation (1.65)? What (already well-known of course) conservation law is a consequence of Galilean invariance?

(d) Show that G is in fact the generator of the Galilean boosts, in the sense

$$[\mathbf{r}_i, G]_{PB} = \delta\mathbf{r}_i \quad \text{and} \quad [\mathbf{p}_i, G]_{PB} = \delta\mathbf{p}_i$$

REFERENCES

There are many texts on classical mechanics. Two recent editions of the most famous ones are

[1] L. D. LANDAU AND E. M. LIFSHITZ, *Mechanics*, (third edition), Butterworth-Heinemann, 1996, translated by J. B. Sykes and J. S. Bell. Paperback

[2] H. GOLDSTEIN, C. P. POOLE AND J. L. SAFKO, *Classical Mechanics* (third edition) Addison-Wesley, 2002

The reference to Noether's theorem is

[3] E. NOETHER, Nachr. v. d. Ges. d. Wiss. zu Göttingen, (1918) 234

An English translation by M. A. Tavel can be found at "Noether, Amalie Emmy," at Contributions of Women in Physics, <http://www.physics.ucla.edu/~cwp>.

CHAPTER 2

Fundamentals of Quantum Mechanics

2.1 THE SUPERPOSITION PRINCIPLE

2.1.1 The Double-Slit Experiment

The **superposition principle** lies at the foundation of quantum mechanics. One way to see why that is so comes from analyzing the double-slit interference experiment for electrons.

In a double-slit experiment light waves of wavelength λ are incident on an opaque barrier that contains two small slits separated by a distance d. Light from both slits then falls on a screen a distance L farther downstream. Provided $L \gg d$, the angle θ from each slit to a point x on the screen, measured from the beam direction, is nearly the same for both slits, and the distance from x to the two slits differs by $d \sin \theta$. The intensity of the waves falling on the screen can be determined by adding the amplitudes of the waves from these two sources—that is Huygens' principle. At a point equidistant from the slits, the waves interfere constructively. At any point on the screen where the distances to the two slits differ by an integral number of wavelengths, there is also constructive interference. Halfway between these maxima, the waves cancel. The result is the characteristic interference fringe pattern.

The condition for constructive interference is $d \sin \theta = n\lambda$. In particular, the first maximum occurs when $\sin \theta = \lambda/d$.

If you record the intensity at the screen on a photographic paper, the pattern will show up as a darkening of the paper proportional to

$$I \sim \cos^2\left(\frac{\pi d \sin \theta}{\lambda}\right) \tag{2.1}$$

The pattern is sketched in Figure 2.3 below.

But the first thing that was discovered about quantum mechanics was that you have to think of a monochromatic beam of light also as a beam of photons with energy $E = \hbar \omega$. The darkening of the photograph is a measurement of the distribution of photons as they hit the screen. If you look very carefully at the photograph you should see that it is made up of a large number of dots, denser where the function (2.1) is bigger.

Now imagine turning down the strength of the beam, so a photon comes along only every so often. It will take longer for the photographic paper to darken—that is, for any given number of dots to appear—but if you wait long enough you should get the same pattern. In principle you could watch the photons arrive one a time on the screen. The intensity is proportional to the *probability* that any particular photon will strike the screen in an interval Δx.

This probability turns out to be *all* the information contained in the intensity distribution. If you close off one slit, the photons will form a single-slit diffraction pattern offset by $d/2$ so that it is centered downstream from the open slit. That is the pattern when all the photons are required to pass through one slit only. If you close the other slit instead, you will see the same pattern shifted by d. I have sketched these patterns in Figure 2.1.

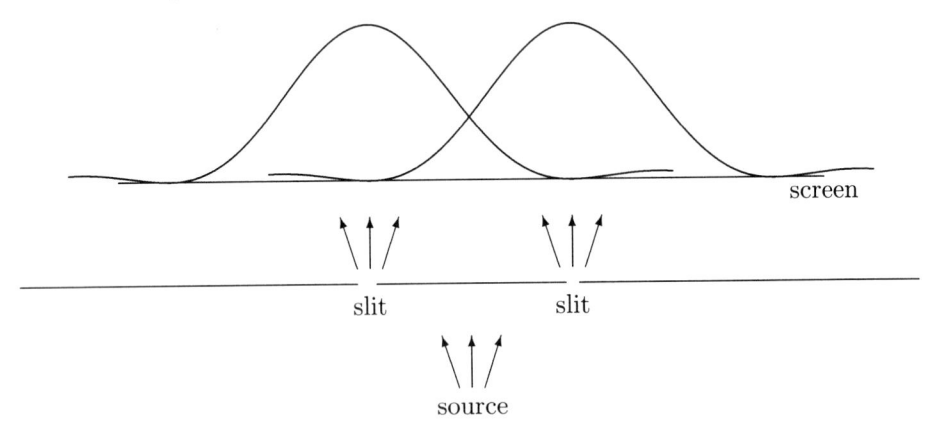

FIGURE 2.1: Photons—or electrons—pass through two slits. If one or the other slit is closed off, the intensity will form one of the two diffraction patterns in the figure, centered over the open slit.

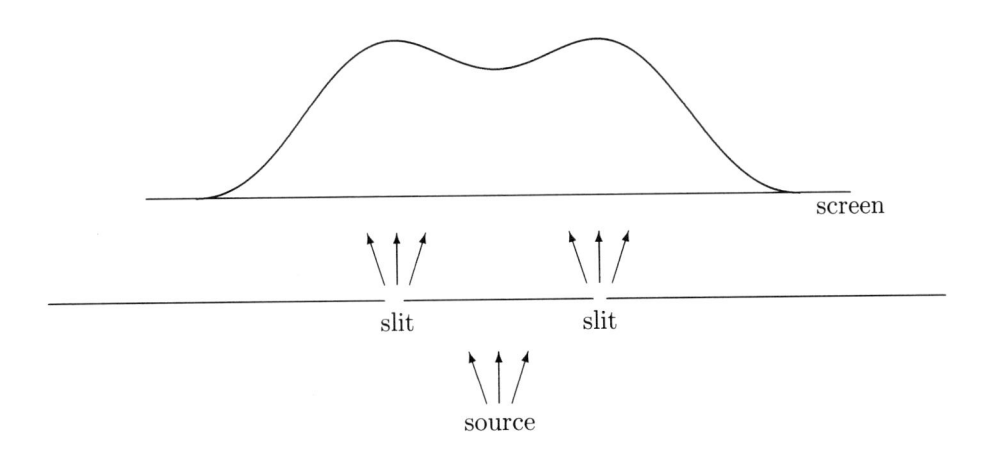

FIGURE 2.2: Photons— or electrons—pass through two open slits. If they were classical particles, the intensity pattern should be just the sum of the two patterns in Figure 2.1.

So if you run the experiment for a while with one slit open, then for the same length of time with the other slit open, you will see on the screen a pattern that is the sum of the two single-slit diffraction patterns, as in Figure 2.2. But if you leave both slits open and cut the intensity of the beam in half, the same number of photons per second—or per day—will pass through each slit, but the pattern on the screen is the double-slit pattern of Figure 2.3, not the sum of the single-slit patterns, as it would be for the particles of classical mechanics.

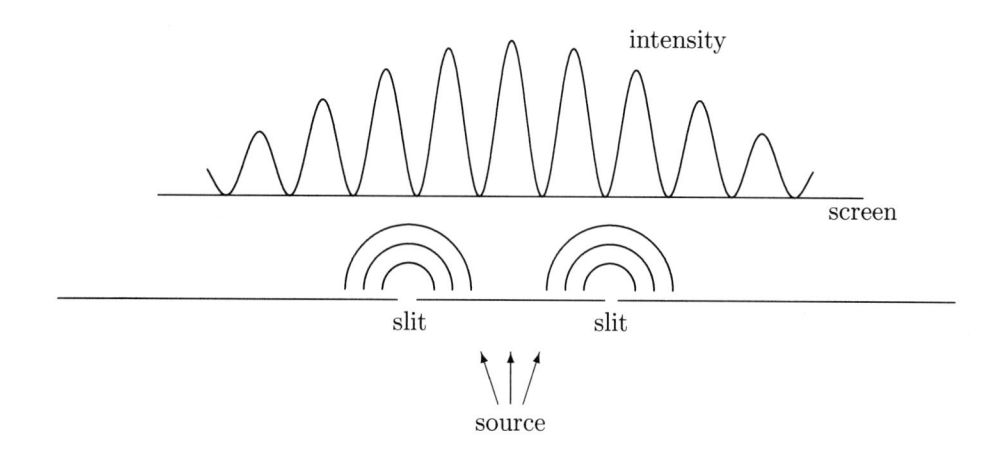

FIGURE 2.3: Instead, when photons—or electrons—pass through the two slit apparatus, with both slits open, the intensity pattern is the classic interference pattern characteristic of waves.

You might think that somehow you could measure which slit any particular photon passes through. A careful analysis shows that this is impossible without destroying the interference pattern. The information in the wave function is that its square is the probability of finding the photon at a particular point on the screen.

Particles like electrons are associated with waves. De Broglie first proposed this idea to explain atomic spectra, and Schrödinger elaborated it soon after. The analysis of a two-slit experiment must hold for electrons also. The square of the wave function is a probability density. The wave function itself is called a "probability amplitude," and it is these amplitudes that add. The amplitude for the particle to be at any point on the screen is the sum of the amplitudes for it to go through each hole. This is an example of the *superposition principle*, applied to electrons.

These quantum mechanical waves share many properties with the more familiar waves of classical mechanics. Like the electric field, the height of a water wave or the pressure of a sound wave, they are amplitudes, and the total effect from several sources is the sum of them. In all these waves, the intensity, or the energy density, is proportional to the square of the magnitude. In quantum mechanics the probability plays the role of the energy density.

> Historical note: The simplest direct proof of the existence of "electron waves" would have been to repeat for electrons the experiment that established the wave nature of light: Thomas Young's 1810 two-slit interference experiment.
>
> In the 1920s it was possible to produce electron beams in vacuum tubes by ionizing a filament with heat and then accelerating the liberated electrons between two

22 Chapter 2 Fundamentals of Quantum Mechanics

electrodes at different voltages. A typical energy for a substantial beam of monoenergetic electrons was of the order 100 eV. (The electrons next to the filament have a lower energy, but for this experiment you need a beam with a dispersion in the energy small compared to the energy itself.) The de Broglie wavelength of such a beam is

$$\frac{\hbar}{p} = \lambda = \frac{2\pi\hbar}{\sqrt{2mE}} \approx 1.2 \text{ Å} \tag{2.2}$$

This wavelength is three or four orders of magnitude shorter than visible light, so to observe the interference pattern you would have to construct slits that are many times closer than needed in the optical experiment. That is clearly impossible, since an Ångstrøm ($= 10^{-8}$ cm) is the order of the spacing between atoms.

A natural version of the double-slit experiment was performed, at first inadvertently, by Davisson and Germer in 1927 at Bell Labs [1]. They scattered a monoenergetic beam of electrons off the surface of a large nickel crystal. A crystal is a regular lattice of atoms, so it acts much like a diffraction grating. Waves reflected off a grating have sharp maxima at angles of the order $\sin\theta \approx \lambda/d$, where d is now the spacing between the grating lines. The crystal is more complicated, being a three-dimensional lattice. Waves from different planes interfere at particular angles to form sharp spots or bands of constructive interference. This is called Bragg scattering, which you might have learned about in the optics part of your introductory physics course, but the details are unimportant here. The important point is that the order of magnitude of the angle of the diffraction maxima is always given by $\sin\theta \approx \lambda/d$. The spacing between the atoms in metallic nickel had already been determined from X-ray scattering. The pattern for electrons whose de Broglie wavelength is about 1 Å was the same as the pattern for X-rays of that wavelength.

The Davisson-Germer experiment was direct confirmation of the de Broglie hypothesis. But by that time, the methods of Schrödinger's wave function had already proved wildly successful in understanding the structure of atoms and molecules.

2.1.2 The Stern-Gerlach Experiment

The Stern-Gerlach experiment illustrates the superposition principle in another way. It has no classical analog, and it cannot be described easily in the wave function language.

A particle's angular momentum in its rest frame is a vector quantity called the **spin s**. An electron at rest has an angular momentum. The earth also has some angular momentum even in its rest frame, but in the earth's case we understand it to be the ordinary orbital motion of the pieces that make up the planet. Not so for an electron or a nucleus. The spin of an elementary particle cannot be ascribed to the motion of its constituent.

An electron at rest also possesses a magnetic moment $\boldsymbol{\mu}$, another vector quantity. $\boldsymbol{\mu}$ and **s** are both pseudovector as, and parallel to each other:

$$\boldsymbol{\mu} = \mu_e \mathbf{s} \tag{2.3}$$

The energy of a magnet in a magnetic field **B** is

$$E = -\boldsymbol{\mu} \cdot \mathbf{B} \tag{2.4}$$

If **B** is constant there is no net force, but otherwise

$$\mathbf{F} = \mathbf{\nabla}(\boldsymbol{\mu} \cdot \mathbf{B}) \qquad (2.5)$$

Consider the magnetic field between two poles of a big magnet, aligned along the z-direction but not quite identical. Then **B** is along the z-axis, and the inhomogeneities in other directions can be ignored.[1]

Now inject into this field a hot "gas" of atoms or molecules and detect them with a screen. The original experiment was done in 1921 by Stern and Gerlach, using silver atoms.[2] (Electrons or other charged particles will not work, since the drift due to the gradient of the magnetic field will be swamped by the Lorentz force.) The motion of the beam particles is so slow that the angular momentum of motion, $\mathbf{r} \times \mathbf{p}$, is negligible. The initial direction of the angular momentum should be random, and classically one expects a continuous "smeared" distribution, corresponding to different projections of the spin **s**, or angular momentum, on the z-axis.

Instead Stern and Gerlach saw two widely separated spots on the detector, corresponding to the two values $+\hbar/2$ and $-\hbar/2$ for s_z. This was the first direct evidence of the quantization of angular momentum, although it was not until 1925 that Uhlenbeck and Goudsmit [3] postulated the existence of the electron spin to explain the Stern-Gerlach experiment and other complicated features of atomic spectra, like the "anomalous" Zeeman effect (see Section 6.1). It was the conceptual simplicity and clarity of the Stern-Gerlach experiment that provided the key.

The Stern-Gerlach experiment has taken on a life of its own as a "thought-experiment" in discussions of the fundamentals of quantum mechanics. It is the simplest since only two states (rather than a continuum) are relevant, and it is possible to prepare a gas of atoms with almost all in the same one of these two states.

If you pass a beam of unpolarized silver atoms or electrons through a Stern-Gerlach apparatus aligned in the z direction it is split into two beams, with s_z being $\pm\hbar/2$ respectively. Suppose only one of the two, say with $s_z = -\hbar/2$, falls on the screen, while the $s_z = +\hbar/2$ beam goes on to be analyzed by a second Stern-Gerlach apparatus, identical to the first. The second time the beam is deflected, but not split. It is made up entirely of $s_z = +\hbar/2$ atoms. We prepared a state with spin along the z-axis in the first experiment, then analyzed it in the second by measuring the spin. Of course one finds $s_z = +\hbar/2$ only. No atoms with spins antiparallel to the z-axis get through.

A more interesting variation is to arrange the second apparatus to measure the spin along the x-axis. The beam, which originally had $s_z = \hbar/2$, is split again into two beams of equal intensity, now with $s_x = \pm\hbar/2$.

Does this mean that half the atoms in the beam with $s_z = \hbar/2$ had their spins parallel to the x-axis, while half had their spins in the opposite direction? That cannot be, for the experiment could be done one atom at a time. Originally the atom had equal probability of $s_z = \pm\hbar/2$, but after passing the first apparatus it is

---

[1] If $\partial B_z/\partial z$ is not identically zero, the field cannot vanish or even be exactly constant in the other two directions because of the constraint $\mathbf{\nabla} \cdot \mathbf{B} = 0$. But the experiment can be arranged so that the x and y components are not important, and I will assume that has been done here.

[2] The original paper is reference [2].

definitely in the $+z$ direction. The only interpretation of the second experiment is that it has equal *probability* for its spin to be aligned in either direction along the x-axis.

But there is more. Complicate the experiment further, and after the $s_z = +\hbar/2$ beam passes through the apparatus aligned in the x direction, let the atoms with $s_x = -\hbar/2$ fall on a screen designed so that only atoms with spins pointing along the positive x-axis get through. This twice-analyzed beam contains only one-fourth of the original particles. What happens if you analyze this beam with a third Stern-Gerlach apparatus? If this third apparatus is also aligned along the x-direction, you just check that those atoms that got through indeed have their spins along the positive x-axis. But suppose the third experiment is again an inhomogeneous field that has a gradient in the z-direction, like the first one. The beam will be split into two parts of equal intensity, containing particles with $s_z = \hbar/2$ and $s_z = -\hbar/2$ in equal parts, *even though after the first Stern-Gerlach apparatus the $-\hbar/2$ component was blocked.*

By "measuring" s_x in between, part of the $s_z = -\frac{1}{2}\hbar$ beam has been regenerated. The Stern-Gerlach devices act like plane polarizers and analyzers for light waves. Measuring s_x destroys any information about s_z that the beam had before. Evidently an atom cannot have a definite value of s_z and s_x at the same time.

I am still using the word "state" rather loosely here. The first experiment is a measurement of s_z and the two possible states are $s_z = \pm\frac{1}{2}\hbar$. The second experiment measures s_x, and the two possible states are $s_x = \pm\frac{1}{2}\hbar$. Evidently "state" means something close to "possible results of a measurement." What kind of mathematical objects have the properties of these states? The analogy with the polarization of electromagnetic waves is the clue: The states in quantum mechanics are vectors.

2.2 THE MATHEMATICAL LANGUAGE OF QUANTUM MECHANICS

2.2.1 Vector Spaces

The mathematics of quantum mechanics is the mathematics of linear vector spaces.[3] The condition of a system is described by a "state vector" $|\psi\rangle$. All possible state vectors of a physical system form a complex vector space. You should have some knowledge of vectors at an introductory level, but you do not have to be an expert. The vector spaces of quantum mechanics are like the ordinary three-dimensional spaces of vectors from introductory physics, the ones we usually write in boldface or with arrows, except that the scalar product is complex, and the dimension is arbitrary.

Much of the material in the next few sections will be review for many of you, but I want to go over it carefully using Dirac's notation. A complex vector space contains **vectors** $|\psi\rangle$ and **scalars** (here complex numbers) α. Vectors can be added and they can be multiplied by scalars: If $|\psi_1\rangle$ and $|\psi_2\rangle$ are vectors, and α and β are

---

[3]The algebraic formulation of quantum mechanics was actually discovered before Schrödinger published his famous wave equation by Heisenberg [4], Dirac [5], Born and Jordan [6], and Born, Heisenberg and Jordan [7]. These papers and others are collected in van der Waerden [8]. Dirac's textbook [9] contains much original work. See footnote 15 in Section 2.7 for references to Schrödinger's original papers.

scalars, then $|\psi\rangle = \alpha|\psi_1\rangle + \beta|\psi_2\rangle$ is also a vector. Addition and scalar multiplication satisfy the distributive and associative laws:

$$\left[\alpha + \beta\right]|\psi\rangle = \alpha|\psi\rangle + \beta|\psi\rangle \tag{2.6a}$$

$$\alpha\left[|\psi_1\rangle + |\psi_2\rangle\right] = \alpha|\psi_1\rangle + \alpha|\psi_2\rangle \tag{2.6b}$$

$$\left[|\psi_1\rangle + |\psi_2\rangle\right] + |\psi_3\rangle = |\psi_1\rangle + \left[|\psi_2\rangle + |\psi_3\rangle\right] \tag{2.6c}$$

$$\alpha\left[\beta|\psi\rangle\right] = \left[\alpha\beta\right]|\psi\rangle \tag{2.6d}$$

A **basis** is a set of vectors $|\psi_n\rangle$ such that any vector can be written

$$|\psi\rangle = \sum_n \alpha_n |\psi_n\rangle \tag{2.7}$$

for some collection of scalars α_n. Usually the word "basis" means that this is no longer true when any of the $|\psi_n\rangle$ is omitted from the list. The α_n are the **components** of $|\psi\rangle$ in the $|\psi_n\rangle$ basis.

> Note: The properties listed so far are true also for the ordinary real three-dimensional vectors of elementary mechanics. In three-dimensional space, there are three orthonormal unit vectors $\hat{\bf n}_x$, $\hat{\bf n}_y$, and $\hat{\bf n}_z$. A general position vector as
>
> $$\mathbf{r} = x\hat{\bf n}_x + y\hat{\bf n}_y + z\hat{\bf n}_z \tag{2.8}$$

Equation (2.7) is the generalization of (2.8).

Scalar Products

In wave mechanics one frequently has to calculate an complex number from two wave functions using a rule like

$$\int_{-\infty}^{\infty} \psi_1(x)^* \psi_2(x) dx \tag{2.9}$$

The generalization of this concept is the scalar product in complex vector spaces. A straightforward mathematical notation for the scalar product would be

$$\Big(|\phi\rangle, |\psi\rangle\Big) \tag{2.10}$$

Unlike the rule

$$\mathbf{a} \cdot \mathbf{b} = \mathbf{b} \cdot \mathbf{a} \tag{2.11}$$

we use for ordinary vectors in mechanics, the scalar product (2.10) is a complex number, and its value depends on the order of the vectors:

$$\Big(|\phi\rangle, |\psi\rangle\Big) = \Big(|\psi\rangle, |\phi\rangle\Big)^* \tag{2.12}$$

26   Chapter 2   Fundamentals of Quantum Mechanics

The rule (2.12) is designed to ensure that $\left(|\psi\rangle, |\psi\rangle\right)$ be real. This scalar product is linear in the second vector, but antilinear in the first vector:

$$\left(|\phi\rangle, \alpha|\psi_1\rangle + \beta|\psi_2\rangle\right) = \alpha\left(|\phi\rangle, |\psi_1\rangle\right) + \beta\left(|\phi\rangle, |\psi_2\rangle\right) \qquad (2.13)$$

But

$$\left(\alpha|\psi_1\rangle + \beta|\psi_2\rangle, |\phi\rangle\right) = \alpha^*\left(|\psi_1\rangle, |\phi\rangle\right) + \beta^*\left(|\psi_2\rangle, |\phi\rangle\right) \qquad (2.14)$$

Instead of these cumbersome braces and commas I will follow the standard physics tradition and use a notation introduced by Dirac. We write

$$\langle\phi|\,\psi\rangle = \langle\psi|\,\phi\rangle^* \qquad (2.15)$$

and if $|\psi\rangle = \alpha|\psi_1\rangle + \beta|\psi_2\rangle$, then

$$\langle\phi|\,\psi\rangle = \alpha\langle\phi|\,\psi_1\rangle + \beta\langle\phi|\,\psi_2\rangle \qquad (2.16)$$

The inner product $\langle\psi|\,\psi\rangle$ is positive definite:

$$\langle\psi|\,\psi\rangle \geq 0 \qquad (2.17)$$

where $\langle\psi|\,\psi\rangle = 0$ if, and only if, $|\psi\rangle = 0$. Here 0 stands for the unique vector such that for any $|\psi\rangle$, $|\psi\rangle + 0 = |\psi\rangle$.

The restriction (2.17) is an independent postulate, and it lets one define a length, or norm, in the vector space:

$$\bigl\||\psi\rangle\bigr\| = +\sqrt{\langle\psi|\,\psi\rangle} \qquad (2.18)$$

The definition (2.18) in turn lets one define limits, infinite series, sequences of vectors, and convergence in the usual way.

Two vectors $|\phi\rangle$ and $|\psi\rangle$ are **orthogonal** if $\langle\phi|\,\psi\rangle = 0$. It is often convenient to choose an orthogonal basis: $\langle\psi_m|\,\psi_n\rangle = 0$ unless $m = n$. If the basis vectors are also normalized, then the vectors $|\psi_m\rangle$ form an **orthonormal** basis:

$$\langle\psi_m|\,\psi_n\rangle = \delta_{mn} \qquad (2.19)$$

It is particularly easy to expand an arbitrary vector in such a basis. Let

$$|\psi\rangle = \sum_n \alpha_n |\psi_n\rangle \qquad (2.20)$$

Then $\alpha_n = \langle\psi_n|\,\psi\rangle$, and

$$\langle\psi|\,\psi\rangle = \sum_n \alpha_n \langle\psi|\,\psi_n\rangle = \sum_n |\alpha_n|^2 \qquad (2.21)$$

For finite dimensional spaces, the sum (2.21) always exists, and so does the expansion (2.20). If the number of dimensions is infinite, this is not automatic. A vector space is called **complete** if the expansion (2.20) always converges to a vector in the space provided $\sum_n |\alpha_n|^2 < \infty$. A complete vector space that has a complex scalar product and a positive definite norm is a **Hilbert space.**[4] Hilbert spaces are exactly the mathematical structures needed to make sense of the superposition principle.

---

[4]Not all vector spaces are Hilbert spaces. The four-vectors one uses in special relativity make up a vector space, but the usual Minkowski product allows for nonzero vectors x such that $x \cdot x = 0$.

2.2.2 The Probability Interpretation

What does all this have to do with physics? The physical states will be represented by a vector in a Hilbert space. Each possible state $|\psi\rangle$ has certain physical properties. If a system is in a state $|\phi\rangle$, the probability that it has the properties of the state $|\psi\rangle$ is $|\langle\psi|\phi\rangle|^2$. This is the mathematical formulation of the superposition principle. The scalar product $\langle\phi|\psi\rangle$ is a **probability amplitude**.

For any observable A (x, p, spin, etc.) there is a list of measurable values for that observable. The list may be infinite or even continuous, but for the moment assume it is discrete, like bound-state energies in simple examples. Call these measurable values λ_i. There is some state $|\lambda_i\rangle$ in which the result of a measurement of A will certainly be λ_i. If the system is in the state $|\lambda_i\rangle$, then the probability for measuring λ_i is unity. The physical requirement is $|\langle\lambda_i|\lambda_j\rangle|^2 = \delta_{ij}$, which implies $\langle\lambda_i|\lambda_j\rangle = \pm\delta_{ij}$. The minus sign is impossible since $\langle\psi|\psi\rangle \geq 0$.

The probability that one measures any other value is zero. Thus

$$\langle\lambda_i|\lambda_j\rangle = \delta_{ij} \tag{2.22}$$

If the system is in any state $|\psi\rangle$, the probability of measuring λ_i is $|\langle\psi|\lambda_i\rangle|^2$. Since there is always some result of a measurement, the sum of these probabilities must be 1:

$$\sum_i |\langle\psi|\lambda_i\rangle|^2 = \sum_i \langle\psi|\lambda_i\rangle\langle\lambda_i|\psi\rangle = \langle\psi|\psi\rangle = 1 \tag{2.23}$$

The collection λ_i contains *all* possible measurable values for A, so the states $|\lambda_i\rangle$ must be a complete orthonormal basis. The probability interpretation of quantum mechanics, together with the superposition principle, requires a Hilbert space, and there is no other way to implement it.

2.2.3 Linear Operators

An **operator** means a function on the vectors in a vector space. A is **linear** if

$$A\left[\alpha|\psi_1\rangle + \beta|\psi_2\rangle\right] = \alpha A|\psi_1\rangle + \beta A|\psi_2\rangle \tag{2.24}$$

Since there will be but one occasion to mention operators that are not linear, usually I will just write "operator" for "linear operator."

For linear operators, it is enough to know what A does to an orthonormal basis $|\psi_n\rangle$:

$$A|\psi_n\rangle = \sum_m |\psi_m\rangle A_{mn} \tag{2.25}$$

where

$$\boxed{A_{mn} = \langle\psi_m|A|\psi_n\rangle} \tag{2.26}$$

These numbers A_{mn} are the **matrix elements** of A in the basis $|\psi_m\rangle$. Then, for any vector $|\psi\rangle = \sum_n \alpha_n|\psi_n\rangle$,

$$A|\psi\rangle = \sum_n A\alpha_n|\psi_n\rangle = \sum_{mn} |\psi_m\rangle A_{mn}\alpha_n \tag{2.27}$$

$\sum_n A_{mn}\alpha_n$ are the components of the vector $A|\psi\rangle$ in the basis $|\psi_m\rangle$. In this way, to every linear operator there corresponds a (possibly infinite) matrix whose elements are A_{mn}.

The sum of two operators is

$$[A+B]|\psi\rangle = A|\psi\rangle + B|\psi\rangle \tag{2.28}$$

The product of two linear operators—this is a new kind of product—is defined in the obvious way: $C = AB$ means that for every vector $|\psi\rangle$, $C|\psi\rangle = A[B|\psi\rangle]$. The matrix elements of AB are

$$(AB)_{kn} = \langle\psi_k|AB|\psi_n\rangle = \sum_m \langle\psi_k|A|\psi_m\rangle B_{mn} = \sum_m A_{km}B_{mn} \tag{2.29}$$

An operator is (isomorphic to) a matrix, and operator composition is ordinary matrix multiplication. But be careful. *The matrix associated with an operator depends on the basis.*

> Mathematical note: Linear operators form a vector space themselves. They can be added
> $$(A+B)|\psi\rangle = A|\psi\rangle + B|\psi\rangle \tag{2.30}$$
> and multiplied by scalars
> $$(\alpha A)|\psi\rangle = \alpha A|\psi\rangle \tag{2.31}$$
> Vector spaces in which there is also a multiplication rule between the vectors are called **algebras**. In particular, the linear operators on a vector space form an algebra under the composition of operators. This is an ordinary associative algebra, but it is not necessarily commutative; in general $AB \neq BA$.

The notation $A = |\phi_1\rangle\langle\phi_2|$ is just a way to define an operator A with the property[5]

$$A|\psi\rangle = |\phi_1\rangle\langle\phi_2|\psi\rangle \tag{2.32}$$

So $A = \sum_i |\phi_i\rangle\langle\psi_i|$ means

$$A|\psi\rangle = \sum_i |\phi_i\rangle\langle\psi_i|\psi\rangle \tag{2.33}$$

The Identity Operator

There is always a unique operator I, the identity:

$$I|\psi\rangle = |\psi\rangle \tag{2.34}$$

for all $|\psi\rangle$. In an orthonormal basis $|\psi_n\rangle$

$$\sum_n |\psi_n\rangle\langle\psi_n|\psi\rangle = |\psi\rangle \tag{2.35}$$

so

$$\boxed{\sum_n |\psi_n\rangle\langle\psi_n| = I} \tag{2.36}$$

---
[5] These forms are sometimes called dyads and dyadics.

Section 2.2 The Mathematical Language of Quantum Mechanics 29

is a way to write the identity operator. Be sure you understand the meaning of (2.36). We will use it over and over again. An orthonormal basis with the propertry (2.36) is complete.

Inverse of an Operator

An operator A may have an **inverse** A^{-1}:

$$AA^{-1} = A^{-1}A = I \qquad (2.37)$$

Products of inverses of operators satisfy a simple theorem:

$$\boxed{(AB)^{-1} = B^{-1}A^{-1}} \qquad (2.38)$$

If there is no solution A^{-1} to equation (2.37), A is **singular**.

In a finite dimensional space, if A is singular, then the determinant of the matrix associated with A according to (2.25) is zero. But the idea of the inverse of an operator exists in spaces of infinite dimensions, though the concept of determinant is not always defined there.

Eigenvectors and Eigenvalues

If $A|\psi\rangle = \lambda|\psi\rangle$, then $|\psi\rangle$ is an **eigenvector** of A, and λ is its **eigenvalue**. If the vector space is finite dimensional, the eigenvalues are solutions to

$$\det(A - \lambda) = 0 \qquad (2.39)$$

but again, the concepts of eigenvectors and eigenvalues are not restricted to finite dimensional vector spaces.

Hermitean Operators

If for all vectors $|\phi\rangle$ and $|\psi\rangle$, $\langle\phi|B|\psi\rangle = \langle\psi|A|\phi\rangle^*$, the operator B is the **Hermitean conjugate** or **adjoint** of A: $B = A^\dagger$. It follows that $A = B^\dagger$ also, and that

$$\langle\phi|A^\dagger|\psi\rangle = \langle\psi|A|\phi\rangle^* \qquad (2.40)$$

In an orthonormal basis,

$$A_{mn} = \langle\psi_m|A|\psi_n\rangle = \langle\psi_n|A^\dagger|\psi_m\rangle^* = A^\dagger_{nm}{}^* \qquad (2.41)$$

The matrix for $A^\dagger$ is the transpose of the complex conjugate of the matrix for A. If $A = A^\dagger$, the operator A is **Hermitean** or **self-adjoint**.[6]

There is another simple theorem:

$$\boxed{(AB)^\dagger = B^\dagger A^\dagger} \qquad (2.42)$$

and three very important ones, which I do not prove here:

1. The eigenvalues of a Hermitean operator are real.

---

[6]Sometimes an operator A is not defined on all the vectors in a Hilbert space, but only on some subset. Then if A satisfies the condition (2.40) for the states on which it is defined, it is technically Hermitian but not self-adjoint. I will not need to distinguish these two definitions.

30 Chapter 2 Fundamentals of Quantum Mechanics

2. The eigenvectors of a Hermitean operator with different eigenvalues are orthogonal.

3. It follows (but not trivially) that the eigenvectors of a Hermitean operator can be chosen to be a complete, orthonormal basis.[7]

We will use these three theorems frequently. They are worth memorizing.

The matrix for a Hermitean operator in a basis of its own eigenvectors has no off-diagonal elements. The eigenvector basis is said to **diagonalize** the operator.

Unitary Operators

Finally, for some linear operator U, let

$$|\phi_1\rangle = U|\psi_1\rangle \quad \text{and} \quad |\phi_2\rangle = U|\psi_2\rangle \tag{2.43}$$

Then if $\langle \phi_1 | \phi_2 \rangle = \langle \psi_1 | \psi_2 \rangle$ for all $|\psi_i\rangle$, U is called a **unitary** operator. Unitary operators are the generalization to complex spaces of rotations; they leave scalar products—lengths and angles—unchanged. If U is unitary,

$$\langle \phi_1 | \phi_2 \rangle = \langle \psi_1 | U^\dagger U | \psi_2 \rangle = \langle \psi_1 | \psi_2 \rangle \tag{2.44}$$

so

$$\boxed{UU^\dagger = U^\dagger U = I} \tag{2.45}$$

The Hermitean conjugate of a unitary operator is its inverse. Observables correspond to Hermitean operators, and symmetry transformations correspond to unitary operators.

2.2.4 Observables

Any observable is automatically associated with a Hermitean linear operator: Let A be an observable, and λ_i the possible values that can be the result of measuring A. Quantum mechanics will restrict what these values might be. The different λ_i are mutually exclusive, so $\langle \lambda_i | \lambda_j \rangle = \delta_{ij}$, and the eigenstates are an orthonormal basis. Therefore one can *construct* the operator

$$A = \sum_n \lambda_n |\lambda_n\rangle\langle\lambda_n| \tag{2.46}$$

with the property that

$$A|\lambda_n\rangle = \lambda_n |\lambda_n\rangle \tag{2.47}$$

By convention, observables have real values. If you want to use a complex-valued observable, treat the real and imaginary parts separately. The λ_n are real, and therefore A is a Hermitean operator with eigenvectors $|\lambda_n\rangle$ and eigenvalues λ_n. If A were the only observable, the construction would have little content. But even the simplest classical systems have more than one observable: Besides x, there is always p, the energy, and so forth. Real systems have many observables, and here is where the subject becomes interesting.

---

[7] When the eigenvalues are continuous, this property is true only in a generalized sense discussed in Section 2.3.

Section 2.2 The Mathematical Language of Quantum Mechanics   31

If A_1 and A_2 are two observables that can be measured simultaneously, like x and y, then a basis is a set of *simultaneous* eigenvectors of A_1 and A_2: In an obvious notation

$$A_1|\lambda_1,\lambda_2\rangle = \lambda_1|\lambda_1,\lambda_2\rangle \qquad A_2|\lambda_1,\lambda_2\rangle = \lambda_2|\lambda_1,\lambda_2\rangle \tag{2.48}$$

In this basis both A_1 and A_2 are diagonal, and the two Hermitean operators commute:

$$[A_1, A_2] = A_1 A_2 - A_2 A_1 = 0 \tag{2.49}$$

Observables that can be measured simultaneously commute. The notation is easily generalized to any number of simultaneously measurable observables. If there are m simultaneously measurable observables A_j, the eigenvectors can be labeled by m eigenvalues:

$$A_j|\lambda_{1n_1},\lambda_{2n_2}\ldots\lambda_{mn_m}\rangle = \lambda_{j n_j}|\lambda_{1n_1},\lambda_{2n_2}\ldots\lambda_{mn_m}\rangle \tag{2.50}$$

The basis vectors are orthonormal and complete.

A collection of observables are "independent" if none of them is a linear combination of the others. Usually there is some maximum number of independent observables that all commute with each other. A set of this number of operators that all commute is a "maximal set" of commuting observables. You can find a different maximal set, but there are always the same number of them.

There will always be some observables that do not commute with each other. For example x and p, or L_x and L_y cannot be measured simultaneously.

Expectation value

The **expectation value** of an observable is the average value of measurements of A when the system is in the state $|\psi\rangle$. By definition, it is the sum of the probabilities that A has each value λ_i multiplied by λ_i. Thus

$$\langle A \rangle = \sum_n \lambda_n |\langle \lambda_n | \psi \rangle|^2 = \sum_{m,n} \langle \psi | \lambda_m \rangle \langle \lambda_m | A | \lambda_n \rangle \langle \lambda_n | \psi \rangle = \langle \psi | A | \psi \rangle \tag{2.51}$$

2.2.5 Examples

Finite Dimensional Spaces

A list of n complex numbers forms a vector in an n-dimensional Hilbert space. It is customary to write the list in a column:

$$|\alpha\rangle = \begin{pmatrix} \alpha_1 \\ \alpha_2 \\ \alpha_3 \\ \ldots \\ \alpha_n \end{pmatrix} \qquad |\beta\rangle = \begin{pmatrix} \beta_1 \\ \beta_2 \\ \beta_3 \\ \ldots \\ \beta_n \end{pmatrix} \tag{2.52}$$

The scalar product is

$$\langle \beta | \alpha \rangle = \sum_{i=1}^{n} \beta_i^* \alpha_i \tag{2.53}$$

Then
$$\langle\alpha|\alpha\rangle = \sum_{i=1}^{n} |\alpha_i|^2 \geq 0 \tag{2.54}$$

The observables are Hermitean $n \times n$ matrices. The action of an operator A on a vector α is
$$A|\alpha\rangle = \begin{pmatrix} \sum_i A_{1i}\alpha_i \\ \sum_i A_{2i}\alpha_i \\ \sum_i A_{3i}\alpha_i \\ \vdots \\ \sum_i A_{ni}\alpha_i \end{pmatrix} \tag{2.55}$$

or with $|\beta\rangle = A|\alpha\rangle$ we have
$$\beta_j = \sum_i A_{ji}\alpha_i \tag{2.56}$$

An obvious basis consists of vectors with a 1 in one row and 0 elsewhere:
$$|\psi_i\rangle = \begin{pmatrix} 0 \\ 0 \\ \vdots \\ 0 \\ 1 \\ 0 \\ \vdots \end{pmatrix} \tag{2.57}$$

The 1 comes in the i-th row. These satisfy
$$\langle\psi_i|\psi_j\rangle = \delta_{ij} \tag{2.58}$$

or equivalently
$$\sum_{i=1}^{n} |\psi_i\rangle\langle\psi_i| = I \tag{2.59}$$

The simplest nontrivial quantum mechanical system has two states, and therefore can describe the Stern-Gerlach experiment discussed in Section 2.1.2. The two states you split the beam into, using a filter aligned in the z direction, can be chosen as the two independent normalized vectors

$$|z+\rangle = \begin{pmatrix} 1 \\ 0 \end{pmatrix} \qquad |z-\rangle = \begin{pmatrix} 0 \\ 1 \end{pmatrix} \tag{2.60}$$

There are only four linearly independent operators:

$$I = \begin{pmatrix} 1 & 0 \\ 0 & 1 \end{pmatrix} \qquad \sigma_x = \begin{pmatrix} 0 & 1 \\ 1 & 0 \end{pmatrix} \qquad \sigma_y = \begin{pmatrix} 0 & -i \\ i & 0 \end{pmatrix} \qquad \sigma_z = \begin{pmatrix} 1 & 0 \\ 0 & -1 \end{pmatrix} \tag{2.61}$$

The vectors $|z\pm\rangle$ are the eigenstates of σ_z, with eigenvalues ± 1. So we can identify them as states with definite s_z, where $s_z = \hbar/2\sigma_z$. Then

$$|x\pm\rangle = \frac{|z+\rangle \pm |z-\rangle}{\sqrt{2}} \tag{2.62}$$

are eigenstates of $s_x = \hbar/2\sigma_x$ with the same eigenvalues. A Stern-Gerlach filter in the x-direction measures s_x. If the beam entering the filter definitely has the value $+\hbar/2$ for s_z, the probability that it has the value $+\hbar/2$ for s_x is, according to the rules stated above,

$$P = |\langle x_+| z_+\rangle|^2 = \frac{1}{2} \tag{2.63}$$

A Stern-Gerlach filter chooses a basis in the vector space!

A Function Space

Most interesting physical systems are infinite dimensional Hilbert spaces, spaces where the observables can take on any of an infinite number of possible values. Common examples of these are function spaces, vector spaces where the vectors are functions, in quantum mechanics often the wave functions themselves.

Here is a simple example: a space that consists of all continuous complex functions of one real variable with period 2π, together with limits of sequences of such continuous functions. The functions have to be "square-integrable," that is,

$$\int_{-\pi}^{\pi} |f(x)|^2 dx < \infty \tag{2.64}$$

Addition and scalar multiplication in this vector space is defined so that the linear combination

$$|\phi\rangle = \alpha|\psi_1\rangle + \beta|\psi_2\rangle \tag{2.65}$$

is the function

$$\phi(x) = \alpha\psi_1(x) + \beta\psi_2(x) \tag{2.66}$$

Define a scalar product on this space:

$$\langle \phi| \psi\rangle = \int_{-\pi}^{\pi} \phi(x)^* \psi(x)\, dx \tag{2.67}$$

The rules (2.67) and (2.66) are enough to ensure that these functions satisfy the axioms of a Hilbert space.

Define an operator D on this space so that $D|\psi\rangle = |\phi\rangle$ means

$$\phi(x) = -i\frac{d}{dx}\psi(x) \tag{2.68}$$

It is easy to check (see Problem 2.4) that D satisfies the requirements in Section 2.2.3 for a Hermitean operator. Its normalized eigenvectors are

$$|e_n\rangle = \frac{1}{\sqrt{2\pi}} e^{inx} \tag{2.69}$$

34 Chapter 2 Fundamentals of Quantum Mechanics

and their corresponding eigenvalues are n ($n = 0, \pm 1, \pm 2, \ldots$), since

$$-i\frac{d}{dx}\frac{1}{\sqrt{2\pi}}e^{inx} = n\frac{1}{\sqrt{2\pi}}e^{inx} \qquad (2.70)$$

The eigenvalues are real and the eigenvectors are orthogonal, as required by the theorems in Section 2.2.3. The states $|e_n\rangle$ form an orthonormal basis:

$$\langle e_m | e_n \rangle = \delta_{mn} \qquad (2.71)$$

Any vector $|\psi\rangle$ can be expanded:

$$|\psi\rangle = \sum_n |e_n\rangle\langle e_n | \psi\rangle \qquad (2.72)$$

or, since these are functions,

$$\psi(x) = \frac{1}{2\pi}\sum_n e^{inx} \int_{-\pi}^{\pi} e^{-int}\psi(t)\,dt \qquad (2.73)$$

Equation (2.73) is a version of Fourier's theorem.

It is tempting to write equation (2.73) as

$$\psi(x) = \int_{-\pi}^{\pi} \left[\frac{1}{2\pi}\sum_n e^{inx}e^{-int}\right]\psi(t)dt \qquad (2.74)$$

What is the function inside the brackets? In a finite dimensional vector space, the components α_i of a vector $|\alpha\rangle$ satisfy

$$\alpha_i = \sum_j \delta_{ij}\alpha_j \qquad (2.75)$$

and the Kronecker delta symbol δ_{ij} is the matrix for the identity operator. It would be useful to think of

$$\frac{1}{2\pi}\sum_n e^{inx}e^{-int} \qquad (2.76)$$

as an identity operator in function space, with a continuous index, but the expression is not a proper function at all. Since $\psi(x)$ is any function, and since for $x \neq t$ $\psi(x)$ cannot depend on $\psi(t)$, equation (2.73) can make sense only if

$$\sum_n e^{inx}e^{-int} = 0 \qquad (x \neq t) \qquad (2.77)$$

but when $x = t$ the infinite sum diverges. Dirac introduced the **Dirac delta function** to unify the notation for finite and infinite dimensional vector spaces.

2.3 CONTINUOUS EIGENVALUES

2.3.1 The Dirac Delta Function

So far we have been pretending that the eigenvalues λ_i are discrete, but some of the most common observables like x and p (but not L) usually have continuous values. A careful theory of such observables exists and is not even very hard. But it is easier to use Dirac's brilliant informal invention, the delta function.

It is informal because it looks like a function and you can use it like a function, but it isn't really a function. It is brilliant because it allows a unification of notation. The Dirac delta function is the continuous generalization of the Kronecker delta δ_{ij}.

You use the delta function as if it were an ordinary function, and you'll never get into trouble until you see meaningless quantities like $\delta^2(x)$ or $1/\delta(x)$. Its properties are

$$\delta(x) = 0 \qquad (x \neq 0) \qquad (2.78a)$$

$$\int_a^b \delta(x)\,dx = 1 \qquad \text{(if 0 is between } a \text{ and } b\text{)} \qquad (2.78b)$$

$$\int_a^b \delta(x)\,dx = 0 \qquad \text{(otherwise)} \qquad (2.78c)$$

Then

$$\int_a^b f(x)\delta(x)\,dx = f(0) \qquad \text{(if 0 is between } a \text{ and } b\text{)} \qquad (2.79a)$$

$$\int_a^b f(x)\delta(x)\,dx = 0 \qquad \text{(otherwise)} \qquad (2.79b)$$

More generally

$$\boxed{\int_a^b f(x)\delta(x-c)\,dx = f(c) \qquad \text{(if } a < c < b\text{)}} \qquad (2.80)$$

The delta function is even: $\delta(-x) = \delta(x)$, and

$$\delta(ax) = \frac{1}{|a|}\delta(x) \qquad (2.81)$$

The step function $\Theta(x)$ defined by[8]

$$\Theta(x) = 1 \quad (x > 0) \qquad (2.82a)$$
$$\Theta(x) = 0 \quad (x < 0) \qquad (2.82b)$$

is closely related to $\delta(x)$.

$$\frac{d}{dx}\Theta(x) = \delta(x) \qquad (2.83)$$

It is not meaningful to talk about the value of either of these "functions" at every particular x. They make sense only inside an integral. Mathematicians call them **distributions**.

---
[8] Some books use $\theta(x)$.

2.3.2 Continuous Observables

One Dimension

With this handy δ function the formulation of quantum mechanics can be extended to continuous observables. The position of a particle in one dimension can have any real value x. There is a continuum of eigenstates $|\psi_{x'}\rangle$:[9]

$$x|\psi_{x'}\rangle = x'|\psi_{x'}\rangle \qquad (2.84)$$

Normalize them to $\langle\psi_x|\psi_{x'}\rangle = \delta(x - x')$. The interpretation must be that if $|\psi\rangle$ is a normalized state with $\langle\psi|\psi\rangle = 1$, then

$$\int_{x_o}^{x_o+\Delta x} |\langle\psi|x\rangle|^2 \, dx \qquad (2.85)$$

is the probability of measuring x between x_o and $x_o + \Delta x$. $|\langle\psi|x\rangle|^2$ is a probability density, not a probability. Nevertheless, in the interest of a unified nomenclature we often just say that $|\langle\psi|x\rangle|^2$ is the probability for a system in the state $|\psi\rangle$ to be "at" x.

A continuous version of the representation (2.36) of the identity operator is

$$I = \int_{-\infty}^{\infty} |\psi_x\rangle\langle\psi_x| \, dx \qquad (2.86)$$

since then

$$I|\psi_{x_o}\rangle = \int_{-\infty}^{\infty} |\psi_x\rangle\langle\psi_x|\psi_{x_o}\rangle \, dx = \int_{-\infty}^{\infty} \delta(x - x_o)|\psi_x\rangle \, dx = |\psi_{x_o}\rangle \qquad (2.87)$$

and so forth.

Three Dimensions

The states of a three-dimensional particle can be labeled $|\psi_{x,y,z}\rangle$, or just $|\psi_\mathbf{r}\rangle$ or even $|\mathbf{r}\rangle$ for short. These are a complete orthonormal set:

$$\langle\psi_{\mathbf{r}'}|\psi_\mathbf{r}\rangle = \delta_3(\mathbf{r}' - \mathbf{r}) \qquad (2.88)$$

with the common notation[10]

$$\delta_3(\mathbf{r}) = \delta(x)\delta(y)\delta(z) \qquad (2.89)$$

The generalization of equation (2.85) to three dimensions is that for a three-dimensional particle in a normalized state $|\psi\rangle$, the probability that it is in a volume V is

$$\int_V |\langle\psi_\mathbf{r}|\psi\rangle|^2 \, d^3r \qquad (2.90)$$

and the integrand

$$\rho(\mathbf{r}) = |\langle\mathbf{r}|\psi\rangle|^2 \qquad (2.91)$$

---

[9] A common notation is to write just $|x\rangle$ instead of $|\psi_x\rangle$, but in this section (only) I want to distinguish the operators and the eigenvalues carefully.

[10] This function is sometimes written $\delta^3(x)$ or $\delta^{(3)}(x)$.

is the probability density. The identity operator is

$$I = \int d^3r |\psi_\mathbf{r}\rangle\langle\psi_\mathbf{r}| \qquad (2.92)$$

The generalization of these ideas to n coordinates is straightforward.

2.3.3 Fourier's Theorem and Representations of $\delta(x)$

Fourier Series Representation of $\delta(x)$

There are many representations of $\delta(x)$. One of them follows from the example at the end of Section 2.2.5. Any function defined in $-\pi < x \le \pi$ satisfies Fourier's theorem in the form

$$\psi(x) = \int_{-\pi}^{\pi} \left[\frac{1}{2\pi} \sum_n e^{inx} e^{-int} \right] \psi(t) dt \qquad (2.74)$$

Since $\psi(x)$ is any function, it cannot depend on $\psi(t)$ for $x \ne t$. Equation (2.74) can make sense only if

$$\sum_n e^{inx} e^{-int} = 0 \qquad (x \ne t) \qquad (2.93)$$

It follows that

$$\frac{1}{2\pi} \sum_n e^{inx} e^{-int} = \delta(x - t) \qquad (2.94)$$

Equation (2.94) is a prototype of a class of ways to write $\delta(x)$.

> Mathematical note: Equation (2.94) looks like sloppy mathematics, but its meaning is straightforward: Let $\psi(t)$ be any function well enough behaved that the integral in equation (2.73) exists for any n. Then equation (2.94) means that if you multiply both sides by $\psi(t)$ and integrate, using (2.79) on the right, and on the left interchanging the order of the sum and the integral, then the result is true.

The Fourier Transform

The Fourier transform theorem is fundamental to our understanding of the Dirac delta function and the relation between coordinates and momenta. It says that

$$\psi(x) = \frac{1}{\sqrt{2\pi}} \int_{-\infty}^{\infty} c(\omega) e^{i\omega x} d\omega \quad \text{where} \quad c(\omega) = \frac{1}{\sqrt{2\pi}} \int_{-\infty}^{\infty} \psi(t) e^{-i\omega t} dt \qquad (2.95)$$

The Fourier transform is a limit of the Fourier series:

> ***Proof of the Fourier transform theorem.*** The functions I called $|e_n\rangle$ in Section 2.2.5 can be defined on an interval of any length L instead of 2π:
>
> $$|e_n\rangle = \frac{1}{\sqrt{L}} e^{2\pi inx/L} \qquad (2.96)$$

38 Chapter 2 Fundamentals of Quantum Mechanics

These are orthonormal with the scalar product

$$\langle \phi | \psi \rangle = \int_{-L/2}^{L/2} \phi(x)^* \psi(x)\, dx \qquad (2.97)$$

instead of (2.67). Now the Fourier expansion takes the form

$$\psi(x) = \sum_n c_n e_n(x) \qquad (2.98)$$

where

$$c_n = \frac{1}{\sqrt{L}} \int_{-L/2}^{L/2} e^{-2\pi i n t/L} \psi(t)\, dt \qquad (2.99)$$

Set

$$\omega_n = \frac{2\pi n}{L} \quad \text{and} \quad \Delta\omega_n = \omega_n - \omega_{n-1} = \frac{2\pi}{L} \qquad (2.100)$$

Then

$$\psi(x) = \frac{1}{2\pi} \sum_n \int_{-L/2}^{L/2} e^{i\omega_n(x-t)} \frac{2\pi}{L} \psi(t)\, dt = \frac{1}{2\pi} \sum_n \Delta\omega_n \int_{-L/2}^{L/2} e^{i\omega_n(x-t)} \psi(t)\, dt \qquad (2.101)$$

In the $L \to \infty$ limit the sum becomes a Riemann integral, with the result (2.95). ∎

From equation (2.95) it follows that for any $f(x)$,

$$f(x) = \frac{1}{2\pi} \int_{-\infty}^{\infty} dz \int_{-\infty}^{\infty} dy\, f(z) e^{i(x-z)y} = \int_{-\infty}^{\infty} \left[\frac{1}{2\pi} \int_{-\infty}^{\infty} e^{i(x-z)y}\, dy \right] f(z)\, dz \qquad (2.102)$$

Again, $f(x)$ is arbitrary, so it cannot depend on $f(z)$ for $z \neq x$. Therefore

$$\int_{-\infty}^{\infty} e^{i(x-z)y}\, dy = 0 \quad (x \neq z) \qquad (2.103)$$

This is one way to derive the very useful representation

$$\boxed{\frac{1}{2\pi} \int_{-\infty}^{\infty} e^{i(x-z)y}\, dy = \delta(x-z)} \qquad (2.104)$$

Other ways to write the Dirac delta function can be found in Section A.1.1 in the appendix.

2.4 CANONICAL COMMUTATORS AND THE SCHRÖDINGER EQUATION

2.4.1 The Correspondence Principle

Identifying the states of physical systems with vectors in a Hilbert space as in Section 2.2 is the correct mathematical statement of the superposition principle. But it does not tell us the structure of any physical system, or how it changes with

Section 2.4 Canonical Commutators and the Schrödinger Equation

time. If we know the classical physics of a system, and we usually do, we know what the observables are, and can insist that classical physics comes out when we average over a macroscopic number of measurements. This is a form of the **correspondence principle**, which, more precisely, I take to mean:[11]

(a) The list of observables in classical mechanics should be included in the list of quantum mechanical observables.

(b) The symmetries of the classical system—e.g. translations and rotations—should hold in the classical limit.

(c) The expectation values of the time dependences of the observables should be the same as in classical mechanics (**Ehrenfest's principle**).

These three rules are not a rigorous program for an axiomatic approach, and I do not want to try to write one down here. But they have proved faithful guides for guessing the correct quantum mechanical rules for the parts of nature we understand. And nature has provided one class of physical systems where the rules are clear and the experiments are precise: the structure of atoms. The forces among electrons and nuclei are electromagnetic, completely understood classically. Experiment confirms that the quantum theory of atoms and molecules, carried out using the rules of this section, is correct.

Point (a) above usually means that the observables are the coordinates r_i of the particles, their conjugate momenta p_i, and functions of them. If we call the operators for these observables R_i and P_i, then from the discussion in Section 2.3.2, generalized to an arbitrary number of coordinates, there is a complete set of states that are eigenstates of all the coordinates:

$$R_i|r_1, r_2, \ldots r_n\rangle = r_i|r_1, r_2, \ldots r_n\rangle \quad (2.105)$$

and another that are eigenstates of all the momenta:

$$P_i|p_1, p_2, \ldots p_n\rangle = p_i|p_1, p_2, \ldots p_n\rangle \quad (2.106)$$

Then one can inquire into the effect of the operators P_i on the eigenstates of the r_i, or equivalently into the scalar products

$$\langle r_1, r_2, \ldots r_n | p_1, p_2, \ldots p_n\rangle \quad (2.107)$$

Most of the physics will follow from the value of the bracket (2.107).

I want to attack this question in a slightly indirect way, by examining the classical limit of the time dependence of observables, and pretend to "discover" how quantum states evolve with time. This approach is really a special case of (b), since time translation is a symmetry, albeit a complicated one.

Begin by defining the **time-translation operator**, $U(t, t_o)$, the operator that relates the state of a system at two different times:

$$|\psi(t)\rangle = U(t, t_o)|\psi(t_o)\rangle \quad (2.108)$$

---

[11] Bohr used the term in a related but slightly different way. By "correspondence principle" he meant that the rules of quantum mechanics should have the same predictions as the rules of classical mechanics in the macroscopic region (e.g. that an electron in a large atomic orbit should emit radiation with the frequency of its classical motion).

40 Chapter 2 Fundamentals of Quantum Mechanics

$U(t, t_o)$ must be unitary in order for the conservation of probability to hold in *any* state.

The time-translation operator has the composition rule

$$U(t_2, t_1)U(t_1, t_o) = U(t_2, t_o) \tag{2.109}$$

and a "boundary condition"

$$U(t, t) = I \tag{2.110}$$

Therefore

$$U(t, t_o)^\dagger = U(t, t_o)^{-1} = U(t_o, t) \tag{2.111}$$

Differentiate $U(t, t_o)$ with respect to t and let $t \to t_o$. Define an operator $K(t_o)$ by

$$\lim_{t \to t_o} \frac{d}{dt} U(t, t_o) = -iK(t_o) \tag{2.112}$$

$K(t)$ tells us what happens when we make a small time translation. Because $U(t, t_o)$ is unitary, K must be Hermitean:

$$0 = \frac{d}{dt} I = \lim_{t \to t_o} \frac{d}{dt} \left[U(t, t_o)^\dagger U(t, t_o) \right] = i \left[K(t_o)^\dagger - K(t_o) \right] \tag{2.113}$$

Any classical observable A is a function of the coordinates q_i, the momenta p_i, and possibly the time t. Its average value is given by equation (2.51). Assume for simplicity that A has no explicit time dependence. Then if the system is in some state $|\psi(t)\rangle$, the time derivative of the expectation value of A at time t_o is

$$\frac{d}{dt}\langle A \rangle \bigg|_{t=t_o} = \lim_{t \to t_o} \frac{d}{dt} \left\langle \psi(t_o) \middle| U(t, t_o)^\dagger A U(t, t_o) \middle| \psi(t_o) \right\rangle$$

$$= i \left\langle \psi(t_o) \middle| K(t_o)A - AK(t_o) \middle| \psi(t_o) \right\rangle = i \langle \psi(t_o) | [K(t_o), A] | \psi(t_o) \rangle \tag{2.114}$$

In particular

$$\frac{d}{dt}\langle p_k \rangle = i \langle [K(t), p_k] \rangle \tag{2.115}$$

and

$$\frac{d}{dt}\langle q_k \rangle = i \langle [K(t), q_k] \rangle \tag{2.116}$$

We can take the correspondence principle to mean that the expectation values of the time dependence of the observables will obey the classical equations of motion, that is, that the time dependence in (2.115) and (2.116) is the time dependence in Hamilton's equations (1.29):

$$\frac{d}{dt}\langle p_k \rangle = -\left\langle \frac{\partial H}{\partial q_k} \right\rangle \quad \text{and} \quad \frac{d}{dt}\langle q_k \rangle = \left\langle \frac{\partial H}{\partial p_k} \right\rangle \tag{2.117}$$

Section 2.4 Canonical Commutators and the Schrödinger Equation 41

Therefore
$$\frac{\partial H}{\partial q_k} = -i[K(t), p_k] \tag{2.118}$$

and
$$\frac{\partial H}{\partial p_k} = i[K(t), q_k] \tag{2.119}$$

Any solution to equations (2.118) and (2.119) will satisfy our rule. One solution is that K is the Hamiltonian itself, and the coordinates and momenta satisfy the following relations:

$$[q_i, q_j] = [p_i, p_j] = 0 \qquad [q_i, p_j] = i\delta_{ij} \tag{2.120}$$

Proof. The proof of this statement is not at all trivial. Here is how it goes: First, suppose H has a term quadratic in the momenta only, as in the free particle problem: $H = \sum_k \alpha_k p_k^2$. Then

$$\frac{\partial H}{\partial p_k} = 2\alpha_k p_k \tag{2.121}$$

while from the commutation rules (2.120)

$$\begin{aligned} i[H, q_k] &= -i \sum_j \alpha_j \left[q_k, p_j^2\right] \\ &= -i \sum_j \alpha_j \left([q_k, p_j]p_j + p_j[q_k, p_j]\right) = 2\alpha_k p_k \end{aligned} \tag{2.122}$$

Similarly, if the Hamiltonian has a term quadratic in the coordinates, as in the harmonic oscillator problem, it follows that

$$\frac{\partial H}{\partial q_k} = -i[H, p_k] \tag{2.123}$$

Next, suppose H has a term cubic in a coordinate:

$$\begin{aligned} \left[p_k, q_j^3\right] &= \left[p_k, q_j^2 q_j\right] = \left[p_k, q_j^2\right]q_j + q_j^2[p_k, q_j] \\ &= -2i\delta_{jk}q_j^2 - iq_j^2\delta_{kj} = -3iq_j^2\delta_{kj} = -i\frac{\partial H}{\partial q_k} \end{aligned} \tag{2.124}$$

Continuing by induction, one can prove that if the coordinates and momenta satisfy (2.120), then the rules (2.118) and (2.119) hold at least for any polynomial in the q_k and for any polynomial in the p_k. The generalization to arbitrary Hamiltonians is straightforward, although not entirely free from ambiguities. ∎

This "demonstration" that $K = H$ solves equations (2.118) and (2.118) is not airtight. But there seems to be no ambiguity in cases of physical interest, and that

42 Chapter 2 Fundamentals of Quantum Mechanics

is all that is needed. Finally, if the theorem is true for the time dependence of the p_k and the q_k, it will hold for the time dependence of any function of them.

To sum up: If

$$\frac{d}{dt}|\psi(t)\rangle = -iH|\psi(t)\rangle \tag{2.125a}$$

$$[q_j, p_k] = i\delta_{ij} \tag{2.125b}$$

then the time dependence of the expectation values of the observables is the same as is classical mechanics.

There are ambiguities in this prescription. For example, if the Hamiltonian has a term that is a product of some p's and some q's, the order might be important in quantum mechanics, but not in classical mechanics. The Hamiltonian for a charged particle in an external electromagnetic field appears to have this defect. Convince yourself that for this system at least there is no ambiguity in the equations of motion.

2.4.2 The Canonical Commutation Relations

While the solutions (2.125) guarantee the correct time dependence for expectation values, they cannot be correct; the two sides of both equations have different dimensions. H has the dimensions of energy, while K has the dimension of inverse time. Similarly, we have set the commutators or q_i and p_j to be dimensionless, while they should have the dimensions of angular momentum. But (2.125) are not the only solutions to equations (2.118) and (2.119)! Both K and the commutation relations can be scaled by a dimensionful constant. Then the tentative rules (2.125) are replaced by

$$\boxed{i\hbar\frac{d}{dt}|\psi\rangle = H\psi} \tag{2.126a}$$

$$\boxed{[q_i, p_j] = i\hbar\delta_{ij}} \tag{2.126b}$$

for some constant $\hbar$ with the dimensions of angular momentum. Equation (2.126a) is **Schrödinger's equation** written in operator form, and equation (2.126b) are the **canonical commutation relations**. The value of $\hbar$ cannot be determined from the correspondence principle.

The time evolution of the expectation value of any observable A (with no explicit time dependence) is now guaranteed to be

$$\hbar\frac{d}{dt}\langle A\rangle = i\langle[A, H(t)]\rangle \tag{2.127}$$

The time derivative of the states, given by equations (2.126), is more fundamental. I have motivated equations (2.126) by the correspondence principle, but they are logically the postulates of quantum mechanics.

For observables like spin that have no close nonrelativistic classical version, we can simply define the Hamiltonian as the operator related to time evolution according to Schrödinger's equation. In relativistic and many body quantum mechanics, the full classical Hamiltonian formalism will again be used, but this time with fields as the coordinates and canonical momenta.

Note on commutators and Poisson brackets: Notice the similarity between the commutator rules in quantum mechanics and the Poisson bracket rules in classical mechanics (see Problems (1.3) and (1.4)):

$$[q_i, q_j] = 0 \quad \text{and} \quad [q_i, q_j]_{PB} = 0 \qquad (2.128a)$$
$$[p_i, p_j] = 0 \quad \text{and} \quad [p_i, p_j]_{PB} = 0 \qquad (2.128b)$$
$$[q_i, p_j] = i\hbar \delta_{ij} \quad \text{and} \quad [q_i, p_j]_{PB} = \delta_{ij} \qquad (2.128c)$$

The rule can be stated as follows: The commutator of any two q_i or p_j is the same as the classical Poisson bracket, multiplied by $i\hbar$. Then, because both commutators and Poisson brackets are Lie algebras, and because the algebraic rule

$$[A, BC] = [A, B]C + B[A, C] \qquad (2.129)$$

holds for both commutators and Poisson brackets, commutators of any two observables (that is, functions of the coordinates and momenta) are $i\hbar$ times the corresponding Poisson bracket.

2.4.3 Planck's Constant

Units

Why do we have to introduce $\hbar$? Why didn't the dimensions come out right in equations equation (2.125)? It is because we insisted that the three dimensions of classical mechanics—length, time, and mass—are independent. That is just a convention. It is as if we had for centuries thought that liters and meters were independent units until somebody discovered a universal conversion factor.

Whether or not it is useful to think of $\hbar$ as having dimensions depends on the context. A similar situation holds in electrostatics, where one can write the force between two charged particles, separated by a distance r, as $F = e_1 e_2 / 4\pi\epsilon_o r^2$. Here e_1 and e_2 are measured in coulombs, r in meters, and ϵ_o in the complicated combination that makes the answer come out in newtons. On the other hand, in cgs units we write $F = e_1 e_2 / r^2$. In that system e^2 has units dyne-cm^2. Whether or not one wants to introduce coulombs, or liters per cubic centimeter, is arbitrary.

Like the speed of light c, $\hbar$ cannot be calculated. It is a conversion factor. Had people known about quantum mechanics earlier, the units of energy might have been chosen as inverse time, and the units of momentum as inverse length. In such a system, $\hbar = 1$. Had people known the speed of light also, and that it was a fundamental quantity, time could have been measured in centimeters, while mass, momentum, and energy could all be in the same energy unit, ergs or electron volts or whatever you like. In such a system, $c = 1$. Later on I will use this system. It is easier than you think to convert back to conventional units.

The Classical Limit

Since $\hbar$ appears on the right hand side of all the commutators, classical mechanics is recovered and all observables commute as $\hbar \to 0$. Sometimes we say that classical

44 Chapter 2 Fundamentals of Quantum Mechanics

mechanics is the $\hbar \to 0$ limit of quantum mechanics. $\hbar$ tells us on what scale classical mechanics may be a good approximation. For ordinary macroscopic phenomena, typical values of dimensionless combinations like $pq/\hbar$ are huge, of the order of the number of atoms in ordinary things, or Avogadro's number.

Numerical Value of $\hbar$

The numerical value of $\hbar$ depends on our anthropomorphic units. In the cgs system the unit of length was originally related to the circumference of the earth through the poles, about 4×10^9 cm. The period of the earth's diurnal rotation was defined as 86400 seconds. And the mass of one cubic centimeter of H_2O at the average atmospheric temperature and pressure at the earth's surface was one gram. With these values, or their modern equivalents, the observed value of $\hbar$ is

$$\boxed{\hbar = 1.0545887\ldots \times 10^{-27}\,\text{erg-sec}} \qquad (2.130)$$

A more useful energy unit is the electron volt, the energy acquired by one elementary electronic charge moving through one volt. Then

$$\boxed{\hbar = 6.582173 \times 10^{-16}\,\text{eV-sec} = 6.582173 \times 10^{-22}\,\text{MeV-sec}} \qquad (2.131)$$

and

$$\boxed{\hbar c = 197.329\,\text{MeV-fermi} = 1973.29\,\text{ev-Ångstrøm}} \qquad (2.132)$$

These conversion factors are all you usually need.

2.5 QUANTUM DYNAMICS

2.5.1 The Time-Translation Operator

The dynamics of a quantum system is contained in the time-translation operator, the operator that takes a state at one time t to another time t_o:

$$|\psi(t)\rangle = U(t, t_o)|\psi(t_o)\rangle \qquad (2.108)$$

$U(t, t_o)$ is related to the Hamiltonian through Schrödinger's equation (2.126a). To see how that works, write (2.126a) as

$$i\hbar \frac{d}{dt} U(t, t_o)|\psi(t_o)\rangle = H(t)U(t, t_o)|\psi(t_o)\rangle \qquad (2.133)$$

Since $|\psi(t_o)\rangle$ can be any state, the time-translation operator itself satisfies the differential identity

$$i\hbar \frac{d}{dt} U(t, t_o) = H(t)U(t, t_o) \qquad (2.134)$$

Together with the boundary condition $U(t_o, t_o) = I$, equation (2.134) can be converted into an integral equation

$$U(t, t_o) = I - \frac{i}{\hbar} \int_{t_o}^{t} H(t')U(t', t_o)dt' \qquad (2.135)$$

Often this form can serve as the basis for an expansion of $U(t, t_o)$ in some small quantity, but in general equation (2.135) is difficult to solve as long as $H(t)$ depends on the time. When H is a constant, as must be the case for an isolated system, the solution is immediate:

$$U(t, t_o) = e^{-iH(t-t_o)/\hbar} \tag{2.136}$$

2.5.2 The Heisenberg Picture

So far x and p have been constant operators, while the time evolution of quantum mechanical systems is contained in the time dependence of the states $|\psi(t)\rangle$. This point of view, called the "Schrödinger picture," leads naturally to the Schrödinger wave equation. Sometimes it is convenient to consider a unitary transformation to the "Heisenberg picture," where the states are constants in time. Operators like x and p, independent of the time in the Schrödinger picture, will depend on the time in the Heisenberg picture. In fact they will obey the classical equations of motion; that is the principal advantage of the Heisenberg picture. Furthermore, the discussion of time-dependent symmetries, like boosts, is somewhat simpler in the Heisenberg picture, so it is a particularly useful device in relativistic quantum mechanics, where one wants to discuss rotations and Lorentz transformations in a unified way.

In the Heisenberg picture the state of a system can be taken to be the same as the Schrödinger picture state at time $t = 0$:

$$|\psi; \mathrm{H}\rangle = |\psi(0)\rangle \tag{2.137}$$

The time dependence of observations goes into the operators. For any observable $A(t)$, define a Heisenberg picture observable $A_\mathrm{H}(t)$ so that at any time the corresponding matrix elements in the two pictures agree.[12]

$$\langle \psi; \mathrm{H}| A_\mathrm{H}(t) |\psi; \mathrm{H}\rangle = \langle \psi(t)| A(t) |\psi(t)\rangle \tag{2.138}$$

The eigenstates of $A_\mathrm{H}(t)$ are now time-dependent vectors even for observables that are constant in the Schrödinger picture. The physics is still the same, since the transformation from $|\psi(t)\rangle$ to $|\psi; \mathrm{H}\rangle$ is a unitary transformation. The pictures are related by a time-dependent change of basis.

From equation (2.137) and the definition (2.108) of the time-translation operator,

$$|\psi(t)\rangle = U(t, 0)|\psi; \mathrm{H}\rangle \tag{2.139}$$

Since $|\psi(t)\rangle$ is an arbitrary state, equations (2.138) and (2.139) imply that

$$A_\mathrm{H}(t) = U(t, 0)^\dagger A(t) U(t, 0) = U(0, t) A(t) U(t, 0) \tag{2.140}$$

In the Heisenberg picture the states are constant. The operators change with time according to the equation of motion

$$\boxed{i\hbar \frac{d}{dt} A_\mathrm{H}(t) = [A_\mathrm{H}(t), H_\mathrm{H}(t)] + i\hbar \dot{A}_\mathrm{H}(t)} \tag{2.141}$$

---

[12] It is sufficient to require this of expectation values.

46 Chapter 2 Fundamentals of Quantum Mechanics

Proof of the Heisenberg Picture Equations of Motion. The rule for the time dependence of operators in the Heisenberg picture follows by differentiation:

$$i\hbar \frac{d}{dt} \langle \psi; H | A_H(t) | \psi; H \rangle = \left\langle \psi; H \left| i\hbar \frac{d}{dt} A_H(t) \right| \psi; H \right\rangle \quad (2.142)$$

Equation (2.138) implies that (2.142) is the same as

$$i\hbar \frac{d}{dt} \langle \psi(t) | A(t) | \psi(t) \rangle = -\langle \psi(t) | H(t) A(t) | \psi(t) \rangle$$
$$+ \langle \psi(t) | A(t) H(t) | \psi(t) \rangle + i\hbar \left\langle \psi(t) \left| \frac{d}{dt} A(t) \right| \psi(t) \right\rangle \quad (2.143)$$

On the right the first two terms are

$$\langle \psi; H | [A_H(t), H_H(t)] | \psi; H \rangle \quad (2.144)$$

and the last is

$$i\hbar \left\langle \psi; H \left| \dot{A}_H(t) \right| \psi; H \right\rangle \quad (2.145)$$

with the definition

$$\dot{A}_H(t) = U(0,t) \left(\frac{d}{dt} A(t) \right) U(t,0) \quad (2.146)$$

In the Schrödinger picture, $A(t)$ is a constant unless there is an explicit time dependence. Since the $|\psi\rangle$ is arbitrary, the Heisenberg picture operator satisfies (2.142). ∎

The last term in equation (2.141) is a complicated object that vanishes when the Schrödinger picture operator is time independent. When Hamiltonian is indeed independent of time, as it must be for an isolated system, then the time translation operator is given explicitly by equation (2.136). If the operator A is also a constant in the Schrödinger picture, like x or p, then equation (2.141) becomes much simpler:

$$\boxed{i\hbar \frac{d}{dt} A_H(t) = [A_H(t), H]} \quad (2.147)$$

and the operators in the two pictures are related by

$$A_H(t) = e^{iHt/\hbar} A e^{-iHt/\hbar} \quad (2.148)$$

In the Schrödinger picture the expectation values of observables obey the classical equations of motion. In the Heisenberg picture the states are constants, and the operators obey the classical equations of motion. For a particle in a potential, for example,

$$\dot{x}_H(t) = \frac{1}{m} p_H(t) \quad (2.149)$$

Since the two pictures are connected by a unitary transformation, commutation rules go over from one to the other unchanged:

$$[x_H(t), p_H(t)] = i\hbar \quad (2.150)$$

Note: That the Heisenberg picture operators obey the classical equation of motion, even if the Hamiltonian is explicitly time dependent, can be shown most elegantly using the Poisson bracket formalism. Compare equation (2.141) to the equation of motion in classical mechanics:

$$\dot{A}(t) = [A, H]_{PB} + \frac{\partial A}{\partial t} \tag{2.151}$$

The correspondence between Poisson brackets and commutators works for time dependence if we write everything in the Heisenberg picture.

2.6 THE UNCERTAINTY PRINCIPLE

With no more than the bare framework introduced above, a fundamental and remarkable property of quantum mechanics can be deduced. The **uncertainty principle** was discovered by Heisenberg [10] even before Dirac developed the whole formalism. The proof is only a few lines.

Let A be the operator for any observable. $\langle A \rangle$ is the expectation value of A in some state $|\psi\rangle$:

$$\langle A \rangle = \langle \psi | A | \psi \rangle \tag{2.152}$$

If $|\psi\rangle$ is an eigenstate of A then the result of measuring A will always be the same. Otherwise the results are scattered about $\langle A \rangle$. If you think about these measurements as a statistical ensemble, $\langle A \rangle$ is the mean of the ensemble and

$$\sigma_A^2 = \left\langle \psi \middle| (A - \langle A \rangle)^2 \middle| \psi \right\rangle \tag{2.153}$$

is the variance. The standard deviation, an intuitively useful measure of the spread of the measurements about $\langle A \rangle$, is σ_A itself.

Let A and B be any two observables, and let $\langle A \rangle$ and $\langle B \rangle$ be their averages in the state $|\psi\rangle$.

$$\Delta A = A - \langle A \rangle \quad \text{and} \quad \Delta B = B - \langle B \rangle \tag{2.154}$$

The **dispersion** of an observable in a state $|\psi\rangle$ is a common name for the variance of the measurements.

$$\sigma_A^2 = \langle \Delta A^2 \rangle \quad \text{and} \quad \sigma_B^2 = \langle \Delta B^2 \rangle \tag{2.155}$$

The **general uncertainty principle** says that if A and B do not commute, then there is a lower bound on the products of the two variances:

$$\boxed{\sigma_A^2 \sigma_B^2 \geq -\frac{1}{4} \langle \psi | [A, B] | \psi \rangle^2} \tag{2.156}$$

The left-hand side is positive because $[A, B]$ is imaginary.

When A and B do not commute, they cannot be measured simultaneously to arbitrary accuracy. The product of the variances has a lower bound given by the inequality (2.156). In particular, they cannot both be zero.

48 Chapter 2 Fundamentals of Quantum Mechanics

Proof of the Uncertainty Principle. The uncertainty principle can be demonstrated as follows: Let C be any linear operator, and $|\psi\rangle$ any state. Let $\phi = C|\psi\rangle$. Then the rule $\langle \phi | \phi \rangle \geq 0$ becomes

$$\langle \psi | C^\dagger C | \psi \rangle \geq 0 \tag{2.157}$$

Next let

$$C = \Delta A + it\Delta B \tag{2.158}$$

for any real number t. Since A and B are Hermitean,

$$C^\dagger = \Delta A - it\Delta B \tag{2.159}$$

Set

$$f(t) = \langle \psi | C^\dagger C | \psi \rangle = \langle \psi | \Delta A^2 + t^2 \Delta B^2 + it[\Delta A, \Delta B] | \psi \rangle \tag{2.160}$$

Now

$$[\Delta A, \Delta B] = [A, B] \tag{2.161}$$

so

$$f(t) = \langle \psi | \Delta A^2 + t^2 \Delta B^2 + it[A, B] | \psi \rangle \tag{2.162}$$

Notice that since

$$\langle \psi | [A, B] | \psi \rangle^* = \left\langle \psi \left| [A, B]^\dagger \right| \psi \right\rangle = -\langle \psi | [A, B] | \psi \rangle \tag{2.163}$$

$\langle \psi | [A, B] | \psi \rangle$ is imaginary, and the last term in equation (2.162) is indeed real. We need $f(t)$ only at its minimum value t_o, i.e. where

$$t_o = -\frac{i}{2} \frac{\langle \psi | [A, B] | \psi \rangle}{\langle \psi | \Delta B^2 | \psi \rangle} \tag{2.164}$$

to get

$$0 \leq f(t_o) = \langle \psi | \Delta A^2 | \psi \rangle + \frac{1}{4} \frac{\langle \psi | [A, B] | \psi \rangle^2}{\langle \psi | \Delta B^2 | \psi \rangle} \tag{2.165}$$

or

$$\langle \psi | \Delta A^2 | \psi \rangle \langle \psi | \Delta B^2 | \psi \rangle \geq -\frac{1}{4} \langle \psi | [A, B] | \psi \rangle^2 \tag{2.166}$$

■

The most important case is a coordinate and its canonically conjugate momentum. From the rule (2.126b) with $i = j$,

$$[q, p] = i\hbar \tag{2.167}$$

equation (2.156) becomes the **Heisenberg Uncertainty Principle**

$$\langle \psi | \Delta p^2 | \psi \rangle \langle \psi | \Delta q^2 | \psi \rangle \geq \frac{\hbar^2}{4} \tag{2.168}$$

In terms of the standard deviations (2.168) is the popular form

$$\boxed{\sigma_p \sigma_q \geq \frac{1}{2}\hbar} \tag{2.169}$$

For a wave, the uncertainty principle is the limit on the precision to which both its location and frequency measured.

2.7 WAVE FUNCTIONS

2.7.1 Wave Functions in Coordinate Space

In one dimension a particle has a single coordinate x, with continuous eigenstates $|\psi_x\rangle$ as in Section 2.3.2. The eigenvalues are all the real numbers. If $|\psi\rangle$ is any normalized state, then the amplitude for the particle to be "at x" is what one means by the value of the wave function at x.

$$\boxed{\psi(x) = \langle \psi_x | \psi \rangle} \tag{2.170}$$

Normalize the eigenstates of x to Dirac delta functions as in Section 2.3.2:

$$\langle \psi_{x'} | \psi_x \rangle = \delta(x' - x) \quad \text{or} \quad \int_{-\infty}^{\infty} |\psi_x\rangle\langle\psi_x| \, dx = I \tag{2.171}$$

If $|\psi\rangle$ is a normalized eigenstate,

$$1 = \langle \psi | \psi \rangle = \int_{-\infty}^{\infty} \langle \psi | \psi_x \rangle \langle \psi_x | \psi \rangle \, dx = \int_{-\infty}^{\infty} |\psi(x)|^2 \, dx \tag{2.172}$$

These rules really mean that

$$\int_{x_1}^{x_2} |\psi(x)|^2 \, dx \tag{2.173}$$

is the probability for finding x between x_1 and x_2. The integrand

$$\rho(x) = \psi(x)^* \psi(x) \tag{2.174}$$

is the **probability density**.

2.7.2 Momentum and Translations

Translations

Momentum is closely related to the translation operator. In classical mechanics a translation $T(a)$ changes a coordinate x to $x + a$, (see Tables 1.1 and 1.2) and leaves a momentum unchanged. In quantum mechanics there is a **translation operator** $T(a)$ that takes the state $|\psi_x\rangle$ into an eigenstate of the position operator $|\psi_{x+a}\rangle$ with eigenvalue $x + a$. The multiplication rule $T(a_1)T(a_2) = T(a_1 + a_2)$ is the same in quantum mechanics as it is in classical mechanics.

Translation should preserve the norms and scalar product of the states, and therefore must be unitary: $T(a)^\dagger T(a) = I$ or $T(a)^\dagger = T(-a)$. So in one dimension the translation operator is

$$T(a)|\psi_x\rangle = e^{i\phi(x)}|\psi_{x+a}\rangle \tag{2.175}$$

with some phase $e^{i\phi(x)}$.

I want to show how to transform the rules about transformations of states into rules about transformations of observables. Compare the matrix element of the operator x between any two states and their translated states:

$$\langle \psi_{x'} | x | \psi_{x''} \rangle = x'' \langle \psi_{x'} | \psi_{x''} \rangle \tag{2.176}$$

50 Chapter 2 Fundamentals of Quantum Mechanics

$$\langle \psi_{x'+a} | x | \psi_{x''+a} \rangle = (x'' + a) \langle \psi_{x'} | \psi_{x''} \rangle = \langle \psi_{x'} | x'' + aI | \psi_{x''} \rangle \quad (2.177)$$

But also
$$\langle \psi_{x'+a} | x | \psi_{x''+a} \rangle = \langle \psi_{x'} | T(-a) x T(a) | \psi_{x''} \rangle \quad (2.178)$$

Since $|\psi_{x'}\rangle$ and $|\psi_{x''}\rangle$ can be any state,
$$T(-a)\, x\, T(a) = x + aI \quad (2.179)$$

Look at small translations:[13] Since $T(a_1)T(a_2) = T(a_1 + a_2)$, it is possible to write the translation operator as
$$T(a) = e^{-iKa} \quad (2.180)$$

for some operator K independent of a; and since $T(a)$ is unitary, K is Hermitean. Expand equation (2.180) in powers of a.
$$T(\epsilon) = e^{-iK\epsilon} = I - iK\epsilon + \cdots \quad (2.181)$$

Then from equation (2.179), to first order,
$$(I + iK\epsilon)\, x\, (I - iK\epsilon) = x + \epsilon + \cdots \quad (2.182)$$

An identity like this holds for any ϵ, so it is true order by order. Comparing the terms on each side first order in ϵ one obtains
$$[x, K] = i \quad (2.183)$$

Equation (2.183) is the first order (sometimes loosely called *infinitesimal*) form of (2.179).

Now consider an eigenstate of momentum $|\psi_{p'}\rangle$. Momentum is invariant under translations, so $T(a)|\psi_{p'}\rangle = |\psi_{p'}\rangle$, and, imitating the steps above, one gets
$$T(-a)\, p\, T(a) = p \quad (2.184)$$

$T(a)$ commutes with p, and the infinitesimal form of (2.184) is $[K, p] = 0$.

The commutation rule (2.183) satisfied by K is the same as that for the momentum up to a factor $\hbar$. We can guarantee the canonical commutation relation by setting
$$p = \hbar K \quad (2.185)$$

so that
$$[x, p] = i\hbar \quad \text{and} \quad [p, p] = 0 \quad (2.186)$$

as required. (Exercise: What would happen if you added a constant: $p = \hbar K + c$?)

Momentum as a Differential Operator on Wave Functions

If $|\psi\rangle$ is a normalized state with wave function $\psi(x)$, what is the wave function of the state $p|\psi\rangle$? In other words, what is $\langle \psi_x | p | \psi \rangle$? To first order, from equation (2.183)

$$\langle \psi_x | T(-\epsilon) | \psi \rangle = \psi(x) + \frac{i\epsilon}{\hbar} \langle \psi_x | p | \psi \rangle \quad (2.187)$$

$$= \langle \psi | \psi_{x+\epsilon} \rangle^* = \langle \psi_{x+\epsilon} | \psi \rangle = \psi(x + \epsilon)$$

---
[13]This is the similar to the technique used in Section 2.4.1 for translations in time.

so
$$\langle \psi_x | p | \psi \rangle = -i\hbar \frac{\psi(x+\epsilon) - \psi(x)}{\epsilon} \tag{2.188}$$

Equation (2.188) is true only to first order in ϵ. Take the $\epsilon \to 0$ limit to get

$$\boxed{\langle \psi_x | p | \psi \rangle = -i\hbar \frac{d\psi(x)}{dx}} \tag{2.189}$$

This is the rule for the momentum operator in elementary wave mechanics.

Momentum Space Wave Functions

The eigenstates of momentum satisfy $p|\psi_{p_o}\rangle = p_o|\psi_{p_o}\rangle$. What are the x-space wave functions for these momentum eigenstates? Set $\psi_p(x) = \langle \psi_x | \psi_p \rangle$. Then

$$-i\hbar \frac{d}{dx} \psi_p(x) = p \psi_p(x) \tag{2.190}$$

The solution to equation (2.190), normalized to $\langle \psi_{p'} | \psi_{p''} \rangle = \delta(p' - p'')$ is

$$\psi_p(x) = \frac{1}{\sqrt{2\pi\hbar}} e^{ipx/\hbar} \tag{2.191}$$

You can define "momentum space wave functions" by $\Phi(p) = \langle \psi_p | \psi \rangle$, the relation between $\Phi(p)$ and the ordinary wave function $\psi(x)$ is

$$\Phi(p) = \langle \psi_p | \psi \rangle = \int_{-\infty}^{\infty} \langle \psi_p | \psi_x \rangle \langle \psi_x | \psi \rangle \, dx = \frac{1}{\sqrt{2\pi\hbar}} \int_{-\infty}^{\infty} e^{-ipx/\hbar} \psi(x) dx \tag{2.192}$$

$\Phi(p)$ is the Fourier transform of $\psi(x)$.

Example: Gaussian wave function

A Gaussian wave function is a simple example. Suppose a one-dimensional particle has the wave function

$$\Psi(x) = A e^{-a^2 x^2 / 2} \tag{2.193}$$

normalized to $\int_{-\infty}^{\infty} |\Psi(x)|^2 \, dx = 1$. Thus[14]

$$|A|^2 \int_{-\infty}^{\infty} e^{-a^2 x^2} \, dx = 1 \tag{2.194}$$

so up to an arbitrary phase, $A = a^{1/2} \pi^{-1/4}$. Then

$$\Phi(p) = \frac{1}{\sqrt{2\pi\hbar}} a^{\frac{1}{2}} \pi^{-\frac{1}{4}} \int_{-\infty}^{\infty} e^{-ipx/\hbar} e^{-a^2 x^2 / 2} \, dx \tag{2.195}$$

Complete the square in the exponent:

$$\Phi(p) = \sqrt{\frac{a}{2\pi^{\frac{3}{2}} \hbar}} \int_{-\infty}^{\infty} e^{-(\frac{1}{2} a^2)(x + ip/a^2 \hbar)^2} e^{-p^2/2a^2 \hbar^2} \, dx$$

$$= \sqrt{\frac{a}{2\pi^{\frac{3}{2}} \hbar}} \frac{\sqrt{2\pi}}{a} e^{-p^2/2a^2 \hbar^2} = \frac{1}{\sqrt{a\hbar\sqrt{\pi}}} e^{-p^2/2a^2 \hbar^2} \tag{2.196}$$

---

[14] See the integrals (A.22) in the appendix.

52 Chapter 2 Fundamentals of Quantum Mechanics

A Gaussian in x-space is also a Gaussian in p-space!

It is instructive to check the uncertainty principle (2.169) for a state whose wave function is (2.193): $\langle\psi|\,x\,|\psi\rangle = \langle\psi|\,p\,|\psi\rangle = 0$, while

$$\langle\Psi|\,x^2\,|\Psi\rangle = \frac{a}{\sqrt{\pi}}\int_{-\infty}^{\infty} x^2 e^{-a^2 x^2}\,dx = \frac{1}{2a^2} \tag{2.197}$$

and similarly

$$\langle\Psi|\,p^2\,|\Psi\rangle = \int_{-\infty}^{\infty} p^2|\Phi(p)|^2\,dp = \frac{1}{2}a^2\hbar^2 \tag{2.198}$$

so that

$$\langle\Psi|\,x^2\,|\Psi\rangle\langle\Psi|\,p^2\,|\Psi\rangle = \frac{1}{4}\hbar^2 \tag{2.199}$$

For a Gaussian (and only for a Gaussian), the "$\geq$" in the uncertainty principle become an equality.

2.7.3 Schrödinger's Wave Equation

Equation (2.189) leads directly to Schrödinger's wave equation for a one-dimensional particle in a potential. If the Hamiltonian is $H = p^2/2m + V(x)$, then the general form (2.126b) of Schrödinger's equation becomes

$$i\hbar\frac{\partial}{\partial t}\psi(x,t) = -\frac{\hbar^2}{2m}\frac{\partial^2}{\partial x^2}\psi(x,t) + V(x)\psi(x,t) \tag{2.200}$$

In three dimensions, the states are $|\psi_\mathbf{r}\rangle$, and the requirement $[r_i, p_j] = i\hbar\delta_{ij}$ leads by a similar argument to

$$\langle\psi_\mathbf{r}|\,\mathbf{p}\,|\psi\rangle = -i\hbar\boldsymbol{\nabla}\psi(\mathbf{r}) \tag{2.201}$$

When the Hamiltonian can be written as $H = \mathbf{p}^2/2m + V(\mathbf{r})$, Schrödinger's equation now appears in its original form, as a partial differential equation for a three-dimensional wave function:[15]

$$\boxed{i\hbar\frac{\partial}{\partial t}\psi(\mathbf{r},t) = -\frac{\hbar^2}{2m}\boldsymbol{\nabla}^2\psi(\mathbf{r},t) + V(\mathbf{r})\psi(\mathbf{r},t)} \tag{2.202}$$

Instead of working in the abstract space with basis vectors $|\psi_\mathbf{r}\rangle$, it is often convenient to identify the vectors with the wave functions. Identify $|\psi\rangle$ with the wave function $\psi(\mathbf{r})$, and define the scalar product $\langle\phi|\,\psi\rangle = \int \phi(\mathbf{r})^*\psi(\mathbf{r})\,d^3r$. I made such an identification in the discussion of Fourier's theorem in Section 2.3.3. The operators on the space become differential or integral operators on the functions. That is,

$$\mathbf{p}\psi(\mathbf{r}) = -i\hbar\boldsymbol{\nabla}\psi(\mathbf{r}) \quad \text{and} \quad r_i\psi(\mathbf{r}) = r_i\psi(\mathbf{r}) \tag{2.203}$$

In the last equation, the symbol r_i stands for an operator on the left but a number on the right. This is the Schrödinger wave function approach, and it is equivalent for simple systems.

---

[15] Schrödinger's original papers are references [11, 12, 13, 14].

In three dimensions the probability to find the particle in some volume V is

$$\int_V \rho(\mathbf{r})d^3r \qquad (2.204)$$

where in analogy to the one-dimensional definition (2.174) the probability density is

$$\rho(\mathbf{r}) = \psi(\mathbf{r})^*\psi(\mathbf{r}) \qquad (2.205)$$

Because $V(\mathbf{r})$ is real as long as the Hamiltonian is Hermitean, Schrödinger's equation conserves probability, but the probability density changes according to the continuity equation

$$\boxed{\frac{\partial \rho(\mathbf{r})}{\partial t} + \nabla \cdot \mathbf{j}(\mathbf{r}) = 0} \qquad (2.206)$$

where

$$\mathbf{j}(\mathbf{r}) = \frac{i\hbar}{2m}\left[\psi(\mathbf{r})\nabla\psi(\mathbf{r})^* - \psi(\mathbf{r})^*\nabla\psi(\mathbf{r})\right] = \frac{\hbar}{m}\operatorname{Im}\psi(\mathbf{r})^*\nabla\psi(\mathbf{r}) \qquad (2.207)$$

is the **probability current**. Since $\mathbf{j}(\mathbf{r}) \to 0$ as $r \to \infty$ the continuity equation (2.206) implies the conservation of probability.

2.7.4 Time-Dependent Free Particle Wave Functions

Wave Packets

Consider a free particle in one dimension. The momentum eigenstates $|p\rangle$ have wave functions

$$\langle x|p\rangle = \psi_p(x) = \frac{1}{(2\pi\hbar)^{1/2}}e^{ipx/\hbar} \qquad (2.208)$$

For a free particle these are also energy eigenstates:

$$H|p\rangle = \frac{p^2}{2m}|p\rangle = E|p\rangle \qquad (2.209)$$

Suppose the state vector is $|p\rangle$ at $t = 0$. Its time evolution (in the Schrödinger picture) is

$$i\hbar\frac{d|\psi(t)\rangle}{dt} = H|\psi(t)\rangle \qquad (2.210)$$

or

$$|\psi(t)\rangle = e^{-iHt/\hbar}|\psi(0)\rangle = e^{-iEt/\hbar}|\psi(0)\rangle \qquad (2.211)$$

So the time-dependent wave function is

$$\psi(x,t) = \frac{1}{(2\pi\hbar)^{1/2}}e^{i(px-Et)/\hbar} \qquad (2.212)$$

Define the wave number $k = p/\hbar$ and the (angular) frequency $\omega = E/\hbar$. For a pure plane wave, ω/k is the phase velocity. The wavelength is $\lambda = 2\pi/k$.

The wave function for a real particle is not a plane wave, but a wave packet

$$\psi(x,t) = \frac{1}{\sqrt{2\pi}}\int_{-\infty}^{\infty} e^{i(kx-\omega t)}f(k)\,dk \qquad (2.213)$$

54 Chapter 2 Fundamentals of Quantum Mechanics

$\psi(x,t)$ will be normalized provided

$$\int_{-\infty}^{\infty} |f(k)|^2 \, dk = 1 \qquad (2.214)$$

Wave and Group Velocities

There appears to be a puzzle. The plane waves represent particles with some momentum p, and they travel with a speed

$$v_p = \frac{E}{p} = \frac{p}{2m} \qquad (2.215)$$

that is half the speed of a mechanical particle with that momentum. The resolution is that the theorem that says $d\langle x\rangle/dt$ is the mechanical velocity does not apply to plane waves—it applies to normalizable wave packets like (2.213). The wave function of a real particle is not infinitely long, but it can be very long. It can have a form like (2.213) with $f(k)$ strongly peaked around some value k_o. In that case it describes a particle whose momentum is always close to $p_o = \hbar k_o$.

Expand the exponent in a power series about k_o:

$$\omega t - kx = \omega_o t - k_o x + (k - k_o)[v_g(k_o)t - x] + \frac{1}{2}(k - k_o)^2 \frac{d^2\omega(k_o)}{dk} t + \cdots \qquad (2.216)$$

where $\omega_o = \omega(k_o)$ and

$$v_g = \frac{d\omega(k_o)}{dk} = \frac{dE}{dp} \qquad (2.217)$$

is the **group velocity**. If the higher terms in the expansion (2.216) can be ignored, then approximately

$$\psi(x,t) \approx \frac{1}{\sqrt{2\pi}} \int_{-\infty}^{\infty} e^{-i(\omega_o t - k_o x + (k-k_o)(v_g t - x) + \cdots)} f(k) dk$$

$$= \frac{1}{\sqrt{2\pi}} e^{-i(\omega_o - k_o v_g)t} \int_{-\infty}^{\infty} e^{ik(x - v_g t) + \cdots} f(k) dk \qquad (2.218)$$

$$= \frac{1}{\sqrt{2\pi}} e^{-i(\omega_o - k_o v_g)t} \Psi(x - v_g t)$$

$\psi(x,t)$ can be approximated by a plane wave contained in an envelope that travels at the group velocity. The **phase velocity** is

$$v_p = \frac{\omega_o}{k_o} = \frac{E}{p} \qquad (2.219)$$

but the group velocity depends on the function $\omega(k)$ or $E(p)$. The mean location of the wave packet moves with the group velocity, not the phase velocity, and $\langle\psi|x|\psi\rangle$ has the value predicted by the general rule (2.117).

For a free nonrelativistic particle, $E = p^2/2m$, and the phase velocity is $v_p = p/2m$, half the speed of a classical particle with this energy and momentum. But the group velocity $v_g = dE/dp = p/m$ as expected.

Note: With relativistic kinematics, $E^2 = p^2c^2 + m^2c^4$, so the phase velocity is $E/p = c^2/v$, which if $v < c$ is faster than the speed of light! But the group velocity is $dE/dp = pc^2/E = v$. The expectation value of the quantum mechanical observable x obeys the classical equation of motion, and the group velocity is less than c.

Spread of the Wave Packet

In this approximation the dispersion of the wave packet is independent of time. This is not generally true, and the packet will begin to spread appreciably when the quadratic term in equation (2.216) becomes significant, that is, when $(p - p_o)^2 t \gg 2m\hbar$. The spread of a Gaussian wave packet is the subject of Problem 2.7. At least at first, $(p - p_o)^2$ is of the order of the original dispersion Δp^2 of the momentum, which itself is of the same order of magnitude as $\hbar^2/\Delta x^2$. The linear approximation is good as long as

$$t \ll \frac{\hbar m}{\Delta p^2} \quad \text{or} \quad t \ll \frac{m(\Delta x^2)}{\hbar} \tag{2.220}$$

PROBLEMS

✓ 2.1. The Spin Operators

This is a problem on vector spaces and measurement in quantum mechanics: There is a piece of the electron angular momentum called *spin*. The components are three observables s_x, s_y, and s_z. The only possible results of an observation of any of these are $\hbar/2$ and $-\hbar/2$. Thus ignoring all its other properties, an electron can be thought of as having only two states $|+\rangle$ and $|-\rangle$, eigenstates of the operator s_z:

$$s_z|+\rangle = \frac{\hbar}{2}|+\rangle \quad \text{and} \quad s_z|-\rangle = -\frac{\hbar}{2}|-\rangle$$

The operators s_x and s_y also have eigenvalues $\pm\hbar/2$, but they are not diagonal in the basis that diagonalized s_z. They are defined by

$$s_x|+\rangle = \frac{\hbar}{2}|-\rangle \quad s_x|-\rangle = \frac{\hbar}{2}|+\rangle \quad \text{and} \quad s_y|+\rangle = i\frac{\hbar}{2}|-\rangle \quad s_y|-\rangle = -i\frac{\hbar}{2}|+\rangle$$

(a) Show that the eigenvalues of s_x and s_y are indeed $\pm\hbar/2$ as claimed. What are the corresponding eigenstates?

(b) Show that $[s_x, s_y] = i\hbar s_z$, and the two relations obtained from this one by cyclic permutation of the subscripts.

(c) Let $\hat{n}$ be a unit vector in the direction of the spherical coordinates θ and ϕ:

$$n_x = \sin\theta\cos\phi \quad n_y = \sin\theta\sin\phi \quad n_z = \cos\theta$$

Show that

$$|\psi\rangle = \cos\frac{\theta}{2}|+\rangle + e^{i\phi}\sin\frac{\theta}{2}|-\rangle$$

is an eigenstate of $\mathbf{s} \cdot \hat{n} = n_x s_x + n_y s_y + n_z s_z$ with eigenvalue $\hbar/2$. What is the eigenvector of $\mathbf{s} \cdot \hat{n}$ with the other eigenvalue?

(d) A beam of electrons goes through a series of "Stern-Gerlach" type measurements as follows:

- The first measurement accepts $s_z = \hbar/2$ electrons and rejects $s_z = -\hbar/2$ electrons.

56 Chapter 2 Fundamentals of Quantum Mechanics

- The second measurement accepts electrons in eigenstates of $\mathbf{s} \cdot \hat{\mathbf{n}}$ ($\hat{\mathbf{n}}$ is in the x-z plane, that is, $\phi = 0$) with eigenvalue $\hbar/2$, and rejects those with eigenvalue $-\hbar/2$.
- The third measurement accepts $s_z = -\hbar/2$ electrons and rejects $s_z = \hbar/2$ electrons.

What is the relative intensity of the final beam to the initial beam? What is the value of θ that maximizes this intensity?

2.2. Expectation Values and Dispersions in Spin Measurements

Let $|\psi\rangle$ be the eigenstate of $\mathbf{s} \cdot \hat{\mathbf{n}}$ with eigenvalue $\hbar/2$ as in Problem 2.1.

(a) What are the expectation values of s_x and s_y in the state $|\psi\rangle$?

(b) Prove that the dispersion [see equation (2.155)] of any observable A is

$$\langle \psi | A^2 | \psi \rangle - \langle \psi | A | \psi \rangle^2$$

(c) What are the the dispersions of s_x and s_y in the state $|\psi\rangle$ above?

(d) What directions $\hat{\mathbf{n}}$ minimize the dispersions of s_x and s_y respectively? What is the minimum value of the product of these dispersions?

2.3. Units and Some Natural Constants

Here are some numerical values:

$$\frac{h}{2\pi} = \hbar = 1.054887 \times 10^{-27} \text{ erg-seconds}$$

$$\text{charge of the proton} = 1.60218 \times 10^{-19} \text{ coulombs}$$

$$\text{speed of light} = c = 2.99792458 \times 10^{10} \frac{\text{cm}}{\text{sec}}$$

$$\text{mass of the electron} = 9.109534 \times 10^{-28} \text{ g}$$

An **electron volt** (eV) is the electrostatic work done on (or by) one electron moving through one volt. An MeV is 10^6 eV. One coulomb moving through one volt does one joule of work.

(a) How many ergs is one eV?

(b) What is the rest-energy mc^2 of an electron in MeV?

(c) A fermi is 10^{-13} cm, and is a widely used unit in nuclear and high-energy physics. Derive the value of $\hbar c$ in MeV-fermi that is given in equation (2.132).

(d) If charge is measured in coulombs, the electrostatic energy of two electrons (charges $-e$ each) a distance r meters apart is

$$V = \frac{e^2}{4\pi\epsilon_o r}$$

where according to Maxwell ϵ_o has the same numerical value as $10^7/4\pi c^2$ (in MKS).[16] What is ϵ_o?

> Note: Don't look up the values of these constants! Compute them from the numbers given, and show how you got your answer!

---
[16] The MKS dimensions of ϵ_o are coulombs^2/newton-meter^2

(e) Since $\hbar c$ has dimensions energy-distance, one can write
$$V = \frac{\alpha \hbar c}{r}$$
where α is dimensionless and therefore thankfully independent of unit conventions. Calculate the value of α from e in coulombs and ϵ_o in MKS units.

(f) Faraday discovered that the amount of electricity needed to liberate one mole of hydrogen, oxygen, or chlorine from water or hydrochloric acid, or the amount needed to electroplate a surface with one mole of zinc, silver, and so forth, were all low integer multiples of about 96,500 coulombs. What is Avogadro's number?

Note: I will usually follow the atomic physics custom of measuring everything in cgs units, so that Coulomb's law, for example, becomes $V = e^2/r$. However both systems (and others) are in common use. The best way to avoid confusion is to replace e^2 (in cgs) by $\alpha \hbar c$ whenever possible. Then expressions can be evaluated consistently in any units.

2.4. A Simple Function Space

A variety of mathematical systems obey the postulates for the kind of vector space introduced in Section 2.2. Particularly important in quantum mechanics are **function spaces**, where the vectors are functions and the scalar product is an integral. These were mentioned briefly in Section 2.7. This problem and Problem 2.5 are about such function spaces.

The collection of all complex functions $f(x)$ of a real variable x with period 2π form a complex vector space. Define the scalar product by

$$\langle f(x) | g(x) \rangle = \int_{-\pi}^{\pi} f^*(x) g(x) \, dx$$

(a) Show that $D = -i d/dx$ is a Hermitean operator on this space. What are its eigenvalues and normalized eigenvectors (see page 33)?

(b) What are the eigenvalues and eigenvectors of $D^2 = -d^2/dx^2$?

(c) Define the reflection operator $\mathcal{R}$ on this space by
$$\mathcal{R} f(x) = f(-x)$$
Show that the commutator $[\mathcal{R}, D] \neq 0$. What is $[\mathcal{R}, D^2]$?

(d) Find the simultaneous eigenvectors of $\mathcal{R}$ and D^2.

2.5. Legendre Polynomials

Legendre polynomials (see Section A.2.2) are orthonormal basis vectors in a function space, as follows: Consider the space of all real functions $f(z)$ defined and regular on the real interval $-1 < z \leq 1$. Let the scalar product be

$$\langle f(z) | g(z) \rangle = \int_{-1}^{1} f(z) g(z) \, dz$$

(a) Show that the differential operator L defined by
$$L = \frac{d}{dz}(z^2 - 1)\frac{d}{dz}$$
is Hermitean with this scalar product.

58 Chapter 2 Fundamentals of Quantum Mechanics

(b) There is an eigenvector (eigenfunction) $f_l(z)$ of the operator L that is a polynomial of degree l for each nonnegative integer l. For each l write

$$f_l(z) = \sum_{n=0}^{l} a_{l,n} z^n$$

and obtain a recursion relation for the coefficients $a_{l,n}$. Compare the coefficients of z^l on both sides of this equation to obtain the eigenvalues λ_l. The $f_l(z)$ are proportional to the Legendre polynomials.

(c) Since the f_l are eigenvectors of a Hermitean operator with different eigenvalues, they must be orthogonal:

$$\int_{-1}^{1} f_l(z) f_{l'}(z) \, dz = 0 \quad (l \neq l')$$

Find f_o, f_1, f_2 and f_3 from the recursion relations for the $a_{l,n}$ (up to the arbitrary normalization constant), and show explicitly that they are orthogonal to each other.

(d) Conventionally, the Legendre polynomials $P_l(z)$ are proportional to these functions, but they are not normalized to unity. Rather $P_l(1) = 1$. It can be shown (see Section A.2.2) that this rule implies

$$\int_{-1}^{1} [P_l(z)]^2 \, dz = \frac{2}{2l+1}$$

Check this rule for the four f_l ($l = 0, 1, 2, 3$) you calculated above. Assuming it is true for all l, find constants c_l such that,

$$\sum_l c_l P_l(z) P_l(z') = \delta(z' - z)$$

2.6. Some Properties of the Rotation Matrices

The real 3×3 matrix $\bar{R}(\hat{\mathbf{n}}, \theta)$ rotates any vector by an angle θ about the direction of the unit vector $\hat{\mathbf{n}}$. We write

$$\bar{R}(\hat{\mathbf{n}}, \theta) = e^{-i\theta \hat{\mathbf{n}} \cdot \bar{\mathbf{J}}} \quad \text{or} \quad \bar{R}(\hat{\mathbf{n}}_i, \theta) = e^{-i\theta \bar{J}_i}$$

where the matrices $\bar{J}_i$ are

$$(\bar{J}_i)_{jk} = -i\epsilon_{ijk}$$

(a) Show that $(\bar{J}_z)^3 = \bar{J}_z$.

(b) Use the result (a) to sum the infinite series that defines the exponent:

$$\bar{R}(\hat{\mathbf{n}}_z, \theta) = \exp(-i\theta \bar{J}_z) = \sum_0^\infty \frac{1}{n!} (-i\theta \bar{J}_z)^n$$

(c) What are the eigenvalues of $\bar{J}_z$?

2.7. Time Evolution of a Free Gaussian Wave Function

A free particle in one dimension has a Gaussian wave function such that the expectation value of x^2 is $b^2/2$ at time $t = 0$. What are the expectation values of x^2 and p^2 at any later time t?

2.8. Complex Potentials and Probability Conservation

Consider the following Schrödinger equation for a particle in three dimensions:

$$i\hbar \frac{\partial \psi(\mathbf{r}, t)}{\partial t} = -\frac{\hbar^2}{2m} \nabla^2 \psi(\mathbf{r}, t) + [V_1(\mathbf{r}) + iV_2(\mathbf{r})] \psi(\mathbf{r}, t)$$

where $V_1(\mathbf{r})$ and $V_2(\mathbf{r})$ are real functions. Show that probability is not conserved unless $V_2 = 0$. Give an expression for the rate at which probability is lost or gained.

2.9. Two-Particle Systems

Classical mechanics

(a) Write the Lagrangian for a system of two three-dimensional particles, with coordinates $\mathbf{r}_1$ and $\mathbf{r}_2$, and masses m_1 and m_2, respectively. Let the potential energy be a function $V(|\mathbf{r}_1 - \mathbf{r}_2|)$ of the magnitude of the distance between the particles only. Write the Hamiltonian as a function of the coordinates $\mathbf{r}_1$ and $\mathbf{r}_2$ and their canonically conjugate momenta $\mathbf{p}_1$ and $\mathbf{p}_2$.

(b) Show that the Hamiltonian is invariant (classically) under spatial translations. What physical quantity is conserved as a consequence of this invariance?

(c) Now introduce new coordinates: the position of the center-of-mass and the relative coordinate, defined by

$$\mathbf{R} = \frac{m_1 \mathbf{r}_1 + m_2 \mathbf{r}_2}{m_1 + m_2} \quad \text{and} \quad \mathbf{r} = \mathbf{r}_1 - \mathbf{r}_2$$

What are the new momenta $\mathbf{P}$ and $\mathbf{p}$ conjugate to $\mathbf{R}$ and $\mathbf{r}$? Write the Hamiltonian as a function of these new variables. What are the physical meanings of the new momenta $\mathbf{p}$ and $\mathbf{P}$?

Quantum mechanics

(d) Write the quantum-mechanical Schrödinger equation for the two-particle wave-function $\psi(\mathbf{r}_1, \mathbf{r}_2)$ in terms of the original coordinates. Then transform by direct substitution (*i.e.*, using the chain rule for partial derivatives) to the coordinates $\mathbf{r}$ and $\mathbf{R}$. Verify that the new equation is the same as the one you would have gotten by applying the canonical rules directly to the Hamiltonian in the second form above.

(e) Finally, show that this Schrödinger equation can be separated into two equations, one each in $\mathbf{r}$ and $\mathbf{R}$.

2.10. The Baker-Hausdorff Lemma

For any operators A and B, and for a real parameter θ, prove that

$$e^{i\theta B} A e^{-i\theta B} = A + i\theta[B, A] - \frac{\theta^2}{2}[B, [B, A]] + \cdots$$

where the n-th term is

$$\frac{1}{n!} i^n \theta^n [B, [B, [B \cdots, [B, A]]]]$$

This is the **Baker-Hausdorff lemma**.[17]

---

[17] Sometimes the Baker-Campbell-Hausdorff lemma.

2.11. Spread of Wave Function

(a) Suppose an electron's wave function had a spread of the order a shortly after the big bang, and still has a comparable spread now. What is the order of magnitude of the minimum possible value of a? (Since this is an order of magnitude calculation, take the current age of the universe to be 10^{10} years.)

(b) Same problem for an object whose mass is one kilogram.

2.12. Another Representation of the Delta Function

A useful representation of the Dirac delta function is

$$\frac{1}{2\pi}\int_{-\infty}^{\infty} e^{i(x-z)y}\, dy = \delta(x-z) \qquad (2.104)$$

The left-hand side of this equation is not a true function because the infinite integral does not converge uniformly in $x-z$. One way to give meaning to an expression like this is to replace it by

$$\delta(x-z,\epsilon) = \frac{1}{2\pi}\int_{-\infty}^{\infty} e^{i(x-z)y}e^{-\epsilon|y|}\, dy$$

where ϵ is a small, positive real number. For $\epsilon > 0$, the function $\delta(x-z,\epsilon)$ is well-behaved. As elsewhere, the physics behind introducing ϵ is that if it is small enough it cannot be observed, so there can be no harm in introducing it. In this context quantities like ϵ are sometimes called **regulators**.

Show that in the limit $\epsilon \to 0$, the function $\delta(x-z,\epsilon)$ has the properties of the Dirac delta function, that is, that

$$\lim_{\epsilon \to 0}\delta(x-z,\epsilon) = 0$$

when $x \neq z$, and that

$$\lim_{\epsilon \to 0}\int_{-\infty}^{\infty}\delta(x-z,\epsilon)\, dx = 1$$

REFERENCES

The Davisson-Germer and Stern-Gerlach experiments can be found in

[1] C. DAVISSON AND L. H. GERMER, Phys. Rev. **30** (1927), 705.

[2] W. GERLACH AND O. STERN, Zeitschrift für Physik **9** (1922), 349.

The electron spin was first suggested by

[3] G. E. UHLENBECK AND S. GOUDSMIT, Naturwiss. **13** (1925), 953.

The algebraic formulation of quantum mechanics was discovered by

[4] W. HEISENBERG, Zeitschrift für Physik **33** (1925), 879.

[5] P. A. M. DIRAC, Proc. Roy. Soc. **A109** (1925), 642.

[6] M. BORN AND P. JORDAN, Zeitschrift für Physik **34** (1925), 858.

[7] M. BORN, W. HEISENBERG, AND P. JORDAN, Zeitschrift für Physik **35** (1925), 557.

These original papers and others are reprinted with commentary and translated into English (when necessary) in

[8] B. L. VAN DER WAERDEN, *Sources of quantum mechanics*, North Holland, 1967.

Dirac's textbook contains much original work:

[9] P. A. M. DIRAC, *Quantum mechanics*, Oxford University Press, 1930-1967. See also Appendix D.

The Uncertainty principle first appeared in

[10] W. HEISENBERG, Zeitschrift für Physik **43** (1927), 172.

Schrödinger's wave equation first appeared in a series of papers:

[11] E. SCHRÖDINGER, Annalen der Physik **79** (1926), 361.

[12] E. SCHRÖDINGER, Annalen der Physik **79** (1926), 489.

[13] E. SCHRÖDINGER, Annalen der Physik **79** (1926), 734.

[14] E. SCHRÖDINGER, Annalen der Physik **80** (1926), 437.

CHAPTER 3

Stationary States

3.1 ELEMENTARY EXAMPLES

3.1.1 States with Definite Energy

An eigenstate of the Hamiltonian is a **stationary state**. Its eigenvalues are its **energy spectrum**.[1] Typically the spectrum of a Hamiltonian contains bound states and unbound states. The bound states have energies where the classical orbits would be confined to a finite region of space, like the planets revolving around the sun. They are generally discrete states that can be normalized to unity. The unbound states occur at higher energies than bound states, energies at which the classical orbits are not confined, like the energy of a comet that comes in from outer space fast enough to escape the sun's gravity entirely. Unbound particles, averaged over a long enough time, have a finite probability to be at any distance from the origin. They cannot be described by normalizable state vectors. For them we use the continuous eigenstate formalism of Section 2.3.

This chapter begins with some examples of the energy spectrum problem in one and three dimensions, followed by two sections on angular momentum. It concludes with the two most important problems in elementary quantum mechanics, the harmonic oscillator and the hydrogen atom. The techniques introduced for these simple idealized systems will be useful later in a variety of more complicated situations.

For constant H, the stationary states contain all the information about a quantum mechanical system. When the Hamiltonian H is not a function of the time, its eigenvectors $|\psi_n\rangle$ form a time-independent basis. The eigenvalues are the possible observable values for the energy:

$$H|\psi_n\rangle = E_n|\psi_n\rangle \tag{3.1}$$

Any state can be expanded in these eigenvectors with time dependent coefficients:

$$|\psi(t)\rangle = \sum_n a_n(t)|\psi_n\rangle \tag{3.2}$$

The time dependence of $|\psi(t)\rangle$ is given by Schrödinger's equation

$$i\hbar \frac{d}{dt}|\psi\rangle = H|\psi\rangle \tag{2.126a}$$

so

$$a_n(t) = a_{on} e^{-iE_n t/\hbar} \tag{3.3}$$

---

[1] The word is a relic from the line spectra emitted by atoms as they decay from one energy eigenstate to another.

In this way the time-dependent Schrödinger equation is reduced to the energy eigenvalue problem. If at time t_o the system is in one of these energy eigenstates, for example, $a_{om} = \delta_{mn}$, then at a later time

$$|\psi(t)\rangle = e^{-iE_n(t-t_o)/\hbar}|\psi_n\rangle \tag{3.4}$$

The state is a constant multiplied by an oscillating phase factor. The energy of the state remains E_n forever.

3.1.2 A Two-State System

The first example is a system with only two states. It is a fair approximation to the ground states of a molecule of ammonia, NH$_3$, and provides a lesson in the power of thinking about the symmetries of even this simplest of all quantum mechanics systems.

The full Hamiltonian for a NH$_3$ molecule is some complicated function of the positions and momenta of the four nuclei and the ten electrons; even that is an approximation, since we should include the constituents of the nuclei, the quantum effects of the electromagnetic field, and so forth. One of the beauties of atomic and molecular systems is that these effects can to a good approximation be swept under the rug. It is not even a bad approximation to ignore the electrons as dynamical variables and describe the system by a potential among the four nuclei.

The three protons (H nuclei) lie in some plane. The N nucleus must be on a line through the center of the triangle formed by the H's, so to a good approximation it can be described by a one-dimensional wave function with a potential energy $V(x)$, some function of the distance x of the N from the plane of the H nuclei. It is convenient to let $V(x) \to 0$ as $x \to \infty$, so that the molecule has "zero energy" when it is completely dissociated. As x decreases, the potential decreases because the electrons attract the nitrogen nucleus; but then it must increase again as the nitrogen nucleus approaches the plane of the hydrogen nuclei, because of the electrostatic repulsion by the protons (see Figure 3.1). Thus $V(x)$ has a minimum at some finite distance from the origin. In fact it must have two minima, one to the left and one to the right of that plane. Classically these are the two stable positions of the N nucleus.

You can mock up this one-dimensional potential and find the energy eigenvalues. But here I want to make an even more drastic approximation; namely that there are only two states, one with the N to the left of the plane of the H's, and one to the right, as illustrated in Figure 3.2. Call these $|\psi_1\rangle$ and $|\psi_2\rangle$.

The Hamiltonian is invariant under reflections about the plane of the H nuclei, which now means simply the interchange $|\psi_1\rangle \leftrightarrow |\psi_2\rangle$. The expectation value of H in either state is the same:

$$\langle\psi_1|H|\psi_1\rangle = \langle\psi_2|H|\psi_2\rangle = E_o \tag{3.5}$$

That symmetry has survived all the drastic approximations.

The potential barrier between the two states is very high, on the scale of the binding energy. But as long as the barrier height is finite these are not eigenstates

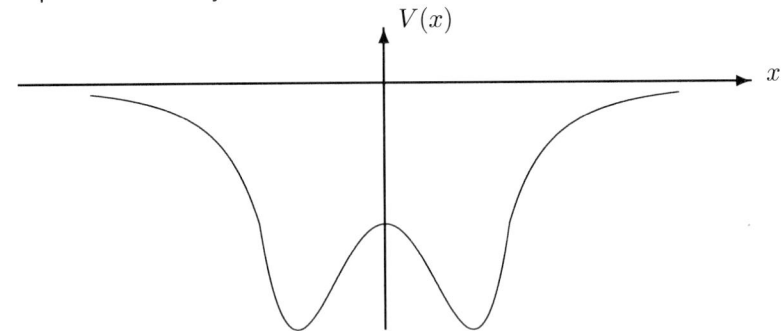

FIGURE 3.1: The nitrogen nucleus sees an attractive potential well on either side of the plane of the hydrogen nuclei, and a high repulsive barrier in between. The potential is symmetric about that plane.

FIGURE 3.2: In the two-state approximation, the nitrogen nucleus is on one or the other side of the plane of the hydrogens.

of the Hamiltonian. There is the quantum mechanical "tunneling" effect, an off-diagonal element of the Hamiltonian,

$$\langle \psi_1 | H | \psi_2 \rangle = \langle \psi_2 | H | \psi_1 \rangle^* = -A \tag{3.6}$$

In matrix form

$$H = \begin{pmatrix} E_o & -A \\ -A & E_o \end{pmatrix} \tag{3.7}$$

A can be chosen real.

E_o is of the order of the electronic energy levels of the molecule, corresponding to frequencies in the visible or infrared. If the potential barrier is high, A will be very small. For the NH_3 molecule, A corresponds to frequencies in the microwave region.

Energy eigenvalue problems are always simplified whenever there is a symmetry transformation that leaves the Hamiltonian invariant. Here that symmetry is the invariance of our problem under $|\psi_1\rangle \leftrightarrow |\psi_2\rangle$, which is what is left of reflections in the plane of the H atoms. Define a reflection operator $\mathcal{R}$:

$$\mathcal{R}|\psi_1\rangle = |\psi_2\rangle \quad \text{and} \quad \mathcal{R}|\psi_2\rangle = |\psi_1\rangle \tag{3.8}$$

The operator $\mathcal{R}$ is unitary, since if $|\phi_i\rangle = \mathcal{R}|\psi_i\rangle$ then $\langle \phi_i | \phi_j \rangle = \langle \psi_i | \psi_j \rangle$. Furthermore, $\mathcal{R}^2 = 1$, so $\mathcal{R}^{-1} = \mathcal{R}$. The operator $\mathcal{R}$ is its own inverse. It follows

that
$$\mathcal{R}^\dagger = \mathcal{R}^{-1} = \mathcal{R} \qquad (3.9)$$

$\mathcal{R}$ is Hermitean as well as unitary, and its eigenvalues, called the **parities** of the eigenstates, are ± 1. Furthermore, $[\mathcal{R}, H] = 0$.

You learned in Section 2.2.4 that if two observables can be measured simultaneously (i.e., if there is a basis of vectors that are eigenvectors of both observables) then the two observables commute. The converse is also true, and the above is the simplest example.

Proof. Let $|\psi_n\rangle$ be an (orthonormal) basis of eigenstates of H: $H|\psi_n\rangle = E_n|\psi_n\rangle$, and suppose $[\mathcal{R}, H] = 0$. Then

$$H\mathcal{R}|\psi_n\rangle = \mathcal{R}H|\psi_n\rangle = E_n\mathcal{R}|\psi_n\rangle \qquad (3.10)$$

The vector $\mathcal{R}|\psi_n\rangle$ is also an eigenstate of H, with the same eigenvalue E_n. If no two $|\psi_n\rangle$ have the same E_n, then $\mathcal{R}|\psi_n\rangle$ must be proportional to $|\psi_n\rangle$:

$$\mathcal{R}|\psi_n\rangle = \lambda|\psi_n\rangle \qquad (3.11)$$

and $|\psi_n\rangle$ is an eigenvector of $\mathcal{R}$.

If $E_n = E_m$ for some $m \neq n$, then $\mathcal{R}|\psi_n\rangle$ must be a linear combination of $|\psi_m\rangle$ and $|\psi_n\rangle$. Any such vector is an eigenvector of H with eigenvalue E_n, so there is a linear combination that is an eigenvector of both H and $\mathcal{R}$. Proceeding in this way by induction, one can easily prove in general that you can always find a basis of eigenvectors of H that are also eigenvectors of $\mathcal{R}$. ∎

Here, the eigenstates of $\mathcal{R}$ are obviously

$$|\psi_\pm\rangle = \frac{|\psi_1\rangle \pm |\psi_2\rangle}{\sqrt{2}} \qquad (3.12)$$

Then

$$H|\psi_\pm\rangle = (E_o \mp A)|\psi_\pm\rangle \qquad (3.13)$$

Since A is small compared to E_o, the splitting between these two states is much smaller than the spacing between the ordinary electronic levels of this molecule.[2] Notice how many important features can be abstracted from symmetry and order-of-magnitude arguments! I shall return more than once to this two-state problem.

Time Evolution of the Two-State System

At any time the state is a linear combination of $|\psi_1\rangle$ and $|\psi_2\rangle$, or equivalently $|\psi_+\rangle$ and $|\psi_-\rangle$. Suppose, for example, at time $t = 0$ the N atom is somehow known to

---

[2] The difference between the energies of the two states is about 9.84×10^{-5} eV. The frequency of the radiation emitted in the transition is 2.4×10^{10} sec^{-1}. The transition is used in the ammonia maser.

be on one side of the molecule: $|\psi(0)\rangle = |\psi_1\rangle$. Then, since the Hamiltonian has no explicit time dependence,

$$\begin{aligned}|\psi(t)\rangle &= \frac{1}{\sqrt{2}} e^{-iHt/\hbar} \left[|\psi_+\rangle + |\psi_-\rangle\right] \\ &= \frac{1}{\sqrt{2}} \left[e^{-i(E_o - A)t/\hbar}|\psi_+\rangle + e^{-i(E_o + A)t/\hbar}|\psi_-\rangle\right] \\ &= e^{-iE_o t/\hbar} \left[[\cos(At/\hbar)]|\psi_1\rangle + [i\sin(At/\hbar)]|\psi_2\rangle\right]\end{aligned} \quad (3.14)$$

The system has completely tunneled through to the state $|\psi_2\rangle$ when $t = \pi\hbar/2A \approx \hbar/A$. This is the first example of the energy-time uncertainty relation, although here "uncertainty" is a misnomer.

3.1.3 One-Dimensional Potential Problems

A single particle, without spin or any other kind of structure, moving in a potential, is an important collection of problems. There is one coordinate x, and one momentum p. The Hamiltonian has the form

$$H = \frac{1}{2m}p^2 + V(x) \quad (3.15)$$

and except in the case of a free particle $[p, H] \neq 0$. The potential $V(x)$ is a function of the coordinate, but not of the momentum for a velocity-independent potential. The operator $V(x)$ is diagonal in the $|x\rangle$ basis. Its matrix elements are

$$\langle x'| V |x\rangle = V(x)\delta(x' - x) \quad (3.16)$$

The possible results of observing the energy are solutions to

$$H|\psi_n\rangle = E_n|\psi_n\rangle \quad (3.17)$$

Take the matrix element of both sides of this equation with $\langle x|$:

$$\left(-\frac{\hbar^2}{2m}\frac{d^2}{dx^2} + V(x)\right)\psi_n(x) = E_n\psi_n(x) \quad (3.18)$$

This is the time-independent Schrödinger equation for a particle in a one-dimensional potential.

Free Particle in One Dimension

Simplest of all is a free particle, $V = 0$. The eigenvalues are all possible positive real numbers $E = p^2/2m$. For each E there are two independent states $|p\rangle$ and $|-p\rangle$. H alone is not a "complete set of observables." The meaning of this phrase is that eigenvalues of H are not enough to specify a basis uniquely. The eigenstates of H are **degenerate**. For a free particle we could label the energy eigenstates uniquely by their eigenvalues of p as well as H, but that will not work when the spectrum is degenerate but $V \neq 0$.

A more powerful approach is to examine the symmetries of the Hamiltonian. When $V(x) = 0$, H is invariant under one-dimensional reflections, which change p to $-p$ but leave H unchanged. Reflection invariance will hold for many nonzero potentials as well.

In Section 2.7.2 I showed how the *operators* x and p behave under a translation:

$$T(-a)\, x\, T(a) = x + aI \tag{2.179}$$

and

$$T(-a)\, p\, T(a) = p \tag{2.184}$$

How do they transform under $\mathcal{R}$? $\mathcal{R}$ changes the states of definite x into ones with a new value of x related to the original by the rule for the classical reflection operator in Table 1.1:

$$\mathcal{R}|x\rangle = |-x\rangle \tag{3.19}$$

$\mathcal{R}$ has all the algebraic properties of the operator with the same name introduced in Section 3.1.2. It is both Hermitean and unitary, an observable with possible eigenvalues ± 1, the **parities** of the eigenstates of $\mathcal{R}$. Furthermore, for any state,

$$x\mathcal{R}|x'\rangle = x|-x'\rangle = -x'|-x'\rangle \tag{3.20}$$

while

$$\mathcal{R}\, x|x'\rangle = \mathcal{R}\, x'|x'\rangle = x'\mathcal{R}|x'\rangle = x'|-x'\rangle \tag{3.21}$$

Therefore there is an operator identity

$$\mathcal{R}\, x\, \mathcal{R} = -x \tag{3.22}$$

and, similarly,

$$\mathcal{R}\, p\, \mathcal{R} = -p \tag{3.23}$$

x and p are odd under reflections. Since the free Hamiltonian is even in x and p, it is invariant: $\mathcal{R}H\mathcal{R} = H$.

The result (3.11) is general: It is possible to find eigenstates of H that are also eigenstates of $\mathcal{R}$. In our example of a free particle, these are just states proportional to $|p\rangle \pm |-p\rangle$. The wave functions of these states are proportional to

$$\langle x|\,p\rangle - \langle x|-p\rangle \sim \sin\frac{px}{\hbar} \quad \text{and} \quad \langle x|\,p\rangle + \langle x|-p\rangle \sim \cos\frac{px}{\hbar} \tag{3.24}$$

Parity can be used to label the energy eigenstates even when $V(x) \neq 0$ as long as it is an even function: $V(-x) = V(x)$. But it is impossible to find simultaneous eigenstates of energy, momentum, and parity. Momentum and parity do *not* commute.

One-Dimensional Infinite Square Well

Consider a particle in one dimension confined to $|x| < a$. The potential is $V = 0$ for $|x| < a$ and $V = \infty$ for $|x| > a$. (If you are not comfortable with infinite potentials, think of $V(x)$ as the limit of a finite-depth well.) $\psi'(x)$ cannot be discontinuous unless there is a delta function in $\psi''(x)$, hence in the potential also. For the finite well, both $\psi(x)$ and $\psi'(x)$ are continuous at $x = \pm a$. In the infinite well limit, $\psi'(x)$

becomes discontinuous also. Since $\psi(x) = 0$ for $|x| > a$, $\psi(\pm a) = 0$ is the boundary condition. In terms of the wave function, the energy eigenvalue equation is

$$-\frac{\hbar^2}{2m}\frac{d^2\psi_n(x)}{dx^2} = E_n \psi_n(x) \tag{3.25}$$

Solutions inside the well are linear combinations of $e^{\pm ipx/\hbar}$, provided they satisfy the boundary condition for arbitrary p. Those that do are (omitting the normalization)

$$\psi_n^-(x) = \sin(n\pi x/2a) \quad (n \text{ even and positive}) \tag{3.26a}$$

and

$$\psi_n^+(x) = \cos(n\pi x/2a) \quad (n \text{ odd and positive}) \tag{3.26b}$$

$V(x)$ is an even function of x, so H is reflection invariant. The eigenstates of H are not degenerate, so they must automatically turn out to be eigenstates of $\mathcal{R}$ also. The energies are

$$\boxed{E_n = \frac{p_n^2}{2m} = \frac{n^2 \pi^2 \hbar^2}{8ma^2}} \tag{3.27}$$

The ground state is ψ_1^+. In simple problems, the state of lowest energy is almost always a state of even parity.

Let us test the uncertainty principle for the ground state. The normalized wave function is[3]

$$\psi_1^+(x) = \frac{1}{\sqrt{a}}\cos(\pi x/2a)\,\Theta(|x| < a) \tag{3.28}$$

The expectation values of both x and p are zero. For p^2,

$$\langle \psi_1 | p^2 | \psi_1 \rangle = 2mE = \frac{\pi^2 \hbar^2}{4a^2} \tag{3.29}$$

while for x^2,

$$\langle \psi_1 | x^2 | \psi_1 \rangle = \frac{1}{a}\int_{-a}^{a} x^2 \cos^2(\pi x/2a)\, dx = \frac{a^2}{\pi^2}\left(\frac{\pi^2}{3} - 2\right) \tag{3.30}$$

so

$$\langle \psi_1 | x^2 | \psi_1 \rangle \langle \psi_1 | p^2 | \psi_1 \rangle = \frac{\hbar^2}{4}\left(\frac{\pi^2}{3} - 2\right) = 1.2899\ldots \frac{\hbar^2}{4} > \frac{\hbar^2}{4} \tag{3.31}$$

as required by equation (2.168).

3.2 THE HARMONIC OSCILLATOR

In one dimension the harmonic oscillator Hamiltonian is

$$H = \frac{p^2}{2m} + \frac{1}{2}m\omega^2 x^2 \tag{3.32}$$

The classical motion is oscillation of the form

$$x = A\cos(\omega t + \phi). \tag{3.33}$$

The harmonic oscillator is the basis for studying vibrational levels of atoms and molecules. It is at the foundation of the second quantization formalism for many body systems and for relativistic quantum fields.

---
[3] The step function $\Theta(x)$ is defined in equation (2.82).

3.2.1 The Spectrum

Set
$$p = \sqrt{m}P \quad \text{and} \quad x = \frac{Q}{\sqrt{m}} \qquad (3.34)$$

Then
$$H = \frac{1}{2}\left(P^2 + \omega^2 Q^2\right) \qquad (3.35)$$

The new variables are also canonically conjugate:
$$[P, Q] = [p, q] = -i\hbar \qquad (3.36)$$

Since the spectrum of H can be found from equation (3.36), it cannot depend on m.

Raising and Lowering Operators

All the information in this problem is contained in the *algebra* of the commutators of the operators.[4] Construct the operator

$$a = \frac{1}{\sqrt{2\omega\hbar}}\left(\omega Q + iP\right) \qquad (3.37)$$

Then
$$Q = \frac{\sqrt{2\omega\hbar}}{2\omega}\left(a + a^\dagger\right) \quad \text{and} \quad P = \frac{\sqrt{2\omega\hbar}}{2i}\left(a - a^\dagger\right) \qquad (3.38)$$

Since a is not Hermitean, $a^\dagger \neq a$. Their commutator is

$$[a, a^\dagger] = 1 \qquad (3.39)$$

The normalization in equation (3.37) was chosen so that the commutator would have this simple value. The Hamiltonian is

$$H = \frac{1}{4}\omega\hbar\left(-(a^\dagger - a)^2 + (a^\dagger + a)^2\right) = \frac{1}{2}\omega\hbar\left(a^\dagger a + a a^\dagger\right) \qquad (3.40)$$

From the commutation relation (3.39) the Hamiltonian can be written

$$H = \hbar\omega\left(a^\dagger a + \frac{1}{2}\right) \qquad (3.41)$$

The commutator of a with $a^\dagger a$ is:[5]

$$[a, a^\dagger a] = [a, a^\dagger]a + a^\dagger[a, a] = a \qquad (3.42)$$

so that
$$[a, H] = \hbar\omega a \qquad (3.43)$$

and
$$[a^\dagger, H] = [H, a]^\dagger = -\hbar\omega a^\dagger \qquad (3.44)$$

---
[4] Section 34 of Dirac's textbook, reference [9] in Chapter 2.
[5] See equation (2.129).

Energy Level Spacing

Let $|\psi_n\rangle$ be the energy eigenstates: $H|\psi_n\rangle = E_n|\psi_n\rangle$. Then

$$Ha^\dagger|\psi_n\rangle = a^\dagger H|\psi_n\rangle + \hbar\omega a^\dagger|\psi_n\rangle$$
$$= E_n a^\dagger|\psi_n\rangle + \hbar\omega a^\dagger|\psi_n\rangle = (E_n + \hbar\omega)\, a^\dagger|\psi_n\rangle \quad (3.45)$$

It follows that if $|\psi_n\rangle$ is an eigenstate of H with energy E_n, then $a^\dagger|\psi_n\rangle$ is an eigenstate of H with energy $E_n + \hbar\omega$. There is a ladder of states $|\psi_n\rangle$ with energies increasing in units of $\hbar\omega$. $a^\dagger$ is called a **raising** operator, since it increases the eigenvalue of the energy operator.

Similarly, a is a *lowering* operator, since

$$Ha|\psi_n\rangle = aH|\psi_n\rangle + [H,a]|\psi_n\rangle = (E_n - \hbar\omega)a|\psi_n\rangle \quad (3.46)$$

Ground State

Let $|\phi_n\rangle = a|\psi_n\rangle$. Equation (3.46) says that either $|\phi_n\rangle$ is proportional to $|\psi_{n-1}\rangle$ or $|\phi_n\rangle = 0$. In either case

$$0 \leq \langle\phi_n|\phi_n\rangle = \langle\psi_n|a^\dagger a|\psi_n\rangle = \left\langle\psi_n\left|\frac{H}{\hbar\omega} - \frac{1}{2}\right|\psi_n\right\rangle = \frac{E_n}{\hbar\omega} - \frac{1}{2} \quad (3.47)$$

so

$$E_n \geq \frac{\hbar\omega}{2} \quad (3.48)$$

The ladder of states cannot go down indefinitely; there is a state $|\psi_o\rangle$ of lowest energy, such that $a|\psi_o\rangle = 0$:

$$0 = a^\dagger a|\psi_o\rangle = \left(\frac{E_o}{\hbar\omega} - \frac{1}{2}\right)|\psi_o\rangle \quad (3.49)$$

The energy of the lowest state is $E_o = \hbar\omega/2$.

The whole spectrum has been found by the simple and elegant argument.[6] For each nonnegative integer there is one state with energy

$$\boxed{E_n = \hbar\omega\left(n + \frac{1}{2}\right)} \quad (3.50)$$

Level-Number Operator

$N = a^\dagger a$ is a particularly simple operator:

$$N|\psi_n\rangle = a^\dagger a|\psi_n\rangle = \left[\frac{H}{\hbar\omega} - \frac{1}{2}\right]|\psi_n\rangle = n|\psi_n\rangle \quad (3.51)$$

N is the level-number operator. It is Hermitean, its eigenvectors are the $|\psi_n\rangle$, and its eigenvalues the level numbers n.

---

[6] There is only one ground state, therefore only one ladder. See footnote 7.

3.2.2 Matrix Elements

The eigenvectors of the Hamiltonian are orthogonal and can be normalized to unity: $\langle \psi_n | \psi_m \rangle = \delta_{mn}$. The state $a^\dagger |\psi_n\rangle$ is proportional to $|\psi_{n+1}\rangle$. What is the coefficient of proportionality? Let

$$a^\dagger |\psi_n\rangle = c_n |\psi_{n+1}\rangle \tag{3.52}$$

Then, from the commutation rule (3.39) and the level-number operator (3.51) we obtain

$$|c_n|^2 = |c_n|^2 \langle \psi_{n+1} | \psi_{n+1} \rangle = \langle \psi_n | aa^\dagger | \psi_n \rangle = \langle \psi_n | a^\dagger a + 1 | \psi_n \rangle$$
$$= (n+1) \tag{3.53}$$

The phases of the constants c_n are arbitrary. Take them to be real and positive. Then

$$a^\dagger |\psi_n\rangle = \sqrt{n+1} |\psi_{n+1}\rangle \tag{3.54}$$

The matrix a looks like

$$a = \begin{pmatrix} 0 & \sqrt{1} & 0 & 0 & \cdots \\ 0 & 0 & \sqrt{2} & 0 & \cdots \\ 0 & 0 & 0 & \sqrt{3} & \cdots \\ \vdots & \vdots & \vdots & \vdots & \vdots \\ 0 & 0 & 0 & \cdots & \sqrt{n} \cdots \\ \cdots & \cdots & \cdots & \cdots & \cdots \end{pmatrix} \tag{3.55}$$

The matrix for $a^\dagger$ is the transpose of (3.55).

The correctly normalized state can be constructed by applying $a^\dagger$ repeatedly to the ground state:

$$|\psi_n\rangle = \frac{1}{\sqrt{n!}} a^{\dagger n} |\psi_0\rangle \tag{3.56}$$

Thus the matrix elements of a and $a^\dagger$ are given by

$$a^\dagger |\psi_n\rangle = \sqrt{n+1} |\psi_{n+1}\rangle \tag{3.57a}$$

$$a |\psi_n\rangle = \sqrt{n} |\psi_{n-1}\rangle \tag{3.57b}$$

The expectation value of the square of the coordinate Q in any energy eigenstate is

$$\langle \psi_n | Q^2 | \psi_n \rangle = \frac{\hbar}{2\omega} \langle \psi_n | (a + a^\dagger)^2 | \psi_n \rangle = \frac{\hbar}{2\omega} \langle \psi_n | a^\dagger a + aa^\dagger | \psi_n \rangle$$
$$= \frac{2}{\hbar \omega} \frac{\hbar}{2\omega} \langle \psi_n | H | \psi_n \rangle = \frac{1}{\omega^2} E_n \tag{3.58}$$

The expectation value of the potential energy is therefore

$$\langle \psi_n | V(Q) | \psi_n \rangle = \frac{1}{2} \omega^2 \langle \psi_n | Q^2 | \psi_n \rangle = \frac{1}{2} E_n \tag{3.59}$$

Similarly

$$\langle\psi_n|\,P^2\,|\psi_n\rangle = -\frac{\hbar\omega}{2}\langle\psi_n|\,(a-a^\dagger)^2\,|\psi_n\rangle = \frac{\hbar\omega}{2}\langle\psi_n|\,a^\dagger a + aa^\dagger\,|\psi_n\rangle \qquad (3.60)$$
$$= \frac{\hbar\omega}{2}\frac{2}{\hbar\omega}\langle\psi_n|\,H\,|\psi_n\rangle = E_n$$

So the expectation value of the kinetic energy is

$$\frac{1}{2}\langle\psi_n|\,P^2\,|\psi_n\rangle = \frac{1}{2}E_n \qquad (3.61)$$

The expectation values of the kinetic and the potential energies are equal, each half the total energy.

3.2.3 The Ground-State Energy

The Hamiltonian is

$$H = \frac{1}{2}\left(P^2 + \omega^2 Q^2\right) = \hbar\omega\left(N + \frac{1}{2}\right) \qquad (3.62)$$

Without the last term, the ground-state energy would be zero. But only *differences* between energies are measurable, and the zero of the energy can be chosen for convenience. The energy of the lowest level came out to be nonzero because the potential energy was chosen to be is zero when $x = 0$. With this convention the classical oscillator has zero energy when it is at rest at the origin.

In quantum mechanics this choice for the zero of the energy scale is not so compelling. Nothing prevents one from shifting the scale so that the energy of the ground state is zero. The Hamiltonian

$$H = \hbar\omega N = \hbar\omega a^\dagger a \qquad (3.63)$$

describes the same physical system, except that the energies of the stationary states are

$$E_n = n\hbar\omega \qquad (3.64)$$

In terms of the original coordinates and momenta, the new Hamiltonian (3.63) is

$$H = \hbar\omega a^\dagger a = \frac{1}{2}\left(P^2 + \omega^2 Q^2 + i\omega(QP - PQ)\right) \qquad (3.65)$$

In classical mechanics $QP - PQ$ vanishes. The revised Hamiltonian is just as good as the old one in classical mechanics, but in quantum mechanics it appears to be different.

The canonical quantization prescription in Section 2.4 seems to be ambiguous. Different forms of the classical Hamiltonian, equivalent because everything commutes in classical mechanics, can result in apparently different quantum theories. In the present example the difference is only formal: the zero of the energy is not measurable. If there were any experimentally testable consequence, one would have to find out which theory was correct from experiment. Here, the two possibilities have the same observable consequences. Sometimes the ambiguity is resolved by requiring some symmetry, like gauge invariance, to hold. In other cases requiring the Hamiltonian to be Hermitean decides the order (there is an interesting example in Section 3.5.1). But in every physically interesting case I know it turns out that this quantization ambiguity is not a physical distinction.

3.2.4 Wave Functions

It is useful to find the wave functions $\psi_n(x) = \langle x| \psi_n\rangle$. Returning to the original coordinate x is

$$x = \sqrt{\frac{\hbar}{2m\omega}}(a + a^\dagger) \qquad (3.66)$$

one can compute its matrix elements in terms of the matrix elements of a and $a^\dagger$ above:

$$\langle \psi_n| x |\psi_{n+1}\rangle = \sqrt{\frac{\hbar}{2m\omega}}\sqrt{n+1} \qquad (3.67a)$$

$$\langle \psi_n| x |\psi_{n-1}\rangle = \sqrt{\frac{\hbar}{2m\omega}}\sqrt{n} \qquad (3.67b)$$

The other matrix elements of x vanish.

The ground state is defined by the property $a|\psi_o\rangle = 0$, or

$$\langle x'| m\omega x + ip |\psi_o\rangle = 0 \qquad (3.68)$$

Now $\langle x'| x |\psi_o\rangle = x'\psi_o(x')$, and $\langle x'| p |\psi_o\rangle = -i\hbar(d/dx')\psi_o(x')$. So (drop the primes)

$$m\omega x\psi_o(x) + \hbar\frac{d\psi_o(x)}{dx} = 0 \qquad (3.69)$$

The solution is[7]

$$\psi_o(x) = \left[\frac{m\omega}{\pi\hbar}\right]^{\frac{1}{4}} e^{-m\omega x^2/2\hbar} \qquad (3.70)$$

According to the discussion at the end of Section 2.7.2, this is the state for which $(\Delta x)^2(\Delta p)^2$ attains its minimum allowed value $\hbar^2/4$.

The remaining energy eigenfunctions can be found by applying the raising operator repeatedly. For instance the next state is

$$|\psi_1\rangle = a^\dagger|\psi_o\rangle = \frac{1}{\sqrt{2m\omega\hbar}}(m\omega x - ip)|\psi_o\rangle \qquad (3.71)$$

or

$$\psi_1(x) = \frac{1}{\sqrt{2m\omega\hbar}}\left(m\omega x\psi_o(x) - \hbar\frac{d\psi_o(x)}{dx}\right) = \sqrt{\frac{2m\omega}{\hbar}}\left[\frac{m\omega}{\pi\hbar}\right]^{\frac{1}{4}} x e^{-m\omega x^2/2\hbar} \qquad (3.72)$$

For arbitrary n set $z = x\sqrt{m\omega/\hbar}$ and use equation (3.56):

$$\psi_n(x) = \sqrt{\frac{1}{2^n n!}}\left[\frac{m\omega}{\pi\hbar}\right]^{\frac{1}{4}}\left(z - \frac{d}{dz}\right)^n e^{-z^2/2} \qquad (3.73)$$

---

[7] Since the solution is unique, equation (3.70) is the proof that the levels in the one-dimensional oscillator are not degenerate.

74 Chapter 3 Stationary States

It is a simple exercise to prove by induction on n the identity

$$\left(z - \frac{d}{dz}\right)^n e^{-z^2/2} = (-1)^n e^{z^2/2} \frac{d^n}{dz^n} e^{-z^2} \tag{3.74}$$

so that

$$\psi_n(x) = \sqrt{\frac{1}{2^n n!}} \left[\frac{m\omega}{\pi\hbar}\right]^{\frac{1}{4}} e^{-m\omega x^2/2\hbar} H_n\left(x\sqrt{\frac{m\omega}{\hbar}}\right) \tag{3.75}$$

where

$$H_n(z) = (-1)^n e^{z^2} \frac{d^n}{dz^n} e^{-z^2} \tag{3.76}$$

are the Hermite polynomials. They are even or odd with n. A few are listed in Section A.2.4 of the appendix.

3.3 SPHERICALLY SYMMETRIC POTENTIALS AND ANGULAR MOMENTUM

3.3.1 Spherical Symmetry

A three-dimensional particle moving in a potential is described by Cartesian coordinates x, y, z or the spherical coordinates r, θ, ϕ. The Hamiltonian is

$$H = \frac{\mathbf{p}^2}{2m} + V(r) \tag{3.77}$$

V is independent of the momentum. Here I will consider only those potentials that are a function of the radial coordinate r alone. In that case H is invariant under rotations.

The Hamiltonian alone is not a complete set of observables. Since the eigenstates of H are degenerate, the states are not specified uniquely by their energies. We need to find enough other observables that commute with H and with each other to be able to name the states unambiguously.

The time-independent Schrödinger wave equation in three dimensions is

$$\left(-\frac{\hbar^2}{2m}\boldsymbol{\nabla}^2 + V(r)\right)\psi(\mathbf{r}) = E\psi(\mathbf{r}) \tag{3.78}$$

In three dimensions there is an interesting new dynamical operator, the angular momentum. I adopt the convention that the angular momentum is $\hbar\mathbf{L}$, so that $\mathbf{L}$ is dimensionless. Thus $\hbar\mathbf{L} = \mathbf{r} \times \mathbf{p}$. This convention, common but not universal, will eliminate a lot of $\hbar$-s from our equations.

The commutator of the momentum with the angular momentum is[8]

$$\hbar[p_i, L_j] = \sum_{kl} \epsilon_{lkj}[p_i, r_l p_k] = i\hbar \sum_k \epsilon_{ijk} p_k \tag{3.79}$$

---

[8]There is a classical mechanics result (Problem 1.6) that for any vector observable

$$[V_i, L_j]_{PB} = \sum_k \epsilon_{ijk} V_k$$

This quantum mechanical commutator is yet another example of the prescription $[A, B] = i\hbar[A, B]_{PB}$.

Section 3.3 Spherically Symmetric Potentials and Angular Momentum

and similarly
$$[r_i, L_j] = i\hbar \sum_k \epsilon_{ijk} r_k \tag{3.80}$$

3.3.2 Orbital Angular Momentum as a Differential Operator

In spherical coordinates (see Section A.3 in the appendix),
$$\nabla^2 = \frac{1}{r^2}\frac{\partial}{\partial r} r^2 \frac{\partial}{\partial r} - \frac{1}{r^2}\mathcal{L}^2 \tag{3.81}$$

where
$$\mathcal{L}^2 = -\left(\frac{1}{\sin\theta}\frac{\partial}{\partial\theta}\sin\theta\frac{\partial}{\partial\theta} + \frac{1}{\sin^2\theta}\frac{\partial^2}{\partial\phi^2} \right) \tag{3.82}$$

The differential operator $\hbar^2 \mathcal{L}^2$ is the same as the operator for the square of the total angular momentum. That is
$$\mathcal{L}^2 = \sum_i [(-i\mathbf{r} \times \boldsymbol{\nabla})_i]^2 \tag{3.83}$$

The proof of equation (3.83) is straightforward:

Proof. Let $\mathbf{p}$ stand for the differential operator $-i\hbar\boldsymbol{\nabla}$. The operator $\mathbf{r} \cdot \mathbf{p}$ picks out the component of $\mathbf{p}$ in the radial direction: $\mathbf{r} \cdot \mathbf{p} = -i\hbar r \partial/\partial r$. Thus

$$\hbar^2 \mathbf{L}^2 = \sum_i (\mathbf{r} \times \mathbf{p})_i (\mathbf{r} \times \mathbf{p})_i = \sum_{ijklm} \epsilon_{ijk}\epsilon_{ilm} r_j p_k r_l p_m$$

$$= \sum_{ijklm} \epsilon_{ijk}\epsilon_{ilm} r_j (r_l p_k - i\hbar\delta_{kl}) p_m$$

$$= \sum_{jklm}(\delta_{jl}\delta_{km} - \delta_{jm}\delta_{kl}) r_j r_l p_k p_m + 2i\hbar \sum_{jm}\delta_{jm} r_j p_m \tag{3.84}$$

$$= r^2 p^2 - \mathbf{r} \cdot (\mathbf{r} \cdot \mathbf{p})\mathbf{p} + 2i\hbar \mathbf{r} \cdot \mathbf{p}$$

$$= r^2 p^2 + \hbar^2 rr \frac{\partial}{\partial r}\frac{\partial}{\partial r} + 2\hbar^2 r\frac{\partial}{\partial r} = r^2 \left(p^2 + \hbar^2 \frac{1}{r^2}\frac{\partial}{\partial r} r^2 \frac{\partial}{\partial r} \right)$$

Thus
$$\mathbf{p}^2 = -\hbar^2 \frac{1}{r^2}\frac{\partial}{\partial r} r^2 \frac{\partial}{\partial r} + \frac{\hbar^2}{r^2}\mathbf{L}^2 \tag{3.85}$$

as claimed. ∎

The differential operator $\mathcal{L}^2$ depends only on the angles. It commutes with the radial coordinate r and with $\partial/\partial r$. Therefore, as long as the potential is a function of r alone,
$$[H, \mathcal{L}^2] = 0 \tag{3.86}$$

H and $\mathbf{L}^2$ can be diagonalized simultaneously.

76 Chapter 3 Stationary States

L_z acting on a wave function is the differential operator $-i\partial/\partial\phi$. $\mathbf{L}^2$ also commutes with L_z, and so does H, since the angle ϕ doesn't appear in H. Thus H, $\mathbf{L}^2$, and L_z can be simultaneously diagonalized.

L_x and L_y also commute with H and with $\mathbf{L}^2$, but because of the special role the z axis plays in the spherical coordinate system, these properties are not so obvious (see Section 3.3.3). But there do not exist simultaneous eigenstates of *all* these observables because in general $[L_i, L_j] \neq 0$. Since $[L_i, \mathbf{L}^2] = 0$, one may choose $\mathbf{L}^2$ and any one of the three components. It is conventional to choose L_z.

Label the stationary states $|E, l, m\rangle$, where

$$H|E, l, m\rangle = E|E, l, m\rangle \tag{3.87a}$$

$$\mathbf{L}^2|E, l, m\rangle = \lambda_l |E, l, m\rangle \tag{3.87b}$$

$$L_z|E, l, m\rangle = m|E, l, m\rangle \tag{3.87c}$$

There is no logical reason not to label the states by the eigenvalue λ_l of $\mathbf{L}^2$, but it is a useful convention to do it another way. For the moment I leave open the relation between λ_l and l.

E will depend on the potential $V(r)$; but the spectrum of $\mathbf{L}^2$ and $\mathcal{L}_z$ follows from spherical symmetry alone. This is an important result, so I will go over the derivation carefully.

3.3.3 The Angular Momentum Commutator Algebra

What are the commutators among the components of $\mathbf{L}$? These commutators are fundamental and will be the only input to the calculation of the spectrum. The result of a short computation is

$$\boxed{[L_i, L_j] = i \sum_k \epsilon_{ijk} L_k} \tag{3.88}$$

Here is the proof:

Proof. Use the definition $\hbar L_i = (\mathbf{r} \times \mathbf{p})_i = \sum_{jk} \epsilon_{ijk} r_j p_k$ twice:

$$\hbar^2 [L_i, L_j] = \sum_{klmn} \epsilon_{ikl} \epsilon_{jmn} [r_k p_l, r_m p_n] \tag{3.89}$$

Use the canonical commutation rules (2.126b) and a generalization of equation (2.129) to write

$$\hbar^2 [L_i, L_j] = i\hbar \left[-\sum_{kmn} \epsilon_{ikm} \epsilon_{jmn} r_k p_n + \sum_{klm} \epsilon_{ikl} \epsilon_{jmk} r_m p_l \right] \tag{3.90}$$

Finally, sum over the repeated subscript using one of the rules from Section

Section 3.3 Spherically Symmetric Potentials and Angular Momentum   77

A.1 of the appendix.

$$\hbar^2[L_i, L_j] = i\hbar \left[-\sum_{kn} (\delta_{in}\delta_{jk} - \delta_{ij}\delta_{nk})\, r_k p_n + \sum_{ml} (\delta_{lj}\delta_{im} - \delta_{ij}\delta_{lm})\, r_m p_l \right]$$

$$= i\hbar\, (r_i p_j - r_j p_i) = i\hbar^2 \sum_k \epsilon_{ijk} L_k \tag{3.91}$$

∎

The rule $[L_i, L_j] = i \sum_k \epsilon_{ijk} L_k$ is the same as the commutators among the "generators" of the rotation matrices in equation (1.62), and it will come up again later. So it is useful to ask a more general question: What are the finite dimensional Hilbert spaces on which the commutator rules

$$\boxed{[J_i, J_j] = i \sum_k \epsilon_{ijk} J_k} \tag{3.92}$$

are defined among three Hermitean operators J_i?

We construct the algebra in the basis of eigenstates $|j, m\rangle$ of $\mathbf{J}^2$ and J_z. Here m is the eigenvalue of J_z, but j is not the eigenvalue of $\mathbf{J}^2$. That is some λ_j, a number that depends on j, like λ_l in equation (3.87b) above. We will choose the definition of j itself shortly according to a standard convention. The first question is, what are the possible values of λ_j and of m? Then, if a system is in an eigenstate of these two operators with eigenvalues λ_j and m, what are J_x and J_y in this basis? These questions can be answered in the following nine easy steps:

Step (i) Define

$$\mathbf{J}^2 = J_x^2 + J_y^2 + J_z^2 \tag{3.93}$$

The matrix $\mathbf{J}^2$ is Hermitean because the J_i are Hermitean.

Step (ii) Show that J_i commutes with $\mathbf{J}^2$:

$$[J_i, \mathbf{J}^2] = \sum_j [J_i, J_j] J_j + \sum_j J_j [J_i, J_j] = 0 \tag{3.94}$$

Step (iii) Define the combinations[9]

$$J_\pm = J_x \pm i J_y \tag{3.95}$$

and solve for J_x and J_y:

$$J_x = \frac{J_+ + J_-}{2} \quad \text{and} \quad J_y = \frac{J_+ - J_-}{2i} \tag{3.96}$$

The new combinations have these commutators with the J_i and with each other:

$$[J_z, J_\pm] = \pm J_\pm \tag{3.97a}$$

---
[9] Some books put an extra $\sqrt{2}$ in this definition.

78 Chapter 3 Stationary States

$$[J_+, J_-] = 2J_z \tag{3.97b}$$

The $J_\pm$ also commute with $\mathbf{J}^2$.

Step (iv) Use equation (3.96) to write $\mathbf{J}^2$ in terms of J_+ and J_-:

$$\mathbf{J}^2 = \frac{1}{2}(J_+ J_- + J_- J_+) + J_z^2 \tag{3.98}$$

Step (v) Now show that $J_\pm$ are raising and lowering operators for the eigenvalues of J_z. Choose a basis $|j, m\rangle$ such that $J_z|j, m\rangle = m|j, m\rangle$. Since $\mathbf{J}^2$ commutes with all three J_i, $\mathbf{J}^2|j, m\rangle = \lambda_j |j, m\rangle$ with the same value of λ_j independent of m. Then[10]

$$\begin{aligned} J_z J_+ |j, m\rangle &= \bigl([J_z, J_+] + J_+ J_z\bigr)|j, m\rangle = (J_+ + J_+ J_z)|j, m\rangle \\ &= J_+(1 + m)|j, m\rangle = (m + 1)J_+ |j, m\rangle \end{aligned} \tag{3.99}$$

so, either $J_+|j, m\rangle = 0$ or else $J_+|j, m\rangle$ is an eigenstate of J_z with eigenvalue $m+1$, i.e. proportional to $|j, m+1\rangle$. So here, as in the harmonic oscillator problem, there is a ladder of states; they are eigenstates of J_z with eigenvalues separated by unity. Each ladder is labeled in some way by j, and all the states in the ladder are eigenstates of $\mathbf{J}^2$ with the same eigenvalue λ_j.

Step (vi) Show that these ladders of states have a top and a bottom:

$$\lambda_j - m^2 = \langle j, m | \mathbf{J}^2 - J_z^2 | j, m \rangle = \langle j, m | J_x^2 + J_y^2 | j, m \rangle \tag{3.100}$$

Since J_x is Hermitean,

$$\langle j, m | J_x^2 | j, m \rangle = \sum_{m'} |\langle j, m | J_x | j, m' \rangle|^2 \geq 0 \tag{3.101}$$

and $\langle j, m | J_y^2 | j, m \rangle \geq 0$ similarly. Therefore

$$m^2 \leq \lambda_j \tag{3.102}$$

The ladders of eigenvalues of J_z are bounded both above and below.

Step (vii) Compute the allowed values of λ_j:

There may be different collections of possible m-values for each j, but for any j they are bounded. Now I can reveal what the label j will be. Conventionally we name the ladders by the value j of the maximum m in the ladder:

$$J_+|j, j\rangle = 0 \tag{3.103}$$

Then use $J_+|j, j\rangle = 0$ and equation (3.97b) to write

$$\begin{aligned} \mathbf{J}^2 |j, j\rangle &= \left[\frac{1}{2}(J_+ J_- + J_- J_+) + J_z^2\right]|j, j\rangle = \left[\frac{1}{2}(J_+ J_- - J_- J_+) + J_z^2\right]|j, j\rangle \\ &= \left(J_z + J_z^2\right)|j, j\rangle = j(j+1)|j, j\rangle \end{aligned} \tag{3.104}$$

---

[10] and similarly $J_z J_- |j, m\rangle = (m-1)|j, m\rangle$.

Section 3.3 Spherically Symmetric Potentials and Angular Momentum

Since for given j, the value of λ_j is independent of m,

$$\mathbf{J}^2|j,m\rangle = j(j+1)|j,m\rangle \tag{3.105}$$

for all m. Let $m_{\min}$ be the lowest eigenvalue of J_z in a ladder whose highest value is j.

$$J_-|j,m_{\min}\rangle = 0 \tag{3.106}$$

Then $\mathbf{J}^2|j,m_{\min}\rangle = j(j+1)|j,m_{\min}\rangle$, but also

$$\begin{aligned}\mathbf{J}^2|j,m_{\min}\rangle &= \left[\frac{1}{2}(J_+J_- + J_-J_+) + J_z^2\right]|j,m_{\min}\rangle \\ &= \left[\frac{1}{2}(-J_+J_- + J_-J_+) + J_z^2\right]|j,m_{\min}\rangle \\ &= (m_{\min}^2 - m_{\min})|j,m_{\min}\rangle\end{aligned} \tag{3.107}$$

The lowest eigenvalue satisfies

$$m_{\min}(m_{\min} - 1) = j(j+1) \tag{3.108}$$

$m_{\min} = j+1$ cannot be a solution to the original problem, since $m_{\min} \le j$. Therefore

$$m_{\min} = -j \tag{3.109}$$

The ladder goes from $m = -j$ to $m = j$.

Step (viii) Find the allowed values of j:

The J_z ladders run from $-j$ to j in integer steps. Therefore, $j - (-j) = 2j$ is an integer. The allowed values of j are positive integers or half-integers:

$$j = 0, \frac{1}{2}, 1, \frac{3}{2}, 2, \ldots \tag{3.110}$$

Here is the origin of a feature of the microscopic world that people accustomed to a classical description of nature once found so strange: $\mathbf{J}^2$ can have any value in classical mechanics, but only certain discrete values are allowed by the rules of quantum mechanics.

Step (ix) Finally, find the matrix elements of $J_\pm$:

We will need these, or the matrix elements of J_x and J_y that follow from them, frequently. For some numbers a_{jm},

$$J_+|j,m\rangle = a_{jm}|j,m+1\rangle \tag{3.111}$$

or

$$\begin{aligned}|a_{jm}|^2 &= \langle j,m|\,J_-J_+\,|j,m\rangle \\ &= \frac{1}{2}\langle j,m|\,J_-J_+ + J_+J_-\,|j,m\rangle + \frac{1}{2}\langle j,m|\,J_-J_+ - J_+J_-\,|j,m\rangle \\ &= \langle j,m|\,\mathbf{J}^2 - J_z^2\,|j,m\rangle - \langle j,m|\,J_z\,|j,m\rangle = j(j+1) - m(m+1)\end{aligned} \tag{3.112}$$

The phase of a_{jm} is arbitrary. It is conventional to take these matrix elements to be real and positive. Thus

$$\boxed{J_+|j,m\rangle = \sqrt{j(j+1) - m(m+1)}|j, m+1\rangle} \qquad (3.113a)$$

$$\boxed{J_-|j,m\rangle = \sqrt{j(j+1) - m(m-1)}|j, m-1\rangle} \qquad (3.113b)$$

These are the rules for *all* the finite-dimensional representations generated by three Hermitean operators with the angular momentum algebra.

3.3.4 Classification of the States

Finally, let us return to the original problem, how to classify the eigenstates of the Hamiltonian for a three-dimensional particle in a spherically symmetric potential. Because H, $\mathbf{L}^2$, and L_z all commute, they can be simultaneously diagonalized, and the states can be chosen as in equations (3.87). The eigenstates of $\mathbf{L}^2$ come in ladders $|l, m\rangle$ with $l \geq m \geq -l$ in integer steps. All these $2l + 1$ states can be obtained from $|l, l\rangle$ by repeatedly applying the lowering operator L_-.

Not only do they all have the same value of $\mathbf{L}^2$ but also the same value of the energy, since L_- and L_+ commute with H. For instance, suppose the state with $m = l - 1$ is also an energy eigenstate. Then

$$H|E_{l,l-1}, l, l-1\rangle = \frac{1}{\sqrt{2l}} H L_- |E_{l,l-1}, l, l\rangle$$

$$= \frac{1}{\sqrt{2l}} L_- H |E_{l,l-1}, l, l\rangle = E_l |E_{l,l-1}, l, l-1\rangle \qquad (3.114)$$

Continuing down the ladder in this fashion, you see that all the states have the same eigenvalue E_l of the Hamiltonian. The energy spectrum is degenerate.

3.4 SPHERICALLY SYMMETRIC POTENTIALS: WAVE FUNCTIONS

3.4.1 Spherical Coordinates and Spherical Harmonics

Consider a particle in three dimensions with the Hamiltonian

$$H = \frac{\mathbf{p}^2}{2m} + V(\mathbf{r}) \qquad (3.115)$$

where V is local and depends on the radial coordinate only. In classical mechanics, this is the "central potential problem."

The energy eigenstates can be chosen eigenstates of the square of the orbital angular momentum and its z-component also. Label them $|E, l, m\rangle$ as in Section 3.3. All the states with the same E and l have the same energy, and the wave functions are

$$\psi_{Elm}(\mathbf{r}) = \langle \mathbf{r} | Elm \rangle \qquad (3.116)$$

For every E and l there are $2l + 1$ states, each with different m.

Spherical Coordinates

In spherical coordinates the Schrödinger equation (3.78) is

$$\left[-\frac{\hbar^2}{2m}\left(\frac{1}{r^2}\frac{\partial}{\partial r}r^2\frac{\partial}{\partial r} - \frac{\mathbf{L}^2}{r^2}\right) + V(r)\right]\psi_{Elm}(r,\theta,\phi) = E\psi_{Elm}(r,\theta,\phi) \quad (3.117)$$

The eigenvalue equation (3.87c) becomes

$$-i\frac{\partial}{\partial \phi}\psi_{Elm}(\mathbf{r}) = m\psi_{Elm}(\mathbf{r}) \quad (3.118)$$

The solutions have the form

$$\psi(r,\theta,\phi)F(r,\theta)e^{im\phi} \quad (3.119)$$

and m must be an integer in order that $\psi(\mathbf{r})$ be single-valued. Even though according to equation (3.110) the algebra of the L_i allows half-integral l and m, only integers are allowed here for these simple functions. Orbital angular momentum cannot be a half-integral multiple of $\hbar$.

The states are also eigenstates of $\mathbf{L}^2$ with eigenvalues $l(l+1)$. Therefore in addition to the ϕ-dependence (3.118) $\psi_{Elm}(\mathbf{r})$ satisfies

$$\mathbf{L}^2\psi_{Elm}(\mathbf{r}) = -\left(\frac{1}{\sin\theta}\frac{\partial}{\partial\theta}\sin\theta\frac{\partial}{\partial\theta} + \frac{1}{\sin^2\theta}\frac{\partial^2}{\partial\phi^2}\right)\psi_{Elm}(r,\theta,\phi)$$
$$= l(l+1)\psi_{Elm}(r,\theta,\phi) \quad (3.120)$$

Use (3.119) for $\psi(r,\theta,\phi)$ and cancel the ϕ-dependent factor:

$$\mathbf{L}^2 F(r,\theta) = -\left(\frac{1}{\sin\theta}\frac{\partial}{\partial\theta}\sin\theta\frac{\partial}{\partial\theta} - \frac{m^2}{\sin^2\theta}\right)F(r,\theta) = l(l+1)F(r,\theta) \quad (3.121)$$

Spherical Harmonics

Since there is no reference to r in equation (3.121) the solutions have the form[11]

$$\psi_{Elm}(\mathbf{r}) = F(r,\theta)e^{imr} = R_{Elm}(r)Y_l^m(\theta,\phi) \quad (3.122)$$

where

$$\mathbf{L}^2 Y_l^m(\theta,\phi) = -\left(\frac{1}{\sin\theta}\frac{\partial}{\partial\theta}\sin\theta\frac{\partial}{\partial\theta} + \frac{1}{\sin^2\theta}\frac{\partial^2}{\partial\phi^2}\right)Y_l^m(\theta,\phi)$$
$$= l(l+1)Y_l^m(\theta,\phi) \quad (3.123)$$

The functions $Y_l^m(\theta,\phi)$ are the familiar **spherical harmonics** (See Section A.2.2 in the appendix), conventionally normalized by

$$\int |Y_l^m(\theta,\phi)|^2 d\Omega = 1 \quad (3.124)$$

---

[11] Many authors write Y_{lm} instead of Y_l^m.

82   Chapter 3   Stationary States

For $m = 0$, the spherical harmonics are proportional to the Legendre polynomials. You can check that

$$Y_l^0(\theta, \phi) = \sqrt{\frac{2l+1}{4\pi}} P_l(\cos\theta) \tag{A.105}$$

If $P_l(z)$ is a solution to Legendre's equation (A.69), then so is $P_l(-z)$. It follows that the Legendre polynomials are either even or odd. The leading term in $P_l(z)$ goes like z^l, so $P_l(-z) = (-1)^l P_l(z)$. The polynomials $P_l(\cos\theta)$ are even or odd under $\theta \to \pi - \theta$, a reflection in the x-y plane. The spherical harmonics are even or odd under a reflection through the origin:

$$Y_l^m(\pi - \theta, \phi + \pi) = (-1)^l Y_l^m(\theta, \phi) \tag{3.125}$$

The eigenstates of orbital angular momentum have parity $(-1)^l$.

3.4.2   The Radial Wave Equation

The Schrödinger equation (3.117) can now be written

$$\left[-\frac{\hbar^2}{2Mr^2}\left(\frac{\partial}{\partial r} r^2 \frac{\partial}{\partial r} - l(l+1)\right) + V(r)\right] R_{Elm}(r) Y_l^m(\theta, \phi)$$
$$= E R_{Elm}(r) Y_l^m(\theta, \phi) \tag{3.126}$$

Cancel the Y_l^m to obtain the **radial wave equation**

$$\left[-\frac{\hbar^2}{2Mr^2}\left(\frac{d}{dr} r^2 \frac{d}{dr} - l(l+1)\right) + V(r)\right] R_{El}(r) = E R_{El}(r) \tag{3.127}$$

Evidently $R_{Elm}(r)$ is independent of m, so henceforth I will write it as just $R_{El}(r)$. The normalization (3.124) for the spherical harmonics then implies that for normalized states the radial factors satisfy

$$\int_0^\infty R_{El}(r) r^2 dr = 1 \tag{3.128}$$

Note on the "separation of variables" technique: If you already know that method for solving equations like (3.117), the argument here might seem a bit convoluted. But note that

- These eigenvalues of $\mathbf{L}^2$ and L_z are not just properties of these ordinary differential equations. We knew what they had to be before we started, from the general discussion of the angular momentum algebra.

- The separation of variables method does not prove that these products of functions are all the solutions, but that is what I have shown here. Any solution of the Schrödinger equation can be expanded in functions of the form $R_{El}(r) Y_l^m(\theta, \phi)$.

This is about as far as one can go toward finding the solution of the equation (3.78) or (3.117) without specifying the potential $V(r)$.

Equation (3.127) is much simpler that equation (3.78): It is an ordinary differential equation, not a partial differential equation. If necessary it is possible find a numerical solution.

You can save some writing by setting[12]

$$k^2 = \frac{2mE}{\hbar^2} \quad \text{and} \quad U(r) = \frac{2mV(r)}{\hbar^2} \tag{3.129}$$

and replacing R_{El} by

$$u_{El}(r) = rR_{El}(r) \tag{3.130}$$

Then (the prime means differentiation with respect to r)

$$u''_{El} + \left(k^2 - \frac{l(l+1)}{r^2}\right) u_{El} - U(r) u_{El} = 0 \tag{3.131}$$

Equation (3.131) looks like the one-dimensional Schrödinger equation (3.18) with an additional repulsive potential $\hbar^2 l(l+1)/2mr^2$, except that it has the boundary condition $u(0) = 0$ absent for one-dimensional wave functions. Since the centrifugal term is repulsive and grows with l, the energies of the bound states should increase with l.

Asymptotic Properties of the Wave Function

Some asymptotic properties of the radial wave-functions can be read directly off the equation. Provided the potential falls off fast enough $u_{El}(r) \to e^{\pm ikr}$.[13] If $E > 0$,

$$R_{El}(r) \xrightarrow[r \to \infty]{} \frac{1}{r} e^{ikr} \quad \text{or} \quad R_{El}(r) \xrightarrow[r \to \infty]{} \frac{1}{r} e^{-ikr} \tag{3.132}$$

(or a linear combination of these). When $E < 0$, as for bound states,

$$R_{El}(r) \xrightarrow[r \to \infty]{} \frac{1}{r} e^{-\alpha r} \tag{3.133}$$

where $\alpha = \sqrt{-2mE/\hbar^2}$.

There is also a general rule for the $r \to 0$ limit. In that limit $V(r)$ is negligible [a potential that goes as fast as $1/r^2$ as $r \to 0$ will be unacceptable] and the centrifugal term dominates. There are two possible solutions:

$$R_{El}(r) \sim r^l \quad \text{or} \quad R_{El} \sim r^{-l-1} \tag{3.134}$$

The second is unacceptable for finite wave functions, since $l \geq 0$.

The claim that solutions with $l = 0$ and $R(r) \sim 1/r$ are not acceptable requires some explanation, since they do not appear to violate the normalization condition (3.128). The reason is that $\nabla^2(1/r) = -4\pi \delta_3(\mathbf{r})$. So in fact, functions that are proportional to $1/r$ do not solve the original three-dimensional equation (3.78); there is an extra delta function at the origin.

---

[12] k is imaginary when $E < 0$.
[13] The exact condition is $V(r) <$ constant$/r$ for large enough r.

84   Chapter 3   Stationary States

Note: Dirac delta functions are really instructions for the values of integrals that contain them. The appearance of the delta function in the Laplacian means that if you thought a function going like $1/r$ at the origin was a solution to (3.78), you would discover that the expectation value of the Hamiltonian was not the energy E on the right, but had an extra term.

Free Particle Radial Wave Functions

An elementary but important example is a free particle: $V = 0$. The radial wave function then depends on k and r only through the combination kr. Set $\rho = kr$. Then

$$u_l'' + \left(1 - \frac{l(l+1)}{\rho^2}\right) u_l = 0 \tag{3.135}$$

where now the primes indicates differentiation with respect to ρ.

The solutions to equation (3.135) are proportional to the functions $\rho j_l(\rho)$, where

$$j_l(\rho) = \sqrt{\frac{\pi}{2\rho}} J_{l+\frac{1}{2}}(\rho) \tag{3.136}$$

are the **spherical Bessel functions**. Some of the properties of these and related functions are listed in Appendix A.2.3.

3.5 HYDROGENLIKE ATOMS

3.5.1 The Symmetries

The rapid acceptance of quantum mechanics in the 1920s was due in large part to the dramatic agreement of the one-electron atom spectra with the theoretical prediction. The Hamiltonian of a hydrogen atom, or a one electron ion like He$^+$ or Li^{++} is

$$H = \frac{\mathbf{p}^2}{2m} - \frac{Z^2}{e} r \tag{3.137}$$

Here e is the magnitude of the electron or proton charge, and Z is the atomic number of the nucleus. For hydrogen, $Z = 1$. Since the dimensions of e^2 are the same as the dimensions of the product of $\hbar$ and c it is useful to write it as

$$e^2 = \alpha \hbar c \tag{3.138}$$

α is called the **fine structure constant**, for reasons explained in Chapter 7. α is dimensionless, and its value is about $1/137.036$.

Remember the Runge-Lenz vector in the classical mechanical treatment of the Kepler problem (Problem 1.2).

$$\mathbf{A} = \mathbf{L} \times \mathbf{p} + Ze^2 m \frac{\mathbf{r}}{r} \tag{3.139}$$

The vector $\mathbf{A}$ is a constant of the motion. It points in the direction of the major axis of the orbit, which stays put because the orbits are closed. In classical mechanics the components of $\mathbf{A}$ are three conserved quantities special to the Kepler problem.

There is a minor difficulty in using $\mathbf{A}$ in quantum mechanics. The operator[14]

$$\mathbf{A} = \hbar \mathbf{L} \times \mathbf{p} + Ze^2 m \frac{\mathbf{r}}{r} \tag{3.140}$$

is not Hermitean, since $\mathbf{p}$ and $\mathbf{L}$ do not commute. But the classical theory does not determine the order of the factors p_j and L_k. Here the ordering ambiguity is fixed by the requirement that observables be Hermitean operators. The correct form is

$$\mathbf{A} = \frac{1}{2}(\hbar \mathbf{L} \times \mathbf{p}) - \frac{1}{2}(\mathbf{p} \times \hbar \mathbf{L}) + Ze^2 m \frac{\mathbf{r}}{r} \tag{3.141}$$

You can check that the vector $\mathbf{A}$ defined by (3.141) is conserved, that is, that it commutes with the Hamiltonian (see Problem 3.9).

Besides the three L_i, the three components A_i also commute with H. So in quantum mechanics the Coulomb problem has more symmetries, more observables that commute with the Hamiltonian, than does general motion with spherical symmetry, and the energy levels have more than the $2l+1$ degeneracy.

I will work out the commutator algebra of these six operators, the generalization of the rules (3.88) for the three L_i alone.[15] The commutators can be computed from the known commutators for the p_i and the r_i. They are

$$\boxed{[L_i, L_j] = i \sum_k \epsilon_{ijk} L_k \qquad [L_i, A_j] = i \sum_k \epsilon_{ijk} A_k} \tag{3.142a}$$

$$\boxed{[A_i, A_j] = -2im\hbar^2 H \sum_k \epsilon_{ijk} L_k} \tag{3.142b}$$

Computing (3.142b) can get tedious, but is elementary (see Problem 5.13).

Equations (3.142) are another example of a quantum mechanical system defined on a finite dimensional space. There are five observables that all commute with the Hamiltonian and each other, and the eigenstates of H can be chosen to be eigenstates not only of $\mathbf{L}^2$ and L_z, but also of $\mathbf{A}^2$ and A_z. None of the six operators L_i and A_i changes the energy eigenvalue.

The Coulomb problem has a larger symmetry than the problem of an arbitrary spherically symmetric potential, and so there are more states with the same energy.

Note: Before going into all the algebra, be sure you understand why this is so. First review how $[L_i, H] = 0$ implies degeneracy. For any problem with spherical symmetry, the states $|E, l, m\rangle$ and

$$L_+|E, l, m\rangle = \sqrt{l(l+1) - m(m+1)}|E, l, m+1\rangle \tag{3.143}$$

have the same energy. Why? From equation (3.97b)

$$L_z L_+|E, l, m\rangle = L_+ L_z |E, l, m\rangle + L_+|E, l, m\rangle \tag{3.144}$$

whence

$$L_z L_+|E, l, m\rangle = (m+1) L_+|E, l, m\rangle \tag{3.145}$$

---
[14] Remember I use $\mathbf{r} \times \mathbf{p} = \hbar \mathbf{L}$ in quantum mechanics.
[15] Runge [2], Lenz [3], Pauli [4].

86 Chapter 3 Stationary States

The state $L_+|E,l,m\rangle$ is an eigenstate of L_z with eigenvalue $m+1$. And then from

$$HL_+|E,l,m\rangle = L_+H|E,l,m\rangle = EL_+|E,l,m\rangle \qquad (3.146)$$

This state is also an eigenstate of the Hamiltonian with the same energy, unless it is zero. It cannot *always* be zero, since L_+ is not the zero operator. That's how the degeneracy of energy eigenstates came about.

In a similar fashion, define $A_+ = A_x + iA_y$. Then

$$[L_z, A_+] = A_+ \qquad (3.147)$$

Consider a state $|E,l,m\rangle$. Imitating the argument above for L_+, one arrives at

$$L_z A_+|E,l,m\rangle = (m+1)A_+|E,l,m\rangle \qquad (3.148)$$

That is, $A_+|E,l,m\rangle$ is an eigenstate of L_z with eigenvalue $m+1$, and is also an eigenstate of H with the same energy. In particular, apply A_+ to the top state $|E,l,l\rangle$ of some $|E,l,m\rangle$ ladder. The new state has $m = l+1$, so if $A_+|E,l,m\rangle$ is not zero, it must belong to a *different* ladder. Two ladders, with different l, are degenerate.

To find out what the degeneracy is, we have to work out the problem in more detail.

3.5.2 The Energy Spectrum

From the commutators of the angular momentum algebra we were able to show that the energy spectrum consists of $2l+1$ degenerate states, with the stated eigenvalues of $\mathbf{L}^2$ and L_z. I want to make the same kind of analysis for the larger algebra just described. I will restrict the states to ones with same energy E, and simplify the problem a bit by rescaling $\mathbf{A}$:

$$\mathbf{A} = \sqrt{-2mE\hbar^2}\,\mathbf{M} \qquad (3.149)$$

The six-dimensional algebra of the L_i and the M_i is

$$[L_i, L_j] = i\sum_k \epsilon_{ijk} L_k \qquad [L_i, M_j] = i\sum_k \epsilon_{ijk} M_k \qquad (3.150a)$$

$$[M_i, M_j] = \frac{1}{-2mE\hbar^2}[A_i, A_j] = i\sum_k \epsilon_{ijk} L_k \qquad (3.150b)$$

Finding all the representations of this larger algebra looks a lot harder than finding all the representations of the three L_i. Fortunately, the larger problem can be reduced to the smaller one here, and no new detailed computation is required. Define the combinations $\mathbf{J}$ and $\mathbf{K}$ as follows:

$$\mathbf{J} = \frac{\mathbf{L}+\mathbf{M}}{2} \qquad \mathbf{K} = \frac{\mathbf{L}-\mathbf{M}}{2} \qquad (3.151)$$

Then

$$[J_i, J_j] = i\sum_k \epsilon_{ijk} J_k \qquad [K_i, K_j] = i\sum_k \epsilon_{ijk} K_k \qquad [J_i, K_j] = 0 \qquad (3.152)$$

The operators J_i and the operators K_i independently form standard angular momentum type algebras, and they commute with other. That is all there is to it! Label the states (suppressing E) by $|j, m_j, k, m_k\rangle$, eigenstates of the following operators:

$$\mathbf{J}^2|j, m_j, k, m_k\rangle = j(j+1)|j, m_j, k, m_k\rangle$$
$$\mathbf{K}^2|j, m_j, k, m_k\rangle = k(k+1)|j, m_j, k, m_k\rangle$$
$$J_z|j, m_j, k, m_k\rangle = m_j|j, m_j, k, m_k\rangle$$
$$K_z|j, m_j, k, m_k\rangle = m_k|j, m_j, k, m_k\rangle \quad (3.153)$$
$$J_+|j, m_j, k, m_k\rangle = \sqrt{j(j+1) - m_j(m_j+1)}|j, m_j+1, k, m_k\rangle$$
$$K_+|j, m_j, k, m_k\rangle = \sqrt{k(k+1) - m_k(m_k+1)}|j, m_j, k, m_k+1\rangle$$

where j and k can be any integers or half-integers—there is no reason to exclude the half-integer representations here—and

$$-j \leq m_j \leq j \quad -k \leq m_k \leq k \quad (3.154)$$

For each choice of j and k, there are $(2j+1)(2k+1)$ states. The operators defined by (3.153) on these $(2j+1)(2k+1)$ states $|j, m_j, k, m_k\rangle$ will solve the commutation relations (3.152).

There is one more simplification. The operators $\mathbf{A}$ and $\mathbf{L}$ are orthogonal (classically, $\mathbf{A}$ is in the plane of the orbit, and $\mathbf{L}$ is perpendicular to that plane). Not every solution to the commutator algebra problem is allowed because $\mathbf{L} \cdot \mathbf{M} = 0$, which using (3.151) becomes $\mathbf{J}^2 = \mathbf{K}^2$. So the only states that occur in our problem have $j = k$, and each level has $(2j+1)^2$ states. Set $n = 2j+1$. Then each energy level has n^2 states. Since j may be any nonnegative integer or a half integer, n can be any positive integer.

What is the energy? A little computation gives (see Problem 3.9)

$$\mathbf{A}^2 = 2mE(\hbar^2 \mathbf{L}^2 + \hbar^2) + m^2 Z^2 e^4 \quad (3.155)$$

in agreement with the classical result

$$\mathbf{A}^2 = 2mE\mathbf{L}^2 + m^2 Z^2 e^4 \quad (3.156)$$

as $\hbar \to 0$. In terms of the vector $\mathbf{M}$ equation (3.155) is

$$\mathbf{M}^2 + \mathbf{L}^2 + 1 = -\frac{mZ^2 e^4}{2E\hbar^2} \quad (3.157)$$

Now

$$\mathbf{M}^2 + \mathbf{L}^2 = (\mathbf{J}+\mathbf{K})^2 + (\mathbf{J}-\mathbf{K})^2 = 2(\mathbf{J}^2 + \mathbf{K}^2) = 4\mathbf{J}^2 = 4j(j+1) \quad (3.158)$$

so

$$\boxed{E = -\frac{mZ^2 e^4}{2\hbar^2}\frac{1}{\mathbf{M}^2 + \mathbf{L}^2 + 1} = -\frac{Z^2\alpha^2 mc^2}{2(4j^2+4j+1)} = -\frac{Z^2\alpha^2 mc^2}{2n^2}} \quad (3.159)$$

where $n = 2j + 1$ can be any positive integer.

Equation (3.159) is Bohr's famous formula for the energy levels of one-electron atoms.[16] Notice that $L_z = J_z + K_z$, so that the eigenvalue m of L_z is $m_k + m_j$, which must be between $-2j$ and $2j$. Since $|m|$ cannot exceed l, the allowed values of l in the n-th level are not greater than $2j = n - 1$. A simple counting argument will show that each value of l from 0 through $n-1$ occurs once and only once in the n-th energy level.

3.5.3 The Radial Wave Functions

It will be useful to know the explicit form of the radial wave functions for hydrogenlike atoms. It is possible to discover them by applying the operators of the **L**, **A** algebra, as I did for the harmonic oscillator problem in Section 3.2. But that procedure gets complicated here. Instead, it is important to review the series solution technique.

The wave functions are

$$\psi_{Elm}(\mathbf{r}) = R_{El}(r) Y_l^m(\theta, \phi) \tag{3.160}$$

The first few radial functions for $Z = 1$ are tabulated in Table 3.1.

$$a^{\frac{3}{2}} R_{10}(r) = 2 e^{-r/a}$$

$$a^{\frac{3}{2}} R_{20}(r) = \tfrac{1}{\sqrt{2}} \left(1 - \tfrac{r}{2a}\right) e^{-r/2a} \qquad a^{\frac{3}{2}} R_{21}(r) = \tfrac{1}{2\sqrt{6}} \tfrac{r}{a} e^{-r/2a}$$

$$a^{\frac{3}{2}} R_{30}(r) = \tfrac{2}{3\sqrt{3}} \left(1 - \tfrac{2r}{3a} + \tfrac{2r^2}{27a^2}\right) e^{-r/3a} \qquad a^{\frac{3}{2}} R_{31}(r) = \tfrac{8}{27\sqrt{6}} \tfrac{r}{a} \left(1 - \tfrac{r}{6a}\right) e^{-r/3a}$$

$$a^{\frac{3}{2}} R_{32}(r) = \tfrac{4}{81\sqrt{30}} \left(\tfrac{r}{a}\right)^2 e^{-r/3a}$$

$$a^{\frac{3}{2}} R_{40}(r) = \tfrac{1}{4} \left(1 - \tfrac{3r}{4a} + \tfrac{r^2}{8a^2} - \tfrac{r^3}{192a^3}\right) e^{-r/4a}$$

$$a^{\frac{3}{2}} R_{41}(r) = \tfrac{\sqrt{5}}{16\sqrt{3}} \tfrac{r}{a} \left(1 - \tfrac{r}{4a} + \tfrac{r^2}{80a^2}\right) e^{-r/4a}$$

$$a^{\frac{3}{2}} R_{42}(r) = \tfrac{1}{64\sqrt{5}} \left(\tfrac{r}{a}\right)^2 \left(1 - \tfrac{r}{12a}\right) e^{-r/4a} \qquad a^{\frac{3}{2}} R_{43}(r) = \tfrac{1}{768\sqrt{35}} \left(\tfrac{r}{a}\right)^3 e^{-r/4a}$$

TABLE 3.1: Hydrogen atom radial wave functions for $n \leq 4$

They can be calculated as follows: Make the substitutions suggested in equations (3.129) and (3.130). Then

$$\frac{d^2}{dr^2} u_{El} - \left(\kappa^2 + \frac{l(l+1)}{r^2}\right) u_{El} + \frac{2Zmc\alpha}{\hbar r} u_{El} = 0 \tag{3.161}$$

---

[16] In 1885 Balmer [1] pointed out that the frequencies of the four known hydrogen spectral lines are proportional to $(m^2 - 4)/m^2$ for $m = 3, 4, 5 \cdots$. Since the spectrum was first identified by Balmer, the energy level formula is often called the Balmer formula.

Section 3.5 Hydrogenlike Atoms

where $\kappa^2 = -k^2$ and $u_{El}(r) = rR_{El}(r)$. $u_{El}(r)$ has to be normalizable, so that

$$\int_0^\infty u_{El}(r)^2 \, dr = 1 \tag{3.162}$$

Change variables from E to n, where

$$E = -\frac{Z^2\alpha^2 mc^2}{2n^2} = -\frac{Z^2\hbar^2}{2mn^2a^2} \tag{3.163}$$

$a = \hbar/mc\alpha$ has the dimensions of length, and is called the **Bohr radius**.

The radial equation (3.161) now reads

$$\frac{d}{dr^2}u(r) + \left(-\frac{Z^2}{(na)^2} - \frac{l(l+1)}{r^2} + \frac{2Z}{ar}\right)u(r) = 0 \tag{3.164}$$

Change variables from r to $\rho = Zr/na$ (now the primes mean differentiation with respect to ρ):

$$u''(\rho) + \left(-1 - \frac{l(l+1)}{\rho^2} + \frac{2n}{\rho}\right)u(\rho) = 0 \tag{3.165}$$

First, factor out the known behavior of u as $\rho \to 0$. Set

$$u(\rho) = \rho^{l+1}g(\rho) \tag{3.166}$$

Then

$$g''(\rho) + \frac{2(l+1)}{\rho}g'(\rho) + \left(\frac{2n}{\rho} - 1\right)g(\rho) = 0 \tag{3.167}$$

For large ρ, $u'' \to u$, so $u \to \rho^m e^{\pm\rho}$ for any m. The plus sign is unacceptable. It leads to functions that cannot be normalized. Therefore set $g(\rho) = e^{-\rho}f(\rho)$. Then $f(\rho)$ should be a polynomial satisfying

$$\rho f''(\rho) + 2(l+1-\rho)f'(\rho) + 2(n-l-1)f(\rho) = 0 \tag{3.168}$$

The function $f(\rho)$ has a series expansion:

$$f(\rho) = \sum_{m=0}^\infty a_m \rho^m \tag{3.169a}$$

$$f'(\rho) = \sum_{m=0}^\infty m a_m \rho^{m-1} \tag{3.169b}$$

$$f''(\rho) = \sum_{m=0}^\infty m(m-1) a_m \rho^{m-2} \tag{3.169c}$$

so

$$\sum_m a_m \left[(m(m-1) + 2m(l+1))\rho^{m-1} + 2(n-l-1-m)\rho^m\right] = 0 \tag{3.170}$$

The coefficient of each power of ρ must separately vanish:

$$a_{m+1} = -2a_m \frac{n-l-1-m}{(m+1)(m+2l+2)} \tag{3.171}$$

From this recursion relation all the a_m can be computed in terms of a_o; this last coefficient must be determined by the normalization condition.

If the series is infinite, then for large m, $a_{m+1}/a_m \to 2/m$; it behaves like $e^{2\rho}$, and the solution to the original equation behaves like e^ρ. This is the unacceptable, nonnormalizable solution. Therefore the series must terminate for some m. The condition is

$$n - l - 1 = m \tag{3.172}$$

for some nonnegative integer m. The number n must be a positive integer, starting with $n = 1$, and the allowed energy levels are again those discovered in equation (3.159).

$$E_n = -\frac{Z^2\alpha^2 mc^2}{2n^2} \quad (n = 1, 2, 3 \ldots) \tag{3.173}$$

The functions $R_{nl}(r)$ can be gotten from the recursion relation for the a_m and the normalization condition. They are related to the associated Laguerre polynomials. Some of the radial eigenfunctions are listed in Table 3.1 and plotted in Figures 3.3 and 3.4.

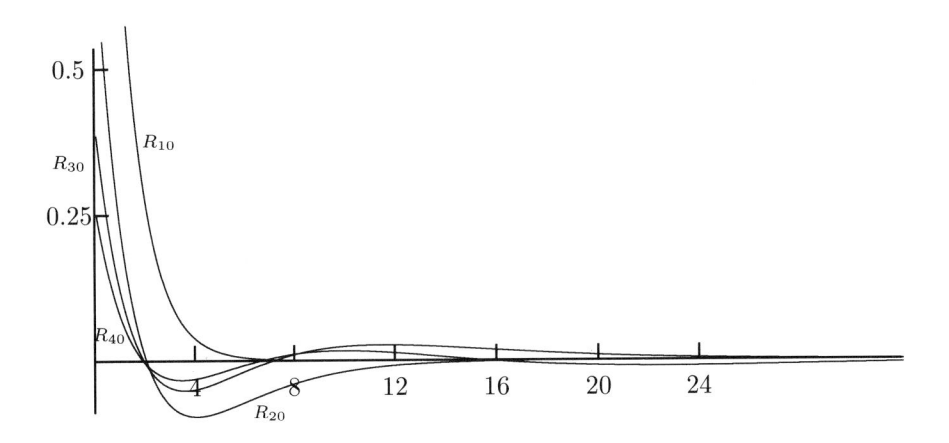

FIGURE 3.3: The radial functions $a^{\frac{3}{2}}R_{n0}(r/a)$, for $n = 1, 2, 3, 4$

In each level there is a $2l + 1$ angular momentum ladder of states for each nonnegative l less than n. The total number of states in each level is

$$\sum_{l=0}^{n-1}(2l+1) = n^2 \tag{3.174}$$

The radial wave function method completely reproduces the results of the algebraic method. But it does not explain at all why the problem has the extra degeneracy.

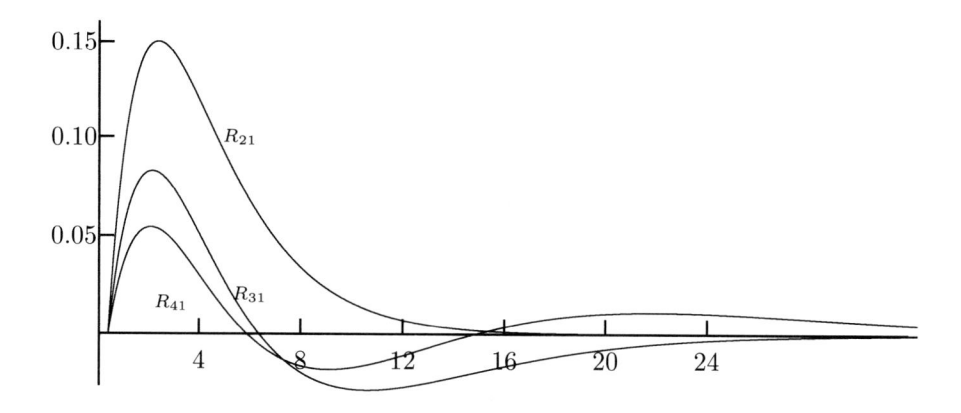

FIGURE 3.4: The radial functions $a^{\frac{3}{2}} R_{n1}(r/a)$, for $n = 2, 3, 4$

PROBLEMS

3.1. One-Dimensional Square Well

Consider a particle moving in a one-dimensional attractive square well (Figure 3.5):

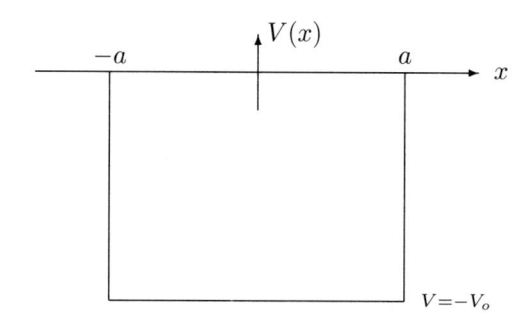

$$V(x) = -V_o \quad (|x| < a)$$
$$V(x) = 0 \quad (|x| > a)$$

FIGURE 3.5: A one-dimensional square well potential

(a) Describe the bound-state energy eigenfunctions and the values of the energy without using reflection symmetry.

(b) Next define a reflection operator $\mathcal{R}$ by $\mathcal{R}\psi(x) = \psi(-x)$. Show that the commutator $[\mathcal{R}, H] = 0$ (H is the Hamiltonian), and that the energy eigenfunctions can also be chosen to be eigenfunctions of $\mathcal{R}$. Show that the only possible eigenvalues of $\mathcal{R}$ (the *parity* of the state) are ± 1.

(c) Which of the bound states have even parity, and which have odd parity? What is the parity of the lowest state? What is the condition that there be an odd-parity bound state? What is the condition that there be an even-parity bound state?

3.2. One-Dimensional Delta-Function Potential

Find the bound-state energy levels, if any, of a one-dimensional particle in the potential
$$V(x) = -a\delta(x) \qquad a > 0$$
This problem should take under half a page!

3.3. Oscillator Strength Sum Rule

(a) Consider a one-dimensional particle with Hamiltonian of the form
$$H = \frac{p^2}{2m} + V(x)$$
Show that
$$[H, x] = -i\hbar \frac{p}{m}$$

(b) Suppose that $|\psi_n\rangle$ are a complete set of energy eigenstates with energies E_n:
$$H|\psi_n\rangle = E_n|\psi_n\rangle$$
Use the result of part (a) to evaluate
$$\sum_n (E_n - E_k)|\langle\psi_k|x|\psi_n\rangle|^2 = \; ?$$
for any $|\psi_k\rangle$. The sum is over all n. This is an important sum rule in atomic spectroscopy. Hint: evaluate $\big[[H, x], x\big]$.

3.4. Double Delta-Function Potential

A one-dimensional particle of mass m moves in an attractive potential
$$V(x) = -V_o b\big[\delta(x - b) + \delta(x + b)\big]$$
where $b > 0$ and $V_o > 0$, as in Figure 3.6.

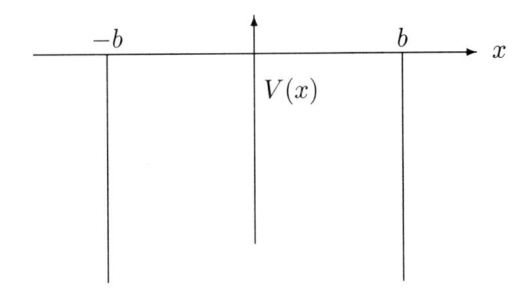

FIGURE 3.6: A one-dimensional potential has Dirac delta functions at $x = \pm b$.

(a) What is the form of the wave function for a bound state with even parity?
(b) Find the expression that determines the bound-state energies for even parity states, and determine graphically how many even parity bound states there are.

(c) Solve for the even parity bound state analytically in the limit $mV_o b^2/\hbar^2 \ll 1$.
(d) Repeat parts (a) and (b) for odd parity. For what values of V_o and b are there bound states?
(e) Find the even and odd parity state binding energies for $mV_o b^2/\hbar^2 \gg 1$. Explain physically why these energies move closer and closer together as $b \to \infty$.

3.5. Infinite Well with Delta Function
A one-dimensional particle is confined to the region $|x| \leq a$, with a delta-function potential at the origin:

$$V(x) = \beta E_o a\, \delta(x) \qquad (-a \leq x \leq a)$$
$$V(x) = \infty \qquad \text{(elsewhere)}$$

Here E_o is the ground state energy, and β is a dimensionless parameter.
(a) Find a (transcendental) equation for the bound-state energies.
(b) Expand your answer in the small number β through second order, that is, write

$$E_n = E_n^{(0)} + \beta E_n^{(1)} + \beta^2 E_n^{(2)} \cdots$$

3.6. Exponential Wave Function
The normalized wave function of a one-dimensional particle is

$$\langle x | \psi \rangle = \psi(x) = N \left[e^{\kappa x} \Theta(-x) + e^{-\kappa x} \Theta(x) \right]$$

for some $\kappa > 0$. Here $\Theta(x)$ is the step function, and N is real and positive.
(a) What is N?
(b) What is the expectation value of x^2?
(c) What is the momentum space wave function $\langle p | \psi \rangle$?
(d) What is the expectation value of p^2?
(e) Verify the expectation values of x^2 and p^2 satisfy the uncertainty principle inequality.
(f) Suppose the Hamiltonian is

$$H = \frac{p^2}{2m} - a\delta(x)$$

as in Problem 3.2, for some real positive a. Choose κ to be the correct value for the energy eigenstate you found in that problem. Compute the expectation value $\langle \psi | H | \psi \rangle$ and show that it agrees with the bound-state energy.

3.7. Coupled Oscillators
Three particles in one dimension with masses m_i are connected by two identical springs, each with natural length a and spring constant β. Let the two particles on the ends have the same mass ($m_1 = m_3 = M$). The particle whose coordinate is x_2 is the middle, and has a different mass ($m_2 \neq M$), so the Hamiltonian is

$$H = \frac{p_1^2}{2M} + \frac{p_2^2}{2m_2} + \frac{p_3^2}{2M} + \frac{\beta}{2}\left[(x_2 - x_3 - a)^2 + (x_1 - x_2 - a)^2\right]$$

Find the energy levels in the center of mass frame. (Hint: Rewrite H as a sum of three independent oscillators, that is, find its normal modes.)

94 Chapter 3 Stationary States

3.8. Quantum Virial Theorem
(a) Let A be any well-defined Hermitean operator, and let $|\psi\rangle$ be a normalized bound state with definite energy E:
$$H|\psi\rangle = E|\psi\rangle$$
What is the expectation value of the commutator $[A, H]$ in the state $|\psi\rangle$?

(b) Let $H = \mathbf{p}^2/2m + V(r)$ be the Hamiltonian for a three-dimensional particle in a local, spherically-symmetric potential. Show that
$$[H, \mathbf{r}] = -i\hbar \frac{\mathbf{p}}{m} \quad \text{and} \quad [H, \mathbf{p}] = i\hbar \boldsymbol{\nabla} V(r)$$
Then, using the result of part (a) with $A = \mathbf{r} \cdot \mathbf{p}$, prove that
$$2\langle\psi|T|\psi\rangle = \left\langle\psi\left|r\frac{\partial V}{\partial r}\right|\psi\right\rangle$$
where T is the kinetic energy and $|\psi\rangle$ is an energy eigenstate. This is the quantum version of the virial theorem.

(c) If $V(r)$ is proportional to r^m, show that for an energy eigenstate with energy E,
$$E = \left(1 + \frac{m}{2}\right)\langle\psi|V|\psi\rangle$$

3.9. Runge-Lenz Vector
In the classical mechanics of the Kepler problem, you studied the Runge-Lenz vector
$$\mathbf{A} = (\mathbf{L} \times \mathbf{p}) + \frac{km\mathbf{r}}{r}$$
and found an expression for $\mathbf{A}^2$ (see Problem 1.2).

(a) Show that in quantum mechanics, where L_i, p_i and r_i are operators, that
$$\mathbf{A} = (\hbar \mathbf{L} \times \mathbf{p}) + \frac{Ze^2 m \mathbf{r}}{r}$$
is not Hermitean. Redefine $\mathbf{A}$ as in equation (3.141):
$$\mathbf{A} = \frac{1}{2}(\hbar \mathbf{L} \times \mathbf{p}) - \frac{1}{2}(\mathbf{p} \times \hbar \mathbf{L}) + \frac{Ze^2 m \mathbf{r}}{r}$$
This new $\mathbf{A}$ reduces to the classical observable in the limit $\hbar \to 0$, since in that limit all operators commute. Show that the redefined $\mathbf{A}$ is Hermitean.

(b) Derive equation (3.155) for the quantum mechanical $\mathbf{A}^2$ in terms of E, Ze^2, $\mathbf{L}^2$, and so forth.

(c) Show that the components of $\mathbf{A}$, defined as the Hermitean operators (3.141), indeed commute with the Hamiltonian (3.137).

Suggestion: Begin by making a catalog of commutators like $[p_i, r^n]$ and $[\mathbf{p}^2, r^n]$. Then if you're careful the calculation should not take much more than a page.

3.10. Harmonic Oscillator Wave Functions
This problem is worked out in many elementary quantum mechanics textbooks. Try to do it first on your own.

(a) Write the time-independent Schrödinger equation for the one-dimensional harmonic oscillator: $V(x) = m\omega^2 x^2/2$.

(b) How do the wave functions behave asymptotically for large x?
(c) Factor out this high x behavior, and write the differential equation satisfied by the remaining factor. (Hint: Change variables to $y = cx$, where c is chosen to eliminate the dimensionful factors.)
(d) Express the solution as a power series of the form

$$H(y) = \sum_{n=0}^{\infty} a_n y^n$$

and show that if this series does not terminate, the wave functions will not be normalizable.
(e) What is the condition that the functions $H(y)$ be polynomials (i.e., that the series above has only a finite number of terms)? Use this condition to calculate the allowed energies.
(f) Compute explicitly the normalized wave functions for the two lowest energies.

3.11. Spherical Square Well

Consider a particle in an attractive spherical square well as in Figure 3.7. The potential is

$$V(\mathbf{r}) = -V_0 \Theta(a - r)$$

Here $V_0 > 0$. a is also a positive number, called the **range** of the potential. Θ is the step function. The eigenvalue condition is that the radial wave function and its derivative must be continuous (why?) at $r = a$.[17]

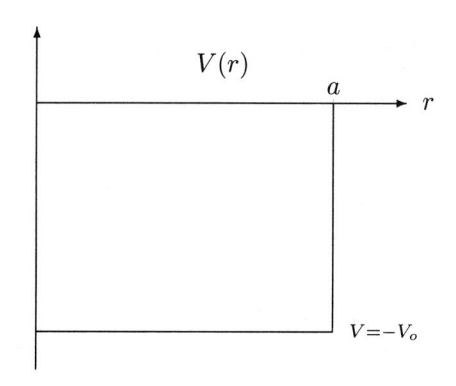

FIGURE 3.7: An attractive spherical square well potential

(a) Write down a transcendental equation for the energies of the $l = 0$ states.
(b) For fixed range a, what is the minimum value of the depth V_0 of the well for which there is a bound state?
(c) As V_0 is increased, at what values do more $l = 0$ and $l = 1$ bound states begin to appear?
(d) At what V_0 do $l = 2$ bound states begin to appear?

---

[17]Some properties of the spherical Bessel functions and related functions are listed in Section A.2.3 of the appendix.

(e) The one-dimensional square well always has a bound state. That is not true in three dimensions, even though the equation for $u_o = rR_o$ is the same as in the one-dimensional case. What is the difference between the one- and the three-dimensional cases?

3.12. Nuclear Forces I

As a first approximation, the force between a neutron and a proton can be described by an attractive square-well potential, like the one in Problem 3.11. The range of the force between the neutron and the proton is known to be of the order of the Compton wavelength of the pion, $\hbar/mc$, where m is the pion mass, about 139 MeV.[18]

(a) If this simple square-well is an adequate description of the neutron-proton system, what is the minimum depth the potential must have? What upper bound can be put on this depth from the knowledge that there are not *two* neutron-proton bound states? Remember to use the "reduced mass" (see Problem 2.9). The neutron and proton have approximately equal masses (about 940 MeV).[19]

(b) Now turn the problem around, and ask what is V_o for a given E. Improve on these estimates using the known binding energy 2.22 MeV for the deuteron. (This is a numerical computation. Try to invent an efficient numerical or graphical method, but do not spend too much time on it.)

3.13. Delta-Shell Potential

Consider a particle moving in a three-dimensional attractive delta-shell potential:

$$V(r) = -V_o a \delta(r - a)$$

Here again $V_o > 0$, and $\delta(r - a)$ is the ordinary one-dimensional Dirac delta function.[20] For the analogous one-dimensional problem, there is always one and only one bound state.

(a) Write the radial equation for this problem (it is similar to the one-dimensional potential in Problem 3.2).

(b) First specialize to $l = 0$. What is the minimum value of V_o needed for there to be a bound state? How many bound states can there be for $V > V_o$?

(c) For general l, show that there is a bound state provided $\lambda > 2l + 1$, where λ is the dimensionless quantity $\lambda = 2mV_o a^2/\hbar^2$.

> Do not spend too much time on this part—it is easy to get stuck. You may need some properties of the spherical Bessel and Hankel functions, and remember the Wronskian!

3.14. Nuclear Forces II

(a) Take the delta-shell approximation in the previous problem as a model for the neutron-proton system. If the range a is again the pion Compton wavelength (see Problem 3.12), what is the minimum value of V_o (in Mev) for which there is a bound state?

---

[18] Introductory texts often define the Compton wavelength with an extra factor of 2π, the way Compton did. But $\hbar/mc$ is a better value experimentally for the range of nuclear forces.

[19] Do not convert electron volts into ergs or joules. The only dimensional numbers you need to work in these units are $c = 2.9979525 \times 10^{10}$ cm per second and $\hbar c = 197.3286$ MeV-fermi. A fermi is 10^{-13} centimeters. It is usually adequate to approximate these numbers by 3×10^{10} and by 197.

[20] Do not confuse it with $\delta_3(\mathbf{r})$.

(b) What value for V_o is required for the energy of the bound state to be -2.22 Mev?

3.15. Three-Dimensional Harmonic Oscillator

The Hamiltonian for the three-dimensional harmonic oscillator is

$$H = \frac{\mathbf{p}^2}{2m} + \frac{m\omega^2 \mathbf{r}^2}{2}$$

(a) Find the energy levels by solving the Schrödinger equation in Cartesian coordinates, as follows:
Write H as

$$H = H_x + H_y + H_z$$

where each H_i depends on r_i and p_i, but not on any of the other coordinates or momenta. The wave function can be chosen to be an eigenstate of each of the H_i separately, so first find the spectrum E_i of H_i. Then this wave function is an eigenstate of H also, with energy

$$E = E_1 + E_2 + E_3$$

What are the possible energy levels, and what is the degeneracy of each level?

(b) Now solve the same problem in spherical coordinates. Since H is spherically symmetric, you can choose eigenstates of $\mathbf{L}^2$ and L_z. Write the Schrödinger equation for the radial wave function $R(r)$ and for $u(r) = rR(r)$.
As $r \to 0$, $u(r)$ is proportional to r^{l+1}. As $r \to \infty$, $u(r)$ is proportional to

$$u(r) \to \sim r^n e^{-\beta r^\gamma}$$

for some constants n, β, and γ. By inspecting the radial wave equation, find the values of β and γ. Convince yourself that n is not determined by the differential equation.
Next write

$$u_{El}(r) = r^{l+1} e^{-\beta r^\gamma} f_{El}(r)$$

and find the equation satisfied by $f(r)$. Expand $f(r)$ in a power series

$$f_{El}(r) = \sum_k a_k r^k$$

(the a_k depend on E and l) and obtain a recursion relation among the a_k.
Prove that all the odd a_k vanish. Show that if the series does not terminate, then $u_{El}(r)$ increases without limit as $r \to \infty$. Therefore the series must terminate for some even integer k. This is the eigenvalue condition.[21] What are the allowed values of the energy? Find the degeneracy of each level directly from your solution.

3.16. Hydrogen Atom—Ground State Wave Functions

(a) Consider an electron in the eigenstate of the hydrogen atom labeled $|E, l, m\rangle$. Express the expectation value of $1/r$ in terms of a (the Bohr radius) n, l, and m. (Hint: Use the virial theorem, Problem 3.8). What is the expectation value of the kinetic energy in the n-th energy level?

---

[21] The $f(r)$ are "associated Laguerre polynomials" of some order, but this information is not much use.

98 Chapter 3 Stationary States

(b) Consider an electron in the ground state of a hydrogen atom. What is the probability that the electron be found at a distance from the nucleus greater than would be allowed classically for an electron of the same energy?

(c) Compute the expectation values of x^2 and p_x^2 in the ground state of the hydrogen atom, and verify that their product satisfies the uncertainty principle.

3.17. Hydrogen Atom—Spectral Lines

Background: The spectra of radiation emitted when an excited hydrogen atom decays into a lower level are named after the spectroscopists who first detected the lines and measured their frequencies. Radiation from decays into the ground state is called the Lyman series. Decays into the $n = 2$ level are the Balmer series; into the $n = 3$ and $n = 4$ levels the Paschen and Brackett series respectively. Decays into the $n = 1$ level from $n = 2, 3, 4 \ldots$ are called the Lyman-α line, the Lyman-β line, the Lyman-γ line, etc.

(a) What are the wavelengths of the Lyman-α, the Balmer-α and the Paschen-α lines?

(b) What are the wavelengths of the shortest lines (initial state is $n \to \infty$) in each of these series?

(c) Which series do you think was discovered first (and why)?

(d) The frequencies in the Lyman series get closer and closer together as $n \to \infty$. Observing the lines in a gas of atoms, one does not see absolutely sharp lines, since the individual atoms are in motion, and therefore radiation emitted from each atom is slightly Doppler-shifted. Assume that the average speed of an atom is given by the kinetic energy $(3/2)\,kT$, and take T to be room temperature, about 300 Kelvin. Boltzmann's constant is 8.62×10^{-5} eV per degree. At what energy level does the Doppler shift of the Lyman series lines become comparable to the spacing between the lines? What is the answer if $T = 900$ Kelvin?

(e) Show that the radiation emitted by a transition between adjacent levels has the same frequency as the classical rotation frequency of an electron with the same energy in a circular orbit, in the limit $n \to \infty$.

3.18. Momentum in Spherical Coordinates

Write the classical Hamiltonian for a particle in a static, spherically symmetric potential, using spherical coordinates.

(a) On the space of 3-dimensional wave functions in spherical coordinates with the scalar product

$$\int \psi^*(r, \theta, \phi) \psi(r, \theta, \phi) d^3r$$

show that the operator $i\partial/\partial r$ is not Hermitean.

(b) Show that

$$p_r = -i\hbar \left(\frac{\partial}{\partial r} + \frac{1}{r} \right)$$

is the "correct" definition of p_r in the sense that if you use it for p_r in the Hamiltonian, written in spherical coordinates, it gives the same answer as writing the Laplacian in spherical coordinates; and that it is Hermitean.

(c) With the replacement $\mathbf{p} = -i\hbar \nabla$, show that in spherical coordinates.

$$\frac{1}{r}\mathbf{r} \cdot \mathbf{p} = -i\hbar \frac{\partial}{\partial r}$$

(d) While in classical mechanics p_r is just $(1/r)\mathbf{r} \cdot \mathbf{p}$, this cannot be the correct quantum mechanical substitution. Why? Show that the symmetrized version

$$\frac{1}{2}\left[\frac{1}{r}\mathbf{r}\cdot\mathbf{p} + \mathbf{p}\cdot\mathbf{r}\frac{1}{r}\right]$$

is satisfactory, and equivalent to the "correct" differential operator above.

3.19. Expectation Values of Powers of $1/r$ – I

Discussion: Small corrections to the hydrogen atom levels (see Chapter 7) will require the expectation values $\langle nlm|\, r^{-k}\, |nlm\rangle$, in the unperturbed bound states, of several powers of $1/r$, where r is the electron's radial coordinate. Here $|nlm\rangle$ are the bound states of the unperturbed, nonrelativistic, spinless hydrogen atom.

For given l, the Schrödinger equation for the radial function is

$$H_l R_{nl}(r) = E_{nl} R_{nl}(r)$$

with

$$H_l = \frac{1}{2m}\left[p_r^2 + \frac{\hbar^2 l(l+1)}{r^2}\right] - \frac{\alpha\hbar c}{r}$$

Here, p_r is the differential operator (see Problem 3.18)

$$p_r = -i\hbar\left(\frac{\partial}{\partial r} + \frac{1}{r}\right)$$

The wave functions are normalized to

$$\int_0^\infty |R_{nl}(r)|^2\, r^2\, dr = 1$$

It is possible to get a closed form for the radial wave functions using the "generating function" for the associated Laguerre Polynomials, and then to compute formulas for various powers of $1/r$. These methods are complicated and not very transparent. You computed the expectation value of $1/r$ itself in Problem 3.16. This problem and the following ones will show you how to compute some other powers.

Problem: Compute

$$\left\langle nlm\left|\frac{1}{r^2}\right|nlm\right\rangle$$

The trick here is to realize that there are normalizable solutions to the radial wave equation for each nonnegative M whether or not l is an integer[22]. Write the radial wave equation, hold M fixed, and think of it as an equation with a parameter l that can have any real value! Then *differentiate* with respect to l. Use the fact that the differential operator H_l defined above is Hermitean for any real number l to show that[23]

$$\int_0^\infty R_{nl}(r)\left(H_l - E_{nl}\right)\left(\frac{dR_{nl}(r)}{dl}\right) r^2\, dr = 0$$

and thus calculate

$$\left\langle nlm\left|\frac{1}{r^2}\right|nlm\right\rangle = \frac{1}{a^2(l+1/2)n^3}$$

---

[22] Of course if l is not an integer, then the solution is not directly related to the original problem.
[23] This trick is a variant on the Feynman-Hellmann theorem.

3.20. Expectation Values of Powers of $1/r$ – II

Obtain a general recursion relation for expectation values of inverse powers of r in the hydrogen atom bound states. You can do this any way you want to, but I suggest the following steps:

Step (i) The radial wave function $u(r) = rR_{nl}(r)$ satisfies

$$u''(\rho) + \left(-1 - \frac{l(l+1)}{\rho^2} + \frac{2n}{\rho}\right)u(\rho) = 0 \quad (3.165)$$

where n and ρ are defined by

$$E_n = -\frac{\alpha^2 mc^2}{2n^2} \quad \text{and} \quad \rho = \frac{r}{na}$$

Here a is the Bohr radius and E_n is the energy eigenvalue.

Step (ii) For an arbitrary integer k, multiply this equation by $\rho^k u'(\rho)$ and integrate from zero to infinity. Integrate by parts until you have an expression for

$$\int_0^\infty \rho^{k-1}(u')^2 d\rho$$

as a sum of expectation values of powers of ρ. Use

$$\frac{d}{d\rho}(u')^2 = 2u'u'' \quad \text{and} \quad \frac{d}{d\rho}(u)^2 = 2u'u$$

and assume that l is big enough so that the integrals all converge at $\rho \to 0$, and so that all the end-point terms vanish.

Step (iii) Similarly, go back to the equation in Step (i), multiply it by $\rho^{k-1}u(\rho)$, and integrate from zero to infinity. Integrate by parts to get another expression for $\int_0^\infty \rho^{k-1}(u')^2 d\rho$ as a sum of expectation values of powers of ρ.

Step (iv) Combine the last two results to eliminate the integrals over $(u')^2$ and obtain a relation among three powers of ρ.

3.21. Expectation Values of Powers of $1/r$ – III

(a) From the result of Problems 3.19 and 3.20, obtain the expectation values

$$\left\langle nlm \left| \frac{1}{r} \right| nlm \right\rangle, \quad \left\langle nlm \left| \frac{1}{r^3} \right| nlm \right\rangle, \quad \text{and} \quad \left\langle nlm \left| \frac{1}{r^4} \right| nlm \right\rangle$$

(b) Find

$$\langle nlm| r |nlm \rangle \quad \text{and} \quad \left\langle nlm \left| r^2 \right| nlm \right\rangle$$

3.22. Coherent States

(a) Consider a one-dimensional harmonic oscillator. Define the state

$$|\psi\rangle = e^{\beta a^\dagger}|0\rangle$$

where $|0\rangle$ is the ground state and β an arbitrary complex number. Use the Baker-Hausdorff lemma, Problem 2.10, to show that $|\psi\rangle$ is an eigenstate of the lowering operator a with eigenvalue β. a is not Hermitean, so its eigenvalues do not have to be real.

(b) Let $|n\rangle$ be the usual energy eigenstates. Find the coefficients $\langle n|\psi\rangle$, and thus find the eigenstate $|\beta\rangle$ of a, proportional to $|\psi\rangle$, that is normalized to unity.

(c) Since a is not Hermitean, its eigenvectors with different eigenvalues need not be orthogonal. Let $|\beta\rangle$ and $|\gamma\rangle$ be two normalized eigenstates of a. Compute $|\langle\gamma|\beta\rangle|^2$.

(d) What is the expectation value of the energy in the state $|\beta\rangle$?

(e) Compute the expectation values and the dispersions (see Problem 2.1) of x and p in the state $|\beta\rangle$. Is there a simple way to understand the result?

REFERENCES

The algebraic solution to the harmonic oscillator can be found in Chapter 34 of Dirac's book, reference [9] in Chapter 2.

The formula for the hydrogen spectrum was discovered by

[1] J. BALMER, Annalen der Physik und Chemie **25** (1885), 80.

The Runge-Lenz vector was invented for classical mechanics by

[2] C. RUNGE, *Vektoranalysis*, vol. 1, S. Hirzel, 1919.

Lenz used it to try to calculate electromagnetic perturbations of the Bohr orbits with "old quantum theory" methods:

[3] W. LENZ, Zeitschrift für Physik **24** (1924), 197.

The application of the Runge-Lenz vector to the hydrogen atom problem in quantum mechanics is originally due to Pauli:

[4] W. PAULI, Zeitschrift für Physik **36** (1926), 336.

CHAPTER 4

Symmetry Transformations on States

4.1 INTRODUCTION

Isolated nonrelativistic systems always have the space-time symmetries listed in Section 1.2: Translations in time and space, rotations, Galilean boosts, reflection, and time inversion. Reflection and time inversion are discrete; the others are continuous. In relativistic quantum mechanics, Galilean boosts are replaced by Lorentz boosts, and there is a new discrete symmetry, charge conjugation. There may also be "internal" symmetries, like isotopic spin.

4.1.1 Symmetries and Transformations

This and the next chapter are about symmetry transformations, especially rotations, in some detail. I will show that systems invariant under rotations or translations must possess a conserved total angular momentum operator **J** and a total momentum operator **P**, and if time translation is a symmetry then there is a constant energy operator H also. Symmetries constrain the form of the Hamiltonian.

Symmetry principles are not at all obvious. They are invariances of the laws of nature, not properties of the physical states we observe. We induce them from observations and they predict consequences that can be tested: There is no observation that will determine the origin of the coordinates, the direction of the z-axis, the zero of the clock, or the frame of reference at absolute rest. You cannot prove such statements by experiments. At best you can claim that the symmetry principles have never yet been observed to be violated.

Be careful to distinguish the symmetries from transformations. When we say a system has a symmetry, we mean that its laws of motion are invariant under a transformation.

Sometimes it is useful to define a transformation on a physical system whether or not it is invariant under that transformation. For example, you can define translations on a particle moving in a potential $V(\mathbf{r})$, although because the coordinate appears explicitly the Hamiltonian is not invariant under translations, and the point $\mathbf{r} = 0$ plays a special role. You can rotate the coordinates and momenta of particles in a constant external electric field, but the physics isn't invariant because the field defines a special direction. These systems are either simplified models or, like the second example, systems that are not truly isolated. On the other hand a two-particle Hamiltonian like

$$H = \frac{p_1^2}{2m_1} + \frac{p_2^2}{2m_2} + V\left(|\mathbf{r}_1 - \mathbf{r}_2|\right) \tag{4.1}$$

is invariant under both translations and rotations.

A transformation is an operator on the vector space of states. For instance, if $T(\mathbf{a})$ is a translation by $\mathbf{a}$, and $|\mathbf{r}\rangle$ is a coordinate eigenstate, then up to a phase $T(\mathbf{a})|\mathbf{r}\rangle$ should be the state $|\mathbf{r}+\mathbf{a}\rangle$. The preservation of probability under these transformations requires them to be unitary operators.[1]

4.1.2 Groups of Transformations

Some Terminology

A **group** is a collection of mathematical objects together with an operation, usually called the group multiplication, on any ordered pair of them. For space-time transformations, or for linear operators in vector spaces, group multiplication is the composition of transformations. AB means first do B and then do A. Group multiplication is associative, but not necessarily commutative: That means $(AB)C = A(BC)$, but AB is not always the same as BA. The product of any two group elements is a group element, there must be an identity, and each element must have an inverse. These are all the requirements for a group. The integers are a group under addition (zero is the identity), but not multiplication, because not every integer has an inverse.

The groups I will be talking about are groups of transformations on quantum mechanical states, groups of unitary operators on a vector space. Groups were invented to study symmetry transformations, and that is the use we put them to here. The collection of all symmetry transformations on a physical system forms a group. The group of all symmetry transformations has many **subgroups** (subsets of a group that themselves form a group), and usually we want to focus our attention on one of these subgroups. The identity I is in all subgroups.

The reflection operator $\mathcal{R}$ and the identity I form the smallest nontrivial group, one with just two elements. Its "multiplication table" is $\mathcal{R}^2 = I^2 = I$ and $\mathcal{R}I = I\mathcal{R} = \mathcal{R}$. But most of the transformations that concern us are **continuous groups**. They have an infinite number of elements, and are labeled by continuous real parameters like a distance or an angle. The mathematics of continuous groups is a vast subject. While a physicist would do well to learn as much about groups as possible, you will not need most of it to understand quantum mechanics. For us, words like "group" and "subgroup" will usually just be a useful terminology. I will try to demonstrate carefully the few technical properties I will need.

These continuous groups, like the collection of rotations or translations, are called **Lie groups**. If $g(\alpha)$ is a one-parameter subgroup, and g_o any group element, then the group multiplication rule for $g(\alpha)g_o$ has to be differentiable with respect to α in order for the continuous group to qualify as a Lie group. All the continuous groups we shall encounter are Lie groups.

Lie groups are characterized by the number of **dimensions** they have, the number of real parameters needed to describe the elements of the group. The translation group for example is a three-dimensional Lie group, since any translation $T(\mathbf{a})$ is specified by the three real numbers a_x, a_y, and a_z. Since translations all commute with each other there is not much more to say about them.

---

[1] There is one exception that we will come to in Section 5.3.2.

104   Chapter 4   Symmetry Transformations on States

The rotation group is three-dimensional. The big group made up of rotations, translations, and all possible products of these, called the "translation-rotation group," is therefore six-dimensional.

One-Dimensional Subgroups

In complicated Lie groups all the elements do not commute. But they have the following property: It is always possible to choose the way the transformations are parametrized so that the elements characterized by one parameter alone commute among themselves. For example, rotations do not all commute with one another. (Try rotating a book around two directions in some order, and then in the opposite order.) Nevertheless all rotations $R(\hat{\boldsymbol{n}}_x, \theta)$ about the x-axis by some angle θ do commute:

$$R(\hat{\boldsymbol{n}}_x, \theta_1) R(\hat{\boldsymbol{n}}_x, \theta_2) = R(\hat{\boldsymbol{n}}_x, \theta_1 + \theta_2) = R(\hat{\boldsymbol{n}}_x, \theta_2) R(\hat{\boldsymbol{n}}_x, \theta_1) \tag{4.2}$$

and likewise for rotations about any other axis.

The key to understanding continuous groups is to look at these transformations in a single direction, like rotations about the z-axis or translations in the x-direction, that all commute with each other. For these one-dimensional commutative subgroups, it is always possible to choose the parameterization so that if $g(a)$ is a transformation, then

$$g(a_1) g(a_2) = g(a_1 + a_2) \tag{4.3}$$

For translations the parameter a is the displacement. For rotations about some axis, a is the angle. For Galilean boosts, a is the velocity. We will see later how to choose a for Lorentz boosts.

These one-dimensional subgroups are important because they let us define generators. Consider a one-dimensional subgroup with elements $g(a)$ and the multiplication rule (4.3). Define an operator G by

$$g(1) = e^{-iG} \tag{4.4}$$

Then because of the rule (4.3) all the transformations along this direction can be written

$$g(a_i) = e^{-i a_i G} \tag{4.5}$$

For example, translations in the z-direction are[2]

$$T(a\hat{\boldsymbol{n}}_z) = e^{-iaP_z/\hbar} \tag{4.6}$$

On a quantum mechanical vector space the $g(a)$ are always unitary operators. Then since g is unitary, $g^\dagger = g^{-1}$, or $G = G^\dagger$. G, called the **generator** of the transformations $g(a)$, is Hermitean, and therefore an observable. In the example (4.6), the Hermitean operator P_z is the z-component of the momentum, as well as the generator (in units of $\hbar$) of translations in the z direction.

---
[2] The notation is defined in Section 1.2. The role of the momentum operator in translations was explained in Section 2.7.2.

The translation group is particularly simple. Since translations in different directions commute, any general translation can be written explicitly as a function of the three parameters a_i and the generators P_i:

$$T(\mathbf{a}) = e^{-i\mathbf{a}\cdot\mathbf{P}/\hbar} \tag{4.7}$$

The factor $\hbar$ is there so that $\mathbf{P}$ will have the usual meaning.

Rotations are more complicated. The next several sections are devoted to a detailed analysis of the rotation group in quantum mechanics. The generators of rotations will turn out to be components of the angular momentum.[3] The methods will also apply to any collection of continuous transformations, and in particular to Lorentz transformations in relativistic theories and to internal symmetries like isotopic spin in nuclear physics.

4.1.3 Classical and Quantum Symmetries

Mathematical Note: Here is a short digression for the mathematically inclined reader: Since quantum mechanics, not classical mechanics, is the fundamental theory, for every classical symmetry transformation g^C, there should correspond a quantum symmetry transformation g^Q, but not necessarily the other way around.

Furthermore, the multiplication rule for the classical transformations must be the same as the rule for the corresponding quantum transformations. The classical transformations are *functions* of the quantum transformations, with the property that

$$g^C(g_1^Q)g^C(g_2^Q) = g^C(g_1^Q g_2^Q) \tag{4.8}$$

The correspondence is a homomorphism, but not necessarily an isomorphism. That means there may be more that one quantum mechanical symmetry corresponding to the classical identity, but not the other way around. If E is the collection of all g^Q such that $g^C(g^Q) = 1$, the classical group is the factor group $\{g^Q\}/E$. Do not worry if this jargon is unfamiliar to you.

Here is a simple, if trivial, example: Multiply all states by a common phase; $|\psi\rangle \to e^{i\phi}|\psi\rangle$. Then all matrix elements of all operators are unchanged, so this is certainly a quantum mechanical symmetry, but it corresponds to the classical identity for any ϕ. Nor is this the only example. You probably already know that the quantum rotation group is larger than the classical one, and there are quantum mechanical rotations that correspond to the classical identity transformation!

4.2 THE ROTATION GROUP AND ALGEBRA

4.2.1 Representations of Groups

Representations

For any Lie group with elements g_i, a collection of operators $D(g_i)$ that obey the

---

[3]Generators are defined in a related way for continuous classical symmetries also. These are observables G such that under an infinitesimal transformation an observable A changes by

$$\delta A = A(\epsilon) - A(0) = \epsilon[A, G]_{PB}$$

The classical generators for translations and rotations are discussed in Problems 1.5 and 1.6. If the Hamiltonian H is invariant under the continuous symmetry, $[H, G]_{PB} = 0$, then G is conserved (see Problem 1.4). The Poisson bracket formalism makes these formal analogies between classical mechanics and quantum mechanics more transparent.

group multiplication rule
$$D(g_1)D(g_2) = D(g_1 g_2) \tag{4.9}$$
and is called a **representation** of the group.

It often helps to think of the group itself as a collection of matrices, rather than an abstraction defined by the multiplication rules,. We can say that for the three-dimensional rotation group, the real 3×3 matrices $\bar{R}(\hat{\boldsymbol{n}}, \psi)$ introduced in Section 1.2.4 *are* the rotation group. The group multiplication rule is matrix multiplication. Mathematicians call this group $SO(3)$. The symbol $O(n)$ means the collection of real orthogonal $n \times n$ matrices. $SO(n)$ means the same thing, except the matrices must also be unimodular.

A representation of the rotation group is any collection of linear operators on some vector space that satisfy

$$D\left[\bar{R}(\hat{\boldsymbol{n}}_1, \psi_1)\right] D\left[\bar{R}(\hat{\boldsymbol{n}}_2, \psi_2)\right] = D\left[\bar{R}(\hat{\boldsymbol{n}}_1, \psi_1) \bar{R}(\hat{\boldsymbol{n}}_2, \psi_2))\right] \tag{4.10}$$

The 3×3 matrices $\bar{R}$ themselves are a representation, the **defining representation**.

Reducible and Irreducible Representations

From two representations you can always construct a third: Suppose $D^{(1)}(\bar{R})$ and $D^{(2)}(\bar{R})$ are matrix representations of dimensions d_1 and d_2 respectively. Construct the $d_1 + d_2$ dimensional matrices

$$D(\bar{R}) = \begin{pmatrix} D^{(1)}(\bar{R}) & 0 \\ 0 & D^{(2)}(\bar{R}) \end{pmatrix} \tag{4.11}$$

In equation (4.11) the off-diagonal zeros stand for $d_1 \times d_2$ and $d_2 \times d_1$ dimensional zero matrices respectively. Obviously $D(\bar{R})$ is a matrix representation if $D^{(1)}(\bar{R})$ and $D^{(2)}(\bar{R})$ are representations.

A representation like (4.11), with off-diagonal blocks of zeros, is called **reducible**. A representation that cannot be written in this kind of diagonal form is **irreducible**. So if you want to know what are all the representations of some group, it is enough to list all the irreducible representations. The others are **direct sums** of the irreducible ones, and in some basis they have the form of equation (4.11).

To catalog all representations, or even all irreducible representations, might seem a long and complicated project, but it is not. The global structure of continuous groups is almost completely determined by their structure for small parameters.

4.2.2 Representations of the Generators of Rotations

Translations

Most of what you will ever want to know about groups of continuous symmetries is contained in the commutator algebra of the generators. For example the translations can be parametrized

$$T(\mathbf{a}) = e^{-i\mathbf{a}\cdot\mathbf{P}/\hbar} \tag{4.7}$$

where the generators P_i all commute: Thus $[P_i, P_j] = 0$ leads to

$$[T_i(a), T_j(a)] = 0 \qquad (4.12)$$

for all values of a. From the commutators of the P_i you learn the complete (here trivial) structure of the translations: Translations all commute. It makes no difference in which order you translate the position of a particle. The algebra of translations is just the algebra of the addition of three-dimensional vectors, and the commutation rule is the same as $\mathbf{a} + \mathbf{b} = \mathbf{b} + \mathbf{a}$.

Rotations

Not so for rotations. But as I have emphasized, rotations about the same direction *do* commute. For the 3×3 matrices $\bar{R}(\hat{\mathbf{n}}, \psi)$ rotations by any angle about the same axis $\hat{\mathbf{n}}$ have the form[4]

$$\bar{R}(\hat{\mathbf{n}}, \psi) = e^{-i\psi \bar{J}(\hat{\mathbf{n}})} \qquad (4.13)$$

for some matrix $\bar{J}(\hat{\mathbf{n}})$.

Since they commute, all the representation matrices $D(\hat{\mathbf{n}}, \psi)$ about $\hat{\mathbf{n}}$ also can be written in exponential form using a single real parameter. In any representation define $D[\bar{J}(\hat{\mathbf{n}})]$ as the matrix that goes in the exponent, the same place where $\bar{J}(\hat{\mathbf{n}})$ goes in equation (4.13):

$$D[\bar{R}(\hat{\mathbf{n}}, \psi)] = e^{-i\psi D[\bar{J}(\hat{\mathbf{n}})]} \qquad (4.14)$$

In particular, for rotations about one of the coordinate axes[5]

$$D(\hat{\mathbf{n}}_i, \psi) = e^{-i\psi D(\bar{J}_i)} \qquad (4.15)$$

since $D(\bar{J}_i) = D[\bar{J}(\hat{\mathbf{n}}_i)]$. Technically it is bad notation to use the same letter D for two things, but it is conventional. It is bad because $D(\bar{J})$ is not the same function as $D(\bar{R})$. The advantage here will be that we will be able to label the representation of the rotation and the generators the same way.

But be careful: For a representation of the generators, it is not true that $D(J_1)D(J_2) = D(J_1 J_2)$. What we should say is that to every representation $D^{(i)}(\bar{R})$ there corresponds a representation $\tilde{D}^{(i)}(\bar{J})$, defined by equation (4.14), but I will not bother to make the distinction in the notation. It is standard and should not be confusing.

4.2.3 Generators in an Arbitrary Direction

The matrix representations for the generators of rotations about the three coordinate axes, $D(\bar{J}_i)$ are known from Section 3.3.3. Rotations about any other axis $\hat{\mathbf{n}}$ are generated by some matrices $D[\bar{J}(\hat{\mathbf{n}})]$. These are related in a simple way:

$$D[\bar{J}(\hat{\mathbf{n}})] = \sum_i n_i D(\bar{J}_i) = \hat{\mathbf{n}} \cdot D(\bar{\mathbf{J}}) \qquad (4.16)$$

The proof is made by examining small rotations:

---

[4]See Section 4.1.2.
[5]For rotations, I will frequently use $D(\hat{\mathbf{n}}, \psi)$ as shorthand for $D[\bar{R}(\hat{\mathbf{n}}, \psi)]$.

Proof. The theorem is true for the original defining matrices $\bar{R}$ [see equation (1.59)]. It follows that

$$\bar{R}(\hat{n}, \psi) = e^{-i\psi\hat{n}\cdot\bar{J}} = e^{-i\psi n_x \bar{J}_x} e^{-i\psi n_y \bar{J}_y} e^{-i\psi n_z \bar{J}_z} + O(\psi^2)$$
$$= \bar{R}(\hat{n}_x, n_x\psi)\bar{R}(\hat{n}_y, n_y\psi)\bar{R}(\hat{n}_z, n_z\psi) + O(\psi^2) \quad (4.17)$$

Since the $D(\hat{n}, \psi)$ have the same multiplication rule, these can be written as the *same* functions of the $D(\bar{J}_i)$: For small ψ,

$$D(\hat{n}, \psi) = D(\hat{n}_x, n_x\psi) D(\hat{n}_y, n_y\psi) D(\hat{n}_z, n_z\psi) + O(\psi^2)$$
$$= e^{-in_x\psi D(\bar{J}_x)} e^{-in_y\psi D(\bar{J}_y)} e^{-in_z\psi D(\bar{J}_z)} + O(\psi^2) \quad (4.18)$$

or

$$I - i\psi D\left[\bar{J}(\hat{n})\right] = I - in_x\psi D(\bar{J}_x) - in_y\psi D(\bar{J}_y) - in_z\psi D(\bar{J}_z) + \cdots \quad (4.19)$$

whence

$$D\left[\bar{J}(\hat{n})\right] = D(\hat{n}\cdot\bar{J}) = n_x D(\bar{J}_x) + n_y D(\bar{J}_y) + n_z D(\bar{J}_z) = \hat{n}\cdot D(\bar{J}) \quad (4.20)$$

I used an expansion in ψ to derive the rule (4.20), but the rule doesn't involve ψ at all! It is an exact rule, not an approximation for small ψ. $D[\bar{J}(\hat{n})]$ is a linear function of the three $\mathbf{J}_i$. ∎

So you don't have to figure out the matrices $D[\bar{J}(\hat{n})]$ for all $\hat{n}$. *To find all the unitary representation matrices in a representation, it is enough to find the three generators of rotations about the coordinate axes.*

4.2.4 Commutators of the Generators

The Fundamental Theorem

A rotation matrix $\exp\left(-i\psi\hat{n}\cdot\bar{J}\right)$ is represented by

$$D\left(e^{-i\psi\hat{n}\cdot\bar{J}}\right) = e^{-i\psi\hat{n}\cdot D(\bar{J})} \quad (4.21)$$

The representation matrices for the generators don't preserve the ordinary products of the $\bar{J}_i$. But the *commutator algebra* of the $D(\bar{J}_i)$ is the same as that of the $\bar{J}_i$ themselves. That is

$$[D(\bar{J}_i), D(\bar{J}_j)] = D\left([\bar{J}_i, \bar{J}_j]\right) \quad (4.22)$$

The commutator on the right hand side is known:[6]

$$D\left([\bar{J}_i, \bar{J}_j]\right) = i\sum_k \epsilon_{ijk} D\left(\bar{J}_k\right) \quad (4.23)$$

---
[6]Equation (1.62).

This is a very strong constraint on the representations $D(\hat{n}, \theta)$. They have to be defined by a rule like (4.21), where the generators satisfy

$$[D(\bar{J}_i), D(\bar{J}_j)] = i\sum_k \epsilon_{ijk} D(\bar{J}_k) \tag{4.24}$$

All possible solutions to this constraint were found in Section 3.3.3. Here is a proof of this important result:

Proof. Write the product of two unitary matrices as

$$e^{\psi A} e^{\psi B} = e^{\psi(A+B) + \psi^2 X + \cdots} \tag{4.25}$$

If A and B commuted, X would be zero, so X must be proportional to the commutator of A and B. To find the proportionality constant, expand both sides of equation (4.25) in powers of ψ:

$$\left(I + \psi A + \frac{1}{2}\psi^2 A^2 + \cdots\right)\left(I + \psi B + \frac{1}{2}\psi^2 B^2 + \cdots\right)$$
$$= I + \psi(A+B) + \psi^2 X + \frac{1}{2}\psi^2(A^2 + AB + BA + B^2) + \cdots \tag{4.26}$$

where the unwritten terms are still higher order in ψ. Thus

$$X = \frac{1}{2}[A, B] \tag{4.27}$$

so that

$$e^{\psi A} e^{\psi B} = e^{\psi A + \psi B + \frac{1}{2}\psi^2 [A, B] + \cdots} \tag{4.28}$$

The coefficient of ψ^2 in the exponent is indeed a commutator.

Apply equation (4.28) to a product of small rotations about the x and y axis:

$$\bar{R}(\hat{n}_x, \psi)\bar{R}(\hat{n}_y, \psi) = e^{-i\psi \bar{J}_x} e^{-i\psi \bar{J}_y} = e^{-i\psi \bar{J}_x - i\psi \bar{J}_y - \frac{1}{2}\psi^2 [\bar{J}_x, \bar{J}_y] + \cdots}$$
$$= e^{-i\psi \bar{J}_x - i\psi \bar{J}_y - \frac{i}{2}\psi^2 \bar{J}_z + \cdots} \tag{4.29}$$

where in the last form I used the explicit commutator $[\bar{J}_x, \bar{J}_y] = i\bar{J}_z$, a special case of equation (1.62). The rotation axis of the product has a small component in the z direction.

From the definition of a representation, therefore,

$$D\left(\bar{R}(\hat{n}_x, \psi)\bar{R}(\hat{n}_y, \psi)\right) = D\left(e^{-i\psi \bar{J}_x - i\psi \bar{J}_y - \frac{i}{2}\psi^2 \bar{J}_z + \cdots}\right)$$
$$= e^{-i\psi D(\bar{J}_x) - i\psi D(\bar{J}_y) - \frac{i}{2}\psi^2 D(\bar{J}_z) + \cdots} \tag{4.30}$$

also. Now use the rule (4.28) for the representation matrices:

$$D\left(\bar{R}(\hat{n}_x, \psi)\bar{R}(\hat{n}_y, \psi)\right) = D\left[\bar{R}(\hat{n}_x, \psi)\right] D\left(\bar{R}(\hat{n}_y, \psi)\right) = e^{-i\psi D(\bar{J}_x)} e^{-i\psi D(\bar{J}_y)}$$
$$= e^{-i\psi D(\bar{J}_x) - i\psi D(\bar{J}_y) - \frac{1}{2}\psi^2 [D(\bar{J}_x), D(\bar{J}_y)] + \cdots} \tag{4.31}$$

Compare the coefficient of ψ^2 in equations (4.29) and (4.31) to conclude

$$[D(\bar{J}_x), D(\bar{J}_y)] = iD(\bar{J}_z) = D\left([\bar{J}_x, \bar{J}_y]\right) \qquad (4.22)$$

∎

Lie Algebras

The collection of matrices $\bar{J}_i$ and their linear combinations constitute a **Lie algebra**, which is the name for a vector space with a multiplication rule that is neither associative nor commutative, but has the properties of the commutator, namely an anticommutativity rule,

$$[J_i, J_j] = -[J_j, J_i] \qquad (4.32)$$

and, in place of the associativity rule, the **Jacobi identity**:

$$[[A, B], C] + [[C, A], B] + [[B, C], A] = 0 \qquad (4.33)$$

Commutator algebras are not the only Lie algebras—ordinary vector cross products and Poisson brackets are others (see Problem 1.3).

To every Lie group there corresponds a Lie algebra. Suppose g_i are elements of a continuous group. Then a representation is a collection of matrices with the property

$$D(g_1)D(g_2) = D(g_1 g_2) \qquad (4.34)$$

The properties I have shown here for rotations are easily generalized to any Lie group:

$$g = e^{-i \sum_i \theta_i J_i} \qquad (4.35)$$

for some *finite* number of generators J_i. The group element is determined by the real numbers θ_i. If you write the representation in the same way

$$D(g) = e^{-i \sum_i \theta_i D(J_i)} \qquad (4.36)$$

then the new matrices $D(J_i)$, which we call the representations of the generators, all have the same commutator algebra:

$$[D(J_i), D(J_j)] = D\left([J_i, J_j]\right) \qquad (4.37)$$

The representation of a commutator is the commutator of the representation. Equation

$$[J_i, J_j] = i \sum_k \epsilon_{ijk} J_k \qquad (3.92)$$

is the rule for the simplest non-trivial Lie algebra, and the $2j+1$ dimensional operators found in Section 3.3.3 are its representations.

The Converse of the Theorem

The converse is not quite true. Given a closed collection of commutation relations,

$$[X_i, X_j] = i \sum_k c_{ijk} X_k \qquad (4.38)$$

one can always construct the matrices $\exp(-i\sum_k \psi_k X_k)$, and they will indeed form a group, with the generators satisfying the original algebra. But it is not necessarily the group you started with. For small values of the parameters, the exponentiated matrices satisfy the group rules, but they need not do so globally.[7]

So to be a representation of a given group, the generators must satisfy the commutation relations; but this is only a necessary condition, not a sufficient condition. The correct converse theorem is more difficult, and I will not demonstrate it here.

4.2.5 Explicit Form of the Finite Dimensional Representations

Now it is possible to write explicity the irreducible, finite dimensional representations of the rotation group. These are collections of matrices whose multiplication rule is the same as the rule for the 3×3 matrices $\bar{R}$. For any such collection there are also matrices $D(\bar{J}_i)$ that satisfy the commutation relations

$$[D(\bar{J}_i), D(\bar{J}_j)] = i \sum_k \epsilon_{ijk} D(\bar{J}_k) \tag{4.24}$$

Equations (4.24) are the same commutation relations that the L_i satisfy, so the problem of finding all solutions is exactly the one solved in Section 3.3.3. There is one irreducible representation for each nonnegative integer or half-integer j, and it acts on a $2j+1$ dimensional space. Conventionally, these are labeled $D^{(j)}(\bar{J}_i)$. Once you know three Hermitean operators $D^{(j)}(\bar{J}_i)$, they can be exponentiated to find the unitary $D^{(j)}(\hat{n}, \theta)$.

From Section 3.3.3 you know how the $D^{(j)}(J_i)$ acts on the $2j+1$ states $|j, m\rangle$, for $-j \leq m \leq j$. The states are eigenstates of $D^{(j)}(\bar{J}_z)$

$$D^{(j)}(\bar{J}_z)|jm\rangle = m|jm\rangle \tag{4.39}$$

From equations (3.113)

$$D^{(j)}(\bar{J}_\pm)|jm\rangle = \sqrt{j(j+1) - m(m \pm 1)}|j, m \pm 1\rangle \tag{4.40}$$

where

$$D^{(j)}(\bar{J}_\pm) = D^{(j)}(\bar{J}_x) \pm i D^{(j)}(\bar{J}_y) \tag{4.41}$$

Put another way, the matrix elements of the operators $D^{(j)}$ are

$$D^{(j)}(\bar{J}_z)_{m'm} = m \delta_{m'm} \tag{4.42}$$

and

$$D^{(j)}(\bar{J}_\pm)_{m'm} = \sqrt{j(j+1) - m(m \pm 1)} \delta_{m', m \pm 1} \tag{4.43}$$

Furthermore,

$$\sum_i D^{(j)}(\bar{J}_i) D^{(j)}(\bar{J}_i)|j, m\rangle = j(j+1)|j, m\rangle \tag{4.44}$$

There is one representation of every integer dimension.

---

[7] Consider for instance the one-dimensional "matrices" e^{ia} and e^a. They both have the same trivial generator algebra—everything commutes—but one is the group of rotations in two dimensions, and the other the group of one-dimensional translations.

112 Chapter 4 Symmetry Transformations on States

These matrices would still be a representation if we changed the phases of some of the states. By choosing the rules (4.42) and (4.43), we make the convention that the $D^{(j)}(\bar{J}_z)$ are diagonal and the matrix elements of $D^{(j)}(\bar{J}_+)$ are real and positive. With these rules the $2j+1$ dimensional representation is unique, and any other irreducible representation of the same dimension must be related to this conventional one by a unitary transformation.[8]

The finite dimensional rotation operators are obtained by exponentiating the $D^{(j)}(\bar{J}_i)$.

$$D^{(j)}(\bar{R}) = \exp\left(-i\theta \sum_i n_i D^{(j)}(\bar{J}_i)\right) \quad (4.45)$$

The collection of matrix elements of the $D^{(j)}(\bar{R})$ with the conventional phases are the **Wigner matrices**.

Of course this does not prove that all the matrices with the form (4.45) are really representations of the rotation group, only that all the representations must be among them. You will see shortly that only the ones with integer j are really representations of $SO(3)$.

4.2.6 Summary

The last two sections contain a lot of mathematical words, which I will now feel free to use. Much of it is just nomenclature: symmetries and transformations, groups and subgroups, discrete groups and continuous groups. I have mentioned Lie groups, and the idea of generators of Lie groups, and Lie algebras. I introduced the notion of a representation, and distinguished reducible from irreducible representations. You should know the definitions of all these words.

There was one important theorem, which for rotations reads as follows: If $D(\hat{n}, \theta)$ are a representation of the rotation group, then

$$D(\hat{n}, \theta) = e^{-i\theta D[\bar{J}(\hat{n})]} \quad (4.46)$$

This rule defines the operators $D[\bar{J}(\hat{n})]$. They, in turn, can be written in terms of just three operators

$$D\left[\bar{J}(\hat{n})\right] = \sum_i n_i D(\bar{J}_i) \quad (4.47)$$

where $D(\bar{J}_i)$ is shorthand for $D[\bar{J}(\hat{n}_i)]$. And finally, the operators $D(\bar{J}_i)$ satisfy the same commutator algebra as the $\bar{J}_i$:

$$[D(\bar{J}_i), D(\bar{J}_j)] = i \sum_k \epsilon_{ijk} D(\bar{J}_k) \quad (4.48)$$

Because you know the most general solution to this commutator algebra, you know all the representations of the rotation group. A similar derivation works for any Lie group.

---

[8] In particular, the matrices $\bar{J}_i$ and the matrices $D^{(1)}(\bar{J}_i)$ (constructed in Section 4.3.6 below) are related by a unitary transformation, or change of basis. See Problem 4.2.

4.3 SPIN AND ROTATIONS IN QUANTUM MECHANICS

4.3.1 Rotations and Spinless Particles

The unitary operators R that rotate every state depend on the angle and axis of the rotation, and together they are a representation of the rotation group:

$$R(\bar{R}_1)R(\bar{R}_2) = R(\bar{R}_1\bar{R}_2) \tag{4.49}$$

As a first example consider a single particle in three dimensions, of the type we studied in Section 3.3. The states $|\mathbf{r}\rangle$ are a basis, so to find out what the rotation operators do to any state it is enough to see what they do to the states $|\mathbf{r}\rangle$: The rotated states must be located at the point one gets by rotating the vector $\mathbf{r}$. That is the geometrical meaning of rotation, and it guarantees that the operators satisfy the group rule. Thus

$$R(\bar{R})|\mathbf{r}\rangle = |\mathbf{r}'\rangle \tag{4.50}$$

where

$$r'_i = \sum_j \bar{R}_{ij} r_j \tag{4.51}$$

The space of the states is infinite-dimensional, so the operators $R(\bar{R})$ form an infinite-dimensional representation of the rotations.

What are the generators in this representation? The method should by now be familiar. If $|\psi'\rangle = R(\bar{R})|\psi\rangle$ is a rotated state, its wave function $\psi'(\mathbf{r}) = \langle \mathbf{r}| \psi' \rangle$ satisfies

$$\psi'(\mathbf{r}') = \psi(\mathbf{r}) \tag{4.52}$$

Since I called the rotation matrices just R instead of $D(\bar{R})$, I will call the generators J_i. These are the operators on the space of quantum mechanical states.

$$\langle \mathbf{r}| e^{-i\epsilon \hat{\mathbf{n}} \cdot \mathbf{J}} |\psi\rangle = \langle \bar{R}^{-1}\mathbf{r}| \psi \rangle = \psi(\bar{R}^{-1}\mathbf{r}) = \psi(\mathbf{r} - \epsilon \hat{\mathbf{n}} \times \mathbf{r}) \tag{4.53}$$

Expand both sides to first order:

$$\psi(r) - i\epsilon \sum_i n_i \langle \mathbf{r}| J_i |\psi\rangle = \psi(\mathbf{r}) - \epsilon \sum_{ijk} n_i \epsilon_{ijk} r_j \frac{\partial \psi}{\partial r_k} \tag{4.54}$$

So

$$\sum_i n_i \langle \mathbf{r}| J_i |\psi\rangle = -i \sum_{ijk} n_i r_j \epsilon_{ijk} \frac{\partial \psi}{\partial r_k} = -i \sum_i n_i (\mathbf{r} \times \boldsymbol{\nabla})_i \psi(\mathbf{r}) = \sum_i n_i \langle \mathbf{r}| L_i |\psi\rangle \tag{4.55}$$

Since $\hat{\mathbf{n}}$ can be in any direction,

$$\hbar \langle \mathbf{r}| L_i |\psi\rangle = -i\hbar (\mathbf{r} \times \boldsymbol{\nabla})_i \psi(\mathbf{r}) = \langle \mathbf{r}| (\mathbf{r} \times \mathbf{p})_i |\psi\rangle \tag{4.56}$$

$\hbar \mathbf{L} = \mathbf{r} \times \mathbf{p}$ is the orbital angular momentum associated with the motion of $\mathbf{r}$. These operators L_i satisfy the same commutation relations as the $\bar{J}_i$ [see equation (3.88)]. It is not a coincidence. On the states of a simple point particle, the operators L_i are the generators of the rotations. In general, the generators of rotations $\hbar \mathbf{J}$

will be the total angular momentum, and will be conserved if the Hamiltonian is rotationally invariant.

This representation of the rotation operators is not irreducible. To see its irreducible components, change the basis from the states $|\mathbf{r}\rangle$ to states $|Elm\rangle$ as in equations (3.87). E can be the eigenvalue of any spherically symmetric Hamiltonian. In the $|Elm\rangle$ basis the angular momentum operators are in block diagonal form (4.11), where each block is a $2l+1$ dimensional irreducible representation. The full representation is an infinite stack of finite dimensional ones. The operators L_i mix up the $2l+1$ states $|Elm\rangle$ only with each other, not states with a different l or E. The same is true for the rotation matrices themselves, since

$$R = e^{-i\psi\hat{\mathbf{n}}\cdot\mathbf{J}} \tag{4.57}$$

4.3.2 Spin

The states of most particles cannot be described by their position $\mathbf{r}$ alone; and, experimentally, the orbital angular momentum $\hbar\mathbf{L} = \mathbf{r}\times\mathbf{p}$ is not conserved. An electron or a proton has angular momentum even when $\mathbf{p}=0$. A complete set of observables for these particles consists of $\mathbf{r}$ and another, discrete, spin quantum number m. A basis is now labeled $|\mathbf{r}, m\rangle$ where for an electron or proton m takes on two values. An electron at $\mathbf{r}$ can have its spin "up" or "down." Here is how it works: Under rotations, such states transform with the rule

$$R(\hat{\mathbf{n}}, \theta)|\mathbf{r}, m\rangle = \sum_{m'} |\mathbf{r}', m'\rangle D_{m'm}(\hat{\mathbf{n}}, \theta) \tag{4.58}$$

for some matrices $D(\hat{\mathbf{n}}, \theta)$, where again the components of $\mathbf{r}'$ are

$$r'_i = \sum_j \bar{R}_{ij} r_j \tag{4.59}$$

If $\mathbf{r}' = \bar{R}(\hat{\mathbf{n}}_2, \theta_2)\mathbf{r}$ and $\mathbf{r}'' = \bar{R}(\hat{\mathbf{n}}_1, \theta_1)\mathbf{r}'$, then

$$\begin{aligned} R(\hat{\mathbf{n}}_1, \theta_1) R(\hat{\mathbf{n}}_2, \theta_2)|\mathbf{r}, m\rangle &= \sum_{m'} R(\hat{\mathbf{n}}_1, \theta_1)|\mathbf{r}', m'\rangle D_{m'm}(\hat{\mathbf{n}}_1, \theta_1) \\ &= \sum_{m'} |\mathbf{r}'', m''\rangle [D(\hat{\mathbf{n}}_2, \theta_2) D(\hat{\mathbf{n}}_1, \theta_1)]_{m''m} \end{aligned} \tag{4.60}$$

The rotated states are a linear combination of the two states $|\mathbf{r}', m\rangle$. The two dimensional matrices $D(\hat{\mathbf{n}}, \theta)$ are a representation of the rotations, and therefore according to the theorem of Section 4.2.4 they can be written

$$D(\hat{\mathbf{n}}, \theta) = e^{-i\theta\hat{\mathbf{n}}\cdot\mathbf{s}} \tag{4.61}$$

for some matrices $s_i = D(\bar{J}_i)$ that satisfy the fundamental commutation relations

$$[s_i, s_j] = i\sum_k \epsilon_{ijk} s_k \tag{4.62}$$

The rotation operator R does two things to these states. It mixes up the discrete index m with a representation matrix of the appropriate dimension, and it changes $\mathbf{r}$ to $\mathbf{r}'$ just as for spinless particles. You can write it as a product:

$$R(\hat{\mathbf{n}}, \theta) = e^{-i\theta \hat{\mathbf{n}} \cdot \mathbf{J}} = e^{-i\theta \hat{\mathbf{n}} \cdot \mathbf{L}} e^{-i\theta \hat{\mathbf{n}} \cdot \mathbf{s}} \tag{4.63}$$

where

$$e^{-i\theta \hat{\mathbf{n}} \cdot \mathbf{L}} |\mathbf{r}, m\rangle = |\mathbf{r}', m\rangle \tag{4.64}$$

and

$$e^{-i\theta \hat{\mathbf{n}} \cdot \mathbf{s}} |\mathbf{r}, m\rangle = \sum_{m'} |\mathbf{r}, m'\rangle D_{m'm}(\hat{\mathbf{n}}, \theta) \tag{4.65}$$

L and **s** commute. $\hbar \mathbf{s}$ is the **spin angular momentum** (**spin** for short), and $\hbar \mathbf{L}$ is the **orbital angular momentum**. In an isolated system with rotational symmetry the Hamiltonian commutes with $\mathbf{J}$, but not with $\mathbf{L}$ or $\mathbf{s}$ separately. $\hbar \mathbf{J} = \hbar \mathbf{L} + \hbar \mathbf{s}$ is the total angular momentum, as well as the generator of rotations.

Wave Functions for Particles with Spin

For spinless particles, the wave function $\psi(\mathbf{r})$ was $\langle \mathbf{r} | \psi \rangle$. For a particle with spin define the wave function by

$$\psi_m(\mathbf{r}) = \langle \mathbf{r}, m | \psi \rangle \tag{4.66}$$

$\psi(\mathbf{r})$ has as many components as there are possible values of m.

When the Hamiltonian depends on the spin index, the Schrödinger equation includes both differential operators on $\mathbf{r}$ and matrices operating on the spin index. $\psi(\mathbf{r})$ is a column vector function of $\mathbf{r}$, sometimes called a **spinor**, especially when s is half-integral.

What happens to $\psi_m(\mathbf{r})$ under rotations? A component of the rotated state's wave function is (watch carefully)

$$\psi'_m(\mathbf{r}) = \langle \mathbf{r}, m | R | \psi \rangle = \langle \psi | R^\dagger | \mathbf{r}, m \rangle^* = \langle \psi | e^{i\theta \hat{\mathbf{n}} \cdot \mathbf{L}} e^{i\theta \hat{\mathbf{n}} \cdot \mathbf{s}} | \mathbf{r}, m \rangle^*$$
$$= \sum_{m'} D_{mm'}(\hat{\mathbf{n}}, \theta) \langle \bar{R}^{-1} \mathbf{r}, m' | \psi \rangle = \sum_{m'} D_{mm'}(\hat{\mathbf{n}}, \theta) \psi_{m'}(\mathbf{r}'') \tag{4.67}$$

where $r_i = \sum_j \bar{R}_{ij} r''_j$. In equation (4.67) I used the property that the matrices $D(\hat{\mathbf{n}}, \theta)$ are a unitary representation:

$$D_{m'm}(\bar{R}^{-1}) = D_{m'm}(\bar{R})^\dagger = D_{mm'}(\bar{R})^* \tag{4.68}$$

The dependence of the wave function on the coordinate transforms the same way it does under rotations of $\mathbf{r}$ alone. The spin index is transformed by ordinary matrix multiplication.

$$s_i |\mathbf{r}, m\rangle = \sum_{m'} |\mathbf{r}, m'\rangle D_{m'm}(\bar{J}_i) \tag{4.69}$$

or, in terms of wave functions,

$$\langle \mathbf{r}, m | s_i | \psi \rangle = \sum_{m'} D_{mm'}(\bar{J}_i) \psi_{m'}(\mathbf{r}) \tag{4.70}$$

Let us look at some examples.

4.3.3 The Spin-Zero Representation

This is the trivial one dimensional representation. There is only one state, $|0,0\rangle$, with $j = m = 0$, so that

$$D^{(0)}(\bar{J}_i) = 0 \quad \text{and} \quad D^{(0)}(R) = 1 \tag{4.71}$$

4.3.4 The Spin-Half Representation

I have used the argument of Section 3.3.3 backward. If the $D^{(j)}(\bar{R})$ obey the group multiplication rule, then the $D^{(j)}(\bar{J}_i)$ must obey the angular momentum commutator algebra, but I pointed out in Section 4.2.4 that it does not have to work the other way around. All true matrix representations of this algebra can be found by exponentiating the $D^{(j)}(\bar{J}_i)$, but some of the exponentiated forms might not really be representations of $SO(3)$, the 3×3 rotations matrices. This is what happens for half-integral j.

From equations (4.42) and (4.43) with $j = 1/2$, one obtains the nonvanishing matrix elements of this smallest nontrivial representation. They are

$$D^{(\frac{1}{2})}(\bar{J}_z)_{\frac{1}{2},\frac{1}{2}} = \frac{1}{2} \quad D^{(\frac{1}{2})}(\bar{J}_z)_{-\frac{1}{2},-\frac{1}{2}} = -\frac{1}{2} \tag{4.72}$$

$$D^{(\frac{1}{2})}(\bar{J}_+)_{\frac{1}{2},-\frac{1}{2}} = \sqrt{\frac{1}{2} \cdot \frac{3}{2} - (-\frac{1}{2}) \cdot \frac{1}{2}} = 1 \tag{4.73}$$

$D^{(\frac{1}{2})}(\bar{J}_-)$ is the Hermitean conjugate of $D^{(\frac{1}{2})}(\bar{J}_+)$. There is a standard notation for these 2×2 matrices:

$$\boxed{D^{(\frac{1}{2})}(\bar{J}_i) = \frac{1}{2}\sigma_i} \tag{4.74}$$

The σ_i are the three **Pauli matrices**:[9]

$$\boxed{\sigma_x = \begin{pmatrix} 0 & 1 \\ 1 & 0 \end{pmatrix} \quad \sigma_y = \begin{pmatrix} 0 & -i \\ i & 0 \end{pmatrix} \quad \sigma_z = \begin{pmatrix} 1 & 0 \\ 0 & -1 \end{pmatrix}} \tag{4.75}$$

It is easy to verify that

$$\left[\frac{1}{2}\sigma_i, \frac{1}{2}\sigma_j\right] = i\sum_k \epsilon_{ijk} \frac{1}{2}\sigma_k \tag{4.76}$$

The Pauli matrices obey some simple properties worth committing to memory:

$$\sigma_i = \sigma_i^\dagger = \sigma_i^{-1} \qquad (\sigma_i)^2 = I$$

$$\det \sigma_i = -1 \qquad \text{Tr}\,\sigma_i = 0$$

$$(\hat{\mathbf{n}} \cdot \boldsymbol{\sigma})^2 = 1 \qquad \sigma_i\sigma_j = \delta_{ij} + i\sum_k \epsilon_{ijk}\sigma_k$$

$$\sigma_i\sigma_j + \sigma_j\sigma_i = 2\delta_{ij} \qquad [\sigma_i, \sigma_j] = 2i\sum_k \epsilon_{ijk}\sigma_k \tag{4.77}$$

---

[9] Matrix representations for rotations and their generators are conventionally written with the rows and columns labeled by m is *decreasing* order, highest first.

Then,
$$\sum_i \left(D^{(\frac{1}{2})}(\bar{J}_i)\right)^2 = \frac{1}{4}\sigma_x^2 + \frac{1}{4}\sigma_y^2 + \frac{1}{4}\sigma_x^2 = \frac{3}{4} = \frac{1}{2}\cdot\left(\frac{1}{2}+1\right) \qquad (4.78)$$

as required.

Making use of the properties (4.77) you can exponentiate these matrices:
$$D^{(\frac{1}{2})}(\hat{n},\theta) = e^{-i\frac{1}{2}\theta\hat{n}\cdot\boldsymbol{\sigma}}$$
$$= I - \frac{i}{2}\theta\hat{n}\cdot\boldsymbol{\sigma} - \frac{1}{2!}\left(\frac{\theta}{2}\right)^2(\hat{n}\cdot\boldsymbol{\sigma})^2 + i\frac{1}{3!}\left(\frac{\theta}{2}\right)^3(\hat{n}\cdot\boldsymbol{\sigma})^3 + \cdots$$
$$= I\cos(\theta/2) - i\hat{n}\cdot\boldsymbol{\sigma}\sin(\theta/2)$$

The two-dimensional "representation" of the rotation matrices is
$$D^{(\frac{1}{2})}(\hat{n}_z,\theta) = \begin{pmatrix} e^{-i\theta/2} & 0 \\ 0 & e^{i\theta/2} \end{pmatrix}$$
$$D^{(\frac{1}{2})}(\hat{n}_x,\theta) = \begin{pmatrix} \cos(\theta/2) & -i\sin(\theta/2) \\ -i\sin(\theta/2) & \cos(\theta/2) \end{pmatrix} \qquad (4.79)$$
$$D^{(\frac{1}{2})}(\hat{n}_y,\theta) = \begin{pmatrix} \cos(\theta/2) & -\sin(\theta/2) \\ \sin(\theta/2) & \cos(\theta/2) \end{pmatrix}$$

Now there is a surprise. The matrix representing a rotation by 2π about any axis is
$$D^{(\frac{1}{2})}(\hat{n},2\pi) = I\cos\pi - i\sigma_i\sin\pi = -I \qquad (4.80)$$

These matrices are not representations of the rotation group after all, since a rotation by 2π is the identity! You have to rotate by 4π in order to get back to the identity.

We cannot just throw this representation out, since states of angular momentum $\hbar/2$ exist in nature. Here is an example of the possibility raised in Section 4.1.3 that the quantum mechanical symmetry group may be larger than the classical one. The quantum mechanical rotation group is not $SO(3)$, the collection of all real 3×3 unimodular orthogonal matrices. Rather it is $SU(2)$, the group of 2×2 unimodular unitary matrices. Technically, the matrices $\bar{R}$ are a representation of the $SU(2)$ matrices, but not the other way around, since the image of both two dimensional matrices I and $-I$ are the three-dimensional identity matrix. But physicists often call the half-integral representations "double-valued" representations of $SO(3)$. Both of these groups have the same commutator algebra for the generators.

If a state $|\psi\rangle$ that transforms under rotations according to one of these half-integral representations be rotated by 360°, the rotated state is $-|\psi\rangle$. The expectation value of any observable is unchanged, and the classical result remains correct for expectation values.

4.3.5 Euler Angles

The general rule for the finite dimensional representations of the angular momentum operators is given by equations (4.39) and (4.40). Exponentiating these rules to get the representations $D^{(j)}(\bar{R})$ of the rotations for $j > 1/2$ is just a question of some computation, but it must get harder and harder as j increases. The most efficient way to proceed is to parametrize the $D^{(j)}(\bar{R})$ by the Euler angles instead of by $\hat{n}$ and θ.

A famous theorem due to Euler says that any rotation matrix $\bar{R}(\hat{n}, \theta)$ can be written uniquely in the form

$$\bar{R}(\hat{n}, \theta) = \bar{R}(\hat{n}_z, \alpha)\bar{R}(\hat{n}_y, \beta)\bar{R}(\hat{n}_z, \gamma) \tag{4.81}$$

With a little thought this theorem should be geometrically obvious. The explicit relation between the Euler angles α, β, and γ on the one hand and θ and the two real angles needed to specify $\hat{n}$ on the other is not so obvious and is left as an exercise.[10]

The Euler angles are useful because in the standard representation $D^{(j)}(\bar{J}_z)$ is diagonal, and so the representation of a rotation about the z-axis is

$$D^{(j)}{}_{m'm}(\hat{n}_z, \theta) = \langle j, m' | e^{-i\theta J_z} | j, m \rangle = e^{-im\theta}\delta_{m'm} \tag{4.82}$$

For notational brevity I shall write

$$\bar{R}(\alpha, \beta, \gamma) = \bar{R}(\hat{n}_z, \alpha)\bar{R}(\hat{n}_y, \beta)\bar{R}(\hat{n}_z, \gamma) \tag{4.83}$$

and

$$D^{(j)}(\alpha, \beta, \gamma) = D^{(j)}\left(\bar{R}(\alpha, \beta, \gamma)\right) \tag{4.84}$$

Then

$$D^{(j)}{}_{m',m}(\alpha, \beta, \gamma) = \langle j, m' | e^{-i\alpha J_z} | j, m' \rangle\langle j, m' | e^{-i\beta J_y} | j, m \rangle\langle j, m | e^{-i\gamma J_z} | j, m \rangle$$
$$= e^{-i(\alpha m' + \gamma m)} d^{(j)}{}_{m'm}(\beta) \tag{4.85}$$

where $d^{(j)}(\beta)$ is defined as

$$d^{(j)}(\beta) = D^{(j)}(\hat{n}_y, \beta) = e^{-i\beta D^{(j)}(\bar{J}_y)} \tag{4.86}$$

The problem is reduced to computing the representations of rotations about the y-axis. $d^{(j)}(\beta)$ is chosen about the y-axis rather that the x-axis because with our phase conventions $D^{(j)}(\bar{J}_y)$ is always imaginary, and then $d^{(j)}(\beta)$ is real.

---

[10] See Problem 4.3.

4.3.6 The Spin-One Representation

I will look at one more special case, $j = 1$, and then turn to the general formula. For the three-dimensional spin-one representation

$$D^{(1)}(\bar{J}_z) = \begin{pmatrix} 1 & 0 & 0 \\ 0 & 0 & 0 \\ 0 & 0 & -1 \end{pmatrix} \tag{4.87a}$$

$$D^{(1)}(\bar{J}_+) = \begin{pmatrix} 0 & \sqrt{2} & 0 \\ 0 & 0 & \sqrt{2} \\ 0 & 0 & 0 \end{pmatrix} \tag{4.87b}$$

$$D^{(1)}(\bar{J}_-) = D^{(1)}(\bar{J}_+)^\dagger \tag{4.87c}$$

$D^{(1)}(\bar{J}_y)$ is needed to compute $d^{(1)}(\beta)$:

$$D^{(1)}(\bar{J}_y) = \frac{1}{2i}\left(D^{(1)}(\bar{J}_+) - D^{(1)}(\bar{J}_-)\right) = \frac{1}{2i}\begin{pmatrix} 0 & \sqrt{2} & 0 \\ -\sqrt{2} & 0 & \sqrt{2} \\ 0 & -\sqrt{2} & 0 \end{pmatrix} \tag{4.88}$$

All three of the $D^{(1)}(\bar{J}_i)$ have eigenvalues 1, 0, and -1. Therefore,[11]

$$D^{(1)}(\bar{J}_i)^3 = D^{(1)}(\bar{J}_i) \tag{4.89}$$

Expand the exponential in a power series, using equation (4.89) repeatedly, and collect terms to get

$$\begin{aligned} d^{(1)}(\beta) &= D^{(1)}(e^{-i\beta \bar{J}_y}) \\ &= I - iD^{(1)}(\bar{J}_y)\sin\beta - D^{(1)}(\bar{J}_y)^2(1 - \cos\beta) \end{aligned} \tag{4.90}$$

In contrast to the spin-half representation, the spin 1 matrices are a faithful representation of the rotation group. Rotations by 2π are the identity.

4.3.7 Arbitrary j

For $j > 1$, the method of explicit exponentiation gets cumbersome. The general formula was discovered by Wigner. You can find a derivation, not really very complicated, in Section B.1.2 of the appendix. The result is

$$d^{(j)}{}_{m'm}(\beta)$$

$$= (-1)^{m'-m} \sum_\mu (-1)^\mu \frac{\sqrt{(j+m)!(j-m)!(j+m')!(j-m')!}}{\mu!(j+m-\mu)!(j-\mu-m')!(m'+\mu-m)!}$$

$$\times \left(\cos(\beta/2)\right)^{2j+m-m'-2\mu} \left(\sin(\beta/2)\right)^{2\mu+m'-m} \tag{4.91}$$

The summation goes over all values of μ such that all the factors in the denominator are factorials of nonnegative integers. Check that this formula agrees with the special cases we worked out above.

---
[11] The rule (4.89) is particular to the spin-one representation.

4.4 ADDITION OF ANGULAR MOMENTA

4.4.1 Spin and Orbital Angular Momentum

Instead of the states $|\mathbf{r}, m\rangle$ defined in Section 4.3.2 you can choose a basis of eigenstates of $\mathbf{L}^2$, L_z, s_z, and some other observable (like the energy) that have the form $|n, l, m_l, m_s\rangle$, where

$$\mathbf{L}^2|n,l,m_l,m_s\rangle = l(l+1)|n,l,m_l,m_s\rangle \qquad \mathbf{s}^2|n,l,m_l,m_s\rangle = s(s+1)|n,l,m_l,m_s\rangle$$

$$L_z|n,l,m_l,m_s\rangle = m_l|n,l,m_l,m_s\rangle \qquad s_z|n,l,m_l,m_s\rangle = m_s|n,l,m_l,m_s\rangle \tag{4.92}$$

The wave functions for these new basis states are two-component spinors that look like

$$\langle \mathbf{r}, m_{s'}|\, n,l,m_l,m_s\rangle = \delta_{m_{s'}m_s} R_{nlm_s}(r) Y_l^{m_l}(\theta,\phi) \tag{4.93}$$

If the form of the Hamiltonian is $H = \mathbf{p}^2/2m + V(r)$, then $\mathbf{L}$ and $\mathbf{s}$ all commute with H, and the states in equation (4.92) are energy eigenstates. In this simple situation the energy would be independent of the spin quantum number m_s. A central potential Hamiltonian does not mix up spin and orbital angular momentum.

The real world is more complicated. Rotational invariance requires only that the Hamiltonian be invariant under total rotations:

$$H = R^\dagger H R \tag{4.94}$$

or equivalently

$$[J_i, H] = 0 \tag{4.95}$$

The total angular momentum $\mathbf{J} = \mathbf{L} + \mathbf{s}$ is conserved, but $\mathbf{L}$ and $\mathbf{s}$ are not necessarily conserved separately. There are terms in H like $\mathbf{s} \cdot \mathbf{p}$ or $\mathbf{s} \cdot \mathbf{L}$ that contain the spin operator explicitly. These violate orbital and spin angular momentum conservation, but commute with the components of $\mathbf{J}$:

$$\begin{aligned}[J_i, \mathbf{L}\cdot\mathbf{s}] &= \sum_j [(L+s)_i, L_j s_j] = \sum_j ([L_i, L_j s_j] + [s_i, L_j s_j]) \\ &= \sum_j ([L_i, L_j]s_j + L_j[s_i, s_j]) = i\sum_{jk} \epsilon_{ijk}(L_k s_j + L_j s_k) = 0\end{aligned} \tag{4.96}$$

The hydrogen atom Hamiltonian will have this property when magnetic forces are included.

The energy eigenstates are always eigenstates of $\mathbf{J}^2$ and J_z, and $\mathbf{L}^2$ and $\mathbf{s}^2$, but not the components of $\mathbf{L}$ or $\mathbf{s}$. The wave functions with definite $\mathbf{L}^2$, L_z and s_z are often easier to write down immediately using the spherical harmonics. So we will need to construct the $\mathbf{J}^2$ and J_z eigenstates in terms of the $\mathbf{L}^2$, L_z and s_z ones. This problem is usually called the addition of angular momenta.

$\mathbf{L}^2$ commutes with the three J_i even though its individual components L_i do not, so there are eigenstates of $\mathbf{J}^2$ and J_z that are also eigenstates of $\mathbf{L}^2$. All one needs to do is to combine the finite number of states with a fixed value of l into $\mathbf{J}^2$ eigenstates.

4.4.2 Two Simple Examples

Angular Momentum Zero + Spin Half

Let us include the spin quantum number for an electron in a hydrogen atom. The states of the type (4.92) are eigenstates of $\mathbf{L}^2$, L_z, s_z, and

$$H_o = \frac{\mathbf{p}^2}{2m} - \frac{\alpha \hbar c}{r} \qquad (4.97)$$

with eigenvalues $l(l+1)$, m_l, m_s, and $E_n = -\alpha^2 mc^2/2n^2$ respectively.

What are the eigenstates of $\mathbf{J}^2$? There are two $1S$ states, $|n, l, m_l, m_s\rangle = |1, 0, 0, 1/2\rangle$ and $|n, l, m_l, m_s\rangle = |1, 0, 0, -1/2\rangle$. L_i is zero on either of these states, so[12]

$$J_i \left| 1, 0, 0, \pm \frac{1}{2} \right\rangle = s_i \left| 1, 0, 0, \pm \frac{1}{2} \right\rangle \qquad (4.98)$$

These two states transform under $\mathbf{J}$ like the two-dimensional spin-half representation. They are eigenstates of $\mathbf{J}^2$ and J_z with eigenvalues $\frac{3}{4}$ and $\pm\frac{1}{2}$ respectively.

Angular Momentum One + Spin Half

S states are trivial because $\mathbf{L}^2 = 0$ acting on them. The six $2P$ states are more complicated. These are not eigenstates of $\mathbf{J}^2$, but are still eigenstates of $J_z = L_z + s_z$:

$$J_z \left| 2, 1, m_l, \pm \frac{1}{2} \right\rangle = \left(m_l \pm \frac{1}{2} \right) \left| 2, 1, m_l, \pm \frac{1}{2} \right\rangle \qquad (4.99)$$

Here are the six $2P$ states arranged in four rows each with a definite eigenvalue[13] m for J_z.

$$\begin{aligned}
m &= +\frac{3}{2}: & &\left|1, +\frac{1}{2}\right\rangle \\
m &= +\frac{1}{2}: & &\left|0, +\frac{1}{2}\right\rangle \quad \text{and} \quad \left|1, -\frac{1}{2}\right\rangle \\
m &= -\frac{1}{2}: & &\left|0, -\frac{1}{2}\right\rangle \quad \text{and} \quad \left|-1, +\frac{1}{2}\right\rangle \\
m &= -\frac{3}{2}: & &\left|-1, -\frac{1}{2}\right\rangle
\end{aligned} \qquad (4.100)$$

What does $\mathbf{J}^2$ do to the $m = 1/2$ states? Use the rule (3.98) and $J_i = L_i + s_i$ to write

$$\mathbf{J}^2 \left| 0, \frac{1}{2} \right\rangle$$
$$= \left[\frac{1}{2} \left((L_+ + s_+)(L_- + s_-) + (L_- + s_-)(L_+ + s_+) \right) + J_z^2 \right] \left| 0, \frac{1}{2} \right\rangle \qquad (4.101)$$

---

[12] The spectroscopic notation S, P, D, F, and so forth, always refers only to the *orbital* part of the angular momentum.

[13] For brevity I omit the irrelevant quantum numbers n and l here. The labels are $|m_l, m_s\rangle$.

The components of **L** act only on the first label, and the components of **s** on the second. After a little algebra

$$\mathbf{J}^2 \left|0, \frac{1}{2}\right\rangle = \frac{11}{4}\left|0, \frac{1}{2}\right\rangle + \sqrt{2}\left|1, -\frac{1}{2}\right\rangle \tag{4.102}$$

and similarly

$$\mathbf{J}^2 \left|1, -\frac{1}{2}\right\rangle = \sqrt{2}\left|0, \frac{1}{2}\right\rangle + \frac{7}{4}\left|1, -\frac{1}{2}\right\rangle \tag{4.103}$$

On these states $\mathbf{J}^2$ is a 2×2 matrix:

$$\mathbf{J}^2 = \begin{pmatrix} 11/4 & \sqrt{2} \\ \sqrt{2} & 7/4 \end{pmatrix} \tag{4.104}$$

The eigenstates are linear combinations of $|1, -1/2\rangle$ and $|0, 1/2\rangle$. The eigenvalues are the to the characteristic equation:

$$0 = \det \begin{pmatrix} 11/4 - \lambda & \sqrt{2} \\ \sqrt{2} & 7/4 - \lambda \end{pmatrix} = \left(\lambda - \frac{15}{4}\right)\left(\lambda - \frac{3}{4}\right) \tag{4.105}$$

The roots of equation (4.105), $\frac{15}{4} = \frac{3}{2}(\frac{3}{2} + 1)$ and $\frac{3}{4} = \frac{1}{2}(\frac{1}{2} + 1)$, are among the allowed values $j(j+1)$. Of course this had to be so, but it is comforting to see it come out in an explicit computation. There are linear combinations of the $2P$ states that have total angular momentum $\frac{1}{2}$, and others that have total angular momentum $\frac{3}{2}$. The eigenstates can be found by diagonalizing the matrix (4.104).

4.5 CLEBSCH-GORDAN COEFFICIENTS

4.5.1 Definition of the Coefficients

There is a more efficient method for finding the eigenstates of $\mathbf{J}^2 = (\mathbf{L} + \mathbf{s})^2$ than to diagonalize a $2j + 1$ dimensional matrix. The algorithm is explained in this section. It works for adding any two angular momenta; the orbital and spin angular momentum of an electron, as in Section 4.4.2, or the total angular momentum of two different particles. The matrix elements of the new basis states in terms of the old, or the old in terms of the new, are called the Clebsch-Gordan coefficients.

The key to solving this problem is that the states come in $2j+1$ dimensional ladders with the usual properties of angular momentum eigenstates. Let

$$\mathbf{J} = \mathbf{J}_1 + \mathbf{J}_2 \tag{4.106}$$

where $\mathbf{J}_1$ and $\mathbf{J}_2$ commute:

$$[J_{1i}, J_{2j}] = 0 \tag{4.107}$$

and are independently generators of angular momentum algebras:

$$[J_{1i}, J_{1j}] = i \sum_k \epsilon_{ijk} J_{1k} \quad \text{and} \quad [J_{2i}, J_{2j}] = i \sum_k \epsilon_{ijk} J_{2k} \tag{4.108}$$

$\mathbf{J}_1$ and $\mathbf{J}_2$ are vector operators under the total angular momentum:

$$[J_i, J_{1j}] = i\sum_k \epsilon_{ijk} J_{1k} \quad \text{and} \quad [J_i, J_{2j}] = i\sum_k \epsilon_{ijk} J_{2k} \qquad (4.109)$$

and therefore $[J_i, J_j] = i\sum_k \epsilon_{ijk} J_k$ also.

The "old" basis states—the states in equation (4.92) are an example—are $|\psi_{m_1 m_2}\rangle$, with the following properties:[14]

$$\mathbf{J}_1^2 |\psi_{m_1 m_2}\rangle = j_1(j_1+1)|\psi_{m_1 m_2}\rangle \qquad (4.110\text{a})$$
$$\mathbf{J}_2^2 |\psi_{m_1 m_2}\rangle = j_2(j_2+1)|\psi_{m_1 m_2}\rangle \qquad (4.110\text{b})$$
$$J_{1z} |\psi_{m_1 m_2}\rangle = m_1 |\psi_{m_1 m_2}\rangle \qquad (4.110\text{c})$$
$$J_{2z} |\psi_{m_1 m_2}\rangle = m_2 |\psi_{m_1 m_2}\rangle \qquad (4.110\text{d})$$

The "new" basis states are $|\phi_{jm}\rangle$. Since $[J_i, \mathbf{J}_1^2] = [J_i, \mathbf{J}_2^2] = 0$, the "new" states are also eigenstates of $\mathbf{J}_1^2$ and $\mathbf{J}_2^2$, but not necessarily of J_{1z} or J_{2z}. Here are the properties of these states:

$$\mathbf{J}_1^2 |\phi_{jm}\rangle = j_1(j_1+1)|\phi_{jm}\rangle \qquad (4.111\text{a})$$
$$\mathbf{J}_2^2 |\phi_{jm}\rangle = j_2(j_2+1)|\phi_{jm}\rangle \qquad (4.111\text{b})$$
$$\mathbf{J}^2 |\phi_{jm}\rangle = j(j+1)|\phi_{jm}\rangle \qquad (4.111\text{c})$$
$$J_z |\phi_{jm}\rangle = m|\phi_{jm}\rangle \qquad (4.111\text{d})$$

The Clebsch-Gordan coefficients are the projections of the first type of state onto the states of the second type. I will start with some simple examples for low values of j_1 and j_2.

4.5.2 Spin Half + Spin Half

Consider first adding the spins of two spin-half particles. They might be two electrons, or the electron and proton that make up a hydrogen atom. $\mathbf{s}$ is the total spin.

$$\mathbf{s} = \mathbf{s}_1 + \mathbf{s}_2 \qquad (4.112)$$

The eigenvalues of $\mathbf{s}_1^2$ and $\mathbf{s}_2^2$ are always $\frac{3}{4}$. There are four states, labeled in the "old" basis by

$$\left|\psi_{\frac{1}{2},\frac{1}{2}}\right\rangle \quad \left|\psi_{-\frac{1}{2},\frac{1}{2}}\right\rangle \quad \left|\psi_{\frac{1}{2},-\frac{1}{2}}\right\rangle \quad \left|\psi_{-\frac{1}{2},-\frac{1}{2}}\right\rangle \qquad (4.113)$$

We want to find the states $|\phi_{s,m}\rangle$ such that

$$s_z |\phi_{s,m}\rangle = m|\phi_{s,m}\rangle \quad \text{and} \quad \mathbf{s}^2 |\phi_{s,m}\rangle = s(s+1)|\phi_{s,m}\rangle \qquad (4.114)$$

---

[14] When there may be some ambiguity, the states $|\psi_{m_1 m_2}\rangle$ and $|\phi_{jm}\rangle$ below can be written $\left|\psi_{m_1 m_2}^{j_1 j_2}\right\rangle$ and $\left|\phi_{jm}^{j_1 j_2}\right\rangle$. Another common notation is $|j_1, j_2; m_1, m_2\rangle$ and $|j_1, j_2; j.m\rangle$. Usually I will suppress all other quantum numbers except the angular momenta, and suppress the j's also when they are clear from the context. There is no completely standard notation.

There is an immediate simplification. Since $s_z = s_{1z} + s_{2z}$, the states $|\psi_{m_1 m_2}\rangle$ are automatically eigenstates of s_z:

$$s_z |\psi_{m_1 m_2}\rangle = (s_{1z} + s_{2z})|\psi_{m_1 m_2}\rangle = (m_1 + m_2)|\psi_{m_1 m_2}\rangle \tag{4.115}$$

The transformation to the new basis will mix up states with the same $m_1 + m_2$ only.

The new basis states are angular momentum eigenstates, so they come in ladders of length $2s + 1$, where m goes from $-s$ to s in each ladder. The state $|\psi_{\frac{1}{2},\frac{1}{2}}\rangle$ has $m = 1$, so it has to belong to a ladder with $s \geq 1$. But $|\psi_{\frac{1}{2},\frac{1}{2}}\rangle$ is the only state with $m = 1$, and there is no state with higher m. It follows, and *here is the centerpiece of the construction*, that $|\psi_{\frac{1}{2},\frac{1}{2}}\rangle$ is at the top of an $s = 1$ triplet, and we can identify

$$|\psi_{\frac{1}{2},\frac{1}{2}}\rangle = |\phi_{1,1}\rangle \tag{4.116}$$

There must be two more states in an $s = 1$ triplet, having $m = 0$ and -1 respectively. The $s = 1$, $m = 0$ state is some linear combination of the two states with $m = 0$. We can find it using the lowering operator s_-. Compute $s_- |\phi_{11}\rangle$ in the new basis:

$$s_- |\phi_{1,1}\rangle = \sqrt{s(s+1) - m(m-1)} |\phi_{1,0}\rangle = \sqrt{2} |\phi_{10}\rangle \tag{4.117}$$

and also in the old basis:

$$s_- |\phi_{1,1}\rangle = (s_{1-} + s_{2-}) |\psi_{\frac{1}{2},\frac{1}{2}}\rangle = |\psi_{-\frac{1}{2},\frac{1}{2}}\rangle + |\psi_{\frac{1}{2},-\frac{1}{2}}\rangle \tag{4.118}$$

so

$$|\phi_{10}\rangle = \frac{1}{\sqrt{2}} \left[|\psi_{-\frac{1}{2},\frac{1}{2}}\rangle + |\psi_{\frac{1}{2},-\frac{1}{2}}\rangle \right] \tag{4.119}$$

Apply s_- again to get the bottom state in the ladder: The result of course is

$$|\phi_{1,-1}\rangle = |\psi_{-\frac{1}{2},-\frac{1}{2}}\rangle \tag{4.120}$$

These $s = 1$ states span a three-dimensional subspace of the original four-dimensional space. Their phases are related according to the conventions in Section 4.2.5.

There is one remaining independent state. This last state is an angular momentum ladder all by itself, a ladder of length 1, so it must have $s = 0$ and $m = 0$. The $s = 0$ state has to be orthogonal to the state $|\phi_{10}\rangle$, since both are eigenstates of $\mathbf{s}^2$ with different eigenvalues, and eigenstates of Hermitean operators with different eigenvalues must be orthogonal. Up to an irrelevant phase, these properties define the $s = 0$ state uniquely:

$$|\phi_{0,0}\rangle = \frac{1}{\sqrt{2}} \left[|\psi_{-\frac{1}{2},\frac{1}{2}}\rangle - |\psi_{\frac{1}{2},-\frac{1}{2}}\rangle \right] \tag{4.121}$$

4.5.3 Spin Half + Angular Momentum One

Next return to the problem discussed in Section 4.4.2, a spin-half electron in a P-state ($l = 1$). The total angular momentum is $\mathbf{J} = \mathbf{L} + \mathbf{s}$. There are six states in all. In the "old" basis, these are three spin-up states

$$\left|\psi_{1,\frac{1}{2}}\right\rangle \quad \left|\psi_{0,\frac{1}{2}}\right\rangle \quad \left|\psi_{-1,\frac{1}{2}}\right\rangle \tag{4.122}$$

and three spin-down states

$$\left|\psi_{1,-\frac{1}{2}}\right\rangle \quad \left|\psi_{0,-\frac{1}{2}}\right\rangle \quad \left|\psi_{-1,-\frac{1}{2}}\right\rangle \tag{4.123}$$

The state with highest $m = m_1 + m_2$ has $m = 3/2$. The six states must divide into four $j = 3/2$ states and two $j = 1/2$ states. Here the top state must have $j = 3/2$, since there is no state with higher m:

$$\left|\phi_{\frac{3}{2},\frac{3}{2}}\right\rangle = \left|\psi_{1,\frac{1}{2}}\right\rangle \tag{4.124}$$

The other states with $j = 3/2$ can be found by applying $J_- = L_- + s_-$:

$$J_-\left|\phi_{\frac{3}{2},\frac{3}{2}}\right\rangle = \sqrt{3}\left|\phi_{\frac{3}{2},\frac{1}{2}}\right\rangle$$
$$(L_- + s_-)\left|\psi_{1,\frac{1}{2}}\right\rangle = \sqrt{2}\left|\psi_{0,\frac{1}{2}}\right\rangle + \left|\psi_{1,-\frac{1}{2}}\right\rangle \tag{4.125}$$

so

$$\left|\phi_{\frac{3}{2},\frac{1}{2}}\right\rangle = \frac{1}{\sqrt{3}}\left[\sqrt{2}\left|\psi_{0,\frac{1}{2}}\right\rangle + \left|\psi_{1,-\frac{1}{2}}\right\rangle\right] \tag{4.126}$$

Continuing,

$$\left|\phi_{\frac{3}{2},-\frac{1}{2}}\right\rangle = \frac{1}{\sqrt{3}}\left[\left|\psi_{-1,\frac{1}{2}}\right\rangle + \sqrt{2}\left|\psi_{0,-\frac{1}{2}}\right\rangle\right] \tag{4.127}$$

and finally,

$$\left|\phi_{\frac{3}{2},-\frac{3}{2}}\right\rangle = \left|\psi_{-1,-\frac{1}{2}}\right\rangle \tag{4.128}$$

The states come out automatically normalized.

The remaining two states must be orthogonal to those in the $j = 3/2$ ladder, since they are eigenstates of the Hermitean operator $\mathbf{J}^2$ with a different eigenvalue, $j = 1/2$. The state with $m = 1/2$ orthogonal to $\left|\phi_{\frac{3}{2},\frac{1}{2}}\right\rangle$ is (the phase is again arbitrary)

$$\left|\phi_{\frac{1}{2},\frac{1}{2}}\right\rangle = \frac{1}{\sqrt{3}}\left[\sqrt{2}\left|\psi_{1,-\frac{1}{2}}\right\rangle - \left|\psi_{0,\frac{1}{2}}\right\rangle\right] \tag{4.129}$$

and the $m = -\frac{1}{2}$ state is

$$\left|\phi_{\frac{1}{2},-\frac{1}{2}}\right\rangle = \frac{1}{\sqrt{3}}\left[\left|\psi_{0,-\frac{1}{2}}\right\rangle - \sqrt{2}\left|\psi_{-1,\frac{1}{2}}\right\rangle\right] \tag{4.130}$$

4.5.4 Spin Half + Angular Momentum l

To describe all the states of one-electron atoms we need to figure out how to combine spin half with any l. The construction is a generalization of the last two examples.

The total angular momentum of the electron is $\mathbf{J} = \mathbf{L} + \mathbf{s}$. $\mathbf{s}^2$ is always $3/4$. The "old" states are $|\psi_{m_l m_s}\rangle$. The "new" states $|\phi_{jm}\rangle$ will have quantum numbers $\mathbf{J}^2$, $\mathbf{L}^2$, and J_z. The state with $m_l = l$ and $m_s = 1/2$ is an eigenstate of J_z with eigenvalue $l + 1/2$, and there are no states with higher eigenvalues. Therefore, there is a $2j + 1 = 2l + 2$ dimensional ladder of eigenstates with $j = l + 1/2$.

Unless $l = 0$, there are two states with $m = l - 1/2$. One linear combination will be in the $j = l + 1/2$ ladder. The remaining one must be the top of a new ladder, this time with $j = l - 1/2f$. The number of states in these two ladders adds up to $2(2l + 1)$, which is the dimension of the original space. It follows that all states fall into one of these two ladders, and the possible j values are only $l \pm 1/2$.

The states in the $j = l + 1/2$ ladder are linear combinations of one spin-up state and one spin-down state:

$$\left|\phi_{l+\frac{1}{2},m}\right\rangle = \alpha_m \left|\psi_{m-\frac{1}{2},\frac{1}{2}}\right\rangle + \beta_m \left|\psi_{m+\frac{1}{2},-\frac{1}{2}}\right\rangle$$

for some constants α_m and β_m such that

$$|\alpha_m|^2 + |\beta_m|^2 = 1 \tag{4.132}$$

Computing these coefficients is a brief algebraic exercise (see below). The result is

$$\alpha_m = \sqrt{\frac{l+m+1/2}{2l+1}} \qquad \beta_m = \sqrt{\frac{l-m+1/2}{2l+1}} \tag{4.133}$$

These are all the Clebsch-Gordan coefficients for spin l + spin half. The coefficients are really determined only up to a phase. The convention is that all the Clebsch-Gordan coefficients are real numbers:

$$\left|\phi_{l+\frac{1}{2},m}\right\rangle = \sqrt{\frac{l+m+1/2}{2l+1}} \left|\psi_{m-\frac{1}{2},\frac{1}{2}}\right\rangle + \sqrt{\frac{l-m+1/2}{2l+1}} \left|\psi_{m+\frac{1}{2},-\frac{1}{2}}\right\rangle \tag{4.134}$$

The states with $j = l - 1/2$ are orthogonal to the states in equation (4.134):

$$\left|\phi_{l-\frac{1}{2},m}\right\rangle = \sqrt{\frac{l-m+1/2}{2l+1}} \left|\psi_{m-\frac{1}{2},\frac{1}{2}}\right\rangle - \sqrt{\frac{l+m+1/2}{2l+1}} \left|\psi_{m+\frac{1}{2},-\frac{1}{2}}\right\rangle \tag{4.135}$$

These coefficients will be useful for constructing the stationary states of a one-electron atom. (See, for instance, Problem 7.18.)

Calculation of α_m and β_m. Apply the lowering operator to any state:[15]

---

[15] Write $\sqrt{j(j+1) - m(m-1)}$ as $\sqrt{(j+m)(j-m+1)}$.

On the one hand,

$$J_-\left|\phi_{l+\frac{1}{2},m}\right\rangle = \sqrt{(j+m)(j-m+1)}\left|\phi_{l+\frac{1}{2},m-1}\right\rangle$$

$$= \sqrt{\left(l+m+\frac{1}{2}\right)\left(l-m+\frac{3}{2}\right)}\left[\alpha_{m-1}\left|\psi_{m-\frac{3}{2},\frac{1}{2}}\right\rangle\right. \quad (4.136)$$

$$\left. + \beta_{m-1}\left|\psi_{m-\frac{1}{2},-\frac{1}{2}}\right\rangle\right]$$

while on the other,

$$J_-\left|\phi_{l+\frac{1}{2},m}\right\rangle = (L_- + s_-)\left[\alpha_m\left|\psi_{m-\frac{1}{2},\frac{1}{2}}\right\rangle + \beta_m\left|\psi_{m+\frac{1}{2},-\frac{1}{2}}\right\rangle\right]$$

$$= \alpha_m\left[\sqrt{\left(l+m-\frac{1}{2}\right)\left(l-m+\frac{3}{2}\right)}\left|\psi_{m-\frac{3}{2},\frac{1}{2}}\right\rangle + \left|\psi_{m-\frac{1}{2},-\frac{1}{2}}\right\rangle\right] \quad (4.137)$$

$$+ \beta_m\sqrt{\left(l+m+\frac{1}{2}\right)\left(l-m+\frac{1}{2}\right)}\left|\psi_{m-\frac{1}{2},-\frac{1}{2}}\right\rangle$$

Since the states on the right-hand sides of equations (4.136) and (4.137) are the same, the spin-up components and the spin-down components must be equal separately. Compare the coefficients of the spin-up states to get

$$\alpha_{m-1}\sqrt{(l+m+\frac{1}{2})(l-m+\frac{3}{2})} = \alpha_m\sqrt{(l+m-\frac{1}{2})(l-m+\frac{3}{2})} \quad (4.138)$$

The spin-down state will give another relation, determining β, but we do not need it, since $|\beta|^2 = 1 - |\alpha^2|$.

Cancel the factor $\sqrt{l-m+3/2}$ in (4.138) to obtain a recursion relation for the α_m:

$$\frac{\alpha_m}{\sqrt{l+m+\frac{1}{2}}} = \frac{\alpha_{m-1}}{\sqrt{l+m-\frac{1}{2}}} \quad (4.139)$$

The ratio is independent of m, since it remains unchanged if you replace m by $m-1$. That means it can be evaluated for all m if it is known for any. When $m = l + 1/2$, there is no contribution from a spin-down state. $\beta_{l+1/2} = 0$ and $\alpha_{l+1/2} = 1$. Thus

$$\alpha_m = \sqrt{\frac{l+m+\frac{1}{2}}{2l+1}} \quad (4.140)$$

∎

4.5.5 The General Rule

The general idea should now be clear. Add j_1 to j_2 to get the highest j. There is a ladder of $2(j_1 + j_2) + 1$ of these states, starting at the top with the single state

128 Chapter 4 Symmetry Transformations on States

$m_1 = j_1$, $m_2 = j_2$. There will be two states with $m = j_1 + j_2 - 1$. One is in this $j = j_1 + j_2$ ladder, so the other must be the top state of a ladder with $j = j_1 + j_2 - 1$, and so on. What is the lowest j?

> Note: Imagine adding the angular momenta of two classical spinning tops with angular momenta $\hbar j_1$ and $\hbar j_2$ respectively. The maximum angular momentum of the whole system is $\hbar j_1 + \hbar j_2$ and the minimum is $|\hbar j_1 - \hbar j_2|$. That is the answer here too.

Suppose $j_1 \geq j_2$. Then the states with each m have j values as follows:

$$
\begin{array}{ccc}
m_1 & m_2 & m = m_1 + m_2 \\
j_1 & j_2 & j_1 + j_2 \\
\\
j_1 - 1 & j_2 & j_1 + j_2 - 1 \\
j_1 & j_2 - 1 & j_1 + j_2 - 1 \\
\\
j_1 - 2 & j_2 & j_1 + j_2 - 2 \\
j_1 - 1 & j_2 - 1 & j_1 + j_2 - 2 \\
j_1 & j_2 - 2 & j_1 + j_2 - 2 \\
\vdots & \vdots & \vdots \\
j_1 - 2j_2 & j_2 & j_1 - j_2 \\
j_1 - 2j_2 + 1 & j_2 - 1 & j_1 - j_2 \\
j_1 - 2j_2 + 2 & j_2 - 2 & j_1 - j_2 \\
\vdots & \vdots & \vdots \\
j_1 & -j_2 & j_1 - j_2
\end{array}
\tag{4.141}
$$

All together, there are $(2j_1 + 1)(2j_2 + 1)$ states. The top state heads a ladder with $2j + 1$ states, where $j = j_1 + j_2$. One of the two states with $m = j_1 + j_2 - 1$ is in that ladder, so the other starts a new ladder with $j = j_1 + j_2 - 1$. Of the three states with $m = j_1 + j_2 - 2$, two are already accounted for, so there is one new ladder with $j = j_1 + j_2 - 2$ and so on. Going all the way down to $j_1 - j_2$, the total number of states in the various ladders (there is one for each m from $j_1 + j_2$ through $j_1 - j_2$) is[16]

$$\sum_{j=j_1-j_2}^{j=j_1+j_2} (2j + 1) = (2j_1 + 1)(2j_2 + 1) \tag{4.142}$$

Since that is all the states there are, every state has been accounted for. The states with $m < j_1 - j_2$ are all in the ladders with $j \geq j_1 - j_2$.

Thus if $j_1 + j_2$ is an integer (half-integer), each integer (half-integer) j from $j_1 + j_2$ down through $|j_1 - j_2|$ occurs once and only once. Three lines of lengths j_1, j_2, and j can form can form a triangle.

---
[16] The sum of the first n odd integers is n^2.

4.5.6 Recursion Relation for the Coefficients

There is no simple form for all the Clebsch-Gordan coefficients (the spin-l, spin-half case was special in that regard) but there is a recursion relation that can be used for computations. Combine j_1 and j_2 to get eigenstates of $\mathbf{J}^2$ and J_z. Call the "old" states $|\psi_{m_1 m_2}\rangle$ and the "new" states $|\phi_{jm}\rangle$. The Clebsch-Gordan coefficients are defined to be real.

$$\langle \psi_{m_1 m_2} | \phi_{jm} \rangle = \langle \phi_{jm} | \psi_{m_1 m_2} \rangle \tag{4.143}$$

Apply $J_\pm = J_{1\pm} + J_{2\pm}$ to $|\phi_{jm}\rangle$ and take the scalar product of each term with $|\psi_{m_1 m_2}\rangle$:

$$\sqrt{j(j+1) - m(m \pm 1)} \langle \psi_{m_1 m_2} | \phi_{j, m \pm 1} \rangle$$
$$= \sqrt{j_1(j_1+1) - m_1(m_1 \mp 1)} \langle \psi_{m_1 \mp 1, m_2} | \phi_{jm} \rangle \tag{4.144}$$
$$+ \sqrt{j_2(j_2+1) - m_2(m_2 \mp 1)} \langle \psi_{m_1, m_2 \mp 1} | \phi_{jm} \rangle$$

From the recursion relation (4.144) you can determine all the coefficients. Start with the "top" coefficient, which must have unit magnitude. We always choose it to be 1.

4.5.7 The Clebsch-Gordan Series

An equivalent recursion relation, in terms of the finite representation matrices $D^{(j)}$, will also be useful. Under a finite rotation,

$$D^{(j_1)}_{m'_1, m}(\hat{\mathbf{n}}, \theta) D^{(\kappa)}_{q', q}(\hat{\mathbf{n}}, \theta) = \left\langle \psi^{j_1, \kappa}_{m'_1, q'} \middle| e^{-i\theta \hat{\mathbf{n}} \cdot \mathbf{J}} \middle| \psi^{j_1, \kappa}_{m, q} \right\rangle \tag{4.145}$$

On the left-hand side, use

$$I = \sum_{j\mu} \left| \phi^{j_1, \kappa}_{j, \mu} \right\rangle \left\langle \phi^{j_1, \kappa}_{j, \mu} \right| \tag{4.146}$$

twice. Since $\exp(-i\theta \hat{\mathbf{n}} \cdot \mathbf{J})$ is diagonal in these states,[17]

$$D^{(j_1)}_{m'_1 m}(\hat{\mathbf{n}}, \theta) D^{(\kappa)}_{q' q}(\hat{\mathbf{n}}, \theta)$$
$$= \sum_{j=|j_1-\kappa|}^{j_1+\kappa} \sum_{\mu m'} \langle \psi_{m'_1, q'} | \phi_{j, \mu} \rangle \langle \phi_{j, m'} | \psi_{m, q} \rangle D^{(j)}_{\mu m'}(\hat{\mathbf{n}}, \theta) \tag{4.147}$$

Equation (4.147), the multiplication rule for representations of the rotation group, is the **Clebsch-Gordan series**. The Clebsch-Gordan coefficients are the coefficients in an expansion of a product of matrix elements of two $D^{(j)}$ in terms of matrix elements of a single $D^{(j)}$. Equations (4.144) and (4.147) are not independent.

---

[17] The superscripts on the state vectors are omitted to reduce the clutter of indices.

Chapter 4 Symmetry Transformations on States

The former can be obtained by expanding the latter in the rotation angles and examining the lowest order terms.

Equation (4.147) can be turned into a more useful form as follows: Multiply both sides by $\langle \psi_{m'_1 q'} | \phi_{j_3, m_3} \rangle$, where $j_1 \leq j_3 \leq \kappa$, and then sum over m'_1 and q':

$$\sum_{m'_1, q'} \langle \psi_{m'_1 q'} | \phi_{j_3, m_3} \rangle D^{(j_1)}_{m'_1 m}(\hat{n}, \theta) D^{(\kappa)}_{q' q}(\hat{n}, \theta)$$

$$= \sum_{m'_1, q'} \sum_{j=|j_1-\kappa|}^{j_1+\kappa} \sum_{\mu m'} \langle \psi_{m'_1 q'} | \phi_{j_3, m_3} \rangle \langle \psi_{m'_1 q'} | \phi_{j, \mu} \rangle \langle \phi_{j, m'} | \psi_{m, q} \rangle D^{(j)}_{\mu m'}(\hat{n}, \theta)$$

$$= \sum_{m'} \langle \phi_{j_3, m'} | \psi_{m, q} \rangle D^{(j_3)}_{m_3 m'}(\hat{n}, \theta) \quad (4.148)$$

The last form follows from the fact that the Clebsch-Gordan coefficients form an orthogonal matrix. Next multiply by

$$D^{(j_1)}_{m m_1}(\hat{n}, \theta)^\dagger = D^{(j_1)}_{m_1 m}(\hat{n}, \theta)^* \quad (4.149)$$

and sum over m and m_1 using the unitarity of the matrix $D^{(j_1)}$:

$$\sum_{m, m'} \langle \phi_{j_3, m'} | \psi_{m, q} \rangle D^{(j_3)}_{m_3 m'}(\hat{n}, \theta) D^{(j_1)}_{m_1 m}(\hat{n}, \theta)^*$$

$$= \sum_{m, m'_1, q'} \langle \psi_{m'_1 q'} | \phi_{j_3, m_3} \rangle D^{(j_1)}_{m'_1 m}(\hat{n}, \theta) D^{(j_1)}_{m m_1}(\hat{n}, \theta)^\dagger D^{(\kappa)}(\hat{n}, \theta)_{q' q} \quad (4.150)$$

or (reinstating the superscripts, and using the real-symmetric property of the coefficients)

$$\sum_{m, m'} \langle \phi^{j_1, \kappa}_{j_3, m'} | \psi^{j_1, \kappa}_{m, q} \rangle D^{(j_3)}_{m_3 m'}(\hat{n}, \theta) D^{(j_1)}_{m_1 m}(\hat{n}, \theta)^*$$

$$= \sum_{q'} \langle \phi^{j_1, \kappa}_{j_3, m_3} | \psi^{j_1, \kappa}_{m_1 q'} \rangle D^{(\kappa)}_{q' q}(\hat{n}, \theta) \quad (4.151)$$

Equation (4.151) is a set of linear equations for the Clebsch-Gordan coefficients in terms of elements of the rotation matrices. Since the equation is homogenous, it determines the ratios of the coefficients only. Their values are fixed by requiring that when $j_3 = m = j_1 + \kappa$, $\langle \phi_{j_3, m} | \psi_{m_1, q} \rangle = \delta_{m_1, j_1} \delta_{q, \kappa}$.

PROBLEMS

4.1. Spin 3/2 Matrices

Write the three 4×4 matrices $D^{(\frac{3}{2})}(\bar{J}_i)$ explicitly. From these matrices calculate

$$\sum_{i=1}^{3} \left(D^{(\frac{3}{2})}(\bar{J}_i) \right)^2$$

4.2. Equivalence of Two Three-Dimensional Representations

Discussion: There is one representation of the rotations and their generators for every nonnegative integer or half-integer j, unique up to a unitary transformation. That is, if $D^{(j)}(\hat{n}, \theta)$ is one of the standard Wigner matrix representations, and if U is any $(2j+1) \times (2j+1)$ unitary matrix, then the matrices

$$D^{(j)}(\hat{n}, \theta)' = U D^{(j)}(\hat{n}, \theta) U^{-1}$$

will also be a representation, and furthermore, if

$$D^{(j)}(\bar{J}_i)' = U D^{(j)}(\bar{J}_i) U^{-1}$$

then

$$D^{(j)}(\hat{n}, \theta)' = \exp\left(-i\theta \sum_i n_i D^{(j)}(\bar{J}_i)'\right)$$

also. In the standard choice of basis, $D^{(j)}(\bar{J}_z)$ is diagonal, and the matrix elements of $D^{(j)}(\bar{J}_\pm)$ are real and positive.

In three dimensions, we have come across two representations of the $\bar{J}_i$. One is built from the three matrices $D^{(1)}(\bar{J}_i)$, and the other from the 3×3 "defining" matrices $\bar{J}_i$. These must be related by some unitary matrix U.

Problem: Review the construction in Section 4.3.6 of the matrices for $D^{(1)}(\bar{J}_z)$ and $D^{(1)}(\bar{J}_\pm)$ in the basis of eigenstates $|m\rangle$ of $D^{(1)}(\bar{J}_z)$, where $m = 1, 0,$ and -1.

(a) What are the three orthonormal states $|\psi_a\rangle$, $a = x, y, z$, such that the matrices for $D^{(1)}(\bar{J}_i)$ are the $\bar{J}_i$ themselves? That is, find $|\psi_a\rangle$ as linear combinations of $|-1\rangle$, $|0\rangle$, and $|1\rangle$ so that

$$D^{(1)}(\bar{J}_i)|\psi_a\rangle = \sum_b |\psi_b\rangle (\bar{J}_i)_{ba}$$

(b) Find the unitary matrix U_{am} such that[18]

$$\sum_{mn} U_{am} D^{(j)}(\bar{J}_i)_{mn} U^\dagger_{nb} = (\bar{J}_i)_{ab}$$

The matrix U is independent of i.

4.3. Euler Angles

Instead of a rotation by an angle ψ about an axis $\hat{n}$, one can also specify a rotation by three Euler angles as defined in Section 4.3.5.

$$\bar{R}(\hat{n}, \psi) = \bar{R}(\hat{n}_z, \alpha) \bar{R}(\hat{n}_y, \beta) \bar{R}(\hat{n}_z, \gamma)$$

Abbreviate the matrices $\bar{R}(\hat{n}, \psi)$ as $\bar{R}(\alpha, \beta, \gamma)$; similarly the rotation operator on states is abbreviated as $R(\alpha, \beta, \gamma)$ and the finite dimensional matrices as

$$D^{(j)}(\alpha, \beta, \gamma) = D^{(j)}(\hat{n}, \psi)$$

---

[18] In the standard basis, the $D^{(1)}$ matrices are usually written with the index *decreasing* from left to right, or top to bottom: $m = 1, 0, -1$. But the matrices $\bar{J}_i$ are customarily written with the index *increasing*: $a = 1, 2, 3$.

Compute both sides of this equation in the $j = \frac{1}{2}$ representation. Then calculate the angle of rotation ψ and the spherical coordinates θ and ϕ of the axis $\hat{\mathbf{n}}$ in terms of the Euler angles.

Hint: Use the rule $\exp(-i\psi\hat{\mathbf{n}} \cdot \boldsymbol{\sigma}) = \cos\psi - i\hat{\mathbf{n}} \cdot \boldsymbol{\sigma} \sin\psi$ and some of the properties of the Pauli matrices listed in (4.77).

This relation between $\hat{\mathbf{n}}$, ψ, and the Euler angles must be a general one, not restricted to the $j = 1/2$ representation. Why?

4.4. Projection Operators and Radial Wave Equations

An electron in an atom has two states for each value of l and m except for $l = 0$, with $j = l \pm 1/2$ respectively. Here $\mathbf{J} = \mathbf{L} + \mathbf{s}$, and m is the eigenvalue of J_z.

(a) What is the value of $\boldsymbol{\sigma} \cdot \mathbf{L}$ operating on a state of definite j? ($\boldsymbol{\sigma}$ is the operator $2\mathbf{s}$.)

(b) Show that

$$\Lambda_+ = \frac{l+1+\boldsymbol{\sigma} \cdot \mathbf{L}}{2l+1} \quad \text{and} \quad \Lambda_- = \frac{l - \boldsymbol{\sigma} \cdot \mathbf{L}}{2l+1}$$

are projection operators for the states $j = l \pm 1/2$, that is, they are the identity on the indicated states, and zero on the others.

(c) For an electron, the most general rotationally invariant Hamiltonian has the form

$$H = \frac{\mathbf{p}^2}{2m} + V_0(r) + V_1(r)\boldsymbol{\sigma} \cdot \mathbf{L}$$

where r is the radial coordinate. Describe the states by a two-component wave function, or Pauli spinor,

$$\psi(\mathbf{r}) = \begin{pmatrix} \psi_{\frac{1}{2}}(\mathbf{r}) \\ \psi_{-\frac{1}{2}}(\mathbf{r}) \end{pmatrix}$$

$\psi(\mathbf{r})$ is a linear combination of a "spin-up" wave function

$$Y_l^{m_l}(\theta, \phi) \begin{pmatrix} 1 \\ 0 \end{pmatrix}$$

and a "spin-down" wave function

$$Y_l^{m_l}(\theta, \phi) \begin{pmatrix} 0 \\ 1 \end{pmatrix}$$

Spinors like these have definite values of l, m_l, and m_s, but not of j and m. The full wave functions are linear combinations of these with coefficients that can depend on r, but not on θ or ϕ.

The wave functions with definite E, J_z, $\mathbf{J}^2$, and $\mathbf{L}^2$ have the form

$$\psi_{m_s}(\mathbf{r}) = \langle \mathbf{r}, m_s | E, l, j, m \rangle = R_{El}^\pm(r) \mathcal{Y}_l^{m\pm}(\theta, \phi)$$

for the two cases $j = l \pm 1/2$ respectively. $\mathcal{Y}_l^{m\pm}(\theta, \phi)$ are two-component Pauli spinors, and the dependence of $\psi_{m_s}(\mathbf{r})$ on θ, ϕ, and m_s is all in these spinors $\mathcal{Y}_l^{m\pm}(\theta,\phi)$. Furthermore,

$$\langle \mathbf{r},m_s|\mathbf{J}^2|\psi\rangle = j(j+1)\psi_{m_s}(\mathbf{r})$$
$$\langle \mathbf{r},m_s|\mathbf{L}^2|\psi\rangle = l(l+1)\psi_{m_s}(\mathbf{r})$$
$$\langle \mathbf{r},m_s|J_z|\psi\rangle = m\psi_{m_s}(\mathbf{r})$$

and so forth. What are the differential equations for $R_l^+(r)$ and $R_l^-(r)$ in terms of $V_0(r)$ and $V_1(r)$?

4.5. Pauli Spinors

(a) What are the spinors $\mathcal{Y}_1^{\frac{1}{2}+}$ and $\mathcal{Y}_1^{\frac{1}{2}-}$ in terms of the $l=1$ spherical harmonics?[19]

(b) There are two spinors with $j=1/2$, having $l=0$ and $l=1$ respectively. The ones with $m=1/2$ are $\mathcal{Y}_0^{\frac{1}{2}+}$ and $\mathcal{Y}_1^{\frac{1}{2}-}$. Write $\mathcal{Y}_0^{\frac{1}{2}+}$ and $\mathcal{Y}_1^{\frac{1}{2}-}$ explicitly in terms of the spherical harmonics. Now let $\hat{n}$ be the unit vector in the direction of the spherical coordinates θ and ϕ. Using symmetry arguments and properties of the Pauli matrices alone (not the explicit form of $\mathcal{Y}_0^{\frac{1}{2}+}$), show that

$$\boldsymbol{\sigma}\cdot\hat{n}\mathcal{Y}_0^{\frac{1}{2}+} = \pm\mathcal{Y}_1^{\frac{1}{2}-}$$

(c) From explicit calculation, find the sign in part (b).

4.6. Addition of Three Angular Momenta

Consider a system described by *three* independent spins or angular momenta. The states are labeled $|\psi_{m_1 m_2 m_3}\rangle$, where

$$(J_i)_z|\psi_{m_1 m_2 m_3}\rangle = m_i|\psi_{m_1 m_2 m_3}\rangle$$

and

$$\sum_i \mathbf{J}_i\cdot\mathbf{J}_i|\psi_{m_1 m_2 m_3}\rangle = j_i(j_i+1)|\psi_{m_1 m_2 m_3}\rangle$$

Let the total angular momentum be

$$\mathbf{J} = \mathbf{J}_1 + \mathbf{J}_2 + \mathbf{J}_3$$

and specialize to $j_1 = j_2 = j_3 = 1$. Let $|\psi\rangle$ be a linear combination of these states such that

$$\mathbf{J}\cdot\mathbf{J}|\psi\rangle = j(j+1)|\psi\rangle \qquad \text{and} \qquad J_z|\psi\rangle = m|\psi\rangle$$

(a) What are all the possible values for j?

(b) For each value of j and m, how many linearly independent eigenstates are there? In other words, how many total angular momentum ladders with the same j are there (for *two* spins the answer to this question is 1.)?

(c) Construct explicitly the linear combination of the states $|\psi_{m_1 m_2 m_3}\rangle$ for which $j=0$.

4.7. Clebsch-Gordan Coefficients

Let $\left|\psi^{j_1 j_2}_{m_1 m_2}\right\rangle$ be $(2j_1+1)(2j_2+1)$ simultaneous eigenstates of two angular momentum operators $\mathbf{J}_1^2$, $J_{1,z}$, $\mathbf{J}_2^2$, and $J_{2,z}$

$$\mathbf{J}_i^2\left|\psi^{j_1 j_2}_{m_1 m_2}\right\rangle = j_i(j_i+1)\left|\psi^{j_1 j_2}_{m_1 m_2}\right\rangle$$

---

[19]See Problem 4.4, part (a).

and
$$J_{i,z}\left|\psi_{m_1 m_2}^{j_1 j_2}\right\rangle = m_i \left|\psi_{m_1 m_2}^{j_1 j_2}\right\rangle$$

Let $\mathbf{J} = \mathbf{J}_1 + \mathbf{J}_2$. There are linear combinations of the above states such that
$$\mathbf{J}^2 \left|\phi_{jm}^{j_1 j_2}\right\rangle = j(j+1)\left|\phi_{jm}^{j_1 j_2}\right\rangle \quad \text{and} \quad J_z \left|\phi_{jm}^{j_1 j_2}\right\rangle = m\left|\phi_{jm}^{j_1 j_2}\right\rangle$$

(a) If $j_2 = 1$, what is the Clebsch-Gordan coefficient $\left\langle \phi_{j_1,j_1}^{j_1,1} \middle| \psi_{j_1,0}^{j_1,1} \right\rangle$?

(b) If $j_2 = 2$, what is the Clebsch-Gordan coefficient $\left\langle \phi_{j_1,j_1}^{j_1,2} \middle| \psi_{j_1,0}^{j_1,2} \right\rangle$?

4.8. Rotations in n Dimensions

Background: The **orthogonal group** in n dimensions, called $O(n)$, is the collection of all real, orthogonal ($R^{-1} = R^T$) $n \times n$ matrices. We have studied $O(3)$ in detail. I surreptitiously used $O(4)$ in determining the energy levels and degeneracy of the hydrogen atom, and you will need to know more about it when we come to Lorentz transformations. It is useful to know some properties of $O(n)$ for general n.

How many independent parameters are needed to describe an orthogonal matrix in an n-dimensional space? Write
$$\bar{R} = e^{-i\theta \bar{J}}$$

$\bar{J}$ is an $n \times n$ matrix, Hermitean and imaginary, therefore antisymmetric. The number of parameters like θ is the number of independent matrices $\bar{J}$ that satisfy these conditions. The diagonal elements of the $\bar{J}$ are 0. Those below the main diagonal are the complex conjugates of those above. The remaining number is $n(n-1)/2$.

This number is 1 for $n = 2$; rotations in a plane are described by a single angle. It is 3 for $n = 3$ and 6 for $n = 4$. That the number of parameters is equal to the number of dimensions in the three-dimensional world we do experiments in is thus accidental, but the coincidence can be confusing.

For general n, it is not convenient to find $n(n-1)/2$ analogs of the three $\bar{J}_i$ in three dimensions. Rather, define two-index objects $\bar{J}_{ij}$, matrices with a $-i$ in the ij place and an i in the ji place:[20]

$$(\bar{J}_{ij})_{kl} = -i(\delta_{ik}\delta_{jl} - \delta_{il}\delta_{jk})$$

$\bar{J}_{ij}$ is the generator of rotations in the ij plane. In four dimensions the independent matrices are $\bar{J}_{12}, \bar{J}_{13}, \bar{J}_{14}, \bar{J}_{23}, \bar{J}_{24}$, and $\bar{J}_{34}$. In three dimensions $\bar{J}_{ij}$ is just $\sum_k \epsilon_{ijk} \bar{J}_k$.

Then if θ_{ij} are real parameters, subject to the condition $\theta_{ij} = -\theta_{ji}$, any $n \times n$ orthogonal matrix $\bar{R}$ can be written as

$$\bar{R} = e^{-i\frac{1}{2}\sum_{ij} \theta_{ij} \bar{J}_{ij}}$$

---

[20] Don't confuse the two subscripts that label the matrices with the indices that label their matrix elements.

The factor 1/2 is there to take care of double-counting. The structure of the group algebra is given by the commutators of the $\bar{J}$'s.

(a) Show that for any n

$$[\bar{J}_{ij}, \bar{J}_{kl}] = \pm i \left(\delta_{jk}\bar{J}_{il} - \delta_{ik}\bar{J}_{jl} - \delta_{jl}\bar{J}_{ik} + \delta_{il}\bar{J}_{jk} \right)$$

What is the sign?

(b) Now concentrate on $O(4)$, with the $\bar{J}_{ik}$ as defined in the previous problem. Define the six matrices $\bar{J}_i$ and $\bar{K}_i$ by

$$\bar{J}_1 = \bar{J}_{23} \qquad \bar{J}_2 = \bar{J}_{31} \qquad \bar{J}_3 = \bar{J}_{12}$$
$$\bar{K}_1 = \bar{J}_{41} \qquad \bar{K}_2 = \bar{J}_{42} \qquad \bar{K}_3 = \bar{J}_{43}$$

The $\bar{J}$'s and the $\bar{K}$'s are the six generators of $O(4)$. Show that

i) $[\bar{J}_i, \bar{J}_j] = i \sum_k \epsilon_{ijk} \bar{J}_k$

ii) $[\bar{J}_i, \bar{K}_j] = i \sum_k \epsilon_{ijk} \bar{K}_k$

iii) $[\bar{K}_i, \bar{K}_j] = i \sum_k \epsilon_{ijk} \bar{J}_k$

4.9. Two-Dimensional Harmonic Oscillator

In Problem 3.15 you found that, like the hydrogen atom, the three-dimensional harmonic oscillator has greater degeneracy than required by rotation invariance alone. There must be more observables that commute with the Hamiltonian than just the three L_i. For the hydrogen atom the extra degeneracy is a consequence of the conservation of the Runge-Lenz vector. What are the extra observables for the harmonic oscillator, and how can we compute the number of states at each energy level? Since the degeneracy computation is a little more complicated than it is for the hydrogen atom, it will be good practice to work out the analogous problem in two dimensions.

The Hamiltonian is

$$H = \frac{1}{2m} \left(p_x^2 + p_y^2 \right) + \frac{1}{2} m\omega^2 \left(x^2 + y^2 \right)$$

(a) Repeat the calculation of part (a) of Problem 3.15 for the two-dimensional oscillator. That is, find the energies and wave functions in Cartesian coordinates, using the solution for the one-dimensional oscillator, and compute the degeneracy of each level.

(b) Define the two-index symmetric tensor

$$Q_{ij} = m^2 \omega^2 r_i r_j + p_i p_j$$

Use the classical equations of motion to show that for the classical oscillator, the Q_{ij} are constants.

(c) Show that this is true in quantum mechanics also, that is, that $[Q_{ij}, H] = 0$. Show that the angular momentum $\hbar L = x p_y - y p_x$ also commutes with H.

The combination $Q_{xx} + Q_{yy}$ is proportional to the Hamiltonian. The two other independent combinations of this symmetric tensor are two new operators. Together with the angular momentum $\hbar L$, these are three observables that commute with the Hamiltonian.

136   Chapter 4   Symmetry Transformations on States

(d) Write Q_{xy} (in classical mechanics) in terms of the energy, the angular momentum, and the angle between the directions at which r is a maximum and the x-axis. Then from the constancy of Q_{xy} show that the orbit is closed.

4.10. Degeneracy of the Two-Dimensional Harmonic Oscillator
Set
$$a_i = \frac{1}{\sqrt{2m\omega\hbar}}(m\omega r_i + ip_i)$$
and define the four quantities
$$A_{ij} = a_i^\dagger a_j$$
It should be obvious that the A_{ij} commute with the Hamiltonian. A general energy eigenstate is
$$|n_x, n_y\rangle = \frac{1}{\sqrt{n_x!n_y!}}\left(a_x^\dagger\right)^{n_x}\left(a_y^\dagger\right)^{n_y}|0,0\rangle$$

a_j lowers n_j by one unit, and $a_i^\dagger$ raises n_i by one unit. Since the energy depends only on the sum $n_x + n_y$ the operators A_{ij} leave the energy unchanged. So these are the extra operators we are looking for. They are a little more convenient than Hermitean operators L and the Q_{ij} defined in Problem 4.9, but they must be linear combinations of them.

(a) Write the four quantities $A_{ij} = a_i^\dagger a_j$ in terms of the Q_{ij} and L defined in Problem 4.9.

(b) Let σ_i be the Pauli matrices. Label the rows and columns by 1 and 2, instead of the conventional $\pm 1/2$. Then define the three operators
$$T_a = \frac{1}{2}\sum_{i=1}^{2}\sum_{j=1}^{2} A_{ij}(\sigma_a)_{ij}$$

Show that the three T_a satisfy the usual angular momentum algebra.

Since the T_a are functions of the A_{ij}, they commute with the Hamiltonian. These three operators are the larger algebra that is responsible for the degeneracy of the levels. In this two dimensional case, it is the same as the familiar angular momentum algebra, or $SU(2)$.

(c) The combination
$$\mathbf{T}^2 = T_x^2 + T_y^2 + T_z^2 = \frac{1}{2}(T_+ T_- + T_- T_+) + T_z^2$$

is an invariant in any representation. Its value is $j(j+1)$. $\mathbf{T}^2$ is a quadratic function of the A_{ij}. In order to compare it to the Hamiltonian, rewrite $\mathbf{T}^2$ as a linear combination of the A_{ij} and the operators
$$A_{ijkl} = a_i^\dagger a_j^\dagger a_k a_l$$

(d) Similarly,
$$\left(\frac{H}{\hbar\omega} - 1\right)^2 = \left(\sum_{i=1}^{2} A_{ii}\right)^2$$

is a quadratic function of the A_{ij}. Write this quantity in terms of the A_{ij} and the A_{ijkl} also. Now write $\mathbf{T}^2$ in terms of $(H - \hbar\omega)^2$. Since $\mathbf{T}^2$ must be $j(j+1)$, where j is some nonnegative integer or half integer, this result gives the allowed energies and degeneracies. Verify that these are indeed the correct energies and their degeneracies for the two-dimensional isotropic oscillator.

The method of this problem can be generalized to the three-dimensional oscillator. There, the symmetry is $SU(3)$ instead of $SU(2)$.

4.11. Spin Correlations

Two spin-half particles are in a state with total spin zero. Let $\hat{\boldsymbol{n}}_a$ and $\hat{\boldsymbol{n}}_b$ be unit vectors in two arbitrary directions. Calculate the expectation value of the *product* of the spin of the first particle along $\hat{\boldsymbol{n}}_a$ and the spin of the second along $\hat{\boldsymbol{n}}_b$. That is, if $\mathbf{s}_a$ and $\mathbf{s}_b$ are the two spin operators, calculate

$$\langle \psi | \mathbf{s}_a \cdot \hat{\boldsymbol{n}}_a \mathbf{s}_b \cdot \hat{\boldsymbol{n}}_b | \psi \rangle$$

Hint: Because the state is spherically symmetric the answer can depend only on the angle between the two directions.

REFERENCES

The classic group theory book for quantum mechanics is

[1] E. WIGNER, *Group Theory and Its Application to the Quantum Mechanics of Atomic Spectra*, Academic Press, 1959, translated by J. J. Griffin. See also Appendix D.

Two books on Lie groups and algebras written for physicists are

[2] R. N. CAHN, *Semi-simple Lie Algebras and their Representations*, Benjamin Cummings, 1984.

[3] H. GEORGI, *Lie Algebras in Particle Physics*, Perseus, 1999

CHAPTER 5

Symmetry Transformations on Operators

5.1 VECTOR OBSERVABLES

5.1.1 Symmetries, Lifetimes, and Selection Rules

This chapter is about how operators transform under symmetry transformations. There will be many applications, but particularly to selection rules for matrix elements, in order to explain why some transition rates or energy level shifts are forbidden or suppressed. Selection rules account for the vast range of observed lifetimes in otherwise similar systems.

The key is that symmetries imply conservation laws, and approximate symmetries imply approximate conservation laws. For an isolated physical system, the amplitude for a transition from an initial state $|i\rangle$ at time t_o to a final state $|f\rangle$ at a later time t is

$$\langle f | U(t, t_o) | i \rangle \tag{5.1}$$

where

$$U(t, t_o) = e^{-iH(t-t_o)/\hbar} \tag{2.136}$$

The transition is forbidden if $\langle f | H^n | i \rangle$ vanish for all powers of H. When the Hamiltonian has an *exact* continuous symmetry there is a conserved observable.

The argument bears repeating here: If H is invariant under translations and rotations then

$$e^{i\mathbf{a}\cdot\mathbf{P}/\hbar} H e^{-i\mathbf{a}\cdot\mathbf{P}/\hbar} = H \quad \text{and} \quad e^{i\theta\hat{\mathbf{n}}\cdot\mathbf{J}} H e^{-i\theta\hat{\mathbf{n}}\cdot\mathbf{J}} = H \tag{5.2}$$

so that

$$[P_i, H] = 0 \quad \text{and} \quad [J_i, H] = 0 \tag{5.3}$$

For any observable Q, if $[Q, H] = 0$ and if $|\psi_i\rangle$ are eigenstates of Q with eigenvalues q_i, then

$$0 = \langle \psi_1 | [Q, H] | \psi_2 \rangle = \langle \psi_1 | QH - HQ | \psi_2 \rangle = (q_1 - q_2) \langle \psi_1 | H | \psi_2 \rangle \tag{5.4}$$

whence $\langle \psi_1 | H | \psi_2 \rangle = 0$ unless $q_1 = q_2$. Therefore, if $|\psi_1\rangle$ and $|\psi_2\rangle$ are states of definite momentum or a definite (component of the) angular momentum with different eigenvalues, the matrix element of the Hamiltonian between them vanishes and transitions between them are forbidden. These are called **selection rules**.

Particle Name	Mass (MeV/c^2)	Spin	Lifetime (seconds)	Decay Mode
Λ^o	1115.683	$\hbar/2$	2.632×10^{-10}	$\Lambda^o \to p + \pi^-$ (63.9%)
Λ^o				$\Lambda^o \to n + \pi^o$ (35.8%)
p	938.272	$\hbar/2$	∞	
n	939.5653	$\hbar/2$	885.7	$n \to p + e^- + \bar{\nu}_e$
e^-	0.511	$\hbar/2$	∞	
π^-	139.570	0	2.6×10^{-8}	$\pi^- \to \mu^- + \nu$
π^o	134.977	0	8.4×10^{-17}	$\pi^o \to \gamma + \gamma$ (98.8%)
π^o				$\pi^o \to e^- + e^+ + \gamma$ (1.2%)
π^o				$\pi^o \to e^- + e^+ (< 10^{-7})$
μ^-	105.658	$\hbar/2$	2.197×10^{-6}	$\mu^- \to e^- + \nu_\mu + \bar{\nu}_e$
τ^-	1777.0	$\hbar/2$	2.906×10^{-13}	$\nu_\tau + X$
$\nu_e,$	$< 3 \times 10^{-6}$	$\hbar/2$	∞	
ν_μ	$< 2 \times 10^{-7}$	$\hbar/2$	∞	
ν_τ	$< 2 \times 10^{-5}$	$\hbar/2$	∞	
γ	0	$\hbar$	∞	

TABLE 5.1: Masses, lifetimes, and decay modes of some elementary particles

Reflect for a moment on Table 5.1, an extract from a table of particle masses and lifetimes.[1] Each stable particle is stable because of a selection rule. Why is a photon stable? It is the particle with lowest mass, so there is no state it can decay into without violating energy and momentum conservation. Why is a neutrino stable? A decay mode like $\nu \to \gamma + \gamma$ might not violate energy conservation if $m_\nu > 0$, but it would not conserve angular momentum. Why is an electron stable? It is the lowest mass state with nonzero electric charge, and electric charge is exactly conserved. Why is a proton stable? A decay like $p \to e^+ + \gamma$ is allowed by energy, momentum, angular momentum, and charge conservation. We invent an answer: A proton has a property called **baryon number** that is conserved, and it is the lowest mass state with nonzero baryon number. Of course it is not trivial that this new rule is consistent with everything else we know about elementary particles. (Or maybe the proton is not stable. The current limit on its lifetime is something like 10^{33} years.)

What accounts for the vast range of lifetimes among the particle that do decay? We can often understand the answer to this question by breaking the Hamiltonian up into pieces, some much smaller than others. If the matrix elements

---

[1] The listed neutrino masses are limits from direct measurements. Actual masses may be much smaller. Mass differences from interference experiments (see Section 6.3) are in the range $10^{-5} \text{eV}^2 \leq \Delta m^2 \leq 10^{-3} \text{eV}$. The τ^- has many decay modes, but all involve a ν_τ.

140 Chapter 5 Symmetry Transformations on Operators

of the large terms vanish, but the matrix elements of some smaller terms do not, there is an *approximate* symmetry. In elementary particle physics for example, the terms in the Hamiltonian that produce neutrinos are very small—they are called the **weak interaction**. In Table 5.1 you see that π^o decay is very fast compared to those involving neutrinos.

Angular momentum conservation provides a particularly rich collection of selection rules. To explain them I will introduce a view of symmetry transformations complementary to the one in Chapter 4, and look at the way transformations act on operators rather than on the states. This viewpoint is closer to the way one talks about symmetry transformations in classical mechanics, where transformations are defined by what they do to the observables. In quantum mechanics it leads directly to the study of selection rules.

Let $|\psi\rangle$ be any state, g some transformation, and $|\psi'\rangle = g|\psi\rangle$ the transformed state. For any observable A, define A' so that

$$\langle \psi'| A' |\psi'\rangle = \langle \psi| A |\psi\rangle \tag{5.5}$$

for all $|\psi\rangle$. The solution is

$$A' = gAg^\dagger \tag{5.6}$$

For a one-parameter subgroup of continuous symmetries $g(a) = \exp(-i\epsilon G)$, the change of A to first order in ϵ is[2,3]

$$\delta A = A' - A = (I - i\epsilon G) A (I + i\epsilon G) - A = i\epsilon[A, G] \tag{5.7}$$

If the Hamiltonian H is invariant under the transformations $g(a)$, then $[H, G] = 0$, which in turn implies that expectation values of the generator G are constants in time, or that $dG/dt = 0$ in the Heisenberg picture (see Section 2.5.2).

5.1.2 Vector Operators Under Rotations

All observables can be arranged into collections that transform under rotations in some simple fashion. The simplest rule of all belongs to **scalars**, observables that are invariant:

$$RSR^\dagger = S \tag{5.8}$$

In terms of the generators,

$$[J_i, S] = 0 \tag{5.9}$$

A classical vector observable is a collection of three numbers V_i, such that under rotations $V_i \to \sum_j \bar{R}_{ij} V_j$. Examples are **r**, **p** and **L**

In quantum mechanics the expectation values of these observables should change by the same rule when the states are rotated: If $|\psi'\rangle = R|\psi\rangle$, then

$$\langle \psi'| V_i |\psi'\rangle = \langle \psi| R^\dagger V_i R |\psi\rangle = \sum_j \bar{R}_{ij} \langle \psi| V_j |\psi\rangle \tag{5.10}$$

---

[2] See equation (4.5).

[3] Once more, this is the same rule as in classical mechanics, with the standard substitution $[A, B] \to i\hbar[A, B]_{PB}$, and with $\hbar G$ identified with the observable that is the classical generator of the transformations.

Since $|\psi\rangle$ is arbitrary, a vector observable satisfies

$$R^\dagger V_i R = \sum_j \bar{R}_{ij} V_j \tag{5.11}$$

or

$$R V_i R^\dagger = \sum_j V_j \bar{R}_{ji} \tag{5.12}$$

Any three operators that transform under rotations according to the rule (5.11) or (5.12) are by definition components of a vector.

> Note: The word **vector** is unfortunately used in two completely different ways. On the one hand, vectors are elements of the algebraic structures called vector spaces that obey the rules listed in Section 2.2, and in quantum mechanics these are usually denoted by Dirac's symbol $|\psi\rangle$. Here we introduce another use of the word, observables that transform in a particular way under rotations. Try not to confuse these two usages.
>
> That the same word is used to describe two very different kinds of mathematical objects is no accident. They are different generalizations of ordinary three-dimensional vectors like **r**, **p**, or **L**.

Next write the rule (5.12) in terms of the generators:

$$e^{-i\theta \hat{\mathbf{n}} \cdot \mathbf{J}} V_i e^{i\theta \hat{\mathbf{n}} \cdot \mathbf{J}} = \sum_j V_j \left(e^{-i\theta \hat{\mathbf{n}} \cdot \bar{\mathbf{J}}}\right)_{ji} \tag{5.13}$$

and expand both sides to first order. The zero order terms cancel. Compare the terms of order θ using $\left(\bar{J}_i\right)_{jk} = -i\epsilon_{ijk}$:

$$-i\theta \sum_k n_k [J_k, V_i] = -i\theta \sum_{jk} V_j n_k (-i\epsilon_{kji}) \tag{5.14}$$

Since $\hat{\mathbf{n}}$ is arbitrary,

$$\boxed{[J_i, V_j] = i \sum_k \epsilon_{ijk} V_k} \tag{5.15}$$

after renaming some indices. Any vector observable transforms this way. Equation (5.15) is a consequence of equations (5.11) or (5.12). The converse is also true, and can be proved using the Baker-Hausdorff lemma (see Problem 2.10). **J** itself is a vector observable.

For any two vectors the combination $\mathbf{V}_1 \cdot \mathbf{V}_2 = \sum_i V_{1i} V_{2i}$ is always a scalar. I have proved this before in special cases, for example, equations (4.95) and (4.96). The Hamiltonian is an example of a scalar observable.

5.1.3 Spherical Components of Vector Observables

The rules I want to establish look simpler in terms of combinations of the V_i that are analogous to the raising and lowering combinations of J_i. Define the three V^q ($q = -1, 0, 1$) by

142   Chapter 5   Symmetry Transformations on Operators

$$V^1 = -\frac{V_x + iV_y}{\sqrt{2}} \qquad V^{-1} = \frac{V_x - iV_y}{\sqrt{2}} \qquad \text{and} \qquad V^0 = V_z \tag{5.16}$$

In terms of these V^q the rules (5.15) become

$$[J_z, V^q] = qV^q \tag{5.17a}$$

$$[J_\pm, V^o] = \mp (V_x \pm iV_y) = \sqrt{2} V^{\pm 1} \tag{5.17b}$$

$$[J_\pm, V^{\pm 1}] = 0 \tag{5.17c}$$

$$[J_\pm, V^{\mp 1}] = \sqrt{2} V^o \tag{5.17d}$$

The peculiar signs and phases in the definitions (5.16) were chosen so that (5.17a) can be summarized as

$$\boxed{[J_i, V^q] = \sum_{q'} V^{q'} D^{(1)}_{q'q}(\bar{J}_i)} \tag{5.18}$$

as you can easily check using equations (4.87). Equation (5.18) is the first-order form of

$$\boxed{R V^q R^\dagger = \sum_{q'} V^{q'} D^{(1)}_{q'q}(\bar{R})} \tag{5.19}$$

Some nomenclature: The form (5.19) is closely analogous to the form (5.12), since $\bar{R}$ are the same operators as $D^{(1)}$, in a different basis. The Cartesian components of a vector operator transform according to (5.12). The combinations V^q transform according to (5.19), and are called the **spherical** components of the same vector.[4] The spherical components of **J** itself are

$$J^o = J_z \qquad J^1 = -\frac{1}{\sqrt{2}} J_+ \qquad J^{-1} = \frac{1}{\sqrt{2}} J_- \tag{5.20}$$

Finally,

$$\mathbf{a}^\dagger \cdot \mathbf{b} = \sum_{\mu=-1}^{1} a^{\mu\dagger} b^\mu \tag{5.21}$$

5.1.4 Selection Rules for Matrix Elements of Vectors

Consider a matrix element of these spherical vector components between eigenstates of $\mathbf{J}^2$ and J_z, of the form $\langle \alpha', j', m' | V^q | \alpha, j, m \rangle$. Here α stands for all the other quantum numbers. Since $D^{(\kappa)}(\bar{J}_z)$ is diagonal, equation (5.18) applied to J_z becomes

$$[J_z, V^q] = qV^q \tag{5.22}$$

Then from

$$\langle \alpha', j', m' | [J_z, V^q] | \alpha, j, m \rangle = (m' - m) \langle \alpha', j', m' | V^q | \alpha, j, m \rangle \tag{5.23}$$

it follows that

$$(m' - m) \langle \alpha', j', m' | V^q | \alpha, j, m \rangle = q \langle \alpha', j', m' | V^q | \alpha, j, m \rangle \tag{5.24}$$

and therefore,

---
[4] This is because the spherical harmonics $Y^l_m(\hat{\mathbf{n}})$ are the spherical components of the unit vector $\hat{\mathbf{n}}$.

Selection Rule 1:

$$\langle \alpha', j', m' | V^q | \alpha, j, m \rangle = 0 \quad \text{unless } m' = m + q \quad (5.25)$$

The remaining rules are a bit more complicated. They depend on the fact that all matrix elements of the form $\langle \alpha', j', m' | V^q | \alpha, j, m \rangle$ are related for given α, α', j, and j'. If you know this matrix element for one collection of m, m', and q, then you know them for all m, m', and q.

It is easy to see (Problem 5.1) that this is true when $j' = j$. All the $3(2j+1)^2$ matrix elements of the form $\langle \alpha', j, m' | V^q | \alpha, j, m \rangle$ are related by factors that depend on m, m', and q, but not on what the vector $\mathbf{V}$ is. You can compute all these matrix elements by inserting commutators $[J_\pm, V^q]$ and $[J_z, V^q]$ between the states and using the identities (5.17a).

It is not necessary that j' and j be the same. For example, all forty-five matrix elements of the three components of $\mathbf{p}$ between the three $2P$ states and five $3D$ states of a spinless hydrogen atom are related (see Problem 5.1).

What happens if you try to compute matrix elements of a vector operator between two states with $j' - j > 1$? In that case m' can go up to $j + 2$ or higher, and then

$$0 = \langle \alpha', j', j+2 | [J_+, V^+] | \alpha, j, j \rangle$$
$$= \langle \alpha', j', j+2 | J_+ | \alpha', j', j+1 \rangle \langle \alpha', j', j+1 | V^+ | \alpha, j, j \rangle$$

Since the factor $\langle \alpha', j', j+2 | J_+ | \alpha', j', j+1 \rangle$ is *not* zero,

$$\langle \alpha', j', j+1 | V^+ | \alpha, j, j \rangle = 0 \quad (5.26)$$

In fact, *all* the matrix elements

$$\langle \alpha', j', m' | V^q | \alpha, j, m \rangle \quad (5.27)$$

are proportional to $\langle \alpha', j', j+1 | V^+ | \alpha, j, j \rangle$ and thus also vanish, leading to

Selection Rule 2a: All matrix elements of a vector operator $\mathbf{V}$ between eigenstates of $\mathbf{J}^2$ with j differing by more than one unit vanish.

There is one more rule: Suppose $j' = j = 0$. Then

$$\sqrt{2} \langle \alpha', 0, 0 | V^o | \alpha, 0, 0 \rangle = \langle \alpha', 0, 0 | J_+ V^- - V^- J_+ | \alpha, 0, 0 \rangle = 0 \quad (5.28)$$

and

$$\sqrt{2} \langle \alpha', 0, 0 | V^\pm | \alpha, 0, 0 \rangle = \langle \alpha', 0, 0 | J_\pm V^o - V^o J_\pm | \alpha, 0, 0 \rangle = 0 \quad (5.29)$$

Selection Rule 2b: A vector operator has no matrix elements between two $j = 0$ states.

Finally, since the Cartesian components V_i are linear combinations of the V^q, these rules hold for them also:

$$\langle \alpha', j', m' | V_i | \alpha, j, m \rangle = 0 \qquad (5.30)$$

unless $j' = j$ or $j' = j \pm 1$, and the "zero-zero" matrix elements of vector operators are forbidden. We will apply these rules frequently, and for many applications one doesn't need the details of the machinery of spherical vector components used to derive them.

Example:
Let $|n, l, m\rangle$ be the standard basis for the hydrogen atom, ignoring spin. Then if **p** is the momentum operator, the first selection rule says

$$\langle 3, 1, 1 | p_z | 2, 1, 0 \rangle = 0 \quad \text{but} \quad \langle 3, 1, 1 | p_z | 2, 1, 1 \rangle \neq 0 \qquad (5.31)$$

and the second rule says that

$$\langle 3, 2, 1 | p_x | 2, 0, 0 \rangle = 0 \quad \text{but} \quad \langle 3, 1, 1 | p_x | 2, 0, 0 \rangle \neq 0 \qquad (5.32)$$

5.2 TENSOR OBSERVABLES

5.2.1 Cartesian Tensor Operators

Introduction

Some observables transform under rotations in a more complicated way than scalars and vectors. For example consider a product of components of two vectors U and V:

$$T_{ij} = U_i V_j \qquad (5.33)$$

Then

$$RT_{ij}R^\dagger = RU_i V_j R^\dagger = RU_i R^\dagger R V_j R^\dagger = \sum_{i'j'} U_{i'} V_{j'} \bar{R}_{i'i} \bar{R}_{j'j} = \sum_{i'j'} T_{i'j'} \bar{R}_{i'i} \bar{R}_{j'j} \qquad (5.34)$$

Any two-index observable T_{ij} that transforms in this way is defined to be a **rank-two Cartesian tensor**:

$$RT_{ij}R^\dagger = \sum_{i'j'} T_{i'j'} \bar{R}_{i'i} \bar{R}_{j'j} \qquad (5.35)$$

T_{ij} doesn't have to be a product of vectors. $x_i p_j + \hbar^2 L_i s_j$ is also a rank-two Cartesian tensor.

The terminology can be generalized. Any observable with n indices that transforms according to the same rule as the product $V_{1i_1} V_{2i_2} V_{3i_3} \cdots V_{ni_n}$ is by definition a rank n Cartesian tensor.

Scalar operators are rank zero tensors, and transform by the spin-zero representation of the rotation group:

$$S \to RSR^\dagger = S = IS \qquad (5.36)$$

Vector operators are rank-one tensors and transform by the defining representation $\bar{R}$ itself:

$$V_i \rightarrow RV_iR^\dagger = \sum_j V_j \bar{R}_{ji} \qquad (5.37)$$

Reducible and Irreducible Combinations

A collection of operators A_i that transform into linear combinations of each other under rotations are said to **induce** a representation of the rotation group, since $RA_iR^\dagger$ is a combination of the A_i. That is,

$$RA_iR^\dagger = \sum_j A_j M_{ji} \qquad (5.38)$$

Spin-zero operators induce the identity representation. Vector operators induce the three-dimensional (spin-one) representation.

The matrices M form a representation of the rotation group, and therefore we know all about them. Since these higher order Cartesian tensors transform by *products* of the rotation matrices, M are representations of the rotation group of dimensions 1, 3, 9, 27, and so forth, and in contrast to the scalar and vector cases these are not irreducible for higher rank tensors. For working out the selection rules for transitions between eigenstates of the type $|j, m\rangle$, reducibility is an inconvenience.

These collections of operators can be organized into subsets that transform irreducibly. The general method is not hard, but it will be enough to see what happens to the nine components of a rank-two tensor.

A symmetric or antisymmetric tensor preserves that symmetry property property under rotations. For instance, if $T_{ij} = \pm T_{ji}$, then it is simple to show (Problem 5.3), using the orthogonality of the matrices $\bar{R}$, that

$$T'_{ij} = RT_{ij}R^\dagger = \pm T'_{ji} \qquad (5.39)$$

For any rank-two tensor T_{ij}, $T_{ij} + T_{ji}$ is also a rank-two tensor, and under rotations its six independent components mix up only among themselves. $T_{ij} - T_{ji}$ is a rank-two tensor, and its three independent components mix up only among themselves. Finally, the "trace" $\sum_i T_{ii}$ is an invariant, a scalar or rank-zero tensor. The nine components of a rank-two Cartesian tensor can be organized into three groups:

$$T_{ij} = \left[\frac{1}{3}\delta_{ij}\sum_k T_{kk}\right] + \left[\frac{1}{2}(T_{ij} - T_{ji})\right] + \left[\frac{1}{2}(T_{ij} + T_{ji}) - \frac{1}{3}\delta_{ij}\sum_k T_{kk}\right] \qquad (5.40)$$

The second term in equation (5.40) has three independent components, and is antisymmetric. These transform like the three components of a vector, a fact made explicit by writing $V_i = \sum_{jk} \epsilon_{ijk}(T_{ij} - T_{ji})$. The first term, the one that looks like a trace, is a scalar (see Problem 5.2).

Any real symmetric matrix can be expanded in the identity plus five traceless matrices. These five traceless components are the ones in the last bracket, and they also mix up only among themselves. They induce the irreducible $j = 2$ representation. If a tensor has the properties

$$T_{ij} = T_{ji} \quad \text{and} \quad \sum_i T_{ii} = 0 \tag{5.41}$$

then the T_{ij} are all linear combinations of five components that transform only among themselves under equation (5.35).

5.2.2 Spherical Tensor Components

In Section 5.1.3 it was useful to define "spherical" combinations of the components of a vector V_i. We do the same here for rank two-tensors.

For a rank-two tensor that is a product of two vectors, like

$$T_{ij} = U_i V_j \tag{5.33}$$

define the "spherical" vector combinations (5.16). Then set

$$T_2^2 = U^+ V^+ = \frac{1}{2}\left[U_x V_x - U_y V_y + i(U_x V_y + V_x U_y)\right] \tag{5.42a}$$

$$T_2^1 = \frac{1}{\sqrt{2}}\left[U^+ V^o + U^o V^+\right] = -\frac{1}{2}\left[U_x V_z + U_z V_x + i(U_y V_z + U_z V_y)\right] \tag{5.42b}$$

$$T_2^o = \frac{1}{\sqrt{6}}\left[U^+ V^- + U^- V^+ + 2U^o V^o\right] = \frac{1}{\sqrt{6}}\left[3U_z V_z - \mathbf{U} \cdot \mathbf{V}\right] \tag{5.42c}$$

$$T_2^{-1} = \frac{1}{\sqrt{2}}\left[U^- V^o + U^o V^-\right] = \frac{1}{2}\left[U_x V_z + U_z V_x - i(U_y V_z + U_z V_y)\right] \tag{5.42d}$$

$$T_2^{-2} = U^- V^- = \frac{1}{2}\left[U_x V_x - U_y V_y - i(U_x V_y + V_x U_y)\right] \tag{5.42e}$$

You can easily check that these combinations satisfy

$$[J_i, T_2^q] = \sum_{q'} T_2^{q'} D_{q'q}^{(2)}(\bar{J}_i) \tag{5.43}$$

and therefore under finite rotations

$$R T_2^q R^\dagger = \sum_{q'} T_2^{q'} D_{q'q}^{(2)}(\bar{R}) \tag{5.44}$$

Any rank-two tensor transforms the same way under rotations as the product of two vectors, and its components have the same commutators with the J_i. So for any Cartesian rank-two tensor, define

$$T_2^2 = \frac{1}{2}\left[T_{xx} - T_{yy} + i\left(T_{xy} + T_{yx}\right)\right] \tag{5.45a}$$

$$T_2^1 = -\frac{1}{2}\left[T_{xz} + T_{zx} + i\left(T_{yz} + T_{zy}\right)\right] \tag{5.45b}$$

$$T_2^o = \frac{1}{\sqrt{6}}\left[3T_{zz} - \sum_{i=1}^{3} T_{ii}\right] \tag{5.45c}$$

$$T_2^{-1} = \frac{1}{2}\left[T_{xz} + T_{zx} - i\left(T_{yz} + T_{zy}\right)\right] \tag{5.45d}$$

$$T_2^{-2} = \frac{1}{2}\left[T_{xx} - T_{yy} - i\left(T_{xy} + T_{yx}\right)\right] \tag{5.45e}$$

Then these components will transform among themselves according to equations (5.43) and (5.44).

Similarly,

$$W_i = (\mathbf{U} \times \mathbf{V})_i = \sum_{jk} \epsilon_{ijk} U_j V_k \tag{5.46}$$

must transform as a vector. For any rank-two Cartesian tensor, the components

$$W_1^+ = -\frac{1}{\sqrt{2}}\left[T_{yz} - T_{zy} + i(T_{zx} - T_{xz})\right] \tag{5.47a}$$

$$W_1^- = \frac{1}{\sqrt{2}}\left[T_{yz} - T_{zy} - i(T_{zx} - T_{xz})\right] \tag{5.47b}$$

$$W_1^o = T_{xy} - T_{yx} \tag{5.47c}$$

satisfy

$$[J_i, W^q] = \sum_{q'} W^{q'} D^{(1)}_{q'q}(\bar{J}_i) \tag{5.48}$$

and

$$R W^q R^\dagger = \sum_{q'} W^{q'} D^{(1)}_{q'q}(\bar{R}) \tag{5.49}$$

5.2.3 Higher Rank Spherical Tensors

Spherical tensors can be defined with any rank κ, with $2\kappa + 1$ components T_κ^q, $-\kappa \leq q \leq \kappa$. The rank κ can be any nonnegative integer. (Operators that transform as half-integral rank tensors can be constructed, but they are not observables.) The operators T_κ^q are defined to be components of a rank-κ spherical tensor provided

$$\boxed{R T_\kappa^q R^\dagger = \sum_{q'} T_\kappa^{q'} D^{(\kappa)}_{q'q}(\bar{R})} \tag{5.50}$$

Equivalently,

$$\boxed{[J_i, T_\kappa^q] = \sum_{q'} T_\kappa^{q'} D^{(\kappa)}_{q'q}(\bar{J}_i)} \tag{5.51}$$

or, in detail,

$$[J_z, T_\kappa^q] = qT_\kappa^q \quad \text{and} \quad [J_\pm, T_\kappa^q] = \sqrt{\kappa(\kappa+1) - q(q\pm1)}\,T_\kappa^{q\pm1} \qquad (5.52)$$

Mathematical note: A Cartesian tensor of rank κ transforms like the product of κ vectors, so it has 3^κ components. The combinations that transform irreducibly, according to (5.50) or (5.51), can be constructed from combinations that have every possible combination of symmetry and antisymmetry in each pair of indices, and are traceless in every pair. For larger κ this becomes a fancy combinatorial problem, and I will not go into it here in any detail. We will rarely need tensors of rank greater than two.

The highest rank spherical tensor can always be constructed from the prototype, a tensor that is the product of κ vectors:

$$T_{i_1 i_2 \ldots i_\kappa} = V_{1\,i_1} V_{2\,i_2} \cdots V_{\kappa\,i_\kappa} \qquad (5.53)$$

Define

$$T_\kappa^\kappa = V_1{}^+ V_2{}^+ \cdots V_\kappa{}^+ \qquad (5.54)$$

and repeatedly commute both sides with J_-. In this way it is easy to generate $2\kappa+1$ combinations T_κ^q that satisfy equation (5.51).

5.2.4 Selection Rules and the Wigner-Eckart Theorem

In order to calculate energy levels and transition rates, we will need to know the matrix elements of various observables between angular momentum eigenstates. Constructing these matrix elements is greatly simplified by organizing the observables into components of spherical tensors. I will show that for matrix elements like $\langle \alpha', j', m' | T_\kappa^q | \alpha, j, m \rangle$ there exist the following generalizations of the rules we found for vector operators in Section 5.1.4.[5]

(i) $\langle \alpha', j', m' | T_\kappa^q | \alpha, j, m \rangle$ vanishes unless $m' = m + q$.

(ii) $\langle \alpha', j', m' | T_\kappa^q | \alpha, j, m \rangle$ vanishes unless j, j', and κ can form a triangle.

The first rule can be demonstrated without the machinery of the whole Wigner-Eckart theorem. Recall that

$$[J_z, T_\kappa^q] = \sum_{q'} T_\kappa^{q'} D_{q'q}^{(\kappa)}(\bar{J}_z) = qT_\kappa^q \qquad (5.55)$$

so that

$$\begin{aligned} \langle \alpha', j', m' | [J_z, T_\kappa^q] | \alpha, j, m \rangle &= (m' - m)\langle \alpha', j', m' | T_\kappa^q | \alpha, j, m \rangle \\ &= \langle \alpha', j', m' | qT_\kappa^q | \alpha, j, m \rangle \end{aligned} \qquad (5.56)$$

Therefore, $T_\kappa^q = 0$ unless $m' = m + q$. This is the first rule.

---

[5] The states are eigenstates of total angular momentum and its z-component as indicated. α and α' stand for the energy and any other quantum numbers needed to specify the states.

Section 5.2 Tensor Observables

Matrix elements of vectors between states of definite j and m are all related by matrix elements of commutators like $[J_\pm, V^q]$ (see Section 5.1.4). The same is true for spherical tensors of any rank. The Wigner-Eckart theorem says that the matrix elements

$$\langle \alpha', j', m' | T^q_\kappa | \alpha, j, m \rangle \tag{5.57}$$

are proportional to the Clebsch-Gordan coefficients

$$\left\langle \phi^{j\kappa}_{j',m'} \middle| \psi^{j\kappa}_{m,q} \right\rangle \tag{5.58}$$

and that the proportionality constant is *independent* of m', m, and κ. An old atomic spectroscopy convention defines the proportionality constant in terms of a so-called **reduced matrix element**, written with a double-line bracket, according to the rule

$$\boxed{\langle \alpha', j', m' | T^q_\kappa | \alpha, j, m \rangle = \frac{1}{\sqrt{2j'+1}} \left\langle \phi^{j\kappa}_{j',m'} \middle| \psi^{j\kappa}_{m,q} \right\rangle \langle \alpha', j' || T_\kappa || \alpha, j \rangle} \tag{5.59}$$

The factor $\sqrt{2j'+1}$ is conventional, but not universal.

The principal application of the Wigner-Eckart theorem is the generalization of the selection rules we derived for vector operators to tensors of any rank. The matrix elements vanish whenever the indicated Clebsch-Gordan coefficients are zero, leading to:

Selection Rule 1: $\langle \alpha', j', m' | T^q_\kappa | \alpha, j, m \rangle = 0$ unless $m' = m + q$.

Selection Rule 2: $\langle \alpha', j', m' | T^q_\kappa | \alpha, j, m \rangle = 0$ unless j, j', and κ form a triangle.

The rules we discovered for vectors in Section 5.1.4 are special cases. Furthermore, a consequence of the theorem is that for given α, α', j, j', and κ, if you know the matrix element of a tensor for some m', q, and m, then you know it for all other m', q, and m.

It is not hard to see where these rules come from. Construct a new state by operating on an angular momentum eigenstate with a component of a spherical tensor:

$$|\psi\rangle = T^q_\kappa | \alpha, j, m \rangle \tag{5.60}$$

The result of applying the rotation $R(\bar{R})$ to this state is

$$R(\bar{R})|\psi\rangle = R(\bar{R}) T^q_\kappa | \alpha, j, m \rangle = R(\bar{R}) T^q_\kappa R(\bar{R})^\dagger R(\bar{R}) | \alpha, j, m \rangle$$
$$= \sum_{q', m'} D^{(\kappa)}(\bar{R})_{q'q} D^{(j)}(\bar{R})_{m'm} T^{q'}_\kappa | \alpha, j, m' \rangle \tag{5.61}$$

By analogy, consider a state $|\alpha, j, m\rangle$. Its matrix element with another such state must vanish unless j and m are the same. (The two states are orthogonal if they are eigenstates of either $\mathbf{J}^2$ or J_z with different eigenvalues.)

$$\langle \alpha', j', m' | \alpha, j, m \rangle \sim \delta_{j'j} \delta_{m'm} \tag{5.62}$$

150 Chapter 5 Symmetry Transformations on Operators

Suppose that the angular momentum $\mathbf{J}$ of the state on the right is the sum of two angular momenta: $\mathbf{J} = \mathbf{J}_1 + \mathbf{J}_2$. The states can be labeled by the eigenvalues of $\mathbf{J}_i^2$ and $(\mathbf{J}_z)_i$. The matrix element

$$\langle \alpha', j', m' | \alpha, j_1, m_1, j_2, m_2 \rangle \tag{5.63}$$

can be nonzero only when j' is among the values of the total angular momentum possible for the state on the right, that is, when $|j_1 - j_2| \leq j' \leq j_1 + j_2$ and $m' = m_1 + m_2$; equivalently, whenever $\left\langle \phi_{j',m'}^{j_1,j_2} \middle| \psi_{m_1 m_2}^{j_1,j_2} \right\rangle$ can be nonzero. The only relevant property of the state $|\alpha, j_1, m_1, j_2, m_2\rangle$ is the way it transforms under rotations:

$$R(\bar{R})|\alpha, j_1, m_1, j_2, m_2\rangle = \sum_{m_1', m_2'} |\alpha, j_1, m_1', j_2, m_2'\rangle D^{(j_1)}(\bar{R})_{m_1' m_1} D^{(j_2)}(\bar{R})_{m_2' m_2} \tag{5.64}$$

Compare the expressions (5.64) and (5.61). Under a rotation the state $|\psi\rangle$ transforms just like a state whose angular momentum is the sum of j and κ. It therefore has nonvanishing matrix elements

$$\langle \alpha, j', m' | \psi \rangle = \langle \alpha, j', m' | T_\kappa^q | \alpha, j, m \rangle \tag{5.65}$$

only when j' is among the states occuring in the sum of j and κ, and when $m' = m + q$. You can find a complete proof of the Wigner-Eckart theorem along these lines at the end of Section B.2 in the appendix. Another version is presented below.

Proof of the Theorem. Recall the linear recursion relation for the Clebsch-Gordan coefficients derived from the Clebsch-Gordan series:

$$\sum_{m,m'} \left\langle \phi_{j_3,m'}^{j_1,\kappa} \middle| \psi_{m,q}^{j_1,\kappa} \right\rangle D_{m_3 m'}^{(j_3)}(\hat{\mathbf{n}}, \theta) D_{m_1 m}^{(j_1)}(\hat{\mathbf{n}}, \theta)^*$$

$$= \sum_{q'} \left\langle \phi_{j_3,m_3}^{j_1,\kappa} \middle| \psi_{m_1 q'}^{j_1,\kappa} \right\rangle D_{q'q}^{(\kappa)}(\hat{\mathbf{n}}, \theta) \tag{4.151}$$

There is a similar equation for the matrix elements of spherical tensor components. From the defining equation (5.50) for a rank κ tensor

$$\langle \alpha', j_3, m_3 | R(\hat{\mathbf{n}}, \theta) T_\kappa^q R(\hat{\mathbf{n}}, \theta)^{-1} | \alpha, j_1, m_1 \rangle$$

$$= \sum_{q'} \langle \alpha', j_3, m_3 | T_\kappa^{q'} | \alpha, j_1, m_1 \rangle D_{q'q}^{(\kappa)}(\hat{\mathbf{n}}, \theta) \tag{5.66}$$

Now

$$R^{-1}|\alpha, j_1, m_1\rangle = \sum_m |\alpha, j_1, m\rangle D_{m_1 m}^{(j_1)}(\hat{\mathbf{n}}, \theta)^* \tag{5.67}$$

and

$$\langle \alpha', j_3, m_3 | R(\hat{\mathbf{n}}, \theta) = \sum_{m'} \langle \alpha', j_3, m' | D_{m_3 m'}^{(j_3)}(\hat{\mathbf{n}}, \theta) \tag{5.68}$$

So

$$\sum_{m'm} \langle \alpha', j_3, m'| T_\kappa^q |\alpha, j_1, m\rangle D^{(j_3)}_{m_3 m'}(\hat{\boldsymbol{n}}, \theta) D^{(j_1)}_{m_1 m}(\hat{\boldsymbol{n}}, \theta)^*$$

$$= \sum_{q'} \left\langle \alpha', j_3, m_3 \middle| T_\kappa^{q'} \middle| \alpha, j_1, m_1 \right\rangle D^{(\kappa)}_{q'q}(\hat{\boldsymbol{n}}, \theta) \quad (5.69)$$

The tensor matrix elements

$$\left\langle \alpha', j_3, m_3 \middle| T_\kappa^{q'} \middle| \alpha, j_1, m_1 \right\rangle \quad (5.70)$$

satisfy the same homogeneous equation as the Clebsch-Gordan coefficients

$$\left\langle \phi^{j_1,\kappa}_{j_3,m_3} \middle| \psi^{j_1,\kappa}_{m_1 q'} \right\rangle \quad (5.71)$$

for any α, α', j_1 and j_3. The solutions can differ by an overall constant, but the ratios of tensor matrix elements are determined by the ratios of Clebsch-Gordan coefficients. Thus

$$\langle \alpha', j', m'| T_\kappa^q |\alpha, j, m\rangle = \frac{1}{\sqrt{2j'+1}} \left\langle \phi^{j\kappa}_{j',m'} \middle| \psi^{j\kappa}_{m,q} \right\rangle \langle \alpha', j'|| T_\kappa ||\alpha, j\rangle \quad (5.59)$$

The last factor, the so-called reduced matrix element, is the overall factor that depends the tensor T_κ but is independent of m, m', and q. ■

In addition to the selection rules above, the Wigner-Eckart theorem will be of detailed assistance in calculating atomic decay rates in Chapter 11. The projection theorem, discussed in Section B.1.3 in the appendix, is another application.

5.3 DISCRETE SYMMETRIES

How do observables transform under some of the other symmetries listed in Section 1.2? Translations have no structure: They all commute. The Galilean boosts on the other hand are complicated mathematically, and they do not imply interesting selection rules. In any case, we will eventually replace them with the Lorentz form and study those in some detail.

The two discrete symmetries are worth attention. Reflections are similar to rotations in that they imply selection rules on matrix elements of observables. Time inversion symmetry has different consequences.

5.3.1 Reflections and Parity

Isolated systems are invariant under reflections, as well as under rotations and translations, as long as they are subject to electromagnetic, nuclear, or gravitational forces but not the parity-violating weak forces. I have already introduced the reflection operator $\mathcal{R}$, which acts on eigenstates of the position operator as

$$\mathcal{R}|\mathbf{r}\rangle = |-\mathbf{r}\rangle \quad (5.72)$$

and on momentum eigenstates as

$$\mathcal{R}|\mathbf{p}\rangle = |-\mathbf{p}\rangle \tag{5.73}$$

Since $\mathcal{R}$ is a symmetry it is unitary, and since $\mathcal{R}\mathcal{R} = I$, it is also Hermitean. $\mathcal{R}$ commutes with rotations:

$$\mathcal{R}R(\hat{\mathbf{n}}, \theta)\mathcal{R}^{-1} = R(\hat{\mathbf{n}}, \theta) \tag{5.74}$$

but not with space translations:

$$\mathcal{R}T(\mathbf{a})\mathcal{R}|\mathbf{r}\rangle = \mathcal{R}T(\mathbf{a})|-\mathbf{r}\rangle = T(-\mathbf{a})|\mathbf{r}\rangle \quad \text{or} \quad \mathcal{R}T(\mathbf{a})\mathcal{R} = T(-\mathbf{a}) \tag{5.75}$$

In terms of the generators,

$$[\mathcal{R}, \mathbf{J}] = 0 \quad \text{and} \quad \mathcal{R}\mathbf{P}\mathcal{R} = -\mathbf{P} \tag{5.76}$$

From equation (5.72), $\mathcal{R}\mathbf{r}\mathcal{R}|\mathbf{r}_o\rangle = -\mathbf{r}_o|\mathbf{r}_o\rangle$, so $\mathcal{R}\mathbf{r}\mathcal{R} = -\mathbf{r}$. Both $\mathbf{r}$ and $\mathbf{p}$ change sign.

Stationary states can be chosen eigenstates of $\mathcal{R}$ if reflection is a symmetry. The possible eigenvalues of $\mathcal{R}$, the parity, are ± 1. For an electron in a atom, states of definite $\mathbf{L}^2$ are parity eigenstates, with eigenvalue $(-1)^l$, according to equation (3.125).[6]

Transformation of Observables under Reflections

The transformation laws for vectors and tensors lead to selection rules and ultimately to the Wigner-Eckart theorem. There is a similar and much simpler analysis for reflections.

The energy is a scalar under reflections, that is, $\mathcal{R}H\mathcal{R} = H$. The vector observables $\mathbf{r}$ and $\mathbf{p}$ are odd under reflections: $\mathcal{R}\mathbf{r}\mathcal{R} = -\mathbf{r}$, and so forth. Scalars that transform like the Hamiltonian under $\mathcal{R}$, and vectors that transform like $\mathbf{r}$, are ordinary scalars and vectors. On the other hand,

$$\mathcal{R}\mathbf{L}\mathcal{R} = \mathcal{R}\mathbf{r}\mathcal{R} \times \mathcal{R}\mathbf{p}\mathcal{R} = \mathbf{L} \tag{5.77}$$

$\mathbf{L}$ is even under reflection, and the total angular momentum $\mathbf{J}$ is also even under reflections, since reflections commute with rotations. Scalars like $\mathbf{L} \cdot \mathbf{r}$ that are odd under reflections, and vectors that are even under reflections, are called **pseudoscalars** and **pseudovectors** respectively. The notion can be extended to tensors of any rank: A rank n tensor that transforms like a product of n ordinary vectors is an ordinary tensor. One that transforms oppositely under reflections is a pseudotensor.

There is another nomenclature that is used for vectors: Pseudovectors are sometimes called **axial vectors** and vectors odd under reflections are **polar vectors**.

There are parity selection rules, like the Wigner-Eckart selection rules for rotations. If $|\alpha'\rangle$ and $|\alpha\rangle$ are parity eigenstates with parities Π' and Π respectively, and if A is an operator even (odd) under reflections ($\mathcal{R}A\mathcal{R} = \pm A$), then

$$\langle \alpha'|A|\alpha\rangle = \pm\langle \alpha'|\mathcal{R}A\mathcal{R}|\alpha\rangle = \pm \Pi'\Pi\langle \alpha'|A|\alpha\rangle \tag{5.78}$$

---
[6] Ordinary molecular sugar is known not to be in a reflection-invariant state, even though the laws of molecular physics are surely symmetric. How is this possible?

so the matrix element vanishes unless $\pm \Pi' \Pi = 1$. This is the parity selection rule, and it is very powerful. For example, for the spinless hydrogen atom, matrix elements like $\langle n', l, m | z | n, l, m \rangle$ must vanish, even though for $l > 0$ they are not forbidden by the rotation selection rule.

5.3.2 Reversal of the Direction of Motion

This symmetry, often called **time reversal**, is the last of the classical space-time symmetries. It corresponds to the peculiar fact that the laws of physics are invariant if one measures time by a clock that is running backward. It is not an obvious symmetry, since the past and future seem unmistakably different to us, and there is some indirect evidence that time-reversal is not really a symmetry of nature, but the electromagnetic, gravitational and nuclear forces are certainly invariant under time inversion.

Suppose we try to analyze $\mathcal{T}$ the way we analyzed $\mathcal{R}$: $\mathcal{T}$ changes the direction of the momentum, but not the position:

$$\mathcal{T} |\mathbf{r}\rangle = |\mathbf{r}\rangle \quad \text{and} \quad \mathcal{T} |\mathbf{p}\rangle = |-\mathbf{p}\rangle \tag{5.79}$$

In terms of the operators,

$$\mathcal{T} \mathbf{r} \mathcal{T}^{-1} = \mathbf{r} \quad \text{and} \quad \mathcal{T} \mathbf{p} \mathcal{T}^{-1} = -\mathbf{p} \tag{5.80}$$

and of course $\mathcal{T} H \mathcal{T}^{-1} = H$ if $\mathcal{T}$ is a good symmetry. It follows that $\mathbf{L} = \mathbf{r} \times \mathbf{p}$ is odd under $\mathcal{T}$. This is true for any kind of angular momentum, since the sense of the angular momentum changes if time is measured by $-t$:

$$\mathcal{T} \mathbf{J} \mathcal{T}^{-1} = -\mathbf{J} \tag{5.81}$$

On the other hand, rotations and translations are both unchanged by time inversion:

$$\mathcal{T} R(\hat{\mathbf{n}}, \theta) \mathcal{T}^{-1} = R(\hat{\mathbf{n}}, \theta) \quad \text{and} \quad \mathcal{T} T(\mathbf{a}) \mathcal{T}^{-1} = T(\mathbf{a}) \tag{5.82}$$

This discussion, which closely parallels the development for reflections, is inconsistent. Consider the representations of these continuous transformations in terms of the generators:

$$R(\hat{\mathbf{n}}, \theta) = e^{-i\theta \hat{\mathbf{n}} \cdot \mathbf{J}} \tag{5.83a}$$

$$T(\mathbf{a}) = e^{-i\mathbf{a} \cdot \mathbf{P}/\hbar} \tag{5.83b}$$

Consider also the canonical commutation relation

$$[x_i, p_j] = i\hbar \delta_{ij} \tag{5.83c}$$

and the fundamental angular momentum commutators:

$$[J_i, J_j] = \sum_k i\epsilon_{ijk} J_k \tag{5.83d}$$

154   Chapter 5   Symmetry Transformations on Operators

Multiply both sides of all four of these equations (5.83) on the left by $\mathcal{T}$ and on the right $\mathcal{T}^{-1}$. Using equations (5.80), (5.81), and (5.82), equations (5.83) turn into

$$R(\hat{\mathbf{n}}, \theta) = e^{i\theta \hat{\mathbf{n}} \cdot \mathbf{J}} \tag{5.84a}$$

$$T(\mathbf{a}) = e^{i\mathbf{a} \cdot \mathbf{P}/\hbar} \tag{5.84b}$$

$$[x_i, p_j] = -i\hbar \delta_{ij} \tag{5.84c}$$

$$[J_i, J_j] = -\sum_k i\epsilon_{ijk} J_k \tag{5.84d}$$

Each of these has a wrong sign. Equations (5.84) are inconsistent with equations (5.83). There is no unitary operator $\mathcal{T}$ with the desired properties!

What exactly is the problem? Each of the "wrong" equations (5.84) would be a correct equation if on the right-hand side, the i could be replaced by $-i$. I argued in Section 4.1.1 that if U is any symmetry, and if $U|\psi_i\rangle = |\phi_i\rangle$, then the probability for a system in a state $|\psi_1\rangle$ to have the properties of the state $|\psi_2\rangle$ is unchanged by U, and therefore

$$\langle \psi_1 | \psi_2 \rangle = \langle \phi_1 | \phi_2 \rangle \tag{5.85}$$

Actually the condition (5.85) is sufficient, but not necessary. The requirement that probabilities be unchanged is the weaker condition

$$|\langle \psi_1 | \psi_2 \rangle|^2 = |\langle \phi_1 | \phi_2 \rangle|^2 \tag{5.86}$$

If U is a linear operator, the probability amplitudes must indeed be equal. But $\mathcal{T}$ need not be linear. There is another possibility, called an **antilinear operator**, that can happen only for discrete symmetries like $\mathcal{T}$. Continuous symmetries must be linear operators.

Any operator that preserves probabilities as in equation (5.86) must be either linear or antilinear. The theorem is proved in some textbooks, but the proofs, mostly copied from Wigner's original, tend to be incomplete.[7] I shall not try to demonstrate here that these are the only two possibilities, but just define **antilinear** and show that it works for $\mathcal{T}$.

Like a linear transformation, an antilinear transformation has the property that

$$\mathcal{T}\left(|\psi_1\rangle + |\psi_2\rangle\right) = \mathcal{T}|\psi_1\rangle + \mathcal{T}|\psi_2\rangle \tag{5.87}$$

but for any complex number α one requires

$$\mathcal{T}\alpha|\psi\rangle = \alpha^* \mathcal{T}|\psi\rangle \tag{5.88}$$

instead of $\mathcal{T}\alpha|\psi\rangle = \alpha \mathcal{T}|\psi\rangle$. An antilinear operator turns ordinary numbers into their complex conjugates.

---

[7] See Chapter 26 in Wigner's book, reference [1] in Chapter 4, for a discussion and references to earlier papers.

Let $|\psi_n\rangle$ be an orthonormal basis, and let $|\phi_n\rangle = \mathcal{T}|\psi_n\rangle$. If $\langle\phi_n|\phi_m\rangle = \delta_{mn}$, as we assume, then $\mathcal{T}$ is **antiunitary** as well. Let

$$|\psi_a\rangle = \sum_n \alpha_n |\psi_n\rangle \quad \text{and} \quad |\psi_b\rangle = \sum_m \beta_m |\psi_m\rangle \tag{5.89}$$

be any states; set $|\phi_a\rangle = \mathcal{T}|\psi_a\rangle$ and $|\phi_b\rangle = \mathcal{T}|\psi_b\rangle$. Then

$$|\phi_a\rangle = \sum_n \alpha_n^* |\phi_n\rangle \quad \text{and} \quad |\phi_b\rangle = \sum_m \beta_m^* |\phi_m\rangle \tag{5.90}$$

The scalar product is

$$\langle\phi_a|\phi_b\rangle = \sum_{mn} \alpha_n \beta_m^* \langle\phi_n|\phi_m\rangle = \langle\psi_a|\psi_b\rangle^* = \langle\psi_b|\psi_a\rangle \tag{5.91}$$

The scalar product of the transformed states is the complex conjugate of the original. Probabilities will still be preserved, which is the only physical requirement for a symmetry.

Now all the contradictions disappear. Apply $\mathcal{T}$ on the left, and $\mathcal{T}^{-1}$ on the right to both sides of equations (5.83), and remember to use $\mathcal{T}i\mathcal{T}^{-1} = -i$. The results are now consistent. For example,

$$\begin{aligned}\mathcal{T}[x_i, p_j]\mathcal{T}^{-1} &= [\mathcal{T}x_i\mathcal{T}^{-1}, \mathcal{T}p_j\mathcal{T}^{-1}] = [x_i, -p_j] = -[x_i, p_j] \\ &= -i\hbar\delta_{ij} = \mathcal{T}i\hbar\delta_{ij}\mathcal{T}^{-1}\end{aligned} \tag{5.92}$$

Time Development of Time-Reversed States

If $\mathcal{T}$ commutes with the Hamiltonian,

$$H\mathcal{T}|\psi(t)\rangle = \mathcal{T}i\hbar\frac{d|\psi(t)\rangle}{dt} = -i\hbar\frac{d\mathcal{T}|\psi(t)\rangle}{dt} \tag{5.93}$$

As expected, $\mathcal{T}|\psi(t)\rangle$ obeys Schrödinger's equation with t replaced by $-t$.

Time-Reversed Wave Functions

In any particular basis $\mathcal{T}$ can be written (exercise!) in the form $\mathcal{T} = UK$, where U is an ordinary unitary operator, and K is the antilinear identity: On a basis vector $|\psi\rangle$,

$$K\alpha|\psi\rangle = \alpha^*|\psi\rangle \tag{5.94}$$

so $K^2 = I$. Be careful with this representation. The operator U depends on the basis. What are the wave functions of time-reversed states?

Spin Zero: Consider a spinless particle first. The states $|\mathbf{r}\rangle$ make up a basis. The wave functions are $\psi(\mathbf{r}) = \langle\mathbf{r}|\psi\rangle$, or

$$|\psi\rangle = \int d^3r |\mathbf{r}\rangle\langle\mathbf{r}|\psi\rangle = \int d^3r \psi(\mathbf{r})|\mathbf{r}\rangle \tag{5.95}$$

Then

$$\mathcal{T}|\psi\rangle = \int d^3r \psi(\mathbf{r})^* \mathcal{T}|\mathbf{r}\rangle = \int d^3r \psi(\mathbf{r})^* |\mathbf{r}\rangle \tag{5.96}$$

For spin-zero particles, the time-reversed wave function is $\psi(\mathbf{r})^*$.

$\mathcal{T}^2 = \mathcal{T}\mathcal{T}$ is an ordinary unitary operator. On states like (5.95), $\mathcal{T}^2$ is the identity:

$$\mathcal{T}^2|\psi\rangle = \mathcal{T}\int \psi(\mathbf{r})^* d^3r |\mathbf{r}\rangle = \int \psi(\mathbf{r}) d^3r |\mathbf{r}\rangle = |\psi\rangle \tag{5.97}$$

Spin Half: Next consider spin-half states. A complete basis consists of the states $|\mathbf{r}, m\rangle$. The wave functions are two-row spinors: $\psi_m(\mathbf{r}) = \langle \mathbf{r}, m| \psi\rangle$. Since $\mathcal{T}$ changes the sign of the spin, $\mathcal{T}|\mathbf{r}, m\rangle$ must be proportional to $|\mathbf{r}, -m\rangle$. What is the phase? In the standard basis, the behavior of the spin operator $\mathbf{s} = \frac{1}{2}\boldsymbol{\sigma}$ on these states is

$$\mathbf{s}|\mathbf{r}, m\rangle = \frac{1}{2}\sum_n \boldsymbol{\sigma}_{nm}|\mathbf{r}, n\rangle \tag{5.98}$$

so

$$\mathcal{T}\mathbf{s}\mathcal{T}^{-1}|\mathbf{r}, m\rangle = -\frac{1}{2}\sum_n \boldsymbol{\sigma}^*_{nm}|\mathbf{r}, n\rangle \tag{5.99}$$

Write $\mathcal{T} = UK$. U is an ordinary unitary operator that mixes up the spin indices only:

$$U|\mathbf{r}, m\rangle = \sum_k |\mathbf{r}, k\rangle U_{km} \tag{5.100}$$

Then

$$\mathcal{T}\mathbf{s}\mathcal{T}^{-1}|\mathbf{r}, m\rangle = UKsKU^\dagger |\mathbf{r}, m\rangle$$
$$= \frac{1}{2}\sum_{ijn} U^*_{mi}\sigma^*_{ji}U_{nj}|\mathbf{r}, n\rangle = \frac{1}{2}\sum_n \left(U\boldsymbol{\sigma}^* U^\dagger\right)_{nm}|\mathbf{r}, n\rangle \tag{5.101}$$

By comparing equations (5.99) and (5.101) one learns that U is some 2×2 matrix such that $U^\dagger \sigma_i U = -\sigma_i^*$ for all three Pauli matrices. In the standard basis the solution is unique up to a phase: $U = \sigma_y$. The time-reversed wave function is $\langle \mathbf{r}, m| \mathcal{T} |\psi\rangle = \sum_n (\sigma_y)_{mn} \psi_n(\mathbf{r})^*$. This result is especially useful in the application of time-reversal symmetry to scattering problems.

Note: You may wonder how σ_y came to be singled out, since there should be nothing special about the y-direction. This fact can be traced to the convention that s_y is imaginary, while s_x and s_z are real. Had we chosen other phases in equations (3.113) time inversion on spin-half states would have been represented by a different Pauli matrix.

On spin-half states, $\mathcal{T}^2$ is no longer the identity. Instead (remember $\sigma_y^2 = I$),

$$\mathcal{T}^2|\mathbf{r}, m\rangle = \sum_n \mathcal{T}(\sigma_y)_{nm}|\mathbf{r}, n\rangle = \sum_n (\sigma_y)^*_{nm}\mathcal{T}|\mathbf{r}, n\rangle = -\sum_n (\sigma_y)_{nm}\mathcal{T}|\mathbf{r}, n\rangle = -|\mathbf{r}, m\rangle \tag{5.102}$$

Equation (5.102) is an instance of an important general theorem.

A Superselection Rule

While $\mathcal{T}$ is an antilinear operator, $\mathcal{T}^2$ is an ordinary linear operator, since $\mathcal{T}^2\alpha|\psi\rangle = \alpha\mathcal{T}^2|\psi\rangle$. Classically, $\mathcal{T}^2 = 1$, just like $\mathcal{R}^2$. If you change the sense of a clock, then do it again, you should be back where you started. But there is a famous theorem that says T^2 is always -1 on states of half-integer angular momentum.

Proof. For each component of the angular momentum,

$$T J_i T^{-1} = -J_i \tag{5.103}$$

Therefore,

$$T J_+ T^{-1} = -J_- \tag{5.104}$$

Label the states $|\alpha, j, m\rangle$ as usual. T must change the sign of the spin quantum number:

$$T|\alpha, j, m\rangle = c_m |\alpha, j, -m\rangle \tag{5.105}$$

for some constants c_m of absolute value unity. Now

$$T J_+ |\alpha, j, m\rangle = c_{m+1} \sqrt{j(j+1) - m(m+1)} |\alpha, j, -m-1\rangle \tag{5.106}$$

while

$$J_- T |\alpha, j, m\rangle = c_m \sqrt{j(j+1) - m(m+1)} |\alpha, j, -m-1\rangle \tag{5.107}$$

Thus $c_m = -c_{m+1}$. The sign of c_m alternates. The c_m do not have to be real, so

$$c_m = (-1)^{j-m} e^{i\delta} \tag{5.108}$$

for some phase δ. Notice $(-1)^{j-m}$ is real whether j is an integer or a half-integer. Then

$$T^2 |\alpha, j, m\rangle = (-1)^{j-m} e^{-i\delta} T |\alpha, j, -m\rangle = (-1)^{2j} |\alpha, j, -m\rangle \tag{5.109}$$

T^2 is $+1$ for integer j, -1 for half-integer j. ∎

There is a simple application when j is half-integral. If $[T, H] = 0$, then $[T^2, H] = 0$, and the energy eigenstates can be chosen to be eigenstates T^2. Since T commutes with H, $T|\psi\rangle$ has the same energy as $|\psi\rangle$. If $T|\psi\rangle = |\psi\rangle$, then $T^2|\psi\rangle = T|\psi\rangle = -|\psi\rangle$, which is a contradiction. So $T|\psi\rangle \neq |\psi\rangle$. There must be another state with the same energy as $|\psi\rangle$.

When T is a symmetry, half-integer spin energy eigenstates always come in degenerate pairs! This rule is called the **Kramers Degeneracy** in atomic physics.

5.3.3 Identical Particles

All electrons are exactly alike. So are all protons, all photons, all U^{235} nuclei and all iron atoms in their ground state. They are *indistinguishable*: If a system contains two identical particles of the same species (electrons, protons, etc.), then its physical properties do not depend on which particle is called number one and which number two. No experiment can tell which is the first particle and which the second. Exchanging two particles is a symmetry of nature. The states that occur should be a basis for representations of the algebra of these permutations.

As an example, an electron is described by the eigenstates of some complete set of commuting observables, such as $|x, y, z, m_s\rangle$ or $|E, l, j, m\rangle$; A two-electron state is described by two sets of such coordinates, for example, by

$$|x_1, y_1, z_1, m_1, x_2, y_2, z_2, m_2\rangle \quad \text{or} \quad |E_1, l_1, j_1, m_1, E_2, l_2, j_2, m_2\rangle$$

Denote the labels generically by $|\alpha_1, \alpha_2\rangle$, and define the permutation operator P by

$$P|\alpha_1, \alpha_2\rangle = |\alpha_2, \alpha_1\rangle \tag{5.110}$$

where α stands for any complete set of one-particle eigenvalues. It is not completely trivial to show that equation (5.110) is independent of the basis chosen, but it is true. Like the reflection operator, the operator P is its own inverse: $P^2 = I$. The eigenvalues of P can be only ± 1.

The statement that the two particles are indistinguishable means that P is a symmetry of the system. That is

$$PHP = H \tag{5.111}$$

Since P commutes with H the eigenstates of H can be chosen eigenstates of P also.

So far the discussion parallels that of reflections and parity. But here nature is particularly simple. *All* the states with the same number of particles of the same type are eigenstates of P with the same sign. Eigenstates with the other sign do not occur.

For complex systems with more than two identical particles, there is a permutation operator P_{ij} for each pair. There exist representations of the permutation group of mixed symmetry, even under exchange of some pairs of indices, and odd under others. But those do not occur in nature. For a given type of particle, the states are either odd or even for *all* the P_{ij}.

Particles like photons, pi-mesons, and deuterons are even under P. Electrons, protons, and neutrons are odd under P. Objects that are odd under this permutation symmetry are said to obey Fermi-Dirac statistics; those that are even obey Bose-Einstein statistics. Colloquially, the world is made up of **fermions** and **bosons**. There is a general rule, arbitrary in nonrelativistic quantum mechanics, that all half-integral spin systems are fermions and all integral spin system systems are bosons. The origin of this "spin-statistics theorem," first proved by Pauli, Fierz, Lüders, Zumino and others, lies in the mathematical properties of the Lorentz transformations.[8]

If $P|\psi\rangle = -|\psi\rangle$, the quantum numbers of a pair of particles cannot all be the same. For example, if two electrons have the same E, l, j, and m, then a two-electron state with both electrons having the same properties is

$$|\psi\rangle = |E_1, l_1, j_1 m_1, E_1, l_1, j_1, m_1\rangle \tag{5.112}$$

$|\psi\rangle$ can be odd under P only if it is zero. This rule is the **Pauli exclusion principle**. It accounts for the structure of atoms and nuclei, and for the periodic system of the elements.

---

[8]See Section 13.4.

There is no such restriction of particles with the plus sign. Any number of them can be in the same state. Bosons exhibit condensation properties at low temperatures, in which a macroscopic number of particles can be in the same state, as in superfluid helium-4. Lasers are possible because photons are bosons.

In equation (5.110) I wrote states as we would without regard to their indistinguishability, and then asserted that only those states symmetric or antisymmetric under permutations actually exist in nature. This notation is straightforward for systems of two or three identical particles, but becomes cumbersome as the number increases. It would be better to let a single vector stand for each physically distinct state, rather than deal with a lot of different symbols that describe the same physical state. In relativistic systems the notation here is totally unsuitable, since the possibility of producing pairs of particles means that the number of particles in a state is not fixed. I will introduce the more elegant notation of creation and destruction operators in Chapter 11 when we discuss photons. You may already have encountered it in the study of systems of many particles.

5.4 INTERNAL SYMMETRIES: ISOSPIN

Symmetry transformations—transformations on a quantum mechanical vector space that leave the Hamiltonian unchanged—imply conservation laws. Symmetry transformations that form a non-Abelian group imply degeneracies, several states that all have the same energy.[9] My main example has been the $2j+1$ levels of an angular momentum ladder. Here the group, $SO(3)$ in classical mechanics, becomes $SU(2)$ in quantum mechanics. The degeneracy of the energy levels of the spinless hydrogen atom was another example. There the group is $O(4)$, rotations in four dimensions.

The converse is also true. An observed degeneracy is always the consequence of some transformation group that is a symmetry. If this transformation is not one of the space-time symmetries that we know from classical mechanics, it is called an **internal symmetry**. Isospin is the most famous example.

Nucleons

The neutron is almost identical to the proton. It has the same spin, similar nuclear forces—strong, and with a range of the order of one fermi—and almost exactly the same mass. Only the electric charge, and therefore its electromagnetic interactions, are different.

The electromagnetic interactions are small. We can write the Hamiltonian as $H = H_S + H'$, where H' contains all the electromagnetic and weak forces and any others of order α or smaller. H_S has the strong nuclear forces, the ones that dominate when neutrons and protons get to within about one fermi apart. Isospin is a symmetry of H_S. In the approximation that H' is zero, the neutron and proton are two different states of the same particle, the **nucleon**, connected by a set of transformations that commute with H_S, and these two states, called $|n\rangle$ and $|p\rangle$, have the same mass.

---

[9] A group all of whose elements commute is an **Abelian**, or commutative group. Translations and multiplication by a universal phase are examples. Other groups, like rotations, are non-commutative or **non-Abelian**.

160 Chapter 5 Symmetry Transformations on Operators

The name of this internal symmetry is **isospin**.[10] We already know all about it mathematically, since the algebra of the symmetry transformations turns out to be identical with the algebra of rotational symmetry, or $SU(2)$. $|p\rangle$ and $|n\rangle$ form an isospin one-half doublet, and they are the basis for the fundamental two-dimensional representation. With the identification $|p\rangle = |1/2\rangle$ and $|n\rangle = |-1/2\rangle$, we can just take over wholesale our knowledge about rotations on spin-half particles. Under isospin rotations

$$U(\hat{\boldsymbol{n}}, \theta)|i\rangle = \sum_{j=-\frac{1}{2}}^{\frac{1}{2}} |j\rangle D^{(\frac{1}{2})}(\hat{\boldsymbol{n}}, \theta)_{ji} \tag{5.113}$$

with $i = \pm 1/2$. Here

$$D^{(\frac{1}{2})}(\hat{\boldsymbol{n}}, \theta)_{ji} = e^{-i\theta\hat{\boldsymbol{n}}\cdot\boldsymbol{\tau}/2} \tag{5.114}$$

τ_i are the Pauli matrices—they are denoted here by a different letter just to remind us that these transformations act on a different space than real spin. Define the three generators I_M by

$$U(\hat{\boldsymbol{n}}, \theta) = e^{-i\theta\hat{\boldsymbol{n}}\cdot\mathbf{I}} \tag{5.115}$$

then the three Hermitean operators I_M satisfy the angular momentum algebra, and the states are eigenstates of I_3. The proton and neutron are isospin-up and isospin-down states of the nucleon, eigenstates of $I_3 = \tau_3/2$ with eigenvalues $\pm 1/2$.

It is useful to construct isospinors, two-component objects like the Pauli spinors defined in equation (4.66), but with no space-time dependence. For a general nucleon state $|N\rangle = N_{\frac{1}{2}}|p\rangle + N_{-\frac{1}{2}}|n\rangle$, construct the isospinor

$$N = \begin{pmatrix} N_{+\frac{1}{2}} \\ N_{-\frac{1}{2}} \end{pmatrix} \tag{5.116}$$

Then under an isospin rotation,

$$N \to D^{(\frac{1}{2})}(\hat{\boldsymbol{n}}, \theta) N \tag{5.117}$$

or in components

$$N_i \to \sum_j \left[D^{(\frac{1}{2})}(\hat{\boldsymbol{n}}, \theta) \right]_{ij} N_j \tag{5.118}$$

For nucleons I_3 is a linear function of the electric charge:

$$Q = I_3 + \frac{1}{2} \tag{5.119}$$

Nuclei
Nuclei of atomic weight A are made up of A neutrons and protons. The forces that bind them are principally in H_S, so they come in isomultiplets with some isospin I, where I is one of those that can be made by combining A isodoublets. Each

---
[10] It used to be called isotopic spin until people realized it had little to do with isotopes. The idea is apparently due originally to Heisenberg [1]. A translation can be found in Brink [2].

multiplet has $2I+1$ different states, all of (approximately) the same mass. The rule for the electric charge is a generalization of equation (5.119):

$$Q = I_3 + \frac{A}{2} \tag{5.120}$$

Here are some examples: Two nucleons can form states with isospins zero and one. There are no observed pp or nn bound states, so the only $A=2$ nucleus is the deuteron, an isosinglet with ordinary spin one. The two $A=3$ nuclei, ^{3}He, a rare isotope of helium, and tritium, ^{3}H, have almost the same mass or binding energy, and both have ordinary spin half. These two nuclei form an isodoublet. The three $A=14$ nuclei ^{14}O, ^{14}N, and ^{14}C, all spin zero and with approximately equal masses, form an isotriplet.

> Note: Of these three $A=14$ nuclei, ^{14}N has the lowest mass, by a fraction of an MeV, and it is the only one that is stable. The ^{14}O nucleus in a neutral atom captures one of the electrons orbiting nearby, emits a neutrino to conserve angular momentum and energy, and becomes ^{14}N. ^{14}O has a half-life of about 71 seconds. ^{14}C is also unstable, and beta-decays, with a lifetime of about 5700 years, into the stable ^{14}N nucleus. Nevertheless ^{14}C does occur naturally in our atmosphere, where it is continually made in collisions between high-energy cosmic rays and atmospheric nitrogen. ^{14}C is the isotope that is used in radiocarbon dating.

Pi Mesons

Hadron is the name given to all the particles that enter into H_S. Excluded are the photon and all of the **leptons** namely the three charged spin-half particles e^-, μ^-, and τ^-, the neutrinos, and the antiparticles of all these. These particles have interactions in the electromagnetic and weak Hamiltonian, but not in H_S.

The three π mesons are the lowest mass hadrons. They all have zero spin. The π^+ and π^- have exactly the same mass, about 139 MeV (see Table 5.1 in section 5.1.) because they are each other's antiparticle, while the π^o, a few MeV lighter, is its own antiparticle like the photon. They are grouped into an isotriplet that transforms like the spin-one representation of rotations:

$$\begin{pmatrix} \pi^+ \\ \pi^o \\ \pi^- \end{pmatrix} \to D^{(1)}(\hat{\mathbf{n}}, \theta) \begin{pmatrix} \pi^+ \\ \pi^o \\ \pi^- \end{pmatrix} \tag{5.121}$$

Here π^q stands for $\langle q| \psi \rangle$, where $q = 1, 0, -1$. The relation (5.120) still holds provided $A=0$ for these particles.

Example: Proton-deuteron scattering

Beyond the existence of isomultiplets, isospin invariance predicts ratios of scattering and production cross sections. Let $|i\rangle$ and $|f\rangle$ be the initial and final states in some such process. They are related by

$$|f\rangle = U(t_f, t_i)|i\rangle \tag{5.122}$$

where $U(t_f, t_i)$ is the time-translation operator introduced in Section 2.5.1. Since

$$U(t_f, t_i) = e^{-iH(t_f - t_i)} \tag{5.123}$$

162 Chapter 5   Symmetry Transformations on Operators

the operator $U(t_f, t_i)$ will commute with isospin when H does. The amplitude $\langle f | U(t_f, t_i) | i \rangle$ is diagonal in the $|I, I_3\rangle$ basis and independent of I_3. As a consequence, there are restrictions among the rates or cross sections.

Consider scattering a proton off a deuteron. An high enough energy, it is possible to produce a pi meson and either ^{3}He or a tritium nucleus:

$$\text{p} + \text{d} \rightarrow {}^3\text{H} + \pi^+ \tag{5.124a}$$

$$\text{p} + \text{d} \rightarrow {}^3\text{He} + \pi^o \tag{5.124b}$$

The final state is made out of isospin one and isospin half, so it is a combination of states with $I = 3/2$ and $I = 1/2$. Suppressing all the other quantum numbers—coordinates and spin—these states are $|\phi_{II_3}\rangle$, where

$$\mathbf{I}^2 |\phi_{IM}\rangle = I(I+1)|\phi_{IM}\rangle \quad \text{and} \quad I_3 |\phi_{IM}\rangle = M |\phi_{IM}\rangle \tag{5.125}$$

Since the nucleon is an isodoublet, while the deuteron is an isoscalar, the total isospin in the initial state is $1/2$, and $I_3 = 1/2$ also. Since

$$\left\langle \phi_{\frac{3}{2}\frac{1}{2}} \middle| U(t_f, t_i) \middle| \text{p}, \text{d} \right\rangle = 0 \tag{5.126}$$

one finds by the Clebsch-Gordan construction

$$\left\langle \pi^+, {}^3\text{H} \middle| U(t_f, t_i) \middle| \text{p}, \text{d} \right\rangle = \sum_{I=\frac{1}{2}}^{\frac{3}{2}} \left\langle \pi^+, {}^3\text{H} \middle| \phi_{I\frac{1}{2}} \right\rangle \left\langle \phi_{I\frac{1}{2}} \middle| U(t_f, t_i) \middle| \text{p}, \text{d} \right\rangle$$

$$= \left\langle \pi^+, {}^3\text{H} \middle| \phi_{\frac{1}{2}\frac{1}{2}} \right\rangle \left\langle \phi_{\frac{1}{2}\frac{1}{2}} \middle| U(t_f, t_i) \middle| \text{p}, \text{d} \right\rangle \tag{5.127}$$

$$= -\sqrt{\frac{2}{3}} \left\langle \phi_{\frac{1}{2}\frac{1}{2}} \middle| U(t_f, t_i) \middle| \text{p}, \text{d} \right\rangle$$

and similarly

$$\left\langle \pi^o, {}^3\text{He} \middle| U(t_f, t_i) \middle| \text{p}, \text{d} \right\rangle = \sqrt{\frac{1}{3}} \left\langle \phi_{\frac{1}{2}\frac{1}{2}} \middle| U(t_f, t_i) \middle| \text{p}, \text{d} \right\rangle \tag{5.128}$$

The rates, or cross sections, are proportional to the squares of the amplitudes:

$$\frac{\left| \left\langle \pi^+, {}^3\text{H} \middle| U(t_f, t_i) \middle| \text{p}, \text{d} \right\rangle \right|^2}{\left| \left\langle \pi^o, {}^3\text{He} \middle| U(t_f, t_i) \middle| \text{p}, \text{d} \right\rangle \right|^2} = 2 \tag{5.129}$$

in good agreement with experiment.

Antiparticles

All particles have antiparticles, identical to them in mass and spin but having opposite values of additive quantum numbers like electric charge, baryon number, and the components of isospin. Here the isospinor notation comes in handy. If the nucleon N is described by a two component object transforming as in equation (5.117), the antinucleon $\bar{N}$ transforms like $N^\dagger$:

$$\bar{N} \rightarrow \bar{N} D^{(\frac{1}{2})}(\hat{\mathbf{n}}, \theta)^\dagger \tag{5.130}$$

or in components

$$\bar{N}_i \to \sum_j \bar{N}_j \left[D^{(\frac{1}{2})}(\hat{\boldsymbol{n}},\theta)^\dagger\right]_{ji} = \left[D^{(\frac{1}{2})}(\hat{\boldsymbol{n}},\theta)^*\right]_{ij} \bar{N}_j \qquad (5.131)$$

since $D^{(\frac{1}{2})}$ is a unitary matrix. The antinucleons undergo isospin transformations represented by matrices that are the complex conjugates of the usual ones! These matrices are certainly a representation, since

$$D_1^* D_2^* = (D_1 D_2)^* \qquad (5.132)$$

The representation matrix for a rotation about a given axis is

$$D^{(\frac{1}{2})}(\hat{\boldsymbol{n}}_a,\theta)^* = e^{+i\theta \tau_a^*/2} \qquad (5.133)$$

so the generators are $-\tau_a^*/2$. In particular,

$$I_3 \left|\pm 1/2\right\rangle = \mp \frac{1}{2} \left|\pm 1/2\right\rangle \qquad (5.134)$$

which is the value we expect for the antinucleons. The minus sign guarantees that $\bar{p}$ and $\bar{n}$ have values for I_3 opposite from p and n.

This simple notation makes it easy to combine nucleon and antinucleon states into combinations with definite isospin. Instead of a two-component object, the wave functions for nucleon-antinucleon pairs are elements of a 2×2 matrix

$$\psi_{ij} = N_i \bar{N}_j \qquad (5.135)$$

The most general nucleon-antinucleon state is a linear combination

$$\psi = \sum_{ij} \bar{N}_i M_{ij} N_j \qquad (5.136)$$

for some matrix M.

As a mnemonic, one can label the components by the particles they stand for:

$$N = \begin{pmatrix} p \\ n \end{pmatrix} \quad \text{and} \quad \bar{N} = \begin{pmatrix} \bar{p} & \bar{n} \end{pmatrix} \qquad (5.137)$$

The matrix whose components are $N_i \bar{N}_j$ is

$$\begin{pmatrix} p\bar{p} & p\bar{n} \\ n\bar{p} & n\bar{n} \end{pmatrix} \qquad (5.138)$$

These states have baryon number zero. How do they transform under isospin rotations? For an isospin rotation about the a-th axis, N and $\bar{N}$ transform (to first order in θ) as

$$\delta N = -i\frac{1}{2}\theta \tau_a N \quad \text{and} \quad \delta \bar{N} = i\frac{1}{2}\theta \bar{N} \tau_a \qquad (5.139)$$

so for the general combination defined in equation (5.136),

$$\delta\psi = i\frac{1}{2}\theta \bar{N}[\tau_a, M] N \qquad (5.140)$$

The combination
$$\bar{N}N = p\bar{p} + n\bar{n} \tag{5.141}$$
is an invariant—this is the isospin-zero combination of two isospin-half particles. Notice the plus sign in equation (5.141). In a standard basis the spin-zero combination of two spin-half particles is
$$\frac{\left|+\tfrac{1}{2}, -\tfrac{1}{2}\right\rangle - \left|-\tfrac{1}{2}, +\tfrac{1}{2}\right\rangle}{\sqrt{2}} \tag{5.142}$$

The antinucleons do not transform in the standard way.

The other three combinations are proportional to
$$\psi_b = \bar{N}\tau_b N \tag{5.143}$$
whose transformation law is
$$\delta\psi_b = -\theta \sum_c \epsilon_{abc}\psi_c \tag{5.144}$$
just like the Cartesian components of a vector. In particular, the combination $p\bar{p} - n\bar{n}$ is the three-component of an isovector, orthogonal to the invariant combination. If $\pi^\pm$ and π^o are isovector particles, and σ an isoscalar particle, the matrix
$$\Pi = \begin{pmatrix} (\sigma + \pi^o)/\sqrt{2} & \pi^+ \\ \pi^- & (\sigma - \pi^o)/\sqrt{2} \end{pmatrix} \tag{5.145}$$
transforms like the one in equation (5.138).

PROBLEMS

5.1. Matrix Elements of Vector Operators
(a) In the notation of Section 5.1.4, compute
$$\frac{\langle \alpha', j, j-1 | V^- | \alpha, j, j \rangle}{\langle \alpha', j, j | V^o | \alpha, j, j \rangle}$$
and
$$\frac{\langle \alpha', j, j-1 | V^o | \alpha, j, j-1 \rangle}{\langle \alpha', j, j | V^o | \alpha, j, j \rangle}$$
for arbitrary j.

(b) Let **V** be any vector operator, and label the states in the notation of Section 5.1.4. Show that
$$\langle \alpha', j+1, j | V^o | \alpha, j, j \rangle$$
is proportional to
$$\langle \alpha', j+1, j+1 | V^+ | \alpha, j, j \rangle$$
and the ratio of these two matrix elements is independent of **V**. What is the ratio?

5.2. Rank-Two Tensors
Show that if T is a second-rank Cartesian tensor, then $\sum_i T_{ii}$ is a scalar, and

$$V_i = \sum_{jk} \epsilon_{ijk} T_{jk}$$

are components of a vector, as claimed in Section 5.2.1.

5.3. Symmetric and Antisymmetric Tensors
Show that if $T_{ij} = \pm T_{ji}$, then $T'_{ij} = RT_{ij}R^\dagger = \pm T'_{ji}$ as claimed in equation (5.39).

5.4. Rotation Matrices in Rotated Coordinates

Discussion: Let $\hat{\boldsymbol{n}}$ be a unit vector pointing in the direction whose spherical coordinates are θ, ϕ. The unit vector $\hat{\boldsymbol{n}}_z$ can be rotated into $\hat{\boldsymbol{n}}$ by a rotation around the y-axis by θ followed by a rotation around the z-axis by ϕ. That is,

$$\hat{\boldsymbol{n}} = \bar{R}(\hat{\boldsymbol{n}}_z, \phi)\bar{R}(\hat{\boldsymbol{n}}_y, \theta)\hat{\boldsymbol{n}}_z$$

In the Euler angle notation, $\hat{\boldsymbol{n}} = \bar{R}(\phi, \theta, 0)\hat{\boldsymbol{n}}_z$.

Consider a state in the $2j+1$ dimensional representation, that is, one that rotates as

$$D^{(j)}(\hat{\boldsymbol{n}}, \theta)|j, m\rangle = \sum_{m'} |j, m'\rangle D^{(j)}(\hat{\boldsymbol{n}}, \theta)_{m'm}$$

The state $D^{(j)}(\phi, \theta, 0)|j, m\rangle$ ought to have the same properties as $|j, m\rangle$, but with respect to the rotated operators. In particular, matrix elements of J_z in the $|j, m\rangle$ basis should be the same as matrix elements of $\hat{\boldsymbol{n}} \cdot \mathbf{J}$ in the $D^{(j)}(\phi, \theta, 0)|j, m\rangle$ basis. As a consequence,

$$D^{(j)}(\bar{R})J_z D^{(j)}(\bar{R})^{-1} = \hat{\boldsymbol{n}} \cdot \mathbf{J}$$

so for *any* j,

$$D^{(j)}(\bar{R})D^{(j)}(\bar{J}_z)D^{(j)}(\bar{R})^{-1} = \sum_i n_i D^{(j)}(\bar{J}_i)$$

On the left, θ and ϕ parametrize $\bar{R}$; on the right, they parametrize the components n_i of the vector $\hat{\boldsymbol{n}}$. This result should be geometrically obvious; it is not hard to prove in general, but I shall not do it here.

(a) Verify this general result explicitly in the $j = 1$ representation. That is, show that

$$D^{(1)}(\bar{R})D^{(1)}(\bar{J}_z)D^{(1)}(\bar{R})^{-1} = \sum_i n_i D^{(1)}(\bar{J}_i)$$

where

$$D^{(1)}(\bar{R}) = D^{(1)}(\hat{\boldsymbol{n}}_z, \phi)D^{(1)}(\hat{\boldsymbol{n}}_y, \theta)$$

The matrices on the right were worked out in Section 4.3.6.

(b) Suppose a beam of spin-one particles is passed through a Stern-Gerlach filter that selects only those particles whose z-component of the spin is $+\hbar$. Now measure the spin along an axis that points in the direction whose spherical components are θ and ϕ, as above. What are the respective probabilities of measuring the spin of a particle along this new direction as $\hbar$, 0, and $-\hbar$ respectively? Compute the sum of these three probabilities.

5.5. Decomposition of Rank-Two Cartesian Tensors
Define the following rank-two Cartesian tensors:
$$T_{ij} = r_i r_j \quad \text{and} \quad T'_{ij} = p_i p_j$$

(a) Evaluate the expectation value of T_{ij} in the ground state of the three-dimensional harmonic oscillator.
(b) Same for T'_{ij}.

> Hint: Expand T_{ij} and T'_{ij} in terms that transform as components of spherical tensors of definite rank (but it isn't necessary to compute the T^q_κ in terms of the T_{ij}) and use the second Wigner-Eckart theorem selection rule.

5.6. A Wigner-Matrix Sum Rule
(a) Evaluate
$$\sum_{m'=-j}^{j} m' \left| d^{(j)}(\beta)_{mm'} \right|^2$$

(b) Verify your result for $j = 1/2$ and $j = 1$.

5.7. Fundamental Properties of Rank-Two Spherical Tensors
Let T_2 be a rank-two tensor that is the product of two vectors:
$$T_{ij} = U_i V_j$$

The five equations (5.42) define the "spherical" components T^q_2 in terms of the combinations U^q and V^q.

(a) Show that as long as T^q_2 is a sum of terms $U^m V^n$ such that $m + n = q$, the rule
$$[J_z, T^q_2] = \sum_{q'} T^{q'}_2 D^{(2)}_{q'q}(\bar{J}_z)$$
is satisfied.

(b) Complete the demonstration of equation (5.43) by showing explicitly that
$$[J_-, T^q_2] = \sum_{q'} T^{q'}_2 D^{(2)}_{q'q}(\bar{J}_-)$$

5.8. More on the Three-Dimensional Oscillator
The Hamiltonian is[11]
$$H = \frac{1}{2m}\mathbf{p}^2 + \frac{m\omega^2 \mathbf{r}^2}{2} = \frac{1}{2m}(Q_{xx} + Q_{yy} + Q_{zz})$$

The three components of the angular momentum $\mathbf{L} = \mathbf{r} \times \mathbf{p}$ are not the only conserved operators. All the components of the tensor
$$Q_{ij} = m^2 \omega^2 r_i r_j + p_i p_j$$

---
[11] This problem and the next four go together. They explain how to use $SU(3)$ to understand the spectrum of the three-dimensional oscillator.

are also conserved. The proof of this statement is identical to the one for two dimensions in Problem 4.9. In classical mechanics the fact that the Q_{ij} are constant shows that the orbits are closed. In quantum mechanics the extra conserved observables imply a symmetry group for this Hamiltonian larger than the rotations, which in turn implies a degeneracy greater than $2l+1$.

(a) What are the commutators $[L_i, Q_{jk}]$ in terms of the L_i and Q_{jk} (here the total angular momentum is $\hbar \mathbf{L} = \mathbf{r} \times \mathbf{p}$)?

(b) Five linear combinations of the Cartesian tensor Q_{ij} form the components Q_2^q of a rank-two spherical tensor. The normalization is arbitrary, so take

$$Q_2^o = 3Q_{zz} - 2mH$$

Use the rule

$$[L_i, Q_2^q] = \sum_{q'} Q_2^{q'} D^{(2)}(\bar{J}_i)_{q'q}$$

and the result of part (a) to construct the other four components of Q_2^q (use the combinations $L_\pm$) in terms of the Cartesian tensor components.

(c) Consider the $2l+1$ states in the N-th energy level with angular momentum l. Label these as $|N, l, m\rangle$. Let M be the expectation value of Q_2^o in the top state of the ladder:

$$M = \langle N, l, l | Q_2^o | N, l, l \rangle$$

Compute[12]

$$\langle N, l, m | Q_{xx} - Q_{yy} | N, l, l \rangle$$

in terms of M for all m.

Note: You have shown that for the isotropic oscillator, the three L_i and the five independent combinations of

$$Q_{ij} - \frac{1}{3}\delta_{ij} \sum_k Q_{kk}$$

all commute with the Hamiltonian. They have the algebra of $SU(3)$, though this is not obvious yet. The analogous statement for the hydrogen atom was that the three L_i and the three components of the Runge-Lenz vector commute with the Hamiltonian, and have the six-dimensional algebra of $O(4)$. We were able to work out the dimensions of the representations of $O(4)$, and thus find the degeneracy of the hydrogen problem. Here, we need to know something about the representations of $SU(3)$ to find the degeneracy of the oscillator.

5.9. The Three-Dimensional Unitary Unimodular Matrices

Read the mathematical note in Appendix C on $SU(3)$ and its representations.

(a) If $U = e^{-i\Lambda}$ is a unitary matrix, you know that Λ must be Hermitean. Show in general that if a matrix U is finite-dimensional and its determinant is unity, then the trace of Λ is zero.

---

[12]Hint: Write $Q_{xx} - Q_{yy}$ in terms of the Q_2^q, and then compute, twice, matrix elements of the commutator $[J_+, Q_2^q]$ for the appropriate q.

(b) There are $SU(2)$ **subalgebras** in $SU(3)$, that is, collections of three matrices T_i that satisfy $[T_i, T_j] = i \sum_k \epsilon_{ijk} T_k$. Show that $\bar{F}_1$, $\bar{F}_2$, and $\bar{F}_3$ are an $SU(2)$ subalgebra.

(c) For some matrix U, the three matrices $\bar{F}_4$, $\bar{F}_5$, and U form an $SU(2)$ subalgebra. What is U?

(d) For some matrix V, the three matrices $\bar{F}_6$, $\bar{F}_7$, and V form an $SU(2)$ subalgebra. What is V?

(e) The $SU(3)$ structure constants f_{ijk} are analogs of ϵ_{ijk} for $SU(2)$. They are defined by

$$[\bar{F}_i, \bar{F}_j] = i \sum_k f_{ijk} \bar{F}_k$$

The matrices $\bar{F}_i$ are $\bar{F}_i = \lambda_i/2$, with the eight 3×3 matrices λ_i defined in equations (C.3). The structure constants have the property

$$f_{ijk} = -f_{jik}$$

Why?

(f) It is also true but not so obvious that the f_{ijk}, like the ϵ_{ijk}, are antisymmetric under interchange of *any* pair of indices. Use $\text{Tr}(\lambda_i \lambda_j) = 2\delta_{ij}$ and the cyclic property of the trace to show that $f_{ijk} = -f_{ikj}$ also.

5.10. Representations of SU(3)

(a) Write out all the eigenstates of F_3 and F_8, and their eigenvalue pairs (μ, ν), in the representation of $SU(3)$ obtained by combining two fundamental representations, as at the end of the mathematical note in Section C of the appendix.

(b) For any N, what is the dimension of the representation whose top pair of eigenvalues are $(N/2, N/\sqrt{12})$? Check that these are also the dimensions of the three-dimensional oscillator energy levels.[13]

5.11. SU(3) and the Energy Levels of the Three-Dimensional Oscillator

(a) As in the two-dimensional case, Problem 4.10, it is easier to study the operators $A_{ij} = a_i^\dagger a_j$ than the Hermitian combinations in Problem 5.8. Since these A_{ij} do not change the sum $n_x + n_y + n_z$, they all commute with the Hamiltonian. Define the eight operators

$$F_a = \frac{1}{2} \sum_{i=1}^{3} \sum_{j=1}^{3} A_{ij} (\lambda_a)_{ij}$$

where

$$a_i = \frac{1}{\sqrt{2m\omega\hbar}} (m\omega r_i + i p_i)$$

and λ_a are the eight matrices defined in Problem 5.9. Write out these eight operators in terms of the three L_i and five combinations of the Q_{ij}.

(b) Show that the commutator algebra of these eight F_a is $SU(3)$, *i.e.* that

$$[F_a, F_b] = i \sum_c f_{abc} F_c$$

---

[13] $SU(3)$ has representations other than these symmetric ones. Those will not occur in the present problem, since the oscillator states are proportional to $(a_x^\dagger)^{n_x} (a_y^\dagger)^{n_y} (a_z^\dagger)^{n_z} |0\rangle$. The order of the $a_i^\dagger$ is irrelevant.

with the f_{ijk} as in Problem 5.9. The proof is the same as part (b) in Problem 4.10 except for the number of dimensions.

(c) Let F_a be operators obeying the $SU(3)$ algebra in any representation, that is,

$$[F_a, F_b] = i \sum_c f_{abc} F_c$$

Show that $\sum_a (F_a)^2$ commutes with all eight F_a [in analogy to $\mathbf{J}^2$ in $SU(2)$].

(d) Show that

$$\sum_a (F_a)^2 = \frac{1}{2}(T_+ T_- + T_- T_+)$$
$$+ \frac{1}{2}(U_+ U_- + U_- U_+) + \frac{1}{2}(V_+ V_- + V_- V_+) + F_3^2 + F_8^2$$

here

$$T_\pm = F_1 \pm iF_2 \qquad U_\pm = F_4 \pm iF_5 \qquad V_\pm = F_6 \pm iF_7$$

Apply this form to the state with highest eigenvalues $\mu = N/2$ and $\nu = N/\sqrt{12}$ to find the eigenvalue of $\sum_a (F_a)^2$ on all the states of the representation. [Hint: Review the derivation in $SU(2)$ of $\mathbf{J}^2 = j(j+1)$.]

(e) In part (a) above you identified the combinations of $A_{ij} = a_i^\dagger a_j$ for the three-dimensional oscillator that are the F_a on some representation. Now you know the possible dimensions of an irreducible representation, and the eigenvalue of $\sum_a (F_a)^2$ on those states. Write out $\sum_a (F_a)^2$ in terms of the A_{ij} and the $A_{ijkl} = a_i^\dagger a_j^\dagger a_k a_l$, as you did in Problem 4.10 for the two-dimensional oscillator.

(f) Write $H - \frac{3}{2}\hbar\omega$ and $(H - \frac{3}{2}\hbar\omega)^2$ as a combination of these A_{ijkl} and A_{ij}, and then express the constant $\sum_a (F_a)^2$ as a polynomial in $H - \frac{3}{2}\hbar\omega$. In this way find the value of the Hamiltonian on the energy level with the degeneracy you have calculated by considering one of these symmetric irreducible representations. Verify that you have correctly explained the degeneracy *and* the level structure of the three-dimensional isotropic oscillator.

5.12. Antiparticle Representation of Spin Half

The standard spin-half representation of $SU(2)$ is

$$D^{(\frac{1}{2})}(\hat{\mathbf{n}}, \theta) = \exp\left(-i\frac{\theta}{2}\hat{\mathbf{n}} \cdot \boldsymbol{\sigma}\right)$$

The matrices σ_i are the Pauli matrices, and the generators are

$$D^{(\frac{1}{2})}(\mathbf{J}_i) = \frac{1}{2}\sigma_i$$

There is another two-dimensional representation, useful in describing antiparticles.[14] The matrices are

$$\bar{D}^{(\frac{1}{2})}(\hat{\mathbf{n}}, \theta) = D^{(\frac{1}{2})}(\hat{\mathbf{n}}, \theta)^*$$

---

[14] See, for example, the discussion following equation (5.130) in Section 5.4.

170 Chapter 5 Symmetry Transformations on Operators

(a) Show that the generators are

$$\bar{D}^{(\frac{1}{2})}(\bar{J}_i) = -\frac{1}{2}\sigma_i^*$$

and that these matrices indeed satisfy the angular momentum commutation relations.

(b) You know that there can only be one representation of rotations, or $SU(2)$, in any dimension, so the representation $\bar{D}^{(\frac{1}{2})}$ must be related to the standard one by a change of basis. What is the transformation from one basis to the other, that is, find a unitary transformation U such that

$$U\bar{D}^{(\frac{1}{2})}(\hat{\mathbf{n}},\theta)U^{-1} = D^{(\frac{1}{2})}(\hat{\mathbf{n}},\theta)$$

or equivalently

$$U\sigma_i^* U^{-1} = -\sigma_i$$

5.13. Quantum Runge-Lenz Algebra

Prove equation (3.142b) for the commutators among components of the Runge-Lenz vector. Now that you know something about transformation properties of operators, the proof is relatively short:

(a) From symmetry arguments alone, show that the commutator of two components of the Runge-Lenz vector $\mathbf{A}$ must be proportional to $\mathbf{L} = \mathbf{r} \times \mathbf{p}$, that is,

$$[A_i, A_j] = F(\mathbf{r},\mathbf{p})\, i\sum_k \epsilon_{ijk}L_k$$

where $F(\mathbf{r},\mathbf{p})$ is some *scalar* function of $\mathbf{r}$ and $\mathbf{p}$.

(b) Compute

$$[A_i, \mathbf{A}^2]$$

in terms of $F(\mathbf{r},\mathbf{p})$ from this expression for the commutator.

(c) Now compute the same commutator from the expression (3.155) for $\mathbf{A}^2$ derived from its definition and compare with the previous result to obtain the scalar $F(\mathbf{r},\mathbf{p})$.

REFERENCES

The idea of isospin symmetry was originally due to Heisenberg.

[1] W. HEISENBERG, Zeitschrift für Physik **77** (1932), 1.

There is a translation in

[2] D. M. BRINK, *Nuclear forces*, Pergamon, 1965.

Wigner's discussion of time inversion, with references to the original papers, can be found in Chapter 26 of reference [1] in Chapter 4.

CHAPTER 6

Interlude

6.1 EXTERNAL MAGNETIC FIELDS

Before we plunge into the techniques and applications of quantum mechanics, I want to discuss a few loosely related subjects that require the technical background obtained in previous chapters. First, the effect of external magnetic fields on energy eigenstates of an electron. Next there is a section on the density matrix, and another on a new example of quantum interference, oscillations among types of neutrinos. Finally I want to examine some of the questions mentioned in Section 2.2.2 about the probability interpretation of quantum mechanics.

6.1.1 Natural Units

First it is time to revisit the question of units. From here on $\hbar$ and c will be dimensionless and equal to one. You will see that using such a system will save a lot of writing, and it is not hard to put the usual units back in when you need to compute quantities numerically.

Don't worry that any information is getting lost. It isn't. Both $\hbar$ and c are conversion factors. Dimensional analysis with separate units for length, mass, and time is useful when you are sure the equations cannot contain factors of $\hbar$ or c. From now on, that kind of dimensional analysis only indicate where the factors of $\hbar$ and c would go if you put them back in. Students sometimes find these natural units a little hard to get used to at first, but with practice you will appreciate how efficient it is.

With $\hbar$ and c set to unity, energy, momentum, and mass have the same units, while time and distance have the reciprocal units. There is only one independent unit. We could take it to be a unit of energy, length, mass, time, or anything else. It is not necessary to make the choice now—we will make whatever choice is convenient each time we want a number with units.

If you need the security of imagining that our equations are written with some underlying unit in mind, take that unit to be the electron volt(eV), the magnitude of the energy acquired by an electron moving through a potential difference of one volt, about 1.60218×10^{-12} erg. Mass too is measured in electron volts. The electron mass is about $511,000$ eV. Most of the coming applications will be about atoms, electrons, molecules, nuclei, and so forth, and for these systems the eV or multiples like the MeV are the most convenient units. Equations (2.131) and (2.132) provide the conversion factors to conventional units.

6.1.2 Gauge Invariance

With $q = -e$ and $c = 1$, the classical Hamiltonian for an electron in a magnetic field (1.26) is obtained from the **minimal coupling** prescription, which says to replace $\mathbf{p}$ by $\mathbf{p} + e\mathbf{A}(\mathbf{r})$.[1] If the only interaction of an electron is with an external magnetic field, the full Hamiltonian is

$$H = \frac{(\mathbf{p} + e\mathbf{A})^2}{2m} \qquad (6.1)$$

Here, $\mathbf{p}$ is the *canonical* momentum, not $m\mathbf{v}$.

In classical mechanics electromagnetic interactions are invariant under gauge transformations of the form

$$\mathbf{A}(\mathbf{r}) \to \mathbf{A}'(\mathbf{r}) = \mathbf{A}(\mathbf{r}) + \boldsymbol{\nabla}\Lambda(\mathbf{r}) \qquad (6.2)$$

where $\Lambda(\mathbf{r})$ is an arbitrary scalar function of $\mathbf{r}$. Even though the form of the Hamiltonian depends on $\mathbf{A}$, the results of observations are obviously gauge-independent because the equations of motion (1.33) depend only on the invariant combinations $\mathbf{E}$ and $\mathbf{B}$. In classical electrodynamics the vector potential $\mathbf{A}(\mathbf{r})$ is just a mathematical convenience.

But quantum mechanics requires the canonical formalism from the start, so you might well wonder how a Hamiltonian like (10.85) can predict results in quantum mechanics that depend only on the electric and magnetic fields. The way the minimal coupling prescription guarantees gauge invariance is more complicated here than in classical mechanics, but more interesting.

If $\psi(\mathbf{r})$ is any function of $\mathbf{r}$, and $\mathbf{A}'(\mathbf{r}) = \mathbf{A}(\mathbf{r}) + \boldsymbol{\nabla}\Lambda(\mathbf{r})$, then

$$\left[-i\boldsymbol{\nabla} + e\mathbf{A}'(\mathbf{r})\right] e^{-ie\Lambda(\mathbf{r})} \psi(\mathbf{r}) = e^{-ie\Lambda(\mathbf{r})} \left[-i\boldsymbol{\nabla} + e\mathbf{A}(\mathbf{r})\right] \psi(\mathbf{r}) \qquad (6.3)$$

In particular, if

$$\frac{1}{2m}\left[-i\boldsymbol{\nabla} + e\mathbf{A}(\mathbf{r})\right]^2 \psi(\mathbf{r}) = E\psi(\mathbf{r}) \qquad (6.4)$$

then

$$\frac{1}{2m} e^{-ie\Lambda(\mathbf{r})} \left[-i\boldsymbol{\nabla} + e\mathbf{A}(\mathbf{r})\right]^2 \psi(\mathbf{r}) = E e^{-ie\Lambda(\mathbf{r})} \psi(\mathbf{r}) \qquad (6.5)$$

From equation (6.3), used twice, equation (6.5) is the same as

$$\frac{1}{2m}\left[-i\boldsymbol{\nabla} + e\mathbf{A}'\mathbf{r})\right]^2 e^{-ie\Lambda(\mathbf{r})} \psi(\mathbf{r}) = E e^{-ie\Lambda(\mathbf{r})} \psi(\mathbf{r}) \qquad (6.6)$$

Therefore, if $\psi(\mathbf{r})$ satisfies the time-independent Schrödinger equation (6.4) for $\mathbf{A}(\mathbf{r})$ in some gauge, then in some other gauge the transformed wave function

$$\psi'(\mathbf{r}) = e^{-ie\Lambda(\mathbf{r})} \psi(\mathbf{r}) \qquad (6.7)$$

satisfies the Schrödinger equation *with the same energy E.*

A gauge transformation changes the wave function as well as the vector potential, but the energy levels are the same in both gauges. The state vectors differ

---
[1] By convention $e > 0$, so the charge of an electron is $-e$.

by an **r**-dependent phase, so the probability that the particle be in d^3r is the same in both gauges: $|\psi(\mathbf{r})|^2 = |\psi'(\mathbf{r})|^2$. The probability that **p** be in d^3p is not gauge invariant; but the operator **p** depends on the gauge, and is not an observable in that sense. It is true, but not totally obvious, that there are no observable gauge-dependent quantities.

While the physics must be gauge invariant, it is still necessary to choose a gauge in order to define $\mathbf{A}(\mathbf{r})$ uniquely. In quantum mechanics you get into a lot of trouble in a gauge that depends on the time explicitly, since then you have to face the complications of time-dependent Hamiltonians even for static magnetic fields. The only sensible gauge for our purposes is the transverse gauge,

$$\nabla \cdot \mathbf{A}(\mathbf{r}) = 0 \tag{6.8}$$

and I usually will fix **A** in the transverse gauge from now on.[2]

6.1.3 Constant Magnetic Fields and Landau Levels

Start with the case $\mathbf{E} = 0$ and $\mathbf{B}$ = constant. A possible vector potential satisfying the transverse gauge condition (6.8) is the **symmetric gauge**

$$\mathbf{A}(\mathbf{r}) = \frac{1}{2}\mathbf{B} \times \mathbf{r} \tag{6.9}$$

Take the constant magnetic field in the z direction, $\mathbf{B} = B\hat{n}_z$, so that

$$\mathbf{A} = -\frac{1}{2}yB\hat{n}_x + \frac{1}{2}xB\hat{n}_y \tag{6.10}$$

The Hamiltonian (6.1) is (in this discussion ignore the spin of the electron)

$$H = H_{xy} + H_z \tag{6.11}$$

with

$$H_z = \frac{p_z^2}{2m} \quad \text{and} \quad H_{xy} = \frac{1}{2m}\left(p_x - \frac{1}{2}eBy\right)^2 + \frac{1}{2m}\left(p_y + \frac{1}{2}eBx\right)^2 \tag{6.12}$$

H_z is the kinetic energy of the motion in the z-direction, and $[H_{xy}, H_z] = 0$. $E_z = p_z^2/2m$ can have any positive value. What is the spectrum for fixed E_z? Write H_{xy} in the form

$$H_{xy} = H_1 + H_2 \tag{6.13}$$

where

$$H_1 = \frac{1}{2m}\left(p_x^2 + p_y^2\right) + \frac{m\omega^2}{2}\left(x^2 + y^2\right) \quad \text{and} \quad H_2 = \frac{eB}{2m}L_z \tag{6.14}$$

with

$$\omega = \frac{|eB|}{2m} \quad \text{and} \quad L_z = xp_y - yp_x \tag{6.15}$$

---

[2] The choice of gauge is less restricted with the path integral methods introduced in Chapter 10.

Here ω is the Larmor frequency. Since H_1 is invariant under rotations about the z-axis, it commutes with L_z, and so we have managed to write H as a sum of three commuting terms. One is the kinetic energy in the z-direction. The second looks like the Hamiltonian for a two-dimensional harmonic oscillator, and the third is an energy proportional to what looks like the component of the angular momentum parallel to the magnetic field.[3] Choose the eigenstates to be eigenstates of p_z (with any real number eigenvalue), of L_z (with any integer eigenvalue k), and of H_1.

One could construct eigenstates of the two-dimensional oscillator in the usual way:

$$|n_x, n_y\rangle = \frac{1}{\sqrt{n_x! n_y!}} \left(a_x^\dagger\right)^{n_x} \left(a_y^\dagger\right)^{n_y} |0,0\rangle \tag{6.16}$$

Then

$$H_1 |n_x, n_y\rangle = E_N |n_x, n_y\rangle \tag{6.17}$$

where

$$E_N = \omega \left(N + 1\right) \tag{6.18}$$

and $N = n_x + n_y$. The degeneracy of each level is $N + 1$.

There is a more convenient way to do this. Define the combinations

$$a_\pm^\dagger = \frac{1}{\sqrt{2}} \left(a_x^\dagger \pm i a_y^\dagger\right) \tag{6.19}$$

and their conjugates

$$a_\pm = \frac{1}{\sqrt{2}} \left(a_x \mp i a_y\right) \tag{6.20}$$

so that

$$\left[a_-, a_+^\dagger\right] = 0 \tag{6.21}$$

and

$$H_1 = \omega \left(a_x^\dagger a_x + a_y^\dagger a_y + 1\right) = \omega \left(a_+^\dagger a_+ + a_-^\dagger a_- + 1\right) \tag{6.22}$$

H_1 is again a sum of two independent oscillators, so its eigenstates are proportional to

$$\left(a_+^\dagger\right)^{n_+} \left(a_-^\dagger\right)^{n_-} |0,0\rangle \sim |n_+, n_-\rangle \tag{6.23}$$

Now N in equation (6.18) is $N = n_+ + n_-$.

Since the operators $a_x^\dagger$ and $a_y^\dagger$ are linear combinations of $\mathbf{r}$ and $\mathbf{p}$, they transform as components of vectors under rotations:

$$\left[L_i, a_j^\dagger\right] = i \sum_k \epsilon_{ijk} a_k^\dagger \tag{6.24}$$

In particular

$$\left[L_z, a_x^\dagger\right] = i a_y^\dagger \quad \text{and} \quad \left[L_z, a_y^\dagger\right] = -i a_x^\dagger \tag{6.25}$$

---

[3] But it is not! L_z depends on the gauge. It is conserved in the transverse gauge, but not in general.

so
$$\left[L_z, a_\pm^\dagger\right] = \pm a_\pm^\dagger \tag{6.26}$$

In fact
$$L_z = a_+^\dagger a_+ - a_-^\dagger a_- \tag{6.27}$$

$a_+^\dagger$ is a raising operator for L_z, and $a_-^\dagger$ is a lowering operator for L_z. On the other hand, $a_+^\dagger$ and $a_-^\dagger$ both raise the eigenvalue of H_1 by ω.

All the states can be constructed starting from the ground state, $n_+ = n_- = 0$. Any eigenstate is proportional to
$$\left(a_+^\dagger\right)^{n_+} \left(a_-^\dagger\right)^{n_-} |0, 0\rangle \tag{6.28}$$

These states are linear combinations of the states (6.16) with $n_x + n_y = n_+ + n_-$. The point of using the $|n_+, n_-\rangle$ basis is that these states are also eigenstates of L_z, and therefore of the Hamiltonian H_{xy} defined in (6.13) as well.

Now change the notation slightly and label the states by their eigenvalues of H_1 and L_z:
$$H_1|N, k\rangle = (N+1)\omega|N, k\rangle \quad \text{and} \quad L_z|N, k\rangle = \frac{eB}{|eB|}\omega k|N, k\rangle \tag{6.29}$$

then
$$a_+^\dagger|N, k\rangle \sim |N+1, k+1\rangle \quad \text{and} \quad a_-^\dagger|N, k\rangle \sim |N+1, k-1\rangle \tag{6.30}$$

Any eigenstate of H_{xy} is proportional to
$$\left(a_+^\dagger\right)^{n_+} \left(a_-^\dagger\right)^{n_-} |0, 0\rangle \sim |n_+ + n_-, n_+ - n_-\rangle \tag{6.31}$$

and its energy is
$$E_{n,k} = \omega\left[n_+ + n_- + 1 + \frac{eB}{|eB|}(n_+ - n_-)\right] \tag{6.32}$$

If $eB > 0$, then[4]
$$H_{xy} = \omega\left(2a_+^\dagger a_+ + 1\right) \tag{6.33}$$

and the energy of motion in the xy plane is
$$\boxed{E_{n,k} = \omega(2n_+ + 1) = \omega_c\left(n_+ + \frac{1}{2}\right)} \tag{6.34}$$

independent of n_-. Here $\omega_c = 2\omega$ is the cyclotron frequency, the frequency for the classical revolution of a charged particle in a constant magnetic field. Each energy level is characterized by a nonnegative integer n_+, and can have any nonnegative value for n_-. It is infinitely degenerate. These equally spaced levels of a charged particle in a magnetic field, called **Landau levels**, have many applications in the physics of solids.[5]

---
[4] If $eB < 0$, the roles of a_+ and a_- are interchanged.
[5] See Chapter XVI in Landau and Lifshitz [1].

6.1.4 Magnetic Moment

Orbital Magnetic Moment

For small magnetic fields, it is useful to write the Hamiltonian out as

$$H = H_o + \frac{e}{2m}(\mathbf{p} \cdot \mathbf{A} + \mathbf{A} \cdot \mathbf{p}) + \frac{e^2}{2m}\mathbf{A}^2 \qquad (6.35)$$

As long as the field is small, the $\mathbf{A}^2$ term can be neglected. To first order in the dimensionless parameter e, the interaction term in the Hamiltonian is

$$H_{\text{mag}} = \frac{e}{2m}(\mathbf{p} \cdot \mathbf{A} + \mathbf{A} \cdot \mathbf{p}) \qquad (6.36)$$

In the transverse gauge (6.8) $\mathbf{p}$ and $\mathbf{A}$ commute:

$$\langle \mathbf{r} | \mathbf{p} \cdot \mathbf{A}(\mathbf{r}) | \psi \rangle = -i \nabla \cdot [\mathbf{A}(\mathbf{r}) \psi(\mathbf{r})] = -i \mathbf{A}(\mathbf{r}) \cdot \nabla \psi(\mathbf{r}) = \langle \mathbf{r} | \mathbf{A}(\mathbf{r}) \cdot \mathbf{p} | \psi \rangle \qquad (6.37)$$

so

$$H_{\text{mag}} = \frac{e}{m}\mathbf{A}(\mathbf{r}) \cdot \mathbf{p} = \frac{e}{2m}(\mathbf{r} \times \mathbf{p}) \cdot \mathbf{B} = \mu_B \mathbf{L} \cdot \mathbf{B} \qquad (6.38)$$

The positive number $\mu_B = e/2m$ is the **Bohr magneton**. The orbital magnetic moment of the electron is $\boldsymbol{\mu} = -\mu_B \mathbf{L}$.

The units of $\mathbf{B}$ are mass squared or energy squared, so the magnetic moment has units of inverse mass or inverse energy. Physicists often measure macroscopic magnetic fields in gauss, a cgs unit. The units of μ_B are

$$\boxed{\mu_B = 5.79 \times 10^{-9} \frac{\text{eV}}{\text{gauss}} = 9.29 \times 10^{-21} \frac{\text{erg}}{\text{gauss}}} \qquad (6.39)$$

Spin Magnetic Moment

This is only part of the story. The electron also has a spin $\mathbf{s}$, and the total magnetic moment has a term proportional to the spin. It is tempting just to replace $\mathbf{L}$ in the expression for $\boldsymbol{\mu}$ by $\mathbf{J} = \mathbf{L} + \mathbf{s}$, and write $\boldsymbol{\mu} = -\mu_B \mathbf{J}$; but this is experimentally wrong.

There is no reason for the contribution of the spin to the magnetic moment to be $-\mu_B \mathbf{s}$. That would be true if the electron were a rigid uniformly charged sphere, and the magnetic moment were due to the motion of the parts of the electron. But the electron is no such thing; it is a "point" particle, and there is no sense asking what is inside it. The term in $\boldsymbol{\mu}$ proportional to $\mathbf{s}$ cannot be calculated using a nonrelativistic theory. We write

$$\boldsymbol{\mu} = \boldsymbol{\mu}_{\text{orb}} + \boldsymbol{\mu}_{\text{spin}} = -\mu_B \mathbf{L} - g\mu_B \mathbf{s} \qquad (6.40)$$

The number g is the **gyromagnetic ratio**.[6] This point was a source of some confusion in the Bohr-Sommerfeld "Old Quantum Theory," and the contribution

---
[6] Some books use the term "gyromagnetic ratio" to mean the ratio of the magnetic moment to the spin (in units of $\hbar$). In that terminology, the electron's gyromagnetic ratio has the dimensions of magnetic moment and is what I call $ge\hbar/2m$.

to the magnetic moment from the difference between g and 1 is still occasionally called the **anomalous magnetic moment**.

Experimentally, for the electron g is almost exactly 2. This is a prediction of the Dirac equation, and will be explained in Chapter 12. In fact, $g = 2$ is the first term of an expansion in α that can be obtained in relativistic quantum electrodynamics. Schwinger [3] computed the next term to be α/π; the computation has been carried out at least to sixth order in e, with the result

$$\frac{g}{2} = 1 + \frac{\alpha}{2\pi} - 0.328478445 \left(\frac{\alpha}{\pi}\right)^2 + 1.18311 \left(\frac{\alpha}{\pi}\right)^3 + \cdots \tag{6.41}$$

At the nonrelativistic level one must take g from experiment; its observed value agrees with (6.41) to uncanny accuracy.[7] To say that the electron is a "point" particle does not mean that the probability of finding an electron vanishes except at one point; quantum mechanics tells us otherwise. It means that the $1/r$ factors in the electric and magnetic interactions are correct down to arbitrarily small r. This is not true for a proton, and its gyromagnetic ratio is nowhere near 2. For nucleons (neutrons and protons) of mass M one defines the **nuclear magneton** by

$$\mu_N = \frac{e}{2M} \tag{6.42}$$

The nucleon magnetic moment is $\boldsymbol{\mu} = \mu_N \mathbf{L} + g\mu_N \mathbf{s}$ (the proton charge is $+e > 0$). For the proton, $g = 2 \times 2.8$, while for the neutron, $g = -2 \times 1.91$.

For the electron, as long as you are making computations to lowest order in α, you might as well take $g = 2$. Then equation (6.40) becomes

$$\boldsymbol{\mu} = -\mu_B(\mathbf{L} + 2\mathbf{s}) = -\mu_B(\mathbf{L} + \boldsymbol{\sigma}) \tag{6.43}$$

and the interaction Hamiltonian is

$$H_{\text{mag}} = -\boldsymbol{\mu} \cdot \mathbf{B} = \mu_B(\mathbf{L} \cdot \mathbf{B} + \boldsymbol{\sigma} \cdot \mathbf{B}) \tag{6.44}$$

6.1.5 The Hydrogen Atom in a Magnetic Field

Let us apply these rules to the problem of finding the hydrogen atom energy levels, but now with the complication that the atom is in a constant external field in the z-direction.

When $\mathbf{B}$ is zero, the Hamiltonian of an electron in a hydrogen atom was $H_o = \mathbf{p}^2/2m - \alpha/r$ with energies $E^o_{nlm_lm_s} = -\alpha^2 m/2n^2$ as in Section 3.5. The states can be labeled $|n, l, m_l, m_s\rangle$, where m_l and m_s are eigenvalues of L_z and s_z.

Now turn on the magnetic field. H_{mag} is diagonal provided we take $\mathbf{B}$ along the z-axis:

$$\begin{aligned} H|nlm_lm_s\rangle &= \left[\frac{\mathbf{p}^2}{2m} - \frac{\alpha}{r} + H_{\text{mag}}\right]|nlm_lm_s\rangle \\ &= \left[-\frac{\alpha^2 m}{2n^2} + \mu_B B(m_l + 2m_s)\right]|nlm_lm_s\rangle \end{aligned} \tag{6.45}$$

---
[7] See Mohr and Taylor [4] and Kinoshita and Lindquist [5], respectively, for recent experimental and theoretical values.

The eigenvalues are the energy levels of a hydrogen atom in this external field. The energies of the states for the $n = 1$ and $n = 2$ levels are listed in Table 6.1.

$n = 1$	$\lvert 1, 0, 0, \pm\tfrac{1}{2}\rangle$	$E = E_1^o \pm \mu_B B$
$n = 2$	$\lvert 2, 0, 0, \pm\tfrac{1}{2}\rangle$	$E = E_2^o \pm \mu_B B$
$n = 2$	$\lvert 2, 1, 1, \tfrac{1}{2}\rangle$	$E = E_2^o + 2\mu_B B$
$n = 2$	$\lvert 2, 1, 1, -\tfrac{1}{2}\rangle$	$E = E_2^o$
$n = 2$	$\lvert 2, 1, 0, \pm\tfrac{1}{2}\rangle$	$E = E_2^o \pm \mu_B B$
$n = 2$	$\lvert 2, 1, -1, \tfrac{1}{2}\rangle$	$E = E_2^o$
$n = 2$	$\lvert 2, 1, -1, -\tfrac{1}{2}\rangle$	$E = E_2^o - 2\mu_B B$

TABLE 6.1: $n = 2$ Energies of the hydrogen atom in a weak magnetic field

The states originally in the angular momentum ladders are no longer degenerate, since H is not rotationally invariant. But there is still some degeneracy remaining. This splitting of spectral lines by a weak external magnetic field is called the **Zeeman effect** [6].

6.2 THE DENSITY MATRIX

6.2.1 Definition

In a real experiment you do not usually prepare a system over and over again in exactly the same state. You study a large number of independent versions of the same experiment. That is what happens when you heat up a gas and look at intensities of the various spectral lines, or when you scatter a beam of particles off a target and study the distribution of the scattered particles. The experiment is done on a number of particles or atoms with some distribution of quantum states. You may know this distribution exactly—you may actually have succeeded in preparing them all in the same quantum state—or more likely you know some features of the distribution, perhaps that the angular momentum is distributed isotropically, or that the energy is an equilibrium thermal distribution. The density matrix is a convenient way to describe such experiments.

Imagine N independent versions of the same system, an atom or nucleus perhaps, each in some quantum state $|\psi_i\rangle$. What is the average value of measurements of an observable A over the whole ensemble? It is the average of the quantum expectation values over the ensemble. Let $|\phi_a\rangle$ be any orthogonal basis. Then:

$$\langle A \rangle = \frac{1}{N} \sum_{i=1}^{N} \langle \psi_i | A | \psi_i \rangle = \sum_a \langle \phi_a | \rho A | \phi_a \rangle = \mathrm{Tr}\,(\rho A) \qquad (6.46)$$

where

$$\rho = \frac{1}{N} \sum_i |\psi_i\rangle\langle\psi_i| \qquad (6.47)$$

If $|\phi_a\rangle$ are the orthonormal eigenstates of some Hermitean operator, then the probability to find the system in a particular $|\phi_a\rangle$ is

$$\frac{1}{N}\sum_i |\langle\psi_i|\phi_a\rangle|^2 = \text{Tr}\,\rho P_a \tag{6.48}$$

where $P_a = |\phi_a\rangle\langle\phi_a|$ is the projection operator into the state $|\phi_a\rangle$.

The operator ρ is the **density matrix**. It contains all the information about results of measurements on the ensemble, and in particular, the expectation value of any observable is given by equation (6.46). But ρ does not uniquely determine the states of the individual particles. More than one collection of $|\psi_i\rangle$ has the same density matrix.

Some Basic Properties

In an orthonormal basis the matrix elements of the density matrix are

$$\rho_{ba} = \frac{1}{N}\sum_i \langle\psi_i|\phi_a\rangle\langle\psi_i|\phi_b\rangle^* \tag{6.49}$$

Therefore ρ is Hermitean.

The trace of ρ itself is the expectation value of the identity operator, hence

$$\text{Tr}\,\rho = 1 \tag{6.50}$$

If all the individual particles are actually in the same state $|\psi\rangle$, then

$$\rho_{ba} = \frac{1}{N}\sum_i \langle\psi_i|\phi_b\rangle^*\langle\psi_i|\phi_a\rangle = \frac{1}{N}\sum_i \langle\psi|\phi_b\rangle^*\langle\psi|\phi_a\rangle \tag{6.51}$$

and

$$(\rho^2)_{ba} = \frac{1}{N^2}\sum_{i,j,c} \langle\psi|\phi_b\rangle^*\langle\psi|\phi_c\rangle\langle\psi|\phi_c\rangle^*\langle\psi|\phi_a\rangle = \rho_{ba} \tag{6.52}$$

so that $\text{Tr}\,(\rho^2) = \text{Tr}\,\rho = 1$. But in general $\text{Tr}\,(\rho^2) \neq 1$. For example, let $|\phi_a\rangle$ be an eigenstate of ρ itself: $\rho|\phi_a\rangle = \lambda_a|\phi_a\rangle$. The eigenvalue a cannot be negative:

$$\lambda_a = \rho_{aa} = \frac{1}{N}\sum_i |\langle\psi_i|\phi_a\rangle|^2 \geq 0 \tag{6.53}$$

The density matrix is positive semidefinite. It follows that

$$\text{Tr}\,(\rho^2) = \sum_a (\rho_{aa})^2 = \sum_a \lambda_a^2 \leq \left(\sum_a \lambda_a\right)^2 = 1 \tag{6.54}$$

and so

$$0 \leq \text{Tr}\,(\rho^2) \leq 1 \tag{6.55}$$

The equality can hold only when the mixture is in fact a pure state. The density matrix evolves with time according to

$$i\frac{\partial\rho}{\partial t} = \frac{1}{N}\sum_i \left[\left(i\frac{\partial}{\partial t}|\psi_i\rangle\right)\langle\psi_i| + |\psi_i\rangle\left(i\frac{\partial}{\partial t}\langle\psi_i|\right)\right] = [H,\rho] \tag{6.56}$$

6.2.2 Example: Thermodynamic Equilibrium

Suppose a system has a large number N of distinguishable, but otherwise identical, particles. That is a fair description of ordinary matter at ordinary temperature—each particle is in a different place and the overlap of the wave functions is not important. Let ϵ_i be the energies of the single particle stationary states. The number of ways there can be n_i particles in the state with energy ϵ_i is given by the multinomial distribution:

$$A = \frac{N!}{\prod_i n_i!} \tag{6.57}$$

The collection of particles is in thermal equilibrium when each way of distributing the particles is equally likely, subject to the constraints that the total number N and the total energy E is fixed. The most likely configuration is the one that maximizes A:

$$\frac{\partial A}{\partial n_i} = 0 \tag{6.58}$$

subject to

$$\sum_i n_i = N \quad \text{and} \quad \sum_i n_i \epsilon_i = E \tag{6.59}$$

The answer to this well-known problem is[8]

$$n_i \sim e^{-\epsilon_i/kT} \tag{6.60}$$

The density matrix for a mixture like this is

$$\rho = \frac{1}{Z} e^{-H/kT} \tag{6.61}$$

Z is the partition function, and the normalization $\text{Tr } \rho = 1$ requires

$$Z = \text{Tr } e^{-H/kT} \tag{6.62}$$

As a simple example, consider an ensemble of electrons in a constant magnetic field, and suppose they are at rest so the only relevant quantum number is the spin. The Hamiltonian for each particle is

$$H = g\mu_B \mathbf{s} \cdot \mathbf{B}_o \tag{6.63}$$

One basis consists of two eigenstates, with the spin parallel or antiparallel to the magnetic field, with energy eigenvalues

$$\epsilon_i = \pm \frac{g}{2} \mu_B B_o \tag{6.64}$$

and in that basis ρ is diagonal, with diagonal elements

$$\rho_{ii} = \frac{1}{Z} e^{\mp g\mu_B B_o/2kT} \tag{6.65}$$

---

[8]The result is much more general than this schematic derivation—see any good book on statistical mechanics.

The partition function here is

$$Z = 2\cosh\frac{g\mu_B B_o}{2kT} \tag{6.66}$$

The expectation value of the spin along the direction of the magnetic field is

$$\left\langle \mathbf{s}\cdot\frac{\mathbf{B}_o}{B_o}\right\rangle = \text{Tr}\left(\rho\mathbf{s}\cdot\frac{\mathbf{B}_o}{B_o}\right) = -\frac{1}{2}\tanh\frac{g\mu_B B_o}{2kT} \tag{6.67}$$

and you can check that $\text{Tr}\,\rho^2 < 1$.

6.2.3 Example: Spin-Half Systems

The formalism, applied to the spin index only, is especially useful for spin-half particles like electrons. Here the density matrix is a 2 × 2 traceless Hermitean matrix. The most general form is

$$\rho = \frac{1}{2}(I + \mathbf{P}\cdot\boldsymbol{\sigma}) \tag{6.68}$$

ρ is completely specified by three real numbers. The expectation value of a component of the spin is

$$\langle s_i\rangle = \text{Tr}\,(s_i\rho) = \frac{1}{2}P_i \tag{6.69}$$

or

$$\mathbf{P} = 2\langle\mathbf{s}\rangle = \langle\boldsymbol{\sigma}\rangle \tag{6.70}$$

For this reason the vector $\mathbf{P}$ is called the **polarization** of the mixture. $|\mathbf{P}| = 1$ for a pure state, and $|\mathbf{P}| < 1$ otherwise. If $\mathbf{P} = 0$ the mixture is unpolarized.

In this two-dimensional space the Hamiltonian can be written in terms of four real numbers as

$$H = \frac{1}{2}(H_o I + \mathbf{H}\cdot\boldsymbol{\sigma}) \tag{6.71}$$

The time evolution of the polarization is given by

$$\frac{dP_i}{dt} = \frac{d\langle\sigma_i\rangle}{dt} = \text{Tr}\left(\sigma_i\frac{d\rho}{dt}\right) = -i\text{Tr}\,(\sigma_i[H,\rho]) = \sum_{jk}\epsilon_{ijk}H_j P_k \tag{6.72}$$

or

$$\boxed{\frac{d\mathbf{P}}{dt} = \mathbf{H}\times\mathbf{P}} \tag{6.73}$$

Constant Hamiltonian

Equation (6.73) holds whether $\mathbf{H}$ is a constant or not. The special case of a constant Hamiltonian—an electron in a constant magnetic field for example—has many applications and is worth working out in detail:

Let $\mathbf{P}_o = \mathbf{P}(0)$, the initial polarization, and let θ be the angle between $\mathbf{P}_o$ and $\mathbf{H}$. Set up an orthogonal (but not orthonormal) coordinate system and expand the polarization at any time as

$$\mathbf{P}(t) = A\mathbf{H} + B\mathbf{P}_o\times\mathbf{H} + C\mathbf{H}\times(\mathbf{P}_o\times\mathbf{H}) \tag{6.74}$$

182 Chapter 6 Interlude

(If **H** is along the z-axis, and if **P**(0) is along the x-axis, these are vectors along the usual coordinate axes.) Then from the time dependence (6.73) the coefficients satisfy (with $\omega_o = |\mathbf{H}|$)

$$\dot{A} = 0 \qquad \dot{B} = -C\omega_o^2 \qquad \dot{C} = B \qquad (6.75)$$

Therefore, together with the initial conditions,

$$A = \frac{1}{\omega_o^2} \mathbf{P}_o \cdot \mathbf{H} \qquad B(t) = -\frac{1}{\omega_o} \sin \omega_o t \qquad C(t) = \frac{1}{\omega_o^2} \cos \omega_o t \qquad (6.76)$$

$\mathbf{P}(t)$ can be written in terms of a unit vector $\hat{\mathbf{H}} = \mathbf{H}/\omega_o$ in the direction of **H**:

$$\mathbf{P}(t) = \hat{\mathbf{H}} \, \mathbf{P}_o \cdot \hat{\mathbf{H}} - \mathbf{P}_o \times \hat{\mathbf{H}} \sin \omega_o t + \hat{\mathbf{H}} \times \left(\mathbf{P}_o \times \hat{\mathbf{H}} \right) \cos \omega_o t \qquad (6.77)$$

Since

$$\mathbf{P}(t) \cdot \hat{\mathbf{H}} = \mathbf{P}_o \cdot \hat{\mathbf{H}} = |\mathbf{P}_o| \cos \theta \qquad (6.78)$$

is independent of the time, the polarization vector precesses about the constant magnetic field with angular frequency ω_o, keeping the angle θ between $\mathbf{P}(t)$ and **H** constant also. Of course this is obvious from equation (6.73) with constant **H**. The expansion in (6.77) will be useful below.

Suppose that at $t = 0$ the particle definitely has its spin along some direction. The system is in a pure state and $\mathbf{P}_o$ is a unit vector. The probability at a later time for its spin along $\mathbf{P}_o$ to be $\pm 1/2$ is Tr $(P_\pm \rho)$, where according to equation (6.48), $P_\pm$ is the projection matrix into one of the two eigenstates of $\boldsymbol{\sigma} \cdot \mathbf{P}_o$:

$$P_\pm = \frac{1}{2}(1 \pm \boldsymbol{\sigma} \cdot \mathbf{P}_o) \qquad (6.79)$$

At time t the probability for the spin to have flipped to the other state is

$$\text{Tr}\,(P_- \rho) = \frac{1}{4} \text{Tr}\left[(1 - \boldsymbol{\sigma} \cdot \mathbf{P}_o)(1 + \boldsymbol{\sigma} \cdot \mathbf{P}(t))\right]$$
$$= \frac{1}{2}(1 - \mathbf{P}_o \cdot \mathbf{P}) = \sin^2 \theta \sin^2 \frac{\omega_o t}{2} \qquad (6.80)$$

When $\mathbf{P}_o$ is parallel or antiparallel to **H**, then $\sin^2 \theta = 0$; the electron is in an eigenstate of the Hamiltonian and stays there. For any other angle, the probability for transition to the other state oscillates, reaching a maximum when $t = \pi/\omega_o$. If the polarization is at right angles to **H**, then the initial state is totally depleted at this maximum.

There are many applications of the density matrix formalism for spin half. The famous magnetic resonance formula is derived below. Others are in Problems 6.7 and 9.10.

6.2.4 Spin Magnetic Resonance

An electron at rest, or in the ground state of an atom, is placed in a magnetic field $\mathbf{B}_o$ pointing in the z-direction. The energies of the two spin states will be split by $\omega_o = g\mu_B B_o$. The state with $s_z = -1/2$ has the lower energy, and at $t = 0$ the electron is definitely in that lower energy state. Suppose that after $t = 0$, the electron is also subjected to a circularly polarized electromagnetic wave, such that the magnetic field at the location of the electron is

$$\mathbf{B} = \mathbf{B}_1(\hat{\mathbf{n}}_x \cos \omega t + \hat{\mathbf{n}}_y \sin \omega t) \tag{6.81}$$

These are the basic ingredients of magnetic resonance experiments. Without the field $\mathbf{B}_1$ the spin eigenstates would be stationary. The oscillating field induces transitions between them. What it the probability for finding the electron in the upper state at a later time?

The density matrix formalism provides an elegant solution. The time dependence of the polarization is given by equation (6.73). The Hamiltonian matrix has the form (6.71), with $H_o = 0$ and

$$\mathbf{H} = \omega_o \hat{\mathbf{n}}_z + \omega_1 \left(\cos \omega t \, \hat{\mathbf{n}}_x + \sin \omega t \, \hat{\mathbf{n}}_y \right) \tag{6.82}$$

The trick is to make this problem look like the problem with a constant Hamiltonian by transforming to a rotating reference frame. Instead of $P_i(t) = \mathbf{P}(t) \cdot \hat{\mathbf{n}}_i$ and $H_i(t) = \mathbf{H}(t) \cdot \hat{\mathbf{n}}_i$, expand these vectors using components in the rotating coordinate system:

$$P'_i(t) = \mathbf{P}(t) \cdot \hat{\mathbf{n}}'_i(t) \quad \text{and} \quad H'_i(t) = \mathbf{H}(t) \cdot \hat{\mathbf{n}}'_i(t) \tag{6.83}$$

with

$$n'_i(t) = \sum_j \bar{R}(t)_{ij} n_j \tag{6.84}$$

$\bar{R}(t) = \bar{R}(\hat{\mathbf{n}}_z, \omega t)$ is the real 3×3 rotation matrix about the z-axis by ωt. In the rotating frame $\mathbf{H}'$ is has constant components: $\mathbf{H}' = \omega_o \hat{\mathbf{n}}'_z + \omega_1 \hat{\mathbf{n}}'_x$. At $t = 0$, the polarization vector is $\mathbf{P}_o = -\hat{\mathbf{n}}_z$. Since the rotation is around the z-axis, $\mathbf{P}'_o = \mathbf{P}_o \cdot \hat{\mathbf{n}}'_z = -\hat{\mathbf{n}}_z$ also.

Write the components of the general formula (6.73) in the rotating basis:

$$\frac{d\mathbf{P}}{dt} \cdot \hat{\mathbf{n}}'_i(t) = \sum_{jk} \epsilon_{ijk} H'_j P'_k \tag{6.85}$$

But also[9]

$$\frac{d\mathbf{P}}{dt} \cdot \hat{\mathbf{n}}'_i(t) = \frac{d}{dt} \left[\mathbf{P} \cdot \hat{\mathbf{n}}'_i(t) \right] + \boldsymbol{\omega} \times \mathbf{P} \cdot \hat{\mathbf{n}}'_i(t) \tag{6.86}$$

The vector $\boldsymbol{\omega} = \omega \hat{\mathbf{n}}_z$ is the same in the original and in the rotating coordinates. Therefore the components of $\mathbf{P}$ in the rotating frame are

$$\frac{d}{dt} \left[\mathbf{P} \cdot \hat{\mathbf{n}}'_i(t) \right] = \sum_k \epsilon_{ijk} H'_j P'_k - \sum_k \epsilon_{ijk} \omega_j P'_k \tag{6.87}$$

or

$$\frac{d}{dt} P'_i = \left[(\mathbf{H}' - \boldsymbol{\omega}) \times \mathbf{P}' \right]_i \tag{6.88}$$

---

[9] This form follows from the rule (1.41), or by explicit calculation.

184 Chapter 6 Interlude

Since $\mathbf{H}' - \boldsymbol{\omega}$ is a constant vector, $\mathbf{P}'(t)$ precesses about the direction of $\mathbf{H}' - \boldsymbol{\omega}$ with an angular frequency equal to its magnitude. The probability for flipping the spin is given by equation (6.80), with appropriate changes in the meaning of θ and ω_o:

$$\omega_o \to \Omega = |\mathbf{H}' - \boldsymbol{\omega}| = \sqrt{(\omega - \omega_o)^2 + \omega_1^2}$$

$$\cos\theta \to \mathbf{P}_o \cdot \frac{\mathbf{H}' - \boldsymbol{\omega}}{|\mathbf{H}' - \boldsymbol{\omega}|} = \frac{\omega - \omega_o}{\Omega} \qquad (6.89)$$

$$\sin\theta \to \sqrt{1 - \left(\frac{\omega - \omega_o}{\Omega}\right)^2} = \frac{\omega_1}{\Omega}$$

The probability that the spin is flipped after time t is

$$\frac{\omega_1^2}{\Omega^2}\sin^2\left(\frac{\Omega t}{2}\right) \qquad (6.90)$$

These are **Rabi oscillations** and equation (6.90) is Rabi's formula.

> Note: In this calculation the magnetic field is treated classically, but in principle a resonance experiment is the emission or absorption of photons by the electron. (The correct dynamics of photon-electron interactions is complicated—the quantum theory of photons is the subject of Chapter 11.) The energy of one photon in the oscillating field is ω, and at resonance maximum $\omega = \omega_o$ this is the energy difference between the two states of the electron in the large constant magnetic field. But there is substantial absorption of photons in a range of energies on either side of ω_o. This is another example of the energy-time uncertainty principle. Convince yourself that as the period for one oscillation goes up, the spread in energy (the width of the resonance curve) goes down.

The Wave Function:
Even though all the physical information about this magnetic resonance system is contained in the density matrix, it will be useful later to have the solution to the Schrödinger equation

$$i\frac{\partial}{\partial t}\psi(t) = H\psi(t) \qquad (6.91)$$

Here

$$\psi(t) = \begin{pmatrix} a_+(t) \\ a_-(t) \end{pmatrix} \qquad (6.92)$$

is the two-component wave function, with $a_\pm(t)$ the amplitudes for the spin to be up or down, respectively, along the z-axis. The Hamiltonian is given by equations (6.71) and (6.82) above. The most general solution (see Problem 6.7) is

$$\boxed{\psi(t) = c_+ e^{-i\omega\sigma_z t/2} e^{-i\Omega t/2}\begin{pmatrix} \cos\frac{\Theta}{2} \\ \sin\frac{\Theta}{2} \end{pmatrix} + c_- e^{-i\omega\sigma_z t/2} e^{i\Omega t/2}\begin{pmatrix} \sin\frac{\Theta}{2} \\ -\cos\frac{\Theta}{2} \end{pmatrix}} \qquad (6.93)$$

with Ω as in (6.89) and

$$\cos\Theta = \frac{\omega_o - \omega}{\Omega}$$
$$\sin\Theta = \frac{\omega_1}{\Omega} \tag{6.94}$$

6.3 NEUTRINO INTERFERENCE

6.3.1 Neutrinos

This section is a detour into the physics of neutrinos, an elegant example of a simple quantum system.

Neutrinos are neutral, extremely light spin-half fermions, whose interactions with other particles are very weak. Soon after neutrons were discovered in 1932 by James Chadwick, and long before neutrinos were detected, Pauli postulated them to explain the missing energy and angular momentum in neutron decay

$$n \to p + e + \bar{\nu} \tag{6.95}$$

In the 1940s the μ^- particle was discovered, produced by cosmic rays in the earth's atmosphere. High energy cosmic ray protons shake off showers of π^- mesons in the atmosphere. Eventually the π^- decays into a μ^- and a neutrino.[10]

The μ^- and the e^- are alike in all respects but their masses. Each has an **antiparticle**, the e^+ (positron) and the μ^+, a particle with the same mass but the opposite charge. Together with the neutrinos, and their antiparticles the antineutrinos, these are the **leptons**. Because the electron is by definition a "particle," the neutrino that is emitted with it in β-decay is an antineutrino. The number of leptons, minus the number of antileptons, seems to be conserved. Leptons do not feel the strong, short-range nuclear forces. Charged leptons have electromagnetic interactions with other particles, but neutrinos do not.

The neutrinos produced along with electrons turned out to be *different* particles from those produced by a μ^-, and they are now called ν_e and ν_μ respectively. Neutrinos that are produced in beta decay can change a neutron back into a proton:

$$\nu_e + n \to p + e^- \tag{6.96}$$

but a ν_μ will not do that. Rather, if the energy is high enough, the process is

$$\nu_\mu + n \to p + \mu^- \tag{6.97}$$

That is how it was verified that ν_μ and ν_e are indeed different particles.

In 1977 a third charged lepton was discovered. This lepton is called τ^-. It is almost twice as heavy as a proton, and has its own neutrino, a ν_τ.

The mass Fermi calculated for the ν_e was consistent with zero, and no experiment since has ever been able to detect directly a nonzero mass for any of the three neutrinos. It was long thought that they might be exactly massless, and like the photon move always at the speed of light.

---

[10] The masses and lifetimes of these particles are listed in Table 5.1.

The effective interaction between neutrinos and other particles is proportional to Fermi's constant G, about $1.027 \times 10^{-5} m_p^{-2}$ (m_p is the mass of the proton.). G is measured in ordinary β-decays like neutron decay. G is so small that a single neutrino will pass through the entire earth with only one chance in 10^{-15} that it interacts at all. Nevertheless with a large enough detector, and with enough patience, neutrinos have been detected.

The rules for neutrino interaction are now as well established as those for electromagnetic interactions, but any detailed discussion would take us too far afield. One peculiar property is that only the left-handed states—those with negative helicity, with the spin antiparallel to the momentum—interact at all. Therefore failure to respect parity conservation is a feature of all weak interaction processes.

6.3.2 Neutrino Mixing

For a long time there was no compelling reason to doubt that neutrinos are massless. But there was no compelling reason to believe it either, or for that matter to believe that they all have the same mass. More startlingly, there is no reason the particles produced in decays, the ones called ν_μ, ν_e and ν_τ, should be stationary states at all! Some time ago the suggestion was made that these neutrinos might be linear combinations of states with definite energy. Here's how it works:

I want to discuss free neutrinos, like those produced in some decay process, like β-decay or μ decay, with some definite momentum and spin, so we can suppress momentum and spin in the discussion. They will not change since there are no interactions. These three types of neutrinos make up a three-dimensional vector space, with ν_e, ν_μ, and ν_τ as an orthonormal basis.

To keep the problem simple, I will imagine there are just two types of neutrino. This is enough to exhibit the essential features of what is really a three-neutrino problem, and in fact is a reasonable first approximation to the physics of the processes described below. Let $|\nu_1\rangle$ and $|\nu_2\rangle$ be the neutrino states with definite energy, normalized and orthogonal. (They don't have to be exactly orthogonal, but that is a separate complication.) Then

$$\langle \nu_i | \nu_j \rangle = \delta_{ij} \tag{6.98}$$

Then all the neutrino states are vectors in a two-dimensional vector space. Suppose the τ and μ neutrino states, when they are produced at $t = 0$, are some linear combinations of these two, and further that they are also normalized and orthogonal. Then

$$|\nu_\mu\rangle = \cos\theta |\nu_1\rangle + \sin\theta |\nu_2\rangle \tag{6.99a}$$
$$|\nu_\tau\rangle = -\sin\theta |\nu_1\rangle + \cos\theta |\nu_2\rangle \tag{6.99b}$$

The angle θ is called a **mixing angle**.

Suppose a ν_μ is produced at $t = 0$. That is what happens when a $\mu^\pm$ decays in the atmosphere. What is the probability that at a later time t it will be detected as a ν_τ?

The dynamics of this system is formally the same as that of any two-state system, such as the electron spin in a magnetic field. In the basis of mass eigenstates the Hamiltonian is diagonal by definition:

$$H_{diag} = \begin{pmatrix} E_1 & 0 \\ 0 & E_2 \end{pmatrix} \tag{6.100}$$

In this basis (6.100) the top row is ν_1 and the bottom row is ν_2. Write H in Pauli matrix notation (even though the states have nothing to do with spin):

$$H_{diag} = E_o - \omega\sigma_3 \tag{6.101}$$

where

$$\omega = \frac{1}{2}\text{Tr}\,(\sigma_3 H) = \frac{E_2 - E_1}{2} \tag{6.102}$$

E_o is an overall constant energy that has no observable consequences, so redefine the Hamiltonian to be a traceless matrix $H = \omega\sigma_3$.

In the basis $|\nu_\mu\rangle$ and $|\mu_\tau\rangle$ of states with definite interaction properties the Hamiltonian is

$$H = U H_{diag} U^\dagger \tag{6.103}$$

for a unitary matrix U whose elements can be read off the definition (6.99) of the mixing angle. With the E_o term subtracted off,

$$H = -\omega \begin{pmatrix} \cos 2\theta & -\sin 2\theta \\ -\sin 2\theta & -\cos 2\theta \end{pmatrix} = -\omega\left[\cos 2\theta\,\sigma_3 - \sin 2\theta\,\sigma_1\right] \tag{6.104}$$

In this basis (6.104) the top row is ν_μ and the bottom row is ν_τ.

At time t the system is described by a two-component vector

$$\psi(t) = \begin{pmatrix} \psi_{\nu_\mu}(t) \\ \psi_{\nu_\tau}(t) \end{pmatrix} = e^{-iHt}\psi(0) = e^{-i\omega\hat{\mathbf{n}}\cdot\boldsymbol{\sigma} t}\psi(0) = \left[\cos\omega t - i\sin\omega t\,\hat{\mathbf{n}}\cdot\boldsymbol{\sigma}\right]\psi(0) \tag{6.105}$$

where $n_x = \sin 2\theta$, $n_y = 0$, and $n_z = \cos 2\theta$. In particular, if at $t = 0$ the neutrino is in a pure ν_μ state, the probability that it will be detected at a later time as a ν_τ is

$$P_\tau(t) = \left|\left(e^{-iHt}\right)_{21}\right|^2 = \left|\sin 2\theta \sin \omega t\right|^2 \tag{6.106}$$

The oscillation frequency ω is half the energy difference between the stationary levels.

6.3.3 Neutrino Oscillations and the Mass Splitting

What is the period of the oscillation? Call the masses of the neutrinos with definite energy m_1 and m_2. When the neutrino is produced in the atmosphere by μ particle decay, it has some momentum p that can be computed from momentum conservation. The neutrinos are very relativistic—no experiment has ever detected a difference between their speeds and c. It is legitimate to expand the energy difference in the small numbers m_i/p:

$$E_1 - E_2 = \sqrt{p^2 + m_1^2} - \sqrt{p^2 + m_2^2} = \frac{\Delta m^2}{2p} + \cdots \tag{6.107}$$

where $\Delta m^2 = m_2^2 - m_1^2$. The probability to find a ν_τ at time t is

$$P_\tau(t) = \sin^2 2\theta \sin^2 \frac{\Delta m^2 t}{4p} \tag{6.108}$$

One does not really measure the time that the neutrino is detected, but rather the distance L from the production point. Since the neutrinos move at (almost) the speed of light,

$$P_\tau(t) \approx \sin^2 2\theta \sin^2 \left(\frac{\Delta m^2 L}{4E}\right) \tag{6.109}$$

It has become the custom to measure Δm^2 in eV^2/c^4, L in kilometers and the energy in GeV (1 GeV = 10^9eV). In these funny units (exercise!)

$$\boxed{P(t) = \sin^2 2\theta \sin^2 \left(1.27 \frac{\Delta m^2 L}{E}\right)} \tag{6.110}$$

Typically, a neutrino undergoes maximum "oscillation" when

$$1.27 \frac{(\Delta m)^2 L}{E} \approx \frac{\pi}{2} \tag{6.111}$$

By the mid 1990s there were hints that some ν_μ neutrinos were missing in cosmic rays. Unambiguous evidence came in June 1998. At that time a collaboration working at the Kamiokande neutrino detection facility in Japan—originally built to look for possible rare decays of protons—reported a large asymmetry between the number of μ type neutrinos coming down from the zenith and coming up through the earth [15]. Presumably the ν_μ are all made in the atmosphere, so the difference is due to the different distances from the source to the detector – neutrinos interact so weakly with ordinary matter that the result cannot be explained by absorption in the earth. There is no other plausible conclusion than that they oscillated into another type of neutrino somewhere in between.

The rate of ν_μ neutrinos coming downward agreed with predictions from the known properties of neutrinos and cosmic rays. But half of the neutrinos coming directly up through the earth were missing! For the neutrinos coming down from the atmosphere $L = L_{\text{down}}$ is 10 or 15 km, and so

$$1.27 \frac{\Delta m^2 L_{\text{down}}}{E} \ll 1 \tag{6.112}$$

The ones coming up were produced in the atmosphere on the other side of the earth. For them

$$1.27 \frac{\Delta m^2 L_{\text{up}}}{E} \geq \frac{\pi}{2} \tag{6.113}$$

where $L_{\text{up}} \approx 13,000$ km. The incident neutrino energies vary, but are of the order 1 GeV. The neutrino mass difference lies roughly in the range 10^{-1}eV $\geq \Delta m \geq 10^{-2}$eV. The mixing angle is consistent with "maximal" mixing $\theta = \pi/4$.

6.3.4 Solar Neutrinos

A long series of experiments starting as far back as 1968 has been designed to look for neutrinos from the sun. The nuclear fusion processes that produce solar energy are well understood. A variety of sequences of nuclear reactions, with different intermediate steps, all amount to some variation on 4p $\to$ ^{4}He $+ 2e^+ + 2\nu_e$. In contrast to the neutrinos of cosmic ray origin discussed above, solar neutrinos are all the ν_e type and their energies are a few MeV

The result of decades of observation had been the "solar neutrino puzzle": Only a fraction of the ν_e that ought to be there, as calculated from detailed models of the nuclear interactions in the sun, have been seen. It is now clear that these "missing" neutrinos are not really missing. They have turned into one of the other neutrino types. Neutrino oscillations have solved the solar neutrino puzzle.

6.3.5 Neutrino Oscillations in Matter

The Hamiltonian for the interaction of neutrinos with other matter is proportional to G, so the *rate* of collisions is proportional to G^2. It is also proportional at low energies to the neutrino energy. A proton in the sun presents to a neutrino an effective cross section of the order $\approx G^2 M_p E_\nu$, which for a 10 MeV neutrino is about 10^{-45} cm^2. You can work out that the probability for a neutrino produced at the sun's center to scatter before it escapes is about 6×10^{-11}.

Again, to keep the argument simple, suppose that only the two kinds of neutrino, ν_e and ν_τ, are important. The description of the oscillations in free space is as above, with the ν_e replacing ν_μ. They are related to the mass eigenstates by

$$|\nu_e\rangle = \cos\theta |\nu_1\rangle + \sin\theta |\nu_2\rangle \tag{6.99a'}$$
$$|\nu_\tau\rangle = -\sin\theta |\nu_1\rangle + \cos\theta |\nu_2\rangle \tag{6.99b'}$$

The Hamiltonian in vacuum is

$$H = -\omega \begin{pmatrix} \cos 2\theta & -\sin 2\theta \\ -\sin 2\theta & -\cos 2\theta \end{pmatrix} = \omega \left[\cos 2\theta \sigma_3 - \sin 2\theta \sigma_1 \right] \tag{6.104}$$

plus terms proportional to the identity matrix. The mixing angle and $\omega = \Delta m^2/4E_\nu$ are of course not the same as in the ν_μ - ν_τ case.

Now there is another effect in matter, a coherent effect from all the atoms. This is the extra energy from the interaction Hamiltonian itself. When the neutrino passes through matter the interaction with the particles in the background adds a constant to the energy, proportional to the first power of Fermi's constant G as well as to the density of background particles. It will be different for ν_e and ν_τ.

This energy is diagonal in the ν_e, ν_τ basis:

$$H' = \begin{pmatrix} V_e & 0 \\ 0 & V_\tau \end{pmatrix} \tag{6.115}$$

Most of the interactions are the same for ν_e and ν_τ, and contribute the same energy to V_e and V_τ, and so will not contribute to the relative energy. But there is one term in ν_e-e scattering that has no ν_τ analog. The ν_e and the electron can annihilate

190   Chapter 6   Interlude

into a W^- meson, which then decays immediately back into the original pair. The amplitude for this term can be computed exactly from electro weak theory, and it is[11]

$$V_e = \sqrt{2}GN \tag{6.116}$$

N is the density of electrons. There is a new effective Hamiltonian

$$H = -\omega\left[\cos 2\theta \sigma_3 - \sin 2\theta \sigma_1\right] + \frac{GN}{\sqrt{2}}\sigma_3 \tag{6.117}$$

plus terms proportional to the identity matrix.

So in matter the states ν_1 and ν_2 are no longer energy eigenstates. The eigenstates will be some new combinations ν_{im} related to the interaction eigenstates by some new mixing angle θ_m:

$$|\nu_e\rangle = \cos\theta_m|\nu_{1m}\rangle + \sin\theta_m|\nu_{2m}\rangle \tag{6.118a}$$
$$|\nu_\tau\rangle = -\sin\theta_m|\nu_{1m}\rangle + \cos\theta_m|\nu_{2m}\rangle \tag{6.118b}$$

As a neutrino moves through matter with varying density, the new eigenstates will change with time also.

The computation is identical to the one in Section 6.3. In the instantaneously diagonal basis the Hamiltonian is

$$H_m = \omega_m \sigma_3 \tag{6.119}$$

where

$$\omega_m = \sqrt{\left(\omega\cos 2\theta - \frac{GN}{\sqrt{2}}\right)^2 + (\omega\sin 2\theta)^2} \tag{6.120}$$

and the mixing angles are given by

$$\tan 2\theta_m = \frac{\omega\sin 2\theta}{\omega\cos 2\theta - GN/\sqrt{2}} = \frac{\sin 2\theta}{\cos 2\theta - 2\sqrt{2}GNE_\nu/\Delta m^2} \tag{6.121}$$

If the electron density is a constant, then from equation (6.106) with the obvious substitutions, the probability for a ν_e turning into a ν_τ is

$$P = \sin^2 2\theta_m \sin^2 \omega_m L \tag{6.122}$$

The formula for the mixing angle in matter has a maximum ($\theta_m = \pi/2$) when the denominator in (6.121) vanishes, that is, when

$$\omega\cos 2\theta = GN/\sqrt{2} \quad \text{or} \quad N = \frac{\Delta m^2 \cos 2\theta}{2\sqrt{2}E_\nu G} \tag{6.123}$$

so there is a value of the electron density at which the oscillations are large whatever the value of the mixing angle in vacuum, provided only that $\Delta m^2 \neq 0$. In that case the ν_e will be all converted into ν_τ when

$$\frac{\Delta m^2}{4E}\sin 2\theta = \frac{\pi}{2L} \tag{6.124}$$

---

[11] If you do not know any elementary particle physics, just take this on faith—it should be obvious that there is *some* such effect of order G. The point here is how it can modify the oscillation formula.

This enhancement of the detection rate, due to the coherent interaction between neutrinos and bulk matter, is called the *MSW* effect [16, 17]. It is very likely an important factor in explaining the observation of solar neutrinos.

6.4 MEASUREMENTS IN QUANTUM MECHANICS

6.4.1 Wave-Function Collapse

Suppose you measure the spin of an electron along some direction. Before the measurement there was some chance of getting either value. After the spin is measured, you know which value it has, and if you can contrive to repeat the measurement very quickly, you will certainly get the same answer. The wave function has "collapsed." A measurement seems to be different from any other physical event.

Even with complete knowledge of a quantum state it is impossible to predict exactly the outcome of an experiment. The rules tell us how to compute probabilities, but not why a particular measurement gets one result rather than the other. Yet without exception they describe the results of experiment correctly.

In the early days of quantum mechanics some people speculated that because a wave function "collapsed" when it was "measured," the projection of the state vector onto some direction in the vector space has something to do with human thought. An electron was in a linear combination of the two states until someone decided to look at it—then the wave-function collapsed. This is not a widely held view today. The more accepted view is that a measurement occurs when a microscopic system interacts with a macroscopic system. In the case of our electron, the spin is "measured" when the electron leaves a trail on a photographic plate or sets off a counter. The difference is that the macroscopic system—the detector—obeys the laws of classical mechanics.

Most physicists do not believe that macroscopic systems really obey physical laws fundamentally different from electrons and atoms. Macroscopic systems have so many quantum states that are in practice indistinguishable that they appear to obey the classical rules with great accuracy. Linear combinations of states with very different properties become extremely unlikely. An electron can be in a linear combination of spin-up and spin-down, but Schrödinger's cat was never in a state that is in a linear combination of alive and dead. Exactly how this works has never been explained to everyone's satisfaction, but that does not mean it is wrong.

6.4.2 The EPR Paradox

Is a state vector really all there is to know about a physical system? In 1935 Einstein, Podolsky, and Rosen [7] described a thought-experiment that seemed to suggest that all the possible properties were "really there" all the time, even though that information is not supposed to be in the state vector. Here is a simplified version due to Bohm that was elaborated later by Bell:[12]

Suppose a spin-zero particle, like a neutral π^o meson, decays at rest into an electron and a positron, both spin-half particles. The two particles come out in opposite directions. If you measure the spin of one particle, you know that the spin of the second is equal and opposite. This is true in either quantum or classical

---
[12]Bohm [8] Section 22.16.

mechanics. The result of *either* measurement alone is also purely random, but once you measure the spin of the electron, you still know the result of measuring the spin of the positron, even though it may be far away at the time of the measurement. This is the Einstein-Podolsky-Rosen "paradox."

Imagine measuring the spins of both particles along the direction of the electron, taken to be the z-axis. The two-particle state is labeled by the two spin quantum numbers $|m_a m_b\rangle$. The correct combination can be computed using the Clebsch-Gordan coefficients for combining two spin-half particles:[13]

$$|\psi\rangle = \frac{|+-\rangle - |-+\rangle}{\sqrt{2}} \qquad (6.125)$$

where $|\pm\pm\rangle$ is shorthand for $|\pm\frac{1}{2}, \pm\frac{1}{2}\rangle$.

If you measure the spin of the electron, the probability is 50% that you get spin up, and 50% that you get spin down. There is no way to tell in advance which; the state $|\psi\rangle$ does not contain that information. Nevertheless it completely describes the system.

The same is true if you measure the spin of the positron. Before the electron spin is measured, the outcome of measuring the positron spin is unknown, but as soon as the electron spin is measured, the result of measuring the positron spin becomes certain, no matter how far apart they are.

Information appears to travel faster than the speed of light, in some reference frame violating the principle that effects happen after causes, not before. But does measuring the electron spin really "cause" the outcome of a measurement of the positron spin? Not in the usual sense; someone measuring the electron spin cannot send a faster-than-light message to someone else measuring the positron spin. It is only afterward that they can compare notes and see that there has been the correlation required by angular momentum conservation.

The EPR experiment suggested another solution that does not violate the causality principle: Our description of quantum mechanics is incomplete, and the state vector is not all there is to know about a system. In this view there is an underlying, deterministic classical reality to which we have no access. When we make a measurement, we sample this underlying classical system, and the probability that seems fundamental in quantum mechanics is a consequence of this statistical sampling. This underlying level contains parameters that cannot be observed in our macroscopic experiments.

The idea that what we see are statistical averages over these parameters is called a **hidden variables** theory. They are hidden in the sense that we do not know anything about them, but no fundamental principle says we can never know. In this view the statistical nature of quantum mechanical predictions is due to the complicated random behavior of this classical substructure. You might think that the electron and positron were "really" inside the π^o meson all the time, bound by some force, moving in complicated orbits with the spins changing this way and that, always obeying classical mechanics rules and always conserving momentum and angular momentum. At some instant the two particles fly apart. The components of the spins along the direction of motion are always opposite. In this case the

---
[13] See Section 4.5.2.

hidden variables are the values of the electron momentum, its position, and its spin at the instant of decay—in the macroscopic world random and unmeasurable quantities.

For some years, versions of the hidden variable idea were possible alternatives to the orthodox interpretation of quantum mechanics. People like de Broglie who thought nature must be deterministic like classical mechanics were especially fond of hidden variables. While never very popular, hidden variable theories seemed a possible version of quantum mechanics until 1964, when John Bell showed that the predictions of a large class of hidden variable theories are different from those of quantum mechanics, and must therefore be excluded.

6.4.3 Bell's Inequality

Bell's idea [9, 10] was to consider the decay of a spin-zero system into two spin-half particles, as in the EPR thought experiment, but to work out correlations between the two spin measurements even when the measurements are not made along the same axis. That is, measure the spins not along the directions of motion, as in the EPR experiment, but along arbitrary directions, as in Figure 6.1.

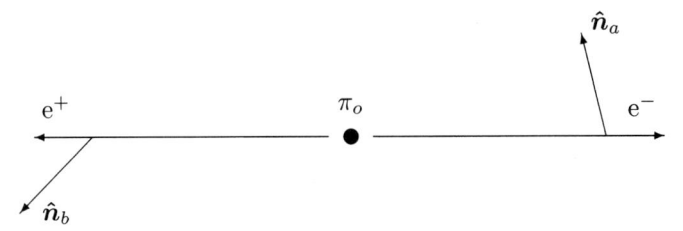

FIGURE 6.1: Schematic diagram of Bell's version of Bohm's version of the EPR experiment. In the EPR experiment the electron and positron spins, $\hat{n}_a$ and $\hat{n}_b$, are parallel.

How could the result of such measurements be a consequence of classical mechanics alone? Classical mechanics does not predict that the angular momentum along any axis is always an integer multiple of $\hbar/2$, but it is not forbidden, so let us assume this is true in the unknown classical mechanics for some unknown reason. In fact, let us measure only the *sign* of the spins along some directions. For each decay, observe $A(\hat{n}_a)$, the sign of the spin of the electron along some direction $\hat{n}_a$, and $B(\hat{n}_b)$, the sign of the positron spin along some other direction. Bell's theorem is about the correlation in a hidden variable theory between these two quantities as a function of the two directions.

Suppose you measure the average of the *product* of $A(\hat{n}_a)$ and $B(\hat{n}_b)$ in a large number of trials. Spherical symmetry requires that this average value depend only on the angle between the two directions, so one can define the average as

$$E(\theta) = \langle A(\hat{n}_a)B(\hat{n}_b)\rangle \tag{6.126}$$

where $\hat{n}_a \cdot \hat{n}_b = \cos\theta$.

If you measure the two spins along the same axis, the values must be opposite: If $A(\hat{n}_a) = 1$, then $B(\hat{n}_a) = -1$, and if $A(\hat{n}_a) = -1$, then $B(\hat{n}_a) = 1$. The product

194 Chapter 6 Interlude

is -1 for each measurement. Similarly, the spins along opposite directions must always be equal: $A(\hat{n}_a)B(-\hat{n}_a) = 1$. Whether or not there is an underlying classical theory, angular momentum conservation alone requires $E(0) = -1$ and $E(\pi) = 1$. $E(\theta)$ is some function that goes from -1 to $+1$ as $0 \leq \theta \leq \pi$. A simple example of such a function is

$$E(\theta) = \frac{2\theta}{\pi} - 1 \tag{6.127}$$

but there is no reason for the actual result to be linear like this.

Quantum Mechanics

What does quantum mechanics predict for $E(\theta)$? In the simple case where both particles have spin half, the state is

$$|\psi\rangle = \frac{|+-\rangle - |-+\rangle}{\sqrt{2}} \tag{6.125}$$

The correlation is the expectation value of the product $(\boldsymbol{\sigma})_a \cdot \hat{n}_a (\boldsymbol{\sigma})_b \cdot \hat{n}_b$. From Problem 4.11,

$$E(\theta) = \langle\psi|\,\boldsymbol{\sigma}_a \cdot \hat{n}_a \boldsymbol{\sigma}_b \cdot \hat{n}_b\,|\psi\rangle = -\cos\theta \tag{6.128}$$

Any decay of a spinless particle into two spin-half particles must have this correlation if it obeys the laws of quantum mechanics. Notice that $E(0) = -1$ and $E(\pi) = 1$ as required.

Classical Hidden Variable Description

Now suppose instead there is really a classical description, but all you know about the probabilities for different outcomes is that they depend somehow on the hidden variable (or variables) λ; and that our experiment samples all values of λ with an unknown probability $p(\lambda)$, subject to

$$\int p(\lambda)d\lambda = 1 \tag{6.129}$$

Then there are some functions $A(\hat{n}_a, \lambda)$ and $B(\hat{n}_b, \lambda)$ that are the results of these two measurements for any particular value of the hidden variables λ. Furthermore, for a given λ, these functions must be opposite for a given direction: $B(\hat{n}_b, \lambda) = -A(\hat{n}_b, \lambda)$. If $\hat{n}_a = \hat{n}_z$ and $\hat{n}_b \cdot \hat{n}_a = \cos\theta_b$, the average value of the product is

$$E(\theta_b) = \int A(\hat{n}_a, \lambda)B(\hat{n}_b, \lambda)p(\lambda)d\lambda \tag{6.130}$$

and similarly for some other direction $\hat{n}_c$

$$E(\theta_c) = \int A(\hat{n}_a, \lambda)B(\hat{n}_c, \lambda)p(\lambda)d\lambda \tag{6.131}$$

Then since $B(\hat{n}_b, \lambda) = -A(\hat{n}_b, \lambda)$, and since $A(\hat{n}_b, \lambda)^2 = 1$

$$E(\theta_b) - E(\theta_c) = \int A(\hat{n}_a, \lambda) \left[B(\hat{n}_b, \lambda) - B(\hat{n}_c, \lambda)\right] p(\lambda)\,d\lambda$$

$$= -\int A(\hat{n}_a, \lambda) \left[A(\hat{n}_b, \lambda) - A(\hat{n}_c, \lambda)\right] p(\lambda)d\lambda \tag{6.132}$$

$$= -\int A(\hat{n}_a, \lambda)A(\hat{n}_b, \lambda) \left[1 - A(\hat{n}_b, \lambda)A(\hat{n}_c, \lambda)\right] p(\lambda)\,d\lambda$$

The quantity in square brackets cannot be negative. Therefore

$$|E(\theta_b) - E(\theta_c)| \leq \int \left|A(\hat{n}_a, \lambda)A(\hat{n}_b, \lambda)\right| \left[1 - A(\hat{n}_b, \lambda)A(\hat{n}_c, \lambda)\right] p(\lambda) \, d\lambda \quad (6.133)$$

Since $|A(\hat{n}_a, \lambda)A(\hat{n}_b, \lambda)| = 1$ and $1 - A(\hat{n}_b, \lambda)A(\hat{n}_c, \lambda) \geq 0$ equation (6.133) becomes

$$\begin{aligned}|E(\theta_b) - E(\theta_c)| &\leq 1 - \int A(\hat{n}_b, \lambda)A(\hat{n}_c, \lambda)p(\lambda) \, d\lambda \\ &= 1 + \int A(\hat{n}_b, \lambda)B(\hat{n}_c, \lambda)p(\lambda) \, d\lambda\end{aligned} \quad (6.134)$$

Take $\hat{n}_b$ and $\hat{n}_c$ in the x-z plane, so that the angle between these two directions is $\theta_{bc} = \theta_b - \theta_c$. Then the integral in equation (6.134) is $E(\theta_{bc})$, and

$$|E(\theta_b) - E(\theta_c)| \leq 1 + E(\theta_b - \theta_c) \quad (6.135)$$

or

$$\boxed{|E(\theta_b) - E(\theta_c)| - E(\theta_b - \theta_c) \leq 1} \quad (6.136)$$

This is Bell's inequality. It is a constraint on the function $E(\theta)$. As an illustration, check that it is satisfied by the simple example (6.127). A linear $E(\theta)$ does not violate the inequality.

Does the quantum mechanical result satisfy Bell's inequality for all angles? Let $\theta_b = \theta$ and $\theta_c = 2\theta$. Bell's inequality requires

$$|E(\theta) - E(2\theta)| - E(\theta) \leq 1 \quad (6.137)$$

or, with $E(\theta) = -\cos\theta$,

$$|\cos\theta - \cos 2\theta| + \cos\theta \leq 1 \quad (6.138)$$

If θ lies between 0 and $\pi/2$, the inequality is violated. But the quantum result is confirmed in a vast number of experiments. The hidden variable model is untenable.[14,15]

PROBLEMS

6.1. Electron on a Ring

(a) Write the classical Hamiltonian for a free electron in terms of spherical coordinates and their canonical momenta. Suppose the electron is confined to a circular ring of radius R. The ring lies in the x-y plane, and the z-axis passes through the center of the ring. Impose this constraint on p_r and p_θ. What is the Hamiltonian with this constraint?

Then treat this expression as a quantum mechanical Hamiltonian. What are the allowed energy levels?

---

[14] A series of experiments in the early 1980s explicitly tested Bell's inequality using the correlation between photon polarizations in a two-photon atomic decay. See Aspect, Grangier, and Roger [11, 12] and Aspect, Dalibard and Roger [13].

[15] The derivation of Bell's inequality here is not difficult, but it is not very transparent either. Mermin has tried to explain the essence of Bell's inequality in a series of interesting papers, all reprinted in [14].

(b) Now let there be an external magnetic field $\mathbf{B} = \mathbf{\nabla} \times \mathbf{A}$, constant in time but not necessarily in space. Take it to be in the z-direction, perpendicular to the plane of the ring: $\mathbf{B} = B_o \hat{\mathbf{n}}_z$. From Section 1.26 the Hamiltonian is

$$H = \frac{1}{2m}\left(\mathbf{p} + \frac{e}{c}\mathbf{A}\right)^2$$

The components of $\mathbf{p}$ are still constrained as in part (a). Decompose $\mathbf{A}(\mathbf{r})$ into its spherical components:

$$\mathbf{A} = A_r \hat{\mathbf{n}}_r + A_\theta \hat{\mathbf{n}}_\theta + A_\phi \hat{\mathbf{n}}_\phi$$

where the $\hat{\mathbf{n}}_q$ are vectors of unit length in the indicated directions. In this notation the components of $\mathbf{p}$ along the three orthogonal directions are proportional to, but not all equal to, the canonical momenta p_r, p_θ, and p_ϕ. Show that A_r and A_θ vanish. Then the constraints require p_r and p_θ to vanish (why?). Write the Hamiltonian in terms of L_z and A_ϕ.

In the transverse gauge

$$\mathbf{A}(\mathbf{r}) = \frac{1}{2}\mathbf{B} \times \mathbf{r} \qquad (6.10)$$

is the correct form of the vector potential. What are the cylindrical-coordinate components A_ρ, A_z, and A_ϕ? What are the quantum energy levels now?

(c) Only differences between energy levels can be measured. Could you use the energy levels of a ring in a constant magnetic field, as in this problem, to measure the strength of the field? (Explain why or why not.)

6.2. Field of a Magnetic Monopole

(a) Let the spherical components of the vector potential be given by

$$A_r = A_\theta = 0 \quad \text{and} \quad A_\phi = -g\frac{1-\cos\theta}{r\sin\theta}$$

Show that the magnetic field $\mathbf{B}(\mathbf{r}) = \mathbf{\nabla} \times \mathbf{A}(\mathbf{r})$ is the field of a magnetic monopole at the origin everywhere except along the negative z-axis.

(b) Show that in the vicinity of the z-axis, the magnetic field in part (a) has a singularity proportional to $\Theta(-z)\delta(x)\delta(y)$. What is the constant of proportionality?[16]

(c) Show that the divergence of this magnetic field is identically zero, as required by Maxwell's equations.

Note: This is the form of the vector potential for a magnetic monopole. Dirac showed that it is impossible to devise a form for $\mathbf{A}(\mathbf{r})$ that does not have such a singularity but still satisfies $\mathbf{\nabla} \cdot (\mathbf{\nabla} \times A) = 0$ everywhere.

(d) Consider an electron confined to a ring as in Problem 6.1, but the center of the ring is on the z-axis at an arbitrary distance z_o from the origin. Let there be a magnetic monopole with magnetic charge g at the origin. What are the quantum energy levels now?

---

[16] $\Theta(x)$ is the step function.

(e) Your answer to (d) should be a function of R and z_o. As the ring is moved adiabatically along the z-axis, the n-th level will shift. What is its value as $z_o \to +\infty$? In that limit the monopole is again infinitely far away, and the allowed energy differences must again be those of the electron-in-a-ring in the absence of an external field calculated in Problem 6.1. Use this fact to obtain a relation between g and e.

6.3. Measurement of $g - 2$

(a) An electron is injected into a cyclotron where there is a constant magnetic field **B** along the z-axis. Initially the electron is moving in the x-direction. What is the orbit of the electron? What is the direction of the electron's momentum **p** as a function of time? (Use classical mechanics.)

(b) Now suppose that at $t = 0$ the electron (gyromagnetic ratio g) is known to be polarized in the x-direction; that is, $|\psi(0)\rangle$ is an eigenstate of s_x with eigenvalue $+1/2$. What is the period of precession of the expectation value of the spin?

(c) Suppose the electron is ejected from the cyclotron at time $t > 0$. Show that the angle between $\langle\psi|\,\mathbf{s}\,|\psi\rangle$ and **p** as a function of t is proportional to $g - 2$. This is how the corrections to the $g = 2$ rule can be measured directly.

6.4. Problem on Landau Levels

An electron moves in a constant magnetic field B along the z-axis. In this problem ignore the electron spin.

(a) Write the classical Hamiltonian equations of motion in the transverse gauge, equation (6.10), and show that the general solution has the form

$$x = x_o + R\cos(\omega_c t - \phi_o)$$
$$y = y_o + R\sin(\omega_c t - \phi_o)$$
$$z = z_o + v_z t$$

with six arbitrary parameters x_o, y_o, z_o, v_z, R, and ϕ_o. Here ω_c is the cyclotron frequency $|eB|/m$. What is the energy of the motion in terms of these parameters?

(b) The classical motion is along a helix whose axis is fixed at $x = x_o$, $y = y_o$. In the quantum description of this problem, show that x_o and y_o are conserved. In other words $[x_o, H] = [y_o, H] = 0$.

> Hint: Write x_o and y_o in terms of the raising and lowering operators defined in equation (6.19). The answer is
>
> $$x_o = \frac{1}{2\sqrt{m\omega}}\left(a_-^\dagger + a_-\right)$$
>
> $$y_o = \frac{i}{2\sqrt{m\omega}}\left(a_-^\dagger - a_-\right)$$
>
> where here $\omega = \omega_c/2$ is the Larmor frequency.

(c) What are the expectation values $\langle x_o \rangle$ and $\langle y_o \rangle$ in the energy eigenstates $|n_+, n_-\rangle$ defined in equation (6.23)? The state

$$|\psi\rangle = \frac{|0,0\rangle + |0,1\rangle}{\sqrt{2}}$$

is also an energy eigenstate, since the energy is independent of n_-. What are the expectation values $\langle x_o \rangle$ and $\langle y_o \rangle$ in this state?

(d) Compute the dispersions [see equation (2.155)] of x_o and y_o in the energy eigenstates $|n_+, n_-\rangle$. In which states are the product of the dispersions the minimum?

(e) Compute the commutator $[x_o, y_o]$. Since this commutator is not zero, these two observables cannot be measured simultaneously, even though both are constants of the motion. Use the generalized uncertainty principle to get a rigorous minimum for the product

$$\left\langle (x_o - \langle x_o \rangle)^2 \right\rangle \left\langle (y_o - \langle y_o \rangle)^2 \right\rangle$$

Does the product achieve this minimum value in any of the states $|n_+, n_-\rangle$?

6.5. Landau Levels in Another Gauge

This is another problem about the Landau levels for a (spinless) electron in a constant magnetic field $\mathbf{B} = B\hat{\mathbf{n}}_z$. Instead of the "symmetric" gauge

$$\mathbf{A} = -\frac{1}{2} yB\hat{\mathbf{n}}_x + \frac{1}{2} xB\hat{\mathbf{n}}_y \tag{6.10}$$

take the vector potential to be in the "Landau Gauge"

$$\mathbf{A}' = \mathbf{A} + \nabla \left(\frac{xyB}{2} \right) = xB\hat{\mathbf{n}}_y$$

(a) Find the stationary energies in this gauge. For fixed E_z, what is the degeneracy of each level?

(b) Write down the general form for the wave functions in this gauge.

6.6. Neutron in Axial Magnetic Field

A neutron, which is a neutral, spin-half particle with mass M and magnetic moment $-g\mu_N \mathbf{s}$, is confined to a circular ring of radius R_o. The ring lies in the x-y plane, centered on the origin. A long solenoid around the z-axis produces a constant magnetic field pointing in the z-direction at all points on the ring. There is also a wire along the z-axis producing an azimuthal field at the ring. The total magnetic field at the ring is

$$\mathbf{B} = B_o \hat{\mathbf{n}} = B_1 \hat{\mathbf{n}}_z + B_2 \hat{\mathbf{n}}_\phi = B_1 \hat{\mathbf{n}}_z + B_2 \left(\cos \phi \hat{\mathbf{n}}_y - \sin \phi \hat{\mathbf{n}}_x \right)$$

The state of the particle can be described by its spin along any direction and its angle ϕ around the ring. The Hamiltonian is

$$H = \frac{L^2}{2MR_o} + \frac{\omega_o}{2} \hat{\mathbf{n}}(\phi) \cdot \boldsymbol{\sigma}$$

where $L = -i\partial/\partial\phi$ is the z component of the angular momentum, $\omega_o = g\mu_N B_o$, and

$$\omega_o \hat{\mathbf{n}}(\phi) = \omega_o \left[\cos\theta \hat{\mathbf{n}}_z + \sin\theta \left(\cos\phi \hat{\mathbf{n}}_y - \sin\phi \hat{\mathbf{n}}_x \right) \right]$$
$$= \omega_1 \hat{\mathbf{n}}_z + \omega_2 \left(\cos\phi \hat{\mathbf{n}}_y - \sin\phi \hat{\mathbf{n}}_x \right)$$

The field makes the angle θ with the z-axis:

$$\frac{B_2}{B_1} = \frac{\omega_2}{\omega_1} \tan\theta$$

What are all possible values of the energy?

If $\mathbf{B} = 0$, this is the same as part (a) of Problem 6.1, and the energy levels are doubly degenerate. Does the magnetic field split the degenerate levels?

PROBLEMS

6.7. Electron Magnetic Resonance

For some later applications, it will be useful to work out the spin resonance formula explicitly, in terms of the two component wave functions rather than in the density matrix formalism of Section 6.2.3. Suppose an electron in the ground state of hydrogen, or a free electron at rest, is placed in a magnetic field $\mathbf{B}_o$ pointing in the z-direction. The energy eigenstates are the eigenstates of spin along the z-axis. Let $|\psi_-\rangle$ be the ground state, and $|\psi_+\rangle$ be the upper state. The energy difference between the two states is

$$\omega_o = 2\mu_B B_o$$

where μ_B is the Bohr magneton.

(a) First let there also be a small constant magnetic field along the x-axis also: $\mathbf{B}_1 = B_1 \hat{\mathbf{n}}_x$. Set $\omega_1 = 2\mu_B B_1$. Write the two component wave function as

$$\psi(t) = \begin{pmatrix} a_+(t) \\ a_-(t) \end{pmatrix}$$

where $a_\pm(t)$ are the amplitudes for the electron to be an eigenstate of s_z with eigenvalue $\pm 1/2$. What is the general formula for $\psi(t)$ in terms of $\psi(0)$? Write the answer in terms of $\bar{\omega} = 2\mu_B \bar{B}$, where $\bar{B}$ is the magnitude of the field, and the angle θ that $\bar{B}$ makes with the z-axis.

If the electron is definitely in the ground state at $t = 0$, what is the probability that it will be in the upper state at a later time?

(b) Now let there also be a circularly polarized electromagnetic wave, such that the magnetic field at the location of the electron is

$$\mathbf{B} = \mathbf{B}_1(\hat{\mathbf{n}}_x \cos \omega t + \hat{\mathbf{n}}_y \sin \omega t)$$

The magnetic field still makes an angle θ with the z-axis, but now precesses with angular frequency ω. Write down the general form of the wave function $\psi(t)$ in this case. This is of course the spin resonance problem described at the end of Section 6.2, an important time-dependent problem unusual in that it can be solved exactly. The trick is to set[17]

$$|\phi(t)\rangle = e^{i\omega\sigma_z t/2}|\psi(t)\rangle$$

which should reduce the problem to the constant field case in part (a). The answer is equation (6.93).

(c) Finally, if the electron is definitely in the ground state at $t = 0$, what is the probability that it will be in the upper state at a later time? The answer should be the same as equation (6.90).

6.8. EMR with Partially Polarized Electrons

(a) In a real observation of spin magnetic resonance you might know that all the electrons are in an eigenstate of spin along the large constant field $\mathbf{B}_o$, but not all of them are in the lower state. In that case the density matrix for the mixture at $t = 0$ is defined by a polarization vector $\mathbf{P}_o$ antiparallel to $\mathbf{B}_o$, but $|\mathbf{P}|_o < 1$. Repeat the calculation at the end of Section 6.2.3 in this case, leading to a generalization of the formula (6.90).

(b) Describe what happens if $\mathbf{P}(0) < 1$ and is pointing in an arbitrary direction.

---

[17] This amounts to a transformation of the states to a rotating coordinate system.

REFERENCES

Landau levels were introduced in Chapter XVI of

[1] L. D. LANDAU AND E. M. LIFSHITZ, *Quantum mechanics: Nonrelativistic theory*, Pergamon, Oxford, 1965-1977, translated by J. B. Sykes and J. S. Bell. See also Appendix D.

For a more recent review of Landau levels and the quantum Hall effect see

[2] R. E. PRANGE AND S. M. GIRVIN, *The quantum Hall effect*, Springer Verlag, 1990.

Schwinger's correction to the gyromagnetic ratio is in

[3] J. SCHWINGER, Phys. Rev. **73** (1948), 416L.

For more recent experimental and theoretical values, respectively, see

[4] P. J. MOHR AND B. N. TAYLOR, Rev. Modern Phys. **72** (2000), 351.

[5] T. KINOSHITA AND W. B. LINDQUIST, Phys. Rev. Letters **47** (1981), 1573.

The splitting of spectral lines in a magnetic field was discovered by

[6] P. ZEEMAN, Phil. Mag. **43** (1897), 226.

The EPR thought-experiment was proposed in

[7] A. EINSTEIN, B. PODOLSKY, AND N. ROSEN, Phys. Rev. **47** (1935), 777.

The version described in the text is on page 611 of

[8] D. BOHM, *Quantum theory*, Prentice Hall, 1951.

Bell's inequality appears in

[9] J. BELL, Physics **1** (1964), 195.

[10] J. BELL, Rev. Modern Phys. **38** (1966), 447.

The direct experimental tests of Bell's inequality are in

[11] A. ASPECT, P. GRANGIER, AND G. ROGER, Phys. Rev. Letters **47** (1981), 460.

[12] A. ASPECT, P. GRANGIER, AND G. ROGER, Phys. Rev. Letters **49** (1982), 91.

[13] A. ASPECT, J. DALIBARD, AND G. ROGER, Phys. Rev. Letters **49** (1982), 1804.

Mermin's papers are reprinted in

[14] N. DAVID MERMIN, *Boojums all the way through*, Cambridge, 1990.

The experiment that first reported atmospheric neutrino oscillations was

[15] Y. FUKUDA ET AL., Phys. Rev. Letters **81** (1998), 1562.

The references for the MSW effect are

[16] L. WOLFENSTEIN, Phys. Rev. **D17** (1978), 2369.

[17] S. P. MIKHEYEV AND A. Y. SMIRNOV, Nuovo Cimento **9C** (1986), 17.

CHAPTER 7

Approximation Methods for Bound States

7.1 BOUND-STATE PERTURBATION THEORY

Bound-state perturbation theory is a procedure to deal with problems that cannot be solved exactly but are close to ones that can.[1] Not all problems can be treated this way, and even when it is possible, it is sometimes an art to figure out how to do it. Nevertheless, the method has led to a complete understanding of the structure of simple atoms and molecules and the effects of weak external electric and magnetic fields on them.

The hydrogen atom will continue to be the principle example. Perturbation methods make it possible to compute its energy levels to phenomenal accuracy. I will discuss perturbations due to external electromagnetic fields, spin-orbit coupling, relativistic corrections, and the effect of the proton size and spin. The method is applicable to more complicated atoms, molecules, nuclei, elementary particles, and many more complicated systems

7.1.1 The Perturbation Expansion

Let E_n^o be the exact spectrum of a Hamiltonian H_o:

$$H_o|\psi_n^o\rangle = E_n^o|\psi_n^o\rangle \tag{7.1}$$

Perturbation theory is a way to compute, by successive approximations, the spectrum of $H = H_o + H'$:

$$H|\psi_n\rangle = E_n|\psi_n\rangle \tag{7.2}$$

when the E_n are close to the unperturbed energies E_n^o, and when H' is "small" in the sense that expansions in powers of H' are expansions in a small quantity.

Proving that the expansions converge is often impossible. For real problems, we will be grateful when they turn out to be expansions in increasing powers of a small number (like the fine structure constant α) and hope for the best. Fortunately the results agree remarkably with experiment.

Define

$$\Delta_n = E_n - E_n^o \tag{7.3}$$

Δ_n is the energy shift of the state $|\psi_n^o\rangle$ due to the perturbation H'. Then multiply the identity $H - H_o = H'$ on the right by $|\psi_n\rangle$ and on the left by $\langle\psi_n^o|$, and use the

---

[1]Originally due to Rayleigh, the method was first applied to wave mechanics by Schrödinger, reference [13] in Chapter 2.

Section 7.1 Bound-State Perturbation Theory

facts that these are eigenstates of H and H_o respectively. Then

$$\langle \psi_n^o | H - H_o | \psi_n \rangle = \langle \psi_n^o | H' | \psi_n \rangle$$
$$\langle \psi_n^o | E_n - E_n^o | \psi_n \rangle = \langle \psi_n^o | H' | \psi_n \rangle \tag{7.4}$$

and so

$$\Delta_n = \frac{\langle \psi_n^o | H' | \psi_n \rangle}{\langle \psi_n^o | \psi_n \rangle} \tag{7.5}$$

To compute Δ_n to some order you need $|\psi_n\rangle$ to one order lower only. In zeroth order, E_n is just E_n^o, and you make an error of higher order only setting $|\psi_n\rangle \to |\psi_n^o\rangle$ in the last line of equation (7.4). So the first-order expression for the energy shift is

$$\boxed{\Delta_n^1 = \langle \psi_n^o | H' | \psi_n^o \rangle = H'_{nn}} \tag{7.6}$$

The lowest order approximation to the energy shift is the expectation value of the perturbation in the unperturbed eigenstate. In equation (7.6) and in all that follows, H'_{mn} are the matrix elements of H' is the $|\psi_n^o\rangle$ basis:

$$H'_{mn} = \langle \psi_m^o | H' | \psi_n^o \rangle \tag{7.7}$$

To proceed further, you need a formula suitable for expanding the state vector $|\psi_n\rangle$ in powers of H' or of its matrix elements. Write

$$(H - E_o^n + E_o^n - H_o)|\psi_n\rangle = H'|\psi_n\rangle$$
$$(H_o - E_n^o)|\psi_n\rangle = (\Delta_n - H')|\psi_n\rangle \tag{7.8}$$

Resist the temptation to solve equation (7.8) by multiplying both sides by $(H_o - E_n^o)^{-1}$. That is illegal here, since $H_o - E_n^o$ has no inverse. But it is almost right. The thing to do is to exclude the state $|\psi_n^o\rangle$ on which $H_o - E_n^o$ is zero, and impose the boundary condition that $|\psi_n^o\rangle = |\psi_n\rangle$ when $H' = 0$. A solution to equation (7.8) that satisfies these rules is

$$|\psi_n\rangle = |\psi_n^o\rangle + \sum_{m \neq n} \frac{|\psi_m^o\rangle\langle\psi_m^o|\Delta_n - H'|\psi_n\rangle}{E_m^o - E_n^o} \tag{7.9}$$

Proof. The proof that equation (7.9) satisfiies equation (7.8) is brief:

$$(H_o - E_n^o)|\psi_n\rangle = (H_o - E_n^o) \sum_{m \neq n} \frac{|\psi_m^o\rangle\langle\psi_m^o|\Delta_n - H'|\psi_n\rangle}{E_m^o - E_n^o}$$
$$= \sum_{m \neq n} (E_m^o - E_n^o) \frac{|\psi_m^o\rangle\langle\psi_m^o|\Delta_n - H'|\psi_n\rangle}{E_m^o - E_n^o} \tag{7.10}$$
$$= \sum_m |\psi_m^o\rangle\langle\psi_m^o|\Delta_n - H'|\psi_n\rangle = (\Delta_n - H')|\psi_n\rangle$$

in agreement with equation (7.8). In the last line, the sum over $m = n$ was added, since

$$(H_o - E_n^o)|\psi_n^o\rangle = 0 \tag{7.11}$$

∎

204   Chapter 7   Approximation Methods for Bound States

Equations (7.4) and (7.9) can be solved together by successive iterations. Write

$$|\psi_n\rangle = |\psi_n^0\rangle + |\psi_n^1\rangle + |\psi_n^2\rangle + \cdots$$
$$E_n = E_n^o + E_n^1 + E_n^2 + \cdots \quad (7.12)$$

Here E_n^k and $|\psi_n^k\rangle$ are proportional to the k-th power of the perturbation H'. Sometimes it helps to write

$$H' = \lambda H'$$
$$|\psi_n\rangle = |\psi_n^0\rangle + \lambda|\psi_n^1\rangle + \lambda^2|\psi_n^2\rangle + \cdots \quad (7.13)$$
$$E_n = E_n^o + \lambda E_n^1 + \lambda^2 E_n^2 + \cdots$$

Compare terms with the same number of powers of λ, and then set $\lambda \to 1$ at the end. In zeroth order

$$|\psi_n\rangle = |\psi_n^o\rangle \quad \text{and} \quad E_n = E_n^o \text{ or } \Delta_n = 0 \quad (7.14)$$

To get the next approximation, replace $|\psi_n\rangle$ by $|\psi_n^o\rangle$ on the right-hand side of equation (7.9):

$$|\psi_n^0\rangle + |\psi_n^1\rangle = |\psi_n^o\rangle + \sum_{m \neq n} |\psi_m^o\rangle \frac{\langle \psi_m^o | \Delta_n - H' | \psi_n^o \rangle}{E_m^o - E_n^o}$$
$$= |\psi_n^o\rangle - \sum_{m \neq n} \frac{H'_{mn}|\psi_m^o\rangle}{E_m^o - E_n^o} \quad (7.15)$$

With this result Δ_n can be computed through second order:

$$\boxed{\Delta_n^1 + \Delta_n^2 = H'_{nn} - \sum_{m \neq n} \frac{|H'_{mn}|^2}{E_m^o - E_n^o} + \cdots} \quad (7.16)$$

Equation (7.16) has an interesting corollary: If the expectation value of H' in the ground state vanishes, then the effect of any perturbation is to depress the ground-state energy. The Stark effect in Section 7.2 on the $n = 1$ state of hydrogen will provide an example of this effect.

The form for $|\psi_n\rangle$ through second order can be obtained by putting $|\psi_n^1\rangle$ and Δ_n^1 into the general expression for $|\psi_n\rangle$ (see Problem 7.1).

Normalization of the Perturbed States

Since equation (7.8) is homogeneous, the solution is not unique. To any solution you can add a solution to the homogeneous equation

$$(H_o - E_n^o)|\psi_n\rangle = 0 \quad (7.17)$$

The solution written in equation (7.9) makes a particular choice.

$$\langle \psi_n^o | \psi_m^o \rangle = \delta_{mn} \quad (7.18)$$

Then the eigenstates of H are orthogonal:

$$\langle \psi_n | \psi_m \rangle = 0 \quad (m \neq n) \tag{7.19}$$

but they are not normalized to unity. Instead

$$\langle \psi_n^o | \psi_n \rangle = \langle \psi_n^o | \psi_n^o \rangle = 1 \tag{7.20}$$

The normalization is

$$\begin{aligned}\langle \psi_n | \psi_n \rangle &= 1 + \sum_{m \neq n} \langle \psi_n | \psi_m^o \rangle \frac{\langle \psi_m^o | \Delta_n - H' | \psi_n \rangle}{E_m^o - E_n^o} \\ &= 1 + \sum_{m \neq n} |\langle \psi_n | \psi_m^o \rangle|^2 \end{aligned} \tag{7.21}$$

It is possible to force the perturbed states to come out normalized by putting a constant in the definition (7.9) of $|\psi_n\rangle$, but that seems to me unnecessarily complicated.

Two Restrictions on the Perturbation Method

The method has two technical restrictions:

- First, there must be a one-to-one correspondence between the eigenstates $|\psi_n^o\rangle$ of the unperturbed Hamiltonian H_o and the eigenstates $|\psi_n\rangle$ of the full Hamiltonian H. That is, if one replaces H' by $\lambda H'$, the solution is presumed smooth in the $\lambda \to 0$ limit. In particular, $E_n \to E_n^o$ as $\lambda \to 0$. This is not always true. For a three-dimensional particle, for example, you cannot take $H_o = \mathbf{p}^2/2m$ and $H' = -\alpha/r$, since the spectrum of H_o is all the positive numbers (the possible energies of a free particle) while the spectrum of $H = H_o + H'$ includes bound-state energies as well.

- There is a second technical restriction. We will usually work with a subset of all the states, and those must be collections of states such that no two have the same unperturbed energy: $E_n^o \neq E_m^o$ unless $n = m$. Furthermore, we will want to include only those connected by H', those for which $\langle \psi_m^o | H' | \psi_n^o \rangle \neq 0$. These need not be all the states. For example if H' and H_o both conserve total angular momentum, there will always be the $(2j+1)$ fold degeneracy. Two states in one of these ladders will have the same unperturbed energy, so we study states of a single m value only. This restriction is not a problem as long as $H'_{mn} = \langle \psi_m^o | H' | \psi_n^o \rangle = 0$ whenever $E_m = E_n$. We will see that this is usually the case, but not always.

The reason for the first restriction is obvious. The reason for the second restriction will be revealed shortly.

7.1.2 Example: Harmonic Oscillator

It is always comforting to check a new method by an example where the answer is known. Let us look at the one-dimensional harmonic oscillator of Section 3.2. To the Hamiltonian

$$H_o = \frac{p^2}{2m} + \frac{m\omega^2 x^2}{2} = \frac{\omega}{2}(aa^\dagger + a^\dagger a) \tag{7.22}$$

add a perturbation
$$H' = \frac{1}{2}\lambda x^2 \tag{7.23}$$

According to equation (7.6), the first-order perturbation of the n-th level is

$$\Delta_n^{(1)} = H'_{nn} = \frac{\lambda}{2}\langle\psi_n^o|x^2|\psi_n^o\rangle = \frac{\lambda}{4m\omega}\langle\psi_n^o|(a^2 + a^{\dagger 2} + 2a^\dagger a + 1)|\psi_n^o\rangle \tag{7.24}$$

Since the raising and lowering operators have no diagonal elements,

$$\Delta_n^1 = \left(n + \frac{1}{2}\right)\frac{\lambda}{2m\omega} \tag{7.25}$$

The exact eigenvalues are $E_n = (n + \frac{1}{2})\omega'$, where

$$\omega'^2 = \omega^2\left(1 + \frac{\lambda}{m\omega^2}\right) \tag{7.26}$$

Expand ω' in powers of $\lambda/m\omega$:

$$\omega' = \omega\sqrt{1 + \frac{\lambda}{m\omega^2}} = \omega + \frac{\lambda}{2m\omega} + \cdots \tag{7.27}$$

The eigenvalues are

$$E_n = \left(n + \frac{1}{2}\right)\omega' \tag{7.28}$$

in agreement with (7.25). It is not hard to check that the term second order in λ is correct also.

7.2 STATIC EXTERNAL ELECTRIC FIELDS

The **Stark effect** [1] is the shift of atomic energy levels when the atoms are placed in a static electric field. Let the perturbing field be $\mathbf{E} = |\mathbf{E}|\hat{n}_z$. The electrostatic potential is $\phi = -Ez$, and the extra energy is

$$H' = e\mathbf{E}\cdot\mathbf{r} = e|\mathbf{E}|z \tag{7.29}$$

For hydrogen atoms, spin can be ignored in calculating the Stark effect, since the electron's spin does not enter H'. The energy levels are described by the state vectors $|n, l, m\rangle$. The unperturbed energies are

$$E^{(o)}{}_{nlm} = -\frac{\alpha^2 m}{2n^2} \tag{7.30}$$

Because z is odd under reflections, $\langle\psi|z|\psi\rangle = 0$ if $|\psi\rangle$ is a state of definite parity. In particular, $\langle 100|z|100\rangle = 0$. There is no first-order effect on the ground state. The perturbed energies are given through second order by equation (7.16):

$$E_{nlm} = E^{(o)}{}_{nlm} + e|\mathbf{E}|z_{nlm,nlm} + e^2\mathbf{E}^2\sum_{n'l'm'\neq nlm,}\frac{|z_{n'l'm',nlm}|^2}{E^{(o)}{}_{nlm} - E^{(o)}{}_{n'l'm'}} + \cdots \tag{7.31}$$

The second-order term lowers the ground-state energy.

In the absence of an electric field $-e\mathbf{r}$ is the dipole moment operator; more generally for a constant electric field the dipole moment $\mathbf{d}$ is related to the energy by $E = -\mathbf{E} \cdot \mathbf{d}$, or

$$d_i = -\frac{\partial}{\partial E_i} E \tag{7.32}$$

If $\mathbf{d}$ is nonzero when $\mathbf{E} \to 0$ the atom is said to have a permanent dipole moment. Since the n-th order in the perturbation expansion (7.31) is proportional to the n-th power of the external field the permanent dipole moment is given completely by the first-order term. A hydrogen atom in the ground state has no permanent dipole moment.

7.2.1 Perturbation of the First Excited Level

Something new happens in the $n = 2$ level. There are four $n = 2$ states, three $2P$ states and one $2S$ state. These are all parity eigenstates, so the diagonal matrix elements $\langle 2lm| z |2lm \rangle$ vanish.

There seems to be no first-order effect. But because all four states have the same energy, there is a zero in the denominator of the second-order term. What has gone wrong?

The assumption that no two states connected by H' were degenerate does not hold for the spinless hydrogen atom, violating the second restriction mentioned on page 205. The matrix element $\langle 210| z |200 \rangle$ does not vanish even though the $2S$ and $2P$ states with $m = 0$ have the same energy. The existence of this matrix element does not violate reflection invariance (parity conservation). The two states have opposite parity. Nor does it violate either of the two selection rules derived from the Wigner-Eckart theorem.

The way to get around this difficulty is to choose a new basis that partially diagonalizes H'. H' does not have to completely diagonal. It is necessary just to choose a basis such that all states $|\phi\rangle$ with the same energy as $|\psi\rangle$ satisfy $\langle \phi| H' |\psi \rangle = 0$. The states $|200\rangle$ and $|210\rangle$ are unsuitable. Instead, we should find linear combinations of them that diagonalize z, and then proceed safely with the standard formulation of perturbation theory in the new basis. This observation is the content of "degenerate perturbation theory."

The problem encountered here will usually not occur if H_o and H' are both invariant under some symmetry like rotations, since then H' will not connect states with the same unperturbed energy. The hydrogen atom is a special case, since the spectrum of H_o has the "accidental" degeneracy whose origin you studied in some detail back in Section 3.5. The constancy of the Runge-Lenz vector is a symmetry of an idealized approximation that is not a symmetry of nature.

The $2P$ states with $m = \pm 1$ do not connect z to any other $n = 2$ state, so there is no first-order Stark effect for them. Consider the two-dimensional space spanned by $|2S\rangle$ and $|2P\rangle$ with the index $m = 0$ suppressed.

In this basis, H' restricted to the $n = 2$, $m = 0$ states looks like

$$H' = eE \begin{pmatrix} 0 & z_{SP} \\ z_{PS} & 0 \end{pmatrix} \tag{7.33}$$

208 Chapter 7 Approximation Methods for Bound States

while
$$H_o = -\frac{\alpha^2 m}{8}\begin{pmatrix} 1 & 0 \\ 0 & 1 \end{pmatrix} \tag{7.34}$$

The wave functions are
$$\psi_{nlm}(\mathbf{r}) = R_{nl}(r)Y_l^m(\theta, \phi) \tag{7.35}$$

and the matrix element is
$$z_{SP} = z_{PS} = \int_0^\infty R_{21}(r)R_{20}(r)Y_1^o(\theta,\phi)Y_o^o(\theta,\phi)zr^2\,dr d\Omega \tag{7.36}$$

This is a simple two-state problem, like the one in Section 3.1.2. The eigenvalues are $\pm\langle 2P|\,z\,|2S\rangle$, and the eigenvectors are
$$|\psi_\pm\rangle = \frac{|210\rangle \pm |200\rangle}{\sqrt{2}} \tag{7.37}$$

Now $\langle\psi_+|\,H'\,|\psi_-\rangle = 0$, and the standard perturbation theory expression works in the new basis without difficulty. No selection rule requires the diagonal matrix element to vanish, since $|\psi_\pm\rangle$ are not eigenstates of $\mathbf{J}^2$ or $\mathcal{R}$. The unperturbed energies are equal, and the energy shift to first order is
$$\mp\Delta = \langle\psi_\pm|\,H'\,|\psi_\pm\rangle = \mp eE\langle 210|\,z\,|200\rangle \tag{7.38}$$

Equation (7.38) is the first-order Stark effect shift for atomic hydrogen. Because of the degeneracy of the unperturbed levels, the atom acts as if it has an intrinsic dipole moment.

To compute Δ set $z = r\cos\theta$, and write
$$\Delta = e|\mathbf{E}|\int_0^\infty R_{21}(r)R_{20}(r)r^3 dr \iint Y_1^o(\theta,\phi)Y_o^o(\theta,\phi)\cos\theta d\Omega \tag{7.39}$$

The angular integral is
$$\iint Y_1^o(\theta,\phi)Y_o^o(\theta,\phi)\cos\theta d\Omega = \frac{1}{\sqrt{3}} \tag{7.40}$$

The radial functions are listed in Table 3.1.
$$\int_0^\infty R_{21}R_{20}r^3\,dr = \frac{a}{8\cdot\sqrt{3}}\int_0^\infty (2-\rho)e^{-\rho}\rho^4\,d\rho = -3\sqrt{3}a \tag{7.41}$$

The perturbation is $\Delta = -3e|\mathbf{E}|a$. The splitting between the levels is 2Δ.

7.2.2 Polarizability of the Ground State

Electric fields induce a polarization in the atom even when its intrinsic dipole moment vanishes. The term in the energy quadratic in the field has the form
$$E = -\frac{1}{2}\sum_{ij}\alpha_{ij}E_i E_j \tag{7.42}$$

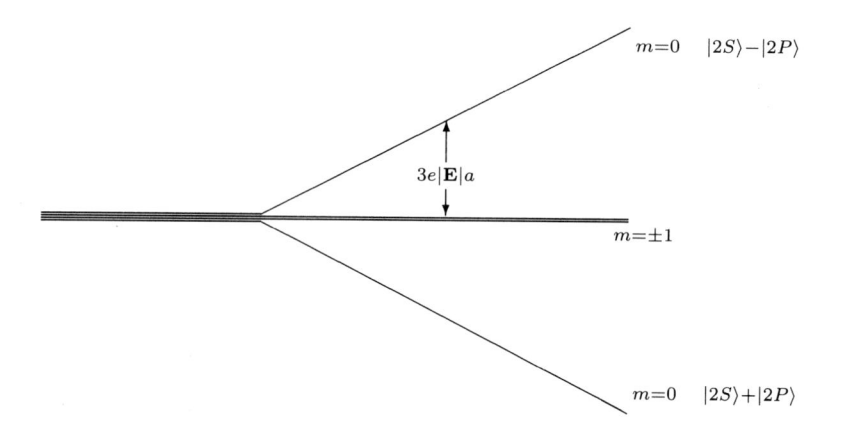

FIGURE 7.1: The Stark effect in the $n=2$ levels of the hydrogen atom. As the electric field is increased, two of the four energy levels are shifted in opposite directions, while the other two remain unchanged and degenerate.

so from the definition (7.32) there is an "induced dipole moment"

$$d_i = \sum_j \alpha_{ij} E_j \qquad (7.43)$$

The symmetric tensor α is the **polarizability tensor**. Because the ground state of the hydrogen atom is an S state, α is diagonal with all the diagonal elements equal, and the energy of the ground state is

$$E_{100} = E^{(o)}{}_{100} - \frac{1}{2}\alpha\mathbf{E}^2 + \cdots = E^{(o)}{}_{100} + e^2\mathbf{E}^2 \sum_{n'l'm' \neq 100} \frac{|z_{n'l'm',100}|^2}{E^{(o)}{}_{100} - E^{(o)}{}_{n'l'm'}} + \cdots \qquad (7.44)$$

The scalar α is the polarizability of the atom in its ground state. The second-order correction to the energy levels gives the theoretical prediction for this experimentally well-measured number α.

One approach is to evaluate as many terms in the sum for which you have the strength or the time, and hope that this is a fair approximation to the sum. Since the selection rules require all the intermediate states to have $l=1$ and $m=0$ the energy shift through second order is

$$\Delta^{(2)} = -2a\mathbf{E}^2 \sum_{n=2}^{\infty} \frac{n^2}{n^2-1} |\langle n10| z |100\rangle|^2 \qquad (7.45)$$

Approximating the sum by its first term we get

$$\Delta^{(2)} = -\frac{8}{3}a\mathbf{E}^2 |\langle 210| z |100\rangle|^2 = -\frac{2^{18}}{3^{11}}a^3\mathbf{E}^2 = -1.4798\ldots a^3\mathbf{E}^2 \qquad (7.46)$$

Since every term is negative, $\Delta^{(2)}$ in equation (7.46) is an upper bound. You can do better by including a few more terms (see Problem 7.9).

It is possible to get a lower bound also: Since $E_n \geq E_2$ in the sum that occurs in equation (7.45),

$$\Delta^{(2)} \geq e^2 \mathbf{E}^2 \sum_{n=2}^{\infty} \frac{|\langle n10|\, z\, |100\rangle|^2}{E_1 - E_2} = \frac{e^2 \mathbf{E}^2}{E_1 - E_2} \sum_{n=1}^{\infty} |\langle n10|\, z\, |100\rangle|^2 \qquad (7.47)$$

$$= -\frac{8}{3} a \mathbf{E}^2 \langle 100|\, z^2\, |100\rangle = -\frac{8}{3} a^3 \mathbf{E}^2$$

Exact Solution

Amazingly, it is possible to sum the series in (7.45) and compute the polarizability exactly.[2] I cannot resist presenting this result here briefly, even though the method is not generalizable to a very wide class of problems.

We developed the perturbation method starting from the form of Schrödinger's equation (7.8).

$$(H_o - E_o^o)|\psi_o\rangle = (\Delta - H')|\psi_o\rangle \qquad (7.48)$$

Here $|\psi_o\rangle$ is the exact ground state. Its energy is $E = E_o^o + \Delta$, where $E_o^o = -\alpha^2 m/2$. The unperturbed Hamiltonian is $H_o = \mathbf{p}^2/2m - \alpha/r$ and H' is given by equation (7.29). Let $|\psi_o^n\rangle$ be the n-th term in the perturbation series expansion of $|\psi_o\rangle$. Then

$$\left(H_o - E_o^o\right)\left(|\psi_o^0\rangle + |\psi_o^1\rangle\right) = (\Delta^1 - H')\,|\psi_o^0\rangle = -H'|\psi_o^0\rangle \qquad (7.49)$$

is the same as (7.48) correct through first order. In the last form I used the fact that for this problem $\Delta^1 = 0$.

Next suppose $|\psi_o^1\rangle = B|\psi_o^0\rangle$ for some operator B. Then

$$(H_o - E_o^o)\, B|\psi_o^0\rangle = -H'|\psi_o^0\rangle \qquad (7.50)$$

or[3]

$$[H_o, B]|\psi_o^0\rangle = -H'|\psi_o^0\rangle \qquad (7.51)$$

Then the energy shift is given to second order by equation (7.4):

$$\Delta^2 = \langle\psi_o^0|\, H'\, |\psi_o^1\rangle = \langle\psi_o^0|\, H'B\, |\psi_o^0\rangle \qquad (7.52)$$

Try to find an operator B that satisfies equation (7.51) and is a function of the coordinates alone: $\langle \mathbf{r}|\, B\, |\psi\rangle = B(\mathbf{r})\psi(\mathbf{r})$. In that case

$$\langle \mathbf{r}|\, H_o B\, |\psi_o^0\rangle = -\frac{1}{2m}\nabla^2 B(\mathbf{r})\psi_{100}(\mathbf{r}) - \frac{\alpha}{r} B(\mathbf{r})\psi_{100}(\mathbf{r})$$

$$= -\frac{1}{2m}\left[\left(\nabla^2 B(\mathbf{r})\right)\psi_{100}(\mathbf{r}) + 2\nabla B(\mathbf{r}) \cdot \nabla\psi_{100}(\mathbf{r}) + B(\mathbf{r})\nabla^2\psi_{100}(\mathbf{r})\right]$$

$$- \frac{\alpha}{r} B(\mathbf{r})\psi_{100}(\mathbf{r}) \qquad (7.53)$$

---

[2] Dalgarno and Lewis [2], Schwartz [3], Merzbacher [4] page 461.
[3] Equation (7.49) does not define $|\psi_o^1\rangle$ uniquely. We also need the normalization condition (7.19), which here reads $\langle\psi_o^0|\, B\, |\psi_o^0\rangle = 0$.

so $B(\mathbf{r})$ is a solution to

$$(\nabla^2 B(\mathbf{r}))\psi_{100}(\mathbf{r}) + 2\nabla B(\mathbf{r}) \cdot \nabla \psi_{100}(\mathbf{r}) = 2me|\mathbf{E}|z\psi_{100}(\mathbf{r}) \tag{7.54}$$

The ground-state wave function $\psi_{100}(\mathbf{r})$ is a function of the radial coordinate only. Write $B(\mathbf{r})$ also in spherical coordinates, expand it in spherical harmonics, and compare coefficients. Only the term in $P_1(\cos\theta)$ survives. Thus $B(\mathbf{r}) = b(r)\cos\theta$, where

$$\frac{1}{r^2}\frac{d}{dr}r^2\frac{d}{dr}b(r) - \frac{2}{r^2}b(r) + 2b'(r)\frac{R'_{10}(r)}{R_{10}(r)} = 2me|\mathbf{E}|r \tag{7.55}$$

Equation (7.55) has the solution

$$b(r) = -me|\mathbf{E}|ar\left(a + \frac{r}{2}\right) \tag{7.56}$$

So the ground-state wave function, correct to first order in the perturbation, is

$$\langle \mathbf{r}| \psi_o \rangle = \psi_{100}(\mathbf{r}) - me|\mathbf{E}|az\left(a + \frac{r}{2}\right)\psi_{100}(\mathbf{r}) \tag{7.57}$$

This wave function is indeed normalized as required by equation (7.20):

$$\langle \psi_o^0 | \psi_o^1 \rangle = -meEa\int z\left(a + \frac{r}{2}\right)|\psi_{100}(\mathbf{r})|^2 d^3r = 0 \tag{7.58}$$

so the energy, correct to second order, is given by equation (7.4) as

$$\Delta^1 + \Delta^2 = \langle \psi_o^0 | H' | \psi_o^1 \rangle = -me^2\mathbf{E}^2 a\int z^2\left(a + \frac{r}{2}\right)|\psi_{100}(\mathbf{r})|^2 d^3r$$

$$= -\frac{1}{3}me^2\mathbf{E}^2 a\left[a\langle \psi_o^0 | r^2 | \psi_o^0\rangle + \frac{1}{2}\langle \psi_o^0 | r^3 | \psi_o^0\rangle\right] \tag{7.59}$$

$$= -\frac{1}{3}me^2\mathbf{E}^2 a\left[3a^3 + \frac{15}{4}a^3\right] = -\frac{9}{4}\mathbf{E}^2 a^3$$

Remarks

Even when there is no external field, the degenerate levels of a hydrogen atom are split slightly by spin effects, (see Section 7.3) so that there remains a small correction to the treatment here of the degeneracy problem. But it will be a good approximation as long as the electric field is strong enough that this fine structure is small on the scale of the first-order energy shifts calculated here. Then the fine structure splitting is a small correction that can calculated as a perturbation.

A more serious problem is that as z gets large, the perturbing potential (7.29) grows linearly and therefore gets arbitrarily big no matter how small $|E|$ is. As a consequence, the energy for the problem is not really finite, and the perturbation series cannot really converge.

Nevertheless, the first-order and second-order approximations in this section agree very well with experiment. How is this possible? As elsewhere, the answer is that the mathematics is only an approximation to the physics. A real constant

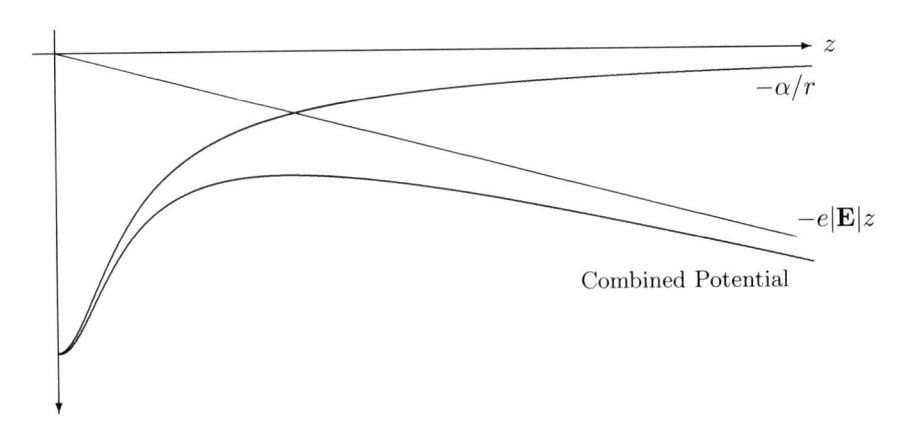

FIGURE 7.2: The total potential energy is shown as the sum of the Coulomb potential and the potential of a constant electrostatic field.

electric field does not go on forever, and these divergent terms are not really there. Still, you might imagine an electric field strong enough so that for some large enough z the perturbing electric field force $e|\mathbf{E}|$ is bigger in magnitude than the attractive Coulomb force $-\alpha/r^2$, as in Figure 7.2. In that case it would be energetically possible for the bound electron to tunnel through and escape the atom, although for typical electric fields the tunneling time for the ground state is greater than the age of the universe.

7.3 FINE STRUCTURE OF THE HYDROGEN ATOM

There are also small effects for a completely isolated atom. The same perturbation formalism applies to both situations. The mathematics does not know whether you turned on a perturbation that originally was not there, or just decided to include a small effect that had been neglected.

7.3.1 The Spin-Orbit Coupling

The electron has a spin magnetic moment. There is a term in the Hamiltonian for the energy of this little moving magnet in the electric field of the nucleus. A classical magnet would precess. Is it possible to guess the interaction from a classical analog? The answer is that the form is the same, but in detail the classical answer would depend on the structure of the magnet.

The whole system is rotationally invariant, so $\mathbf{J} = \mathbf{L} + \mathbf{s}$ commutes with H, even though $[\mathbf{L}, H]$ and $[\mathbf{s}, H]$ are not separately zero. H depends only on the scalars you can make out of the operators describing the spinning electron, namely, $\mathbf{r}^2$, $\mathbf{p}^2$. $\mathbf{L}^2$, and $\mathbf{L} \cdot \mathbf{s}$.

Let $V(r)$ be the potential energy of the electron due to electrostatic field of the nucleus. For the hydrogen atom, $V(r) = -\alpha/r$. There is no interaction energy between a magnetic moment at rest and a static electric field, but there is one if the magnet is moving. If the magnet—here an electron—is moving slowly enough, you can obtain the interaction energy by transforming to the frame in which the electron is at rest. In that frame the proton is moving with velocity $-\mathbf{v}$.

The magnetic field of a point charge q moving with velocity $\mathbf{u}$ is can be obtained from the Biot-Savart law. If the moving charge is at the origin, the magnetic field at some point $\mathbf{r}$ is

$$\mathbf{B} = q\frac{\mathbf{u} \times \mathbf{r}}{r^3} = \mathbf{u} \times \mathbf{E} \tag{7.60}$$

where $\mathbf{E}$ is the electrostatic field of the charge. As long as the electrostatic potential $\phi(\mathbf{r})$ is spherically symmetric, $\mathbf{E} = -\boldsymbol{\nabla}\phi(r)$, and the magnetic field seen by the electron in its rest frame is

$$\mathbf{B} = -\mathbf{v} \times \mathbf{E} = \mathbf{v} \times \boldsymbol{\nabla}\phi(r) = \mathbf{v} \times \frac{\mathbf{r}}{r}\frac{\partial \phi}{\partial r} = -\frac{\mathbf{r} \times \mathbf{p}}{mr}\frac{\partial \phi}{\partial r} = -\frac{\mathbf{L}}{mr}\frac{\partial \phi}{\partial r} \tag{7.61}$$

The magnetic energy becomes a spin-orbit coupling:

$$H_{LS} = -\boldsymbol{\mu} \cdot \mathbf{B} = -\frac{ge}{2m}\frac{\mathbf{L} \cdot \mathbf{s}}{mr}\frac{\partial \phi(r)}{\partial r} = \frac{g}{2m^2 r}\mathbf{L} \cdot \mathbf{s}\frac{dV(r)}{dr} \tag{7.62}$$

But this is not right. The r-dependence is correct, and the $\mathbf{L} \cdot \mathbf{s}$ factor is correct, but the observed coefficient is half that in equation (7.62), for complex atoms as well as hydrogen. The experimental result for an electron in *any* spherically symmetric potential is[4]

$$H_{LS} = \frac{g}{4m^2 r}\frac{dV(r)}{dr}\mathbf{L} \cdot \mathbf{s} \tag{7.63}$$

For the hydrogen atom $dV(r)/dr = \alpha/r^2$, and with $g = 2$,

$$\boxed{H_{LS} = \frac{\alpha}{2m^2}\frac{1}{r^3}\mathbf{L} \cdot \mathbf{s}} \tag{7.64}$$

For point particles like the electron, equation (7.64), together with the gyromagnetic ratio $g = 2$, is predicted by a correct relativistic treatment using the Dirac equation. But there is no classical explanation.

There was really no reason to expect the classical argument to give the correct result. The argument was flawed, since it used a transformation to an accelerated reference frame. The factor of two that distinguishes equation (7.62) from equation (7.63) is sometimes said to be due to the "Thomas precession." An accelerated magnetic moment should precess, thus in effect reducing the time-averaged moment along any axis. There are nonrelativistic "derivations" of the Thomas precession, but I believe they all assume some property of a classical electron designed to get the right answer. The correct derivation can be found in Section 12.4.3.

---

[4]For complex atoms, the energy levels of the valence electron(s) can often be reproduced by some effective spherically symmetric potential $V(r)$. In that case the fine structure is given correctly by equation (7.63). This expression is the origin of the **Landé interval rule** [5, 6].

7.3.2 Correction to Energy Levels

Now put H_{LS} from equation (7.64) into the perturbation theory machinery and calculate to first order. Because of the factor $\mathbf{L}\cdot\mathbf{s}$, H_{LS} is not diagonal in the basis $|\psi_{m_l m_s}^{nl}\rangle$, but rather in the basis $|\Phi_{jm}^{nl}\rangle$. One set of basis states can be expressed in terms of the other with the Clebsch-Gordan coefficients. The new basis still diagonalizes H_o, since H_o is independent of the spin: The first-order energy shift is

$$\langle \Phi_{jm}^{nl}| H_{LS} |\Phi_{jm}^{nl}\rangle = \frac{\alpha}{2m^2}\left\langle \Phi_{jm}^{nl}\left|\frac{1}{r^3}\mathbf{L}\cdot\mathbf{s}\right|\Phi_{jm}^{nl}\right\rangle \tag{7.65}$$

$\mathbf{L}\cdot\mathbf{s}$ commutes with all the components of $\mathbf{J}$, so it is diagonal in the $|\Phi_{jm}^{nl}\rangle$ basis. Therefore,

$$\langle \Phi_{jm}^{nl}| H_{LS} |\Phi_{jm}^{nl}\rangle = \frac{\alpha}{2m^2}\left\langle \Phi_{jm}^{nl}\left|\frac{1}{r^3}\right|\Phi_{jm}^{nl}\right\rangle\left\langle \Phi_{jm}^{nl}\left|\mathbf{L}\cdot\mathbf{s}\right|\Phi_{jm}^{nl}\right\rangle \tag{7.66}$$

The expectation value of $1/r^3$ is independent of j and m. From Problem 3.21

$$\left\langle \Phi_{jm}^{nl}\left|\frac{1}{r^3}\right|\Phi_{jm}^{nl}\right\rangle = \frac{\alpha^3 m^3}{n^3 l(l+1)(l+\frac{1}{2})} \tag{7.67}$$

The expectation value of $\boldsymbol{\sigma}\cdot\mathbf{L}$ is l in the states with $j = l+1/2$ and is $-l-1$ in the states with $j = l-1/2$. The first-order energy shift due to the spin-orbit coupling is

$$\boxed{\Delta_{LS} = \langle \Phi_{jm}^{nl}| H_{LS} |\Phi_{jm}^{nl}\rangle = \frac{\alpha^4 m}{4n^3 l(l+1)(l+\frac{1}{2})}\binom{l}{-l-1}} \tag{7.68}$$

The top row is $j = l+1/2$ and the bottom row is $j = l-1/2$.

Note on $l = 0$:
I have not really derived equation (7.68) for $l = 0$ states. In that case the expression is $0/0$. A careful derivation comes from approximating the Dirac equation up to terms of the order $\alpha^4 m$ in the energy. There is no legitimate way to derive the result in nonrelativistic quantum mechanics.

In the nonrelativistic limit of the Dirac equation the factor $1/r^3$ that occurs in equation (7.65) is replaced by a function slightly less divergent at the origin (see Section 12.4.3). As a consequence, the expectation value of this function is really finite for S states, so the spin-orbit coupling term is zero when $l = 0$. But there is another term, called the **Darwin term**,

$$H_3 = -\frac{1}{8m^2}\boldsymbol{\nabla}^2 V(r) \tag{7.69}$$

where for the hydrogen atom $V(r) = -\alpha/r$. Because $\boldsymbol{\nabla}^2(1/r) = -4\pi\delta_3(\mathbf{r})$, the Darwin term is a contact term, and affects S states only. Its contribution to the energy is

$$E_3 = \frac{\alpha^4 m}{2n^3}\delta_{l0} \tag{7.70}$$

Therefore, for any l we can write the sum of the spin-orbit energy and the Darwin term energy when $j = l + 1/2$ as

$$\Delta_{LS} = \frac{\alpha^4 m}{4n^3(l+1)(l+\frac{1}{2})} \tag{7.71}$$

for any l. There are no S states with $j = l - 1/2$. These terms will be derived correctly in Section 12.4.3.

7.3.3 The Relativistic Kinetic Energy Correction

Equation (7.68) still does not agree with experiment! The magnetic field due to a moving charge is a first-order relativistic effect. To include all effects of the same order, you need take into account also the relativistic change in the electron's kinetic energy. This correction is the same order in α as the spin-orbit correction. The relativistic kinetic energy is

$$T = E_{\text{rel}} - m \tag{7.72}$$

with

$$E_{\text{rel}}^2 = m^2 + p^2 \tag{7.73}$$

so that a better value for T is

$$T = \sqrt{m^2 + p^2} - m = \frac{p^2}{2m} - \frac{p^4}{8m^3} + \cdots \tag{7.74}$$

The first term is the usual nonrelativistic kinetic energy. The second term can be tacked on as a perturbation H_{kin}. Then to this order it is all right to set p equal to the nonrelativistic momentum—corrections will be even higher order. Evaluate H_{kin} as follows, with $T_o = p^2/2m$:

$$H_{\text{kin}} = -\frac{1}{2m}\frac{p^4}{4m^2} = -\frac{1}{2m}T_o^2 = -\frac{1}{2m}(H_o - V)^2$$

$$= -\frac{1}{2m}(H_o^2 - H_o V - V H_o + V^2) \tag{7.75}$$

Since for the hydrogen atom $\langle \Phi_{jm}^{nl} | V | \Phi_{jm}^{nl} \rangle = 2E_n^o$,

$$\langle \Phi_{jm}^{nl} | (H_o^2 - V H_o - H_o V) | \Phi_{jm}^{nl} \rangle = -3(E_n^o)^2 \tag{7.76}$$

and, since (see Problem 3.20)

$$\langle \Phi_{jm}^{nl} | V^2 | \Phi_{jm}^{nl} \rangle = \frac{\alpha^2}{a^2 n^3(l+\frac{1}{2})} \tag{7.77}$$

the expectation value of V^2 is

$$\langle \Phi_{jm}^{nl} | V^2 | \Phi_{jm}^{nl} \rangle = \frac{\alpha^2}{a^2 n^3 (l+\frac{1}{2})} \tag{7.78}$$

The energy shift is

$$\Delta_{\text{kin}} = \langle \Phi_{jm}^{nl} | H_{\text{kin}} | \Phi_{jm}^{nl} \rangle = -\frac{1}{2m}\left(-\frac{3\alpha^4 m^2}{4n^4} + \frac{\alpha^2}{a^2 n^3 (l+\frac{1}{2})}\right) \qquad (7.79)$$
$$= \frac{\alpha^4 m}{2n^4}\left(\frac{3}{4} - \frac{n}{l+\frac{1}{2}}\right)$$

7.3.4 The Fine Structure of the Hydrogen Atom

The total energy shift is the sum of (7.68) and (7.79):

$$\Delta_{FS} = \Delta_{\text{kin}} + \Delta_{LS} = \frac{\alpha^4 m}{2n^4}\left[\frac{n}{2l(l+1)(l+\frac{1}{2})}\left(\begin{array}{c}l\\-l-1\end{array}\right) + \frac{3}{4} - \frac{n}{l+\frac{1}{2}}\right] \qquad (7.80)$$

For fixed l, equation (7.80) written in terms of j instead of l works out to be

$$\Delta_{FS} = \frac{\alpha^4 m}{2n^4}\left(\frac{3}{4} - \frac{2n}{2j+1}\right) \qquad (7.81)$$

for both $j = l + 1/2$ and $j = l - 1/2$.

The value of Δ_{FS} depends on n and j, but surprisingly not on l. The $2P_{\frac{1}{2}}$ and the $2P_{\frac{3}{2}}$ levels are split, but the $2S_{\frac{1}{2}}$ and the $2P_{\frac{1}{2}}$ levels are still degenerate, even though this degeneracy is not required by rotational symmetry. There is no

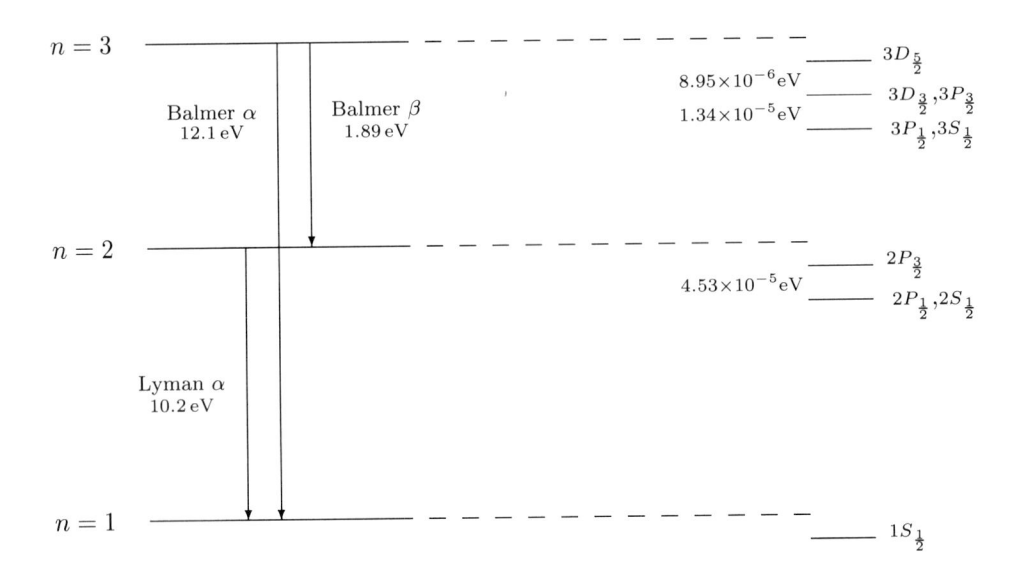

FIGURE 7.3: Schematic diagram of the fine structure splitting in the lowest three levels of atomic hydrogen (not to scale)

real symmetry of nature that requires the $2S_{\frac{1}{2}}$ and the $2P_{\frac{1}{2}}$ states to have the same energy, and there is a small effect that *does* split them. It is called the Lamb shift, and can be computed in relativistic quantum field theory. The Dirac equation (Chapter 12) predicts that the energy levels depend on j only to all orders.

For any n, the fine structure splitting between the $j = \frac{1}{2}$ and the $j = \frac{3}{2}$ states is

$$\Delta = \frac{\alpha^4 m}{4n^3} \tag{7.82}$$

For $n = 2$, the splitting is about 4.5283×10^{-5} eV. The fine structure splitting for the lowest levels of the hydrogen atom is illustrated in Figure 7.3.

7.3.5 External Magnetic Field Again

Now we can compute the effect of a weak external magnetic field when the fine structure is taken into account. Here I confine myself to fields weak compared to the fine structure splitting. The more general problem, fields weak compared to the gross level splitting but large compared to the fine structure splitting, is left as an exercise (see Problem 7.18).

Redefine the "unperturbed" part of the Hamiltonian by writing

$$H_o = \frac{\mathbf{p}^2}{2m} - \frac{\alpha}{r} + H_{FS} \tag{7.83}$$

so that the energy levels computed in Section 7.3.4 are the unperturbed levels, and treat an external magnetic field in the z-direction as the perturbation. Set $g = 2$, since the correction to g is higher order in α:

$$H' = \mu_B B (L_z + 2s_z) \tag{7.84}$$

$L_z + 2s_z$ is not diagonal in the $\left|\psi_{jm}^{nl}\right\rangle$ basis, but its diagonal elements can be computed using the Clebsch-Gordan coefficients.[5]

Now compute the effect of H' in the $n = 2$ levels, between states of definite j and m. The matrix elements of H' between eigenstates $\left|\psi_{m_l,m_s}^{nl}\right\rangle$ of L_z and s_z are

$$\left\langle \psi_{m_l,m_s}^{2l} \right| H' \left| \psi_{m_l,m_s}^{2l} \right\rangle = \mu_B B (m_l + 2m_s) \tag{7.85}$$

Then using

$$\left|\phi_{\frac{3}{2},\frac{3}{2}}^{21}\right\rangle = \left|\psi_{1,\frac{1}{2}}^{21}\right\rangle \qquad \left|\phi_{\frac{1}{2},\frac{1}{2}}^{20}\right\rangle = \left|\psi_{0,\frac{1}{2}}^{20}\right\rangle \tag{7.86a}$$

$$\left|\phi_{\frac{1}{2},\frac{1}{2}}^{21}\right\rangle = \frac{\left|\psi_{0,\frac{1}{2}}^{21}\right\rangle - \sqrt{2}\left|\psi_{1,-\frac{1}{2}}^{21}\right\rangle}{\sqrt{3}} \tag{7.86b}$$

one gets

$$\left\langle \phi_{\frac{3}{2},\frac{3}{2}}^{21} \right| H' \left| \phi_{\frac{3}{2},\frac{3}{2}}^{21} \right\rangle = 2\mu_B B \qquad \left\langle \phi_{\frac{1}{2},\frac{1}{2}}^{20} \right| H' \left| \phi_{\frac{1}{2},\frac{1}{2}}^{20} \right\rangle = \mu_B B \tag{7.87a}$$

$$\left\langle \phi_{\frac{1}{2},\frac{1}{2}}^{21} \right| H' \left| \phi_{\frac{1}{2},\frac{1}{2}}^{21} \right\rangle = \frac{1}{3}\mu_B B \tag{7.87b}$$

---
[5] m_j is the eigenvalue of $J_z = L_z + s_z$.

218 Chapter 7 Approximation Methods for Bound States

You could continue in this fashion, computing the remaining states in each ladder with the Clebsch-Gordan coefficients. But since μ is a vector operator, all its matrix elements of the form $\langle \phi^{nl}_{jm'} | \mu_i | \phi^{nl}_{jm} \rangle$ are related by coefficients that depend only on i, m', and m.[6] It follows that

$$\langle \phi^{nl}_{jm} | \mu_z | \phi^{nl}_{jm} \rangle \sim \langle \phi^{nl}_{jm} | J_z | \phi^{nl}_{jm} \rangle = m \tag{7.88}$$

The proportionality constant can be read off equation (7.87). The $n = 2$ level, for example, is split as follows:

$$2S_{\frac{1}{2}}: \quad E = -\frac{\alpha^2 m}{8} - \frac{5\alpha^4 m}{128} + 2m_j \mu_B B$$

$$2P_{\frac{1}{2}}: \quad E = -\frac{\alpha^2 m}{8} - \frac{5\alpha^4 m}{128} + \frac{2}{3} m_j \mu_B B \tag{7.89a}$$

$$2P_{\frac{3}{2}}: \quad E = -\frac{\alpha^2 m}{8} - \frac{\alpha^4 m}{128} + \frac{4}{3} m_j \mu_B B$$

This use of the Wigner-Eckart theorem is a special case of the projection theorem, Section B.1.3.

7.3.6 The Hyperfine Structure of the Hydrogen Atom

So far I have taken H_o to be the Hamiltonian of an electron moving in the electrostatic potential of the nucleus. The nucleus also has spin. Like the electron, it is a little dipole, and produces a magnetic field. The correction due to the motion of the electron in the magnetic field of the nucleus is the **hyperfine structure**.

The proton, also a spin-half particle, has a magnetic moment is $\boldsymbol{\mu}_p = g_p \mu_N \mathbf{s}_p$, where $\mathbf{s}_p$ is the proton's spin. Its gyromagnetic ratio (measured, for example, in nuclear magnetic resonance experiments) is about 2×2.793, and $\mu_p = e/2M$. The proton's mass M is about $1836\,m$.

The total angular momentum of the electron and the proton is

$$\mathbf{F} = \mathbf{L} + \mathbf{s} + \mathbf{s}_p \tag{7.90}$$

Only the total $\mathbf{F}$ is a constant of the motion, because the whole atom is rotationally symmetric, but not the electron or the proton separately.

The magnetic field due to a magnetic moment $\boldsymbol{\mu}_p$ is $\mathbf{B} = \boldsymbol{\nabla} \times \mathbf{A}$, where

$$\mathbf{A} = -\boldsymbol{\mu}_p \times \boldsymbol{\nabla} \frac{1}{r} \tag{7.91}$$

so the magnetic field of the nucleus is

$$\mathbf{B} = -\boldsymbol{\nabla} \times \left(\boldsymbol{\mu}_p \times \boldsymbol{\nabla} \frac{1}{r} \right) = \boldsymbol{\mu}_p \cdot \boldsymbol{\nabla} \left(\boldsymbol{\nabla} \frac{1}{r} \right) - \boldsymbol{\mu}_p \boldsymbol{\nabla}^2 \frac{1}{r} \tag{7.92}$$

$\mathbf{B}$ enters both the orbital and the spin parts of the electron electromagnetic interaction. The orbital part is obtained from the minimal coupling rule:

$$H = \frac{1}{2m} \left(\mathbf{p} + e\mathbf{A}(\mathbf{r}) \right)^2 = \frac{\mathbf{p}^2}{2m} + H'_{\text{orbital}} \tag{7.93}$$

---
[6] See Section 5.1.4 and Problem 5.1.

Section 7.3 Fine Structure of the Hydrogen Atom

so that[7]

$$H'_{\text{orbital}} = \frac{e}{m}\mathbf{p} \cdot \boldsymbol{\nabla}\left(\frac{1}{r}\right) \times \boldsymbol{\mu}_p = -\frac{g_p e^2}{2Mm}\frac{1}{r^3}\mathbf{p} \cdot \mathbf{r} \times \mathbf{s}_p = \frac{g_p \alpha}{2Mm}\frac{1}{r^3}\mathbf{s}_p \cdot \mathbf{L} \quad (7.94)$$

plus terms higher order in e^2.

The interaction of the proton's magnetic field with the electron spin (the proton magnetic field is effectively constant over the dimension of the electron) is

$$H'_{\text{spin}} = -\boldsymbol{\mu} \cdot \mathbf{B} = \frac{ge}{2m}\mathbf{s} \cdot \mathbf{B}$$
$$= \frac{ge}{2m}\frac{g_p e}{2M}\left[(s_p \cdot \boldsymbol{\nabla})(\mathbf{s} \cdot \boldsymbol{\nabla})\frac{1}{r} - \mathbf{s}_p \cdot \mathbf{s}\boldsymbol{\nabla}^2\frac{1}{r}\right] = \frac{gg_p\alpha}{4mM}\sum_{ij}s_{ei}s_{pj}T_{ij}(\mathbf{r}) \quad (7.95)$$

The last factor is a symmetric rank-two Cartesian tensor T_{ij} that acts on the coordinate only:

$$T_{ij} = \left(\partial_i\partial_j - \delta_{ij}\boldsymbol{\nabla}^2\right)\left(\frac{1}{r}\right) = \frac{1}{3}\delta_{ij}\sum_k T_{kk} + \left(T_{ij} - \frac{1}{3}\delta_{ij}\sum_k T_{kk}\right) \quad (7.96)$$

Using

$$\sum_k T_{kk} = -2\boldsymbol{\nabla}^2\left(\frac{1}{r}\right) \quad (7.97)$$

it can be written in the form of equation (5.40):

$$T_{ij} = -\frac{2}{3}\delta_{ij}\boldsymbol{\nabla}^2\left(\frac{1}{r}\right) + \left(\partial_i\partial_j - \frac{1}{3}\delta_{ij}\boldsymbol{\nabla}^2\right)\left(\frac{1}{r}\right) \quad (7.98)$$

Therefore,

$$H'_{\text{spin}} = H'_{\text{contact}} + H'_{\text{dipole}} \quad (7.99)$$

where

$$H'_{\text{contact}} = \frac{gg_p\alpha}{4mM}\frac{8\pi}{3}\mathbf{s}_e \cdot \mathbf{s}_p\delta_3(\mathbf{r}) \quad (7.100)$$

and

$$H'_{\text{dipole}} = \frac{gg_p\alpha}{4mM}\sum_{ij}s_{ei}s_{pj}\left(\partial_i\partial_j - \frac{1}{3}\delta_{ij}\boldsymbol{\nabla}^2\right)\left(\frac{1}{r}\right)$$
$$= \frac{gg_p\alpha}{4mM}\frac{1}{r^5}\sum_{ij}s_{ei}s_{pj}\left(3r_ir_j - \delta_{ij}r^2\right) \quad (7.101)$$

Hyperfine Splitting of the Ground State

In the $n = 1$ level $\mathbf{L} = 0$ and $\mathbf{F} = \mathbf{s}_p + \mathbf{s}$. The four states can be labeled either by the electron spin quantum number $m_s = \pm\frac{1}{2}$ and the proton spin quantum number $m_p = \pm\frac{1}{2}$, or else by F and m_F. The orbital term does not contribute in

---

[7]Since the hyperfine correction is very small, the $\mathbf{A}^2$ term can be neglected, as can relativistic corrections.

an S-state, nor, because of one of the selection rules, does the piece that transforms under spatial rotations like a rank-two tensor. Only the contact term is left. The expectation value of H_{HFS} in the $1S$ state is

$$\frac{gg_p\alpha}{4mM} \frac{8\pi}{3} \mathbf{s}_e \cdot \mathbf{s}_p \int \delta_3(\mathbf{r})|\psi_{100}(r)|^2 \, d^3r \tag{7.102}$$

The hyperfine splitting is proportional the value of the ground-state wave function at the origin.

$$|\psi_{100}(0)|^2 = \frac{1}{\pi a^3} = \frac{m^3\alpha^3}{\pi} \tag{7.103}$$

H_{HFS} has off-diagonal elements in the basis $|n, l, j, m_j, m_F\rangle$, so technically "degenerate perturbation theory" is called for here. The basis that diagonalizes H_{HFS} is made up of the eigenstates of $\mathbf{F}^2$ and F_z. The possible values of the total angular momentum F are 1 and 0. The expectation value of $2\mathbf{s} \cdot \mathbf{s}_p$ in states of fixed $\mathbf{F}^2$ is $1/2$ for $F = 1$ and $-3/2$ for $F = 0$. Thus,

$$\boxed{\Delta_{HFS} = \frac{\alpha^4}{3} \frac{m}{M} gg_p m \begin{pmatrix} \frac{1}{2} \\ -\frac{3}{2} \end{pmatrix} \quad \text{for} \quad F = \begin{pmatrix} 1 \\ 0 \end{pmatrix}} \tag{7.104}$$

The splitting is

$$\Delta = 2\Delta_{HFS} = \frac{2}{3} \times \frac{1}{137.036^4} \times \frac{1}{1836} \times 2 \times 2 \times 2.78 \times 511{,}000\,\text{eV}$$
$$= 5.85009 \times 10^{-6}\,\text{eV} \tag{7.105}$$

The hyperfine splitting and the fine structure splitting are the same order in α, but the hyperfine splitting is smaller by a factor of order m/M.

The hyperfine transition is the last signal from most of the atoms that make up much of the dark matter in space. Its wavelength is

$$\lambda = \frac{2\pi}{\Delta} = \frac{2\pi \times 1.97 \times 10^{-5}\,\text{eV-cm}}{5.86 \times 10^{-6}\,\text{eV}} \approx 21.2\,\text{cm} \tag{7.106}$$

This is the famous 21-cm line that was first observed using radio astronomy techniques in 1951. It pervades the universe, varying with its source in intensity and Doppler shift. It was used to map for the first time the spiral arms of our own galaxy.[8]

7.4 OTHER ATOMS

7.4.1 The Ground State of Helium

The helium atom is a good introduction to the many-electron systems. I will begin with a drastic oversimplification, and ignore not only magnetic effects but also the electrostatic force between the electrons. Then this is a problem to which you already know the solution. With no extra effort we can include all two-electron

---
[8] Purcell and Ewen [8], Muller and Oort [7].

Section 7.4 Other Atoms

ions such as the negative hydrogen atom or the positive lithium ion as well as the neutral helium atom.

Let the charge of the nucleus be Ze. For helium, $Z = 2$. The Hamiltonian contains the energy of each electron separately in the Coulomb field of the nucleus.

$$H_o = H_1 + H_2 \tag{7.107}$$

where

$$H_i = \frac{p_i^2}{2m} - \frac{Ze^2}{r_i} \tag{7.108}$$

The eigenstates of the Hamiltonian (7.107) are

$$|n_1, l_1, m_{1l}, m_{1s}, n_2, l_2, m_{2l}, m_{2s}\rangle \tag{7.109}$$

with wave functions

$$\psi_{m_{1s} m_{2s}}(\mathbf{r}_1, \mathbf{r}_2) = R_{n_1 l_1}(r_1) Y_{l_1}^{m_{1l}}(\theta_1, \phi_1) R_{n_2 l_2}(r_2) Y_{l_2}^{m_{2l}}(\theta_2, \phi_2) \chi_{m_{1s}, m_{2s}} \tag{7.110}$$

$\chi_{m_{1s}, m_{2s}}$ is any properly normalized function of the spin indices. The functions $R_{nl}(r)$ are the hydrogen atom radial wave functions from Section 3.5.3. For low n these can be read from Table 3.1 with α in the hydrogen case replaced by $Z\alpha$. The energies are

$$E = -\frac{Z^2 \alpha^2 m}{2n_1^2} - \frac{Z^2 \alpha^2 m}{2n_2^2} \tag{7.111}$$

Electrons are fermions, and these two-electron states have to change sign when you exchange all their properties. It is useful to combine the two spins first, since the states with definite total spin have definite symmetry under exchange of the spin quantum number. Let $\mathbf{s} = \mathbf{s}_1 + \mathbf{s}_2$ be the total spin, and choose the basis of eigenstates of $\mathbf{s}^2$ and s_z, instead of s_{1z} and s_{2z}. Suppressing the space indices, write down the $|\phi_{sm}\rangle$ states in terms of the $|\psi_{m_1, m_2}\rangle$ states making use of the Clebsch-Gordan coefficients for combining two spin-half systems:

$$|\phi_{1,1}\rangle = \left|\psi_{\frac{1}{2},\frac{1}{2}}\right\rangle \tag{7.112a}$$

$$|\phi_{1,0}\rangle = \frac{1}{\sqrt{2}}\left(\left|\psi_{\frac{1}{2},-\frac{1}{2}}\right\rangle + \left|\psi_{-\frac{1}{2},\frac{1}{2}}\right\rangle\right) \tag{7.112b}$$

$$|\phi_{1,-1}\rangle = \left|\psi_{-\frac{1}{2},-\frac{1}{2}}\right\rangle \tag{7.112c}$$

$$|\phi_{0,0}\rangle = \frac{1}{\sqrt{2}}\left(\left|\psi_{\frac{1}{2},-\frac{1}{2}}\right\rangle - \left|\psi_{-\frac{1}{2},\frac{1}{2}}\right\rangle\right) \tag{7.112d}$$

The three spin-one states are symmetric under the interchange of m_1 and m_2; the singlet spin-zero state is antisymmetric. So if the states have definite total spin, they have definite symmetry under exchange of their spin indices. To guarantee total antisymmetry, the triplet states must be antisymmetric and the singlet state symmetric when $\mathbf{r}_1$ and $\mathbf{r}_2$ are interchanged.

7.4.2 The Perturbation Method for the Helium Atom

Now add the Coulomb repulsion between the electrons. One approach is to take H_o as above, and the interelectron repulsion as a perturbation, although it is not obvious that H' is smaller than H_o in any useful sense.

$$H' = +\frac{e^2}{|\mathbf{r}_1 - \mathbf{r}_2|} \tag{7.113}$$

Without H', an eigenstate of H would be the product of ground-state wave functions for each electron as in equation (7.110) with $n_1 = n_2 = 1$. Since this state is symmetric in $\mathbf{r}_1$ and $\mathbf{r}_2$ it must be a spin singlet state. I will suppress the spin index here with the understanding that the spin is zero for the ground state.

The wave function for this spin-singlet ground $|\psi_o\rangle$ is

$$\psi_o(\mathbf{r}_1, \mathbf{r}_2) = \psi_{100}(\mathbf{r}_1)\psi_{100}(\mathbf{r}_2) \tag{7.114}$$

where

$$\psi_{100}(\mathbf{r}) = \sqrt{\frac{Z^3}{\pi a^3}} e^{-Zr/a} \tag{7.115}$$

Then the expectation value of H_o is $2Z^2 E_1^o$, where $E_1^o = -\alpha^2 m/2$ is the unperturbed energy of an electron in the Coulomb potential of a single proton, about -13.6 ev.

These are the zero order energies for the heliumlike atoms and ions. Here is a table. For comparison I have listed the experimental energies and the energies $Z^2 E_1^o$ of a single electron H atom, or He^+ and Li^{++} ions.

	Zero Order	Experiment
H^-	-27.21	-14.35
He	-108.85	-78.99
Li^+	-244.90	-198.1

TABLE 7.1: Zero order energies of some two-electron ions, compared with one-electron ions and with experiment (in electron volts)

While the unperturbed energies are the same order of magnitude as the experimental result, the difference seems too big for low order perturbation theory to work well. Nevertheless it is instructive to pursue the perturbation method a bit further. The first-order energy shift is

$$\Delta^{(1)} = \langle \psi | H' | \psi \rangle = e^2 \frac{Z^6}{\pi^2 a^6} \iint \frac{e^{-2Zr_1/a} e^{-2Zr_2/a}}{|\mathbf{r}_1 - \mathbf{r}_2|} d^3 r_1 \, d^3 r_2 \tag{7.116}$$

The six-dimensional integral in equation (7.116) is straightforward to calculate analytically. From Section A.1.3 in the appendix, the Fourier transform of $1/r$ is

$$\frac{1}{r} = \frac{4\pi}{(2\pi)^3} \int \frac{1}{k^2} e^{i\mathbf{k}\cdot\mathbf{r}} d^3 k \tag{A.35}$$

The first-order energy shift (7.116) is

$$\Delta^{(1)} = e^2 \frac{Z^6}{a^6} \frac{1}{2\pi^4} \int \frac{1}{k^2} \left(\int e^{-2Zr/a} e^{i\mathbf{k}\cdot\mathbf{r}} d^3r \right)^2 d^3k \qquad (7.117)$$

The integral in parentheses is equation (A.37a). Then with the help of equation (A.31) also

$$\Delta^{(1)} = \frac{128 e^2}{\pi^2} \left(\frac{Z}{a}\right)^8 \int \frac{1}{k^2 \left[(2Z/a)^2 + k^2\right]^4} d^3k$$

$$= \frac{4Ze^2}{\pi a} \int_0^\infty \frac{1}{(1+x^2)^4} dx = \frac{4Ze^2}{\pi a} \frac{5\pi}{32} = -\frac{5}{4} Z E_o \qquad (7.118)$$

Since E_o is negative, $\Delta^{(1)}$ increases the predicted energies, as expected. The numerical results are shown in Table 7.2.

	First Order	Experiment	Single Electron
H⁻	−10.20	−14.35	−13.61
He	−74.83	−78.99	−54.43
Li⁺	−193.88	−198.1	−122.45

TABLE 7.2: First-order energies of some two-electron ions, compared with one-electron ions and with experiment (in electron volts)

It turns out to be a much better approximation than one had any right to expect. The approximation of equation (7.118) differs from the experimental result by only about 4 eV for all three of these ions. But while the error is small, it is too big to be accounted for by the effects we have neglected. The second-order term is significant.

For the negative hydrogen ion, the answer is of no help at all. The predicted energy is *higher* than the energy of a single-electron atom. If −10.2 eV were the correct answer, the H⁻ ion would not be stable, since it could decay into a hydrogen atom and a free electron. Experimentally, the ion, while weakly bound, is stable, though not by much.

7.5 THE VARIATIONAL METHOD

7.5.1 The General Method

The variational, or Rayleigh-Ritz method is a nonperturbative way to the approximate calculations of eigenvalues.[9] Most methods for finding the energy levels of atoms and molecules are variational schemes. Here is the basic idea.

Let $|\psi_n\rangle$ be an eigenstate of H, and let $|\psi\rangle = |\psi_n\rangle + |\delta\psi\rangle$ be some nearby state, normalized to $\langle\psi|\psi\rangle = 1$. Precisely this means that if $|\psi_n\rangle$ are some complete

---

[9]The original applications are in Section 88 of Rayleigh [10], and in Ritz [9].

224 Chapter 7 Approximation Methods for Bound States

orthonormal set of eigenstates of H, then

$$|\psi\rangle = |\psi_n\rangle + |\delta\psi\rangle = N|\psi_n\rangle + \sum_{m \neq n} \epsilon_m |\psi_m\rangle \qquad (7.119)$$

for some "small" numbers ϵ_m. N is chosen so the normalization comes out correctly. Let E be the expectation value of H in $|\psi\rangle$. The variational method works so well because the change of E due to small ϵ_m vanishes to first order:

$$\begin{aligned} \delta E &= \langle\psi| H |\psi\rangle - \langle\psi_n| H |\psi_n\rangle \\ &= \langle\psi_n| H |\delta\psi\rangle + \langle\delta\psi| H |\psi_n\rangle + \langle\delta\psi| H |\delta\psi\rangle \\ &\approx E_n \Big[\langle\psi| \, \delta\psi\rangle + \langle\delta\psi| \, \psi\rangle \Big] \end{aligned} \qquad (7.120)$$

In the last form I left off the term that is second order in the ϵ_n. Since the states are normalized, $1 = \langle\psi| \psi\rangle$, and

$$0 = \delta\langle\psi| \psi\rangle = \langle\psi| \, \delta\psi\rangle + \langle\delta\psi| \, \psi\rangle \qquad (7.121)$$

and therefore (again up to second order)

$$\boxed{\delta E = E_n \, \delta\langle\psi_n| \psi_n\rangle = 0} \qquad (7.122)$$

The expectation value of the Hamiltonian is an extremum when $|\psi\rangle$ is an energy eigenstate.

If you can guess a state that is near an eigenstate, the expectation value of the Hamiltonian is near the eigenvalue. If you make a good guess $|\psi\rangle$ for $|\psi_n\rangle$, then $\langle\psi| H |\psi\rangle$ is an even better guess for E_n. and if the error in the state is first order in some small numbers ϵ_i, the error in the expectation value of the energy will be second order. If $|\psi\rangle$ depends on some parameters, the value E_n for the parameters that extremize $\langle\psi| H |\psi\rangle$ is the best E_n among all of them. With enough parameters it is often possible to do very well, and even one parameter is frequently surprisingly good.

The ground state is an important special case. Let H be the Hamiltonian, and $|\psi_n\rangle$ an orthonormal set of eigenstates: $H|\psi_n\rangle = E_n|\psi_n\rangle$. Let $|\psi_o\rangle$ be the ground state, and suppose it is not degenerate. Then for any state $|\psi\rangle$,

$$\langle\psi| H |\psi\rangle = \sum_n E_n |\langle\psi| \psi_n\rangle|^2 \geq \sum_n E_1^o |\langle\psi| \psi_n\rangle|^2 = E_1^o \langle\psi| \psi\rangle = E_1^o \qquad (7.123)$$

The ground-state energy is a true minimum.

Here is an elementary example: Consider the three-dimensional harmonic oscillator, $H = T + V$, where

$$T = \frac{\mathbf{p}^2}{2m} \quad \text{and} \quad V = \frac{1}{2}m\omega^2 r^2 \qquad (7.124)$$

Suppose you didn't know the exact ground-state energy, and guessed incorrectly a spherically symmetric trial wave function of the form

$$\langle \mathbf{r}| \psi(a)\rangle = \psi(a; \mathbf{r}) = R(a, r) Y_0^0(\theta, \phi) \qquad (7.125)$$

with
$$R(a;r) = \frac{2}{a^{3/2}} e^{-r/a} \tag{7.126}$$

$R(a;r)$ has the form of the correctly normalized solution to another problem, with one free parameter. I chose it because we know the expectation value of the kinetic energy:

$$\langle \psi(a) | T | \psi(a) \rangle = \frac{1}{2ma^2} \tag{7.127}$$

while

$$\langle \psi(a) | V | \psi(a) \rangle = \frac{m\omega^2}{2} \int_0^\infty R(a;r)^2 r^4 \, dr = \frac{3}{2} m\omega^2 a^2 \tag{7.128}$$

The estimate for the energy is

$$E(a) = \langle \psi(a) | H | \psi(a) \rangle = \frac{1}{2ma^2} + \frac{3}{2} m\omega^2 a^2 \tag{7.129}$$

There is a minimum when

$$a^4 = \frac{1}{3m^2\omega^2} \tag{7.130}$$

and with a given by (7.130)

$$E(a) = \sqrt{3}\,\omega = 1.732\ldots\omega \tag{7.131}$$

Not bad for such a crude guess! The correct result is $1.5\,\omega$.

7.5.2 The Helium Atom

Ground State

For the helium atom we want to "guess" a trial wave function $\psi(\mathbf{r}_1, \mathbf{r}_2)$ that might be a good approximation to the correct ground-state wave function.

There is some art in choosing a good collection of trial functions, and physical intuition helps. For the spin singlet states, $|\psi\rangle$ must be symmetric in $\mathbf{r}_1$ and $\mathbf{r}_2$. We can improve on the product $\psi_{100}(\mathbf{r}_1)\psi_{100}(\mathbf{r}_2)$ that we used as the unperturbed wave function as follows: Each electron does not really feel the Coulomb potential of the nucleus except when it is much closer to the nucleus than the other electron. When it is farther away, the nuclear charge is screened by the second electron and feels a potential with charge $Z - 1$. A better approximation is a ground-state-like wave function but with $Z = \sigma$, where σ is somewhere between Z and $Z - 1$.[10] Let σ be arbitrary, and then vary it to get the minimum energy. In other words, choose trial functions of the form

$$\psi(\mathbf{r}_1, \mathbf{r}_2) = \frac{\sigma^3}{\pi a^3} e^{-\sigma(r_1+r_2)/a} \tag{7.132}$$

The value of σ that minimizes $E(\sigma) = \langle \psi | H | \psi \rangle$ is the best approximation to E_1^o among all the $E(\sigma)$.

---

[10] Of course there are even better approximations—this is just a first step, but one easy to deal with analytically.

226 Chapter 7 Approximation Methods for Bound States

Write $H = H_1 + H_2 + H'$, with H_i as in equation (7.108), H' as in equation (7.113), and compute $E(\sigma)$. Use

$$\int \psi_{100}(\mathbf{r}_i) \left(\frac{p_i^2}{2m} - \frac{\sigma e^2}{r_i} \right) \psi_{100}(\mathbf{r}_i) \, d^3r = E_1^o(\sigma) = \sigma^2 E_1^o \qquad (7.133)$$

and

$$-(\sigma e^2) \int \psi_{100}(\mathbf{r}_i)^2 \frac{1}{r} d^3r = 2\sigma^2 E_1^o \qquad (7.134)$$

to obtain

$$\langle \psi | H_1 | \psi \rangle = (2\sigma Z - \sigma^2) E_1^o \qquad (7.135)$$

and the same for H_2.

According to equation (7.118), the value of $\langle \psi | H' | \psi \rangle$ is

$$\langle \psi | H' | \psi \rangle = -\frac{5}{4} \sigma E_1^o \qquad (7.136)$$

So

$$E(\sigma) = \langle \psi | H | \psi \rangle = \left(4Z\sigma - 2\sigma^2 - \frac{5}{4}\sigma \right) E_1^o \qquad (7.137)$$

The value of σ that minimizes $E(\sigma)$ is $\sigma = Z - 5/16$. As expected, σ is between Z and $Z - 1$. The best value of the energy is

$$\boxed{E = 2\left(Z - \frac{5}{16} \right)^2 E_1^o} \qquad (7.138)$$

The numerical values for the first three two-electron ions are summarized in Table 7.3.

	First-Order Perturbation	One-Parameter Variational	Experiment	Single Electron
H$^-$	-10.20	-12.86	-14.35	-13.61
He	-74.83	-77.491	-78.99	-54.42
Li$^+$	-193.88	-196.54	-198.1	-122.45

TABLE 7.3: One parameter variational energies for some two-electron ions, compared with experiment, with first-order perturbation theory, and with the fully ionized energy (in electron volts).

Remember that the error in the energy is second order in the error of the trial function. Even this simple trial function is amazingly good. The close agreement for helium was one of the first detailed confirmations that quantum mechanics was a correct description of atoms. In all three cases, the variational energy is higher than the experimental energy, as it must be for the ground state. The difference between the variational result and experiment is about the same for the three ions shown.

Unfortunately, the result for the negative hydrogen atom is above the energy of an ordinary a one-electron atom. You can do even better with more than one parameter, and, finally, get a H$^-$ energy in the stable region, as observed.[11]

Excited States

For the excited states you can use trial functions in which the two electrons are in different eigenstates of the unperturbed Hamiltonian. You have to construct symmetric and antisymmetric combinations, depending on whether the spin state is singlet or triplet. Thus

$$\psi(\mathbf{r}_1, \mathbf{r}_2) = \frac{1}{\sqrt{2}}\left[\psi_{nlm}(\mathbf{r}_1)\psi_{n'l'm'}(\mathbf{r}_2) \pm \psi_{nlm}(\mathbf{r}_2)\psi_{n'l'm'}(\mathbf{r}_1)\right] \quad (7.139)$$

For helium it is easy to see that unless at least one of these functions is a ground-state function, the expectation value of the Hamiltonian is positive; when both electrons are in a excited state, they are so far from the nucleus that the electrostatic repulsion between them overcomes the attraction of each to the nucleus, and the atom dissociates. So all the excited states can be found starting from trial functions of the form

$$\psi(\mathbf{r}_1, \mathbf{r}_2) = \frac{1}{\sqrt{2}}\left[\psi_{nlm}(\mathbf{r}_1)\psi_{100}(\mathbf{r}_2) \pm \psi_{nlm}(\mathbf{r}_2)\psi_{100}(\mathbf{r}_1)\right] \quad (7.140)$$

For example, we could use the form (7.140) to find a $1s$-$2p$ state with trial functions

$$\psi_{1s}(\mathbf{r}) = \frac{Z_1^{3/2}}{\sqrt{\pi a^3}}e^{-Z_1 r/a} \quad \text{and} \quad \psi_{2p}(\mathbf{r}) = \frac{1}{2\sqrt{6}}\left(\frac{Z_2}{a}\right)^{\frac{5}{2}} r e^{-Z_2 r/2aq} Y_1^m(\theta, \phi) \quad (7.141)$$

and minimize with respect to the two parameters Z_1 and Z_2.

Even without any explicit computations, it should be clear that there is a large splitting between the triplet state and the singlet state. The exclusion principle simulates a spin-dependent interaction, even though the spin-dependent terms that actually appear in the Hamiltonian are only of the order of the fine structure splitting. This is a quantum effect with no classical counterpart.

In the singlet state the wave function is symmetric in $\mathbf{r}_1$ and $\mathbf{r}_2$, and there is a large probability that both electrons be at the same point in space, while in the triplet state the wave function vanishes when $\mathbf{r}_1 = \mathbf{r}_2$ and the two electrons cannot be at the same point. So the expectation value of the repulsion between the electrons is higher in the singlet state than in the triplet state. All other terms, not depending on the correlation between the two electrons, contribute comparably. The triplet state has the lower energy, as a direct consequence of the exclusion rule.

This is the basis for an empirical principle called **Hund's rule**: It says that in many systems the states with the highest spin symmetry, and therefore the most antisymmetric wave functions, tend to have the lowest energy [14]. The most extreme example is a ferromagnet, where the state with a huge number of parallel spins has the lowest energy.

There is an ancient terminology: The more common triplet states are conventionally called **orthohelium**. The singlet states are **parahelium**.

---

[11] The classic papers are by Hylleraas [12, 11]. There is a nice discussion in Chapter 3 of Bethe and Jackiw [13].

7.5.3 The Eigenvalue-Variational Scheme

The variational method is the starting point for a number of numerical schemes for solving bound-state problems.

Suppose you want to solve some complicated energy eigenvalue problem: $H|\psi\rangle = E|\psi\rangle$. You can always choose a complete set of states $|\psi_n\rangle$ that form a basis. For instance, they might be the harmonic oscillator stationary states in the appropriate number of dimensions. Choose as a trial state a linear combination of a finite number of the $|\psi_n\rangle$:

$$|\psi\rangle = \sum_n \beta_n |\psi_n\rangle \qquad (7.142)$$

where $\langle \psi_n | \psi_m \rangle = \delta_{nm}$. Then

$$E(\beta) = \langle \psi | H | \psi \rangle = \frac{\sum_{mn} \beta_m^* \beta_n H_{mn}}{\sum_n |\beta_n|^2} \qquad (7.143)$$

Allow all the β_n to be independent, and think of them as a collection of variational parameters. Generally the coefficients can be complex, so $E(\beta)$ should be extremized with respect to the real and imaginary parts, independently, of each β_n.[12] Since β_n and β_n^* are two independent linear combinations or $\mathrm{Re}\,\beta_n$ and $\mathrm{Im}\,\beta_n$, it is the same thing, and easier, to differentiate

$$\sum_{mn} \beta_m^* \beta_n H_{mn} = \sum_n |\beta_n|^2 E(\beta) \qquad (7.144)$$

with respect to β_i and β_i^* and then set

$$\partial E / \partial \beta_i = 0 \quad \text{and} \quad \partial E / \partial \beta_i^* = 0 \qquad (7.145)$$

to get

$$\sum_m H_{im} \beta_m = E(\beta) \beta_i \quad \text{and} \quad \sum_m H_{mi} \beta_m^* = E(\beta) \beta_i^* \qquad (7.146)$$

Since H is Hermitean, $E(\beta)$ is real and both forms are the same. $E(\beta)$ is an eigenvalue of the matrix H, and β_i are the components of the eigenvector in the basis $|\psi_n\rangle$. The variational method becomes an eigenvalue problem.

The method is even more efficient if you choose the matrix elements H_{mn} to depend on a parameter ω. In a one-dimensional problem you might choose the basis to be eigenstates of

$$p^2 + m^2 \omega^2 x^2 \qquad (7.147)$$

and keep ω as another variational parameter. Then the idea is to work in a small subspace of the full space of states, in that subspace find the eigenvalues of the matrix H_{mn}, choose the lowest one, and minimize that with respect to ω. For many one- and three-dimensional potentials, this scheme will allow you to calculate the eigenvalues to great accuracy with only a little numerical work. Problem 7.19 provides an example.

---

[12] For computational simplicity, you might want to restrict the trial functions to real combinations of the basis vectors.

At the other extreme imagine allowing the trial functions to span the whole space. Then the solution of the variational problem is any solution to the time independent Schrödinger equation. In other words, if *any* variation is allowed

$$\delta \langle \psi | H | \psi \rangle = 0 \tag{7.148}$$

is equivalent to

$$H|\psi\rangle = E|\psi\rangle \tag{7.149}$$

7.6 MOLECULES

Molecules are complicated. They have both nuclei and electrons, sometimes many of both. They usually lack the spherical symmetry that simplifies classifying atomic states.

The secret to any approximation scheme is that the electrons move rapidly in the almost static field of the nuclei, and the nuclei move slowly in some kind of average field of the swiftly moving electrons, and in the electrostatic field of the other nuclei. The first approximation is to hold the nuclear coordinates fixed and compute the electron wave functions as if they were moving in a fixed potential. Then the motion of the nuclei can be computed using the potential provided by the electron energy.[13]

7.6.1 The Born-Oppenheimer Approximation

Here is how the program is put into practice: Let $\mathbf{R}_i$ be the nuclear coordinates of the N nuclei and $\mathbf{r}_i$ the coordinates of the n electrons. $\mathbf{R}$ and $\mathbf{r}$ are shorthand for all the nuclear and electron coordinates. Normally these will be vectors in a $3N$ or $3n$ dimensional space. Fix the $\mathbf{R}_i$ and solve for the electronic motion as if $\mathbf{R}_i$ were known parameters. In some range of $\mathbf{R}_i$ there will be electronic bound states with energies $E(\mathbf{R})$. The idea is to compute $E(\mathbf{R})$ and then let the nuclei move in this potential. The scheme, called the **Born-Oppenheimer** method, works because the nuclear excitations are much smaller than the electronic binding energy. Born and Oppenheimer showed that the neglected terms are indeed small because the mass ratio is small.[14]

Let $H_e(\mathbf{R})$ be all the terms in the full Hamiltonian except the nuclear kinetic energy. For fixed $\mathbf{R}$ the Schrödinger equation

$$H_e(\mathbf{R})|\psi_n(\mathbf{R})\rangle = E_n(R)|\psi_n(\mathbf{R})\rangle \tag{7.150}$$

defines the electron energies $E_n(\mathbf{R})$ and a complete set of electron eigenstates. The total energy is $H_e(\mathbf{R})$ plus the nuclear kinetic energy. The electron wave functions are

$$\psi_n(\mathbf{r}, \mathbf{R}) = \langle \mathbf{r} | \psi_n \rangle \tag{7.151}$$

---

[13]In what follows I will ignore the spins of the particles and consider only the Coulomb forces. Spin and magnetic effects provide small corrections just as they do in atoms. In realistic calculations for complex molecules it is sometimes better to consider only the valence electrons, moving rapidly in the fields of the atomic cores, which are made up of the nuclei and the more tightly bound electrons.

[14]Born and Oppenheimer [15]. There is also a large recent literature. A good place to start is Moody, Shapere and Wilczek [16].

230   Chapter 7   Approximation Methods for Bound States

In the electron subsystem $\mathbf{R}$ is just a parameter. If the nuclei were infinitely heavy and fixed at $\mathbf{R} = \mathbf{R}_o$, the full wave functions would be

$$\Psi_n(\mathbf{r},\mathbf{R}) = \langle \mathbf{r},\mathbf{R}|\,\mathbf{R},n\rangle = \psi_n(\mathbf{r},\mathbf{R})\delta_3\left(\mathbf{R}-\mathbf{R}_o\right) \qquad (7.152)$$

Expand $\Psi(\mathbf{r},\mathbf{R})$ in the states $|\mathbf{R},n\rangle$ appropriate to each $\mathbf{R}$. That is, write

$$\Psi(\mathbf{r},\mathbf{R}) = \langle \mathbf{r},\mathbf{R}|\,I\,|\Psi\rangle = \sum_n \phi_n(\mathbf{R})\psi_n(\mathbf{r},\mathbf{R}) \qquad (7.153)$$

which can be obtained using

$$I = \iiint \sum_n |\mathbf{R}',\mathbf{r}\rangle\langle \mathbf{R}',\mathbf{r}|\,\mathbf{R},n\rangle\langle \mathbf{R},n|\,d^3\mathbf{R}'\,d^3\mathbf{R}\,d^3\mathbf{r} \qquad (7.154)$$

for the identity operator. The expansion turns the original problem into a complicated set of coupled equations for the coefficients.

Be careful here. This is a different kind of basis from those in problems solved by separation of variables. The Hamiltonian is not written as a sum of commuting terms.

If $\mathbf{R}$ were really an external parameter, and if the electron were in the instantaneous ground state at some time, it would remain in the ground state as long as $\mathbf{R}$ changes slowly. This is the **adiabatic theorem**. You can find the precise formulation and proof in many books.[15]

Although $\mathbf{R}$ is really a dynamical variable we still assume that the lowest states of the molecule all have the electron in its ground state $|\psi_o\rangle$, that is, approximate the wave function by the first term in the expansion (7.153)[16]

$$\Psi(\mathbf{r},\mathbf{R}) = \phi(\mathbf{R})\psi_o(\mathbf{r},\mathbf{R}) \qquad (7.155)$$

and use $\Psi(\mathbf{r},\mathbf{R})$ as a trial function for a variational estimate. Then the normalization requires that

$$\int |\phi(\mathbf{R})|^2 d^3R = 1 \qquad (7.156)$$

and $\phi(\mathbf{r})$ is effectively the nuclear wave function.

The variational estimate is

$$E = \langle \Psi|\,H\,|\Psi\rangle = \int \Psi(\mathbf{r},\mathbf{R})^* \left(\sum_i \frac{\mathbf{P}_i^2}{2M_i} + H_e(\mathbf{R})\right) \Psi(\mathbf{r},\mathbf{R})\,d^3R\,d^3r$$

$$= \int \Psi(\mathbf{r},\mathbf{R})^* \left(-\sum_i \frac{1}{2M}\nabla_i^2 + E_e(\mathbf{R})\right) \Psi(\mathbf{r},\mathbf{R})\,d^NR\,d^nr \qquad (7.157)$$

The sum is over all the coordinates of all the nuclei. The symbol $\boldsymbol{\nabla}_i$ means $\partial/\partial R_i$. Define[16] $|\boldsymbol{\nabla}_i \psi_o\rangle$ as the electron state whose wave function is

$$\langle \mathbf{r}|\,\boldsymbol{\nabla}_i \psi_o\rangle \equiv \boldsymbol{\nabla}_i \psi_o(\mathbf{r},\mathbf{R}) \qquad (7.158)$$

---

[15] For example pages 744–755 in Messiah [26] or pages 327–330 in Griffiths [25]. Messiah has references to the original proofs.

[16] The $\mathbf{R}$ dependence of $|\psi_n\rangle$ and $|\boldsymbol{\nabla}_i\psi_n\rangle$ is suppressed.

and define the "vector"
$$A_i(\mathbf{R}) \equiv i\langle \psi_o | \nabla_i \psi_o \rangle \tag{7.159}$$

From
$$\nabla_i \int \psi_o(\mathbf{r},\mathbf{R})^* \psi_o(\mathbf{r},\mathbf{R}) d^n r = 0 \tag{7.160}$$

it follows that $A_i(\mathbf{R})$ is real.

Next integrate out the electron wave function to obtain the effective nuclear Hamiltonian. The key to this calculation is the identity (see below)

$$\int \psi_o(\mathbf{r},\mathbf{R})^* P_i^2 \psi_o(\mathbf{r},\mathbf{R}) \phi(\mathbf{R}) = P_i^2 \phi(\mathbf{R}) - A_i(\mathbf{R}) \nabla_i \phi(\mathbf{R}) - \nabla_i A_i(\mathbf{R}) \phi(\mathbf{R})$$
$$- \phi(\mathbf{R}) \int \left(\nabla_i \psi_o(\mathbf{r},\mathbf{R})^* \right) \nabla_i \psi_o(\mathbf{r},\mathbf{R}) d^n r \tag{7.161}$$

Therefore,
$$E = \int \phi(\mathbf{R})^* \left[\sum_i \frac{(P_i - A_i)^2}{2M_i} + E_e(\mathbf{R}) + U(\mathbf{R}) \right] \phi(\mathbf{R}) d^N R \tag{7.162}$$

Since any nuclear trial function $\phi(\mathbf{R})$ is allowed, the variational principle $\delta E = 0$ is equivalent here to Schrödinger's equation with the effective Hamiltonian

$$\boxed{H_{\text{eff}} = \sum_i \frac{(P_i - A_i)^2}{2M_i} + E_e(\mathbf{R}) + U(\mathbf{R})} \tag{7.163}$$

acting on the nuclear coordinates alone. Besides the electronic energy, there seems to be an effective vector potential $\mathbf{A}$ and an effective scalar potential

$$U(\mathbf{R}) = -\int \left(\nabla_i \psi_o(\mathbf{r},\mathbf{R})^* \right) \nabla_i \psi_o(\mathbf{r},\mathbf{R}) d^n r - \sum_i \frac{A_i^2}{2M_i} \tag{7.164}$$

Steps that lead to (7.161):

$$\int \psi_o(\mathbf{r},\mathbf{R})^* \nabla_i^2 \psi_o(\mathbf{r},\mathbf{R}) \phi(\mathbf{R}) = \nabla_i^2 \phi(\mathbf{R})$$
$$+ 2 \left[\nabla_i \phi(\mathbf{R}) \right] \int \psi_o(\mathbf{r},\mathbf{R})^* \left[\nabla_i \psi_o(\mathbf{r},\mathbf{R}) \right] d^n r$$
$$+ \phi(\mathbf{R}) \int \left(\psi_o(\mathbf{r},\mathbf{R})^* \nabla_i^2 \psi_o(\mathbf{r},\mathbf{R}) \right) d^n r \tag{7.165}$$

Now use
$$\nabla_i A_i(\mathbf{R}) = i \int \left(\nabla_i \psi_o(\mathbf{r},\mathbf{R})^* \right) \left(\nabla_i \psi_o(\mathbf{r},\mathbf{R}) \right) d^n r$$
$$+ i \int \psi_o(\mathbf{r},\mathbf{R})^* \nabla_i^2 \psi_o(\mathbf{r},\mathbf{R}) d^n r \tag{7.166}$$

232 Chapter 7 Approximation Methods for Bound States

to write equation (7.165) as

$$\int \psi_o(\mathbf{r},\mathbf{R})^* \boldsymbol{\nabla}_i^2 \psi_o(\mathbf{r},\mathbf{R}) \phi(\mathbf{R})$$
$$= \boldsymbol{\nabla}_i^2 \phi(\mathbf{R}) - 2i\left[\boldsymbol{\nabla}_i \phi(\mathbf{R})\right] A_i(\mathbf{R}) - i\phi(\mathbf{R}) \boldsymbol{\nabla}_i A_i(\mathbf{R})$$
$$+ \phi(\mathbf{R}) \int \left(\boldsymbol{\nabla}_i \psi_o(\mathbf{r},\mathbf{R})^*\right) \boldsymbol{\nabla}_i \psi_o(\mathbf{r},\mathbf{R}) d^n r \qquad (7.167)$$
$$= \boldsymbol{\nabla}_i^2 \phi(\mathbf{R}) - iA_i(\mathbf{R}) \boldsymbol{\nabla}_i \phi(\mathbf{R}) - i\boldsymbol{\nabla}_i \left(A_i(\mathbf{R})\phi(\mathbf{R})\right)$$
$$+ \phi(\mathbf{R}) \int \left(\boldsymbol{\nabla}_i \psi_o(\mathbf{r},\mathbf{R})^*\right) \boldsymbol{\nabla}_i \psi_o(\mathbf{r},\mathbf{R}) d^n r$$

∎

Look at the first term in the effective Hamiltonian (7.163). To simplify the writing, assume that all the M_i are the same and that the parameter space is three dimensional—none of the conclusions will depend on these simplifications. The first term in H_{eff} is proportional to $(\mathbf{P}-\mathbf{A})^2$. Born and Oppenheimer just wrote this as the ordinary kinetic energy of the nuclei. Is the vector potential $\mathbf{A}$ really there?

Even though $\mathbf{A}$ has nothing to do with magnetism it *looks* like the magnetic vector potential introduced in Section 6.1. Try to remove it by a "gauge" transformation (see Section 6.1.2). If you change the electron wave function by

$$\psi_o(\mathbf{r},\mathbf{R}) \rightarrow e^{i\lambda(\mathbf{R})}\psi_o(\mathbf{r},\mathbf{R}) \qquad (7.168\text{a})$$

then from the definition (7.159) the vector potential changes by

$$\mathbf{A}(\mathbf{R}) \rightarrow \mathbf{A}(\mathbf{R}) - \boldsymbol{\nabla}\lambda(\mathbf{R}) \qquad (7.168\text{b})$$

If you also change the nuclear wave function by

$$\phi(\mathbf{R}) \rightarrow e^{-i\lambda(\mathbf{R})}\phi(\mathbf{R}) \qquad (7.168\text{c})$$

The full wave function $\Psi(\mathbf{r},\mathbf{R}) = \phi(\mathbf{R})\psi(\mathbf{r},\mathbf{R})$ remains unchanged, so the physics must be the same. All you need to get rid of $\mathbf{A}$ is this kind of transformation with $\boldsymbol{\nabla}\lambda(\mathbf{R}) = \mathbf{A}(\mathbf{R})$, or

$$\lambda(\mathbf{R}) = \int_{\mathbf{R}_o}^{\mathbf{R}} \mathbf{A}(\mathbf{R}) \cdot \mathbf{R} \qquad (7.169)$$

for any $\mathbf{R}_o$. The function defined in equation (7.169) will work provided $\lambda(\mathbf{R})$ is uniquely defined at each point, that is, provided the line integral in (7.169) be independent of the path. And the condition for that is that the "magnetic field" $\boldsymbol{\nabla}\times\mathbf{A}$ vanish. The vector potential in the effective nuclear Hamiltonian H_{eff} can be safely ignored provided

$$\boldsymbol{\nabla}\times\mathbf{A}(\mathbf{R}) = 0 \qquad (7.170)$$

The effect described here is really elementary, even if it may seem to you a bit complicated on first reading. It is a bit humbling that it went unnoticed for over fifty years. It is best understood as a consequence of the nontrivial phase of the electron wave functions that was discovered in 1983 by Michael Berry. You can learn a little more about Berry's phase in Section 10.3. It would take us too far afield to go further into the general theory of the Born-Oppenheimer method here.

7.6.2 The Hydrogen Molecular Ion

I will work out in detail only the simplest chemical compound, the positive hydrogen ion H_2^+, one electron and two protons.

Let $\mathbf{R}_i$ be the proton coordinates and $\mathbf{r}$ the electron coordinates. $\mathbf{P}_i$ and $\mathbf{p}$ are the corresponding momenta. In the center of mass frame (ignoring the electron mass) the relative coordinate of the protons is $\mathbf{R} = \mathbf{R}_1 - \mathbf{R}_2$, and their momenta are $\mathbf{P} = \mathbf{P}_1 = -\mathbf{P}_2$. The Hamiltonian is

$$H = \frac{\mathbf{P}^2}{M} + \frac{\mathbf{p}^2}{2m} + \alpha \left(\frac{1}{R} - \frac{1}{|\mathbf{r} - \mathbf{R}/2|} - \frac{1}{|\mathbf{r} + \mathbf{R}/2|} \right) \tag{7.171}$$

The piece of the Hamiltonian that acts on the electron coordinates alone, including the Coulomb potential terms, is

$$H_e(\mathbf{R}) = \frac{\mathbf{p}^2}{2m} + \alpha \left(\frac{1}{R} - \frac{1}{|\mathbf{r} - \mathbf{R}/2|} - \frac{1}{|\mathbf{r} + \mathbf{R}/2|} \right) \tag{7.172}$$

If you fix the nuclei to compute the electron wave functions, the latter depend only on the scalar distance R between them. For this simple problem, the "vector potential" can be transformed away by a gauge transformation like (7.168). In fact, in this case the electron states can be chosen to be real, so $\mathbf{A}$ vanishes identically. You need more complicated molecules to see its effect.[17] For the hydrogen molecular ion, the effective nuclear potential is just what Born and Oppenheimer thought it should be:

$$H_{\text{eff}} = \frac{\mathbf{P}^2}{2M} + E_o^e(R) + U(R) \tag{7.173}$$

where

$$H_e(\mathbf{R}) |\psi(\mathbf{r}, R)\rangle = E_o^e(R) |\psi(\mathbf{r}, R)\rangle \tag{7.174}$$

The extra piece $U(R)$ is small.

LCAO Trial Wave Functions

If the electron were bound to one of the two nuclei only, its ground state would be

$$|\psi_1(\mathbf{R})\rangle \quad \text{or} \quad |\psi_2(\mathbf{R})\rangle \tag{7.175}$$

where

$$\langle \mathbf{r} | \psi_i(\mathbf{R}) \rangle = \psi_i(\mathbf{r}, \mathbf{R}) = \frac{1}{\sqrt{\pi a^3}} e^{-r_i/a} \tag{7.176}$$

with

$$r_1 = \left| \mathbf{r} - \frac{\mathbf{R}}{2} \right| \quad \text{and} \quad r_2 = \left| \mathbf{r} + \frac{\mathbf{R}}{2} \right| \tag{7.177}$$

Wave functions like this are called **atomic orbitals**. They are not good trial functions because the molecule should be even or odd under reflections through the plane through the origin perpendicular to $\mathbf{R}$. So choose for trial functions the even and odd **Linear combination of atomic orbitals**:

$$|\psi_\pm(\mathbf{R})\rangle = N_\pm \left(|\psi_1(\mathbf{R})\rangle \pm |\psi_2(\mathbf{R})\rangle \right) \tag{7.178}$$

---

[17] Problem 10.3 is a simple example where the vector potential cannot be ignored.

234 Chapter 7 Approximation Methods for Bound States

As $R \to \infty$, $N_\pm \to 1/\sqrt{2}$, but for finite R the wave functions are not orthogonal. The normalization condition is

$$1 = |N_\pm|^2 2(1 \pm I) \tag{7.179}$$

where I is the overlap integral[18]

$$I = \langle\psi_1(\mathbf{R})|\,\psi_2(\mathbf{R})\rangle = \frac{1}{\pi a^3}\int e^{-(r_1+r_2)/a}d^3r = e^{-R/a}\left[1 + \frac{R}{a} + \frac{1}{3}\left(\frac{R}{a}\right)^2\right] \tag{7.180}$$

Since $|\psi_i(\mathbf{R})\rangle$ are the ordinary hydrogen atom ground-state wave functions about the nuclei,

$$\frac{\mathbf{p}^2}{2m}|\psi_1(\mathbf{R})\rangle = \left[E_1^o + \frac{\alpha}{r_1}\right]|\psi_1(\mathbf{R})\rangle \tag{7.181a}$$

$$\frac{\mathbf{p}^2}{2m}|\psi_2(\mathbf{R})\rangle = \left[E_1^o + \frac{\alpha}{r_2}\right]|\psi_2(\mathbf{R})\rangle \tag{7.181b}$$

where $E_1^o = -\alpha^2 m/2$ is the hydrogen atom ground-state energy. Therefore,

$$H_e|\psi_\pm(\mathbf{R})\rangle = N_\pm\Big(H_e|\psi_1(\mathbf{R})\rangle \pm H_e|\psi_2(\mathbf{R})\rangle\Big)$$
$$= \left(E_1^o + \frac{\alpha}{R}\right)|\psi_\pm(\mathbf{R})\rangle - N_\pm\alpha\left(\frac{1}{r_2}|\psi_1(\mathbf{R})\rangle \pm \frac{1}{r_1}|\psi_2(\mathbf{R})\rangle\right) \tag{7.182}$$

The estimate of the energy with these trial functions is

$$E_\pm(R) = \langle\psi_\pm(\mathbf{R})|\,H_e\,|\psi_\pm(\mathbf{R})\rangle$$
$$= \left(E_1^o + \frac{\alpha}{R}\right) - 2\alpha N_\pm^2\left(\left\langle\psi_1(\mathbf{R})\left|\frac{1}{r_2}\right|\psi_1(\mathbf{R})\right\rangle \pm \left\langle\psi_1(\mathbf{R})\left|\frac{1}{r_2}\right|\psi_2(\mathbf{R})\right\rangle\right) \tag{7.183}$$

The integrals are[18]

$$\left\langle\psi_1(\mathbf{R})\left|\frac{1}{r_2}\right|\psi_1(\mathbf{R})\right\rangle = \frac{1}{R} - \left(\frac{1}{a} + \frac{1}{R}\right)e^{-2R/a} \tag{7.184}$$

and

$$\left\langle\psi_1(\mathbf{R})\left|\frac{1}{r_2}\right|\psi_2(\mathbf{R})\right\rangle = \left(\frac{1}{a} + \frac{R}{a^2}\right)e^{-R/a} \tag{7.185}$$

The estimate of the energy as a fraction of $-E_1^o = \alpha^2 m/2 = \alpha/2a$ is

$$\frac{E_\pm(R)}{|E_1^o|} = \frac{2a}{R} - 1 - \frac{2a}{1 \pm I}\left[\left(\frac{1}{R} - \left(\frac{1}{a} + \frac{1}{R}\right)e^{-2R/a}\right) \pm \left(\left(\frac{1}{a} + \frac{R}{a^2}\right)e^{-R/a}\right)\right] \tag{7.186}$$

---

[18]These integrals can be evaluated easily in elliptic coordinates (See Section A.3.3 in the appendix) or using the Fourier transform as in the evaluation of the integral in equation (7.116).

Because there are only two lengths in this simple problem, this dimensionless ratio is a function of $t = R/a$ alone:

$$\frac{E_{\pm}(t)}{|E_1^o|} = -1 + \frac{2}{t}\left[\frac{(1+t)e^{-2t} \pm \left((1-(2t^2/3))\right)e^{-t}}{1 \pm (1+t+t^2/3)e^{-t}}\right] \quad (7.187)$$

These two functions are plotted in Figure 7.4.

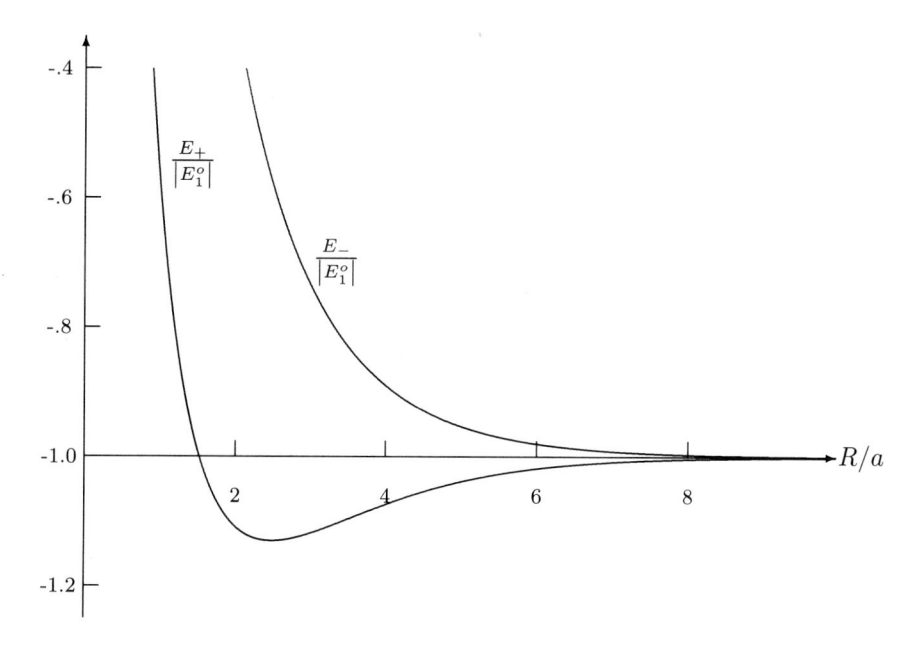

FIGURE 7.4: The hydrogen molecular ion. Simple variational estimate of the electronic energy as a function of the internuclear distance.

The molecule will not have a bound state unless the energy is less than E_1^o. Otherwise it would decay into a free proton plus a free hydrogen atom. For large separations the energy is just E_1^o. For small separations, the Coulomb repulsion between the nuclei dominates. In between, the nuclear repulsion can be overcome by the attraction of the electron, but this effect is important only when the electron is far from either nucleus. In the antisymmetric state the electron spends most of its time near one of the protons, and the odd wave function has no minimum and does not produce a bound state. But in the symmetric state the electron spends more time, on the average, halfway in between, where the attraction is greatest. Our trial symmetric wave function has a minimum at $R = 2.49\,a$, where $E_+ = -1.13\,E_1^o$. The binding energy is $0.13 E_1^o \approx 1.77\,\text{eV}$.

The approximation so far gives the right order of magnitude, but you can do much better with a more complicated trial wave function, for example, by varying the nuclear charge as we did in the helium atom problem. (In the present calculation, the answer depends only on the ratio R/a, so there is no point in varying a as well as R.) An exact numerical solution has been obtained after separating

236 Chapter 7 Approximation Methods for Bound States

the variables in elliptic coordinates, with a result in agreement with observation. The measured binding energy is 2.8 eV, and the equilibrium separation is about $R_o = 2.00\,a$.[19]

Vibrational and Rotational Levels

The eigenvalues of H_e are the *electronic* energy levels, typically, as for atoms, of the order of a few electron volts:

$$E_{\text{electronic}} \approx \alpha^2 m \tag{7.188}$$

Molecules more complex than H_2^+ will have many electronic energy levels, and transitions between them will emit photons with energies in this range.

The total energy also includes the energy of the nuclear motion. From equation (7.173), omitting the small term $U(R)$, the Schrödinger equation for the nuclear wave function is the Schrödinger equation for a particle of mass $M/2$ moving in the potential $E(R)$.

$$\left[-\frac{1}{M}\nabla_\mathbf{R}^2 + E(R) - E\right]\phi(\mathbf{R}) = 0 \tag{7.189}$$

Since the potential is central, the energies are determined by the radial equation

$$\left[-\frac{1}{M}\left(\frac{d^2}{dR^2} - \frac{l(l+1)}{R^2}\right) + E(R)\right]\phi(R) = E\phi(R) \tag{7.190}$$

There is no point trying to solve this approximate equation exactly. You can get an idea of the nature of the spectrum as follows: Since the nuclear distance is always close to the minimum R_o, the *rotational* energy, the dependence on l, can be approximated by replacing the centrifugal term with

$$\frac{l(l+1)}{MR_o^2} \tag{7.191}$$

Then equation (7.189) is the equation for a dumbell or rigid rotator, and the energy levels are

$$E_l = \frac{l(l+1)}{MR_o^2} \tag{7.192}$$

which are of the order

$$E_{\text{rotational}} \approx \frac{1}{MR_o^2} \approx \frac{1}{Ma^2} \approx \frac{m}{M}E_{\text{electronic}} \tag{7.193}$$

or about 10^{-3} eV. Rotational levels like this, proportional to $l(l+1)$ with $l = 0, 1, 2 \ldots$ are typical features of molecular spectra.

It is also possible to excite radial nuclear motions. Set $l = 0$ and approximate equation (7.189) by

$$\left[-\frac{1}{M}\frac{d^2}{dR^2} + \frac{M\omega^2(R-R_o)^2}{4}\right]\phi(R) = E\phi(R) \tag{7.194}$$

---

[19] For this problem the Schrödinger equation can be separated into three ordinary differential equations using elliptic coordinates. I do not believe there is an actual algebraic formula for the energies, but they can be found numerically to arbitrary precision by a well-defined successive approximation scheme. See Wallis and Hulbert [17] or Bates, Ladsham, and Stewart [18].

The frequency ω is determined by the curvature of the electronic potential at the minimum:

$$\omega^2 = \frac{2}{M} \frac{d^2 E(R_o)}{dR^2} \tag{7.195}$$

The order of magnitude of the vibrational levels are

$$E^2_{\text{vibrational}} = \omega^2 \approx \frac{1}{M} \left. \frac{d^2 E(R_o)}{dR^2} \right|_{R \approx a} \approx \frac{1}{M} \frac{E^o_1}{a^2} \frac{d^2}{dt^2} \left(\frac{E_+}{|E^o_1|} \right) \approx \frac{1}{Mma^4} \tag{7.196}$$

or

$$E_{\text{vibrational}} \approx \sqrt{\frac{m}{M}} E_{\text{electronic}} \tag{7.197}$$

Many simple molecules share the qualitative features of the H_2^+ molecular ion. The low-lying vibrational levels resemble those of the harmonic oscillator, so there is roughly a constant spacing between them. The rotational levels, in contrast, have the quadratic $l(l+1)$ spacing. There is a hierarchy of energies:

$$\frac{E_{\text{rotational}}}{E_{\text{vibrational}}} \approx \frac{E_{\text{vibrational}}}{E_{\text{electronic}}} \approx \sqrt{\frac{m}{M}} \tag{7.198}$$

7.7 THE WKB METHOD

The WKB method is a less general way to obtain approximate solutions to the one-dimensional Schrödinger equation, but it has important applications. It is also applicable to three-dimensional problems when they can be reduced to solving a radial wave equation. The WKB method has been particularly successful in attacking one-dimensional tunneling problems and bound-state problems.[20]

Let $\psi(x)$ be a solution to the one-dimensional time-independent Schrödinger equation:

$$\psi''(x) + 2m[E - V(x)] = 0 \tag{7.199}$$

If you write the wave function in the form

$$\psi(x) = e^{iS(x)} \tag{7.200}$$

then $S(x)$ satisfies

$$iS''(x) - (S')^2 + k^2(x) = 0 \tag{7.201}$$

where

$$k^2(x) = 2m[E - V(x)] \tag{7.202}$$

Replacing a linear differential equation with a nonlinear one may not look like progress, but here it is a good beginning for a scheme of successive approximations. For a free particle, $V(x) = 0$, $S'(x) = \pm k$, and $S''(x) = 0$. Take $S''(x) = 0$ as the zero-order approximation to $S(x)$ even when $V(x) \neq 0$:

$$S_o(x) = \pm \int k(x) dx \tag{7.203}$$

---

[20] Wentzel [19], Kramers [20], Brillouin [21, 22].

238 Chapter 7 Approximation Methods for Bound States

These are two possible solutions; the choice depends on the boundary conditions. The constant of integration will be determined by the normalization.

Next insert the expression (7.203) into equation (7.201):

$$S'(x) = \pm\sqrt{k(x)^2 + iS''(x)} \simeq \pm\sqrt{k(x)^2 + iS_o''(x)} = \pm\sqrt{k(x)^2 + ik'(x)} \quad (7.204)$$

provided $k'(x)$ is small compared to $k^2(x)$. Thus the first-order approximation is

$$S_1(x) = \pm \int \sqrt{k^2(x) \pm ik'(x)}\, dx \quad (7.205)$$

The approximation is that $V(x)$ doesn't change much over one de Broglie wavelength—the fractional change in wavelength over one wavelength should be small: $|S'(x)/S(x)| \ll 1$. It is possible to do a little better. Write

$$S_1(x) \approx \pm \int k(x)\,dx + \frac{i}{2}\int \frac{k'(x)}{k(x)}\,dx = \pm \int k(x)\,dx + \frac{i}{2}\log k(x) \quad (7.206)$$

or

$$\boxed{\psi(x) \approx \frac{1}{\sqrt{k(x)}} \exp\left[\pm i \int k(x)\,dx\right]} \quad (7.207)$$

Equation (7.207) is the WKB approximation.

When $E < V(x)$, these formulas still hold with $k(x) \to i\kappa(x)$, with

$$\kappa^2(x) = 2m\bigl[V(x) - E\bigr] \quad (7.208)$$

7.7.1 Turning Points and Connection Formulas

Let a be a "turning point" of the classical motion, a point where $V(a) = E$, and suppose that $V(x)$ has a negative slope at $x = a$, so that $V(x) < a$ for $x > a$ and $V(x) > a$ for $x < a$. Since $k(a) = 0$, the approximation (7.207) breaks down near $x = a$. For $x \ll a$

$$\psi(x) \approx \frac{A}{\sqrt{\kappa(x)}} \exp\left[-\int_a^x \kappa(x')\,dx'\right] + \frac{B}{\sqrt{\kappa(x)}} \exp\left[\int_a^x \kappa(x')\,dx'\right] \quad (7.209)$$

while for $x \gg a$

$$\psi(x) \approx \frac{C}{\sqrt{k(x)}} \exp\left[-i\int_a^x k(x')\,dx'\right] + \frac{D}{\sqrt{k(x)}} \exp\left[i\int_a^x k(x')\,dx'\right] \quad (7.210)$$

The constants A, B, C, and D will be determined by the boundary conditions of the problem at hand; κ and k are defined by equations (7.208) and (7.202).

The WKB approximation breaks down near the turning points of the classical motion, but Schrödinger's equation is perfectly regular there. You need to interpolate between the WKB approximations on either side. In principle this is always possible as long as $V(x)$ is well behaved near $x = a$; you could calculate the solution numerically in the region near the turning point and determine C and D in terms of A and B.

A and B are arbitrary, fixed by some conditions far from the turning points. Since the wave function for $x > a$ depends linearly on its value for $x < a$, A and B have to be linear functions of C and D. The connection has the form

$$A = \alpha_1 C + \alpha_2 D \quad \text{and} \quad B = \alpha_3 C + \alpha_4 D \qquad (7.211)$$

for some constants α_i. The α_i are dimensionless, so they cannot depend on the slope of the potential near $x = a$; the connection formulas are universal linear equations.

The one-dimensional time-independent Schrödinger equation (7.199) has real parameters, so $\psi(x)^*$ is another solution, with different boundary conditions. You can get this solution from equations (7.209) and (7.210) simply by replacing A and B with A^* and B^*, and replacing C and D with D^* and C^* respectively. Since the connection coefficients are universal,

$$B = \alpha_3^* D + \alpha_4^* C \quad \text{and} \quad A = \alpha_1^* D + \alpha_2^* C \qquad (7.212)$$

from which it follows that $\alpha_1^* = \alpha_2$ and $\alpha_3^* = \alpha_4$.

A further relation among these coefficients follows from current conservation. From the time-dependent Schrödinger equation, it follows that the probability current

$$j(x) = \frac{1}{m} \operatorname{Im} \psi(x)^* \psi'(x) \qquad (7.213)$$

is a constant.[21] Equate the constant on both sides of the turning point to get

$$\frac{1}{m}\left[|C|^2 - |D|^2\right] = \frac{2}{m}\operatorname{Im}(B^*A) = \frac{2}{m}\operatorname{Im}(\alpha_2^*\alpha_4)\left[|C|^2 - |D|^2\right] \qquad (7.214)$$

Therefore,

$$2\operatorname{Im}(\alpha_2^*\alpha_4) = 1 \qquad (7.215)$$

So the connection formulas (7.211) depend on three real parameters:

$$A = r_1 e^{i\theta_1} C + r_1 e^{-i\theta_1} D \quad \text{and} \quad B = r_3 e^{i\theta_3} C + r_3 e^{-i\theta_3} D \qquad (7.216)$$

subject to the constraint

$$r_1 r_3 \sin(\theta_1 - \theta_3) = \frac{1}{2} \qquad (7.217)$$

I do not know how to go any further than equation (7.216) in an elementary way. But to the extent that $V(x)$ can be approximated by a straight line between the regions where (7.209) and (7.210) are good approximations, there is an analytical solution. The next section contains the derivation of the connection coefficients. For the reader in a hurry, all you need for applications are the relations (7.246) and (7.248).

---

[21] See Section 2.7.3.

240 Chapter 7 Approximation Methods for Bound States

7.7.2 The Linear Approximation

Near the turning point approximate the potential by[22]

$$V(x) = E - \beta(x - a) \qquad (7.218)$$

with $\beta > 0$. Equation (7.218) describes a potential that decreases linearly as x increases through $x = a$. Schrödinger's equation with this potential is

$$-\frac{1}{2m}\psi''(x) - \beta(x-a)\psi(x) = 0 \qquad (7.219)$$

Change variables to

$$t = (2m\beta)^{\frac{1}{3}}(x-a) \qquad (7.220)$$

so that

$$\psi''(t) + t\psi(t) = 0 \qquad (7.221)$$

Positive t

When $x > a$,

$$k^2 = (2m\beta)^{\frac{2}{3}} t \qquad (7.222)$$

So

$$\frac{dk}{dx} = \frac{1}{2}(2m\beta)^{\frac{2}{3}} t^{-\frac{1}{2}} \qquad (7.223)$$

and the condition $k'(x) \ll k^2(x)$ for the validity of the representation (7.210) is

$$t \gg \left(\frac{1}{2}\right)^{\frac{2}{3}} \qquad (7.224)$$

An analytic solution to equation (7.221) can be found using some standard nineteenth-century analysis. Set

$$z = \frac{2}{3} t^{\frac{3}{2}} = \frac{2}{3}\frac{k^3}{2m\beta} \qquad (7.225)$$

in the region where $t > 0$. Then

$$\int_a^x k(x')dx' = \int_0^t t'^{\frac{1}{2}} dt' = z \qquad (7.226)$$

In terms of z the differential equation (7.221) becomes

$$z\psi''(z) + \frac{1}{3}\psi'(z) + z\psi(z) = 0 \qquad (7.227)$$

where the primes mean differentiation with respect to the argument indicated. Next, set

$$\psi(z) = z^{\frac{1}{3}} F(z) \qquad (7.228)$$

---
[22] Results similar to these can also be obtained in the unusual case that the turning points are not linear.

so that
$$z^2 F''(z) + zF'(z) + \left(z^2 - \frac{1}{9}\right) F(z) = 0 \qquad (7.229)$$

Equation (7.229) is Bessel's equation of order $\pm 1/3$. $\psi(t)$ is $z^{\frac{1}{3}}$ times a linear combination of $J_{\frac{1}{3}}(z)$ and $J_{-\frac{1}{3}}(z)$. These Bessel functions of fractional order are singular at $z = 0$, but $t = 0$ is not a singular point of equation (7.221), and the singularity in the Bessel functions will disappear when the transformations that led from equation (7.203) to (7.229) are undone.

Negative t

In this region
$$\kappa^2 = -(2m\beta)^{\frac{2}{3}} t \qquad (7.230)$$
so here instead of (7.225) define
$$z = \frac{2}{3}(-t)^{\frac{3}{2}} > 0 \qquad (7.231)$$
Then
$$\int_a^x \kappa(x')dx' = -\frac{2}{3}(-t)^{\frac{3}{2}} = -z \qquad (7.232)$$
and the differential equation (7.221) is
$$z\psi''(z) + \frac{1}{3}\psi'(z) - z\psi(z) = 0 \qquad (7.233)$$
Again, set
$$\psi(z) = z^{\frac{1}{3}} F(z) \qquad (7.228)$$
$F(z)$ satisfies
$$z^2 F''(z) + zF'(z) - \left(z^2 + \frac{1}{9}\right) F(z) = 0 \qquad (7.234)$$

The solutions to equation (7.221) are linear combinations of the two functions $z^{\frac{1}{3}} J_{\frac{1}{3}}(iz)$ and $z^{\frac{1}{3}} J_{-\frac{1}{3}}(iz)$.

Airy Functions

The asymptotic forms of these Bessel functions are described in Section A.2.5 of the appendix.[23] You can find the general solution to the linear approximation by hooking these forms for large z together and using the behavior of the Bessel functions near $t = 0$, as discussed in Section A.2.5. Finally, these forms can be compared to the WKB approximations (7.209) and (7.210), and if the various regions do indeed overlap you can get a single form over a large range of x.

---

[23] Most students of quantum mechanics will be satisfied to take results like these on faith. Others need to know where every result comes from. These appendices are included to satisfy the curiosity of those in the second group while saving them the trouble of searching through the literature.

242 Chapter 7 Approximation Methods for Bound States

I will just quote the result. The linear combinations of solutions we want are the **Airy functions** Ai(t) and Bi(t) of negative argument, defined as

$$\text{Ai}(-t) = \frac{1}{3}\sqrt{t}\left[J_{\frac{1}{3}}\left(\frac{2}{3}t^{\frac{3}{2}}\right) + J_{-\frac{1}{3}}\left(\frac{2}{3}t^{\frac{3}{2}}\right)\right] \tag{7.235}$$

$$\text{Bi}(-t) = \sqrt{\frac{x}{3}}\left[J_{-\frac{1}{3}}\left(\frac{2}{3}t^{\frac{3}{2}}\right) - J_{+\frac{1}{3}}\left(\frac{2}{3}t^{\frac{3}{2}}\right)\right] \tag{7.236}$$

The large t limits of the Airy functions follow from the various asymptotic forms of the Bessel functions:

$$\text{Ai}(-t) = \frac{1}{\sqrt{\pi}}t^{-\frac{1}{4}}\cos\left(z - \frac{\pi}{4}\right) \qquad (t \to \infty) \tag{7.237a}$$

$$\text{Ai}(-t) = \frac{1}{2}\frac{1}{\sqrt{\pi}}|t|^{-\frac{1}{4}}e^{-z} \qquad (t \to -\infty) \tag{7.237b}$$

and

$$\text{Bi}(-t) = -\frac{1}{\sqrt{\pi}}t^{-\frac{1}{4}}\sin\left(z - \frac{\pi}{4}\right) \qquad (t \to \infty) \tag{7.238a}$$

$$\text{Bi}(-t) = \frac{1}{\sqrt{\pi}}|t|^{-\frac{1}{4}}e^{z} \qquad (t \to -\infty) \tag{7.238b}$$

where $z = \frac{2}{3}|t|^{\frac{3}{2}}$ as above. A general solution to equation (7.219) is

$$\psi(x) = C_A \text{Ai}(-t) + C_B \text{Bi}(-t) \tag{7.239}$$

For large positive x, this is

$$\psi(x) \xrightarrow[x \to +\infty]{} \frac{1}{\sqrt{\pi}}(2m\beta)^{-\frac{1}{6}}\frac{1}{\sqrt{k(x)}}$$
$$\times \left[C_A \cos\left(\int_a^x k(x')dx' - \frac{\pi}{4}\right) - C_B \sin\left(\int_a^x k(x')dx' - \frac{\pi}{4}\right)\right] \tag{7.240}$$

while for large negative x, it is

$$\psi(x) \xrightarrow[x \to -\infty]{} \frac{1}{\sqrt{\pi}}(2m\beta)^{-\frac{1}{6}}\frac{1}{\sqrt{\kappa(x)}}$$
$$\times \left[\frac{1}{2}C_A \exp\left(\int_0^a \kappa(x')dx'\right) + C_B \exp\left(-\int_0^a \kappa(x')dx'\right)\right] \tag{7.241}$$

Explicit form for the Coefficients

Presume that there is a range of positive t for which equations (7.240) and (7.210) are both good approximate forms of the same function. Replacing (7.210) with

$$\psi(x) \xrightarrow[x \to +\infty]{} C'\frac{1}{\sqrt{k(x)}}\cos\left(\int_a^x k(x')dx' - \frac{\pi}{4}\right) + D'\frac{1}{\sqrt{k(x)}}\sin\left(\int_a^x k(x')dx' - \frac{\pi}{4}\right) \tag{7.242}$$

Identify

$$C' = \frac{1}{\sqrt{\pi}}(2m\beta)^{-\frac{1}{6}}C_A \quad \text{and} \quad D' = -\frac{1}{\sqrt{\pi}}(2m\beta)^{-\frac{1}{6}}C_B \quad (7.243)$$

$$A = \frac{1}{\sqrt{\pi}}(2m\beta)^{-\frac{1}{6}}C_B \quad \text{and} \quad B = \frac{1}{2}\frac{1}{\sqrt{\pi}}(2m\beta)^{-\frac{1}{6}}C_A \quad (7.244)$$

The two asymptotic regions are connected by

$$C' = 2B \quad \text{and} \quad D' = -A \quad (7.245)$$

In terms of the original coefficients defined in equation (7.210),

$$D = Be^{-i\pi/4} + \frac{1}{2}Ae^{i\pi/4} \quad (7.246a)$$

$$C = Be^{i\pi/4} + \frac{1}{2}Ae^{-i\pi/4} \quad (7.246b)$$

or

$$A = Ce^{i\pi/4} + De^{-i\pi/4} \quad (7.247a)$$

$$B = \frac{1}{2}\left(Ce^{-i\pi/4} + De^{i\pi/4}\right) \quad (7.247b)$$

Equations (7.247) are a particular case of the general rules (7.216) and (7.217).

I have taken the potential to have a negative slope at the turning point $x = a$. There is also the connection formulas in the other case. If $V(x)$ is increasing through a turning point $x = a$, then equation (7.210) or equivalently equation (7.242) holds for $x \ll a$ and equation (7.209) for $x \gg a$. You can repeat the analysis with the opposite sign for β in equation (7.218), or simply change the sign of the argument of the Airy function. The result is

$$D = \frac{1}{2}Be^{-i\pi/4} + Ae^{i\pi/4} \quad (7.248a)$$

$$C = \frac{1}{2}Be^{i\pi/4} + Ae^{-i\pi/4} \quad (7.248b)$$

Equations (7.246) and (7.248) are the famous **connection formulas**.

7.7.3 Bound States

As a simple example consider a potential well with a single minimum. I will examine an energy eigenfunction with energy E such that $V(x) > E$ for $x < a$ and $x > b$, while $E > V(x)$ for $a < x < b$. Call these regions I, II, and III, starting from the left, as in Figure 7.5.

The wave function has the form

$$\psi(x) \approx \frac{B}{\sqrt{\kappa(x)}} \exp\left[\int_a^x \kappa(x')dx'\right] \quad (7.249)$$

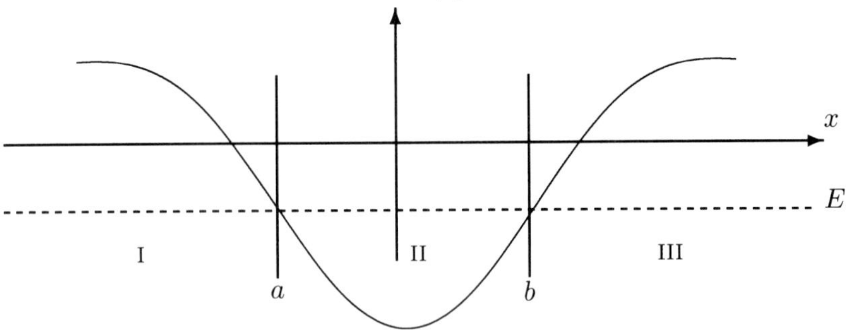

FIGURE 7.5: A Single well potential

in region I, and

$$\psi(x) \approx \frac{A}{\sqrt{\kappa(x)}} \exp\left[-\int_a^x \kappa(x')dx'\right] \qquad (7.250)$$

in region III.[24] In between, in region II, the wave function looks like

$$\psi(x) \approx \frac{C}{\sqrt{k(x)}} \exp\left[-i\int_a^x k(x')dx'\right] + \frac{D}{\sqrt{k(x)}} \exp\left[i\int_a^x k(x')dx'\right] \qquad (7.210)$$

or equivalently

$$\psi(x) \approx \frac{C}{\sqrt{k(x)}} e^{-i\Phi} \exp\left[-i\int_b^x k(x')dx'\right] + \frac{D}{\sqrt{k(x)}} e^{i\Phi} \exp\left[i\int_b^x k(x')dx'\right] \qquad (7.251)$$

where

$$\Phi = \int_a^b k(x')dx' \qquad (7.252)$$

The connection formulas (7.246) and (7.248) become

$$C = Be^{i\pi/4} \quad \text{and} \quad D = Be^{-i\pi/4} \qquad (7.253)$$

and

$$Ce^{-i\Phi} = Ae^{-i\pi/4} \quad \text{and} \quad De^{i\Phi} = Ae^{i\pi/4} \qquad (7.254)$$

respectively. These two rules are inconsistent unless $e^{2i\Phi} = e^{i\pi}$:

$$\Phi = \left(n + \frac{1}{2}\right)\pi \qquad (7.255)$$

Then

$$C = iD = e^{i\pi/4}B \quad \text{and} \quad A = (-1)^n B \qquad (7.256)$$

---

[24]The terms with the other sign in the exponent must vanish in order for the wave function to be normalizable.

In particular, equation (7.210) for the wave function in the middle region is

$$\psi(x) \approx \frac{2B}{\sqrt{k(x)}} \cos\left[\int_a^x k(x')dx' - \frac{\pi}{4}\right] \qquad (7.257)$$

Equation (7.255) is the WKB approximation to the eigenvalue condition. If $V(x) = V(-x)$, so that $a = -b$, it is easy to see that the solutions are even or odd functions of x.

Example: The harmonic oscillator

Apply the WKB method to the one-dimensional harmonic oscillator potential

$$V(x) = \frac{1}{2}m\omega^2 x^2 \qquad (7.258)$$

as in Section 3.2. For an energy E, the classical turning points are

$$x_{\pm} = \pm\sqrt{\frac{2E}{m\omega^2}} \qquad (7.259)$$

The phase defined in equation (7.252) is

$$\Phi = \int_{x_-}^{x_+} \sqrt{2m[E - V(x)]}\,dx = \int_{x_-}^{x_+} \sqrt{2mE - (m\omega x)^2}\,dx = \frac{E\pi}{\omega} \qquad (7.260)$$

The eigenvalue condition (7.255) is then

$$E = \left(n + \frac{1}{2}\right)\omega \qquad (7.261)$$

The WKB approximation to all the energy levels is fortuitously exact for the harmonic oscillator. The approximate wave functions, of course, differ from the correct ones.

A similar analysis can be made for other potential wells. The calculation for a quartic potential is the subject of Problem 7.15.

Example: A double well

A double well with a high barrier in the middle (illustrated in Figure 7.6) is a useful approximation to many physical systems. In Section 3.1.2 I discussed an application to the NH_3 molecule. There I made the drastic approximation that there were effectively only two states, but a better job is possible since the distance between the minima and the curvature near the minima can be obtained from other features of the molecular spectrum. The double well has also been used to study simple molecules like H_2 or N_2.

Let $\psi_i(x)$ be the WKB approximation to the wave function in the wave func-

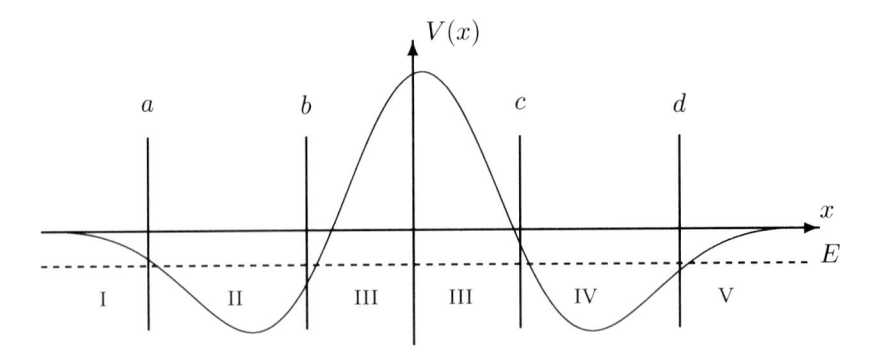

246 Chapter 7 Approximation Methods for Bound States

FIGURE 7.6: A double well potential illustrating the five separate WKB-approximation regions

tion in the region i:

$$\psi_I(x) \approx \frac{B_1}{\sqrt{\kappa(x)}} \exp\left[\int_a^x \kappa(x')dx'\right] \tag{7.262a}$$

$$\psi_{II}(x) \approx \frac{C_2}{\sqrt{k(x)}} \exp\left[-i\int_a^x k(x')dx'\right] + \frac{D_2}{\sqrt{k(x)}} \exp\left[i\int_a^x k(x')dx'\right] \tag{7.262b}$$

$$\psi_{III}(x) \approx \frac{A_3}{\sqrt{\kappa(x)}} \exp\left[-\int_b^x \kappa(x')dx'\right] + \frac{B}{\sqrt{\kappa(x)}} \exp\left[\int_b^x \kappa(x')dx'\right] \tag{7.262c}$$

$$\psi_{IV}(x) \approx \frac{C_4}{\sqrt{k(x)}} \exp\left[-i\int_c^x k(x')dx'\right] + \frac{D_4}{\sqrt{k(x)}} \exp\left[i\int_c^x k(x')dx'\right] \tag{7.262d}$$

$$\psi_V(x) \approx \frac{A_5}{\sqrt{\kappa(x)}} \exp\left[-\int_d^x \kappa(x')dx'\right] \tag{7.262e}$$

Applying the appropriate connection formula at each of the four turning points gives

$$C_2 = e^{i\pi/4}B_1 \quad D_2 = e^{-i\pi/4}B_1 \tag{7.263a}$$

$$A_3 = \sin\Phi_2 B_1 \quad B_3 = 2\cos\Phi_2 B_1 \tag{7.263b}$$

$$A_3 = 2\cos\Phi_4 e^{\Phi_3} A_5 \quad B_3 = \sin\Phi_4 e^{-\Phi_3} A_5 \tag{7.263c}$$

$$C_4 = e^{i\Phi_4}e^{-i\pi/4}A_5 \quad D_4 = e^{-i\Phi_4}e^{i\pi/4}A_5 \tag{7.263d}$$

where

$$\Phi_2 = \int_a^b k(x')dx' \tag{7.264a}$$

$$\Phi_3 = \int_b^c \kappa(x')dx' \tag{7.264b}$$

$$\Phi_4 = \int_c^d k(x')dx' \tag{7.264c}$$

All but one of the coefficients are determined. Eliminate A_3 and B_3 from equations (7.263b) and (7.263c) to obtain

$$2B_1 \cos\Phi_2 = \sin\Phi_4 e^{-\Phi_3} A_5$$
$$B_1 \sin\Phi_2 = 2\cos\Phi_4 e^{\Phi_3} A_5 \tag{7.265}$$

Equations (7.265) are consistent provided

$$\tan\Phi_2 \tan\Phi_4 = 4e^{2\Phi_3} \tag{7.266}$$

Since the three phase factors depend on the energy, equation (7.266) is the eigenvalue condition.

In all the applications cited, the physical system has a reflection symmetry: $V(x) = V(-x)$ and therefore $\Phi_4 = \Phi_2$. In this case

$$\tan\Phi_2 = \pm 2e^{\Phi_3} \tag{7.267}$$

and $A_5 = \pm B_1$. The solutions are either symmetric or antisymmetric.

In many physical applications Φ_3 is a large number. In that case there are two solutions for every integer n, one for each sign in equation (7.267), of the form

$$\Phi_2 = \left(n + \frac{1}{2}\right)\pi \mp \frac{1}{2}e^{-\Phi_3} \pm \frac{1}{24}e^{-3\Phi_3} + \cdots \tag{7.268}$$

As the quadratic minima on each side get more and more separated, $\Phi_3 \to \infty$, and the energy levels become the same as the ordinary harmonic oscillator levels, but doubly degenerate.

7.7.4 Tunneling through a Barrier

An α particle, or ^{4}He nucleus, is bound to a heavy nucleus by a short-range nuclear force like the square well with a wall that is high, but not infinitely high. If the particle can tunnel through the potential barrier, it gets repelled by the Coulomb force between it and the parent nucleus and escapes. It was realized quite early that tunneling, totally forbidden by the rules of classical mechanics, is the explanation for α decay. The first quantitative calculation of quantum tunneling was the explanation of α decay.[25] Let us apply the WKB method to tunneling through a barrier. Consider a one-dimensional particle moving in a potential with some energy $E > 0$.

---
[25] Gamow [23], Gurney and Condon [24].

248 Chapter 7 Approximation Methods for Bound States

The potential vanishes as $x \to \pm\infty$, but exceeds the energy E somewhere between two turning points a and b. Let the particle be incident on the barrier from the left. For $x < a$, the WKB approximation to the wave function has the form (7.209):

$$\psi(x) \approx \frac{C}{\sqrt{k(x)}} \exp\left[-i\int_a^x k(x')dx'\right] + \frac{D}{\sqrt{k(x)}} \exp\left[i\int_a^x k(x')dx'\right] \quad (7.269)$$

D is the coefficient of the incident wave, while C is the coefficient of the wave reflected off the barrier. In the middle region, $a < x < b$ the wave function has the form (7.209):

$$\psi(x) \approx \frac{A}{\sqrt{\kappa(x)}} \exp\left[-\int_a^x \kappa(x')dx'\right] + \frac{B}{\sqrt{\kappa(x)}} \exp\left[\int_a^x \kappa(x')dx'\right]$$

$$= \frac{A}{\sqrt{\kappa(x)}} e^{-\Phi} \exp\left[-\int_b^x \kappa(x')dx'\right] + \frac{B}{\sqrt{\kappa(x)}} e^{\Phi} \exp\left[\int_b^x \kappa(x')dx'\right] \quad (7.270)$$

with

$$\Phi = \int_a^b \kappa(x)dx \quad (7.271)$$

Finally, when $x > b$, the wave function has the form (7.209) again, but if the wave is incident from the left, the boundary condition is that there is only an outgoing term:

$$\psi(x) \approx \frac{F}{\sqrt{k(x)}} \exp\left[i\int_a^x k(x')dx'\right] \quad (7.272)$$

F is the coefficient of the transmitted wave.

Apply the connection formula (7.248) at $x = a$ and (7.247) at $x = b$ to get

$$F = \frac{4D}{4e^\Phi + e^{-\Phi}} \quad (7.273)$$

The **transmission coefficient**, the probability for tunneling through the barrier, is

$$T = \frac{|F|^2}{|D|^2} = \frac{16}{(4e^\Phi + e^{-\Phi})^2} \xrightarrow[\Phi \gg 1]{} e^{-2\Phi} \quad (7.274)$$

PROBLEMS

7.1. State Vectors in Second Order

Continue the expansion started in equation (7.9) and write the perturbed states through second order in the perturbing Hamiltonian H', i.e.

$$|\psi_n\rangle = |\psi_n^o\rangle + \sum_{m \neq n} \frac{|\psi_m^o\rangle H'_{mn}}{E_n^o - E_m^o} + ?$$

7.2. Oscillator with Cubic Perturbation

Consider a force that increases quadratically with distance from the origin as a perturbation to the one-dimensional harmonic oscillator; that is, $H = H_o + H'$, where

$$H_o = \frac{p^2}{2m} + \frac{1}{2}m\omega^2 x^2 \quad \text{and} \quad H' = -\beta x^3$$

Find the shifts of the first two energy levels through the lowest nonvanishing order in perturbation theory.

7.3. Harmonic Oscillator Perturbation Problem

Compute the next term in the expansion (7.27) and verify that it agrees with the perturbation result.

7.4. Infinite Well with Delta Function Revisited

Consider a one-dimensional particle confined to the region $-a \leq x \leq a$. This is the infinite square well potential. Let there be a delta function perturbation at the origin:

$$H' = \beta E_o a \, \delta(x)$$

where E_o is the unperturbed ground-state energy, and β is a dimensionless parameter. Find the correction to the unperturbed energy levels, through second order in β, and compare your answer with the expansion of the exact solution (see Problem 3.5).

7.5. A Two-State System

Approximate the ammonia molecule NH$_3$ by a simple two-state system, as in Section 3.1.2. The three H nuclei are in a plane, and the N nucleus is at a fixed distance either above or below the plane of the H's. Each is approximately a stationary state with some energy E_o. But there is a small amplitude for transition from up to down. Thus $H = H_o + H'$, where

$$H_o = \begin{pmatrix} E_o & 0 \\ 0 & E_o \end{pmatrix} \quad \text{and} \quad H' = \begin{pmatrix} 0 & -A \\ -A & 0 \end{pmatrix}$$

where $|A| \ll |E_o|$.

(a) What are the exact eigenvalues of H?

(b) Now suppose the molecule is in an electric field, which distinguishes the two states. The new Hamiltonian is $H_o + H' + H''$, where

$$H'' = \begin{pmatrix} \epsilon_1 & 0 \\ 0 & \epsilon_2 \end{pmatrix}$$

What are the exact energy levels now?

(c) What answer does perturbation theory give in the lowest nonvanishing order when $|\epsilon_i| \ll |A|$? When $|\epsilon_i| \gg |A|$? Compare both results to the exact answer.

7.6. External Electric Quadrupole Field

An electron in a hydrogen atom potential is subject to a small external electric field $\mathbf{E}(\mathbf{r})$:

$$\mathbf{E} = -2Gx\hat{n}_x + 2Gy\hat{n}_y$$

in addition to the Coulomb field of the nucleus. G is some constant. (In this problem ignore the electron spin.)

(a) What is the extra piece of the Hamiltonian H' due to this electric field?

(b) What are the nonvanishing matrix elements of H' in the five $n = 3$ D states $(l = 2)$?

> Note: These matrix elements are not all independent; they have the same radial integral, so you need compute that only once. You can either compute the angular parts explicitly or just compute one of them and then use the fact that $x^2 - y^2$ is a sum of spherical components of a rank-two tensor, together with the explicit form Wigner-Eckart theorem (Section 5.2.4).
>
> To use the Wigner-Eckart theorem, you will need the Clebsch-Gordan coefficients for combining two spin-two systems into one of angular momentum 2. Some of these coefficients are given by
>
> $$\sqrt{7}\left|\phi_{22}^{22}\right\rangle = \sqrt{2}\left|\psi_{20}^{22}\right\rangle + \sqrt{2}\left|\psi_{02}^{22}\right\rangle - \sqrt{3}\left|\psi_{11}^{22}\right\rangle$$
>
> $$\sqrt{14}\left|\phi_{21}^{22}\right\rangle = \sqrt{6}\left|\psi_{2,-1}^{22}\right\rangle + \sqrt{6}\left|\psi_{-1,2}^{22}\right\rangle - \left|\psi_{10}^{22}\right\rangle - \left|\psi_{01}^{22}\right\rangle$$
>
> $$\sqrt{14}\left|\phi_{20}^{22}\right\rangle = \left|\psi_{1,-1}^{22}\right\rangle + \left|\psi_{-1,1}^{22}\right\rangle + 2\left|\psi_{2,-2}^{22}\right\rangle + 2\left|\psi_{-2,2}^{22}\right\rangle - 2\left|\psi_{00}^{22}\right\rangle$$
>
> The states on the right are labeled by $\left|\psi_{m_1 m_2}^{j_1 j_2}\right\rangle$. The states on the left are labeled by $\left|\phi_{jm}^{j_1 j_2}\right\rangle$. You don't have to derive these coefficients, but should check that J_+ in the top state yields zero, while J_- on that state yields the next one down, and so forth. If you really want to understand this subject, you should compute the angular integrals directly and verify that they agree with the Wigner-Eckart rules.

(c) Because the Hamiltonian is no longer invariant under rotations, these five energy levels need not be degenerate. Diagonalize H' among these five 3D states.

(d) How many distinct levels are there now? What are the perturbed energies to first order in perturbation theory?

7.7. Finite Size of the Nucleus

There is a correction to atomic energy levels due to the finite size of the nucleus. To get an idea of the size of this correction for atomic hydrogen, neglect spin, and assume the proton to be a sphere with uniform charge density of radius $R = 1/m_\pi$, where $m_\pi = 140\,\text{MeV}$ is the π meson mass, and with total charge e. R is the Compton wavelength of the pion (see Problem 3.12). How much does the energy of an electron in this field differ from its energy in a pure $1/r$ potential?[26] Compute the effect of this correction to first order in perturbation theory on the ground state and on the $2S$ state of the atom. In your computation ignore terms of order R/a relative to the leading term. What is the difference in electron volts? How does it compare with the fine structure splitting for $n = 2$?

> Note: Isotopes are nuclei with the same charge but different masses. They have the same number of protons but different numbers of neutrons. The volume of a nucleus is, roughly, proportional to the total number of nucleons, the atomic weight A. The radii of two isotopes therefore differ by some

---
[26] Be sure that both H and H_o have the same value as $r \to \infty$ so that H really approaches H_o when the perturbation is small.

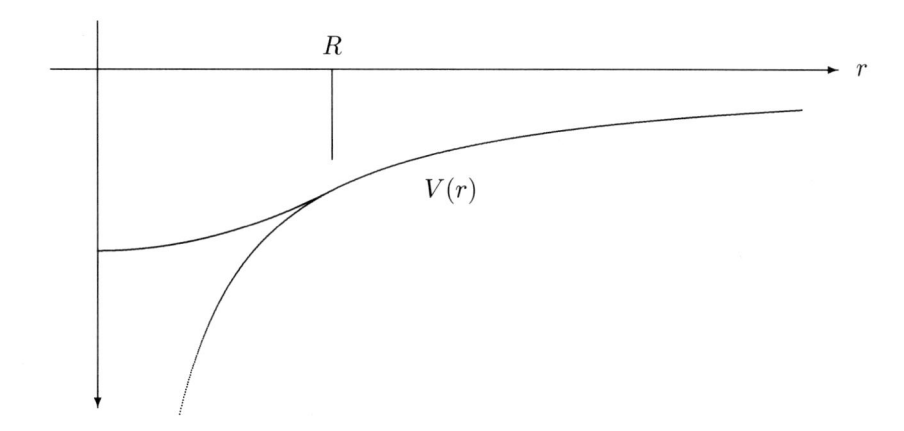

FIGURE 7.7: Effect of the finite size of the nucleus. The potential differs from a pure Coulomb potential only in the small volume $r \leq R$. The lower curve is the Coulomb potential. The upper curve is the potential corrected for $r \leq R$. The perturbation is the difference between the two, and vanishes for $r > R$.

number proportional to $A^{1/3}$. There is a small but detectable difference in the energy levels of two isotopes, which can be computed by the method of this problem. In that context the effect you just calculated is the **isotope effect**.

7.8. **Dipole and Quadrupole Interaction of a Spin-3/2 Nucleus**
   (a) A nucleus with spin j has (only) $2j+1$ states $|j,m\rangle$, which all have the same energy, and transform according to the $D^{(j)}$ representation under rotations. Suppose a spin-3/2 nucleus at the origin is placed in an inhomogeneous electric field whose potential is $V(\mathbf{r})$. Is there an electric dipole interaction? Why or why not?
   (b) From rotational invariance, the quadrupole interaction must have the form
   $$H' = c\,\mathbf{s}\cdot\boldsymbol{\nabla}\mathbf{s}\cdot\boldsymbol{\nabla}V(0)$$
   where $\mathbf{s}$ is the nuclear spin operator, for some constant c related to the quadrupole moment of the nucleus. Since V is an external potential, $\nabla^2 V(\mathbf{r}) = 0$. Use this fact to show that one can write
   $$H' = \frac{c}{3}\sum_{ij}(3s_is_j - \delta_{ij}\mathbf{s}^2)\frac{\partial}{\partial r_i}\frac{\partial}{\partial r_j}V(\mathbf{r})\bigg|_{r=0}$$
   (c) Show that one can choose a coordinate system that eliminates the off-diagonal elements, so that in the new coordinate system
   $$H' = c\sum_i (s_i)^2 \frac{\partial V(\mathbf{r})}{\partial r_i^2}\bigg|_{r=0}$$

   Hint: Be careful here. This is not a transformation of the quantum-mechanical states; the idea is so show that there exists an orthogonal matrix M such that if $r'_i = \sum_j M_{ij}r_j$, then H' has the stated form.

(d) Write H' as
$$H' = a\left(3s_z^2 - \mathbf{s}^2\right) + b\left(s_+^2 + s_-^2\right)$$
and express a and b in terms of the constant c and $V(\mathbf{r})/r_i^2$. What are the eigenvalues and the corresponding eigenstates? Is there any degeneracy?

7.9. The Quadratic Stark Effect

This problem is about the spinless hydrogen atom in an external electric field.
(a) Fill in the missing steps in equation (7.46).
(b) In an external constant electric field of strength E compute the contributions from the $n = 3$ and $n = 4$ levels respectively to the second-order energy of the ground state.
(c) Compute the electric field in the atom due to a bare proton (i.e., an H^+ ion) along the z-axis at a distance of ten Bohr radii from the atom's center, and approximate it by a constant field equal to its value at the center. What is the splitting between the $n = 2$ levels due to the first-order Stark effect?[27] Compare this splitting to the second-order shift of the ground state you computed above. Assume that most of the contribution to the second-order splitting comes from the states considered above; is the second-order effect really negligible?
(d) What happens if the proton is only two Bohr radii away?

7.10. Some Magnetic Field Effects

(a) *The earth's magnetic field*: The earth's magnetic field near the surface is of the order 0.5 gauss. How does the size of the splitting of the $2P$ states of atomic hydrogen due to the earth's magnetic field compare with the fine structure splitting? What about the fields in excess of 100,000 gauss produced in the laboratory?
(b) *The quadratic Zeeman effect*: Estimate the magnitude of the quadratic term in the Hamiltonian for a constant magnetic field [see equation (6.35)] in atomic hydrogen. The energy shift due to this term can have the opposite sign from the linear term. How large a magnetic field do you need before this energy shift just equals, in magnitude, the first-order Zeeman effect?

The quadratic effect is called **diamagnetism**; is diamagnetism important for the earth's magnetic field?
(c) Consider the effect of a constant external magnetic field of strength B on the $n = 2$ levels of a hydrogen atom, in the limit of *strong* magnetic fields. This is the "Paschen-Back" limit. The fine structure splitting is small compared to the splitting due to the magnetic field, so that the perturbation expansion of Section 7.3.5 is not correct.

What is the strength, in gauss, of an external magnetic field that makes the splitting between the highest and lowest $n = 2$ levels the same size as the $n = 2$ fine structure splitting?
(d) Compute the corrections to the energy levels you found in the previous part, adding in the effects that cause the fine structure splitting as a perturbation.

7.11. The van der Waals Effect

Background: The interaction between two atoms can be approximated by the energy of the nuclei fixed at a distance R. This is a version of the Born-Oppenheimer approximation. Then it becomes a two-body problem with potential $V(R)$. For large R, $V(R) \to 0$ as $R \to \infty$, since in that limit the electrostatic force is completely shielded. In the other limit, the Coulomb repulsion between

---

[27] See Section 7.2.

the nuclei of charges Ze and $Z'e$ dominates: $V(R) \to ZZ'e^2/R$ as $R \to 0$. In between, the energy $V(R)$ may be determined by "solving" for the electronic energy states. This is impossible to do exactly except in the simplest cases.

The van der Waals effect is the first correction to $V = 0$ as $R \to \infty$. It tells us how close the atoms have to be before they begin to interact, and therefore, for example, the first corrections to the ideal gas law. The sign is also crucial. If it is negative, then there must be a minimum, and the atoms may bind. If it is positive, they probably will not.

Now consider two hydrogen atoms with their nuclei nailed down a distance R apart. The "unperturbed" Hamiltonian is

$$H_o = \frac{\mathbf{p}_1^2}{2m} + \frac{\mathbf{p}_2^2}{2m} - \frac{e^2}{r_1} - \frac{e^2}{r_2}$$

The eigenstates of H_o are the usual hydrogen atom states $|n_1, l_1, m_1, n_2, l_2, m_2\rangle$, with energies $E_1^o + E_2^o$, where $E_i^o = -\alpha^2 m/2n_i^2$. n_i is the principal quantum number of the i-th atom. This is the zero of the potential in the discussion in the paragraphs above. The rest of the potential is the perturbation H':

$$H' = e^2 \left(\frac{1}{R} - \frac{1}{|\mathbf{R} - \mathbf{r}_1|} - \frac{1}{|\mathbf{R} + \mathbf{r}_2|} + \frac{1}{|\mathbf{R} + \mathbf{r}_2 - \mathbf{r}_1|} \right)$$

Here $\mathbf{r}_1$ and $\mathbf{r}_2$ point from each of the nuclei to "its" electron, and $\mathbf{R}$ points from the first nucleus to the second. See Figure 7.8.

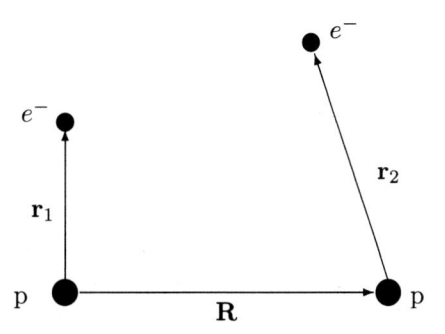

FIGURE 7.8: The van der Waals effect between two hydrogen atoms. $\mathbf{R}$ points from the first proton to the second, while $\mathbf{r}_1$ and $\mathbf{r}_2$ point from each proton to one of the electrons.

For large R expand H' in powers of $1/R$. The $1/R$ and $1/R^2$ terms cancel, so the leading term is cubic. A little algebra yields

$$H' \to -e^2 (\mathbf{r}_1 \cdot \nabla)(\mathbf{r}_2 \cdot \nabla) \frac{1}{R} = e^2 \left(\frac{\mathbf{r}_1 \cdot \mathbf{r}_2}{R^3} - 3 \frac{(\mathbf{r}_1 \cdot \mathbf{R})(\mathbf{r}_2 \cdot \mathbf{R})}{R^5} \right)$$

This looks like the interaction between two electric dipoles $-e\mathbf{r}_1$ and $-e\mathbf{r}_2$.

Usually, parity forbids atoms to have an intrinsic electric dipole moment, so first-order perturbation theory gives zero for the perturbation, and the leading effect is second order. This second-order effect is hard to compute exactly, but it is easy to see that it goes like $1/R^6$. That is the usual form of the van der

254 Chapter 7 Approximation Methods for Bound States

Waals interaction. It can be computed for two hydrogen atoms in the ground (1S) state, and the answer is claimed to be

$$-6.5 \frac{e^2}{a} \left(\frac{a}{R}\right)^6$$

The sign shows that two hydrogen atoms in the ground state attract, so a bound molecule is probable. Second-order perturbation theory always depresses the ground state.

But when the two hydrogen atoms are both in the $n=2$ states, something different happens. You know from Section 7.2 that for atomic hydrogen there is an electric dipole moment to *first* order because of the degeneracy of the $2S$ and $2P$ states. The interaction is H' above, and there *will* be a contribution to first order in perturbation theory.[28]

Problem: Two hydrogen atoms, both in the $n=2$ state, are close to each other with their nuclei a fixed distance R apart. Using perturbation theory to first order in this two-electron system, find the eigenvalues of the energy that vanish as $1/R^3$ as $R \to \infty$. Neglect spin and the identity of the electrons.

Hint: There are sixteen states: Each atom can be in the $2S$ state or in one of the three $2P$ states. They all have the same unperturbed energy, and H' certainly connects some of them. You could just try to diagonalize this 16×16 matrix, and find the 16 eigenvalues. But the problem is simplified by exploiting the remaining axial symmetry.

Take the line joining the two nuclei as the z-axis, and write H' in terms of spherical components of vectors. Label the states $|l_1, m_1, l_2, m_2\rangle$ (omit the principal quantum numbers, which are both here always 2). The total L_z of each state is $m_1 + m_2$. The matrix elements of H' vanish unless $m_1 + m_2$ are the same on both sides. This number can have each of the five values ± 2, ± 1, and 0. It will be enough to work out the energies for the nonnegative values of m.

7.12. Hyperfine Structure of the Hydrogen Atom

In Section 7.3.6 you can find the several terms that contribute to the hyperfine structure splitting of the hydrogen atom energy levels, but I worked it out there in detail only for the ground state.

The perturbing Hamiltonian has three parts:

$$H'_{HFS} = H'_{\text{contact}} + H'_{\text{dipole}} + H'_{\text{orbital}}$$

with

$$H'_{\text{contact}} = \frac{g_e g_p \alpha}{4mM} \frac{8\pi}{3} \mathbf{s}_e \cdot \mathbf{s}_p \delta(\mathbf{r})$$

$$H'_{\text{dipole}} = \frac{g_e g_p \alpha}{4mM} \frac{1}{r^5} \sum_{ij} s_{ei} s_{pj} \left(3 r_i r_j - \delta_{ij} r^2\right)$$

---

[28] You should recognize the form of H' form as proportional to the $q=0$ component of a rank-two spherical tensor made up of linear combinations of $r_{1i} r_{2j}$. Here you cannot use the full power of the Wigner-Eckart theorem, since by setting $\mathbf{R} = R \hat{\mathbf{n}}_z$ we have abandoned complete spherical symmetry. Nevertheless it is still possible to exploit the remaining axial symmetry.

$$H'_{\text{orbital}} = \frac{g_p \alpha}{2Mm} \frac{1}{r^3} \mathbf{s}_p \cdot \mathbf{L}$$

To first order in perturbation theory, find a formula for the energy shifts of *all* the bound-state levels of a hydrogen atom due to the term in the Hamiltonian that comes from the interaction of the electron with the nucleus' magnetic field. This is the general formula for the hyperfine splitting. Work it out any way you want, but I suggest following the steps outlined here.

(a) For the S states only, the contact term enters. Review the derivation of the shift in the ground states. Let $\mathbf{F} = \mathbf{J} + \mathbf{s}_p$ be the total angular momentum, where $\mathbf{J} = \mathbf{L} + \mathbf{s}_e$. The eigenvalues of $\mathbf{F}^2$ are $F(F+1)$, which for S states can be 1 or 0. As a function of F and the principal quantum number n, what are the energies of the levels (to first order in H'_{HFS}) for all the other S states. Hint: $R_{n0}(0)$ is proportional to $n^{-3/2}$.

(b) To compute the dipole term contribution, you need matrix elements of the symmetric tensor

$$Q_{ij} = \frac{1}{r^5}(3r_i r_j - \delta_{ij} r^2)$$

between states of the same n and l. Since these are linear combinations of components of a rank-two spherical tensor, one needs to compute only one of these matrix elements. Then the others are determined by the Wigner-Eckart theorem. The same is true for any other such tensor, in particular for

$$K_{ij} = \frac{3}{2}(L_i L_j + L_j L_i) - \delta_{ij} L^2$$

that is,

$$\langle nlm' | Q_{ij} | nlm \rangle = c \langle nlm' | K_{ij} | nlm \rangle$$

The constant c is independent of m' and m, so you can find it from any matrix element of any component. What is the constant c? Express the answer in terms of the expectation value of $1/r^3$. Hint: Use the normalization of the Legendre polynomals and the recursion relation

$$x P_l(x) = \frac{(l+1)P_{l+1} + l P_{l-1}(x)}{2l+1}$$

and choose $m' = m = 1$ and $i = j = 3$.

(c) Set $g = 2$, and write the expectation value of the dipole term in the form

$$\langle nljFm_F | H'_{\text{dipole}} | nljFm_F \rangle$$
$$= C_1 \left\langle nl0 \left| \frac{1}{r^3} \right| nl0 \right\rangle \langle nljFm_F | C_2 \mathbf{s}_p \cdot \mathbf{L} + C_3 \mathbf{s}_p \cdot \mathbf{s}_e | nljFm_F \rangle \quad (7.275)$$

for some three constants C_i.

(d) Write the term $\langle nljFm_F | H'_{\text{orbital}} | nljFm_F \rangle$ in a similar way.

(e) Apply the projection theorem (Section B.1.3 in the appendix) and the rule

$$\langle nljFm_F | \mathbf{s}_p \cdot \mathbf{J} | nljFm_F \rangle = \frac{1}{2}\left[F(F+1) - j(j+1) - \frac{3}{4} \right]$$

to express $\langle nljFm_F | \mathbf{s}_p \cdot \mathbf{L} | nljFm_F \rangle$ and $\langle nljFm_F | \mathbf{s}_p \cdot \mathbf{s}_e | nljFm_F \rangle$ in terms of i, j, and l.

256   Chapter 7   Approximation Methods for Bound States

(f) Now put it all together to obtain a formula for the first-order energy shifts

$$\langle nljFm_F | H' | nljFm_F \rangle \qquad (7.276)$$

Use the rule

$$\left\langle nlm \left| \frac{1}{r^3} \right| nlm \right\rangle = \frac{\alpha^3 m^3}{n^3 l(l+\frac{1}{2})(l+1)} \qquad (7.277)$$

from Problem 3.21

(g) Show that your rule is valid for S states also.

7.13. Two Elementary Variational Method Problems

(a) Consider a one-dimensional harmonic oscillator. The Hamiltonian is

$$H = \frac{p^2}{2m} + \frac{m\omega^2 x^2}{2}$$

The ground-state energy is $E = \hbar\omega/2$. Pretend you do not know this and try a variational estimate. Estimate the ground-state wave function by

$$\langle x | \psi \rangle = \psi(x) = N e^{-\beta |x|}$$

Determine N by the normalization condition. Find the value of β that minimizes $\langle \psi | H | \psi \rangle$. What is the energy? (Watch out: the derivative of the trial function has a discontinuity.) How close do you get to the true ground-state energy?

(b) Similarly, consider the hydrogen atom without spin. The Hamiltonian is

$$H = \frac{\mathbf{p}^2}{2m} - \frac{\alpha}{r}$$

The true ground-state energy is $E_o = -\alpha^2 m/2$. Since the ground state is an S state the wave function must be spherically symmetrical. Suppose you could not solve this problem exactly either. Estimate the ground-state wave function with a Gaussian:

$$\psi(\mathbf{r}) = N e^{-r^2/b^2}$$

Compute N so that $\psi(\mathbf{r})$ is correctly normalized. Compute the expectation value of H in this state, and find the value of b which minimizes it. The best estimate is $\langle \psi | H | \psi \rangle$ with this optimum value of b. How much does it differ from the correct answer?

7.14. Variational Estimate for a Quartic Potential

(a) Consider a one-dimensional particle moving in a quartic potential: $V(x) = \lambda x^4$. Show the energies depend on λ and the mass only through the combination

$$E_n = \bar{E}_n \left(\frac{\lambda}{4m^2} \right)^{\frac{1}{3}}$$

where $\bar{E}$ is dimensionless.

(b) Estimate the ground-state energy using a trial function of the form

$$\psi(x) = N e^{-a^2 x^2/2}$$

(c) Estimate the energy of the first excited state using a trial function of the form[29]
$$\psi(x) = Nxe^{-a^2x^2/2}$$

7.15. WKB Estimate for a Quartic Potential
Estimate the ground-state energy, and the energies of the first and second excited states, for the quartic potential of Problem 7.14, using the WKB method.

7.16. Variational Method for the Stark Effect on the Ground State of Atomic Hydrogen
Consider the hydrogen atom, without spin, in a weak electric field of strength E_o along the z-axis. The first-order perturbation vanishes. In Problem 7.9 you wrote the formula for the second-order contribution, and computed the first two terms. While this is not a bad approximation, these are only the first two terms in an infinite sum. Since every term in the sum is negative, the result of Problem 7.9 at least gives an upper bound for the energy.

A variational estimate provides a different bound. Take as a trial function
$$\psi(\mathbf{r}) = Ne^{-r/a}\left(1 + \frac{\beta}{a}z\right)$$
with β as a variational parameter. Compute an upper bound to this energy using the variational method, as follows:
(a) Compute the normalization constant N as a function of β.
(b) Write $H = T + V_o + H'$ with
$$T = \frac{\mathbf{p}^2}{2m} \qquad V_o = -\frac{e^2}{r} \qquad H' = -eE_oz$$
and compute separately the expectation values
$$T(\beta) = \langle\psi|T|\psi\rangle \qquad V_o(\beta) = \langle\psi|V_o|\psi\rangle \qquad H'(\beta) = \langle\psi|H'|\psi\rangle$$
(c) Find the value of β that minimizes $E(\beta) = T(\beta) + V_o(\beta) + H'(\beta)$.
(d) Find the minimum value of the energy $E(\beta)$. [Keep only terms to second order in E_o.] This is also an upper bound (why?). Is it better or worse that the bound you obtained in Problem 7.9?

7.17. Born-Oppenheimer Approximation—A Simple Model
Here is a solvable toy model in one dimension for a "molecule" like the H_2^+ ion, containing one light "electron" of mass m connected to two heavy "protons" with mass M by two identical springs, each with natural length a and spring constant β. The Hamiltonian is
$$H = \frac{p_1^2}{2M} + \frac{p_2^2}{2m} + \frac{p_3^2}{2M} + \frac{\beta}{2}\left[(x_2 - x_3 - a)^2 + (x_2 - x_1 + a)^2\right]$$

(a) Fix the distance between the two protons to be R, and find the energy spectrum $E_n(R)$ of the electron as a function of R.
(b) Now treat $E_n(R)$ as a potential between the protons, and solve for the proton energies E_{mn}. Do the calculation in the proton center of mass frame. The result is the "Born-Oppenheimer approximation" for this problem.

---
[29] The numerical values of the three lowest energies, to eight significant figures, are $\bar{E}_o = 1.060362\ldots$, $\bar{E}_1 = 3.799673\ldots$, and $\bar{E}_2 = 7.455703\ldots$.

258 Chapter 7 Approximation Methods for Bound States

(c) The energy levels of this model can be found exactly (see Problem 3.7). What is the error in the Born-Oppenheimer method? Show that the error is small when $M \gg m$.

7.18. Zeeman Effect for Intermediate Magnetic Fields

Compute the combined effect of the fine structure splitting and the splitting caused by an external constant magnetic field on the atomic hydrogen atom energy levels. Take the gyromagnetic ratio of the electron to be exactly 2.

First, calculate the splitting to first order when the magnetic energy is small compared to the fine structure splitting. I did this calculation for the $n = 2$ levels in Section 7.3.5. Next, compute the splitting to first order when the magnetic energy is large compared to the fine structure splitting. Try to obtain the results in these two limits valid for all states.

Finally, compute the combined shifts exactly for the $n = 2$ levels when the Zeeman splitting and the fine structure splitting energies are comparable, and compare with the perturbation theory results in the two limits.

Some Background: The full Hamiltonian is

$$H = H_{\text{coul}} + H_{\text{mag}} + H_{FS}$$

The first term is

$$H_{\text{coul}} = \frac{\mathbf{p}^2}{2m} - \frac{\alpha}{r}$$

The magnetic perturbation is

$$H_{\text{mag}} = -\boldsymbol{\mu} \cdot \mathbf{B}$$

$\mathbf{B}$ is the external magnetic field. Take it to be along the z-axis. $\boldsymbol{\mu}$ is the electron magnetic moment:

$$\boldsymbol{\mu} = -\mu_B(\mathbf{L} + 2\mathbf{s}) = -\mu_B(\mathbf{J} + \mathbf{s})$$

where $\mu_B = e/2m$.

The last term is the fine structure perturbation, which in turn has two parts, the spin-orbit part and the relativistic correction to the kinetic energy of the electron:

$$H_{FS} = H_{LS} + H_{\text{kin}}$$

where

$$H_{LS} = \frac{g\alpha}{4mr^3}\mathbf{L}\cdot\mathbf{s} \quad \text{and} \quad H_{\text{kin}} = -\frac{1}{2m}\left(\frac{\mathbf{p}^2}{2m}\right)^2$$

A complete set of states is either $\left|\psi^{n,l}_{m_l m_s}\right\rangle$ or $\left|\phi^{n,l}_{j,m}\right\rangle$. Each set is a linear combination of the other. The Clebsch-Gordan coefficients that allow one to pass between the two kinds of basis states are given in equations (4.134) and (4.135).

The states $\left|\psi^{nl}_{m_l m_s}\right\rangle$ are eigenstates of $H_{\text{coul}} + H_{\text{mag}}$.
The states $\left|\phi^{nl}_{jm}\right\rangle$ are eigenstates of $H_{\text{coul}} + H_{FS}$.

Suggested Outline: Here is a suggested outline of the computation:

(a) *Matrix elements:* Make a catalog of matrix elements of H_{mag}:

$$\left\langle \psi^{nl}_{m_l m_s} \middle| H_{\text{mag}} \middle| \psi^{nl}_{m_l m_s} \right\rangle = -\mu_B B(m_l + 2m_s)$$

$$\left\langle \phi_{jm}^{nl} \middle| H_{\text{mag}} \middle| \phi_{jm}^{nl} \right\rangle = -\mu_B B m - \mu_B B \left\langle \phi_{jm}^{nl} \middle| S_z \middle| \phi_{jm}^{nl} \right\rangle = ?$$

$$\left\langle \phi_{l+\frac{1}{2},m}^{n,l} \middle| H_{\text{mag}} \middle| \phi_{l-\frac{1}{2},m}^{n,l} \right\rangle = ?$$

(b) *Weak field limit:* In the limit $H_{\text{mag}} \to 0$ (i.e., $\mathbf{B} \to 0$) compute

$$\Delta_{mag} = \left\langle \phi_{jm}^{nl} \middle| H_{\text{mag}} \middle| \phi_{jm}^{nl} \right\rangle$$

as a function of l and m to first order in perturbation theory, for both $j = l \pm 1/2$. Verify that the results of Section 7.3.5 are a special case.

(c) *Strong field limit:* In this case take $H_o = H_{\text{coul}} + H_{\text{mag}}$, and $H' = H_{FS}$. Compute the first-order perturbation

$$\Delta_{FS} = \left\langle \psi_{m_l m_s}^{nl} \middle| H_{FS} \middle| \psi_{m_l m_s}^{nl} \right\rangle$$

as a function of n, l, and m_l for both $m_s = 1/2$ and $m_s = -1/2$.

(d) *Intermediate fields—states with highest m:* When neither limit is valid, take $H_o = H_{\text{coul}}$ and $H' = H_{FS} + H_{\text{mag}}$. While H_{coul} is diagonal in either basis, H' is generally now diagonal in neither. However, the state $\left| \phi_{l+\frac{1}{2},l+\frac{1}{2}}^{nl} \right\rangle$ is the same as the state $\left| \psi_{l\frac{1}{2}}^{nl} \right\rangle$. Verify that your results for the top states of a $j = l + 1/2$ ladder in the weak and strong field limits are indeed the same.

(e) *The 2P states:* In general, the states mix. Work out the $n = 2$ states only. Then the states that mix are those with $m = \pm 1/2$. Measure all energies in units of the fine structure splitting Δ between the $2P_{\frac{3}{2}}$ and the $2P_{\frac{1}{2}}$ states:

$$\Delta = \frac{\alpha^4 m}{32}$$

It might be convenient to take for the zero of the energy scale the energy of the lowest $n = 2$ state in the absence of a magnetic field. (The $2S$ states do not enter the calculation of the P state energies. Why not?)

Consider the $2P$ states with $m = +1/2$. (The calculation for $m = -1/2$ is entirely analogous.) Take the basis to be the two $m = 1/2f$ states, with $j = 3/2$ or $1/2$. Then $H' = H_{\text{mag}} + H_{FS}$ is a 2×2 matrix. Diagonalize the matrix and find the eigenvalues. Expand your answer for small $\mu_B B/\Delta$ and compare with the weak field limit. Then expand your answer for small $\Delta/\mu_B B$ and compare with the strong field limit. Make a rough graph of the energy as a function of the magnetic field.

7.19. Numerical Estimate of Energy Eigenvalues

This problem is in part a numerical project, so you will have to know how to use a computer to do some simple computations. Use the ideas outlined in Section 7.5.3 to find the ground-state energy of a one-dimensional quartic potential to five or six significant figures. The Hamiltonian is $H = T + V$, where

$$T = \frac{p^2}{2m} \quad \text{and} \quad V = \lambda x^4$$

but since the energies scale as[30]

$$E_n = \bar{E}_n \left(\frac{\lambda}{4m^2} \right)^{\frac{1}{3}}$$

---

[30] See Problem 7.14.

Chapter 7 Approximation Methods for Bound States

it is enough to set $2m = \lambda = 1$ and take
$$H = p^2 + x^4$$

You estimated this energy in Problem 7.14, using a trial function with one parameter. Do the calculation any way you want to, but here is a possible outline:

(a) Choose a basis $|\psi_n\rangle$ of eigenstates of a harmonic oscillator potential with $2m = 1$: $H(\omega) = T + V(\omega)$ where
$$T = p^2 \quad \text{and} \quad V(\omega) = \frac{\omega^2}{4} x^2$$

Make a table of all the nonvanishing matrix elements of x^2, T, and x^4 between the eigenstates. To do that, write x^2 in terms of the raising and lowering operators, and use rules like
$$\langle \psi_m | T + V(\omega) | \psi_n \rangle = \omega \left(n + \frac{1}{2} \right) \delta_{mn}$$
and
$$\langle \psi_m | x^4 | \psi_n \rangle = \sum_i \langle \psi_m | x^2 | \psi_i \rangle \langle \psi_i | x^2 | \psi_n \rangle$$

(b) Since the quartic potential is even, matrix elements H_{0n} will be nonzero only if n is even. Take the trial function to be an arbitrary combination of $|\psi_0\rangle$ and $|\psi_2\rangle$. For any fixed ω, the best combination is an eigenvalue of
$$M = \begin{pmatrix} H_{00} & H_{20} \\ H_{20} & H_{22} \end{pmatrix}$$

The estimate for the lowest is the lowest solution to the characteristic equation
$$\det(M - E(\omega)) = 0$$
or
$$E(\omega)^2 - E(\omega)(H_{00} + H_{22}) + H_{00} H_{22} - H_{20}^2 = 0$$

For each ω you can find the roots of this quadratic equation analytically, then vary ω to get the lowest energy. Or you can just program a computer to do the whole computation numerically. You should get an answer somewhat lower than the one-parameter estimate in Problem 7.14. There is also a second, higher, eigenvalue. That is an estimate of the energy of the first excited state.

Note: This is really a two parameter trial function. You have taken $|\psi\rangle = \alpha |\psi_0\rangle + \beta |\psi_1\rangle$. Here α and β are not independent, since the method assumes the trial states are normalized. The second parameter is ω, buried in the definition of the $|\psi_n\rangle$.

(c) Now go on to three, four, five parameters, and so forth, until you either get bored or achieve the desired precision. The only new thing you need to do analytically is learn a general formula for the characteristic equation:

Mathematical Note: The characteristic equation, whose roots are the eigenvalues, has the form
$$E(\omega)^n + a_1 E(\omega)^{n-1} + a_2 E(\omega)^{n-2} + \cdots a_{n-1} E(\omega) + a_n = 0$$

The coefficients depend on the eigenvalues only, so they are linear combinations of products of the eigenvalues, which can be written in an invariant way as traces of powers of the matrix M. The trick is to realize that these functions are independent of the dimension, so we can discover them one by one, writing the characteristic equation in $2, 3, 4 \ldots$ dimensions. The matrix M satisfies its characteristic equation (this is the **Cayley-Hamilton theorem**). From the definition, a_1 is always $-\text{Tr } M$. In two dimensions,

$$M^2 + a_1 M + a_2 = 0$$

Then (take the trace)

$$a_2 = -\frac{1}{2} \left[\text{Tr } M^2 + a_1 \text{Tr } M \right]$$

and continuing to three, four, and more dimensions,

$$a_3 = -\frac{1}{3} \left[\text{Tr } M^3 + a_1 \text{Tr } M^2 + a_2 \text{Tr } M \right]$$
$$a_4 = -\frac{1}{4} \left[\text{Tr } M^4 + a_1 \text{Tr } M^3 + a_2 \text{Tr } M^2 + a_3 \text{Tr } M \right]$$

and so forth. Of course a_n always turns out to be $(-1)^n$ times the determinant of M.

In each dimension, calculate the needed powers of M, calculate their traces, use these to find the a_i and then find the roots of the characteristic equation. The more of this you can manage to automate, the easier it gets to go to higher dimensions.

REFERENCES

Bound-state perturbation methods were first applied to quantum mechanics by Schrödinger, reference [13], Chapter 2.

Atomic level splitting by an electric field was discovered by

[1] J. STARK, Akad. Wiss. Berlin **40** (1913), 932.

The closed form for the polarizability of a hydrogen atom can be found in

[2] A. DALGARNO AND J. T. LEWIS, Proc. Roy. Soc. **A233** (1956), 70.

[3] C. SCHWARTZ, Annals of Physics **6** (1959), 156.

[4] E. MERZBACHER, *Quantum mechanics*, Wiley, 1998, page 461. See also Appendix D.

The Landé interval rule was introduced in

[5] A. LANDÉ, Zeitschrift für Physik **15** (1923), 189.

[6] A. LANDÉ, Zeitschrift für Physik **19** (1923), 112.

The interstellar 21 cm radiation was discovered simultaneously by

[7] C. A. MULLER AND J. H. OORT, Nature **168** (1951), 357.

[8] E. M. PURCELL AND H. I. EWEN, Nature **168** (1951), 356.

The original applications of the variational method are in

[9] W. RITZ, Journal Reine Angew. Math. **135** (1908), 1.

[10] LORD RAYLEIGH, *Theory of sound*, MacMillan (Dover), 1937, 1945, Section 88.

The classic papers applying the variational method to the helium atom energy levels are

[11] E. A. HYLLERAAS, Zeitschrift für Physik **65** (1920), 209.

[12] E. A. HYLLERAAS, Zeitschrift für Physik **54** (1929), 347.

Two-electron atoms and ions are discussed in Chapter 3 of

[13] H. BETHE AND R. JACKIW, *Intermediate quantum mechanics*, Benjamin, 1968. See also Appendix D.

Hund's rule was promulgated in

[14] F. HUND, Zeitschrift für Physik **33** (1925), 345.

The original paper on separating fast and slow coordinates in molecules is

[15] M. BORN AND J. R. OPPENHEIMER, Annalen der Physik **84** (1927), 457.

There is also a large recent literature. A good place to start is

[16] J. MOODY, A. SHAPERE, AND F. WILCZEK, in Shapere and Wilczek, reference [10] in Chapter 10, p. 160.

The energy levels in the H_2^+ ion were computed in

[17] R. F. WALLIS AND H. HULBERT, J. Chem. Phys. **22** (1954), 774.

[18] D. R. BATES, K. LADSHAM, AND A. L. STEWART, Phil. Trans. Roy. Soc. **A246** (1953), 215.

The WKB method was first worked out among others by

[19] G. WENZEL, Zeitschrift für Physik **38** (1926), 518.

[20] H. KRAMERS, Zeitschrift für Physik **39** (1926), 828.

[21] L. BRILLOUIN, J. Phys. Radium **7** (1926), 353.

[22] L. BRILLOUIN, Comptes Rendus **183** (1926), 24.

α decay was first explained in

[23] G. GAMOW, Zeitschrift für Physik **81** (1928), 204.

[24] R. W. GURNEY AND E. U. CONDON, Phys. Rev. **33** (1927), 127.

The adiabatic theorem is proved in

[25] D. GRIFFITHS, *Introduction to Quantum mechanics*, Prentice Hall, (1995), pages 327-330. See also Appendix D.

[26] A. MESSIAH, *Quantum mechanics*, North-Holland Wiley, (1958), pages 744-755. Also Dover (2000). See also Appendix D.

CHAPTER 8

Potential Scattering

8.1 INTRODUCTION

In 1930 E. O. Lawrence built the first cyclotron in Berkeley. Ever since, we have been building bigger and cleverer accelerators to probe the structure of matter at smaller and smaller distances. Richard Feynman was fond of remarking that scattering experiments are like learning how a fine watch works by slamming it into a wall as hard as possible. Nevertheless, much of what we know about the structure of atoms, nuclei, and subnuclear particles comes from collision experiments.

I will begin developing scattering theory in the context of a spinless particle scattering off a fixed potential. The generalization to spin and to two particles interacting through a nonrelativistic potential energy term is straightforward. Collisions that change the state of the target, and collisions involving photons and other relativistic collision problems, will need the more general methods presented in Chapter 9.

8.1.1 Kinematics of Scattering

Probability Current Density
A beam of particles, the "projectile," is scattered off a fixed target. The number of particles crossing a square centimeter per second is called the flux density or current density. It is the particle density ρ multiplied by the velocity.

$$\mathbf{j} = \rho \mathbf{v} \tag{8.1}$$

The total flux of particles through any surface S is

$$J = \int \mathbf{j}(\mathbf{r}) \cdot \hat{\boldsymbol{n}} dS \tag{8.2}$$

For a beam of particles described by a quantum-mechanical state vector or wave function, $\mathbf{j}$ is the probability current density.

A particle scattering off a nonrelativistic potential has a wave function $\psi(\mathbf{r})$ that satisfies Schrödinger's equation:

$$i\frac{\partial \psi}{\partial t} = -\frac{1}{2m}\boldsymbol{\nabla}^2 \psi + V\psi \tag{8.3}$$

The probability density itself according to equation (2.91) is

$$\rho(\mathbf{r}, t) = \psi^*(\mathbf{r}, t)\psi(\mathbf{r}, t) \tag{8.4}$$

so with the definition

$$\mathbf{j} = -\frac{i}{2m}(\psi^*\boldsymbol{\nabla}\psi - \psi\boldsymbol{\nabla}\psi^*) \tag{8.5}$$

it follows that

$$\frac{\partial \rho}{\partial t} + \boldsymbol{\nabla} \cdot \mathbf{j} = 0 \tag{8.6}$$

This is the continuity equation for the probability current density $\mathbf{j}$ from Section 2.7.3.

Cross Sections

Experiments measure the number of collisions per second. That number is proportional to the intensity of the beam. It is more convenient to talk about the target's effective area, or **cross section** σ. The number of scattering events per second is the flux density times the cross section.

There may be partial cross sections σ_i for different possible final states, perhaps states with different spins or even different particle types.

$$\sigma = \sum_i \sigma_i \tag{8.7}$$

The cross section for scattering into the solid angle $d\Omega$ is a special kind of partial cross section called the **differential** cross section $d\sigma$. If there are several channels,

$$\sigma_{tot} = \sum_i \int \frac{d\sigma_i}{d\Omega} d\Omega \tag{8.8}$$

$d\sigma_i$ is the partial cross section into the particle types i and the solid angle $d\Omega$.

If the same particle comes out as goes in, without changing the target, the scattering is called **elastic**. The kinetic energy of the projectile is conserved in elastic scattering.

8.1.2 Scattering and Wave Functions

The incident particles are in an eigenstate of the momentum $\mathbf{k}$ and the kinetic energy $E = \mathbf{k}^2/2m$ (for a nonrelativistic particle). Let us always normalize the states to

$$< \mathbf{k}'|\mathbf{k} > = \delta_3(\mathbf{k}' - \mathbf{k}) \tag{8.9}$$

so that the wave function is

$$\phi(\mathbf{r}, t) = \frac{1}{(2\pi)^{\frac{3}{2}}} e^{i\mathbf{k}\cdot\mathbf{r}} e^{-i\omega t} \tag{8.10}$$

You can check from the definitions (8.4) and (8.5) that $\phi(\mathbf{r}, t)$ describes a plane wave with probability density

$$\rho = \frac{1}{(2\pi)^3} \tag{8.11}$$

and current density

$$\mathbf{j} = \frac{\mathbf{v}}{(2\pi)^3} \tag{8.12}$$

266 Chapter 8 Potential Scattering

Consider scattering off a fixed potential $V(\mathbf{r})$ that vanishes as $r \to \infty$. For the problem to make sense, it is sufficient, but not necessary, to require $V = 0$ for $r > R$ for some R. The precise condition, which I do not prove here, is that $r^{1+\epsilon}V(\mathbf{r}) \to 0$ for some $\epsilon > 0$. The scattering particles are in an eigenstate of the full Hamiltonian:

$$\left(-\frac{1}{2m}\nabla^2 + V(\mathbf{r})\right)\psi(\mathbf{r}) = E\psi(\mathbf{r}) \tag{8.13}$$

The problem is to solve the time-independent wave equation only, as in the bound-state problem, but now with $E > 0$.

Simplify the notation by setting $U(\mathbf{r}) = 2mV(\mathbf{r})$ and $k^2 = 2mE$. Then Schrödinger's equation (8.13) reads

$$(\nabla^2 + k^2)\psi(\mathbf{r}) = U(\mathbf{r})\psi(\mathbf{r}) \tag{8.14}$$

When $U = 0$ there is no scattering, and $\psi(\mathbf{r})$ is the plane wave $\phi(\mathbf{r})$. Write $\psi(\mathbf{r})$ as this plane wave plus a correction $\chi(\mathbf{r})$ due to the presence of the potential:

$$\psi(\mathbf{r}) = \frac{1}{(2\pi)^{\frac{3}{2}}}e^{i\mathbf{k}\cdot\mathbf{r}} + \chi(\mathbf{r}) = \phi(\mathbf{r}) + \chi(\mathbf{r}) \tag{8.15}$$

A boundary condition on these functions is that $\chi(\mathbf{r})$ is negligible far from the scattering region. Since $(\nabla^2 + k^2)\phi(\mathbf{r}) = 0$, that means $(\nabla^2 + k^2)\chi(\mathbf{r}) \to 0$ as $r \to \infty$. In spherical coordinates the Laplacian is the differential operator[1]

$$\nabla^2 = \frac{1}{r^2}\frac{\partial}{\partial r}\left(r^2\frac{\partial}{\partial r}\right) + \frac{1}{r^2\sin\theta}\frac{\partial}{\partial\theta}\left(\sin\theta\frac{\partial}{\partial\theta}\right) + \frac{1}{r^2\sin^2\theta}\frac{\partial^2}{\partial\phi^2} \tag{8.16}$$

As $r \to \infty$, only the first term survives, and $\chi(\mathbf{r})$ satisfies

$$\left[\frac{1}{r^2}\frac{\partial}{\partial r}\left(r^2\frac{\partial}{\partial r}\right) + k^2\right]\chi(\mathbf{r}) \to 0 \tag{8.17}$$

The condition (8.17) implies that

$$\chi(\mathbf{r}) \xrightarrow[r\to\infty]{} \frac{f(\theta,\phi)}{(2\pi)^{\frac{3}{2}}}\frac{e^{\pm ikr}}{r} \tag{8.18}$$

for some function $f(\theta,\phi)$. Since $\psi(\mathbf{r})$ should be the plane wave $\phi(\mathbf{r})$ plus a scattered wave that expands outward, as in figure 8.1, one must choose the plus sign for the outgoing wave (8.18). With the minus sign, our equations would describe a beam of particles coming in radially from infinity with just the right phases to combine into a plane wave. The outward radial current density in (8.18) is

$$v\frac{|f(\theta,\phi)|^2}{(2\pi)^3 r^2} \tag{8.19}$$

The current through an infinitesimal solid angle $d\Omega$ is

$$v\frac{|f(\theta,\phi)|^2}{(2\pi)^3}d\Omega \tag{8.20}$$

---
[1] See Section 3.3.2 or Section A.3 in the appendix.

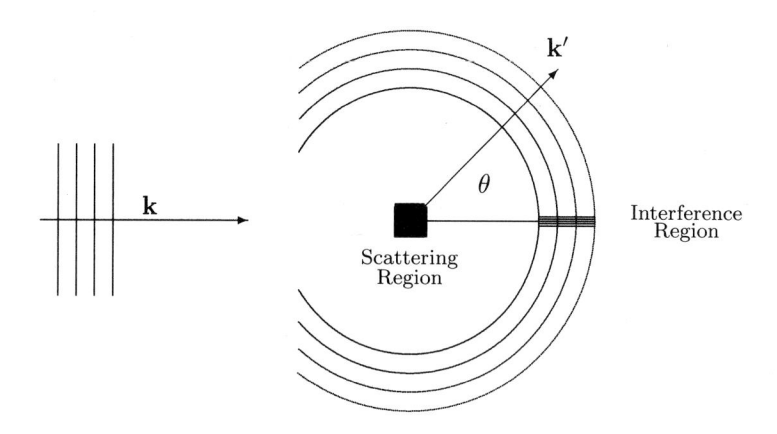

FIGURE 8.1: A projectile with momentum **k** scatters off a target. θ is the scattering angle, and **k'** is the final state momentum. The state is the sum of an incident plane wave and an outgoing spherical wave that interfere in the forward direction.

You can check this more formally by computing $\int \mathbf{j}(\mathbf{r}) \cdot \hat{\mathbf{n}}\, dS$ from the wave function in equation (8.18) through an infinitesimal area dS of a large sphere.

Asymptotically, the full wave function looks like

$$\psi(\mathbf{r}) \to \frac{1}{(2\pi)^{\frac{3}{2}}} \left[e^{i\mathbf{k}\cdot\mathbf{r}} + f(\theta,\phi)\frac{e^{ikr}}{r} \right] \quad (8.21)$$

as $r \to \infty$. The cross section into $d\Omega$ is the total rate for scattering into $d\Omega$ divided by the incident current density $v(2\pi)^{-3}$:

$$\frac{d\sigma}{d\Omega} = \frac{v|f(\theta,\phi)|^2/(2\pi)^3}{v/(2\pi)^3} = |f(\theta,\phi)|^2 \quad (8.22)$$

From equation (8.22) you can obtain the differential cross section if you can solve the time-independent Schrödinger equation (8.13) subject to the constraint (8.18). The total cross section is

$$\sigma_{tot} = \int |f(\theta,\phi)|^2 d\Omega \quad (8.23)$$

Wave Function Normalization

A plane wave has a finite particle density, and since the cross section depends on the ratio of the incoming to the outgoing density the cross section will come out finite even though I have used nonnormalizable wave functions. These nonnormalizable states are not really members of the vector space of all states. If using the Dirac delta function normalization bothers you, use the following trick: Imagine instead the whole experiment inside a large but finite box, so that the wave functions are properly normalized. As long as the box is large on any scale having to do with the experiment, in the region where you need to know the wave functions they are indistinguishable from plane waves. You can check that the size of the quantization box cancels out in the formula for the cross section. This is a standard trick you can find in several textbooks.

Wave Packets

The normalization problem can be avoided by constructing a normalized wave packet that describes a particle of approximately definite momentum at an approximately definite location approaching the scattering center at approximately the right speed. Then one can examine what happens to the wave packet after the scattering. At $t = 0$ a normalized state has a form like

$$\phi(\mathbf{r}, 0) = \frac{1}{(2\pi)^{\frac{3}{2}}} \int g(\mathbf{k}) e^{i\mathbf{k}\cdot\mathbf{r}} d^3k \qquad (8.24)$$

The function $g(\mathbf{k})$ is chosen to have the states properties. At any later time,

$$\phi(\mathbf{r}, t) = \frac{1}{(2\pi)^{\frac{3}{2}}} \int g(\mathbf{k}) e^{i\mathbf{k}\cdot\mathbf{r}} e^{-i\omega(\mathbf{k})t} d^3k \qquad (8.25)$$

Without violating the uncertainty principle, it is possible to construct normalizable wave packets that have only a small spread in momentum but are still approximately localizable. Quantitatively, the wave packets need to be large compared to the range of the potential, and large enough that they do not spread appreciably during the experiment, yet small compared to the distance a classical particle would travel during the experiment. While we want to talk about times that are very long compared to the time it takes for the particle to traverse the range of the potential, we must be careful to avoid arbitrarily long times.

In this wave packet language the solution to Schrödinger's equation is the original superposition of plane waves plus a superposition of spherical waves that move outward from the potential. A packet constructed from incident plane wave plus a scattered wave, that is, solutions of the form $\psi(\mathbf{r})$ in equation (8.15), will indeed have all the correct time-dependent properties. But since the precise shape of the wave packet is unimportant, it is much easier to dispense with the packets completely and study scattering for plane waves. There is a large literature on the subject if you are interested.[2] For a physicist, it is usually sufficient that a sensible use of nonnormalizable plane wave states gives sensible answers. I do not pretend to have proved to you that this idealization is really correct, but I know of no physical insight that is gained from pursuing the wave packet idea any further.

---

[2] Scattering is an ancient subject. Two classic books on scattering are Goldberger and Watson [1] and Newton [2]. There is also a good discussion in Gottfried [3].

8.2 THE SCATTERING AMPLITUDE

8.2.1 Equation for the Scattering Amplitude

The function $f(\theta, \phi)$ is the **scattering amplitude**. If you can find a Green's function $G(\mathbf{r}, \mathbf{r}')$ that satisfies

$$\left[\nabla^2 + k^2\right] G(\mathbf{r}, \mathbf{r}') = \delta_3(\mathbf{r}' - \mathbf{r}) \tag{8.26}$$

then the function

$$\chi(\mathbf{r}) = \int G(\mathbf{r}, \mathbf{r}') U(\mathbf{r}') \psi(\mathbf{r}') d^3 r' \tag{8.27}$$

provides a formal solution to

$$(\nabla^2 + k^2)\psi(\mathbf{r}) = U(\mathbf{r})\psi(\mathbf{r}) \tag{8.14}$$

or equivalently, since $(\nabla^2 + k^2)\phi(\mathbf{r}) = 0$,

$$\left(\nabla^2 + k^2\right) \chi(\mathbf{r}) = U(\mathbf{r})\psi(\mathbf{r}) \tag{8.28}$$

$G(\mathbf{r}, \mathbf{r}')$ is not uniquely defined by equation (8.26). The outgoing wave condition as $r \to \infty$ requires

$$\boxed{G(\mathbf{r}, \mathbf{r}') = -\frac{e^{ik|\mathbf{r}-\mathbf{r}'|}}{4\pi|\mathbf{r} - \mathbf{r}'|}} \tag{8.29}$$

Then

$$\psi(\mathbf{r}) = \frac{1}{(2\pi)^{\frac{3}{2}}} \left[e^{i\mathbf{k}\cdot\mathbf{r}} - \frac{(2\pi)^{\frac{3}{2}}}{4\pi} \int \frac{e^{ik|\mathbf{r}'-\mathbf{r}|}}{|\mathbf{r}' - \mathbf{r}|} U(\mathbf{r}')\psi(\mathbf{r}') d^3 r' \right] \tag{8.30}$$

A similar function, with a minus sign in the exponential, satisfies equation (8.18), but would describe an incoming scattered wave.

Equation (8.30) is an integral form of Schrödinger's equation that incorporates the outgoing scattered wave condition.

The cross section can be read off the behavior of ψ for large enough r. Take r so large that it is outside the range of the potential, so only $r' \leq r$ contributes to the integral. In that region expand in powers of r'/r.

$$|\mathbf{r}' - \mathbf{r}| = \sqrt{(r^2 - 2\mathbf{r} \cdot \mathbf{r}' + r'^2)} = r - \mathbf{r} \cdot \mathbf{r}'/r + \cdots \tag{8.31}$$

Then for large r

$$\psi(\mathbf{r}) \to \frac{1}{(2\pi)^{\frac{3}{2}}} \left[e^{i\mathbf{k}\cdot\mathbf{r}} - \frac{(2\pi)^{\frac{3}{2}}}{4\pi} \int \frac{e^{ikr}}{r} e^{-i k \mathbf{r}\cdot\mathbf{r}'/r} U(\mathbf{r}')\psi(\mathbf{r}') d^3 r' \right] \tag{8.32}$$

The scattering amplitude is

$$\boxed{f(\theta, \phi) = -\frac{(2\pi)^{\frac{3}{2}}}{4\pi} \int e^{-i\mathbf{k}'\cdot\mathbf{r}'} U(\mathbf{r}')\psi(\mathbf{r}') \, d^3 r' = -2\pi^2 \langle \phi_{\mathbf{k}'} | U | \psi_{\mathbf{k}} \rangle} \tag{8.33}$$

where $\mathbf{k}' = k\mathbf{r}/r$ is a vector of magnitude k (the kinetic energy does not change in elastic scattering) but in the direction of the angles θ and ϕ.

8.2.2 The Born Series

Equations (8.32) and (8.33) are a set of coupled integral equations reminiscent of the coupled equations for the wave function and energy shift in bound-state perturbation theory. Equation (8.33) is a good starting point for successive approximations to $f(\theta, \phi)$ in powers of the potential. Define a plane wave in any direction $\mathbf{k}$ by

$$\phi_\mathbf{k}(\mathbf{r}) = \frac{e^{i\mathbf{k}\cdot\mathbf{r}}}{(2\pi)^{\frac{3}{2}}} \tag{8.34}$$

Let $\psi_\mathbf{k}(\mathbf{r})$ be the solution to the Schrödinger equation when $\phi_\mathbf{k}(\mathbf{r})$ is the incident wave:

$$\psi_\mathbf{k}(\mathbf{r}) = \phi_\mathbf{k}(\mathbf{r}) + \chi(\mathbf{r}) \tag{8.35}$$

To get the first **Born approximation** simply replace $\psi_\mathbf{k}(\mathbf{r})$ with $\phi_\mathbf{k}(\mathbf{r})$ in the expression (8.33):

$$\boxed{f^{(1)}(\theta, \phi) = -\frac{1}{4\pi} \int d^3r\, e^{-i\mathbf{k}'\cdot\mathbf{r}} U(\mathbf{r}) e^{i\mathbf{k}\cdot\mathbf{r}} = -2\pi^2 \langle \phi_{\mathbf{k}'} | U(r) | \phi_\mathbf{k} \rangle} \tag{8.36}$$

The Born approximation to the differential cross section is

$$\frac{d\sigma}{d\Omega} = (2\pi)^4 m^2 \left| \langle \phi_{\mathbf{k}'} | V | \phi_\mathbf{k} \rangle \right|^2 \tag{8.37}$$

The general coupled iteration scheme is

$$\psi^{(n)}(\mathbf{r}) = \frac{1}{(2\pi)^{\frac{3}{2}}} \left[e^{i\mathbf{k}\cdot\mathbf{r}} - 2\pi^2 \int d^3r'\, \frac{e^{ik|\mathbf{r}'-\mathbf{r}|}}{4\pi|\mathbf{r}'-\mathbf{r}|} U(r') \psi^{(n-1)}(\mathbf{r}') \right] \tag{8.38a}$$

$$f^{(n)}(\theta, \phi) = -\frac{(2\pi)^{\frac{3}{2}}}{4\pi} \int d^3r'\, e^{-i\mathbf{k}'\cdot\mathbf{r}'} U(r') \psi^{(n-1)}(\mathbf{r}') \tag{8.38b}$$

On the right-hand sides of equations (8.38) the angles θ and ϕ are buried in the definition of $\mathbf{k}'$.

8.2.3 Spherically Symmetric Potentials

The first Born approximation is especially simple for spherically symmetric potentials. In that case

$$f^{(1)}(\theta, \phi) = -\frac{1}{4\pi} \int e^{i(\mathbf{k}-\mathbf{k}')\cdot\mathbf{r}} U(r)\, d^3r \tag{8.39}$$

In this approximation the scattering amplitude depends on the energy and the scattering angle only through the combination $\mathbf{q} = \mathbf{k} - \mathbf{k}'$, the momentum transferred to the target. Its magnitude is $q = |\mathbf{k}-\mathbf{k}'| = 2k\sin(\Theta/2)$ where Θ is the scattering angle, the angle between $\mathbf{k}$ and $\mathbf{k}'$. With $\mathbf{q}\cdot\mathbf{r} = qr\cos\theta$,

$$\begin{aligned} f^{(1)}(\theta, \phi) &= -\frac{1}{4\pi} \int e^{i\mathbf{q}\cdot\mathbf{r}} U(\mathbf{r})\, d^3r \\ &= \frac{1}{2} \int_0^\infty r^2 dr \int_{-1}^1 e^{iqr\cos\theta} U(r) d\cos\theta = -\frac{1}{q} \int_0^\infty r\sin(qr) U(r) dr \end{aligned} \tag{8.40}$$

This is the general rule for spherically symmetric potentials.

Example: The Yukawa potential

As an example, take V(r) to be an attractive Yukawa potential:

$$V(r) = -V_o a \frac{e^{-r/a}}{r} \qquad (8.41)$$

a is the "range" of the potential. Yukawa first suggested this potential to describe scattering of neutrons off protons. It is a fair model for low energy nucleon-nucleon forces, with a roughly the Compton wavelength of the pion $2\pi/m_\pi$. The Born approximation to the scattering is[3]

$$f^{(1)}(\theta, \phi) = \frac{2mV_o a}{q} \int_0^\infty \sin(qr) e^{-r/a} dr = \frac{2mV_o a}{q} \frac{q}{(q^2 + (1/a)^2)} \qquad (8.42)$$

Example: Coulomb scattering

Consider the scattering of an electron off a point nucleus with charge Ze.

$$U(r) = 2mV(r) = -2m(Ze^2)/r \qquad (8.43)$$

The scattering amplitude is

$$f^{(1)}(\theta, \phi) = \frac{2mZe^2}{q} \int_0^\infty \sin(qr) dr \qquad (8.44)$$

The integral is ambiguous because a $1/r$ potential does not exactly satisfy the mathematical conditions of our method. But the method is good for any potential falling off more rapidly than $1/r$, and physically it cannot make any difference if we introduce a regulator, multiplying $V(r)$ by $e^{-\epsilon r}$ and take the limit $\epsilon \to 0$ at the end. Equation (8.44) is replaced by

$$f^{(1)}(\theta, \phi) = \frac{2mZe^2}{q} \lim_{\epsilon \to 0} \int_0^\infty e^{-\epsilon r} \sin(qr) dr \qquad (8.45)$$

For $\epsilon > 0$ this is the Yukawa amplitude (8.42) with $V_o a \to Ze^2$ and $a \to 1/\epsilon$. Therefore

$$f^{(1)}(\theta, \phi) = \lim_{\epsilon \to 0} 2mZe^2 \frac{1}{q^2 + \epsilon^2} = \frac{2mZe^2}{q^2} \qquad (8.46)$$

independent of ϵ. The differential cross section is

$$\boxed{\frac{d\sigma}{d\Omega} = \frac{4m^2 Z^2 e^4}{q^4} = \frac{4m^2 Z^2 e^4}{16 k^4 \sin^4(\Theta/2)}} \qquad (8.47)$$

which is exactly Rutherford's classical formula for Coulomb scattering.

Equation (8.47) is a strange result! Why should the quantum mechanical cross section agree with the classical one? And even more strange, why should the Born approximation be exact?

---
[3] Cf. equation (A.28) in the appendix.

8.2.4 The Optical Theorem

In the early nineteenth century the wave idea of light idea was supported by Young's double-slit interference experiment, but it was still controversial. Fresnel worked out the mathematics in more detail, and computed many of the complicated edge effects that we call diffraction. Poisson pointed out that Fresnel's method led to the absurd prediction that behind a totally absorbing black sphere there should be a white spot exactly at the center. That is the point where all the "wavelets" interfere destructively. Arago managed to construct a sphere whose diameter was so precise that the error was less than the wavelength of visible light, and observed the predicted spot, laying to rest all doubts about the wave theory of light.[4] In quantum mechanical scattering it might appear that the total flux of particles is the flux in the incident wave plus the flux in the scattered wave. Together that sum is greater that the original flux in the incident wave alone. Where did the new flux of particles come from? It comes from the quantum analog of the Poisson-Arago white spot, the interference between the incident and scattered waves in the $\theta = 0$ direction. The quantitative formulation of both effects is called the **optical theorem**. A careful analysis shows that the scattering amplitude has an imaginary absorptive part in the forward direction, corresponding exactly to the depletion of the incident beam needed to conserve probability. The mathematical statement of this relation is the optical theorem in quantum mechanics:

$$\sigma_{tot} = \frac{4\pi}{k} \operatorname{Im} f(0) \tag{8.48}$$

The key to the demonstration of the optical theorem is the conservation of probability. Here is a derivation using the conservation of the probability current density.

Demonstration of the optical theorem. The probability current density (8.5) is

$$\mathbf{j} = \frac{1}{m} \operatorname{Im} (\psi^* \boldsymbol{\nabla} \psi) \tag{8.49}$$

Since ψ is a stationary solution to Schrödinger's equation, there is no net probability flux out of a sphere S

$$\operatorname{Im} \int (\psi^* \boldsymbol{\nabla} \psi) \hat{\mathbf{n}} \cdot dS = \operatorname{Im} \int \psi^* \frac{\partial \psi}{\partial r} dS = 0 \tag{8.50}$$

The same is true for the free wave-function ϕ, which is a solution to another Schrödinger equation.

$$\operatorname{Im} \int \phi^* \frac{\partial \phi}{\partial r} dS = 0 \tag{8.51}$$

Write $\psi = \phi + \chi$, and insert equation (8.51) into equation (8.50):

$$0 = \operatorname{Im} \int \left[\phi^* \left(\frac{\partial \chi}{\partial r}\right) + \chi^* \left(\frac{\partial \phi}{\partial r}\right) + \chi^* \left(\frac{\partial \chi}{\partial r}\right) \right] r^2 d\Omega \tag{8.52}$$

---
[4] See for example Sommerfeld [4], page 215; or Hecht and Zajac [5], page 373.

The condition (8.18) says that in the large r limit

$$\frac{\partial \chi}{\partial r} \to ik \frac{e^{ikr}}{(2\pi)^{\frac{3}{2}} r} f(\theta, \phi) \tag{8.53}$$

and so

$$\int \chi^* \left(\frac{\partial \chi}{\partial r}\right) r^2 d\Omega \to \frac{ik}{(2\pi)^3} \int |f(\theta, \phi)|^2 d\Omega = \frac{ik\sigma_{tot}}{(2\pi)^3} \tag{8.54}$$

The imaginary part of any number is minus the imaginary part of its complex conjugate. In particular,

$$\operatorname{Im} \psi^* \left(\frac{\partial \phi}{\partial r}\right) = -\operatorname{Im} \psi \left(\frac{\partial \phi^*}{\partial r}\right) \tag{8.55}$$

Therefore,

$$\begin{aligned}
\frac{k\sigma_{tot}}{(2\pi)^3} &= -\operatorname{Im} \int \left(\phi^*\left(\frac{\partial \chi}{\partial r}\right) + \chi^*\left(\frac{\partial \phi}{\partial r}\right)\right) r^2 d\Omega \\
&= -\operatorname{Im} \int \left(\phi^*\left(\frac{\partial \chi}{\partial r}\right) - \chi^*\left(\frac{\partial \phi}{\partial r}\right)\right) r^2 d\Omega \\
&= -\operatorname{Im} \int \left(\phi^*\left(\frac{\partial \psi}{\partial r}\right) - \psi\left(\frac{\partial \phi^*}{\partial r}\right)\right) r^2 d\Omega \\
&= -\operatorname{Im} \int (\phi^* \boldsymbol{\nabla} \psi + \psi \boldsymbol{\nabla} \phi^*) \cdot \hat{n} dS = -\operatorname{Im} \int (\phi^* \boldsymbol{\nabla}^2 \psi - \psi \boldsymbol{\nabla}^2 \phi^*) d^3 r \\
&= -\operatorname{Im} \int \phi^*(r) U(r) \psi(r) d^3 r = \frac{1}{2\pi^2} \operatorname{Im} f(0)
\end{aligned} \tag{8.56}$$

and the optical theorem (8.48) follows immediately. ∎

The total cross section is simply related to the imaginary part of the forward elastic amplitude.

The rule (8.48) is the optical theorem, and it is much more general than the derivation here implies. The optical theorem (8.48) gets its name from a similar theorem in classical electromagnetism relating the index of refraction to the absorption of light by a medium.

8.2.5 The Refractive Index

Consider a beam of particles described by a plane wave $\phi(z) = \phi_o e^{ikz}$ incident on a "gas" of scatterers between two planes $z = 0$ and $z = dz$. Let there be N scatterers per unit volume, randomly distributed so that there are no correlations between the scattered waves from each scatterer. Take dz large enough that the volume contains many scatterers, but small enough that the phase of the wave does not change significantly in dz. N must be small enough that one can ignore multiple scattering—the incident particle scatters at most once in the slab.

274   Chapter 8   Potential Scattering

Presume further that the scattering is independent of the azimuthal angle. What is the amplitude of the wave at a point on the z axis, with $z = z_o$, some distance from the slab? Besides the incident wave, each scatterer contributes

$$\psi_j = \phi_o \frac{1}{r_j} e^{ikr_j} f(\theta_j) \tag{8.57}$$

where $r_j^2 = x_j^2 + y_j^2 + (z_o - z)^2$, and θ_j is the angle between the z-axis and the line from the observation point to the scatterer. The contribution to the scattered wave at z_o from all the scatterers in the slab is

$$d\psi_{sc} = \phi_o N dz \int_{-\infty}^{\infty} dx \int_{-\infty}^{\infty} dy \, \frac{1}{r} e^{ikr} f(\theta) = 2\pi N dz \phi_o \int_0^{\infty} \frac{1}{r} e^{ikr} f(\theta) \rho d\rho$$

$$= 2\pi N dz \, \phi_o \int_{z_o}^{\infty} e^{ikr} f(\theta) dr \tag{8.58}$$

with $\rho^2 = x^2 + y^2$ and $r^2 = z_o^2 + \rho^2$. The integral in equation (8.58) doesn't converge, but the integrand can be correct only for scatterers a reasonable distance from the observation point. At large distances there will be significant absorption due to multiple scattering. Take care of this absorption with a convergence factor and replace equation (8.58) with

$$d\psi_{sc} = 2\pi N dz \phi_o \int_{z_o}^{\infty} e^{ikr - \epsilon r} f(\theta) dr \tag{8.59}$$

The answer will be independent of ϵ in the limit $\epsilon \to 0$ (this is another example of a regulator!). The convergence factor permits integration by parts:

$$\psi_{sc} = 2\pi N dz \phi_o \left. \frac{e^{ikr - \epsilon r}}{ik - \epsilon} \right|_{z_o}^{\infty} - 2\pi N dz \phi_o \frac{1}{ik - \epsilon} \int_{z_o}^{\infty} e^{ikr - \epsilon r} \frac{df(\theta)}{dr} dr$$

$$\xrightarrow[\epsilon \to 0]{} \frac{2\pi i N dz}{k} \phi_o e^{ikz_o} f(0) + \frac{2\pi i N dz}{k} \phi_o \int_{z_o}^{\infty} e^{ikr - \epsilon r} \frac{df(\theta)}{dr} dr \tag{8.60}$$

Only forward scattering, scattering due to particles very close to the z-axis, can affect the wave function at z_o. As in a diffraction problem, all the other wavelets interfere destructively provided the scattering amplitude $f(\theta)$ is effectively constant while r changes over one wavelength. Quantitatively, the condition is

$$\left| \frac{df(\theta)/dr}{f(\theta)} \right| \ll k \tag{8.61}$$

Then the second term in equation (8.60) can be neglected and

$$\psi_{sc} \approx \frac{2\pi i N dz}{k} \phi_o e^{ikz_o} f(0) \tag{8.62}$$

It should be possible to choose z_o so that $z_o - dz \ll dz$, so that the observation point is practically on the surface of our slab, yet still far enough away that the density of scatterers inside the slab can be considered continuous. Then the wave amplitude at dz, including the incident wave, is

$$\psi(dz) = \phi_o \left[1 + \frac{2\pi i N dz}{k} f(0) + \cdots \right] \tag{8.63}$$

Passing through the slab, the incident wave is changed by

$$d\psi(z) = \frac{2\pi i N dz}{k} f(0)\psi(0) \tag{8.64}$$

The effective amplitude can be written as

$$\psi(z) = \psi(0)e^{iknz} \tag{8.65}$$

Then

$$\boxed{n = 1 + \frac{2\pi N}{k^2} f(0)} \tag{8.66}$$

For electromagnetic waves, Re n is the index of refraction. The nomenclature is sometimes used for other particles as well.

If the incident wave passes through a distance z of the scattering material, its intensity is attenuated by

$$|\psi(z)|^2 = |\psi(0)|^2 e^{-2kz\text{Im } n} = |\psi(0)|^2 e^{-(4\pi N z/k)\text{Im } f(0)} \tag{8.67}$$

From the definition of the total cross section this should be

$$|\psi(z)|^2 = |\psi(0)|^2 e^{-N\sigma z} \tag{8.68}$$

so again

$$\sigma = \frac{4\pi}{k} \text{Im } f(0) \tag{8.69}$$

This derivation of the optical theorem, in the context of light scattering in matter, indicates the origin of its name. The depletion of the forward wave is accounted for by the scattering.

8.3 PARTIAL WAVES

When the potential energy is a function of the radial coordinate alone, the eigenstates of the Hamiltonian can be taken to be eigenstates of L_z and L^2 as well. For given $E > 0$, the solutions have the form

$$\psi_{lm}(\mathbf{r}) = R_l(r) Y_l^m(\theta, \phi) \tag{8.70}$$

Again set $U(r) = 2mV(r)$, $u_l(r) = rR_l(r)$, and $k^2 = 2mE$. The radial wave function satisfies

$$\frac{d^2 u_l(r)}{dr^2} - \frac{l(l+1)}{r^2} u_l(r) + k^2 u_l(r) - U(r) u_l(r) = 0 \tag{8.71}$$

Equation (8.71) determines $R_l(r)$ up to an arbitrary normalization constant.

Chapter 8 Potential Scattering

The form of the asymptotic behavior of $u_l(r)$ is known from Section 3.4.2. It is proportional to r^{l+1} as $r \to 0$, and to a linear combination of e^{ikr} and e^{-ikr} as $r \to \infty$. In the bound-state problem, the large r limit was $e^{-\kappa r}$ or $e^{\kappa r}$, and $e^{\kappa r}$ was inadmissible because of the normalization condition. For positive energy solutions there is no such restriction and both solutions are possible.

$R_l(r)$ goes like r^l as $r \to 0$, and like some combination of e^{ikr}/r or e^{-ikr}/r as $r \to \infty$.

8.3.1 Expansion of a Plane Wave in a Legendre Series

How does one expand the scattering states in these eigenstates of energy and angular momentum? If $\mathbf{k}$ is along the z-axis, $\mathbf{k} = k\hat{n}_z$, then

$$\mathbf{k} \cdot \mathbf{r} = kz = kr\cos\theta \tag{8.72}$$

In this case e^{ikz} has an expansion in Legendre polynomials.

$$e^{ikz} = e^{ikr\cos\theta} = \sum_l a_l(r) P_l(\cos\theta) \tag{8.73}$$

For a free particle, the solutions to the radial equation are all proportional to the spherical Bessel functions: $R_l(r) = A_l j_l(kr)$. So for some constants A_l

$$e^{i\rho z} = \sum_l A_l j_l(\rho) P_l(z) \tag{8.74}$$

The constant coefficients can be evaluated in a variety of ways. The answer is

$$\boxed{e^{ikr\cos\theta} = \sum_l i^l (2l+1) j_l(kr) P_l(\cos\theta)} \tag{8.75}$$

Here is a simple derivation, using the Rodrigues expansion

$$P_l(z) = \frac{1}{2^l l!} \frac{d^l}{dz^l} (z^2 - 1)^l \tag{A.101}$$

for Legendre polynomials:

Derivation of the Expansion (8.75). Invert the expansion:

$$A_l j_l(\rho) = \frac{2l+1}{2} \int_{-1}^{1} P_l(z) e^{i\rho z} dz = \frac{2l+1}{2} \sum_n \int_{-1}^{1} i^n \frac{\rho^n z^n}{n!} P_l(z) dz \tag{8.76}$$

From the explicit Rodrigues formula for the Legendre polynomials you can get the coefficient of z^l in $P_l(z)$:

$$P_l(z) = \frac{(2l)!}{2^l (l!)^2} z^l + O(z^{l-1}) \tag{8.77}$$

The remaining terms are combinations of $P_{l'}(z)$ for $l' < l$. Therefore,

$$z^n = \frac{2^n(n!)^2}{(2n)!} P_n(z) + \cdots \tag{8.78}$$

plus terms in lower order Legendre polynomials. In particular all the terms in equation (8.76) with $n < l$ vanish. Compare the coefficient of the leading power of ρ (see Table A.6 in the appendix) with the known behavior of the spherical Bessel function for small ρ:

$$\lim_{\rho \to 0} j_l(\rho) = \rho^l \frac{2^l l!}{(2l+1)!} \tag{8.79}$$

to get

$$\begin{aligned} A_l \rho^l \frac{2^l l!}{(2l+1)!} &= \frac{2l+1}{2} \sum_n i^n \frac{\rho^n}{n!} \frac{2^n(n!)^2}{(2n)!} \int_{-1}^{1} P_n(z) P_l(z) dz \\ &= i^l \frac{\rho^l}{l!} \frac{2^l(l!)^2}{(2l)!} \end{aligned} \tag{8.80}$$

So the expansion coefficients are

$$A_l = (2l+1)i^l \tag{8.81}$$

as claimed. ■

Arbitrary Direction

The addition theorem for spherical harmonics says that

$$P_l(\cos\Theta) = \sum_m \frac{4\pi}{2l+1} Y_l^m(\theta_1, \phi_1)^* Y_l^m(\theta_2, \phi_2) \tag{B.24}$$

where Θ is the angle between the directions $\hat{\mathbf{n}}_1$ and $\hat{\mathbf{n}}_2$. It follows that the expansion of a plane wave in an arbitrary direction $\mathbf{k}$ is

$$\boxed{e^{i\mathbf{k}\cdot\mathbf{r}} = 4\pi \sum_{l,m} i^l j_l(kr) Y_l^m(\theta, \phi) Y_l^m(\theta_1, \phi_1)^*} \tag{8.82}$$

where $\mathbf{k} = k\hat{\mathbf{n}}_1$.

The functions $j_l(\rho)$ vanish like ρ^l as $\rho \to 0$. For large ρ their behavior is

$$j_l(\rho) \xrightarrow[\rho \to \infty]{} \frac{1}{\rho} \cos\left(\rho - \frac{\pi}{2}(l+1)\right) = \frac{1}{\rho}\sin\left(\rho - \frac{\pi l}{2}\right) \tag{8.83}$$

A few of these spherical Bessel functions are plotted in Figure 8.2.

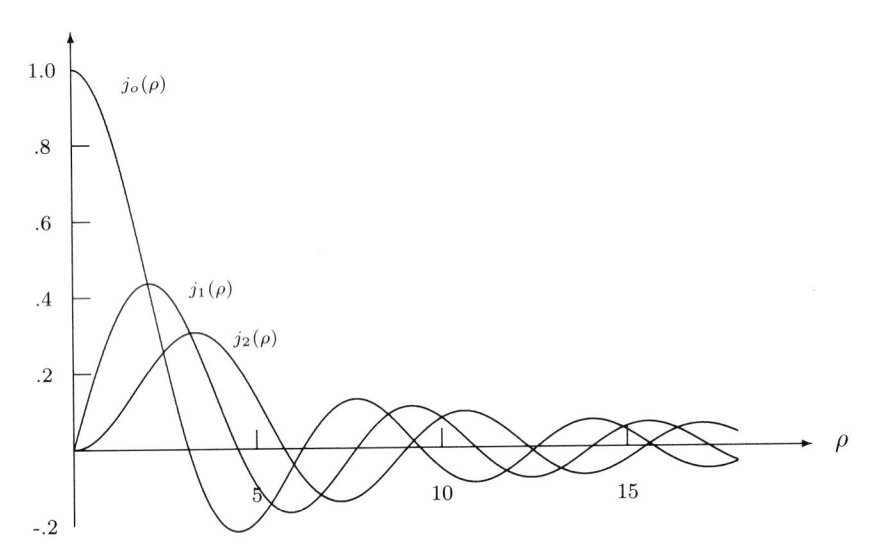

FIGURE 8.2: Spherical Bessel functions for $l = 0, 1, 2$

8.3.2 Partial Wave Expansion of $\psi(\mathbf{r})$

In the general case $V(r)$ does not vanish identically, but is assumed to be negligible for large r, so as $r \to \infty$ the radial wave function $R_l(r)$ approaches a free radial function and is a linear combination of e^{ikr} and e^{-ikr}. For large r, the radial function $R_l(r)$ looks like the $j_l(kr)$ shown in Figure 8.2, but with the phase of the oscillation displaced by δ_l. Write this rule as

$$R_l(r) \to \frac{c_l}{\rho} \sin(\rho - \pi l/2 + \delta_l) \tag{8.84}$$

with $\rho = kr$ as usual.

The constant c_l is arbitrary, but the **phase shift** δ_l is not. $R_l(r)$ must be real, so δ_l is real. The phase shift δ_l summarizes how the l-th partial wave differs from that of a free particle. Thus

$$\lim_{r \to \infty} R_l(r) = \frac{(-i)^{l+1}}{2\rho} c_l \left[e^{i\rho} e^{i\delta_l} - (-1)^l e^{-i\rho} e^{-i\delta_l} \right] \tag{8.85}$$

It is convenient to choose $c_l = e^{i\delta_l}$. Then[5]

$$\lim_{r \to \infty} R_l(r) = \frac{(-i)^{l+1}}{2\rho} \left[e^{i\rho} e^{2i\delta_l} - (-1)^l e^{-i\rho} \right] \tag{8.86}$$

The asymptotic form of the spherical Bessel functions is

$$j_l(\rho) \xrightarrow[\rho \to \infty]{} \frac{(-i)^{l+1}}{2\rho} \left[e^{i\rho} - (-1)^l e^{-i\rho} \right] \tag{8.87}$$

---
[5]This choice is not universal in the literature.

Thus the combination

$$R_l(r) - j_l(\rho) \to \frac{(-i)^{l+1}}{2\rho}\left(e^{2i\delta_l} - 1\right) e^{i\rho} \tag{8.88}$$

has only an outgoing wave as $r \to \infty$. The coefficient of the incoming wave is not changed by the potential.

Next expand $\psi(\mathbf{r})$ and $\phi(\mathbf{r})$, taking the incident plane wave along the z-axis. The form of $\phi(\mathbf{r})$ is

$$\phi(\mathbf{r}) = \frac{1}{(2\pi)^{\frac{3}{2}}} e^{ikz} = \frac{1}{(2\pi)^{\frac{3}{2}}} \sum_l i^l (2l+1) j_l(kr) P_l(\cos\theta) \tag{8.89}$$

The wave function $\psi(\mathbf{r})$ has a combination of incoming and outgoing waves as $r \to \infty$, but the condition (8.88) says that the incoming wave in $\psi(\mathbf{r})$ is identical to that in $\phi(\mathbf{r})$; $\chi(\mathbf{r})$ has an outgoing wave only as $r \to \infty$. Therefore,

$$\psi(\mathbf{r}) = \frac{1}{(2\pi)^{\frac{3}{2}}} \sum_l i^l (2l+1) R_l(r) P_l(\cos\theta) \tag{8.90}$$

where as $r \to \infty$

$$\chi(\mathbf{r}) = \psi(\mathbf{r}) - \phi(\mathbf{r}) \to \frac{1}{(2\pi)^{\frac{3}{2}}} \sum_l i^l (2l+1) \frac{(-i)^{l+1}}{2kr}(e^{2i\delta_l} - 1) e^{ikr} P_l(\cos\theta) \tag{8.91}$$

Finally, from

$$\chi(\mathbf{r}) \to \frac{1}{(2\pi)^{\frac{3}{2}}} \frac{e^{ikr}}{r} f(\theta,\phi) \tag{8.92}$$

it follows that

$$\boxed{f(\theta,\phi) = \sum_l (2l+1) \frac{e^{2i\delta_l} - 1}{2ik} P_l(\cos\theta) = \frac{1}{k} \sum_l (2l+1) e^{i\delta_l} \sin\delta_l P_l(\cos\theta)} \tag{8.93}$$

The total cross section is

$$\boxed{\sigma_{tot} = \int |f(\theta,\phi)|^2 d\Omega = \frac{4\pi}{k^2} \sum_l (2l+1) \sin^2\delta_l} \tag{8.94}$$

Since $P_l(1) = 1$ for all l, the optical theorem

$$\sigma_{tot} = \frac{4\pi}{k} \text{Im } f(0) \tag{8.95}$$

is demonstrated in yet another way.

8.3.3 Calculation of the Phase Shift

With the incident beam chosen along the z-axis, $f(\theta,\phi)$ is independent of the azimuthal angle ϕ. From equation (8.33)

$$f(\theta,\phi) = -2\pi^2 \int \phi_{\mathbf{k}'}(\mathbf{r}')^* U(r') \psi_{\mathbf{k}}(\mathbf{r}') \, d^3r' \tag{8.96}$$

$\mathbf{k}'$ is a vector of magnitude k in the θ,ϕ direction (see Figure 8.1), but $\mathbf{k}$ is in the z-direction. From equation (8.82) and the addition theorem for spherical harmonics (B.24) once again,

$$\phi_{\mathbf{k}'}(\mathbf{r}') = \frac{4\pi}{(2\pi)^{\frac{3}{2}}} \sum_{l,m} i^l Y_l^m(\theta,\phi)^* Y_l^m(\theta',\phi') j_l(kr) \tag{8.97}$$

and

$$\psi_{\mathbf{k}}(\mathbf{r}') = \frac{4\pi}{(2\pi)^{\frac{3}{2}}} \sum_{l,m} i^l Y_l^m(\theta,\phi)^* Y_l^m(0,\phi) R_l(r') \tag{8.98}$$

Therefore the scattering amplitude is

$$f(\theta,\phi) = -4\pi \int \left(\sum_{l,m} (-i)^l Y_l^m(\theta,\phi)^* Y_l^m(\theta',\phi') j_l(kr') \right)$$

$$\times U(r') \left(\sum_{l'} i^{l'} Y_{l'}^o(\hat{\mathbf{n}}) R_{l'}(r') Y_{l'}^o(\hat{\mathbf{n}}_z) \right) r'^2 \, dr' \, d\Omega' \tag{8.99}$$

$$= -\int_0^\infty \left[\sum_l (2l+1) j_l(kr') U(r') R_l(r') P_l(\cos\theta) \right] r'^2 \, dr'$$

Compare equation (8.99) with the expansion (8.93) for the scattering amplitude in terms of the phase shifts to obtain a rule for calculating the phase shifts:

$$e^{i\delta_l} \sin\frac{\delta_l}{k} = -\int_0^\infty j_l(kr) U(r) R_l(r) r^2 \, dr \tag{8.100}$$

The left-hand side of (8.100) is sometimes called the **partial wave scattering amplitude**.

In this way an infinity of one-dimensional equations (8.100), one for each l, is substituted for one three-dimensional partial differential equation or integral equation for $\psi(\mathbf{r})$.

In classical scattering, if b is the impact parameter, the angular momentum is kb, and when b is much larger than the "range" a of the potential, there will not be much scattering. The quantum analog of this observation is that the phase shifts should be insignificant for $l \gg ka$. It is not hard to show that this is in fact the case. So the phase shifts δ_l to fall off rapidly with increasing l. The number of phase shifts contributing significantly to the scattering increases as k increases. For small k there are only a few, and as $k \to 0$ only $l=0$ survives. In that number the scattering is characterized by a single number.

8.4 THE RADIAL WAVE FUNCTION

8.4.1 The Integral Equation

The phase shifts and the radial wave functions can be computed by the method of successive approximations. In zeroth order $R_l(r) = j_l(kr)$. The Born approximation follows immediately from equation (8.100)

$$e^{i\delta_l} \sin \delta_l = -k \int U(r) j_l^2(kr) r^2 dr \qquad (8.101)$$

for the phase shifts.

To do better than the Born approximation, you need to turn the differential equation for $R_l(r)$,

$$\left[\frac{1}{r^2} \frac{d}{dr} r^2 \frac{d}{dr} + k^2 - \frac{l(l+1)}{r^2} \right] R_l(r) = U(r) R_l(r) \qquad (8.102)$$

together with the boundary condition implied by equation (8.88), into an integral equation for R_l. The result is

$$R_l(r) = j_l(kr) - ik \int_0^\infty j_l(kr_<) h_l(kr_>) U(r') R_l(r') r'^2 dr' \qquad (8.103)$$

where $r_>$ (or $r_<$) stand for r or r', whichever is greater (or lesser). Equation (8.103) is in a form that can be solved by successive iterations, providing a series in powers of the potential for both the wave function and the phase shifts. These are usually detailed numerical computations. One simple application is in Problem 8.4.

8.4.2 Partial Wave Green's Functions

Where does the representation (8.103) come from? The integral equation is a partial wave version of

$$\psi(\mathbf{r}) = \frac{1}{(2\pi)^{\frac{3}{2}}} \left[e^{i\mathbf{k} \cdot \mathbf{r}} - \frac{(2\pi)^{\frac{3}{2}}}{4\pi} \int \frac{e^{ik|\mathbf{r}'-\mathbf{r}|}}{|\mathbf{r}'-\mathbf{r}|} U(\mathbf{r}') \psi(\mathbf{r}') d^3 r' \right] \qquad (8.30)$$

To pursue the analogy write the radial wave function as

$$R_l(r) = j_l(kr) + \chi_l(r) \qquad (8.104)$$

where from equation (8.88)

$$\chi_l(r) \xrightarrow[r \to \infty]{} (-i)^l e^{i\delta_l} \sin \delta_l \frac{e^{ikr}}{kr} \qquad (8.105)$$

Then a Green's function $G_l(r, r')$ that is a solution to

$$\left[\frac{1}{r^2} \frac{d}{dr} r^2 \frac{d}{dr} + k^2 - \frac{l(l+1)}{r^2} \right] G_l(r, r') = \frac{\delta(r - r')}{r^2} \qquad (8.106)$$

provides a formula for the l-th scattered wave:

$$\chi_l(r) = \int_0^\infty G_l(r,r') U(r') R_l(r') r'^2 dr' \tag{8.107}$$

The boundary condition is imposed by requiring $G_l(r,r')$ to look like e^{ikr}/r for $r \gg r'$.

For $r \neq r'$, $G_l(r,r')$ is a solution to the homogeneous differential equation, so it is a linear combination of $j_l(kr)$ and $n_l(kr)$, with coefficients that may depend on r'. As $r \to 0$, $G_l(r,r')$ is finite, so for $r < r'$, $G_l(r,r') = f_l(r') j_l(kr)$. The boundary condition as $r \to \infty$ implies that for $r > r'$, $G_l(r,r') = F_l(r') h_l(kr)$, where

$$h_l(\rho) = j_l(\rho) + i n_l(\rho) \tag{8.108}$$

The spherical Hankel functions $h_l(\rho)$ are discussed in Section A.2.3 in the appendix.

Since $G_l(r,r')$ has to be continuous at $r = r'$,

$$G_l(r,r') = C_l j_l(kr) h_l(kr') \quad (r < r') \tag{8.109a}$$
$$G_l(r,r') = C_l h_l(kr) j_l(kr') \quad (r' < r) \tag{8.109b}$$

C_l must be a constant because $G_l(r,r')$ is a function of $r - r'$ and therefore satisfies equation (8.106) with r and r' interchanged.

It is possible to find C_l by plugging the definition (8.109) directly into the equation that defines the Green's function. Here is a more efficient way: Start from the three-dimensional Green's function $G(\mathbf{r},\mathbf{r}')$, the solution to

$$\left[\nabla^2 + k^2 \right] G(\mathbf{r},\mathbf{r}') = \delta_3(\mathbf{r}' - \mathbf{r}) \tag{8.26}$$

$G(\mathbf{r},\mathbf{r}')$ is a function of r, r' and the angle Θ between $\mathbf{r}$ and $\mathbf{r}'$: $\mathbf{r} \cdot \mathbf{r}' = rr' \cos\Theta$, so it has a Legendre function expansion

$$G(\mathbf{r},\mathbf{r}') = \sum_l \frac{(2l+1)}{4\pi} P_l(\cos\Theta) G_l(r,r') \tag{8.110}$$

for some functions $G_l(r,r')$. Put this expansion into equation (8.26), write ∇^2 and $\delta_3(\mathbf{r}-\mathbf{r}')$ in spherical coordinates, and use the addition theorem for spherical harmonics (B.24):

$$\sum_l \frac{(2l+1)}{4\pi} \left(\frac{1}{r^2} \frac{\partial}{\partial r} r^2 \frac{\partial}{\partial r} - \frac{l(l+1)}{r^2} + k^2 \right) P_l(\cos\Theta) G_l(r,r')$$

$$= \sum_{lm} \left(\frac{1}{r^2} \frac{\partial}{\partial r} r^2 \frac{\partial}{\partial r} - \frac{l(l+1)}{r^2} + k^2 \right) Y_l^m(\theta,\phi) Y_l^m(\theta',\phi')^* G_l(r,r')$$

$$= \frac{\delta(r-r')}{r^2} \delta(\cos\theta - \cos\theta') \delta(\phi - \phi') \tag{8.111}$$

where θ and ϕ are the spherical coordinates of $\mathbf{r}$, the similarly for $\mathbf{r}'$. Multiply both sides by $Y_L^M(\theta',\phi')$ and integrate over θ' and ϕ':

$$\left(\frac{1}{r^2} \frac{\partial}{\partial r} r^2 \frac{\partial}{\partial r} - \frac{L(L+1)}{r^2} + k^2 \right) Y_L^M(\theta,\phi) G_L(r,r')$$

$$= \frac{\delta(r-r')}{r^2} Y_L^M(\theta,\phi) \tag{8.112}$$

So $G_l(r, r')$ in (8.110) is indeed the function defined by equation (8.106). The boundary condition at $r \to \infty$ is fixed by the boundary condition on $G(\mathbf{r}, \mathbf{r}')$.

Now suppose $r > r'$, but both of them are very small. Then use the expansion (A.79) in (8.29). The left hand side of equation (8.110) is

$$G(\mathbf{r}, \mathbf{r}') \to -\frac{1}{4\pi} \frac{1}{|\mathbf{r} - \mathbf{r}'|} = -\frac{1}{4\pi} \sum_l \frac{r'^l}{r^{l+1}} P_l(\cos\Theta) \tag{8.113}$$

To evaluate the right-hand side, use the limiting forms from Table A.6 in the representation (8.109):

$$\sum_l \frac{(2l+1)}{4\pi} P_l(\cos\Theta) G_l(r, r')$$

$$\to \sum_l \frac{(2l+1)}{4\pi} P_l(\cos\Theta) C_l \left(-\frac{i(2l-1)!!}{(kr)^{l+1}}\right)\left(\frac{(kr)^l}{(2l+1)!!}\right) \tag{8.114}$$

$$= \frac{1}{4\pi} \frac{1}{ik} \sum_l C_l \frac{r'^l}{r^{l+1}} P_l(\cos\Theta)$$

It follows that $C_l = -ik$ for all l. This completes the proof of the integral equation (8.103) for the radial wave function.

8.4.3 Scattering by an Impenetrable Sphere

Often the radial wave function can be obtained by other means. As an example, consider the potential

$$V(r) = V_o \Theta(a - r) \tag{8.115}$$

in the limit $V_o \to \infty$. The wave function satisfies the free Schrödinger equation for $r > a$ and is zero for $r < a$. For $r > a$, The radial wave function is

$$R_l(r) = a_l j_l(kr) + b_l n_l(kr) \tag{8.116}$$

Since R_l must vanish at $r = a$,

$$a_l j_l(ka) = -b_l n_l(ka) \tag{8.117}$$

Then for large r

$$R_l(r) \to a_l \sin(kr - \pi l/2)/kr - b_l \cos(kr - \pi l/2)/kr$$
$$= e^{i\delta_l} \sin(kr - \pi l/2 + \delta_l)/kr \tag{8.118}$$

so that

$$\tan \delta_l = -b_l/a_l = j_l(ka)/n_l(ka) \tag{8.119}$$

This elementary example illustrates that δ_l goes like k^{2l+1} for small k. In particular, near the scattering threshold $k \to 0$ only the $l = 0$ phase shift survives. The differential cross section is

$$\frac{d\sigma}{d\Omega} = \left| \sum_l (2l+1) \frac{e^{i\delta_l} \sin \delta_l}{k} P_l(\cos\theta) \right|^2 \tag{8.120}$$

Near $k = 0$, the total cross section is

$$\sigma_{tot}(0) = \frac{4\pi}{k^2} \sin^2 \delta_o \qquad (8.121)$$

where

$$\tan \delta_o = j_o(ka)/n_o(ka) = -\tan(ka) \qquad (8.122)$$

and so $\delta_o = -ka$. For small k therefore,

$$\sigma_{tot}(0) \to 4\pi a^2 \qquad (8.123)$$

What a peculiar result! Equation (8.123) is four times the classical cross sectional area of a sphere of radius a.

For general l, write

$$e^{i\delta_l} \sin \delta_l = \frac{1}{\cot \delta_l - i} = \frac{1}{n_l(ka)/j_l(ka) - i} \qquad (8.124)$$

or equivalently

$$e^{2i\delta_l} = \frac{1 + i \tan \delta_l}{1 - i \tan \delta_l} = -\frac{h_l(ka)^*}{h_l(ka)} \qquad (8.125)$$

This is the exact solution to the hard-sphere scattering problem for any l.

For $l = 1$ for instance,

$$h_1(r) = -(\frac{1}{\rho} + \frac{i}{\rho^2})e^{i\rho} = \frac{1}{\rho^2}(\rho + i)e^{i\rho} \qquad (8.126)$$

After a short computation

$$\delta_1 = -ka + \frac{1}{2} \arctan\left(\frac{2ka}{1 - (ka)^2}\right) = -ka + \arctan(ka) \qquad (8.127)$$

PROBLEMS

8.1. Scattering by an Atom

An electron scatters off a hydrogen *atom* in the ground state. Ignore spin and effects due to the indistinguishability of the two electrons. Assume that the potential seen by the scattering electron is

$$V(r) = -\frac{\alpha}{r} + \alpha \int \rho(r') \frac{1}{|\mathbf{r} - \mathbf{r}'|} d^3 r'$$

where $\rho(\mathbf{r}) = |\psi_{100}(\mathbf{r})|^2$, the probability distribution of the bound electron. You will see later that this is the correct interaction potential.

(a) What is the scattering amplitude in Born approximation? What is the scattering length?[6]

---

[6] The **scattering length** a_o is the negative of the scattering amplitude at zero momentum:

$$a_o = -\lim_{k \to 0} f(\theta, \phi) = -\lim_{k \to 0} \frac{\delta_o}{k}$$

(b) Still in Born approximation, compute the differential and total cross section as a function of the energy of the incident electron.

Hint: Change variables in the second term to $\mathbf{r} - \mathbf{r}'$.

8.2. Scattering by an Attractive Square Well
A particle scatters in the three-dimensional potential

$$V(r) = -V_o \Theta(a - r)$$

with $V_o > 0$ as in Problem 3.11. Define k_o by

$$k_o^2 = 2mV_o$$

Inside the well the radial wave functions are proportional to $j_l(k_1 r)$, where $k_1^2 = k_o^2 + k^2$.

(a) Compute $k \cot \delta_o$ explicitly for this potential. What is the scattering length? What is the total threshold ($k = 0$) cross section?

(b) What is the Born approximation to the full scattering amplitude $f(\theta, \phi)$? Compare the Born approximation in the $k \to 0$ limit with the S-wave scattering length you computed in part (a), and confirm that they are equal when $k_o a \ll 1$.

8.3. A Neutron-Proton Scattering Model—I
In Problem 3.11 you computed the binding energy of an attractive square well. If the binding energy $E_B > 0$ is related to κ by

$$\kappa^2 = 2mE_B$$

then (review the derivation) κ is the solution to

$$k' \cot(k' a) = -\kappa$$

where $k'^2 = k_o^2 - \kappa^2$, and $k_o^2 = 2\mu V_o$. In Problem 3.12 you considered the neutron-proton system, and took the range of the potential to be $2\pi/m_\pi$.

There is no bound state unless $V_o > \pi^2/8\mu a^2$ (where μ now is the reduced mass $m_N/2$). What is the lower limit (see Problem 3.12)? How much larger did the value of V_o have to be to make a bound state with $E_B = 2.2$ MeV, the binding energy of the deuteron?

(a) Now you have another way to test whether the square well is a reasonable approximation to the neutron-proton force. With $V_o = 66.3$ MeV, what is the cross section at threshold predicted for neutron-proton scattering? The experimental result is about 23×10^{-24} cm^2. Is this square-well model sensible?

(b) The value $\sim 1/m_\pi$ for the range of this force follows from very general arguments involving the uncertainty principle. Moreover it is confirmed, for example, by how much backwards proton-proton scattering deviates from the Rutherford formula. Is the calculated cross section sensitive to the range of the force? Try varying a over an order of magnitude or so, and for each a adjust V_o to give $E_B = 2.22$ MeV. Can you get closer to the observed threshold cross section?

(c) For some fixed range a, what is the scattering length when the potential has just the minimum value needed to form a bound state?

8.4. Scattering by a Delta-Shell Potential

This problem too can be solved exactly, and is rather simpler than scattering by a square well, Problem 8.2. The potential is

$$V(r) = -V_o a \delta(r-a)$$

with $V_o > 0$.

(a) As in Problem 3.13, define a dimensionless parameter $\lambda = 2mV_o a^2$. Use

$$R_l(r) = j_l(kr) - ik \int_0^\infty j_l(kr_<) h_l(kr_>) U(r') R_l(r') r'^2 dr'$$

and

$$\frac{e^{i\delta_l} \sin \delta_l}{k} = -\int_0^\infty j_l(kr) U(r) R_l(r) r^2 dr$$

to find $e^{i\delta_l} \sin \delta_l / k$ *algebraically* for any l.

(b) What are the scattering length and the threshold cross section?

(c) If the potential is repulsive instead of attractive, what are the upper and lower limits for the scattering length and the threshold cross section?

8.5. A Neutron-Proton Scattering Model—II

(a) Perhaps the delta shell is a better model for neutron-proton scattering than the square well. Repeat the calculations of Problem 8.3 for the delta-shell potential. For the same range $a = 1/m_\pi$, what value of V_o (or λ) will result in a binding energy of 2.22 MeV? What is the scattering length and the total threshold cross section for these values? Try varying a here too and see if you can do better.

(b) Show that the energy of a bound state in this system corresponds to a pole in the scattering amplitude, considered as a function of complex k.

(c) For some fixed range a what is the scattering length when the potential has just the minimum value needed to form a bound state?

8.6. Neutron-Proton Scattering and the Spin of the Deuteron

Discussion: If you did Problems 8.3 and 8.5 correctly, you discovered that it is impossible to find values of the parameters that result in both the correct experimental binding energy and the correct threshold cross section. You would get the same discrepancy for any potential, for example a Yukawa potential of the same range and the strength needed to bind the deuteron.[7]

It is known from corrections to the Rutherford formula in the backward direction that the order of magnitude of the range of the nucleon potential is about one fermi, approximately the pion Compton wavelength. If the strength of the n-p potential is adjusted to give the deuteron binding energy, and has almost any reasonable shape, the threshold cross section will come out to be of the order 3×10^{-24} cm^2.

The conclusion is that our model is wrong. The error, of course, is ignoring the spin of the neutron and proton. In fact there are *two* S-waves, a spin triplet and a spin singlet, with angular momenta one and zero respectively. Which one corresponds to the computations we have just done? It is the one whose angular

---

[7] A proof of this statement uses the effective range formula, which you can read about. The essential ingredients are that bound states occur at poles in scattering amplitudes, and that $k \cot \delta_o$ is an *even* function of k.

momentum is the same as the deuteron. This is how the spin of the deuteron was first discovered.

Problem: If somehow you could arrange the scattering experiment so that the neutron and proton were definitely in a spin-1 state, the cross section at $k = 0$ would indeed be about 23×10^{-24} cm^2. But in a real experiment, the orientations of the nucleon spins are usually random, and the observed cross section is the weighted average. Using the delta-shell model to estimate the threshold cross section in the spin triplet state, estimate the "singlet threshold cross section," the same quantity in an imaginary experiment in which the neutron and proton are arranged to be definitely in the spin singlet state. How does the strength of the singlet potential compare with the strength (parametrized by V_o) of the triplet potential? Can the singlet force be repulsive?

REFERENCES

The classic books on scattering in quantum mechanics are

[1] M. L. GOLDBERGER AND K. M. WATSON, *Collision theory*, Wiley, 1964.

[2] ROGER G. NEWTON, *Scattering theory of waves and particles*, McGraw Hill, 1966.

There is a careful discussion in section 56 of

[3] K, GOTTFRIED, *Quantum Mechanics, Volume I*, Benjamin, 1966. See also Appendix D.

The story of Arago's spot is recounted many places, including

[4] A. SOMMERFELD, *Optics*, Academic Press, 1964, Translated by O. Laporte and P. A. Moldauer.

[5] E. HECHT AND A. ZAJAC, *Optics*, Addison Wesley, 1979.

CHAPTER 9

Transitions

9.1 TRANSITIONS IN AN EXTERNAL FIELD

9.1.1 Time-Dependent Perturbations

Consider a system described by the Hamiltonian $H = H_o + H'(t)$, where H' is small but explicitly time dependent. Let $|\psi_n\rangle$ be eigenstates of H_o with eigenvalues E_n as in bound-state perturbation theory. Let $|\psi(t)\rangle$ be the solution to Schrödinger's equation with the full Hamiltonian H, and expand it in the unperturbed basis:

$$|\psi(t)\rangle = \sum_n c_n(t) e^{-iE_n t} |\psi_n\rangle \qquad (9.1)$$

Then the coefficients satisfy the coupled differential equations[1]

$$i \frac{dc_m(t)}{dt} = c_m(t) H'_{mm}(t) + \sum_{n \neq m} c_n(t) e^{i\omega_{mn} t} H'_{mn}(t) \qquad (9.2)$$

where $\omega_{mn} = E_m - E_n$.

Suppose $|\psi(0)\rangle = |\psi_a\rangle$. Then for small enough positive times, $c_a(t)$ is of order one and all the other $c_m(t)$ are of order H'. For $b \neq a$, to first order in the perturbation H',

$$c_b(t) = -i \int_o^t H'_{ba}(t') e^{i\omega_{ba} t'} dt'. \qquad (9.3)$$

By successive iterations the coefficients can be found to any order, but the first-order result (9.3) is all I will need here.

9.1.2 The Semiclassical Method

Transistions in a Plane Wave

Now specialize to an electron in an atom, originally in the state $|\psi_a\rangle$ at $t = 0$. Suppose the atom is in an oscillating electromagnetic field, a plane wave with the vector potential

$$\mathbf{A}(\mathbf{r}, t) = A_o \hat{\varepsilon} \cos(\mathbf{k} \cdot \mathbf{r} - \omega t) \qquad (9.4)$$

where $\hat{\varepsilon}$ is the polarization vector:

$$\hat{\varepsilon} \cdot \hat{\varepsilon} = 1 \quad \text{and} \quad \hat{\varepsilon} \cdot \mathbf{k} = 0 \qquad (9.5)$$

and A_o must be a real number. If the field is weak then the perturbation is

$$H'(t) = V e^{i\omega t} + V^\dagger e^{-i\omega t} \qquad (9.6)$$

---
[1] See Problem 9.1.

with
$$V = \frac{e}{2m} A_o \mathbf{p} \cdot \hat{\boldsymbol{\varepsilon}} e^{-i\mathbf{k}\cdot\mathbf{r}} \qquad (9.7)$$

From equation (9.3) the first-order amplitude for a transition to some other state $|\psi_b\rangle$ is

$$c_b(t) = -\left[V_{ba} \frac{e^{i(\omega_{ba}+\omega)t} - 1}{\omega_{ba} + \omega} + V_{ba}^\dagger \frac{e^{i(\omega_{ba}-\omega)t} - 1}{\omega_{ba} - \omega}\right] \qquad (9.8)$$

The probability that the system will be in the state $|\psi_b\rangle$ at some later time t is

$$P_b(t) = |c_b(t)|^2 = \left|V_{ba} \frac{e^{i(\omega_{ba}+\omega)t} - 1}{\omega_{ba} + \omega} + V_{ba}^\dagger \frac{e^{i(\omega_{ba}-\omega)t} - 1}{\omega_{ba} - \omega}\right|^2 \qquad (9.9)$$

The expression (9.9) looks complicated. But the second term is large only for states $|\psi_b\rangle$ where $\omega_{ba} \approx \omega$, while the first term is large only when $\omega_{ba} \approx -\omega$. Presumably these two regions do not overlap—in any particular case you can check this if you get suspicious—so only one or the other term at most is important. If only the second term is important, $\omega_{ba} = |\omega_{ba}| > 0$, and

$$P_b(t) \approx |V_{ba}|^2 \left|\frac{e^{i(\omega_{ba}-\omega)t} - 1}{\omega_{ba} - \omega}\right|^2$$

$$= \frac{\alpha}{4m^2} A_o^2 |\langle\psi_b| \mathbf{p} \cdot \hat{\boldsymbol{\varepsilon}} e^{-i\mathbf{k}\cdot\mathbf{r}} |\psi_a\rangle|^2 \left|\frac{\sin(|\omega_{ba}| - \omega)t/2}{|\omega_{ba}| - \omega}\right|^2 \qquad (9.10)$$

What happens if you somehow prepare the atom in the upper state and shine light of the same frequency ω on it? Then $|\psi_b\rangle$ and $|\psi_a\rangle$ exchange roles, ω_{ab} changes sign, and the *first* term in equation (9.9) is large. The rest of the calculation goes through exactly the same, so equation (9.10) is also the formula for the transition from the upper state to the lower one. This is induced, or stimulated emission. The interaction induces the atom to emit energy into the field.

According to equation (9.10), energy can be emitted to or absorbed from any wavelength for a finite time, but the rate is very small outside the frequency region $|\omega - \omega_{ba}| \leq 2\pi/t$. In the limit $t \to \infty$, energy is exchanged with the perturbing potential only at the Bohr frequency for the transition. The electromagnetic field is a classical field here, but the correspondence principle is working behind the scenes to ensure that the transition from "classical field" to "photons" will be a smooth one.

Incoherent Radiation

The transition rate (9.10) depends on the time in a complicated way that may not seem intuitively correct. If you shine ordinary light on a gas of atoms, you expect the light to be absorbed at a constant rate. The difference is that an ordinary electromagnetic wave is not monochromatic. Sunlight, for instance, is made up of a spectrum of radiation with a continuous distribution of frequencies with random phases.

I want to compute the probability for an atom to absorb energy from this incoherent radiation. First translate A_o^2 into an energy density. The energy density of the plane electromagnetic wave is

$$u(\mathbf{r}) = \frac{1}{8\pi}\left[\mathbf{E}(\mathbf{r})^2 + \mathbf{B}(\mathbf{r})^2\right] = \frac{1}{8\pi}\left[\left|\frac{\partial \mathbf{A}}{\partial t}\right|^2 + |\boldsymbol{\nabla}\times\mathbf{A}|^2\right] \qquad (9.11)$$

$$= \frac{1}{8\pi}\left[\omega^2\mathbf{A}^2 + |\mathbf{k}\times\hat{\boldsymbol{\varepsilon}}|^2\mathbf{A}^2\right] = \frac{\omega^2}{4\pi}A_o^2\cos^2(\mathbf{k}\cdot\mathbf{r}-\omega t)$$

The local time dependence is not particularly interesting. What one measures is the time-average of the energy density:

$$u(\mathbf{r}) = u = \frac{\omega^2}{4\pi}A_o^2\,\overline{\cos^2(\mathbf{k}\cdot\mathbf{r}-\omega t)} = \frac{\omega^2}{8\pi}A_o^2 \qquad (9.12)$$

$u(\mathbf{r})$ is a constant in space.

Incoherent radiation is made up of plane waves with different frequencies, spaced so close together that effectively there is a continuous distribution of energy, as in blackbody radiation. Imagine that $d\omega$ is so small that in each small frequency interval $d\omega$ there is a single monochromatic plane wave.

> Note: The assumption that there is only one plane wave in each $d\omega$ is a simplifying trick, and you can amuse yourself by assuming that there might be two, or none, in some of the intervals. The final answer is the same. What is really being assumed here is that the radiation in each interval $d\omega$ is absorbed independently by the atom.
>
> I suppose this cannot really be true for arbitrarily small $d\omega$. The phases of the radiation have to be correlated over some frequency $\delta\omega$ that is negligibly small compared to any other frequency relevant to the problem.

The energy density of the radiation in $d\omega$ is

$$\rho(\omega)d\omega = \frac{\omega^2}{8\pi}A_o^2 \qquad (9.13)$$

To get the *total* probability to excite the atom, due to the effects of all the components of the radiation, replace equation (9.10) by the integral over all frequencies:

$$P(t) = \frac{\alpha}{4m^2}\int_0^\infty \frac{8\pi}{\omega^2}\rho(\omega)|\langle\psi_b|\mathbf{p}\cdot\hat{\boldsymbol{\varepsilon}}e^{-i\mathbf{k}\cdot\mathbf{r}}|\psi_n\rangle|^2\left|\frac{\sin\left[(|\omega_{ba}|-\omega)t/2\right]}{|\omega_{ba}|-\omega}\right|^2 d\omega \qquad (9.14)$$

where I have used equation (9.13) to replace the amplitude of the wave by the energy density. For large t,

$$\frac{\sin^2 xt}{x^2} \to \pi t\delta(x) \qquad (9.15)$$

Then

$$P(t) \xrightarrow[t\to\infty]{} \frac{\alpha}{4m^2}\frac{8\pi^2}{\omega_{ba}^2}\rho(\omega_{ba})|\langle\psi_b|\mathbf{p}\cdot\hat{\boldsymbol{\varepsilon}}e^{-i\mathbf{k}\cdot\mathbf{r}}|\psi_n\rangle|^2 t \qquad (9.16)$$

The complicated time-dependent behavior is washed out when a large number of frequencies are combined incoherently.

The rate of transitions is the time derivative of $P(t)$. It turns out, as it should, to be a constant.

$$\Gamma_b = \frac{\alpha}{4m^2} \frac{8\pi^2}{\omega_{ba}^2} \rho(\omega_{ba}) \left| \langle \psi_b | \mathbf{p} \cdot \hat{\varepsilon} e^{-i\mathbf{k}\cdot\mathbf{r}} | \psi_n \rangle \right|^2 \tag{9.17}$$

For atomic transitions, kr is usually small for the values of r that contribute significantly to the matrix element, and it is legitimate to make a power series expansion of the exponential. When the first term, obtained by setting $\exp i\mathbf{k}\cdot\mathbf{r} \to 1$, is not zero, the computation of the matrix element can be further simplified using $[\mathbf{r}, H_o] = i\mathbf{p}/m$. Then

$$\langle \psi_b | \hat{\varepsilon} \cdot \mathbf{p} e^{i\mathbf{k}\cdot\mathbf{r}} | \psi_a \rangle \approx -im\hat{\varepsilon} \cdot \langle \psi_b | [\mathbf{r}, H_o] | \psi_a \rangle$$
$$= im\omega_{ba} \hat{\varepsilon} \cdot \mathbf{r}_{ba} \tag{9.18}$$

and the rate is

$$\Gamma_b = \frac{4\pi^2 c\alpha}{\hbar} \rho(\omega_{ba}) |\hat{\varepsilon} \cdot \mathbf{r}_{ba}|^2 \tag{9.19}$$

This is the **electric-dipole approximation**. (In the long wavelength limit the interaction can be approximated as $H' = -e\mathbf{E}\cdot\mathbf{r} = -\mathbf{E}\cdot\mathbf{D}$, where $\mathbf{D}$ is the electron dipole moment.)

Isotropic Radiation

Equation (9.19) gives the transition rate in radiation made up of incoherent pieces of different frequencies, but all with the same direction and the same polarization $\hat{\varepsilon}$. Write

$$\hat{\varepsilon}\cdot\mathbf{r} = |\mathbf{r}|\cos\theta \tag{9.20}$$

where θ is the angle between the constant vector $\mathbf{r}$ and the polarization. If the radiation is unpolarized and isotropic, just replace θ by its average value:

$$\cos\theta \to \frac{1}{4\pi} \int \cos\theta\, d\Omega = \frac{1}{3} \tag{9.21}$$

to get

$$\Gamma_b = \frac{4\pi^2 \alpha}{3} \rho(\omega_{ba}) \mathbf{r}_{ab}^2 \tag{9.22}$$

This semiclassical treatment of radiation does not provide a general method for going beyond lowest-order perturbation theory, and is completely incapable of accounting for the decays from excited levels. Still, it provides a useful way of thinking about the classical limit of interactions with photons. I will not explore it further.

9.2 THE TRANSITION MATRIX

Transition theory is the general framework for calculating cross sections and lifetimes. The evolution of quantum states over very long times—in particular, scattering and decays of long-lived excited states—requires methods more powerful than

292 Chapter 9 Transitions

those of potential scattering in Chapter 8 or the semiclassical treatment of Section 9.1. I will go through the formal theory of transitions first, and then, in the next chapter, take a detour through the quantum theory of the electromagnetic field before returning to atoms and molecules.

9.2.1 The Transition Matrix

Consider transitions between eigenstates of some H_o, the asymptotic states. The transitions are caused by H', another piece of the Hamiltonian that is important only for a short time. It is often not so important that H' be small as that it have short range. Long before and long after the scattering event the system should be adequately described by H_o alone.

The full Hamiltonian is $H = H_o + H'$, and as long as the whole system is isolated and energy is conserved, the Hamiltonian is a constant. For a particle in a potential, H_o is $\mathbf{p}^2/2m$. For the scattering of two elementary particles, H_o is the Hamiltonian for the free particles. For the scattering of one atom by another, or the scattering of a photon or electron by an atom, H_o contains the part of the interaction that binds the electrons in the atom, but not the interaction between the projectile and the target.

We pretend to (but never really do) construct normalizable wave packets out of the plane wave states, packets that describe an incident beam plus a scattered beam with reasonably narrow dispersions in both position and momentum. In Chapter 8 that meant choosing time-independent solutions that were an incoming plane wave plus an outgoing spherical wave. In the formalism of this chapter there will be a similar boundary condition.

Suppose that at some time t_o, a time long before the collision, the projectile is far from the target and the system is in an eigenstate $|\phi_a\rangle$ of H_o with energy ω_a. Further, presume the full system to be isolated and therefore time-translation invariant. Then H will have no time dependence, and the state at a later time t is

$$|\psi_a(t)\rangle = e^{-iH(t-t_o)}|\phi_a\rangle \tag{9.23}$$

After the collision, the system will again be in a superposition of eigenstates of H_o. The **transition amplitude** is the amplitude that the system will be in the state $|\phi_b\rangle$ at time t if it was in $|\phi_a\rangle$ at t_o. This fundamental quantity is

$$a_{ba}(t) = \langle \phi_b | \psi_a(t) \rangle = \left\langle \phi_b \left| e^{-iH(t-t_o)} \right| \phi_a \right\rangle \tag{9.24}$$

As $t \to \infty$, $a_{ba}(t)$ ought to become a constant times the oscillating factor $e^{-i\omega_b t}$.

Take the limit not of $a_{ba}(t)$ directly but of its Fourier transform. When $t = t_o$ the state $|\psi_a(t)\rangle$ is an eigenstate of H_o, so it is not exactly an eigenstate of the full H. Other frequencies will contribute. Its Fourier transform is not just a delta function in ω, but some function

$$A_{ba}(\omega) = -i \int_{-\infty}^{\infty} e^{i\omega t} a_{ba}(t)\, dt \tag{9.25}$$

with $a_{ba}(\omega)$ given by the inverse transform

$$a_{ba}(t) = \frac{i}{2\pi} \int_{-\infty}^{\infty} e^{-i\omega t} A_{ba}(\omega)\, d\omega \tag{9.26}$$

Section 9.2 The Transition Matrix

The integral in equation (9.25) does not literally exist, because $a_{ba}(t)$ does not vanish as $t \to \pm\infty$, and the inverse Fourier transform will not converge.

Start with a state that is almost an energy eigenfunction, a wave packet with a narrow energy spread, but not so narrow that it cannot be localized at large negative t. Since we do not care what happened for $t < t_o$, I will replace $a_{ba}(t)$ by 0 for $t < t_o$. This device solves the problem in the $t \to -\infty$ limit, and does not affect the solution as $t \to +\infty$.

There is also a divergence from the behavior of $a_{ba}(t)$ as $t \to +\infty$. This divergence reflects the fact that a real wave packet will eventually spread out unacceptably in the energy. But we plan to measure the scattered particles long before that becomes a problem. So the trick here is to cut off $a_{ba}(t)$ before the dispersion in energy becomes very large, but not before the interaction has long been over.

With a physicist's optimism that the mathematics will make sense at the end, just introduce a regulator to take care of large positive t: Multiply $a_{ba}(t)$ by $\exp(-\epsilon(t - t_o))$, with $\epsilon > 0$. This modification changes the rules mathematically and turns off the interaction in a smooth way, at times in the irrelevantly distant future. At the end of the calculation we will set $\epsilon \to 0$. The exact value of ϵ will never be important as long as $1/\epsilon$ is long on any time scale related to the scattering. Then for $t > t_o$ equation (9.23) is replaced by

$$|\psi_a(t)\rangle = \lim_{\epsilon \to 0} e^{-iH(t-t_o)}|\phi_a\rangle \Theta(t-t_o) e^{-\epsilon(t-t_o)} \tag{9.27}$$

where $\Theta(x)$ is the step function. The Fourier transform (9.25) is now modified to be

$$A_{ba}(\omega) = -i e^{i\omega t_o} \int_{t_o}^{\infty} \left\langle \phi_b \middle| e^{i(\omega - H + i\epsilon)(t-t_o)} \middle| \phi_a \right\rangle dt \tag{9.28}$$

The integrand is well behaved at both ends, and $A_{ba}(\omega)$ can be computed as if H were a number.

Proof. Since H is Hermitean, it has a complete set of eigenstates $|\lambda\rangle$ such that

$$H|\lambda\rangle = E_\lambda |\lambda\rangle \quad \text{and} \quad I = \sum_\lambda |\lambda\rangle\langle\lambda|$$

Then

$$A_{ba}(\omega) = -i e^{i\omega t_o} \sum_\lambda \int_{t_o}^{\infty} \left\langle \phi_b \middle| e^{i(\omega - H + i\epsilon)(t-t_o)} \middle| \lambda \right\rangle \langle \lambda | \phi_a \rangle dt$$

$$= -i e^{i\omega t_o} \sum_\lambda \int_0^{\infty} \left\langle \phi_b \middle| e^{i(\omega - \omega_\lambda + i\epsilon)t} \middle| \lambda \right\rangle \langle \lambda | \phi_a \rangle dt$$

$$= e^{i\omega t_o} \sum_\lambda \left\langle \phi_b \middle| \frac{1}{\omega - \omega_\lambda + i\epsilon} \middle| \lambda \right\rangle \langle \lambda | \phi_a \rangle$$

$$= e^{i\omega t_o} \left\langle \phi_b \middle| (\omega - H + i\epsilon)^{-1} \middle| \phi_a \right\rangle$$

∎

294 Chapter 9 Transitions

Thus
$$A_{ba}(\omega) = e^{i\omega t_o}\langle \phi_b| G(\omega) |\phi_a\rangle = e^{i\omega t_o} G_{ba}(\omega) \tag{9.29}$$

where the **Green's operator** $G(\omega)$ is
$$G(\omega) = \frac{1}{\omega - H + i\epsilon} \tag{9.30}$$

Even though the operator $H - \omega$ has no inverse for positive *real* ω, $G(\omega)$ has an inverse for real ω and any finite real ϵ. The sign of the $i\epsilon$ term guarantees that the system develops forward in time. Equation (9.30) is a formal expression for the inverse of $H - \omega$ with the appropriate boundary conditions. (With a minus sign before the term $i\epsilon$ it is the Fourier transform of a solution to Schrödinger's equation describing a state developing backward in time from a definite eigenstate of H_o at $t = t_o$.)

The amplitude that the system be in the state $|\phi_b\rangle$ at time t is
$$a_{ba}(t) = \frac{i}{2\pi}\int_{-\infty}^{\infty} e^{-i\omega(t-t_o)} G(\omega)_{ba}\, d\omega \tag{9.31}$$

The Green's operator is useful because it satisfies an important identity. Define
$$G^o(\omega) = \frac{1}{\omega - H_o + i\epsilon} \tag{9.32}$$

The unperturbed Hamiltonian H_o is diagonal in the states $|\phi_a\rangle$, and, as in bound-state perturbation theory, that the matrix elements of H' between eigenstates of H_o are assumed computable. $G^o(\omega)$ is "known" in the sense that its matrix elements between the $|\phi_a\rangle$ are known.

> Mathematical Note: $G^o(\omega)$ is the Fourier transform, up to a constant, of the Green's functions we introduced in equation (8.26) and elsewhere in Chapter 8. Think of a rule like
> $$(\nabla^2 + k^2)G(\mathbf{r},\mathbf{r}') = \delta\left(\mathbf{r} - \mathbf{r}'\right)$$
> as an operator equation between states $|\mathbf{r}\rangle$ and $|\mathbf{r}'\rangle$. The right-hand side is the identity. If the operator had an inverse, $G(\mathbf{r},\mathbf{r}') = \langle \mathbf{r}'|(\nabla^2 + k^2)^{-1}|\mathbf{r}\rangle$ would be the solution. This expression can be given meaning by replacing k^2 with $k^2 + i\epsilon$. Choosing this sign for the $i\epsilon$ term is the same as choosing the outgoing wave boundary condition.

$G(\omega)$ can be developed in a power series in H'. From the pair of definitions (9.30) and (9.32)
$$G^o(\omega)^{-1} - G(\omega)^{-1} = \omega - H_o + i\epsilon - \omega + H - i\epsilon = H' \tag{9.34}$$

so that
$$G(\omega) - G^o(\omega) = G^o(\omega)\left(G^o(\omega)^{-1} - G(\omega)^{-1}\right)G(\omega) = G^o(\omega)H'G(\omega) \tag{9.35}$$

Thus
$$\boxed{G(\omega) = G^o(\omega) + G^o(\omega)H'G(\omega)} \tag{9.36}$$

or equivalently

$$G(\omega) = G^o(\omega) + G(\omega)H'G^o(\omega) \qquad (9.37)$$

Equation (9.36) or (9.37) is the starting point for perturbation methods in transition theory. From either one, successive iteration leads to the series expansion

$$G(\omega) = G^o(\omega) + G^o(\omega)H'G^o(\omega) + G^o(\omega)H'G^o(\omega)H'G^o(\omega) + \cdots \qquad (9.38)$$

suggesting the definition of the **transition matrix** $T(\omega)$:

$$T(\omega) = H'G(\omega)G^o(\omega)^{-1} \qquad (9.39)$$

Then

$$G(\omega) = G^o(\omega) + G^o(\omega)T(\omega)G^o(\omega) \qquad (9.40)$$

and, from equation (9.36)

$$T(\omega) = H' + H'G(\omega)H' \qquad (9.41)$$

An expansion for T in powers of the perturbation H' follows from (9.41). The first approximation $T(\omega) = H'$ is the **Born approximation**, and the whole series is the **Born series**. The series converges only when H' is small, but the closed forms hold more broadly. Equation (9.41) also provides an integral equation for $T(\omega)$ that does not require first solving for $G(\omega)$. From equations (9.40) and (9.40) it also follows that

$$T(\omega) = H' + H'G^o(\omega)T(\omega) \qquad (9.42)$$

9.2.2 The Lippmann-Schwinger Equation

Section 9.2.1 provides the framework for computing $G(\omega)$ and $T(\omega)$ and therefore transition rates. The discussion seemed to depend on the use of time-dependent state vectors, although I never actually computed a state as a function of t. There is a popular alternative way to formulate the discussion of the transition matrix that looks more like the potential scattering approach in Chapter 8, in that only time independent solutions to the Schrödinger equation are used from the start. This is the Lippmann-Schwinger method. Rather than derive it from scratch using the formalism, I will show how it follows from the transition matrix defined above. Some questions are easier to deal with in this time-independent language.

Start from the time-dependent state $|\psi_a(t)\rangle$ and expand it in the eigenstates $|\phi_b\rangle$ of H_o, using the representation (9.40):

$$|\psi_a(t)\rangle = \sum_b |\phi_b\rangle\langle\phi_b| \psi_a(t)\rangle = \sum_b |\phi_b\rangle a_{ba}(t)$$

$$= \frac{i}{2\pi} \sum_b \int_{-\infty}^{\infty} e^{-i\omega(t-t_o)} |\phi_b\rangle \left[G^o(\omega) + G^o(\omega)T(\omega)G^o(\omega)\right]_{ba} d\omega \qquad (9.43)$$

296 Chapter 9 Transitions

An arbitrary matrix element of the first term is

$$\frac{i}{2\pi}\sum_b \int_{-\infty}^{\infty} e^{-i\omega(t-t_o)} \langle\phi_c|\phi_b\rangle\langle\phi_b|G^o(\omega)|\phi_a\rangle\, d\omega$$

$$= \frac{i}{2\pi}\sum_b \delta_{ca} \int_{-\infty}^{\infty} e^{-i\omega(t-t_o)} \frac{1}{\omega - \omega_a + i\epsilon}\, d\omega \qquad (9.44)$$

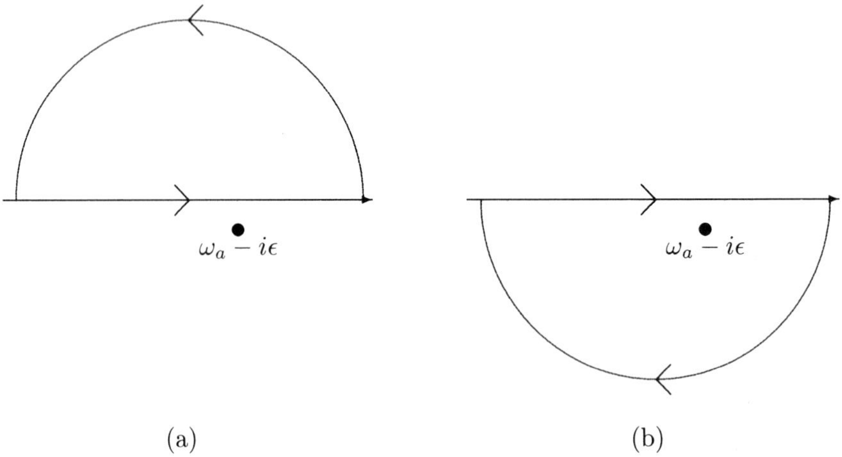

(a)   (b)

FIGURE 9.1: Contours in the complex ω plane for computing the integral in equation (9.44). The only singularity is at $\omega_a - i\epsilon$ as indicated. The original integral is along the real axis. When $t < t_o$ close the contour along a semicircle in the upper half plane, as in (a). When $t > t_o$ close the contour along a semicircle in the lower half plane, as in (b).

The integral can be computed using contour integration. For $t < t_o$, close the contour in the upper half plane. Since there are no singularities in that half plane, the answer is zero by Cauchy's theorem. The operator $G^o(\omega)$ was designed to have just this property. When $t > t_o$, close the integral in the lower half plane. The residue of the simple pole at $\omega_a - i\epsilon$ is $\exp[-i\omega_a(t-t_o)]$. Multiply the residue by $-2\pi i$ to obtain $\delta_{ca}\exp[-i\omega_a(t-t_o)]$ in the limit $\epsilon \to 0$.

The remaining terms in equation (9.43) are

$$|\psi_a(t)\rangle - e^{-i\omega_a(t-t_o)}|\phi_a\rangle$$

$$= \frac{i}{2\pi}\int_{-\infty}^{\infty} e^{-i\omega(t-t_o)} \sum_b |\phi_b\rangle \frac{1}{\omega - \omega_b + i\epsilon}\langle\phi_b|T(\omega)G^o(\omega)|\phi_a\rangle\, d\omega$$

$$= \frac{i}{2\pi}\int_{-\infty}^{\infty} e^{-i\omega(t-t_o)} \sum_b \frac{1}{\omega - H_o + i\epsilon}|\phi_b\rangle \frac{1}{\omega - \omega_a + i\epsilon}\langle\phi_b|T(\omega)|\phi_a\rangle\, d\omega \qquad (9.45)$$

$$= \frac{i}{2\pi} e^{-i\omega_a(t-t_o)} \int_{-\infty}^{\infty} e^{-i(\omega-\omega_a)(t-t_o)} G^o(\omega) \frac{1}{\omega - \omega_a + i\epsilon} T(\omega)|\phi_a\rangle\, d\omega$$

The contribution to the integral from regions where ω is different from ω_a vanishes in the large t limit. Use the representation[2]

$$\lim_{t\to\infty} e^{-i(\omega-\omega_a)(t-t_o)} \frac{1}{\omega - \omega_a + i\epsilon} = -2\pi i \delta(\omega - \omega_a) \qquad (9.46)$$

Then as $t \to \infty$

$$|\psi_a(t)\rangle \to e^{-i\omega_a(t-t_o)}|\phi_a\rangle + e^{-i\omega_a(t-t_o)} G^o(\omega_a) T(\omega_a)|\phi_a\rangle$$
$$= e^{-i\omega_a(t-t_o)}|\psi_a\rangle \qquad (9.47)$$

where the constant state $|\psi_a\rangle$ is a solution to the time independent Schrödinger equation:[3]

$$|\psi_a\rangle = |\phi_a\rangle + G^o(\omega_a) T(\omega_a)|\phi_a\rangle \qquad (9.48)$$

Use the identity (9.42) to write equation (9.48) as

$$\boxed{|\psi_a\rangle = |\phi_a\rangle + G^o(\omega) H'|\psi_a\rangle} \qquad (9.49)$$

Equation (9.49) is the **Lippmann-Schwinger equation** [1].

For potential scattering, (9.49) is an integral equation with the correct boundary conditions for the wave function:

$$\psi(\mathbf{r}) = \phi_a(\mathbf{r}) + \int \langle \mathbf{r}| G^o(\omega) |\mathbf{r}'\rangle V(\mathbf{r}')\psi(\mathbf{r}') d^3r' \qquad (9.50)$$

I leave it as an exercise to show that this is the same as equation (8.30).

Finally, from the Lippmann-Schwinger equation follows an important rule for elements of the transition matrix T. Use equation (9.42) in (9.49) to write

$$T(\omega_a)_{ba} = \langle \phi_b| T(\omega_a) |\phi_a\rangle = \langle \phi_b| H' + H' G^o(\omega) T(\omega_a) |\phi_a\rangle$$
$$= \langle \phi_b|H'\Big(|\phi_a\rangle + G^o(\omega_a) T(\omega_a)|\phi_a\rangle\Big) = \langle \phi_b| H' |\psi_a\rangle \qquad (9.51)$$

9.2.3 Relation to the Scattering Amplitude

When these equations describe the scattering of a single particle by a potential, comparing equation (9.51) to equation (8.33) leads to the identification

$$f(\theta, \phi) = -4\pi^2 m \langle \phi_b| T(\omega_a) |\phi_a\rangle \qquad (9.52)$$

where $|\phi_a\rangle$ is a plane wave state along the z-axis, and $|\phi_b\rangle$ a plane wave state in the direction of the spherical coordinates θ and ϕ.

The Lippmann-Schwinger equation provides a simple way to generalize (9.52) to relativistic particles (and therefore to photons and neutrinos). The coordinate

---

[2] See equation (A.7) in the appendix.
[3] ω_a is the eigenvalue of both H_o and $H = H_o + H'$. The corresponding eigenstates are $|\phi_a\rangle$ and $|\psi_a\rangle$.

298 Chapter 9 Transitions

space version of equation (9.48) (ignoring spin) is

$$\langle \mathbf{r} | \psi_a \rangle = \langle \mathbf{r} | \phi_a \rangle + \int \left\langle \mathbf{r} \left| \frac{1}{\omega_a - H_o + i\epsilon} \right| \phi_b \right\rangle T_{ba}(\omega_a) \, d^3 k_b$$

$$= \langle \mathbf{r} | \phi_a \rangle + \frac{1}{(2\pi)^{\frac{3}{2}}} \int e^{i\mathbf{k}_b \cdot \mathbf{r}} \frac{1}{\omega_a - \omega_b + i\epsilon} T_{ba}(\omega_a) \, d^3 k_b \quad (9.53)$$

Assume the target is spherically symmetric so that T_{ba} is independent of the direction of $\mathbf{k}_b$, and do the angular integrals as in equation (8.40):

$$\int e^{i\mathbf{k}_b \cdot \mathbf{r}} \frac{1}{\omega_a - \omega_b + i\epsilon} T_{ba}(\omega_a) \, d^3 k_b = -\frac{2\pi}{ir} \int_0^\infty k_b \frac{e^{ik_b r} - e^{-ik_b r}}{\omega_a - \omega_b + i\epsilon} T_{ba}(\omega_a) \, dk_b \quad (9.54)$$

T_{ba} vanishes for large k_b, thereby providing the cutoff or regulator that makes the integral finite. Then the integrand can be extended to $\pm\infty$. The term with the negative exponent vanishes, as can be seen by closing the contour in the lower half plane. For the other term, use the representation (A.7) for the Dirac delta function. The result is that for large r,

$$\langle \mathbf{r} | \psi_a \rangle = \langle \mathbf{r} | \phi_a \rangle + \frac{f(\theta)}{r} e^{ik_a r} \quad (9.55)$$

where

$$\boxed{f(\theta) = -4\pi^2 \omega_a T_{ba}} \quad (9.56)$$

This is the relativistic generalization of the identity (9.52).[4]

9.3 SCATTERING AND CROSS SECTIONS

9.3.1 The Scattering Matrix

If a system is in an eigenstate $|\phi_a\rangle$ of H_o at time t_o, then from equations (9.31) and (9.36) the amplitude for it to be in another energy eigenstate $|\phi_b\rangle$ at a much later time t is

$$a_{ba}(t) = \frac{i}{2\pi} \int_{-\infty}^{\infty} e^{-i\omega(t-t_o)} G^o(\omega)_{ba} \, d\omega$$

$$+ \frac{i}{2\pi} \int_{-\infty}^{\infty} e^{-i\omega(t-t_o)} \langle \phi_b | G^o(\omega) T(\omega) G^o(\omega) | \phi_a \rangle \, d\omega \quad (9.57)$$

Since $|\phi_a\rangle$ is an eigenstate of H_o, the first term is

$$\frac{i}{2\pi} \delta_{ba} \int_{-\infty}^{\infty} e^{-i\omega(t-t_o)} \frac{1}{\omega - \omega_a + i\epsilon} \, d\omega \quad (9.58)$$

In the small ϵ limit

$$\frac{i}{2\pi} \int_{-\infty}^{\infty} e^{-i\omega(t-t_o)} G^o(\omega)_{ba} \, d\omega \xrightarrow[\epsilon \to 0]{} \delta_{ba} e^{-i\omega_a(t-t_o)} \quad (9.59)$$

---

[4] You can find the calculation done more carefully in the literature, for example, page 99 of Goldberger and Watson, reference [1] in Chapter 8.

The second term in equation (9.57) is

$$\frac{i}{2\pi}\int_{-\infty}^{\infty} e^{-i\omega(t-t_o)}\frac{1}{(\omega-\omega_b+i\epsilon)}T(\omega)_{ba}\frac{1}{(\omega-\omega_a+i\epsilon)}d\omega \qquad (9.60)$$

Provided $T(\omega)$ is a smooth function of ω, it can be taken outside the integral in the large t limit. As t becomes large, the exponential oscillates faster and faster, and in the limit the integrand vanishes unless the denominators are both zero. I will discuss a little later exactly how smooth $T(\omega)$ needs to be.

The limit $t \to \infty$ and $t_o \to -\infty$ of the expression (9.60) is [using the representation (A.7) again]

$$\frac{i}{2\pi}T(\omega_a)_{ba}e^{-i\omega_a t}e^{i\omega_b t_o}\int_{-\infty}^{\infty}\frac{e^{-i(\omega-\omega_a)t}}{\omega-\omega_a+i\epsilon}\frac{e^{i(\omega-\omega_b)t_o}}{\omega-\omega_b+i\epsilon}d\omega$$

$$\to T(\omega_a)_{ba}e^{-i\omega_a t}e^{i\omega_b t_o}\int_{-\infty}^{\infty}\frac{e^{-i(\omega-\omega_a)t}}{\omega-\omega_a+i\epsilon}\delta(\omega-\omega_b)\,d\omega \qquad (9.61)$$

$$\to -e^{-i\omega_a(t-t_o)}T(\omega_a)_{ba}2\pi i\delta(\omega_a-\omega_b)$$

The order ϵ corrections are no longer needed.

From equations (9.57) and (9.61) we get the rule for the transition amplitude itself in the same limit:

$$e^{i\omega_a(t-t_o)}a_{ba}(t) \to \delta_{ba} - 2\pi i\delta(\omega_a-\omega_b)T(\omega_a)_{ba} = S_{ba} \qquad (9.62)$$

The amplitude $a_{ba}(t)$ approaches a constant times a phase factor oscillating with frequency ω_a. The only time dependence left is the free-particle oscillation contained in the exponential factor on the left. The constant operator on the right is the **S matrix**, or **scattering matrix**.

9.3.2 The Transition Probability

Transition probabilities are the squares of the absolute values of S-matrix elements. For $a \neq b$, this is

$$P_{ba} = |S_{ba}|^2 = 4\pi^2|T_{ba}(\omega_a)|^2\delta(\omega_a-\omega_b)^2 \qquad (9.63)$$

The last factor is infinite because of the continuum normalization. The initial state contains an infinite number of scattering particles, so the probability for the transition increases without limit as $t \to \infty$.

The way to understand this infinity is to back up and compute the rate for large but finite t. The integral in equation (9.60) vanishes in the large t limit unless $\omega \approx \omega_a$. Take t so large that $T_{ba}(\omega_a)$ can be taken outside the integral, and in equation (9.62) replace $\delta(\omega_a-\omega_b)$ by the value of the expression for finite t that it came from in equation (9.61):

$$-2\pi i\delta(\omega_a-\omega_b) \to e^{i\omega_a(t-t_o)}\frac{i}{2\pi}\int_{-\infty}^{\infty} e^{-i\omega(t-t_o)}\frac{1}{(\omega-\omega_b+i\epsilon)(\omega-\omega_a+i\epsilon)}d\omega \qquad (9.64)$$

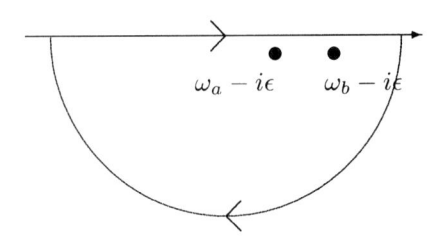

FIGURE 9.2: Contour in the complex ω plane for computing the integral in equation (9.64). The singularities at $\omega_a - i\epsilon$ and $\omega_b - i\epsilon$ are indicated. The original integral is along the real axis. As long as $t > t_o$, you can close the contour along a semicircle in the lower half plane.

Evaluate the integral by closing the contour in the lower half plane. The integrand has simple poles at $\omega_a - i\epsilon$ and $\omega_b - i\epsilon$, so the expression (9.64) is equal to

$$e^{i\omega_a(t-t_o)} \left(\frac{e^{-i\omega_b(t-t_o)}}{\omega_b - \omega_a} + \frac{e^{-i\omega_a(t-t_o)}}{\omega_a - \omega_b} \right) = \frac{e^{-i(\omega_b - \omega_a)(t-t_o)} - 1}{\omega_b - \omega_a} \qquad (9.65)$$

Thus at large but finite t the transition probability is

$$P_{ba} = |T_{ba}(\omega_a)|^2 4\pi^2 \delta(\omega_a - \omega_b)^2 \to |T_{ba}(\omega_a)|^2 \left| \frac{e^{-i(\omega_b - \omega_a)(t-t_o)} - 1}{\omega_b - \omega_a} \right|^2 \qquad (9.66)$$

$$= 2|T_{ba}(\omega_a)|^2 \frac{1 - \cos[(\omega_b - \omega_a)(t-t_o)]}{(\omega_b - \omega_a)^2}$$

P_{ba} indeed increases with the time when $\omega_b = \omega_a$; that is why the probability in the infinite time limit came out to be an energy-conserving delta function multiplied by infinity. The correct *rate* of transitions is constant:

$$\Gamma = \frac{d}{dt} P_{ba} = 2|T_{ba}(\omega_a)|^2 \frac{\sin[(\omega_a - \omega_b)(t-t_o)]}{\omega_b - \omega_a} \xrightarrow[t_o \to -\infty]{t \to \infty} 2\pi |T_{ba}(\omega_a)|^2 \delta(\omega_b - \omega_a) \qquad (9.67)$$

Equation (9.67) is the rule for the transition rate in terms of the transition matrix. (In the last form I have use the representation (A.9) for the Dirac delta function.)

Since the first Born approximation to T_{ba} is H'_{ba}, the first-order approximation to the transition rate is **Fermi's golden rule** for transition rates [2].

$$\boxed{\Gamma = 2\pi |H'_{ba}|^2 \delta(\omega_b - \omega_a)} \qquad (9.68)$$

Fermi actually called this "golden rule number two." His golden rule number one was the formula for second-order transitions.

9.3.3 Cross Sections

If the target is fixed and the states of the incident particle are normalized to Dirac delta functions in momentum, the particle density is $(2\pi)^{-3}$. The flux densities are

$v(2\pi)^{-3}$. The transition rate into a group of final states $|\phi_b\rangle$ with momentum in some range of $\mathbf{k}_b$ (and some collection of spin and other discrete internal quantum numbers if needed) is

$$\Gamma = 2\pi \int |\langle \phi_b | T(\omega_a) | \phi_a \rangle|^2 \delta(\omega_a - \omega_b) d^3 k_b \tag{9.69}$$

The cross section is the transition rate divided by the incident flux density:

$$\sigma = \frac{(2\pi)^4}{v} \int |\langle \phi_b | T(\omega_a) | \phi_a \rangle|^2 \delta(\omega_a - \omega_b) d^3 k_b \tag{9.70}$$

When the kinematics is all nonrelativistic, then $k_b\, dk_b = m\, d\omega_b$. The speed is k_b/m, and

$$\sigma = (2\pi)^4 m^2 \int |\langle \phi_b | T(\omega_a) | \phi_a \rangle|^2 \delta(\omega_a - \omega_b) d\omega_b\, d\Omega_b \tag{9.71}$$

The differential cross section into solid angle $d\Omega_b$ is

$$\boxed{\frac{d\sigma}{d\Omega_b} = (2\pi)^4 m^2 |T_{ba}(\omega_a)|^2} \tag{9.72}$$

The Born approximation to the differential cross section is

$$\boxed{\frac{d\sigma}{d\Omega_b} = (2\pi)^4 m^2 |H'_{ba}|^2} \tag{9.73}$$

Equation (8.37) for potential scattering is a particular instance of equation (9.73).

The Optical Theorem

Equation (9.70) provides a proof of the optical theorem (8.48) more general than the ones we were able to give in the context of potential scattering (section 8.2.4). Start from the definition of the T-matrix

$$T(\omega) = H' + H'G^o(\omega)T(\omega) \tag{9.42}$$

and its Hermitean conjugate

$$T(\omega)^\dagger = H' + T(\omega)^\dagger G^o(\omega)^\dagger H' \tag{9.74}$$

These two equations can be rewritten as

$$T(\omega) = H' + \left[T(\omega)^\dagger - T(\omega)^\dagger G^o(\omega)^\dagger H'\right] G^o(\omega) T(\omega) \tag{9.75a}$$
$$T(\omega)^\dagger = H' + T(\omega)^\dagger G^o(\omega)^\dagger \left[T(\omega) - H' G^o(\omega) T(\omega)\right] \tag{9.75b}$$

whence

$$T(\omega) - T(\omega)^\dagger = T(\omega)^\dagger \left[G^o(\omega) - G^o(\omega)^\dagger\right] T(\omega) \tag{9.76}$$

Therefore, since $G_o(\omega)$ is diagonal in the basis $|\phi_n\rangle$ of eigenstates of H_o, we compute

$$T(\omega)_{ba} - T(\omega)^*_{ab} = \sum_c \langle \phi_b| T(\omega)^\dagger |\phi_c\rangle \left[\frac{1}{\omega - \omega_c + i\epsilon} - \frac{1}{\omega - \omega_c - i\epsilon}\right] \langle \phi_c| T(\omega) |\phi_a\rangle$$

$$= -2i\epsilon \sum_c \frac{1}{(\omega - \omega_c)^2 + \epsilon^2} \langle \phi_b| T(\omega)^\dagger |\phi_c\rangle \langle \phi_c| T(\omega) |\phi_a\rangle$$

$$\xrightarrow[\epsilon \to 0]{} -2\pi i \sum_c \langle \phi_b| T(\omega)^\dagger |\phi_c\rangle \langle \phi_c| T(\omega) |\phi_a\rangle \delta(\omega - \omega_c) \tag{9.77}$$

Set $b = a$ and $\omega = \omega_a$, and use the expression (9.70) for the total cross section:

$$\text{Im } T_{aa} = -\pi \sum_c |\langle \phi_c| T(\omega_a) |\phi_a\rangle|^2 \delta(\omega_c - \omega_a) = -\frac{\pi v}{(2\pi)^4} \sigma_{\text{total}} \tag{9.78}$$

Equation (9.78) is a form of the optical theorem not restricted to elastic scattering or even nonrelativistic kinematics. The scattering amplitude in potential scattering is given in terms of the transition matrix by (9.52). The original form (8.48) of the optical theorem is a special case of (9.78).

The optical theorem is a consequence of the conservation of probability. In potential scattering the optical theorem (8.48) works because the potential energy is real—if it were complex the time-dependent Schrödinger equation would not conserve probability. A real potential means a Hermitean Hamiltonian, which in turn implies a unitary time-evolution operator. The equivalent proof for partial waves in Section 8.3.2 used the fact that the phase shifts are real numbers—again a prerequisite for the total probability flux in the Schrödinger wave function to be conserved. The more general proof given here also depends on probability conservation. It is equivalent to the requirement—buried in all the formulas for T—that the S-matrix is a unitary operator.[5]

9.3.4 Scattering of Electrons by Atoms

The rest of this section is devoted to a few applications that demonstrate the power of the transition matrix formalism and the Lippmann-Schwinger equation, namely, electron scattering by atoms, scattering with recoil, and scattering of identical particles.

Consider first the scattering of an electron by a hydrogen atom. Ignore spin and even ignore the identity of the two electrons involved. The first of these approximations is valid to the extent that the spin-spin and spin-orbit interactions are much smaller that the Coulomb forces. The second will hold to the extent there is little overlap in momentum space between the wave functions of the two electrons, namely at high energies and large momentum transfers.

Let $\mathbf{r}_1$ be the coordinates of the scattering electron, and $\mathbf{r}_2$ those of the bound electron. The initial state is

$$\langle \mathbf{r}_1, \mathbf{r}_2 | \phi_a\rangle = \phi_a(\mathbf{r}_1, \mathbf{r}_2) = \phi_a(\mathbf{r}_1) \psi_{nlm}(\mathbf{r}_2) \tag{9.79}$$

---

[5] See Problem 9.11.

and the final state is

$$\langle \mathbf{r}_1, \mathbf{r}_2 | \phi_b \rangle = \phi_b(\mathbf{r}_1, \mathbf{r}_2) = \phi_b(\mathbf{r}_1)\psi_{n'l'm'}(\mathbf{r}_2) \tag{9.80}$$

where

$$\phi_{a,b}(\mathbf{r}_1) = \frac{1}{(2\pi)^{\frac{3}{2}}} e^{i\mathbf{k}_{a,b}\cdot \mathbf{r}_1} \tag{9.81}$$

n, l, m and n', l', m' are the quantum numbers of the initial and final bound electron state.

For this problem,

$$H_o = \frac{\mathbf{k}_1^2}{2m} + \frac{\mathbf{k}_2^2}{2m} - \frac{\alpha}{r_2} \tag{9.82}$$

so

$$H' = -\frac{\alpha}{r_1} + \frac{\alpha}{|\mathbf{r}_1 - \mathbf{r}_2|} \tag{9.83}$$

Then the Born approximation to the transition matrix is

$$\langle \phi_b | H' | \phi_a \rangle = \frac{1}{(2\pi)^3} \int \psi_{n'l'm'}(\mathbf{r}_2)^* \psi_{nlm}(\mathbf{r}_2) e^{i\mathbf{q}\cdot \mathbf{r}_1} \left(-\frac{\alpha}{r_1} + \frac{\alpha}{|\mathbf{r}_1 - \mathbf{r}_2|} \right) d^3 r_1\, d^3 r_2 \tag{9.84}$$

where $\mathbf{q} = \mathbf{k}_a - \mathbf{k}_b$ is the momentum transferred to the final electron. When the initial and final bound electron are both in the ground state, this is the expression in Problem 8.1, now derived correctly.

Since the bound-state wave functions are orthogonal:

$$\langle \phi_b | H' | \phi_a \rangle$$
$$= \frac{1}{(2\pi)^3} \int e^{i\mathbf{q}\cdot \mathbf{r}} \frac{\alpha}{r} d^3 r \left[-\delta_{n'n}\delta_{l'l}\delta_{m'm} + \int \psi_{n'l'm'}(\mathbf{r}_2)^* \psi_{nlm}(\mathbf{r}_2) e^{i\mathbf{q}\cdot \mathbf{r}_2} d^3 r_2 \right]$$

$$\tag{9.85}$$

Equation (9.85) describes both elastic and inelastic scattering of an electron by a hydrogen atom.

9.3.5 Scattering with Recoil

A real target is not fixed but recoils. After the collision the target is not in the same state as before. Furthermore, the final state can have three or more particles. This is a straightforward generalization that I leave to the problems. I will outline here how to generalize the results of Section 9.3.3.

Consider the scattering of particle 1, the projectile, off particle 2, the target, and let each be in a momentum eigenstate. The initial state is $|\phi_a\rangle = |\mathbf{k}_1, \mathbf{k}_2\rangle$, and the final state $|\phi_b\rangle = |\mathbf{k}'_1, \mathbf{k}'_2\rangle$. For simplicity, ignore any internal quantum numbers like spin, although the formalism can handle them easily. The initial and final wave functions are

$$\langle \mathbf{r}_1, \mathbf{r}_2 | \phi_a \rangle = \phi_a(\mathbf{r}_1, \mathbf{r}_2) = \frac{1}{(2\pi)^3} e^{i(\mathbf{k}_1 \cdot \mathbf{r}_1 + \mathbf{k}_2 \cdot \mathbf{r}_2)} \tag{9.86}$$

and
$$\langle \mathbf{r}_1, \mathbf{r}_2 | \phi_b \rangle = \phi_b(\mathbf{r}_1, \mathbf{r}_2) = \frac{1}{(2\pi)^3} e^{(i\mathbf{k}'_1 \cdot \mathbf{r}_1 + \mathbf{k}'_2 \cdot \mathbf{r}_2)} \tag{9.87}$$

The transition rate into the final state within some range of the final momenta is
$$\Gamma = 2\pi \int |T_{ba}|^2 \, \delta(\omega_b - \omega_a) d^3 k'_1 d^3 k'_2 \tag{9.88}$$

If momentum is conserved, then
$$\bar{H}' = \delta_3(\mathbf{K}' - \mathbf{K}) \langle \phi_b | \bar{H}' | \phi_a \rangle \tag{9.89}$$

where $\mathbf{K} = \mathbf{k}_1 + \mathbf{k}_2$ and $\mathbf{K}' = \mathbf{k}'_1 + \mathbf{k}'_2$. In that case the transition matrix will also have a momentum-conserving delta function:
$$\bar{T}' = \delta_3(\mathbf{K}' - \mathbf{K}) \langle \phi_b | \bar{T}' | \phi_a \rangle \tag{9.90}$$

Even though I have written the matrix elements of $\bar{H}'$ and $\bar{T}'$ using the ordinary Dirac notation, these objects are really only defined between initial and final states with the same total momentum.

The cross section is conventionally defined as the transition rate divided by the flux of the projectile in the target rest frame. The *total* cross section is independent of the frame of reference, but the differential cross section is not, so it is awkward to proceed in a completely general frame. The standard convention is to define the cross section only in a frame of reference where the projectile and target momenta are collinear. The projectile velocity is in the $+z$ direction. The target is either at rest or moving in the $+z$ or $-z$ direction. The cross section into a group of final states is
$$\sigma = \frac{(2\pi)^3}{v_1 + v_2} \Gamma = \frac{(2\pi)^4}{v_1 + v_2} \int |T'_{ba}|^2 \, \delta(\omega_b - \omega_a) d^3 k'_1 d^3 k'_2 \tag{9.91}$$

Because T'_{ba} is proportional to $\delta_3(\mathbf{K}' - \mathbf{K})$ the cross section appears to be proportional to the square of a delta function, as was the transition rate in equation (9.63). The explanation is the same. The states are not normalizable. This infinity could have been avoided by the method of quantizing in a large box, rather than normalizing to Dirac delta functions in momentum.

With our normalization convention the target as well as the projectile beam represents a beam of density $(2\pi)^{-3}$. The state $|\phi_b\rangle$ contains an infinite number of target particles, so there are an infinite number of scatterings even in a finite time interval. It should be clear what to do. The delta function came from integrating
$$\frac{1}{(2\pi)^3} \int e^{i(\mathbf{K} - \mathbf{K}') \cdot \mathbf{R}} d^3 R \tag{9.92}$$

over all space. Restrict the range of integration temporarily to a very large volume V. Then the transition rate from these plane wave states is
$$\bar{\Gamma} = 2\pi \int |T'_{ba}|^2 \delta(\omega_a - \omega_b) dk'_1 dk'_2$$
$$= 2\pi \int |\delta_3(\mathbf{K}' - \mathbf{K}) \bar{T}'_{ba}|^2 \delta(\omega_a - \omega_b) d\mathbf{K}' dk' \tag{9.93}$$
$$\approx 2\pi \int |\bar{T}'_{ba}|^2 \delta(\omega_a - \omega_b) \delta_3(\mathbf{K}' - \mathbf{K}) \frac{V}{(2\pi)^3} d\mathbf{K}' dk'$$

The transition rate *per particle* is $\bar{\Gamma}$ divided by $(2\pi)^{-3}V$, the number of particles in the large volume V: The physical transition rate is therefore not $\bar{\Gamma}$ but

$$\Gamma = 2\pi \int |\bar{T}'_{ba}|^2 \delta(\omega_a - \omega_b)\delta_3(\mathbf{K} - \mathbf{K}') d^3k'_1 d^3k'_2 \quad (9.94)$$

and the cross section (per target particle) is

$$\sigma = (2\pi)^4 \frac{1}{v_1 + v_2} \int |\bar{T}'_{ba}|^2 \delta(\omega_a - \omega_b)\delta_3(\mathbf{K} - \mathbf{K}') d^3k'_1 d^3k'_2$$

$$= (2\pi)^4 \frac{1}{v_1 + v_2} \int |\bar{T}'_{ba}|^2 \delta(\omega_a - \omega_b) d^3k' \quad (9.95)$$

Potential Scattering with Recoil

When two nonrelativistic particles interact via a central local potential $V(r)$, where r is the radial relative coordinate, the matrix element can be evaluated in center of mass and relative coordinates $\mathbf{R}$ and $\mathbf{r}$. Let $\mathbf{k}$ be the relative momentum. Then

$$\langle \phi_b | H' | \phi_a \rangle = \int \phi_b(\mathbf{r}_1, \mathbf{r}_2)^* V(|\mathbf{r}_1 - \mathbf{r}_2|) \phi_a(\mathbf{r}_1, \mathbf{r}_2) d^3r_1 d^3r_2$$

$$= \frac{1}{(2\pi)^6} \int e^{i(\mathbf{K}-\mathbf{K}')\cdot\mathbf{R}} e^{i(\mathbf{k}-\mathbf{k}')\cdot\mathbf{r}} V(r) d^3r d^3R \quad (9.96)$$

$$= \frac{1}{(2\pi)^3} \int e^{i(\mathbf{k}-\mathbf{k}')\cdot\mathbf{r}} V(r) d^3r \, \delta_3(\mathbf{K} - \mathbf{K}')$$

For nonrelativistic scattering, with the interaction given by a potential as in equation (9.96),

$$\omega_b - \omega_a = \frac{k'^2}{2\mu} - \frac{k^2}{2\mu} \quad (9.97)$$

where $\mu = m_1 m_2/(m_1 + m_2)$. Thus

$$\int \delta(\omega_b - \omega_a) d^3k' = \int \delta(\omega_b - \omega_a) \mu k' d\omega_b d\Omega = \mu k' \int d\Omega \quad (9.98)$$

The differential cross section is

$$\frac{d\sigma}{d\Omega} = (2\pi)^4 \frac{\mu k'}{v_1 + v_2} \left| \frac{1}{(2\pi)^3} \int e^{i(\mathbf{k}-\mathbf{k}')\cdot\mathbf{r}} V(r) d^3r \right|^2 \quad (9.99)$$

Equation (9.99) is not restricted to the center of mass frame, but is particularly simple there. In that frame

$$\frac{d\sigma}{d\Omega} = (2\pi)^4 \frac{\mu^2 k'}{k} \left| \frac{1}{(2\pi)^3} \int e^{i(\mathbf{k}-\mathbf{k}')\cdot\mathbf{r}} V(r) d^3r \right|^2 \quad (9.100)$$

9.3.6 Identical Particle Scattering

When the target and projectile are identical particles the formalism goes through unchanged, except that the unperturbed wave functions should have the form

$$\langle \alpha_1, \mathbf{r}_1, \alpha_2, \mathbf{r}_2 | \phi_a \rangle = \frac{1}{\sqrt{2}} (\phi_{1,\alpha_1}(\mathbf{r}_1)\phi_{2,\alpha_2}(\mathbf{r}_2) \pm \phi_{2,\alpha_2}(\mathbf{r}_1)\phi_{1,\alpha_1}(\mathbf{r}_2)) \quad (9.101)$$

where α_1 and α_2 are the discrete internal quantum numbers, if any, of the two particles. With this normalization the initial state is the sum of two states (they do not interfere, being orthogonal) each with a density $(2\pi)^{-3}/2$ for particles moving in the direction of the plane wave $\phi_{1,\alpha}(\mathbf{r})$. Since both projectile particles are the same kind, the total incident flux is the same as before.

The new wrinkle is that to compute the transition rate into a group of final states, you have to add the amplitudes for both particles to scatter into the specified angle, since the two events are indistinguishable.

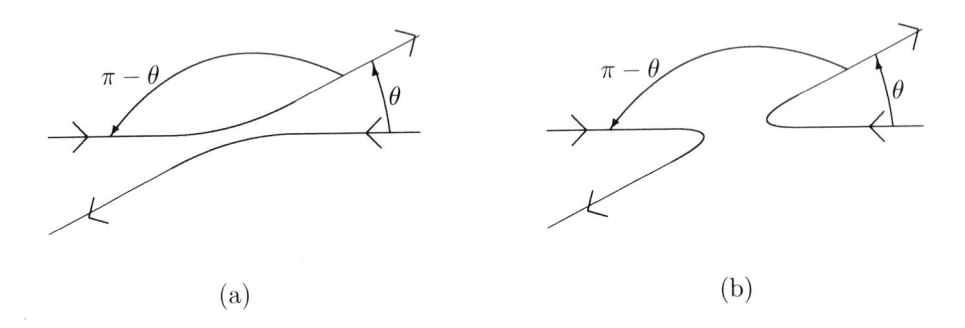

FIGURE 9.3: Schematic diagram of two identical particles scattering in the center of mass frame. Classically only (a) or (b) contributes to the differential cross section, but in quantum mechanics the amplitudes must be added.

The problem is especially simple when H' is independent of the spin. Then the total spin will be conserved, and you can choose states of definite spin, therefore definite particle exchange symmetry also.

The simplest example, the scattering of two electrons, is to take H' to be the Coulomb force, and restricting the calculation to the Born approximation. Then H'_{ba} has four terms. Each must be antisymmetric upon interchanging *all* the electron's properties, spin as well as $\mathbf{r}$. If the initial state has spin one, the states must be antisymmetric in $\mathbf{r}_1 \leftrightarrow \mathbf{r}_2$; for spin zero, symmetric.

In either case the first term is

$$H'_{ba,1} = \frac{1}{2}\frac{1}{(2\pi)^6} \int V(|\mathbf{r}_1 - \mathbf{r}_2|) e^{i(\mathbf{k}_1 - \mathbf{k}'_1)\cdot\mathbf{r}_1} e^{i(\mathbf{k}_2 - \mathbf{k}'_2)\cdot\mathbf{r}_2} d^3r_1 d^3r_2$$
$$= \frac{1}{2}\frac{1}{(2\pi)^3} \int V(r) e^{i(\mathbf{k}-\mathbf{k}')\cdot\mathbf{r}} \delta_3(\mathbf{K}-\mathbf{K}') d^3r$$
(9.102)

The second term is obtained by interchanging $\mathbf{k}'_1$ with $\mathbf{k}'_2$, or equivalently $\mathbf{K}' \to \mathbf{K}'$ and $\mathbf{k}' \to -\mathbf{k}'$:

$$H'_{ba,2} = \pm\frac{1}{2}\frac{1}{(2\pi)^6} \int V(|\mathbf{r}_1 - \mathbf{r}_2|) e^{i(\mathbf{k}_2 - \mathbf{k}'_1)\cdot\mathbf{r}_1} e^{i(\mathbf{k}_1 - \mathbf{k}'_2)\cdot\mathbf{r}_2} d^3r_1 d^3r_2$$
$$= \pm\frac{1}{2}\frac{1}{(2\pi)^3} \int V(r) e^{-i(\mathbf{k}+\mathbf{k}')\cdot\mathbf{r}} \delta_3(\mathbf{K}-\mathbf{K}') d^3r$$
(9.103)

The top sign is for singlet scattering, the bottom sign for triplet scattering. Both terms are proportional to the momentum-conserving delta function. The remaining two terms are the same as the two written, since $\mathbf{r}_1$ and $\mathbf{r}_2$ are just integration variables. In the center of mass frame the two terms differ only by changing the *direction* of the outgoing particle. The transition amplitude is the sum of a "direct" and an "exchange" term.

Equation (9.103) holds for any spin-independent potential. For Coulomb scattering the direct term is

$$T_{\text{direct}}(\omega_a, \theta, \phi) = \frac{1}{(2\pi)^3} \int V(r) e^{i(\mathbf{k}-\mathbf{k}')\cdot\mathbf{r}} \delta_3(\mathbf{K} - \mathbf{K}') d^3r$$
$$= \frac{e^2}{2\pi^2 q^2} = \frac{e^2}{4\pi^2 p^2 (1-\cos\theta)} = \frac{\alpha}{8\pi^2 p^2 \sin^2(\theta/2)} \quad (9.104)$$

while the exchange term is

$$T_{\text{exchange}}(\omega_a, \theta, \phi) = T_{\text{direct}}(\omega_a, \pi - \theta, \phi + 2\pi)$$
$$= \frac{e^2}{16\pi^2 p^2 \cos^2(\theta/2)} = \frac{\alpha}{8\pi^2 p^2 \cos^2(\theta/2)} \quad (9.105)$$

From equation (9.72) the differential cross section is

$$\frac{d\sigma}{d\Omega} = (2\pi)^4 \mu^2 \frac{\alpha^2}{64\pi^4 p^4} \left| \frac{1}{\sin^2(\theta/2)} \pm \frac{1}{\cos^2(\theta/2)} \right|^2 = \frac{\alpha^2 \mu^2}{4p^4} \left| \frac{1}{\sin^2(\theta/2)} \pm \frac{1}{\cos^2(\theta/2)} \right|^2$$
$$= \frac{\alpha^2 \mu^2}{4p^4} \left(\frac{1}{\cos^4(\theta/2)} + \frac{1}{\sin^4(\theta/2)} \pm 2 \frac{1}{\sin^2(\theta/2)} \frac{1}{\cos^2(\theta/2)} \right)$$
$$(9.106)$$

In a real electron-electron scattering experiment, both the target and the projectile are generally unpolarized. The unpolarized cross section is the average of the spin-dependent cross section over spins. In that case three-fourths of the pairs will be in the triplet state statistically, and only one-fourth in the singlet state. The unpolarized cross section is

$$\left(\frac{d\sigma}{d\Omega} \right)_{\text{unpolarized}} = \frac{\alpha^2 \mu^2}{4p^4} \left(\frac{1}{\sin^4(\theta/2)} + \frac{1}{\cos^4(\theta/2)} - \frac{1}{\sin^2(\theta/2)} \frac{1}{\cos^2(\theta/2)} \right) \quad (9.107)$$

9.4 DECAYS OF EXCITED STATES

In atomic decays H_o contains all the electric and magnetic interactions that account for the spectrum of the atom, and H' contains the interaction with the quantum electromagnetic field responsible for the decays. For β-decay, H_o is all the terms that account for the nuclear energy levels; H' is the weak interaction. Were it not for H', the excited states would be stable.

Here the initial state has a discrete energy rather than an energy in the continuous spectrum. In a scattering experiment there is an arbitrarily long time available to prepare the initial state. In the decay of an excited state, the eigenstates of H_o are not stationary states so you cannot prepare a system at time $t = 0$ exactly in

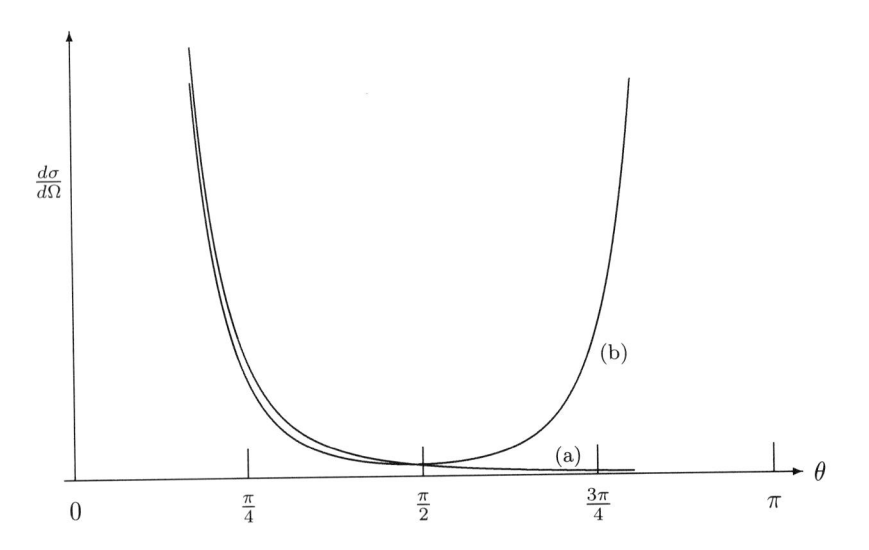

FIGURE 9.4: The electron-electron cross section, treating the particles as distinguishable (a) and as identical particles in an unpolarized beam (b).

an eigenstate of H_o. To prepare an excited state with an absolutely definite energy would require an infinite time, while the state lives for a finite time only.

Nevertheless, it is possible to follow the development of transition theory for scattering closely. The excited state lifetime has to be long compared to the time necessary to prepare it in order to talk sensibly about a decay rate. For an atom excited by absorbing radiation the scale of the times involved in the absorption transition is $1/\omega$, where ω is the energy difference of the transition, the frequency of the radiation. To the extent the lifetime is much longer than $1/\omega$, one can speak sensibly about the system being in the excited state at $t = 0$.

9.4.1 Lowest-Order Transition Rates

Assume that $|\psi(0)\rangle = |\phi_a\rangle$, where $|\phi_a\rangle$ is an eigenstate of H_o but not of H'. At a later time,

$$|\psi(t)\rangle = e^{-iHt}|\phi_a\rangle = \sum_b a_{ba}(t)|\phi_b\rangle \tag{9.108}$$

or

$$a_{ba}(t) = \langle \phi_b | e^{-iHt} | \phi_a \rangle \tag{9.109}$$

Equation (9.109) looks just like equation (9.24) for the scattering problem, except that t_o has been replaced by 0. As in Section 9.2, introduce the exponential factor to ensure convergence as $t \to \infty$. Again the solution with the correct boundary conditions is

$$a_{ba}(t) = \frac{i}{2\pi} \int_{-\infty}^{\infty} e^{-i\omega t} G(\omega)_{ba} d\omega \tag{9.110}$$

where
$$G(\omega) = \frac{1}{\omega - H + i\epsilon} \quad (9.30)$$
and an expansion for the decay amplitude can be obtained from
$$G(\omega) = G^o(\omega) + G^o(\omega)T(\omega)G^o(\omega) \quad (9.40)$$
where
$$T(\omega) = H' + H'G^o(\omega)T(\omega) \quad (9.42)$$
The Born approximation is $T(\omega) = H'$. To that order
$$G(\omega) = G^o(\omega) + G^o(\omega)H'G^o(\omega) \quad (9.111)$$

Repeating the steps in Section 9.3.2 leads to Fermi's golden rule again for decay rate into $|\phi_b\rangle$:
$$\Gamma_{ba} = 2\pi \int |H'_{ba}|^2 \delta(\omega_a - \omega_b) \quad (9.112)$$

In higher order, H' will get replaced by $T(\omega)$. In every order of the perturbation expansion, the transition rate is proportional to an energy-preserving delta function, even though that cannot be an property of the true Γ (because $|\phi_a\rangle$ is not an energy eigenstate).

Generic Example

Here is a template for using the rule. I will use the relativistic relation between momentum and energy since most useful decay problems involve relativistic kinematics. Decays involving electromagnetic radiation are always relativistic, since photons move at speed c. A spinless initial particle of mass M_o and momentum $\mathbf{k}_o$ decays into two spin-zero particles of masses m_1 and m_2.[6] With the δ-function normalization in momentum, the transition rate into some range of final momenta $\mathbf{k}_1$ and $\mathbf{k}_2$ is

$$\bar{\Gamma} = 2\pi \int |\langle \mathbf{k}_1, \mathbf{k}_2| H' |\mathbf{k}_o\rangle|^2 d^3k_1 d^3k_2 \delta(E_a - E_1 - E_2) \quad (9.113)$$

If the Hamiltonian is translationally invariant, the matrix element has the form

$$\langle \mathbf{k}_1, \mathbf{k}_2| H' |\mathbf{k}_o\rangle = f(\mathbf{k}_1, \mathbf{k}_o)\delta_3(\mathbf{k}_1 + \mathbf{k}_2 - \mathbf{k}_o) \quad (9.114)$$

Again the rate looks proportional to the square of a delta function, and again the infinity comes from the nonnormalized plane wave states, ones that describe beams of particles with density $(2\pi)^{-3}$. Instead of integrating over all space, integrate over a finite but arbitrarily large volume V. Then

$$\delta_3(\mathbf{k}_1 + \mathbf{k}_2 - \mathbf{k}_o)^2 \approx \delta_3(\mathbf{k}_1 + \mathbf{k}_2 - \mathbf{k}_o)\frac{V}{(2\pi)^3} \quad (9.115)$$

---

[6]If the particles have spin quantum numbers, transitions from each initial spin to each collection of final spins can be calculated in this way.

310 Chapter 9 Transitions

Equation (9.113) is the rate for decay into *all* the particles in the volume V. The number of particles in V is $V/(2\pi)^3$, so the transition rate per particle, which is independent of normalization conventions, is

$$\Gamma = 2\pi \int |f(\mathbf{k}_1, \mathbf{k}_o)|^2 \, \delta_3(\mathbf{k}_1 + \mathbf{k}_2 - \mathbf{k}_o) \delta(E_1 + E_2 - E_a) d^3k_1 d^3k_2 \qquad (9.116)$$

$$= 2\pi \int |f(\mathbf{k}_1, \mathbf{k}_o)|^2 \, k_1^2 dk_1 d\Omega_1 \delta(E_1 + E_2 - E_a)$$

In the last form, $\mathbf{k}_2$ is determined by the momentum-conservation delta function, and E_1 and E_2 are functions of k_1.

For example, take the initial state to be at rest, so that $E_a = M_o$. Then $\mathbf{k}_1 = -\mathbf{k}_2$, so $k_1 = k_2$, and

$$E_1 + E_2 = \sqrt{k_1^2 + m_1^2} + \sqrt{k_1^2 + m_2^2} \qquad (9.117)$$

Now

$$\delta(E_1 + E_2 - E_a) d(E_1 + E_2) = \delta(E_1 + E_2 - E_a) k_1$$

$$\times \left(\frac{1}{\sqrt{k_1^2 + m_1^2}} + \frac{1}{\sqrt{k_1^2 + m_2^2}} \right) dk_1 = \frac{E_a}{E_1 E_2} \delta(E_1 + E_2 - E_a) k_1 dk_1 \qquad (9.118)$$

So

$$\int \delta(E_1 + E_2 - E_a) k_1 dk_1 = \frac{E_1 E_2}{E_a} \qquad (9.119)$$

and k_1 is fixed by the δ-function. If $f(\mathbf{k}_1)$ is spherically symmetric, the $d\Omega$ integral is just 4π.

This is the right answer for the total decay rate into a group of final states, provided the allowed range of final states energies is much larger than $\hbar\Gamma$. But it is possible to calculate more carefully.

9.4.2 Time Dependence of the Initial State

The amplitude $a_{aa}(t) = \langle \phi_a | \psi(t) \rangle$ starts out at unity when $t = 0$. For short times, the total decay probability per unit time is Γ, so the time average of $(d/dt)|a_{aa}(t)|^2$ must be $-\Gamma$, so long as the amplitude is not significantly different from one. For longer times, you cannot really compute $a_{aa}(t)$ perturbatively.

Instead, develop an iteration scheme for $G(\omega)_{aa}$ as follows: The zeroth-order approximation is

$$G(\omega)_{aa} = G^o(\omega)_{aa} = \frac{1}{\omega - \omega_a + i\epsilon} \qquad (9.120)$$

The pole is at $\omega = \omega_a - i\epsilon$. Insert this value into equation (9.110) to get the time dependence of a free particle:

$$a_{aa}(t) = e^{-i\omega_a t} \qquad (9.121)$$

Section 9.4 Decays of Excited States

To go further, take the diagonal matrix element of equation (9.36), remembering that $G^o(\omega)$ is already diagonal:

$$G(\omega)_{aa} = G^o(\omega)_{aa} + \left[G^o(\omega)H'G(\omega)\right]_{aa} = G^o(\omega)_{aa} + G^o(\omega)_{aa}\sum_b H'_{ab}G(\omega)_{ba} \qquad (9.122)$$

Separate out the term with $b = a$:

$$G(\omega)_{aa} = G^o(\omega)_{aa} + G^o(\omega)_{aa}H'_{aa}G(\omega)_{aa} + G^o(\omega)_{aa}\sum_{b\neq a} H'_{ab}G(\omega)_{ba} \qquad (9.123)$$

For $b \neq a$, $G(\omega)_{ba}$ is of order H', since $G^o(\omega)_{ba}$ vanishes. So through first order

$$G(\omega)_{aa} = G^o(\omega)_{aa} + G^o(\omega)_{aa}H'_{aa}G(\omega)_{aa} + \cdots \qquad (9.124)$$

The solution is

$$G(\omega)_{aa} = \frac{G^o(\omega)_{aa}}{1 - G^o(\omega)_{aa}H'_{aa}} = \frac{1}{\omega - \omega_a + i\epsilon - H'_{aa} + \cdots} \qquad (9.125)$$

With this approximation to the Green's function the amplitude is

$$a_{aa}(t) = e^{-i(\omega_a + H'_{aa})t} \qquad (9.126)$$

The location of the pole is shifted by H'_{aa}, and the amplitude is that of a free particle with the shifted energy. Notice that H'_{aa} is exactly the first-order shift in the energy you would have calculated using bound-state perturbation theory!

The initial state still seems to persist forever. To do better, iterate equation (9.122) once more:

$$G(\omega)_{aa} = G^o(\omega)_{aa} + G^o(\omega)_{aa}H'_{aa}G(\omega)_{aa}$$
$$+ G^o(\omega)_{aa}\sum_{b\neq a} H'_{ab}G^o(\omega)_{bb}\sum_c H'_{bc}G(\omega)_{ca} + \cdots \qquad (9.127)$$

Again, $G(\omega)_{ca}$ is higher order in H' when $c \neq a$. Through second order write

$$G(\omega)_{aa} = G^o(\omega)_{aa} + G^o(\omega)_{aa}H'_{aa}G(\omega)_{aa}$$
$$+ G^o(\omega)_{aa}\sum_{b\neq a} H'_{ab}G^o(\omega)_{bb}H'_{ba}G(\omega)_{aa} + \cdots \qquad (9.128)$$

or

$$G(\omega)_{aa} = \left[\omega - \omega_a + i\epsilon - H'_{aa} - \sum_{b\neq a}|H'_{ba}|^2\frac{1}{\omega - \omega_b + i\epsilon} + \cdots\right]^{-1} \qquad (9.129)$$

The imaginary part of the denominator is in the last term. It can be computed from the identity [equation (A.12) in the appendix]

$$\lim_{\epsilon \to o} \text{Im}\,\frac{1}{x + i\epsilon} = -\lim_{\epsilon \to o}\frac{\epsilon}{x^2 + \epsilon^2} = -\pi\delta(x) \qquad (9.130)$$

Thus

$$G(\omega)_{aa} = \left[(\omega - \omega_a - H'_{aa} - \text{Re}\sum_{b \neq a}|H'_{ba}|^2 \frac{1}{\omega - \omega_b + i\epsilon}\right.$$
$$\left. + i\pi\sum_{b \neq a}|H'_{ba}|^2 \delta(\omega - \omega_a) + \cdots\right]^{-1} \quad (9.131)$$

As a function of complex ω, $G(\omega)_{aa}$ has a pole at a value of ω whose real part, up to second order in H', is

$$\omega' = \omega_a + H'_{aa} + \text{Re}\sum_{b \neq a}|H'_{ba}|^2 \frac{1}{\omega' - \omega_a} + \cdots \quad (9.132)$$

and whose imaginary part to the same order is

$$-\pi\sum_{b \neq a}|H'_{ba}|^2 \delta(\omega - \omega_a) = -\Gamma/2 \quad (9.133)$$

I have now set $\epsilon \to 0$ and assumed H'_{ba} to vary slowly with ω_b.

Both these quantities are familiar. The real part (9.132) is the energy of the initial state computed up to second order in bound-state perturbation theory. The imaginary part (9.133) is half the *total* decay rate as computed in Section 9.4.1. These identifications must continue to hold to any order in the perturbation expansion. Thus

$$G(\omega)_{aa} = \frac{1}{\omega - \omega'_a + i\Gamma/2} \quad (9.134)$$

with $\omega'_a = \text{Re}\,\omega'$ above. Evidently $T(\omega)$ has a singularity in the lower half plane, not at $\omega = \omega_a - i\epsilon$ but at a finite distance below the real axis. The interactions have moved the singularity.

With these approximations, the time dependence of the initial state is

$$a_{aa}(t) = \frac{i}{2\pi}\int_{-\infty}^{\infty} e^{-i\omega t} G(\omega)_{aa}\, d\omega = \frac{i}{2\pi}\int_{-\infty}^{\infty} \frac{e^{-i\omega t}}{\omega - \omega_a' + i\Gamma/2}\, d\omega \quad (9.135)$$

To evaluate the integral, close the contour in the lower half-plane (Figure 9.5). There is a simple pole at $\omega = \omega_a' - i\Gamma/2$. From the residue theorem

$$a_{aa}(t) = e^{-i(\omega_a' - i\Gamma/2)t} = e^{-i\omega_a' t} e^{-\Gamma t/2} \quad (9.136)$$

The first factor is the ordinary time dependence of a stationary state in the Schrödinger picture, with the correct perturbed energy. The probability for finding the system still in the state $|\phi\rangle_a$ at time t is therefore

$$\boxed{|a_{aa}(t)|^2 = e^{-\Gamma t}} \quad (9.137)$$

I suppose equation (9.137) had to be true because of the meaning of Γ, since

$$\frac{d}{dt}|a_{aa}(t)|^2 = -\Gamma |a_{aa}(t)|^2 \quad (9.138)$$

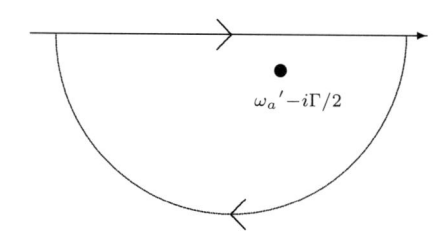

FIGURE 9.5: Contour in the complex ω plane for computing the integral in equation (9.135). The singularity at $\omega_a' - i\Gamma/2$ is indicated. The original integral is along the real axis. As long as $t > t_o$, you can close the contour along a semicircle in the lower half plane.

But a correct derivation is the result of some occasionally subtle approximations.

The Watched Pot

The exponential decay law looks like the statistical exponential decay of an ensemble of independent decaying particles. Precisely for that reason it cannot be exactly right. Suppose you prepare an atom in an excited state at $t = 0$ and measure its energy at time t_1 later. The probability that it is still in the excited state is $\exp(-\Gamma t_1)$. Suppose the result of this measurement is that the system is still in the excited state at t_1. The wave function, which before the measurement was in a superposition of the excited state and the final states, has "collapsed." After another time interval of length t_2, the probability of finding it in the excited state is $\exp(-\Gamma t_2)$ multiplied by the probability that it was in that state at time t_1. The total probability that the atom has not yet decayed is $\exp(-\Gamma t_1)\exp(-\Gamma t_2)$. But that is precisely what you would calculate for the probability that the atom had not decayed by the time $t_1 + t_2$ if you had not made the measurement at time t_1. Far from collapsing a wave function, the measurement at t_1 has no effect at all!

Suppose you wait a finite time t after the system is prepared in the excited state before making a measurement. The probability that it has not decayed is $\exp(-\Gamma t)$. Now divide the time t into N equal subintervals, and make measurements at $t/N, 2t/N, 3t/N, \ldots$. If N is large enough so that t/N is in the range where (9.140) holds, the probability that it is still in the upper state at time t is

$$\left[1 - \beta\left(\frac{t}{N}\right)^2\right]^N \approx 1 - \frac{\beta t^2}{N} \tag{9.139}$$

In the limit $N \to \infty$, the system is observed continuously and it doesn't decay at all. This apparent paradox is sometimes called the **quantum Zeno effect**, but it must be wrong.

Unless a system is in an eigenstate of the Hamiltonian, measuring the energy is supposed to have *some* observable consequence. Let us do the computation a little more carefully. Suppose you measure the energy at a very small time ϵ after the system has been set up.

314   Chapter 9   Transitions

According to equation (9.67) and the discussion preceding it, $P_{ba}(t)$ is linear in t. The probability to find the system still in the excited state at $t = \epsilon$ is $1 - \Gamma\epsilon$. If the result is positive, make another measurement at time 2ϵ. The probability that the system is still found in the upper state is $(1-\Gamma\epsilon)^2 = 1 - 2\Gamma\epsilon + \cdots$. As before, this is the same probability one would get by making just one measurement, at $t = 2\epsilon$, and the conclusion is still that the measurement has no effect on the outcome.

But equation (9.67) is not exactly right, and for very small times it is not right at all. According to the discussion in Section 9.3.2, the rule for the transition probability from state $|\phi_a\rangle$ to state $|\phi_b\rangle$ is

$$P_{ba}(t) = 2|T_{ba}(\omega_a)|^2 \frac{1 - \cos(\omega_b - \omega_a)t}{(\omega_b - \omega_a)^2} \tag{9.66}$$

so when t is very small, the correct formula for the probability is

$$P_{ba}(t) \to |T_{ba}(\omega_a)|^2 t^2 \tag{9.140}$$

and the probability that the system is still in the excited state at $t = \epsilon_1$ is $1 - \beta\epsilon_1^2$ (you can calculate β easily). Then after another interval of length ϵ_2, the probability that it is still in the excited state is

$$\left(1 - \beta\epsilon_1^2\right)\left(1 - \beta\epsilon_2^2\right) = 1 - \beta(\epsilon_1^2 + \epsilon_2^2) + \cdots \tag{9.141}$$

If the first measurement had not been made, the probability after an interval $t = \epsilon_1 + \epsilon_2$ that the atom had not decayed would have been

$$1 - \beta\left(\epsilon_1 + \epsilon_2\right)^2 \tag{9.142}$$

which is closer to unity than the same probability (9.141) when an intermediate measurement is made. Measuring the energy, and discovering the atom still in the upper state, has the expected effect: Later on it is more likely that the atom is still in the excited state.

9.4.3   Distribution of the Final States

The first approximation to H' is

$$\Gamma_b = 2\pi |H'_{ba}|^2 \delta(\omega_b - \omega_a) \tag{9.112}$$

Only the states with the same energy as ω_a occur. This cannot be exactly right, since the $|\phi_b\rangle$ are eigenstates of H_o, and H_o is not exactly conserved. Start again from

$$a_{ba}(t) = \frac{i}{2\pi} \int_{-\infty}^{\infty} e^{-i\omega t} G(\omega)_{ba} d\omega \tag{9.143}$$

and for $b \neq a$

$$G(\omega)_{ba} = G^o(\omega)_{bb} H'_{ba} G(\omega)_{aa} + \cdots \tag{9.144}$$

but now in the last term use

$$G(\omega)_{aa} = \frac{1}{\omega - \omega_a' + i\Gamma/2} \tag{9.134}$$

Section 9.4 Decays of Excited States

instead of the zeroth-order approximation $G(\omega)_{aa} = G^o(\omega)_{aa}$. Then

$$a_{ba}(t) = -\frac{1}{2\pi i}\int_{-\infty}^{\infty} e^{-i\omega t} G(\omega)_{ba}\, d\omega$$

$$= -\frac{1}{2\pi i} H'_{ba} \int_{-\infty}^{\infty} e^{-i\omega t} \frac{1}{\omega - \omega_b + i\epsilon} \frac{1}{\omega - \omega_a' + i\Gamma/2} \quad (9.145)$$

The earlier approximation to equation (9.145) was equivalent to replacing ω_a' by ω_a and $\Gamma/2$ by ϵ in the second factor.

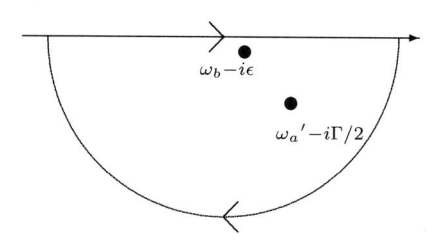

FIGURE 9.6: Contour in the complex ω plane for computing the integral in equation (9.145) as $t \to +\infty$. The singularity at $\omega_a' - i\Gamma/2$ is indicated.

Close the contour in the lower half plane, picking up the poles at $\omega_a - i\epsilon$ and at $\omega_a' - i\Gamma/2$:

$$a_{ba}(t) = H'_{ba} \times \left[e^{-i\omega_b t} e^{-\epsilon t} \frac{1}{\omega_b - \omega_a' + i(\Gamma/2 + \epsilon)} \right.$$

$$\left. - e^{-i\omega_a' t} e^{-\Gamma t/2} \frac{1}{\omega_b - \omega_a + i(\Gamma/2 + \epsilon)} \right] = H'_{ba} e^{-i\omega_b t} \frac{1 - e^{i(\omega_b - \omega_a')t} e^{-\Gamma t/2}}{\omega_b - \omega_a' + i\Gamma/2} \quad (9.146)$$

in the limit $\epsilon \to 0$. Then

$$|a_{ba}(t)|^2 = |H'_{ba}|^2 \frac{1 - 2e^{-\Gamma t/2}\cos(\omega_a' - \omega_b)t + e^{-\Gamma t}}{(\omega_a' - \omega_b)^2 + \Gamma^2/4} \quad (9.147)$$

As $t \to \infty$ the transition rate into $|\phi_b\rangle$ is

$$\frac{d}{dt}|a_{ba}(t)|^2 \to abs H'_{ba}{}^2 \frac{2(\omega_a' - \omega_b)\sin(\omega_a' - \omega_b)t}{(\omega_a' - \omega_b)^2 + \Gamma^2/4} \quad (9.148)$$

For small Γ this expression becomes Fermi's golden rule:

$$|H'_{ba}|^2 \frac{2(\omega_a' - \omega_b)\sin(\omega_a' - \omega_b)t}{(\omega_a' - \omega_b)^2} \to 2\pi |H'_{ba}|^2 \delta(\omega_a - \omega_b) \quad (9.149)$$

From equation (9.147), as $t \to \infty$ the probability that the system is in state $|\phi_b\rangle$ is

$$\boxed{P_{ba} = |H'_{ba}|^2 \frac{1}{(\omega_a' - \omega_b)^2 + \Gamma^2/4}} \quad (9.150)$$

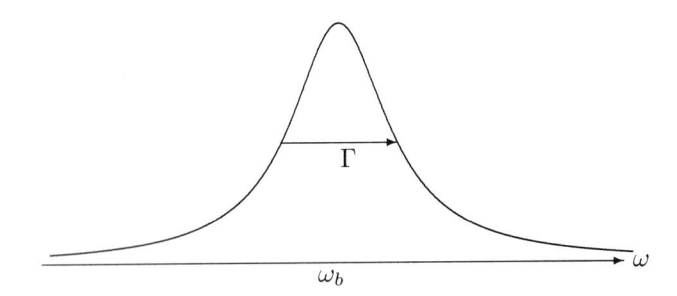

FIGURE 9.7: Typical resonance-curve shape for the energy distribution of final states. ω_b is the "energy" of the decaying state. Γ is the full width at half maximum.

and since

$$\int_0^\infty \frac{d\omega_b}{(\omega'_a - \omega_b)^2 + \Gamma^2/4} = \frac{2\pi}{\Gamma} \tag{9.151}$$

the total probability is unity, as required.

ω_a' is the initial state energy as computed in bound-state perturbation theory. Equation (9.150) is the shape of the final state energy distribution. It is a "resonance curve" like the one in Figure 9.7 as long as H'_{ba} is independent of b in the range where the denominator is significant.

The full width at half maximum is $\hbar\Gamma$. This is the "uncertainty" of the energy of $|\psi_a\rangle$. The state decays according to the exponential decay law; the mean time for decay is

$$\frac{\int t e^{-\Gamma t} dt}{\int e^{-\Gamma t} dt} = 1/\Gamma \tag{9.152}$$

If we let $\Delta t = 1/\Gamma$ and $\Delta E = \hbar\Gamma$, then $\Delta E \Delta t = \hbar$. This is the meaning of the energy-time uncertainty principle. It looks much like the uncertainty principle for $\Delta q \Delta p$, but has a quite different origin.

PROBLEMS

9.1. Time-Dependent Perturbation Theory

A hydrogen atom is in the ground state ($n = 1, l = m = 0$) for $t < 0$. Suppose the atom is placed between the plates of a capacitor, and a weak, spatially uniform but time-dependent decaying field is applied at $t = 0$. The field (for $t > 0$) is

$$\mathbf{E} = \mathbf{E}_o e^{-\gamma t}$$

for some $\gamma > 0$. Take $\mathbf{E}_o$ along the z-axis. What is the probability (to first order in E_o) that the atom will be in each of the four $n = 2$ states as $t \to \infty$? Neglect spin.

9.2. Inelastic Electron-Atom Scattering

Review the scattering of an electron by a hydrogen atom in its ground state (Problem 8.1) still neglecting spin effects and treating the two electrons as dis-

tinguishable. But now suppose the incident energy is high enough to excite the atom into an $n = 2$ state.
  (a) What is the inelastic differential cross section into the $2S$ state, in the first Born approximation?
  (b) What is the total cross section for this channel?

9.3. Electron Magnetic Resonance

Review the spin resonance problem discussed at the end of Section 6.2 and in Problem 6.7. Use the first-order time-dependent perturbation formalism of Section 9.1.1 to compute $|\langle \psi_+ | \psi(t) \rangle|^2$, and compare your answer with the exact result.

9.4. Neutron-Proton Scattering

A general form of the neutron-proton interaction, consistent with translation and rotation invariance, is

$$\begin{aligned} H' &= V_0(r) + \mathbf{s}_p \cdot \mathbf{s}_n\, V_1(r) \\ &+ (\mathbf{s}_p + \mathbf{s}_n) \cdot \mathbf{r}\, V_2(r) + (\mathbf{s}_p + \mathbf{s}_n) \cdot \mathbf{p}\, V_3(r) + (\mathbf{s}_p + \mathbf{s}_n) \cdot \mathbf{L}\, V_4(r) \\ &+ \mathbf{s}_p \cdot \mathbf{r}\, \mathbf{s}_n \cdot \mathbf{r}\, V_5(r) + \mathbf{s}_p \cdot \mathbf{p}\, \mathbf{s}_n \cdot \mathbf{p}\, V_6(r) + (\mathbf{s}_p \cdot \mathbf{r}\, \mathbf{s}_n \cdot \mathbf{p} + \mathbf{s}_n \cdot \mathbf{p}\, \mathbf{s}_p \cdot \mathbf{r})\, V_7(r) \\ &+ (\mathbf{s}_p - \mathbf{s}_n) \cdot \mathbf{r}\, V_8(r) + (\mathbf{s}_p - \mathbf{s}_n) \cdot \mathbf{p}\, V_9(r) + (\mathbf{s}_p - \mathbf{s}_n) \cdot \mathbf{L}\, V_{10}(r) \\ &+ (\mathbf{s}_p \times \mathbf{s}_n) \cdot \mathbf{r}\, V_{11}(r) + (\mathbf{s}_p \times \mathbf{s}_n) \cdot \mathbf{p}\, V_{12}(r) + (\mathbf{s}_p \times \mathbf{s}_n) \cdot \mathbf{L}\, V_{13}(r) \end{aligned}$$

Here $\mathbf{r} = \mathbf{r}_n - \mathbf{r}_p$ is the relative coordinate and $\mathbf{p}$ is the momentum of the neutron in the center of mass. The argument of the potential functions $V_i(r)$ is $r = |\mathbf{r}|$. The list above is not intended to be exhaustive, nor are all the terms necessarily independent.

(a) Assume the interaction is known to be invariant under reflections, but not necessarily under time inversion. Which of the above terms would be forbidden?

(b) Similarly, which of the above terms are forbidden if H' is known to be invariant under time inversion, but not necessarily under reflections?

9.5. $\pi \to e + \nu_e$ decay

Background: This problem and the subsequent ones are exercises in using Fermi's golden rule. They are also exercises in continuum normalization and relativistic kinematics. The subject is one that you may not be very familiar with: decays through the weak interactions.

Some of the matrix elements of the weak Hamiltonian H_W are known; others can be described completely with one or two parameters. Since this is not a treatise on weak interactions, I will simply tell you what H_W to use.

The familiar low-mass particles are stable against decays by the strong and (for the most part) the electromagnetic interactions, because of the conservation laws those interactions obey; they decay by the weak interactions because H_W violates some of those conservation laws. In particular the weak interactions are not reflection invariant: parity conservation does not hold. There is a discussion of particle lifetimes and selection rules in Section 5.1.1.

H_W is proportional to Fermi's constant G, which has the dimensions of inverse mass squared. Its numerical value is about $1.015 \times 10^{-5}\, m_p^{-2}$ (m_p is the mass of the proton.). G is measured in ordinary β-decays like

hadron $\to$ hadron + electron (or positron) + antineutrino (or neutrino)

The most famous example of β decay is neutron decay: $n \to p + e + \bar{\nu}$.

Chapter 9 Transitions

"Hadron" is shorthand for all particles except leptons, photons, and "gauge bosons."[7] The leptons are the negatively charged electron, the μ, and the τ, together with their positively charged antiparticles; and the three neutrinos ν_e, ν_μ, and ν_τ, together with their three antineutrinos. All leptons have spin half.

Neutrinos have very small masses on the scale of other particle masses. For these problems, you can assume that all the neutrino masses vanish, so that their energies and momenta are related by $E = p$. Then neutrinos come in only one spin state: The component of the spin along the direction of their momentum is always $-1/2$. That is, they are always left-handed ($\mathbf{s} \cdot \mathbf{p} = -1/2|p|$). The antineutrinos are right-handed. Any interactions that include neutrinos must violate the reflection and charge-conjugation symmetries.

Table 5.1 is an extract from a table of particle masses and lifetimes.

Consider the decay $\pi^+ \to e^+ + \nu$, starting from a π^+ at rest. Label the initial state by the momentum $\mathbf{p}$ of the pion. Let $\mathbf{q}$ and $\mathbf{k}$ be the momenta of the final state neutrino and positron, respectively. Since the component of the angular momentum of the neutrino along the direction of $\mathbf{q}$ is always $-1/2$, and since the π^+ is a spin-zero meson, angular momentum conservation requires that the component of the positron spin along the direction of $\mathbf{q}$ must be $+1/2$. The spin quantum numbers are fixed, and can be omitted from the state labels. Label the final states by $\mathbf{q}$ and $\mathbf{k}$. Normalize these as usual:

$$\langle \mathbf{p}' | \mathbf{p} \rangle = \delta_3 (\mathbf{p}' - \mathbf{p}) \qquad \langle \mathbf{k}' \mathbf{q}' | \mathbf{k} \mathbf{q} \rangle = \delta_3 (\mathbf{k}' - \mathbf{k}) \delta_3 (\mathbf{q}' - \mathbf{q})$$

Rotational invariance requires the matrix element in the π^+ rest frame to be independent of the direction, so it is a constant. The formula for this constant in terms of other natural constants can be found from the detailed theory. The matrix element is

$$\langle \mathbf{k} \mathbf{q} | H_W | \mathbf{p} \rangle = G f_\pi m_e \sqrt{\frac{1}{8\pi^3 E_e}} \delta_3 (\mathbf{q} + \mathbf{k} - \mathbf{p})$$

where E_e is the final positron energy. The constant f_π has the dimensions of mass, and can be related to the parity-violating part of neutron β-decay. It is measured to be about $93.3 \,\text{MeV}/c^2$.

Problem: Compute the transition rate for $\pi^+ \to e^+ + \nu$ in the π^+ rest frame.

Hint: The matrix element of H_W has a momentum delta function in it, as required by translation invariance. If you blindly square this function you will get infinity. Review the treatment of this issue in Section 9.4.1.

Hint: Do not spend too much time on the numerical computations, but try to do them. You will find it very efficient to use the units $\hbar = c = 1$. Think of the MeV as the basic energy unit. Then mass is also measured in MeV, and both distance and time in MeV^{-1}. The connection to cgs units is given by the factors in equations (2.131) and (2.132).

Note: The positron and the antineutrino enter the weak interactions in a symmetrical way; if the positron were exactly massless, H_W would project out only right-handed positrons (and left-handed electrons), and then the spin-zero π^+ could not decay into such a state at all, and still conserve angular momentum. That's why the matrix element $\langle \mathbf{k} \mathbf{q} | H_W | \mathbf{p} \rangle$ is proportional to the positron mass m_e.

---

[7] See Section 5.4.

9.6. $\pi^+ \to \mu^+ + \nu_\mu$ decay

Background: The μ^- and the μ^+ leptons are heavy versions of the electron and positron. All their interactions – except with gravity – are exactly the same as those of the electron and positron; the *only* observed difference is the mass. For example, the gyromagnetic ratio of the μ is close to two, with the same corrections higher order in α that the electron has.

Therefore, in addition to the decay mode $\pi^+ \to e^+ + \nu$ in Problem 9.5, the π^+ can also decay into a μ-neutrino (ν_μ) and a μ^+. The calculation is exactly the same as in the last problem, except that m_e should be replaced everywhere by m_μ.

Problem: What is the ratio (called the branching ratio) of the $\pi^+ \to e^+ + \nu_e$ rate to the $\pi^+ \to \mu^+ + \nu_\mu$ rate? What is the *total* lifetime of the π^+?

9.7. Λ Decay

The Λ is a neutral spin-half particle, like the neutron but heavier. It would be stable were it not for the weak interactions. The Λ mass and lifetime are given in Table 5.1. It decays 63.9% of the time into a proton and a π^- meson.[8]

(a) What is the decay rate $\Gamma_{p\pi^-}$, in inverse seconds, for this process $\Lambda \to p + \pi^-$?

(b) In the Λ rest frame, the energies of the decay products are fixed by energy and momentum conservation. Show that in that frame

$$E_p = \frac{M_\Lambda^2 + M_p^2 - M_\pi^2}{2M_\Lambda} \quad \text{and} \quad E_\pi = \frac{M_\Lambda^2 + M_\pi^2 - M_p^2}{2M_\Lambda}$$

What is the magnitude p of the momentum of either particle in that frame, in terms of the masses of the two decay products?

(c) The initial state contains a Λ with momentum $\mathbf{p}_\Lambda$ and spin index m_Λ. The final state contains a proton with momentum $\mathbf{p}$ and spin index m_p, and also a π^- with momentum $\mathbf{q}$. The possible values of m_p and m_Λ are $\pm\frac{1}{2}$. The form of the matrix element of H_W between the initial and final state is

$$\langle \mathbf{p} m_p; \mathbf{q} | H_W | \mathbf{p}_\Lambda m_\Lambda \rangle = G f_\Lambda \sqrt{M_\Lambda} \delta_3(\mathbf{p} + \mathbf{q} - \mathbf{p}_\Lambda) \chi_{m_p}^\dagger \mathcal{M} \chi_{m_\Lambda}$$

Here $\chi_{\pm\frac{1}{2}}$ are the Pauli spinors $\chi_{+\frac{1}{2}} = \begin{pmatrix} 1 \\ 0 \end{pmatrix}$ and $\chi_{-\frac{1}{2}} = \begin{pmatrix} 0 \\ 1 \end{pmatrix}$ and $\mathcal{M}$ is a 2×2 matrix. In the Λ rest frame, the most general form of $\mathcal{M}$ consistent with rotational invariance is

$$\mathcal{M} = S + \boldsymbol{\sigma} \cdot \hat{\mathbf{n}} P$$

where $\hat{\mathbf{n}}$ is the proton direction: $\mathbf{p} = p\hat{\mathbf{n}}$, and S and P are some complex numbers satisfying $|S|^2 + |P|^2 = 1$.

Compute the unpolarized decay rate in the Λ rest frame for this process in terms of G, f_Λ, S, P, and the masses.

Hint: You know that if $|\psi_n\rangle$ are an orthonormal basis, then

$$\sum_n |\psi_n\rangle\langle\psi_n| = I$$

---

[8] Most of the remaining decays are into a neutron and a π^0.

320 Chapter 9 Transitions

Similarly, in two dimensions, convince yourself that if χ_m are the two Pauli spinors, then

$$\sum_{m=-\frac{1}{2}}^{\frac{1}{2}} (\chi_m)_i (\chi_m^\dagger)_j = \delta_{ij}$$

and use this fact to show that the sum over both the Λ and proton spins is proportional to Tr $\mathcal{M}\mathcal{M}^\dagger$.

The total rate is the rate into both proton spin states, integrated over all final momenta. *Unpolarized* means averaged over the spin of the initial Λ. You could just pick a Λ spin state, since rotational invariance requires the total rate to be independent of the initial spin. But it is probably simpler (and conceptually more accurate) to average over the initial spins: sum over them and divide by 2.

(d) What is the numerical value of f_Λ in MeV?

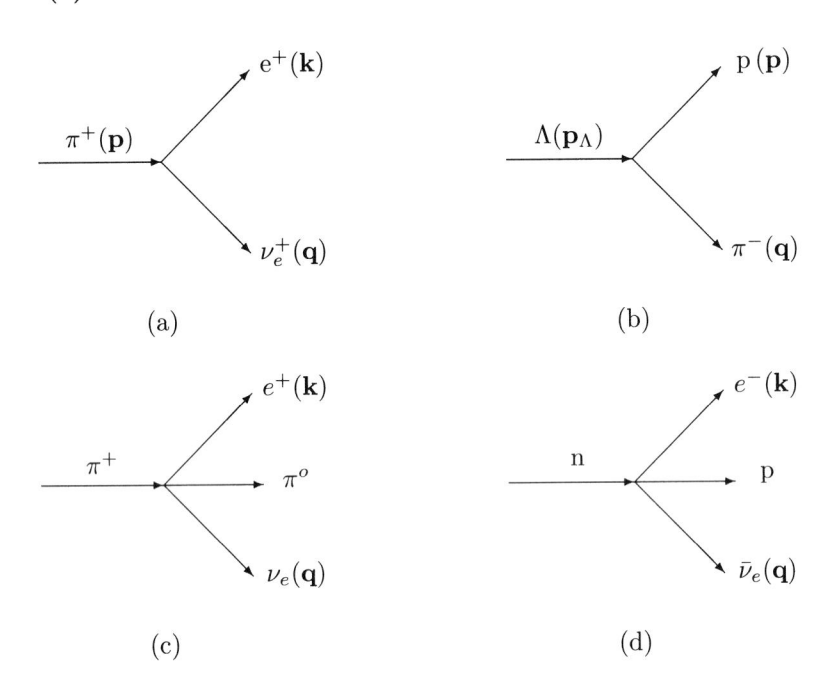

FIGURE 9.8: Schematic diagrams of the weak interaction decay problems: (a) Problem 9.5, (b) Problem 9.7, (c) Problem 9.8, and (d) Problem 9.9.

9.8. π^+ β-Decay

The Fermi constant G is determined unambiguously from β decays of spin-zero systems, for example

$$\pi^+ \to \pi^o + e^+ + \nu$$

The π^o is slightly lighter than the charged pions. The mass difference between the charged and neutral pions is $\Delta = 4.5936$ MeV, small compared to the pion mass itself. If the initial π^+ is at rest, it is a good approximation to let the final π^o to be at rest also, in the way we neglect the recoil of the atom in atomic decays. Think of the π^+ and the π^o as an excited state and the ground state of a very heavy system, like a nucleus. (Can you estimate the maximum error?)

Label the final states by the positron momentum $\mathbf{k}$ and the neutrino momentum $\mathbf{q}$. The angular momentum of the positron is fixed by the momentum of the neutrino. Since the final π^o is "nailed down," energy is conserved but not momentum. (The missing momentum is the π^o momentum.) Normalize the final states to delta functions in the lepton momenta:

$$\langle \mathbf{k}', \mathbf{q}' | \mathbf{k}, \mathbf{q} \rangle = \delta_3(\mathbf{q}' - \mathbf{q})\delta_3(\mathbf{k}' - \mathbf{k})$$

In this problem there is no need for any dimensionful parameters besides the Fermi constant. The matrix element of the weak Hamiltonian is

$$\langle \pi^o(\text{at rest}); \mathbf{k}; \mathbf{q} | H_W | \pi^+(\text{at rest}) \rangle = \frac{G}{\sqrt{32\pi^6}}$$

(a) In contrast to the situation in Problem 9.5, here there is one nontrivial integration left because now there is a three-particle final state. The rate is

$$\Gamma = \int_0^{k_{max}} f(k)dk$$

where k_{max} is the maximum possible positron momentum:

$$(k_{max})^2 = \Delta^2 - m_e^2$$

What is $f(k)$?

Hint: The transition rate is

$$\Gamma = 2\pi \int \delta(E_{e^+} + E_\nu - \Delta)$$
$$\times \left| \langle \pi^o(\text{at rest}); \mathbf{k}; \mathbf{q} | H_W | \pi^+(\text{at rest}) \rangle \right|^2 d^3k \, d^3q$$

E_{e^+} and E_ν are the positron and neutrino energies, respectively. Write

$$d^3q \, d^3k = k^2 \, d\Omega_k q^2 \, d\Omega_q \, dk \, dq$$

The angular integrations are trivial. Use the delta function to do the dq integration.

(b) The integral can be done numerically. If you neglect the positron mass so that $E = k$ for the positron also, it is easy to calculate the integral analytically. Make this approximation, in which case k_{max} is just Δ, and compute Γ.

(c) Can you estimate the error proportional to m_e?

(d) If you did the computation correctly, you found a rate for π^+ β-decay orders of magnitude smaller that the rate for the two-body decay $\pi^+ \to \mu^+ + \nu$ calculated in Problem 9.6. The disparity is entirely due to the values of the final-state momentum integrals.

What is the branching ratio

$$\frac{\Gamma(\pi^+ \to \pi^o + e^+ + \nu)}{\Gamma(\pi^+ \to \mu^+ + \nu)} \quad ?$$

9.9. Neutron β-Decay

Neutron decay into a proton, an antineutrino, and an electron is similar to the β-decay of a π^+. Here the nucleons are spin-half particles, so the transition rate has a spin-dependent piece. Since the neutron and proton are much heavier than the leptons, we can take them to be infinitely heavy (but with a finite energy difference). That is, think of the proton as the ground state of a very heavy system and the neutron as an excited state.

As an exercise, treat all these particles as spinless, so that the momenta are enough to describe the states. The initial state is $|n \text{ (at rest)}\rangle$, and the final state is $|p \text{ (at rest)}; \mathbf{k}; \mathbf{q}\rangle$. As in Problem 9.8 $\mathbf{k}$ and $\mathbf{q}$ are the final state positron and neutrino momenta.

(a) Examining the form of $d\Gamma/dE_e$ near the maximum electron energy is a sensitive way to test whether or not the neutrino is massless. Discuss the behavior of $d\Gamma/dE_e$ near the maximum electron energy E_{max} in the two cases (1) the neutrino is exactly massless and (2) the neutrino has a small mass (much smaller that m_e). You can take $\langle p \text{ (at rest)}; \mathbf{k}; \mathbf{q}| H_W |n \text{ (at rest)}\rangle$ to be a constant over the small energy range considered in this part. How does the power of $E_{max} - E_e$ in $d\Gamma/dE_e$ differ in the two cases?

(b) Now take the neutrino to be massless. The matrix element is[9]

$$\langle p \text{ (at rest)}; \mathbf{k}; \mathbf{q}| H_W |n \text{ (at rest)}\rangle$$
$$= \frac{G}{(2\pi)^3} \left[\frac{1}{E_e E_\nu} \left[(E_e E_\nu + \mathbf{q} \cdot \mathbf{k}) + g_A^2 (3 E_e E_\nu - \mathbf{q} \cdot \mathbf{k}) \right] \right]^{1/2}$$
(9.153)

Show that the quantity in brackets can be written in the form

$$\xi \left[1 + a \mathbf{v}_e \cdot \mathbf{v}_\nu \right]$$

What are ξ and a in terms of g_A?

g_A is a dimensionless constant that can be determined from fitting the correlations between the electron and the antineutrino. Its value is about -1.25.

(c) What is the decay rate $d\Gamma/dE_e$ into a range of electron energies dE_e?

(d) What is the total predicted decay rate of the neutron? Your answer should contain an integral of the same form as in the π^+ β-decay Problem 9.8. In the approximation $m_e \ll \Delta$ there is an analytic form for the decay rate. What is the answer in that approximation? This approximation is not very good here, since m_e/Δ is not so small as in the π^+ case. Can you estimate the correct value of Γ?

9.10. Parity Violation in Λ Decay

(a) Review the discussion of Λ decay in Problem 9.7. Suppose the initial Λ particle is known to have its spin along the z axis. Compute the total rate Γ_{up} for decays into $p + \pi^-$ such that the direction of the final proton is in the upper half plane (that is, $p_z > 0$) in terms of the constants that appear in Problem 9.7, including S and P. Then compute the total rate Γ_{down} in which the final proton direction is in the lower half plane. Check that $\Gamma_{up} + \Gamma_{down}$ is the total rate you computed previously.

---
[9] Taken from the theory, but averaged over neutron spins and summed over final-state spins.

The quantity
$$\alpha = \frac{\Gamma_{\text{up}} - \Gamma_{\text{down}}}{\Gamma_{\text{up}} + \Gamma_{\text{down}}}$$
is called the **asymmetry parameter**. It would be zero if parity were conserved in Λ decay (why?). Calculate the formula for α.

> Hint: It is still convenient to write the sum as a trace of a product of 2×2 matrices. But now you no longer want to average over the spin of the initial Λ. Show that in the present situation
> $$\left(\chi_{\frac{1}{2}}\right)_i \left(\chi_{\frac{1}{2}}^\dagger\right)_j = \frac{1}{2}(I + \sigma_z)_{ij}$$
> and use this result to write the rate as a trace.

(b) For an arbitrary initial Λ spin state, write a formula for the expectation value of (the components of) the proton spin as a function of its direction.

(c) Review the discussion of the density matrix for spin-half systems in Section 6.2.3. In a real experiment the Λ spin state is not known, but often the polarization of the beam can be determined. Write a formula for the expectation value of the proton spin as a function of its direction and of the Λ polarization.

Expand the expectation of the proton spin in an orthogonal basis as follows:
$$F(\hat{\mathbf{n}})\langle \boldsymbol{\sigma}_n \rangle = (\alpha + \mathbf{P} \cdot \hat{\mathbf{n}})\hat{\mathbf{n}} + \beta \mathbf{P} \times \hat{\mathbf{n}} + \gamma \hat{\mathbf{n}} \times (\mathbf{P} \times \hat{\mathbf{n}})$$
where α is the asymmetry parameter in part (a) and β and γ are two other complex constants, functions of S and P. $F(\hat{\mathbf{n}})$ is some function of the proton direction, independent of the Λ polarization. The vector $\mathbf{P}$ is the Λ polarization vector for the mixture, as defined in equation (6.68). Show that
$$\alpha + \beta + \gamma = 1$$

(d) What is the experimental value of the asymmetry parameter α?

9.11. Unitarity of the S Matrix and the Optical Theorem

In Section 9.3.3 the optical theorem was demonstrated by proving that
$$\text{Im } T_{aa} = -\pi \sum_c |\langle \phi_c | T(\omega_a) | \phi_a \rangle|^2 \delta(\omega_c - \omega_a)$$

For spinless particle scattering show that this rule is a consequence of $SS^\dagger = I$, where S is the operator defined in equation (9.62).

REFERENCES

For the Lippman-Schwinger equation see

[1] B. LIPPMANN AND J. SCHWINGER, Phys. Rev. **79** (1950), 469.

The golden rule for transitions first appeared in

[2] E. FERMI, *Nuclear physics*, Univ. of Chicago Press, 1950, page 142.

The relativistic relation between the scattering amplitude and the transition matrix is done carefully on page 99 of reference [1] in Chapter 8.

CHAPTER 10

Further Topics in Quantum Dynamics

10.1 PATH INTEGRATION

There is another way to do quantum mechanics, a way that emphasizes the Lagrangian and the superposition principle rather than the Hamiltonian and the Schrödinger equation. The **path integral** is a generalization of Huygens' principle to de Broglie waves, a generalization of the way we analyzed the double-slit experiment described in Section 2.1. The idea was perfected by Feynman in 1948 following up on earlier hints by Dirac [1, 2, 3].

A transformation is a symmetry if it leaves the action invariant, since the equations of motion are invariant also. The path integral ensures all the symmetries of the theory in a transparent way. (In contrast, the Hamiltonian is not invariant under Lorentz transformations—it is a component of a four-vector.) One of the original glories of the path integral was that it did indeed make relativistic quantum electrodynamics manifestly covariant.

10.1.1 The Propagator as an Integral over Paths

Suppose a one-dimensional particle is in an eigenstate of position $|q_a\rangle$ at time t_a. The amplitude for it to be at q_b at a later time t_b is the **propagator**

$$K(q_b, t_b; q_a, t_a) = \langle q_b | U(t_b, t_a) | q_a \rangle \tag{10.1}$$

where $U(t_b, t_a)$ is the time evolution operator. We already know that to avoid trouble we should make the propagator a **causal** propagator, defining it to be zero when $t_b < t_a$, and for large positive times we should put in a regulating factor $\exp(-\eta)$ and take $\eta \to 0$ at the very end of any calculation. So instead of the definition (10.1) the propagator is defined by

$$K(q_b, t_b; q_a, t_a) = \Theta(t_b - t_a)\langle q_b | e^{-i(H(p,q) - i\eta)(t_b - t_a)} | q_a \rangle \tag{10.2}$$

Assume the Hamiltonian has no explicit time-dependence. (I will assume $t_b > t_a$ so the step function does not have to be written explicitly, and suppress the $i\eta$ until it is needed.) Subdivide the time interval into $N+1$ equal segments, with

$$t_o = t_a \qquad t_{N+1} = t + (N+1)\epsilon = t_b \qquad t_n = t + n\epsilon \tag{10.3}$$

and compute the propagator by inserting a complete set of states at each t_n. That is, use

$$U(t_b, t_a) = e^{-iH(t_b - t_N)} e^{-iH(t_N - t_{N-1})} \cdots e^{-iH(t_1 - t_a)} \tag{10.4}$$

and the identity operator in the form

$$I = \int dq_n |q_n\rangle\langle q_n| \tag{10.5}$$

over and over again to write (all the integrals here go from $-\infty$ to ∞)

$$\begin{aligned}K(q_b, t_b; q_a, t_a) &= \int dq_n\, dq_{n-1} \cdots dq_1 \\ &\times \left\langle q_b \left| e^{-iH(t_b-t_N)} \right| q_N \right\rangle \left\langle q_N \left| e^{-iH(t_N-t_{N-1})} \right| q_{N-1} \right\rangle \cdots \\ &\times \left\langle q_1 \left| e^{-iH(t_1-t_a)} \right| q_a \right\rangle \end{aligned} \tag{10.6}$$

For small enough ϵ, the factors are

$$\begin{aligned}\left\langle q_n \left| e^{-iH(t_n-t_{n-1})} \right| q_{n-1} \right\rangle &\approx \langle q_n | 1 - i\epsilon H | q_{n-1}\rangle + \cdots \\ &= \delta(q_n - q_{n-1}) - i\epsilon \langle q_n | H | q_{n-1}\rangle + \cdots \end{aligned} \tag{10.7}$$

The first term is

$$\delta(q_n - q_{n-1}) = \frac{1}{2\pi} \int e^{ip(q_n - q_{n-1})} dp \tag{10.8}$$

while

$$\langle q_n | p^2 | q_{n-1}\rangle = \frac{1}{2\pi} \int e^{ip(q_n - q_{n-1})} p^2 dp \tag{10.9}$$

and

$$\langle q_n | V(q) | q_{n-1}\rangle = V(q_n)\,\delta(q_n - q_{n-1}) = \frac{1}{2\pi} \int e^{ip(q_n - q_{n-1})} V(q_n)\, dp \tag{10.10}$$

So if the Hamiltonian has the form

$$H = \frac{p^2}{2m} + V(q) \tag{10.11}$$

then through terms that are first order in ϵ

$$\begin{aligned}\left\langle q_n \left| e^{-iH(t_n-t_{n-1})} \right| q_{n-1} \right\rangle &= \frac{1}{2\pi} \int e^{ip(q_n - q_{n-1})} \left[1 - i\epsilon H(p, q_n) dp\right] \\ &\approx \frac{1}{2\pi} \int e^{ip(q_n - q_{n+1})} e^{-i\epsilon H(p, q_n)} dp = \frac{1}{2\pi} \int \exp\left(i\epsilon \left[p\dot{q}_n - H(p, q_n)\right] dp\right) \end{aligned} \tag{10.12}$$

where I have suggestively set $\dot{q}_n = (q_n - q_{n-1})/\epsilon$. The propagator itself is

$$K(q_b, t_b; q_a, t_a) = \int \prod_{n=1}^{N} dq_n \prod_{n=1}^{N+1} \frac{dp_n}{2\pi} \exp\left(i\epsilon \sum_{n=1}^{N+1} (p_n \dot{q}_n - H(p_n, q_n))\right) \tag{10.13}$$

In the limit $N \to \infty$ and $\epsilon \to 0$ in such a way that $\epsilon N \to t_b - t_a$, equation (10.13) is replaced by

$$K(q_b, t_b; q_a, t_a) = \lim_{N\to\infty} \int \prod_{n=1}^{N} dq_n \prod_{n=1}^{N+1} \frac{dp_n}{2\pi} \exp\left(i\epsilon \sum_{n=1}^{N+1} (p_n \dot{q}_n - H(p_n, q_n))\right) \tag{10.14}$$

326 Chapter 10 Further Topics in Quantum Dynamics

or

$$K(q_b, t_b; q_a, t_a) = \int \mathcal{D}q \mathcal{D}p \exp\left[i \int_{t_a}^{t_b} (p\dot{q} - H(p,q))\, dt\right] \tag{10.15}$$

Inside the integral you have to remember the factor $\Theta(t_b - t_a) \exp[-\eta(t_b - t_a)]$ when needed.

This is the Feynman path integral for the propagator. It is a **functional integral** in the sense that it takes functions $p(t)$ and $q(t)$ and turns them into an ordinary number. The integration measure $\mathcal{D}q\mathcal{D}p$ is *defined as* the indicated limit of the finite form (10.13).

In the integral (10.15) the p's and q's are ordinary numbers, no longer operators, and $H(p,q)$ is an ordinary function also. Once you have this expression, you no longer have to be careful with the order of the symbols!

The derivation of the path integral (10.15) goes through unchanged as long as H is a sum of terms, each containing powers only of p or only of q. If there are terms that contain powers of both p and q you can always move all the p's to the left of all the q's. The commutators generate with new terms with fewer factors. We will assume that the operator ordering ambiguity can be resolved correctly in this way—it is not a problem new to path integrals (see the discussion at the end of Section 3.2.3).

When the Hamiltonian has the simple form (10.11), the path integral (10.15) is

$$K(q_b, t_b; q_a, t_a) = \lim_{N \to \infty} \int \prod_{n=1}^{N} dq_n \prod_{n=1}^{N+1} \frac{dp_n}{2\pi} \exp\left[i\epsilon \sum_{n=1}^{N+1} \left(p_n \dot{q}_n - \frac{p_n^2}{2m} - V(q_n)\right)\right] \tag{10.16}$$

The oscillatory integrals converge because of the $e^{-\eta}$ factor from the definition (10.2). A better way to evaluate them is continuing ϵ to a positive imaginary value, completing the square, and using the rule (A.22) for integrating a Gaussian function:

$$\int dp_n \exp\left(i\epsilon p_n \dot{q}_n - i\frac{p_n^2}{2m}\right) = \sqrt{\frac{2m\pi}{i\epsilon}} \exp\left(\frac{i\epsilon m \dot{q}^2}{2}\right) \tag{10.17}$$

so

$$K(q_b, t_b; q_a, t_a) = \lim_{N \to \infty} \left(\frac{m}{2\pi i\epsilon}\right)^{(N+1)/2} \int \prod_{n=1}^{N} dq_n \exp\left[i\epsilon \sum_{n=1}^{N+1} \left(\frac{m\dot{q}_n^2}{2} - V(q_n)\right)\right]$$

$$= \int \mathcal{D}q \exp\left[i \int_{t_a}^{t_b} \left(\frac{m\dot{q}_n^2}{2} - V(q_n)\right) dt\right] \tag{10.18}$$

or

$$K(q_b, t_b; q_a, t_a) = \int \mathcal{D}q \exp\left[i \int_{t_a}^{t_b} L(q, \dot{q}) dt\right] = \int \mathcal{D}q e^{iS[q]} \tag{10.19}$$

Again, the integration measure $\mathcal{D}q$ is defined as the indicated limit. The propagator is given by an integral over all possible paths, weighted by a quantity whose phase is the value of the action along each path.

Even when H is not just $T + V$, the form (10.19) is usually correct. In fact it was once thought that you could always postulate the path integral (10.19) as a substitute for Schrödinger's equation. Feynman even wrote a book with Hibbs [5] deriving all nonrelativistic quantum mechanics from path integrals. But we know now that there are rare yet important cases where the naive version is not quite right. That is why I have derived the path integral carefully from the Hamiltonian and Schrödinger's equation [1, 2, 3].

Note: When the coefficient of p^2 in the Hamiltonian is not a constant, the integration over momentum is more complicated, and the propagator can turn out to be

$$\int \mathcal{D}q \, \exp\left(i \int L_{\text{eff}} \, dt \right) \tag{10.20}$$

with an effective Lagrangian L_{eff} containing some terms not found in the canonical Lagrangian. This is the case when $H = p^2 F(q)$ and it is an important feature of the standard relativistic quantum field theories of elementary particle interactions.

Still it is often useful to take the path integral as the fundamental starting point for quantum mechanics, since an invariant action is what is required to ensure symmetries, including internal symmetries and gauge invariance. The trade-off is that the Hamiltonian may not be what you would have guessed from the canonical rules.[1] The technical point is that conservation of probability—that is, the unitarity of the scattering matrix—is obvious in canonical Hamiltonian formalism but not from the path integral. Path integrals in relativistic quantum mechanics are discussed at length in books on quantum field theory [6, 7, 8].

10.1.2 The Free Particle Propagator

The simplest case is $V(q) = 0$. The action is

$$S[q] = \int_{t_a}^{t_b} \frac{m\dot{q}^2}{2} \, dt \tag{10.21}$$

where $q(t)$ is any path that goes from q_a at t_a to q_b at t_b. Expand $q(t)$ around some particular path $q_o(t)$. For a free particle the Lagrangian depends only on $\dot{q}(t)$, so

$$S[q] = S[q_o] + \int_{t_a}^{t_b} dt \frac{\partial L(\dot{q}_o)}{\partial \dot{q}} \delta\dot{q}(t) + \frac{1}{2} \int_{t_a}^{t_b} dt \frac{\partial^2 L(\dot{q}_o)}{\partial \dot{q}^2} \delta\dot{q}(t)^2 \tag{10.22}$$

There are no higher order terms in the expansion. Now choose $q_o(t)$ to be the classical path itself. The linear term vanishes by Hamilton's principle:

$$S[q] = S[q_o] + \frac{1}{2} \int_{t_a}^{t_b} \frac{\partial^2 L(\dot{q}_o)}{\partial \dot{q}^2} \delta\dot{q}(t)^2 \tag{10.23}$$

---

[1]That is, $H \neq \sum_i p_i \dot{q}_i - L_{\text{eff}}$.

328 Chapter 10 Further Topics in Quantum Dynamics

The classical path is the straight line

$$q_o(t) = \frac{q_b(t-t_a) + q_a(t-t_b)}{t_b - t_a} \tag{10.24}$$

so L is a constant and

$$S[q_o] = \frac{m}{2} \frac{(q_b - q_a)^2}{t_b - t_a} \tag{10.25}$$

So when $t_b > t_a$, the propagator is

$$K(q_b, t_b; q_a, t_a) = \exp(iS[q_o]) \int \mathcal{D}\delta q \, \exp\left[\frac{i}{2} \int_{t_a}^{t_b} \frac{\partial^2 L(\dot{q}_o)}{\partial \dot{q}^2} \delta\dot{q}(t)\right]^2 \tag{10.26}$$

Here $\delta q(t)$ is any path that starts from 0 at t_a and returns to 0 at t_b. Since the classical action vanishes for such paths, the second factor is the propagator for paths with $q_a = q_b = 0$:

$$K(q_b, t_b; q_a, t_a) = K(0, t_b; 0, t_a) \exp(iS[q_o]) \tag{10.27}$$

There are many clever ways to evaluate the first factor. For the free particle, the multiple integral in the definition (10.19) for the propagator can be done easily in closed form (see Problem 10.1).

$$\begin{aligned}
I_N &= \int \prod_{n=1}^{N} dq_n \exp\left(i\epsilon \sum_{n=1}^{N+1} \frac{m\dot{q}^2}{2}\right) \\
&= \int \exp\left[i\frac{m\epsilon}{2}\left(\frac{(q_b - q_N)^2}{\epsilon^2} + \cdots + \frac{(q_1 - q_a)^2}{\epsilon^2}\right)\right] dq_N \, dq_{N-1} \ldots dq_1 \\
&= \frac{1}{\sqrt{N+1}} \left(\frac{2\pi i \epsilon}{m}\right)^{N/2} \exp\left(\frac{im}{2\epsilon(N+1)}(q_b - q_a)^2\right)
\end{aligned} \tag{10.28}$$

so that

$$\begin{aligned}
K(q_b, t_b; q_a, t_a) &= \lim_{N \to \infty} I_N \left(\frac{m}{2\pi i \epsilon}\right)^{(N+1)/2} \\
&= \sqrt{\frac{m}{2\pi i (t_b - t_a)}} \exp\left(\frac{im}{2}\frac{(q_b - q_a)^2}{(t_b - t_a)}\right)
\end{aligned} \tag{10.29}$$

10.1.3 The Harmonic Oscillator

I will work out one more example, the one-dimensional oscillator. The Lagrangian is

$$L = \frac{m}{2}\dot{q}^2 - \frac{m\omega^2}{2}q^2 \tag{10.30}$$

Once again, expand the action about the classical path. Take $t_a = 0$ (since the propagator depends on the times only through the difference $t_b - t_a$). Then

$$\begin{aligned}
S[q] &= S[q_o] + \frac{1}{2}\int_0^{t_b} dt \left(\frac{\partial^2 L(\dot{q}_o)}{\partial \dot{q}^2}[\delta\dot{q}(t)]^2 + \frac{\partial^2 L(tq_o)}{\partial tq^2}[\delta q(t)]^2\right) \\
&= S[q_o] + \frac{m}{2}\int_0^{t_b} dt \left([\delta\dot{q}(t)]^2 - \omega^2 [\delta q(t)]^2\right)
\end{aligned} \tag{10.31}$$

The second term is independent of the end points, so

$$K(q_b, t_b; q_a, 0) = K(0, t_b; 0, 0) \exp(iS[q_o]) \tag{10.32}$$

The classical action comes out to be

$$S[q_o] = \frac{m\omega}{2 \sin \omega t_b} \left[(q_a^2 + q_b^2) \cos \omega t_b - 2 q_a q_b \right] \tag{10.33}$$

and the propagator is

$$K(q_b, t_b; q_a, 0) = K(0, t_b; 0, 0) \exp\left(\frac{im\omega}{2 \sin \omega t_b} \left[(q_a^2 + q_b^2) \cos \omega t_b - 2 q_a q_b \right] \right) \tag{10.34}$$

The coefficient in front depends on t_b but not on the endpoints.

Derivation of (10.33) for the harmonic oscillator. The classical motion is

$$q(t) = Q \sin(\omega t + \phi) \tag{10.35}$$

for some constants Q and ϕ. The boundary conditions are

$$q_b = Q \sin(\omega t_b + \phi) \quad \text{and} \quad q_a = Q \sin \phi \tag{10.36}$$

Then

$$Q \cos \phi = \frac{q_b - q_a \cos \omega t_b}{\sin \omega t_b} \tag{10.37}$$

The classical action is

$$\begin{aligned} S[q_o] &= \frac{m}{2} \int_0^{t_b} \left(\dot{q}(t)^2 - \omega^2 q(t)^2 \right) dt = \frac{m \omega^2 Q^2}{2} \int_0^{t_b} \cos(2\omega t + 2\phi) \, dt \\ &= \frac{m \omega^2 Q^2}{4} \left(\sin(2\omega t_b + 2\phi) - \sin 2\phi \right) \end{aligned} \tag{10.38}$$

Then from the boundary conditions (10.36) and the rule (10.37) for $Q \cos \phi$, a short calculation shows that

$$Q^2 \sin(2\omega t_b + 2\phi) = \frac{2 q_b}{\sin \omega t_b} (q_b \cos \omega t_b - q_a) \tag{10.39a}$$

and

$$Q^2 \sin 2\phi = \frac{2 q_a}{\sin \omega t_b} (q_b - q_a \cos \phi) \tag{10.39b}$$

∎

Examine $K(0, t_b; 0, 0)$ a little more carefully. Let $y(t)$ be any path such that $y_a = y_b = 0$. Then

$$K(0, t_b; 0, 0) = \int \mathcal{D} y \exp \left[i \int_{t_a}^{t_b} \left(\frac{m \dot{y}^2}{2} - \frac{m \omega^2 y^2}{2} \right) dt \right] = \lim_{N \to \infty} \left(\frac{m}{2 \pi i \epsilon} \right)^{(N+1)/2} I_N \tag{10.40}$$

330 Chapter 10 Further Topics in Quantum Dynamics

The form for finite N is no longer needed. In the limit[2]

$$I_N \to \int \prod_{n=1}^{\infty} dy_n \exp\left[i\epsilon \frac{m}{2} \sum_{n=1}^{\infty} (\dot{y}_n^2 - \omega^2 y_n^2)\right]$$

$$= -\int \prod_{n=1}^{\infty} dy_n \exp\left[i\epsilon \frac{m}{2} \sum_{n=1}^{\infty} (y_n \ddot{y}_n + \omega^2 y_n^2)\right] \quad (10.41)$$

$$= \int \prod_{n=1}^{\infty} \prod_{k=1}^{\infty} dy_n\, dy_k \exp\left[i\epsilon \frac{m}{2} \sum_{n=1}^{\infty} \sum_{k=1}^{\infty} (y_k M_{km} y_n)\right]$$

where

$$M = -\left(\frac{d^2}{dt^2} + \omega^2\right) \quad (10.42)$$

The point of this peculiar notation is that since M contains derivatives, it mixes up the different functions $y_m(t)$. To define it more carefully, the derivatives ought to be replaced by finite differences. But the answer is needed only in the $N \to \infty$ limit.

M is a real symmetric matrix, so it can be diagonalized with an orthogonal (real and unitary) matrix. The integral is given by equation (A.25) in the appendix:

$$\int \cdots \int \prod_n dy_n \exp\left(-\sum_i \lambda_i y_i^2\right) = \prod_n \left(\frac{\pi}{\lambda_n}\right)^{1/2} = \frac{\pi^{n/2}}{\sqrt{\det A}} \quad (A.25)$$

In the limit $N \to \infty$

$$C(t_b) \equiv K(0, t_b; 0, 0) = K_o \sqrt{\frac{1}{\det M}} \quad (10.43)$$

The constant K_o is independent of ω.

What is $\det M$? The eigenfunctions are solutions to

$$\left(\frac{d^2}{dt^2} + \omega^2\right) z_n(t) = -\lambda_n z_n(t) \quad (10.44)$$

or

$$z_n(t) = \alpha \sin \omega_n t + \beta \cos \omega_n t \quad (10.45)$$

with $\lambda_n = \omega_n^2 - \omega^2$. The boundary conditions $z_n(0) = z_n(t_b) = 0$ determine ω_n:

$$\omega_n t_b = n\pi \quad \text{or} \quad \lambda_n = \left(\frac{n\pi}{t_b}\right)^2 - \omega^2 \quad (10.46)$$

The determinant is

$$\det M = \prod_{n=1}^{\infty} \left[\left(\frac{n\pi}{t_b}\right)^2 - \omega^2\right] = K_1 \prod_{n=1}^{\infty} \left[1 - \left(\frac{\omega t_b}{n\pi}\right)^2\right] = K_1 \frac{\sin \omega t_b}{\omega t_b} \quad (10.47)$$

---

[2] I use y_n instead of q_n for the dummy variables here as a reminder of the different boundary conditions.

The last line follows from the infinite product representation (A.66) for the $\sin \pi z$. The constant K_1 is independent of ω. If you work it out correctly and multiply by K_o for finite N, the result will remain finite as $N \to \infty$. Therefore,

$$C(t_b) = K(0, t_b; 0, 0) = K_2 \sqrt{\frac{\omega t_b}{\sin \omega t_b}} \tag{10.48}$$

K_2 is another constant independent of ω. It can be determined by requiring that $K(0, t_b; 0, 0)$ agree with the free particle result (10.29) in the limit $\omega \to 0$. The final result is

$$K(q_b, t_b; q_a, 0) = \sqrt{\frac{m\omega}{2\pi i \sin \omega t_b}} \exp\left(\frac{im\omega}{2\sin \omega t_b}\left[(q_a^2 + q_b^2)\cos \omega t_b - 2q_a q_b\right]\right) \tag{10.49}$$

10.1.4 The Euclidean Formalism

We have been integrating integrals like

$$\int \mathcal{D}q \, \exp\left[i \int_{t_a}^{t_b} L(q, \dot{q}) dt\right] \tag{10.50}$$

in equation (10.19) using the rule

$$\int_{-\infty}^{\infty} e^{-ax^2} dx = \sqrt{\frac{\pi}{a}} \tag{10.51}$$

as if they were convergent. The convergence can be enforced by some regulator, but here a better alternative is to rotate the time-integration contour by $\pi/2$ so that it runs along the imaginary axis. That is, set

$$t = -iw \tag{10.52}$$

Then an integral like (10.50) becomes

$$\int \mathcal{D}q \, \exp\left[\int_{it_a}^{it_b} L\left(q, i\frac{dq}{dw}\right) dw\right] = \int \mathcal{D}q \, \exp\left[-\int_{it_a}^{it_b} L_E\left(q, \frac{dq}{dw}\right) dw\right] \tag{10.53}$$

where

$$L_E = -L\left(q, i\frac{dq}{dw}\right) = \frac{1}{2}\left(\frac{dq}{dw}\right)^2 + V(q) \tag{10.54}$$

is the **Euclidean Lagrangian**. The answer comes out in terms of an imaginary time interval

$$(t_b - t_a) = -iT \tag{10.55}$$

Equivalently, for $T > 0$ define the Euclidean propagator

$$K_E(q_b, T; q_a, 0) = \langle q_b | e^{-HT} | q_a \rangle \tag{10.56}$$

332 Chapter 10 Further Topics in Quantum Dynamics

For a particle in a (velocity-independent) potential,

$$K_E(q_b, T; q_a, 0) = \int \mathcal{D}q \exp\left[-\int_0^T L_E(w)dw\right] \tag{10.57}$$

where

$$L_E(w) = -L(iw) = \frac{m}{2}\left(\frac{dq}{dw}\right)^2 + V(q) \tag{10.58}$$

For example, the Euclidean free particle propagator (10.29) is

$$K_E(q_b, T; q_a, 0) = \sqrt{\frac{m}{2\pi T}} \exp\left(-\frac{m}{2}\frac{(q_b - q_a)^2}{T}\right) \tag{10.59}$$

The real time propagator can be found by analytic continuation back to the original variables.[3]

What is the Euclidean propagator good for? Real integrals like the one in (10.53) have no convergence problem, and since the integrand has a sharp peak it is often possible to get a good approximation by expanding the integrand about its maximum. Most of the useful applications are for applications beyond what we can do here, but you will find a couple of examples below.

10.1.5 The Ground-State Energy

Let $\psi_n(q)$ be the energy eigenfunctions, with eigenvalues E_n. Then the Euclidean propagator is

$$K_E(q_b, T; q_a, 0) = \langle q_b | e^{-HT} | q_a \rangle = \sum_n \langle q_b | \psi_n \rangle e^{-E_n T} \langle \psi_n | q_a \rangle$$

$$= \sum_n \psi_n(q_a) \psi_n(q_b)^* e^{-E_n T} \tag{10.60}$$

Since any physical spectrum has a lower bound, all the energies can be chosen positive by choosing the zero point of the energy suitably. Then each term vanishes in the limit $T \to \infty$, the term with the ground-state energy surviving the longest. This observation provides a way to extract the energy of the lowest state.

The harmonic oscillator is an example: The real-time propagator for the harmonic oscillator is

$$K(q_b, t_b; q_a, 0) = \sqrt{\frac{m\omega}{2\pi i \sin \omega t_b}} \exp\left(\frac{im\omega}{2 \sin \omega t_b}\left[(q_a^2 + q_b^2)\cos\omega t_b - 2q_a q_b\right]\right) \tag{10.49}$$

The Euclidean version is

$$K_E(q_b, T; q_a, 0) = \sqrt{\frac{m\omega}{2\pi \sinh \omega T}} \exp\left(-\frac{m\omega}{2\sinh \omega T}\left[(q_a^2 + q_b^2)\cosh\omega T - 2q_a q_b\right]\right) \tag{10.61}$$

---

[3]The Euclidean propagator gets its name from the relativistic version. In relativistic calculations, the Minkowski metric $x^2 + y^2 + z^2 - t^2$ is replaced by the metric of the Euclidean four-dimensional vector space $x^2 + y^2 + z^2 + t^2$, which is often easier to handle.

Since
$$\sinh \omega T \xrightarrow[T\to\infty]{} \frac{1}{2}e^{\omega T} \quad \text{and} \quad \cosh \omega T \xrightarrow[T\to\infty]{} \frac{1}{2}e^{\omega T} \tag{10.62}$$

the leading term in the propagator is

$$K_E(q_b, T; q_a, 0) \xrightarrow[T\to\infty]{} \sqrt{\frac{m\omega}{\pi}} e^{-m\omega q_a^2/2} e^{-m\omega q_b^2/2} e^{-\omega T/2} \tag{10.63}$$

Comparing the limiting form (10.63) with the general rule, (10.60) one gets immediately

$$E_o = \frac{\omega}{2} \quad \text{and} \quad \psi_o(q) = \left(\frac{m\omega}{\pi}\right)^{1/4} e^{-m\omega q_a^2/2} \tag{10.64}$$

in agreement with (3.70) for the ground-state wave function.

The exponential in (10.61) can be expanded in powers of $\exp(-\omega T)$. When multiplied by the factor $\exp(-\omega T/2)$ from the square root in front, each term goes like $\exp[-(n+\frac{1}{2})\omega T]$. The spectrum is

$$E_n = \left(n + \frac{1}{2}\right)\omega \tag{10.65}$$

for $n = 0, 1, 2, \ldots$.[1]

Direct Euclidean Solution

The point of the Euclidean path integral is not to work out a problem in ordinary time and then make an analytic continuation, but to use the imaginary time formalism directly. Consider a potential $V(q)$ slightly more general than the harmonic oscillator, one that has a quadratic minimum $V(q) = 0$ at $q = 0$ but is not necessarily $m\omega^2 q^2/2$ everywhere else. The Euclidean propagator is

$$K_E(q_b, T; q_a, 0) = \sum_n \psi_n(q_a)\psi_n(q_b)^* e^{-E_n T} = \int \mathcal{D}q \exp\left(-\int L_E[q(w)]\right) dw \tag{10.66}$$

where

$$L_E = \frac{m}{2}\left(\frac{dq}{dw}\right)^2 + V(q) \tag{10.67}$$

Expand the action about the path that minimizes it:

$$S_E[q] = \int L_E(w)dw = S[q_o + \delta q]$$
$$= S_E[q_o] + \frac{\delta S_E[q_o]}{\delta \dot q}\delta \dot q + \frac{\delta S_E[q_o]}{\delta q}\delta q + \frac{1}{2}\frac{\delta^2 S_E[q_o]}{\delta \dot q^2}\delta \dot q^2 + \frac{1}{2}\frac{\delta^2 S_E[q_o]}{\delta q^2}\delta q^2 + \cdots \tag{10.68}$$

Integrate the terms in $\delta \dot q^2$ and δq^2 by parts:

$$S_E[q] = S_E[q_o] + \int \left[-m\ddot q_o + V'(q_o)\right]\delta q\, dw + \frac{1}{2}\int \delta q\left[-m\delta\ddot q + V''(q_o)\delta q\right]dw + \cdots \tag{10.69}$$

334   Chapter 10   Further Topics in Quantum Dynamics

The classical path is the one that makes the linear term vanish.

$$m\ddot{q}_o = V'(q_o) \tag{10.70}$$

This is the equation of motion for a particle moving in the potential $-V(q)$, a potential "hill" with a maximum at $q = 0$. The solution is $q_o(w) = 0$ and the classical action vanishes:

$$K_E(q_b, T; q_a, 0) \approx \int dy \exp\left[-\int_0^T y(w)\left(-m\frac{d^2}{dw^2} + V''(0)\right)y(w)dw\right] \tag{10.71}$$

where $y(0) = y(T) = 0$. Again, from the rule (A.25)

$$K_E(q_b, T; q_a, 0) = K_o \frac{1}{\sqrt{\det M}} \tag{10.72}$$

where now the "matrix" M is

$$M = -m\frac{d^2}{dw^2} + V''(0) \tag{10.73}$$

The same argument that led to equation (10.47) above, but with some sign changes, leads here to

$$\lim_{T \to \infty} K_E(q_b, T; q_a, 0) \approx K_o \frac{1}{\sqrt{\sinh V''(0)T/m}} \sim e^{-V''(0)T/2m} \tag{10.74}$$

giving the correct bound-state energy up to cubic terms in the potential.

10.2 PATH INTEGRATION: SOME APPLICATIONS

10.2.1 The Born Series

The propagator (10.19) for a particle in a potential can be expanded in $V(q)$:

$$\begin{aligned}
K(q_b, t_b; q_a, 0) &= \int \mathcal{D}q \exp\left[i\int \left(\frac{1}{2}m\dot{q}^2 - V(q)\right)dt\right] \\
&= \int \mathcal{D}q \exp\left(i\int \frac{1}{2}m\dot{q}^2 dt\right) \exp\left(-i\int V(q)dt\right) \\
&= \int \mathcal{D}q \exp\left(i\int \frac{1}{2}m\dot{q}^2 dt\right)\left[1 - i\int V(q)dt - \frac{1}{2}\left(\int V(q)dt\right)^2 + \cdots\right] \\
&= K_o(q_b, t_b; q_a, 0) + K_1(q_b, t_b; q_a, 0) + K_2(q_b, t_b; q_a, 0) + \cdots
\end{aligned} \tag{10.75}$$

Mathematical Note: Remember that the variables in the path integrals are ordinary numbers, so they all commute. To manipulate the integrals like this, you have to assume everything converges and that you can freely interchange the order of integrations over t and q. As long as these integrals are *defined* as analytic continuations from the Euclidean versions, they converge unambiguously.

Section 10.2 Path Integration: Some Applications 335

The first term is the free particle propagator (10.29):

$$K_o(q_b, t_b; q_a, 0) = \int \mathcal{D}q \, \exp\left(i \int \frac{1}{2} m\dot{q}^2 dt\right)$$

$$= \Theta(t_b) \sqrt{\frac{m}{2\pi i t_b}} \exp\left(\frac{im}{2} \frac{(q_b - q_a)^2}{t_b}\right) \quad (10.76)$$

The term first order in the potential can be evaluated by writing it in the original discrete form:

$K_1(q_b, t_b; q_a, 0)$

$$= -i \lim_{N\to\infty} \left(\frac{m}{2\pi i \epsilon}\right)^{(N+1)/2} \int \prod_{n=1}^{N} dq_n \exp\left[i\epsilon \sum_{n=1}^{N+1} \left(\frac{m\dot{q}_n^2}{2}\right)\right] \sum_{k=1}^{N} \epsilon V(q_k)$$

$$= -i \lim_{N\to\infty} \sum_{k=1}^{N} \epsilon \left(\frac{m}{2\pi i \epsilon}\right)^{k/2} \int \prod_{j=1}^{k-1} dq_j \exp\left[i\epsilon \sum_{n=1}^{k} \left(\frac{m\dot{q}_n^2}{2}\right)\right]$$

$$\times \int V(q_k) dq_k \left(\frac{m}{2\pi i \epsilon}\right)^{(N+1-k)/2} \prod_{j=k+1}^{N} dq_j \exp\left[i\epsilon \sum_{n=k+1}^{N} \left(\frac{m\dot{q}_n^2}{2}\right)\right]$$

$$\to \int_0^{t_b} dt_1 \int_{-\infty}^{\infty} dq_1 K_o(q_b, t_b; q_1, t_1) V(q_1) K_o(q_1, t_1; q_a, 0)$$

$$\to \int_{-\infty}^{\infty} dt_1 \int_{-\infty}^{\infty} dq_1 K_o(q_b, t_b; q_1, t_1) V(q_1) K_o(q_1, t_1; q_a, 0) \quad (10.77)$$

To get the next-to-last line, I rearranged the factors separating those before and after one of the q_k, and in the limit recognized each factor as the free propagator. To get the last line, I use the fact that $K_o(q_b, t_b; q_1, t_1)$ vanishes when $t_b < t_a$.

Now look at the term that is quadratic in the potential. Break up the factors in a similar way:

$$K_2(q_b, t_b; q_a, 0) = -\frac{1}{2} \int \mathcal{D}q \, \exp\left(i \int \frac{1}{2} m\dot{q}^2 dt\right) \left(\int V(q) dt\right)^2$$

$$= -\frac{1}{2} \int_0^{t_b} \int_0^{t_b} dt_1 dt_2 \int_{-\infty}^{\infty} \int_{-\infty}^{\infty} dq_1 dq_2 \quad (10.78)$$

$$\times F(q_b, t_b; q_2, t_2) V(q_2) F(q_2, t_2; q_1, t_1) V(q_1) F(q_1, t_1; q_a, 0)$$

where

$$F(q_b, t_b; q_2, t_2) = K_o(q_b, t_b; q_2, t_2) \quad \text{and} \quad F(q_1, t_1; q_a, 0) = K_o(q_1, t_1; q_a, 0)$$

$$(10.79)$$

but the middle factor depends on the order of the times t_2 and t_1:

$$F(q_2, t_2; q_1, t_1) = K_o(q_2, t_2; q_1, t_1)\Theta(t_2 - t_1) + K_o(q_1, t_1; q_2, t_2)\Theta(t_1 - t_2) \quad (10.80)$$

Inside the integral the two terms in (10.80) are the same:

$$K_2(q_b,t_b;q_a,0) = -\frac{1}{2}\int_{-\infty}^{\infty}\int_{-\infty}^{\infty}dt_1 dt_2 \int_{-\infty}^{\infty}\int_{-\infty}^{\infty}dq_1 dq_2$$

$$\times \Theta(t_2-t_1)K_o(q_b,t_b;q_2,t_2)V(q_2)K_o(q_2,t_2;q_1,t_1)V(q_1)K_o(q_1,t_1;q_a,0)$$

$$-\frac{1}{2}\int_{-\infty}^{\infty}\int_{-\infty}^{\infty}dt_1 dt_2 \int_{-\infty}^{\infty}\int_{-\infty}^{\infty}dq_1 dq_2$$

$$\times \Theta(t_1-t_2)K_o(q_b,t_b;q_1,t_1)V(q_2)K_o(q_1,t_1;q_2,t_2)V(q_1)K_o(q_2,t_2;q_a,0)$$

$$= -\int_{-\infty}^{\infty}\int_{-\infty}^{\infty}dt_1 dt_2 \int_{-\infty}^{\infty}\int_{-\infty}^{\infty}dq_1 dq_2$$

$$\times K_o(q_b,t_b;q_2,t_2)V(q_2)K_o(q_2,t_2;q_1,t_1)V(q_1)K_o(q_1,t_1;q_a,0) \tag{10.81}$$

The pattern should be clear. The n-th order term has n factors of $V(q)$, sandwiched in between $n+1$ free propagators with the appropriate arguments. The coefficient is $(-i)^n$. The factor $1/n!$ from the Taylor expansion is canceled by the $n!$ different ways you can order the factors $V(q_k)$. Visualize this expansion of the propagator as a different kind of "sum over paths." The particle may propagate freely from q_b, t_b to $q_a, 0$. Or it can propagate from q_b, t_b to q_1, t_1, interact, propagate from q_1, t_1 to q_2, t_2, interact, and so forth, any number of times. In each case, the rule is: Write down free propagators, put $-iV[q(t)]$ in between them, and integrate over all possible q's and t's. Finally, add up all the amplitudes with different numbers of interactions.

If at time t_a a particle has the wave function $\psi(x)$, the amplitude for its wave function at a later time t_b is

$$\psi(q_b) = \int dq_a K(q_b,t_b;q_a,t_a)\psi(q_a) \tag{10.82}$$

Then suppose $\phi_b(q_b,t_b)$ is a plane wave, and also an eigenfunction of the Hamiltonian when the potential vanishes:

$$\phi_b(q_b,t_b) = \phi(q_b)\exp(-iE_b t_b) \tag{10.83}$$

Let $\phi_a(q_a,t_a) = \phi(q_a)\exp(-iE_a t_a)$ be another plane-wave eigenstate of H_o. Then I leave it as an exercise to prove that

$$S_{ba} = \lim_{\substack{t_b \to \infty \\ t_a \to -\infty}} \int \phi_b(q_b,t_b)^* K(q_b,t_b;q_a,t_a)\phi_a(q_a,t_a) dq_a\, dq_b \tag{10.84}$$

where S is the scattering matrix defined in Section 9.3.

Note: The expansion above, of which (10.81) is the first few terms, holds only for ordinary potentials. When the potential energy depends on the velocity—as in the important case of a charged particle moving in a magnetic field—you have to do something more complicated.[4]

---
[4] See Feynman and Hibbs [5], pages 189-191.

10.2.2 External Fields and Gauge Invariance

Consider an electron in a general external electromagnetic field with magnetic vector potential $\mathbf{A}(\mathbf{r},t)$ and electrostatic potential $e\phi(\mathbf{r},t)$. The Hamiltonian (ignoring spin interactions) is

$$H = \frac{(\mathbf{p} + e\mathbf{A}(\mathbf{r},t))^2}{2m} - e\phi(\mathbf{r},t) \tag{10.85}$$

This Hamiltonian is not in the form (10.11), so it is not obvious that you can integrate out the momentum and write the propagator in the simple form (10.19), but it is true. Start from the more general phase space path integral (10.15):

$$K(\mathbf{r}_b, t_b; \mathbf{r}_a, t_a) = \int \mathcal{D}\mathbf{r}\mathcal{D}\mathbf{p} \exp\left[i \int_{t_a}^{t_b} \left(\mathbf{p} \cdot \dot{\mathbf{r}} - \frac{(\mathbf{p} + e\mathbf{A}(\mathbf{r},t))^2}{2m} + e\phi(\mathbf{r},t)\right) dt\right] \tag{10.86}$$

Shift the momentum variables from $\mathbf{p}_n$ to $\mathbf{p}'_n = \mathbf{p}_n + e\mathbf{A}(\mathbf{r},t)$. Then $d^3p_n = d^3p'_n$, and the measure of the path integration is unchanged:

$$\begin{aligned}K(\mathbf{r}_b, t_b; \mathbf{r}_a, t_a) &= \int \mathcal{D}\mathbf{r}\mathcal{D}\mathbf{p}' \exp\left[i \int_{t_a}^{t_b} \left((\mathbf{p}' - e\mathbf{A}(\mathbf{r},t)) \cdot \dot{\mathbf{r}} - \frac{\mathbf{p}'^2}{2m} + e\phi(\mathbf{r},t)\right) dt\right] \\ &= \int \mathcal{D}\mathbf{r}\mathcal{D}\mathbf{p} \exp\left[i \int_{t_a}^{t_b} \left(\mathbf{p} \cdot \dot{\mathbf{r}} - \frac{\mathbf{p}^2}{2m} - e\mathbf{A}(\mathbf{r},t) \cdot \dot{\mathbf{r}} + e\phi(\mathbf{r},t)\right) dt\right]\end{aligned} \tag{10.87}$$

Complete the square just as in the steps leading to equation (10.11) and then integrate over $\mathbf{p}$ to get

$$K(\mathbf{r}_b, t_b; \mathbf{r}_a, t_a) = \int \mathcal{D}\mathbf{r} \exp i \int_{t_a}^{t_b} L(\mathbf{r}, \dot{\mathbf{r}}) dt \tag{10.88}$$

where

$$L(\mathbf{r}, \dot{\mathbf{r}}) = \frac{m\dot{\mathbf{r}}^2}{2} - e\mathbf{A}(\mathbf{r},t) \cdot \dot{\mathbf{r}} + e\phi(\mathbf{r},t) \tag{10.89}$$

L is indeed the Lagrangian for a charge $-e$ in a magnetic field.[5] The simple form (10.11) is not completely general, but it holds for an external electromagnetic field.
The propagator is

$$\begin{aligned}K(\mathbf{r}_b, t_b; \mathbf{r}_a, t_a) &= \int \mathcal{D}\mathbf{r} \exp\left[i \int_{t_a}^{t_b} \left(L_o(\mathbf{r}, \dot{\mathbf{r}}) - e\mathbf{A}(\mathbf{r},t) \cdot \dot{\mathbf{r}} + e\phi(\mathbf{r},t)\right) dt\right] \\ &= \int \mathcal{D}\mathbf{r} \exp i S_o[\mathbf{r}] \exp\left[-ie \int_{t_a}^{t_b} \left(\mathbf{A}(\mathbf{r},t) \cdot d\mathbf{r} - \phi(\mathbf{r},t)\right) dt\right]\end{aligned} \tag{10.90}$$

---
[5] See equation (1.15).

The extra factor is an integral of the potentials along the path $\mathbf{r}(t)$. Suppose the potentials are changed by a gauge transformation:

$$\mathbf{A}(\mathbf{r},t) \to \mathbf{A}'(\mathbf{r}) = \mathbf{A}(\mathbf{r},t) + \boldsymbol{\nabla}\Lambda(\mathbf{r},t) \qquad (10.91\text{a})$$

$$\phi(\mathbf{r},t) \to \phi'(\mathbf{r},t) = \phi(\mathbf{r},t) - \frac{\partial}{\partial t}\Lambda(\mathbf{r},t) \qquad (10.91\text{b})$$

In the new gauge the propagator is

$$K'(\mathbf{r}_b,t_b;\mathbf{r}_a,t_a) = \int \mathcal{D}\mathbf{r}\, \exp iS_o[\mathbf{r}]$$
$$\times \exp\left[-ie \int_{\mathbf{r}_a}^{\mathbf{r}_b} \left(\mathbf{A}(\mathbf{r},t)\cdot d\mathbf{r} - \phi(\mathbf{r},t) - \boldsymbol{\nabla}\Lambda(\mathbf{r},t)\cdot\frac{d\mathbf{r}}{dt} - \frac{\partial}{\partial t}\Lambda(\mathbf{r},t)\right)dt\right]$$

$$= \int \mathcal{D}\mathbf{r}\, \exp iS_o[\mathbf{r}] \exp\left[-ie \int_{\mathbf{r}_a}^{\mathbf{r}_b}\left(\mathbf{A}(\mathbf{r},t)\cdot d\mathbf{r} - \phi(\mathbf{r},t) - \frac{d}{dt}\Lambda(\mathbf{r},t)dt\right)\right]$$

$$= K(\mathbf{r}_b,t_b;\mathbf{r}_a,t_a)\exp\left[ie\bigl(\Lambda(\mathbf{r}_a,t_a) - \Lambda(\mathbf{r}_b,t_b)\bigr)\right] \qquad (10.92)$$

If at time t_a the electron wave function is $\psi(\mathbf{r},t_a) = \psi_a(\mathbf{r})$, then at the later time t_b it is

$$\psi(\mathbf{r},t_b) = \int K(\mathbf{r}_b,t_b;\mathbf{r}_a,t_a)\psi_a(\mathbf{r}_a)\,d^3r_a \qquad (10.93)$$

It follows that if $\psi(\mathbf{r},t)$ satisfies Schrödinger's equation in the original gauge, then

$$\psi'(\mathbf{r},t) = e^{-ie\Lambda(\mathbf{r},t)}\psi(\mathbf{r},t) \qquad (10.94)$$

satisfies Schrödinger's equation in the new gauge.

Equation (10.94) is the generalization of the result in Section 6.1.2 for constant magnetic fields. Even though the wave function looks gauge-dependent, there are no observable gauge dependent quantities (see the discussion is Section 6.1.2).

Electromagnetism allows these local gauge transformations, where a change of the magnetic vector potential and the electrostatic potential is accompanied by an **r**-dependent and *t*-dependent phase multiplying the wave function. One consequence is that there are quantum effects that have no classical analog. The most famous was pointed out in 1959 by Aharonov and Bohm [9].

10.2.3 The Aharonov-Bohm Effect

Suppose a beam of electrons with momentum k is moving in the y direction and strikes an impenetrable plane at $y = 0$. There are two small holes in the plane, on the z axis at $x = \pm d/2$, for the beam to pass through. The electrons are detected on a screen some distance $y = L \gg d$ farther downstream. The probability that the electrons strike the screen varies with the distance from the center—this is the two-slit pattern.

Let $\mathbf{r}_a$ be some point on the barrier at $y = 0$ (or near the source, with boundary conditions for the barrier) and $\mathbf{r}_b$ a point on the detecting screen at $y = L$. Then the wave function at $y = L$ is determined by the wave function at $y = 0$ (or at the source, with boundary conditions for the barrier with the slits):

$$\psi(\mathbf{r}_b, t_b) = \int d^3 r_a \int \mathcal{D}\mathbf{r} K(\mathbf{r}_b, t_b; \mathbf{r}_a, t_z) e^{iS_o} \psi(\mathbf{r}_a) \tag{10.95}$$

If the slits are too wide, the beam is not localized enough to produce the interference pattern. If they are too narrow, the approximation that the electron beam has a definite energy fails. Doing the calculation correctly requires working out complicated integrals to get the single-slit diffraction pattern modulating the double-slit interference pattern, and it is not particularly instructive to try.[6] We know that when the slits are wide enough but not too wide the wave function to the right of the barrier is the sum of two spherical wavelets, one coming from each slit:

$$\psi(\mathbf{r}, t) \sim \frac{1}{r_1} e^{i(kr_1 - \omega t)} + \frac{1}{r_2} e^{i(kr_2 - \omega t)} \tag{10.96}$$

r_1 and r_2 are the distances from the observation point to one of the slits. These classical paths are the stationary points of the path integral, and the integrands are in phase only in some region close to the classical paths.

A little algebra and some approximations show that these two terms add up to the double-slit interference pattern

$$|\psi(\mathbf{r}, t)|^2 \sim \frac{4}{r^2} \cos^2 \frac{\pi d \sin\theta}{\lambda} \tag{10.97}$$

Here d is the distance between the slits, r the distance to the point halfway between them, λ the de Broglie wavelength and $\sin\theta = z/L$. The probability that the electrons strike the screen varies with the distance from the center—this is the two-slit pattern.

Now modify the interference experiment so that before they strike the screen the electrons pass through a region where there is no magnetic field, but where the vector potential $\mathbf{A}$ does not vanish. This can be arranged by adding a very long cylindrical solenoid, whose axis is parallel to the z axis, passing through the imaginary triangle you get by connecting the two slits to the center of the interference pattern on the screen, as in Figure 10.1. The solenoid has a radius R_o small enough to fit inside this triangle. Assume the magnet is infinitely long. In that limit there are no edge effects, and the magnetic field is confined to the inside of the cylinder.

Take the origin of coordinates to be where the axis of the solenoid intersects the $z = 0$ plane. Furthermore, let the magnetic field be constant inside the magnet (it is easy to relax this condition) and zero outside. In cylindrical coordinates, $\mathbf{B} = B_o \hat{\mathbf{n}}_z$ for $\rho \leq R_o$ and $\mathbf{B} = 0$ for $\rho \geq R_o$. The vector potential is

$$\mathbf{A}(\mathbf{r}) = \frac{1}{2} B_o \hat{\mathbf{n}}_z \times \mathbf{r} \quad (\rho \leq R_o)$$

$$\mathbf{A}(\mathbf{r}) = \frac{1}{2} \frac{R_o^2 B_o}{\rho} \hat{\mathbf{n}}_\phi \quad (\rho \geq R_o) \tag{10.98}$$

---

[6]But you needn't worry that it doesn't work out for some reason, Aharonov and Bohm did it in Section 4 of their original paper.

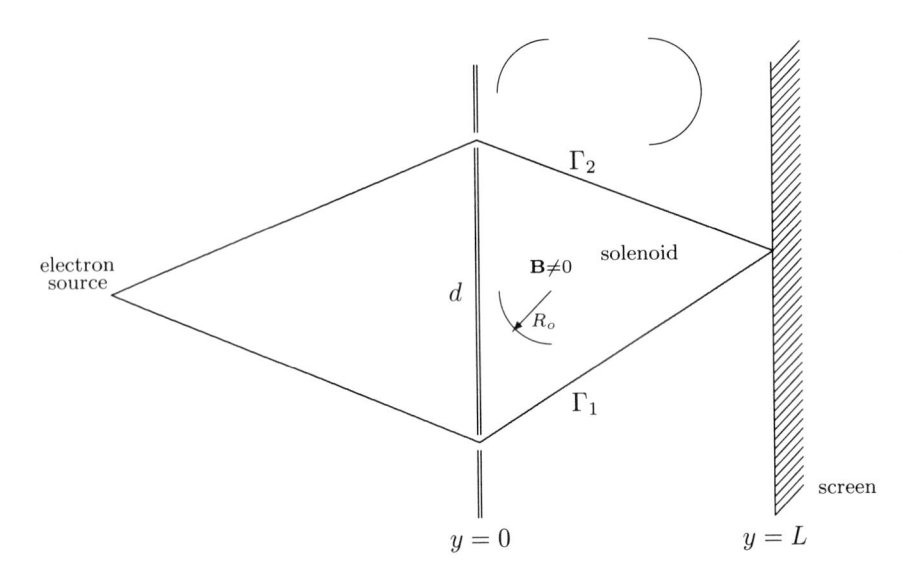

340 Chapter 10 Further Topics in Quantum Dynamics

FIGURE 10.1: Schematic drawing of the Aharonov-Bohm effect. The magnetic field is confined to the infinitely long cylinder of radius R_o, whose axis is perpendicular to the plane of the page.

as you can check using equations (A.174) in the appendix.

Consider each path in the neighborhood of the classical straight line from one of the slits to an observation point. According to the formula for the propagator (10.90), the contribution of that path to the propagator is the contribution of the same path when $\mathbf{A} = 0$, $\exp iS_o[\mathbf{r}]$, multiplied by

$$\exp\left(-ie \int_{\mathbf{r}_a}^{\mathbf{r}_b} \mathbf{A}(\mathbf{r}) \cdot d\mathbf{r}\right) \tag{10.99}$$

Because $\nabla \times \mathbf{A} = \mathbf{B} = 0$, this integral is the same for all the paths in the vicinity of one of the classical paths. But the extra phase is not the same in the vicinity of the other classical path, since you cannot deform one path into the other without intersecting the region where $\mathbf{B} \neq 0$.

Let $\psi_1(\mathbf{r})$ and $\psi_2(\mathbf{r})$ be the contributions to $\psi(\mathbf{r})$ from all the paths near each of the classical paths respectively when $\mathbf{B} = \mathbf{A} = 0$. Approximately, these are the two terms in the double-slit formula (10.96). Now turn on the magnetic field in the solenoid. The wave function at $\mathbf{r}$ is changed to

$$\psi(\mathbf{r}) = \psi_1(\mathbf{r}) \exp\left(-ie \int_{\Gamma_1} \mathbf{A}(\mathbf{r}) \cdot d\mathbf{r}\right) + \psi_2(\mathbf{r}) \exp\left(-ie \int_{\Gamma_2} \mathbf{A}(\mathbf{r}) \cdot d\mathbf{r}\right) \tag{10.100}$$

where Γ_i are the classical straight-line paths from the source through one of the slits to the observation point. The effect is made more obvious by factoring out one of the phases:

$$\psi(\mathbf{r}) = \exp\left(-ie \int_{\Gamma_1} \mathbf{A}(\mathbf{r}) \cdot d\mathbf{r}\right) \left[\psi_1(\mathbf{r}) + \exp\left(-ie \oint_C \mathbf{A}(\mathbf{r}) \cdot d\mathbf{r}\right) \psi_2(\mathbf{r})\right] \tag{10.101}$$

where

$$\oint_C = \int_{\Gamma_2} - \int_{\Gamma_1} \qquad (10.102)$$

The difference between the line integrals is not zero since C is a closed loop that cuts through the solenoid. From Stokes' theorem and the vector potential (10.98), the phase difference is

$$\phi_+ - \phi_- = e\oint_C \mathbf{A}(\mathbf{r}) \cdot d\mathbf{r} = e\pi R_o^2 B_o = e\Phi \qquad (10.103)$$

where Φ is the magnetic flux through the loop. The phase difference (10.103) is in general neither zero nor a multiple of 2π. The interference pattern shifts as you change $\mathbf{B}$. This is the Aharonov-Bohm effect, and it has been observed experimentally.

In classical mechanics the orbits of classical charged particles would be unaffected by the magnetic vector potential in a region where the magnetic field is zero. But the vector potential has observable quantum effects, for example on the phase of a wave function, even when the classical paths are only through a region with $\mathbf{B} = 0$.

Although the Aharonov-Bohm shift depends on $\mathbf{A}$ in a region where $\mathbf{B} = 0$, it is still independent of the gauge. It is a function of the magnetic flux through the loop, a gauge invariant quantity. The vector potential itself is "measurable" in quantum mechanics only up to a gauge transformation.

A Final Note on Path Integration

One reaction to a first encounter with path integration is that it is a difficult way to solve simple problems. Do not be fooled by the elementary nature of the examples here. I have only been able to present an introduction to the subject. Path integration has become a powerful tool in quantum field theory and statistical mechanics.

10.3 BERRY'S PHASE

10.3.1 Origin of the Phase

Let $H(t)$ be a time-dependent Hamiltonian whose instantaneous eigenstates and eigenvalues satisfy

$$H(t)|\psi_n(t)\rangle = E_n(t)|\psi_n(t)\rangle \qquad (10.104)$$

The states $|\psi_n(t)\rangle$ are a basis, but a different one at each time. Suppose $|\psi(0)\rangle = |\psi_o(0)\rangle$. Then if the changes are slow enough the probability to find a different state at a later time vanishes:

$$|\langle\psi_n(t)| \psi(t)\rangle| = 0 \qquad (n \neq 0) \qquad (10.105)$$

That does not mean $|\psi(t)\rangle = |\psi_o(t)\rangle$. The two states might differ by a phase,

$$|\psi(t)\rangle = e^{i\phi}|\psi_o(t)\rangle \qquad (10.106)$$

342 Chapter 10 Further Topics in Quantum Dynamics

From (10.104) and the Schrödinger equation

$$-\dot{\phi}|\psi(t)\rangle + ie^{i\phi}\frac{d}{dt}|\psi_0(t)\rangle = e^{i\phi}E_0(t)|\psi_0(t)\rangle \qquad (10.107)$$

Set

$$\phi = \gamma - \int_{t_o}^{t} E_0(t')dt' \qquad (10.108)$$

If H were constant, the second term above would be the usual time dependence of a stationary state, and γ would be a constant. But in general $\dot{\phi} = \dot{\gamma} - E_o(t)$ and

$$\dot{\gamma} = i\langle \psi_0(t) | \dot{\psi}_0(t) \rangle \qquad (10.109)$$

Since equation (10.104) does not define the instantaneous eigenstates $\psi_n(t)$ uniquely γ and $\dot{\gamma}$ are not unique either. For example, let

$$\left|\tilde{\psi}_0(t)\right\rangle = e^{i\alpha(t)}|\psi_0(t)\rangle \qquad (10.110)$$

where α is some arbitrary real function of t. Then γ is changed to $\tilde{\gamma}$, where

$$\dot{\tilde{\gamma}} = i\langle \tilde{\psi}_0(t) | \dot{\tilde{\psi}}_0(t) \rangle = \dot{\gamma} - \dot{\alpha} \qquad (10.111)$$

It is almost obvious that by choosing the appropriate $\alpha(t)$ it should be possible to get rid of this phase altogether, and for a long time people assumed γ had no physical consequences, that it could always be canceled by another phase factor. But in 1984 Berry [12] pointed out that this is not always true.

Berry considered a system whose Hamiltonian depends on t because it depends on some time-dependent external parameters R_i. One example is an electron in a time-dependent magnetic field **B** (Section 6.2.3). Another is the Born-Oppenheimer approximation (Section 7.6.1), where the nuclear coordinates **R** are external parameters for the electron subsystem, although ultimately they are dynamical coordinates also.

What these examples have in common is that the states are discrete, and that as the parameters R_i are varied, the eigenstates $|\psi_n(\mathbf{R})\rangle$ of the Hamiltonian $H(\mathbf{R})$ do not appear or disappear, and their order does not change. (**R** is shorthand for the collection of parameters R_i.) If at $t = 0$ the system is in one of those states, it should stay in that state as long as the R_i are changed slowly enough. That is the content of the **adiabatic theorem**.[7]

> Note: If the Hamiltonian changes slowly enough, then the eigenstates and eigenvalues don't change very much and a quantum system will stay in the state with the same label. For there to be a substantial probability that it jumps to another state, the Hamiltonian has to change fast enough that its Fourier transform contain significant fraction of its frequencies in the vicinity of the frequency of the level jump. Roughly, the condition is that if a system is in the state a, and if ω_{ab} is the *smallest* energy difference between state a and any other state, then
>
> $$\left|\frac{1}{H_{ba}}\frac{d}{dt}H_{ba}\right| \ll |\omega_{ba}| \qquad (10.112)$$

---
[7]See footnote 15 in Section 7.6 for references.

The simplest example is a problem we have already solved exactly: an electron in a magnetic field precessing around the z-axis, describing a cone with opening angle θ.

10.3.2 Example: Electron in a Precessing Magnetic Field

Consider an electron at rest in a precessing magnetic field: There is a constant magnetic field B_o along the z-direction and a smaller field B_1 rotating in the x-y plane:

$$\mathbf{B} = \mathbf{B}_1(\hat{\mathbf{n}}_x \cos\phi + \hat{\mathbf{n}}_y \sin\phi) \tag{6.81}$$

where $\phi = \omega t$. Set $\omega_o = g\mu_B B_o$ and $\omega_1 = g\mu_B B_1$. The field always makes an angle θ with the z-axis. $\omega_o = \bar{\omega}\cos\theta$ and $\omega_1 = \bar{\omega}\sin\theta$, where

$$\bar{\omega} = g\mu_B \bar{B} = \sqrt{\omega_o^2 + \omega_1^2} \tag{10.113}$$

and $\bar{B}$ is the magnitude of the magnetic field. This is electron magnetic resonance, discussed in Section 6.2.3 and in Problem 6.7.

At any time t the magnetic field is pointing along some direction with spherical coordinates θ and $\phi(t)$. The Hamiltonian is $H(\phi) = \mathbf{H}(\phi) \cdot \boldsymbol{\sigma}/2$, where

$$\mathbf{H} = \omega_o \hat{\mathbf{n}}_z + \omega_1(\cos\phi\,\hat{\mathbf{n}}_x + \sin\phi\,\hat{\mathbf{n}}_y) \tag{6.82}$$

Let $\psi_\pm(t)$ be the instantaneous eigenstates of $H(\phi)$:

$$H(\phi)|\psi_\pm(\phi)\rangle = E_\pm(\phi)|\psi_\pm(\phi)\rangle \tag{10.114}$$

Choose the phases so that the states are single-valued functions of the magnetic field, that is, that $\psi_\pm(\phi + 2\pi) = \psi_\pm(\phi)$.[8]

$$\psi_+(\phi) = \begin{pmatrix} \cos\frac{\theta}{2} \\ \sin\frac{\theta}{2} e^{i\phi} \end{pmatrix} \quad \text{and} \quad \psi_-(\phi) = \begin{pmatrix} \sin\frac{\theta}{2} \\ -\cos\frac{\theta}{2} e^{i\phi} \end{pmatrix} \tag{10.115}$$

These two states are always aligned with the magnetic field. If the electron is initially antiparallel to the magnetic field, $\psi(0) = \psi_-(0)$, then at later times (see Problem 10.5)

$$\psi(t) = i\frac{\omega}{\Omega}\sin\theta \sin\frac{\Omega t}{2} e^{-i\omega t/2}\psi_+(t)$$
$$+ \left[\cos\frac{\Omega t}{2} + i\left(\frac{\bar{\omega} - \omega\cos\theta}{\Omega}\right)\sin\frac{\Omega t}{2}\right] e^{-i\omega t/2}\psi_-(t) \tag{10.116}$$

with

$$\Omega = \sqrt{(\omega_o - \omega)^2 + \omega_1^2} = \sqrt{\bar{\omega}^2 + \omega^2 - 2\omega\bar{\omega}\cos\theta} \tag{10.117}$$

For large ω the wave function flops rapidly back and forth between the two states. But as $\omega \to 0$ the probability for exciting the upper state vanishes like

---

[8]Even with these definitions, the states are not well defined along part of the z-axis. In this example the angle θ is fixed, and the wave functions are well defined for any ϕ. In general there is no way to evade this problem—we will see shortly how to deal with it.

ω^2, and the electron stays in the lower state. This is an example of the adiabatic theorem.

For small ω set $\Omega \approx \bar{\omega} - \omega \cos\theta$. Then

$$\psi(t) \approx e^{i\Omega t/2} e^{-i\omega t/2} \psi_-(t) \approx e^{i\bar{\omega}t/2} e^{-i\omega \cos\theta t/2} e^{-i\omega t/2} \psi_-(t) \qquad (10.118)$$

plus terms of order $\omega/\bar{\omega}$. The first factor is the dynamical phase

$$\exp\left(-i \int_0^t E_-(t')dt'\right) \qquad (10.119)$$

where $E(t) = -\bar{\omega}/2$ is the (here constant) energy of this state. The remaining factor is $\exp i\gamma$, with

$$\gamma = -\frac{\omega t}{2}(\cos\theta + 1) \qquad (10.120)$$

γ is **Berry's phase**, the quantity defined in equation (10.108). After one complete cycle when $t = 2\pi/\omega$,

$$\gamma\left(\frac{2\pi}{\omega}\right) = -\pi(\cos\theta + 1) \qquad (10.121)$$

Equation (10.121) can also be derived directly from the general rule (10.109). Set $\psi_o(t) = \psi_-(\omega t)$ from (10.115). Then from

$$\frac{d}{dt}|\psi_o(t)\rangle = |\psi'_-(\phi)\rangle \frac{d\phi}{dt} \qquad (10.122)$$

with

$$|\psi'_-(\phi)\rangle = \frac{d}{d\phi}|\psi_-(\phi)\rangle = \begin{pmatrix} 0 \\ -ie^{i\phi}\cos\frac{\theta}{2} \end{pmatrix} \qquad (10.123)$$

equation (10.109) can be integrated to give

$$\gamma(2\pi/\omega) = \int_0^{2\pi/\omega} i\langle \psi_-(\phi(t))| \dot{\psi}_-(\phi(t))\rangle \frac{d\phi}{dt} dt$$
$$= \int_0^{2\pi} \langle \psi_-(\phi)| \psi'_-(\phi)\rangle d\phi = -\pi(1 + \cos\theta) \qquad (10.124)$$

The surprising feature of this derivation is that it does not depend at all on the *form* of $\phi(t)$. Berry's phase for a closed loop increases by the same amount no matter how you go around it, provided only that you never go so fast as to invalidate the adiabatic approximation. For that reason it is called a **geometric phase**. It should be clear that you cannot make the result go away by multiplying the state with any smooth, single-valued phase.

10.3.3 The General Formula

There is a general formula for the phase γ as the parameters go around a loop (slowly enough) in parameter space, and a simple criterion for determining when it cannot be transformed away by a phase transformation.

Suppose the Hamiltonian depends on the time through several parameters R_i. I will group them into a "vector" $\mathbf{R}$ even though they may not really be three coordinates of anything. In a molecule, R_i are the $3N$ coordinates of the nuclei. In the spin resonance example, R_i are the components of the magnetic field. Let the instantaneous eigenstates and eigenvalues satisfy

$$H(\mathbf{R})|\psi_n(\mathbf{R})\rangle = E_n(\mathbf{R})|\psi_n(\mathbf{R})\rangle \tag{10.125}$$

There is some phase freedom in the definition of these states. We insist only that they be defined in a way that is smooth in the space of parameters $\mathbf{R}$.

Suppose that the parameters vary with time, and that at $t = 0$ the system is in the state $|\psi_o[\mathbf{R}(0)]\rangle$. Suppose also that the motion is slow enough that the adiabatic approximation is valid, so at a later time

$$|\psi(t)\rangle = e^{i\gamma(t)} \exp\left(-i \int_0^t E_o(\mathbf{R}(t'))dt'\right) |\psi_o(\mathbf{R}(t))\rangle \tag{10.126}$$

There is an obvious generalization of the argument that led to equation (10.124). Define

$$\big|\boldsymbol{\nabla}\psi_o(\mathbf{R})\big\rangle = \boldsymbol{\nabla}\big|\psi_o(\mathbf{R})\big\rangle \tag{10.127}$$

and

$$\mathbf{A}(\mathbf{R}) = i\langle\psi_o(\mathbf{R}(t))|\,\boldsymbol{\nabla}\psi_o(\mathbf{R}(t))\rangle \tag{10.128}$$

Then

$$\dot\gamma = i\Big\langle\psi_o(t)\Big|\dot\psi_o(t)\Big\rangle = i\langle\psi_o[\mathbf{R}(t)]\,|\,\boldsymbol{\nabla}\psi_o[\mathbf{R}(t)]\rangle \cdot \frac{d\mathbf{R}}{dt} = \mathbf{A}(\mathbf{R}) \cdot \frac{d\mathbf{R}}{dt} \tag{10.129}$$

and

$$\gamma(t) = i \int_{R(0)}^{R(t)} \mathbf{A}(\mathbf{R}) \cdot d\mathbf{R} \tag{10.130}$$

If the integral is around a closed path C in parameter space such that $\mathbf{R}(t) = \mathbf{R}(0)$, then from Stokes' theorem

$$\gamma(t) = \int_C \mathbf{A}(\mathbf{R}) \cdot \mathbf{R} = \int \boldsymbol{\nabla} \times \mathbf{A} \cdot d\mathbf{S} \tag{10.131}$$

where the integral of the curl is now over any surface enclosed by C, and the curl is suitably generalized to the appropriate number of dimensions. The integral cannot vanish unless the integral of $\boldsymbol{\nabla} \times \mathbf{A}$ does, and that is a geometrical property of parameter space. This is the general definition of Berry's phase.

You can try to get rid of γ by a "gauge transformation":

$$|\psi_n(\mathbf{R})\rangle \to e^{i\lambda(\mathbf{R})}|\psi_n(\mathbf{R})\rangle \tag{10.132}$$

since then

$$\mathbf{A}(\mathbf{R}) \to \mathbf{A}(\mathbf{R}) - \boldsymbol{\nabla}\lambda(\mathbf{R}) \tag{10.133}$$

But for a closed loop the line integral is invariant under this transformation—it depends only on the "magnetic field" $\boldsymbol{\nabla} \times \mathbf{A}$.

Berry's phase is observable in principle. Imagine taking a coherent beam of particles in some state and passing it through two slits, or splitting it in half with a Stern-Gerlach apparatus. In one branch arrange to vary the external parameters adiabatically around a closed loop and then recombine the two halves. The phases will now be different and can interfere. Each half remembers its history.

There is a useful formula for $\boldsymbol{\nabla} \times \mathbf{A}$. From $\boldsymbol{\nabla}\langle\psi_o(\mathbf{R})|\psi_o(\mathbf{R})\rangle = 0$ it follows that
$$\operatorname{Re}\langle\psi_o(\mathbf{R})|\boldsymbol{\nabla}\psi_o(\mathbf{R})\rangle = 0 \tag{10.134}$$
and therefore
$$(\boldsymbol{\nabla} \times \mathbf{A})_i = -\operatorname{Im}\sum_{jk}\epsilon_{ijk}\boldsymbol{\nabla}_j\langle\psi_o|\boldsymbol{\nabla}_k\psi_o\rangle = -\operatorname{Im}\sum_{jk}\epsilon_{ijk}\langle\boldsymbol{\nabla}_j\psi_o|\boldsymbol{\nabla}_k\psi_o\rangle$$
$$= -\operatorname{Im}\sum_{ij}\sum_{n}\epsilon_{ijk}\langle\boldsymbol{\nabla}_j\psi_o|\psi_n\rangle\langle\psi_n|\boldsymbol{\nabla}_k\psi_o\rangle \tag{10.135}$$

In the last line I inserted a complete set of states. The term with $n=0$ vanishes, being the imaginary part of a real number. To evaluate the remaining terms, differentiate the definition (10.125):
$$(\boldsymbol{\nabla}_i H(\mathbf{R}))|\psi_n(\mathbf{R})\rangle + H(\mathbf{R})|\boldsymbol{\nabla}_i\psi_n(\mathbf{R})\rangle$$
$$= (\boldsymbol{\nabla}_i E_n(\mathbf{R}))|\psi_n(\mathbf{R})\rangle + E_n(\mathbf{R})|\boldsymbol{\nabla}_i\psi_n(\mathbf{R})\rangle \tag{10.136}$$
Take the scalar product with $|\psi_k(\mathbf{R})\rangle$ for $k \neq n$:
$$\langle\psi_k(\mathbf{R})|\boldsymbol{\nabla}_i\psi_n(\mathbf{R})\rangle = \frac{\langle\psi_k(\mathbf{R})|\boldsymbol{\nabla}_i H(\mathbf{R})|\psi_n(\mathbf{R})\rangle}{E_n(\mathbf{R}) - E_k(\mathbf{R})} \tag{10.137}$$
Thus
$$\boldsymbol{\nabla} \times \mathbf{A} = -\operatorname{Im}\sum_{n\neq 0}\frac{\langle\psi_o(\mathbf{R})|\boldsymbol{\nabla}H(\mathbf{R})|\psi_n(\mathbf{R})\rangle \times \langle\psi_n(\mathbf{R})|\boldsymbol{\nabla}H(\mathbf{R})|\psi_o(\mathbf{R})\rangle}{(E_n(\mathbf{R}) - E_o(\mathbf{R}))^2} \tag{10.138}$$

From equation (10.138) one learns that

- Since the states $|\boldsymbol{\nabla}\psi_n(\mathbf{R})\rangle$ do not appear in (10.138), the curl of $\mathbf{A}$ is independent of the way they depend on the parameters $\mathbf{R}$. In particular, $\boldsymbol{\nabla} \times \mathbf{A}(\mathbf{R})$ is unchanged when $|\psi_n(\mathbf{R})\rangle$ is multiplied by an $\mathbf{R}$ dependent phase.

- When another electronic state is degenerate with the ground state for some values of the parameters $\mathbf{R}$, then $\boldsymbol{\nabla} \times \mathbf{A}(\mathbf{R})$ is infinite. The surface integral that is the Berry phase is like the magnetic flux through a finite surface and will not be zero.

Equation (10.138) is easily generalized to parameter spaces of any dimension— I have used three dimensions as an example because the notation for the curl is familiar.[9]

---

[9] You can write down the general formula if you know a little differential geometry jargon. Replace $\boldsymbol{\nabla}$ by the exterior derivative and the cross product by the wedge product and integrate the resulting two-form over the two-dimensional surface whose boundary is the path in parameter space.

10.3.4 Two States near a Degeneracy

Let us return to the example of an electron in a time-dependent magnetic field $\mathbf{B}$, but now allow $\mathbf{B}(t)$ to be an arbitrary, slowly varying, function of the time. The Hamiltonian (with g = 2) is

$$H = \mu_B \mathbf{B}(t) \cdot \boldsymbol{\sigma} \tag{10.139}$$

The parameters R_i are the components of the magnetic field itself. The curl $\boldsymbol{\nabla} \times \mathbf{A}(\mathbf{B})$ can be computed directly from equation (10.138). $\mathbf{A}$ is not the magnetic vector potential whose curl is $\mathbf{B}$; rather it is defined by equation (10.128). The symbol $\boldsymbol{\nabla}$ refers to derivatives with respect to components of the parameters B_i. Thus

$$\boldsymbol{\nabla} H(\mathbf{B}) = \mu_B \boldsymbol{\sigma} \tag{10.140}$$

There are only two states, so if $|\psi_o(\mathbf{B})\rangle$ is the state with spin $-1/2$ along the direction of $\mathbf{B}$, the only state that enters the sum in (10.138) is the other one with spin $+1/2$. The components of $\boldsymbol{\nabla} \times \mathbf{A}(\mathbf{R})$ are

$$(\boldsymbol{\nabla} \times \mathbf{A}(\mathbf{B}))_i$$
$$= -\mathrm{Im} \sum_{jk} \epsilon_{ijk} \frac{\langle \psi_-(\mathbf{B})| \boldsymbol{\nabla}_j H(\mathbf{B}) |\psi_+(\mathbf{B})\rangle \langle \psi_+(\mathbf{B})| \boldsymbol{\nabla}_k H(\mathbf{B}) |\psi_-(\mathbf{B})\rangle}{\left(E_n(\mathbf{B}) - E_o(\mathbf{B})\right)^2} \tag{10.141}$$

The energy difference in the denominator is $E_+(\mathbf{B}) - E_-(\mathbf{B}) = 2\mu_B B$. The states $|\psi_\pm(\mathbf{B})\rangle$ themselves depend on the direction of $\mathbf{B}$ only and are given in equation (10.115).

Suppose $\mathbf{B}$ is along the z-axis: $\mathbf{B} = B\hat{\mathbf{n}}_z$. Then $\langle \psi_+|\sigma_k|\psi_-\rangle$ vanishes for $k = 3$, so $(\boldsymbol{\nabla} \times \mathbf{A})_i = 0$ unless $i = 3$. Since

$$\langle \psi_+|\sigma_1|\psi_-\rangle = 1 \quad \text{and} \quad \langle \psi_+|\sigma_2|\psi_-\rangle = -i \tag{10.142}$$

Thus

$$\boldsymbol{\nabla} \times \mathbf{A}(B\hat{\mathbf{n}}_z) = -\frac{1}{4B^2}\hat{\mathbf{n}}_z \mathrm{Im}\left(\langle \psi_+|\sigma_1|\psi_-\rangle^* \langle \psi_+|\sigma_2|\psi_-\rangle\right.$$
$$\left. - \langle \psi_+|\sigma_2|\psi_-\rangle^* \langle \psi_+|\sigma_1|\psi_-\rangle\right) = -\frac{1}{2B^2}\hat{\mathbf{n}}_z \tag{10.143}$$

Since there is nothing special about the z direction, for general $\mathbf{B}$ the curl of $\mathbf{A}$ must be

$$\boldsymbol{\nabla} \times \mathbf{A}(\mathbf{B}) = -\frac{\hat{\mathbf{B}}}{2|\mathbf{B}|^2} \tag{10.144}$$

In parameter space, where $\mathbf{B}$ is the coordinate, this is the field of a point charge at the origin. For a closed contour C, the integral of $\boldsymbol{\nabla} \times \mathbf{A}$ through any surface is the flux of this charge through the surface:

$$\int_C \mathbf{A}(\mathbf{B}) \cdot d\mathbf{B} = \int_S (\boldsymbol{\nabla} \times \mathbf{A}) \cdot d\mathbf{S} = \pm\frac{1}{2}\Omega \tag{10.145}$$

Ω is the solid angle subtended by C at the origin. The sign depends on the sense in which the curve is traversed.

There is more than one surface whose boundary is the curve C, and if the volume enclosed by any two of them contains the singularity at the origin, they will not give the same answer for the surface integral. Which is correct? The definition (10.115) for $|\psi_-(\mathbf{B})\rangle$ is not unique along the positive z-axis, so the "vector potential" $\mathbf{A}(\mathbf{B})$ that defines γ is not well defined there either.[10] The surface has to be the one that is not penetrated by the singularity.

For example, suppose the magnetic field has constant magnitude and describes a circle with θ fixed at θ_o and $0 \leq \phi \leq 2\pi$, as in the example in Section 10.3.2. The solid angle subtended by this circle (through a surface intersecting the z-axis at negative z) is the area of that part of the unit sphere with $\theta \geq \theta_o$: $\Omega = 2\pi(1+\cos\theta_o)$. The Berry phase is

$$\gamma = \mp \frac{1}{2}\Omega = \mp \pi(1 + \cos\theta) \tag{10.146}$$

in agreement with the result (10.124) from explicit calculation. But the formula (10.145) is much more general.

Berry pointed out that this spin-resonance example describes a more general situation. Suppose $\mathbf{R}$ is a general collection of parameters, and two states are degenerate for some value $\mathbf{R}^*$. Redefine the parameters so that $\mathbf{R}^* = 0$, and shift the energy so that the common energy of the two states is zero when $\mathbf{R} = \mathbf{R}^*$. To first order in $\mathbf{R}$ the other states can be ignored, the average energy is zero, and the Hamiltonian is a 2×2 traceless Hermitean matrix. So define three new coordinates by

$$H = \frac{1}{2}\begin{pmatrix} Z & X - iY \\ X + iY & -Z \end{pmatrix} \tag{10.147}$$

X, Y, and Z are some linear combinations of the components of $\mathbf{R}$. If there are more than three parameters, the other are irrelevant here. Then Berry's phase for motion in a loop in parameter space has the same geometry as the motion in the space of the components of $\mathbf{B}$ in the spin problem. It is the solid angle subtended by a closed loop at the new origin in parameter space, the point $X = Y = Z = 0$ where the two states are degenerate.

Berry's phase and similar geometrical phases have been observed in a large number of experiments. There is a short list in the references at the end of this chapter.

10.3.5 Fast and Slow Coordinates

Molecules are examples of quantum systems where the coordinates can be divided into "fast" coordinates and "slow" coordinates. The electron coordinates depend on the nuclear ones as external parameters. The solution to the electron problem provides an effective potential for the nucleons. The geometrical phase that appears in the dynamics of the electrons is transferred to the nuclear wave functions as a modification of the potential.[11]

---

[10] You could multiply the spinor in (10.115) by $\exp(-i\phi)$, but then $\mathbf{A}$ would be singular along the negative z-axis. There is no way to define it uniquely everywhere—the best you can do is a definition in a patch that excludes some infinitely long line. Mathematically this is the same as the **Dirac string** that is needed to define an ordinary magnetic vector potential for a magnetic monopole.

[11] See equation (7.163).

Section 10.3 Berry's Phase

The origin of this extra term is particularly clear in the path integral formalism. Let $\mathbf{r}$ be the electron coordinates and $\mathbf{R}$ the nuclear coordinates as in Section 7.6.1. The Lagrangian is

$$L = L_N + L_e \tag{10.148}$$

where (take all the nuclear masses equal for simplicity)

$$L_N = \frac{1}{2}M\dot{\mathbf{R}}^2 \quad \text{and} \quad L_e = \frac{1}{2}m\dot{\mathbf{r}}^2 - V(\mathbf{r},\mathbf{R}) \tag{10.149}$$

The propagator is

$$K(\mathbf{r}_b,\mathbf{R}_b,t_b;\mathbf{r}_a,\mathbf{R}_a,t_a) = \int \mathcal{D}\mathbf{R}\,\mathcal{D}\mathbf{r}\, e^{i\int_{t_a}^{t_b} L(\mathbf{r},\mathbf{R})dt}$$
$$= \int \mathcal{D}\mathbf{R}\, e^{i\int_{t_a}^{t_b} L_N(\mathbf{R})dt} \int \mathcal{D}\mathbf{r}\, e^{i\int_{t_a}^{t_b} L_e(\mathbf{r},\mathbf{R})dt} \tag{10.150}$$

The second factor is the electron propagator in the presence of some particular nuclear coordinates $\mathbf{R}$. As in Section 7.6.1, let $|\psi_n(\mathbf{R})\rangle$ be eigenstates of the instantaneous Hamiltonian $H_e(\mathbf{R})$:

$$H_e(\mathbf{R})|\psi_n(\mathbf{R})\rangle = E_n(\mathbf{R})|\psi_n(\mathbf{R})\rangle \tag{10.151}$$

and define

$$|\psi_n(\mathbf{R}),t)\rangle = |\psi_n(\mathbf{R})\rangle e^{-iE_n(\mathbf{R})t} \tag{10.152}$$

Then

$$\int \mathcal{D}\mathbf{r}\, e^{i\int_{t_a}^{t_b} L_e(\mathbf{r},\mathbf{R})dt} = \langle \mathbf{r}_b,\mathbf{R}_b|U(t_b,t_a)|\mathbf{r}_a,\mathbf{R}_a\rangle$$
$$= \sum_{mn} \psi_m(\mathbf{r}_b,\mathbf{R}_b)\langle\psi_m(\mathbf{R}_b)|U(t_b,t_a)|\psi_n(\mathbf{R}_a)\rangle \psi_n(\mathbf{r}_a,\mathbf{R}_a)^* \tag{10.153}$$

$\psi_m(\mathbf{R}_b)$ and $\psi_n(\mathbf{R}_a)$ are not eigenstates of the same Hamiltonian, so the matrix elements of the time-evolution operator are not diagonal.

The Born-Oppenheimer approximation is that $\mathbf{R}$ changes slowly, the adiabatic approximation holds, and $U(t_b,t_a)$ is indeed diagonal. In that approximation, if $|\psi(t_a)\rangle = |\psi_m(\mathbf{R}(t_a))\rangle$, then at a later time

$$|\psi(t)\rangle = e^{i\gamma(t)} \exp\left(-i\int_{t_a}^{t} E_n(\mathbf{R}(t'))dt'\right)|\psi_n(\mathbf{R}(t_a))\rangle \tag{10.154}$$

or

$$i\frac{d}{dt}|\psi(t)\rangle = \left[E_n(\mathbf{R}(t)) - \dot{\gamma}(t)\right]|\psi_n(\mathbf{R}(t))\rangle \tag{10.155}$$

where

$$\dot{\gamma} = i\left\langle \psi_n(\mathbf{R}(t))\middle|\dot{\psi}_n(\mathbf{R}(t))\right\rangle \tag{10.156}$$

Therefore,

$$|\psi(t_b)\rangle = \exp\left[-i\int_{t_a}^{t_b}\left(E_n(\mathbf{R}(t')) + i\left\langle\psi_n(t)\middle|\dot{\psi}_n(\mathbf{R}(t'))dt'\right\rangle\right)\right]|\psi_n(\mathbf{R}(t_a))\rangle \tag{10.157}$$

The exponential above is the matrix element of the time-evolution operator that appears on the right-hand side of equation (10.153), so the full propagator is

$$K(\mathbf{r}_b, \mathbf{R}_b, t_b; \mathbf{r}_a, \mathbf{R}_a, t_a)$$
$$\approx \int \mathcal{D}\mathbf{R}\, e^{i\int_{t_a}^{t_b} L_N(\mathbf{R}) dt} \sum_n \psi_n(\mathbf{r}(t_b), \mathbf{R}(t_b)) \psi_n\left(\mathbf{r}(t_a), \mathbf{R}(t_a)\right)^*$$
$$\times \exp\left[-i \int_{t_a}^{t_b} \left(E_n(\mathbf{R}(t)) - \mathbf{A}_n(\mathbf{R}(t)) \cdot \dot{\mathbf{R}}(t) \right)\right] \quad (10.158)$$

with $\mathbf{A}_n(\mathbf{R}) = i\langle \psi_n(\mathbf{R}) | \boldsymbol{\nabla}\psi_n(\mathbf{R})\rangle$ as above. Let $\phi(\mathbf{R})$ be the nuclear wave function, and approximate the full wave function by a product

$$\Psi(\mathbf{r}, \mathbf{R}) = \phi(\mathbf{R})\psi_o(\mathbf{r}, \mathbf{R}) \quad (7.155)$$

as in Section 7.6.1. The propagator relates the wave function at two different times:

$$\psi\left(\mathbf{r}(t_b), \mathbf{R}(t_b)\right) = \int d\mathbf{R}\, d\mathbf{r}\, K(\mathbf{r}_b, \mathbf{R}_b, t_b; \mathbf{r}_a, \mathbf{R}_a, t_a) \psi\left(\mathbf{r}(t_a), \mathbf{R}(t_a)\right) \quad (10.159)$$

The electron wave function can be integrated out:

$$\phi\left(\mathbf{R}(t_b)\right)$$
$$= \int d\mathbf{R}\, d\mathbf{r}\, K(\mathbf{r}_b, \mathbf{R}_b, t_b; \mathbf{r}_a, \mathbf{R}_a, t_a) \psi(\mathbf{r}(t_a), \mathbf{R}(t_a)) \psi\left(\mathbf{r}(t_b), \mathbf{R}(t_b)\right)^*$$
$$= \left(\int \mathcal{D}\mathbf{R}\, e^{iS_{\text{eff}}}\right) \phi\left(\mathbf{R}(t_a)\right) \quad (10.160)$$

and the effective action for the nuclear coordinates is

$$S_{\text{eff}} = \int_{t_a}^{t_b} L_{\text{eff}}\left(\mathbf{R}(t), \dot{\mathbf{R}}(t)\right) dt = \int_{t_a}^{t_b} \left(\frac{1}{2} M\dot{\mathbf{R}}^2 + A\left(\mathbf{R}(t)\right) \cdot \dot{\mathbf{R}}(t) - E_o\left(\mathbf{R}(t)\right) \right) dt \quad (10.161)$$

Besides the nuclear kinetic energy and the instantaneous electron energy, there is a term coming from the geometric phase factor in the electron sector.

The effective nuclear Hamiltonian can be obtained in the usual way from L_{eff}: Define the "canonical" momenta

$$P_i = \frac{\partial L_{\text{eff}}}{\partial \dot{R}_i} = M\dot{R}_i + A_i(\mathbf{R}) \quad (10.162)$$

The effective Hamiltonian is

$$H_{\text{eff}} = \mathbf{P} \cdot \dot{\mathbf{R}} - L_{\text{eff}} = \frac{(\mathbf{P} - \mathbf{A})^2}{2M} \quad (10.163)$$

This result was obtained in a more elementary (and probably more obscure) way in Section 7.6.1. But there was a small extra term in (7.163) that does not appear here. Order by order, the Lagrangian method is not the same approximation to the full problem as the Hamiltonian approach. A complete discussion of this question is in reference [16] of Chapter 7.

10.3.6 The Aharonov-Bohm Effect Again

Berry pointed out that the Aharonov-Bohm effect (Section 10.2.3) could be thought of as an example of a geometric phase. Consider a long solenoid of radius R_o containing a magnetic field $\mathbf{B}$ as in Figure 10.1. Outside the solenoid the magnetic field vanishes, but $\mathbf{A}$ is given by equation (10.98) and is not zero there.

Put the solenoid along the z-axis, and let an electron be confined to a small three-dimensional box whose center is at $\mathbf{R}$ in the x-y plane. The only important feature is that the box is small enough that all of it is outside the solenoid. When there is no magnetic field, the energy eigenfunctions are $\psi_n(\mathbf{r} - \mathbf{R})$, the usual functions measured relative to the center of the box.

When $\mathbf{A} \neq 0$ the eigenfunctions $\psi(\mathbf{r}; \mathbf{R})$ depend on $\mathbf{A}$ and therefore on $\mathbf{R}$ as a parameter. Make a gauge transformation

$$\mathbf{A}(\mathbf{r}) \to \mathbf{A}'(\mathbf{r}) = \mathbf{A}(\mathbf{r}) + \boldsymbol{\nabla}\Lambda(\mathbf{r}) \tag{10.164}$$

Then if $\psi(\mathbf{r}; \mathbf{R})$ is an energy eigenstate in the original gauge,[12]

$$\psi'(\mathbf{r}; \mathbf{R}) = e^{-ie\Lambda(\mathbf{r})}\psi(\mathbf{r}; \mathbf{R}) \tag{10.165}$$

will be an eigenstate in the new gauge. Since $\boldsymbol{\nabla} \times \mathbf{A} = 0$, the gauge function

$$\Lambda(\mathbf{r}) = -\int_{\mathbf{r}_o}^{\mathbf{r}} \mathbf{A}(\mathbf{r}') \cdot d\mathbf{r}' \tag{10.166}$$

is uniquely defined throughout the box (but not all space) for any $\mathbf{r}_o$. With this choice for the gauge transformation, in the new gauge $\mathbf{A}' = 0$ and $\psi'(\mathbf{r}; \mathbf{R})$ is a solution $\psi_n(\mathbf{r} - \mathbf{R})$ to Schrödinger's equation without the magnetic field. The wave functions for our confined electron can be taken to be

$$\psi(\mathbf{r}; \mathbf{R}) = \exp\left(-ie\int_{\mathbf{R}}^{\mathbf{r}} \mathbf{A}(\mathbf{r}') \cdot d\mathbf{r}'\right) \psi_n(\mathbf{r} - \mathbf{R}) \tag{10.167}$$

For each $\mathbf{R}$ I have chosen the wave function in the center of the box to be the same with or without the vector potential. That guarantees that $\psi(\mathbf{r}; \mathbf{R})$ is a smooth function of the parameters as required.

Now move the box slowly around a circle that encloses the solenoid. The gradient of the wave function with respect to the parameters is

$$\boldsymbol{\nabla}_{\mathbf{R}}\psi(\mathbf{r}; \mathbf{R}) = ie\mathbf{A}(\mathbf{R})\psi(\mathbf{r}; \mathbf{R}) - \exp\left(-ie\int_{\mathbf{R}}^{\mathbf{r}} \mathbf{A}(\mathbf{r}') \cdot d\mathbf{r}'\right) \boldsymbol{\nabla}_{\mathbf{r}}\psi_n(\mathbf{r} - \mathbf{R}) \tag{10.168}$$

and therefore

$$\langle\psi(\mathbf{R})|\,\boldsymbol{\nabla}\psi(\mathbf{R})\rangle = \int \psi(\mathbf{r}; \mathbf{R})^* \boldsymbol{\nabla}_{\mathbf{r}}\psi(\mathbf{r}; \mathbf{R}) d^3r = ie\mathbf{A}(\mathbf{R}) \tag{10.169}$$

(The other term vanishes because the states are normalized for any $\mathbf{R}$.) Except for a numerical factor, the geometric vector $\mathbf{A}$ defined in equation (10.128) in this arrangement is the same as the magnetic vector potential!

---
[12] See Section 10.2.2, especially equation (10.94).

If the box is moved all around the circle, after one cycle the accumulated phase is

$$\gamma = -e \oint \mathbf{A}(\mathbf{R}) \cdot d\mathbf{r} = -e \int (\nabla \times \mathbf{A}) \cdot \hat{\mathbf{n}} dS = -e\Phi \qquad (10.170)$$

where Φ is again the magnetic flux through the loop. This is another way to measure the Aharonov-Bohm phase, by moving one electron all around the solenoid rather than splitting an electron beam, recombining them, and looking for the difference in phases. The Aharonov-Bohm effect is seen to be an example of Berry's phase.

PROBLEMS

10.1. The Free-Particle Propagator

(a) Prove the closed form given for the integral (10.28). Hint: Use induction on N.

(b) Verify that the propagator (10.29) is the same one you would get from the operator expression

$$K(q_b, t_b; q_a, t_a) = \left\langle q_b \left| e^{-iH(t_b - t_a)} \right| q_a \right\rangle$$

for the free one-dimensional Hamiltonian $H = p^2/2m$.

(c) From the explicit form (10.29) of the nonrelativistic free particle propagator in one dimension show that

$$\lim_{t_b \to t_a} K(q_b, t_b; q_a, t_a) = \delta(q_b - q_a)$$

Why does this have to be so?

(d) Show that the free-particle propagator in one dimension (10.76) can be written

$$K_o(q_b, t_b; 0, 0) \sim \int_{-\infty}^{\infty} \frac{e^{-i(\omega t_b - k q_b)}}{k^2/2m - \omega - i\eta} dk \, d\omega$$

Find the constant of proportionality, and thus show that the Born series for the scattering matrix derived in Section 10.2.1 is the same as the one derived in Section 9.3.

10.2. Bound-State Aharonov-Bohm Effect

Consider an electron confined to a ring, as in Problem 6.1, but now instead of a constant magnetic field let the field be confined to an infinitely long cylinder whose axis is the z-axis and whose radius is $R_o < R$. The magnetic field is confined to a region completely inside the ring. (Take the cylinder, which is a solenoid or a magnet, to be infinitely long to avoid edge effects. Then it is possible for the field to be identically zero outside the cylinder.) To be specific, let $\mathbf{B} = B_o \hat{\mathbf{n}}_z$ for $\rho \leq R_o$ and $\mathbf{B} = 0$ for $\rho \geq R_o$, where $\rho^2 = x^2 + y^2$. Then, as in the description of the Aharonov-Bohm effect in Section 10.2.3 the vector potential can be taken to be

$$\mathbf{A}(\mathbf{r}) = \frac{1}{2} B_o \hat{\mathbf{n}}_z \times \mathbf{r} \quad (\rho \leq R_o)$$

$$\mathbf{A}(\mathbf{r}) = \frac{1}{2} \frac{R_o^2 B_o}{\rho} \hat{\mathbf{n}}_\phi \quad (\rho \geq R_o) \qquad (10.98)$$

What are the allowed energies now?

> Note: This problem is a bound state version of the Aharonov-Bohm effect. The energy levels are affected by the vector potential, even though the electron is constrained to be outside the region where there is a nonzero magnetic field.

10.3. Born-Oppenheimer Approximation

Consider a neutron confined to a circular ring in a magnetic field as described in Problem 6.6. In the limit $MR_o\omega_o \gg 1$ the spacing between the two magnetic levels is much greater than the spacing between adjacent rotational levels, and this problem is a candidate for the Born-Oppenheimer approximation. The angle ϕ is a slow coordinate, like the positions of the nuclei in a molecule. The fast coordinate is the spin. Unlike real molecular problems, the fast coordinate here has only two discrete eigenvalues, which are trivial to compute.

(a) First compute the energy the way Born and Oppenheimer might have: Holding ϕ fixed, determine the eigenvalues of the magnetic energy. Then, using that energy as a potential for the circular motion, solve the time independent Schrödinger equation for the energies E_n and the eigenfunctions $\psi_n(\phi)$ of the whole system. Is there any degeneracy?

(b) Calculate the geometrical phase for this problem and redo the energy calculation using the effective Hamiltonian (10.163) or (7.163). What answer do you get now for the energies? Is there any degeneracy?

(c) This toy problem is so simple that it can be solved exactly. Look up your answer to Problem 6.6 and show that in the limit $MR_o\omega_o \gg 1$ it agrees with the result of part (b) above, not part (a).

10.4. Adiabatic Solar Neutrinos

Nuclear reactions near the center of the sun produce the boron isotope ^{8}B, which undergoes beta decay into the beryllium isotope ^{8}Be, a positron, and an electron-type neutrino. The mean energy of the ν_e is 7.2 MeV. On its way through the sun and then on to the earth, the neutrino may oscillate into another neutrino type (suppose only ν_μ is important) according to the formula for neutrino oscillation in matter discussed in Section 6.3.5.

This is a two-state system. The Hamiltonian has two important terms, the Hamiltonian (6.104) for the neutrino in empty space, and the contribution V_e of the electrons in the surrounding matter. V_e is given by equation (6.116), with $N \approx 6 \times 10^{25}$ per cm^3 at the center of the sun, decreasing to zero at the sun's edge, $R_o \approx 700,000$ km.

The ν_e is produced in some linear combination of the two mass eigenstates, and in general it takes some detailed calculation to figure out what that combination has become by the time the neutrino reaches a detector on the earth. But if $\cos 2\theta \approx 1$ and if $\Delta m^2 \lesssim 10^{-5}$ eV, there is a simplification.[13]

(a) Show that with these values of the vacuum mass splitting and mixing angle, the term proportional to the electron density is at least ten times larger than the energies in the vacuum Hamiltonian. Then since the V_e term is diagonal in the ν_e, ν_μ basis, in the approximation that only this V_e term is important the neutrino will be produced in an eigenstate of the Hamiltonian.[14]

---

[13] θ is the vacuum mixing angle, and Δm^2 the difference of the squares of the masses of the vacuum eigenstates. The notation is explained in Section 6.3.

[14] In fact, since it is known that $V_e > 0$, the neutrino will be in the higher of the two states.

(b) The electron density is some smooth function of the sun's radial coordinate r, equal to the stated value at $r = 0$ and vanishing at $r = R_o$. Show that for any reasonable shape for the function $N_e(r)$, the condition (10.112) for the adiabatic approximation is valid as the neutrino travels out from the center. It follows that even though the energy eigenstates at $r = 0$ and at $r = R_o$ may be quite different combinations of $|\nu_e\rangle$ and $|\nu_\mu\rangle$, the electron will remain in a mass eigenstate.

(c) Between the sun and the earth only the vacuum Hamiltonian is nonzero, so the neutrino will remain in an eigenstate all the rest of the way to the earth also, and the probability that it can be detected as a ν_e can be calculated from the mass splitting and vacuum mixing angle alone. What is this probability in terms of Δm^2 and θ?

10.5. Precessing Electron and Instantaneous Eigenstates

For the electron in a precessing magnetic field discussed in Section 10.3.2, if at $t = 0$ the electron is in the spin-down eigenstate of the magnetic field, obtain the projection of the wave function (10.116) on the time-dependent instantaneous eigenstates of the Hamiltonian at later times.

REFERENCES

The idea of path integrals was suggested by Dirac and fully developed by Feynman in 1948.

[1] P. A. M. DIRAC, Physikalisches Z. der Sowjetunion **3** (1933), 64.

[2] R. P. FEYNMAN, Rev. Modern Phys. **20** (1948), 267.

[3] R. P. FEYNMAN, Phys. Rev. **80** (1950), 440.

These papers are reprinted in

[4] J. SCHWINGER (ed.), *Selected papers on quantum electrodynamics*, Dover, 1958.

There is a brief hint in Dirac's textbook, reference [9] in Chapter 2, page 128.

The book on nonrelativistic path integrals is

[5] R. P. FEYNMAN AND A. R. HIBBS, *Quantum mechanics and path integrals*, McGraw Hill, 1965.

There are many books from which you can learn more about path integrals in relativistic quantum field theory, including

[6] M. E. PESKIN AND D. V. SCHROEDER, *An introduction to quantum field theory*, Perseus, 1995.

[7] L. H. RYDER, *Quantum Field Theory*, Cambridge, 1985.

[8] S. WEINBERG, *The quantum theory of fields*, three volumes, Cambridge, 1995.

The Aharonov-Bohm effect was introduced by

[9] Y. AHARONOV AND D. BOHM, Phys. Rev. **115** (1959), 485.

also reprinted in

[10] A. Shapere and F. Wilczek (eds.), *Geometrical phases in physics*, World Scientific, 1989.

The idea can be traced to earlier work:

[11] W. EHRENBERG AND R. E. SIDAY, Proc. Phys. Soc. London **B62** (1949), 8.

Berry's paper appeared in

[12] M. V. BERRY, Proc. Roy. Soc. London **A392** (1984), 45.

Berry's phase has been observed in magnetic resonance experiments by

[13] D. SUTER, G. C. CHINGAS, R. A. HARRIS, AND A. PINES Molecular Physics **61** (1987) 1327.

[14] R. TYCHO, Phys. Rev. Letters **58** (1987) 2281.

in optical fiber experiments with ordinary light (photons)

[15] A. TOMITA AND R. CHIAO, Phys. Rev. Letters **57** (1986) 937.

and using slow neutrons in

[16] T. BITTER AND D. DUBBERS, Phys. Rev. Letters **59** (1987) 251.

How the geometrical phase gets transfered from the fast coordinates to the slow coordinates is explained in reference [16] of Chapter 7. More experimental references can be found in [10] above.

CHAPTER 11

The Quantized Electromagnetic Field

11.1 THE CLASSICAL ELECTROMAGNETIC FIELD HAMILTONIAN

The interaction of the electromagnetic field with an atom or a nucleus or an elementary particle is known exactly, and the rates can be computed to extraordinary accuracy. To do this, we need the quantum theory of the field. The semiclassical methods of Section 9.1 can be used to compute some first-order scattering processes, but they are not powerful enough to describe spontaneous electromagnetic decays at all. The analysis of transitions in the rest of Chapter 9 provides the machinery to do it right. The electric and magnetic fields are no less dynamical observables than the position and momentum of an electron.

Far from an interaction region, these fields will be free fields. The next few sections will develop the machinery for writing down a quantum mechanics of the free electromagnetic fields, that is, to represent them as operators, discover the Hamiltonian as a function of these field operators, and find its spectrum.

11.1.1 Maxwell's Equations and the Transverse Gauge Condition

What are the canonical coordinates and momenta of the free field? This is not entirely trivial, first, because the fields are continuous variables, not discrete ones like the p_i and q_i of Chapter 1. Second, because the six components of **E** and **B** are not independent dynamical variables. They are constrained by Maxwell's equations. First I will review the canonical Hamiltonian formulation of classical electrodynamics, and then use the canonical prescription to pass over to the quantum theory.

Maxwell's equations are two homogenous constraints[1]

$$\nabla \times \mathbf{E} + \frac{\partial \mathbf{B}}{\partial t} = 0 \qquad (11.1\text{a})$$

$$\nabla \cdot \mathbf{B} = 0 \qquad (11.1\text{b})$$

and two equations that determine the fields in terms of their sources, the electric current **j** and the charge density ρ:

$$\nabla \times \mathbf{B} - \frac{\partial \mathbf{E}}{\partial t} = 4\pi \mathbf{j} \qquad (11.2\text{a})$$

$$\nabla \cdot \mathbf{E} = 4\pi \rho \qquad (11.2\text{b})$$

---
[1] Remember that I use nonrationalized units, with $c = 1$.

Section 11.1 The Classical Electromagnetic Field Hamiltonian

Only the fields $\mathbf{E}$ and $\mathbf{B}$ are physically unambiguous, but it is much easier to use the electrostatic potential $\phi(\mathbf{r}, t)$ and the vector potential $\mathbf{A}(\mathbf{r}, t)$, because the potentials have simpler transformation properties and because the interaction is written in terms of the potentials using the minimal coupling prescription introduced in Section 6.1.2. The potentials determine the field strengths:

$$\mathbf{E} = -\boldsymbol{\nabla}\phi - \frac{\partial \mathbf{A}}{\partial t} \tag{11.3a}$$

$$\mathbf{B} = \boldsymbol{\nabla} \times \mathbf{A} \tag{11.3b}$$

The homogeneous equations (11.1) are a consequence of the definitions (11.3).

Electromagnetism is invariant under gauge transformation of the potentials.

$$\mathbf{A} \to \mathbf{A} + \boldsymbol{\nabla}\lambda \qquad \text{and} \qquad \phi \to \phi - \frac{\partial \lambda}{\partial t} \tag{11.4}$$

where $\lambda(x)$ is any scalar field. In principle any gauge will do, but in the Hamiltonian formalism there are a lot of irrelevant complications with a gauge condition that involves time derivatives. In particular we would get into trouble even with a Lorentz invariant gauge condition like

$$\frac{\partial \phi}{\partial t} + \boldsymbol{\nabla} \cdot \mathbf{A} = 0 \tag{11.5}$$

The best choice is the **transverse gauge**.

$$\boldsymbol{\nabla} \cdot \mathbf{A} = 0 \tag{11.6}$$

Note: Fixing the gauge this way means restricting the reference frame. If $\boldsymbol{\nabla} \cdot \mathbf{A}$ holds in one frame, it will not hold in another obtained from the first by a boost. Nevertheless all physical relations involving $\mathbf{E}$ and $\mathbf{B}$ will be unchanged by Lorentz transformations. The path integral formalism provides a technique for quantizing in a Lorentz invariant gauge. But I have to refer you to modern books on quantum electrodynamics. Some are listed at the end of Chapter 10.

Maxwell's equations in terms of the potentials are obtained by using the definitions (11.3) in the equations (11.2).

$$\frac{\partial^2 \mathbf{A}}{\partial t^2} + \boldsymbol{\nabla}\frac{\partial \phi}{\partial t} + \boldsymbol{\nabla} \times (\boldsymbol{\nabla} \times \mathbf{A}) = 4\pi \mathbf{j} \tag{11.7a}$$

$$\frac{\partial}{\partial t}\boldsymbol{\nabla} \cdot \mathbf{A} + \boldsymbol{\nabla}^2 \phi = -4\pi\rho \tag{11.7b}$$

In the transverse gauge (11.6) these simplify to

$$\frac{\partial^2 \mathbf{A}}{\partial t^2} + \boldsymbol{\nabla}\frac{\partial \phi}{\partial t} - \boldsymbol{\nabla}^2 \mathbf{A} = 4\pi \mathbf{j} \tag{11.8a}$$

$$\boldsymbol{\nabla}^2 \phi = -4\pi\rho \tag{11.8b}$$

358 Chapter 11 The Quantized Electromagnetic Field

In this gauge, $\phi(\mathbf{r},t)$ is completely determined by $\rho(\mathbf{r},t)$ at the same instant t. There is no differential equation in t for ϕ, and it is not a dynamically independent field. The solution to Poisson's equation (11.8b) is immediate

$$\phi(x) = \int \frac{\rho(\mathbf{r}')}{|\mathbf{r}-\mathbf{r}'|} d^3 r' \tag{11.9}$$

In the transverse gauge the electrostatic potential is always given by Coulomb's law, and for that reason it is sometimes called the **Coulomb gauge**.

Free Fields

When $\mathbf{j}$ and ρ vanish, $\mathbf{A}$ is a free field. Then from equation (11.8b) and an implied boundary condition at infinity, the electrostatic potential is identically zero. The free-field equations of motion are the three components of the wave equation

$$\frac{\partial^2 \mathbf{A}(\mathbf{r},t)}{\partial t^2} = \mathbf{\nabla}^2 \mathbf{A}(\mathbf{r},t) \tag{11.10}$$

Equation (11.10), the transverse gauge condition, and $\phi = 0$ together are equivalent to Maxwell's equations in the absence of sources.[2]

11.1.2 The Independent Modes

It is easier to disentangle the independent modes of $\mathbf{A}(\mathbf{r},t)$ in terms of its Fourier transform:

$$\mathbf{A}(\mathbf{r},t) = \frac{1}{(2\pi)^{\frac{3}{2}}} \int \mathbf{Q}(\mathbf{k},t) e^{i\mathbf{k}\cdot\mathbf{r}} d^3 k \tag{11.11}$$

Write $\mathbf{k} = \hat{\mathbf{n}}\omega$, where $\hat{\mathbf{n}}$ is a unit vector. The gauge condition (11.6) is

$$\mathbf{\nabla} \cdot \mathbf{A}(\mathbf{r},t) = \frac{1}{(2\pi)^{\frac{3}{2}}} \int i\omega \hat{\mathbf{n}} \cdot \mathbf{Q}(\mathbf{k},t) e^{i\mathbf{k}\cdot\mathbf{r}} d^3 k = 0 \tag{11.12a}$$

while the wave equation (11.10) is

$$\frac{1}{(2\pi)^{\frac{3}{2}}} \int \ddot{\mathbf{Q}}(\mathbf{k},t) e^{i\mathbf{k}\cdot\mathbf{r}} d^3 k = -\frac{1}{(2\pi)^{\frac{3}{2}}} \int \omega^2 \mathbf{Q}(\mathbf{k},t) e^{i\mathbf{k}\cdot\mathbf{r}} d^3 k \tag{11.12b}$$

Therefore,

$$\ddot{\mathbf{Q}}(\mathbf{k},t) + \omega^2 \mathbf{Q}(\mathbf{k},t) = 0 \tag{11.13a}$$

and

$$\hat{\mathbf{n}} \cdot \mathbf{Q}(\mathbf{k},t) = 0 \tag{11.13b}$$

for every $\mathbf{k}$. These two algebraic equations define the independent coordinates.

For each $\mathbf{k}$ the ordinary differential equation (11.13a) has two independent solutions, proportional to $e^{i\omega t}$ and $e^{-i\omega t}$ respectively. Each $\mathbf{Q}(\mathbf{k},t)$ is a linear combination of these two:

$$\mathbf{Q}(\mathbf{k},t) = \mathbf{c}(\mathbf{k},t) + \mathbf{c}'(\mathbf{k},t) = \mathbf{c}(\mathbf{k}) e^{-i\omega t} + \mathbf{c}'(\mathbf{k}) e^{i\omega t} \tag{11.14}$$

---
[2]Since in the tranverse gauge equation (11.10) describes the motion of free fields it is, confusingly, sometimes called the **radiation gauge**.

and
$$\dot{\mathbf{Q}}(\mathbf{k},t) = -i\omega\big[\mathbf{c}(\mathbf{k},t) - \mathbf{c}'(\mathbf{k},t)\big] \tag{11.15}$$

In terms of $\mathbf{Q}$ and $\dot{\mathbf{Q}}$ the vector potential is

$$\mathbf{A}(\mathbf{r},t) = \frac{1}{(2\pi)^{\frac{3}{2}}} \int \big[\mathbf{c}(\mathbf{k},t) + \mathbf{c}'(\mathbf{k},t)\big] e^{i\mathbf{k}\cdot\mathbf{r}} d^3k$$

$$= \frac{1}{(2\pi)^{\frac{3}{2}}} \int \big[\mathbf{c}(\mathbf{k},t)e^{i\mathbf{k}\cdot\mathbf{r}} + \mathbf{c}'(-\mathbf{k},t)e^{-i\mathbf{k}\cdot\mathbf{r}}\big] d^3k \tag{11.16}$$

Both forms of equation (11.16) will be useful. Since $\mathbf{A}(\mathbf{r},t)$ is real (it will be Hermitean in quantum mechanics), then

$$\mathbf{c}'(\mathbf{k},t) = \mathbf{c}(-\mathbf{k},t)^* \tag{11.17}$$

and so

$$\mathbf{A}(\mathbf{r},t) = \frac{1}{(2\pi)^{\frac{3}{2}}} \int \big[\mathbf{c}(\mathbf{k},t)e^{i\mathbf{k}\cdot\mathbf{r}} + \mathbf{c}(\mathbf{k},t)^*e^{-i\mathbf{k}\cdot\mathbf{r}}\big] d^3k \tag{11.18}$$

also.

The three $c_i(\mathbf{k},t)$ are not independent observables because of the transverse gauge condition

$$\hat{\mathbf{n}} \cdot \mathbf{c}(\mathbf{k},t) = 0 \tag{11.19}$$

For each $\mathbf{k}$ define two unit vectors—it does not matter exactly how they are defined, as long they are orthonormal and transverse. $\hat{\varepsilon}_\alpha(\mathbf{k})$ orthogonal to each other and to $\hat{\mathbf{n}}$:

$$\hat{\varepsilon}_\alpha(\mathbf{k})^* \cdot \hat{\varepsilon}_\beta(\mathbf{k}) = \delta_{\alpha\beta} \qquad \text{and} \qquad \hat{\varepsilon}_\alpha(\mathbf{k}) \cdot \hat{\mathbf{n}} = 0 \tag{11.20}$$

Then the general solution to equation (11.19) is

$$\mathbf{c}(\mathbf{k},t) = \sum_{\alpha=1}^{2} \hat{\varepsilon}_\alpha(\mathbf{k},t) c_\alpha(\mathbf{k},t) \tag{11.21}$$

Then

$$\mathbf{A}(\mathbf{r},t) = \frac{1}{(2\pi)^{\frac{3}{2}}} \sum_{\alpha=1}^{2} \int \big[\hat{\varepsilon}_\alpha(\mathbf{k})c_\alpha(\mathbf{k},t)e^{i\mathbf{k}\cdot\mathbf{r}} + \hat{\varepsilon}_\alpha(\mathbf{k})^* c_\alpha(\mathbf{k},t)^* e^{-i\mathbf{k}\cdot\mathbf{r}}\big] d^3k \tag{11.22}$$

and

$$\dot{\mathbf{A}} = -\frac{i}{(2\pi)^{\frac{3}{2}}} \sum_{\alpha} \int \omega \times \big[\hat{\varepsilon}_\alpha(\mathbf{k})c_\alpha(\mathbf{k},t)e^{i\mathbf{k}\cdot\mathbf{r}} - \hat{\varepsilon}_\alpha(\mathbf{k})^* c_\alpha(\mathbf{k},t)^* e^{-i\mathbf{k}\cdot\mathbf{r}}\big] d^3k$$

$$= -i\frac{1}{(2\pi)^{\frac{3}{2}}} \int \omega \big[\mathbf{c}(\mathbf{k},t)e^{i\mathbf{k}\cdot\mathbf{r}} - \mathbf{c}(\mathbf{k},t)^* e^{-i\mathbf{k}\cdot\mathbf{r}}\big] d^3k \tag{11.23}$$

The functions $c_\alpha(\mathbf{k},t)$, two for each $\mathbf{k}$, are almost the independent dynamical variables we are looking for, but they are not quite the canonical p's and q's. The p's and q's are the combinations that make Hamilton's equations come out right.

11.1.3 The Classical Hamiltonian

The Hamiltonian in terms of **E** and **B** is the energy of the electromagnetic field from classical electrodynamics:

$$H = \frac{1}{8\pi} \int \left(\mathbf{E}^2 + \mathbf{B}^2 \right) d^3r \tag{11.24}$$

What are **E** and **B** in terms of the independent modes?
The Fourier transform of the electric field $\mathbf{E}(\mathbf{r},t) = -\dot{\mathbf{A}}(\mathbf{r},t)$ is

$$\mathbf{E}(\mathbf{r},t) = i\frac{1}{(2\pi)^{\frac{3}{2}}} \int \omega \left[\mathbf{c}(\mathbf{k},t)e^{i\mathbf{k}\cdot\mathbf{r}} - \mathbf{c}(\mathbf{k},t)^* e^{-i\mathbf{k}\cdot\mathbf{r}} \right] d^3k \tag{11.25a}$$

$$= i\frac{1}{(2\pi)^{\frac{3}{2}}} \int \omega \left[\mathbf{c}(\mathbf{k},t) - \mathbf{c}(-\mathbf{k},t)^* \right] e^{i\mathbf{k}\cdot\mathbf{r}} d^3k \tag{11.25b}$$

Since $\mathbf{E}(\mathbf{r},t)$ is real,

$$\mathbf{E}(\mathbf{r},t) = -i\frac{1}{(2\pi)^{\frac{3}{2}}} \int \omega \left[\mathbf{c}(\mathbf{k},t)^* - \mathbf{c}(-\mathbf{k},t) \right] e^{-i\mathbf{k}\cdot\mathbf{r}} d^3k \tag{11.25c}$$

also. Similarly (with $\mathbf{k} = \omega \hat{\mathbf{n}}$) write $\mathbf{B}(\mathbf{r},t) = \nabla \times \mathbf{A}(\mathbf{r},t)$ as

$$\mathbf{B}(\mathbf{r},t) = \frac{i}{(2\pi)^{\frac{3}{2}}} \int \omega \hat{\mathbf{n}} \times \left[\mathbf{c}(\mathbf{k},t) + \mathbf{c}(-\mathbf{k},t)^* \right] e^{i\mathbf{k}\cdot\mathbf{r}} d^3k \tag{11.26a}$$

$$= -\frac{i}{(2\pi)^{\frac{3}{2}}} \int \omega \hat{\mathbf{n}} \times \left[\mathbf{c}(\mathbf{k},t)^* + \mathbf{c}(-\mathbf{k},t)e^{-i\mathbf{k}\cdot\mathbf{r}} \right] d^3k \tag{11.26b}$$

The first term in the Hamiltonian (11.24) is

$$\frac{1}{8\pi} \int \mathbf{E} \cdot \mathbf{E}^* d^3r = \frac{1}{8\pi} \frac{1}{(2\pi)^3} \int \omega' \omega e^{i(\mathbf{k}-\mathbf{k}')\cdot \mathbf{r}}$$
$$\times \left[\mathbf{c}(\mathbf{k},t) - \mathbf{c}^*(-\mathbf{k},t) \right] \cdot \left[\mathbf{c}^*(\mathbf{k}',t) - \mathbf{c}(-\mathbf{k}',t) \right] d^3k' \, d^3k \, d^3r \tag{11.27}$$

The integral over **r** is $(2\pi)^3 \delta_3(\mathbf{k}-\mathbf{k}')$. Use the δ-function to integrate over $\mathbf{k}'$. The result is (in the second term change $-\mathbf{k}$ to $\mathbf{k}$ as integration variable)

$$\frac{1}{8\pi} \int \mathbf{E} \cdot \mathbf{E} d^3r = \frac{1}{8\pi} \int \omega^2 \Big[\mathbf{c}(\mathbf{k},t) \cdot \mathbf{c}^*(\mathbf{k},t) + \mathbf{c}^*(\mathbf{k},t) \cdot \mathbf{c}(\mathbf{k},t)$$
$$- \mathbf{c}(\mathbf{k},t) \cdot \mathbf{c}(-\mathbf{k},t) - \mathbf{c}^*(\mathbf{k},t) \cdot \mathbf{c}^*(-\mathbf{k},t) \Big] d^3k \tag{11.28}$$

Similarly[3]

$$\frac{1}{8\pi} \int \mathbf{B} \cdot \mathbf{B} d^3r = \frac{1}{8\pi} \int \omega^2 \Big[\mathbf{c}(\mathbf{k},t) \cdot \mathbf{c}^*(\mathbf{k},t) + \mathbf{c}^*(\mathbf{k},t) \cdot \mathbf{c}(\mathbf{k},t)$$
$$+ \mathbf{c}(\mathbf{k},t) \cdot \mathbf{c}(-\mathbf{k},t) + \mathbf{c}^*(\mathbf{k},t) \cdot \mathbf{c}^*(-\mathbf{k},t) \Big] d^3k \tag{11.29}$$

---

[3] $\hat{\mathbf{n}} \cdot \hat{\mathbf{n}} = 1$ and $\hat{\mathbf{n}} \cdot \mathbf{c}(\mathbf{k},t) = 0$, so that $(\hat{\mathbf{n}} \times \mathbf{c}) \cdot (\hat{\mathbf{n}} \times \mathbf{c}^*) = \mathbf{c} \cdot \mathbf{c}^*$

Section 11.1 The Classical Electromagnetic Field Hamiltonian

When the $\mathbf{E}^2$ and $\mathbf{B}^2$ terms are added, the terms in $\mathbf{c} \cdot \mathbf{c}$ and $\mathbf{c}^* \cdot \mathbf{c}^*$ cancel. The Hamiltonian is

$$H = \frac{1}{4\pi} \int \omega^2 \Big(\mathbf{c}(\mathbf{k},t) \cdot \mathbf{c}^*(\mathbf{k},t) + \mathbf{c}^*(\mathbf{k},t) \cdot \mathbf{c}(\mathbf{k},t) \Big) d^3k$$
$$= \frac{1}{2\pi} \int \omega^2 \mathbf{c}(\mathbf{k},t) \cdot \mathbf{c}^*(\mathbf{k},t) d^3k = \sum_\alpha \int \mathcal{H}_\alpha(\mathbf{k}) \, d^3k \quad (11.30)$$

where [because $\hat{\varepsilon}_\alpha^*(\mathbf{k}) \cdot \hat{\varepsilon}_\beta(\mathbf{k}) = \delta_{\alpha\beta}$]

$$\mathcal{H}_\alpha(\mathbf{k}) = \frac{\omega^2}{2\pi} c_\alpha^*(\mathbf{k},t) c_\alpha(\mathbf{k},t) = \frac{\omega^2}{2\pi} |c_\alpha(\mathbf{k})|^2 \quad (11.31)$$

The time dependence has disappeared. As it should be, H is independent of t.

11.1.4 The Canonical Coordinates

The final result expressed in equations (11.30) and (11.31) is remarkably simple. $\mathcal{H}_\alpha(\mathbf{k})$ is a function of the $c_\alpha(\mathbf{k})$ whose time dependence is proportional to $e^{-i\omega t}$. Thus $\mathcal{H}_\alpha(\mathbf{k})$ is just a harmonic oscillator Hamiltonian, and the canonical coordinates and momenta can be chosen so that harmonic oscillation is the solution to Hamilton's equations for the time dependence. The canonical coordinates and momenta are some real combinations of the complex numbers $c_\alpha(\mathbf{k},t)$.

Define two real observables proportional to the real and imaginary parts of $c_\alpha(\mathbf{k},t)$ by setting

$$c_\alpha(\mathbf{k},t) = \sqrt{\frac{\pi}{\omega^2}} \Big(\omega x_\alpha(\mathbf{k},t) + i p_\alpha(\mathbf{k},t) \Big) \quad (11.32)$$

or

$$x_\alpha(\mathbf{k},t) = \sqrt{\frac{1}{4\pi}} \Big(c_\alpha(\mathbf{k},t) + c_\alpha(\mathbf{k},t)^* \Big) \quad (11.33a)$$

$$p_\alpha(\mathbf{k},t) = -i\sqrt{\frac{\omega^2}{4\pi}} \Big(c_\alpha(\mathbf{k},t) - c_\alpha(\mathbf{k},t)^* \Big) \quad (11.33b)$$

Then

$$\boxed{\mathcal{H}_\alpha(\mathbf{k}) = \frac{1}{2} p_\alpha(\mathbf{k},t)^2 + \frac{1}{2} \omega^2 x_\alpha(\mathbf{k},t)^2} \quad (11.34)$$

These are the desired coordinates and momenta. The field turns out to be equivalent to a collection of oscillators.

You might be worried that there is some problem with observables like $x_\alpha(\mathbf{k})$ that have a continuous label, but the Hamiltonian formalism works in this case also. The Lagrangian for this system must be

$$L(t) = \sum_\alpha \int \left[\frac{1}{2} \dot{x}_\alpha(\mathbf{k},t)^2 - \frac{1}{2} \omega^2 x_\alpha(\mathbf{k},t)^2 \right] d^3k \quad (11.35)$$

362 Chapter 11 The Quantized Electromagnetic Field

Hamilton's principle takes the form

$$0 = \delta \int_{t_1}^{t_2} L(t)dt = \delta \int_{t_1}^{t_2} \sum_\alpha \int \left[\dot{x}_\alpha(\mathbf{k},t)\delta\dot{x}_\alpha(\mathbf{k},t) - \omega^2 x_\alpha(\mathbf{k},t)\delta x_\alpha(\mathbf{k},t)\right] d^3k$$

$$= \delta \int_{t_1}^{t_2} \sum_\alpha \int \left[-\frac{d}{dt}\dot{x}_\alpha(\mathbf{k},t) - \omega^2 x_\alpha(\mathbf{k},t)\right] \delta x_\alpha(\mathbf{k},t) d^3k \quad (11.36)$$

so the Euler-Lagrange equations of motion are

$$\ddot{x}_\alpha(\mathbf{k},t) = -\omega^2 x_\alpha(\mathbf{k},t) \quad (11.37)$$

Then with

$$p_\alpha(\mathbf{k},t) = \dot{x}_\alpha(\mathbf{k},t) \quad (11.38)$$

the Hamiltonian as a function of the coordinates and the momenta is

$$\begin{aligned} H &= \sum_\alpha \int p_\alpha(\mathbf{k},t)\dot{x}_\alpha(\mathbf{k},t) d^3k - L \\ &= \sum_\alpha \int \left[\frac{1}{2}p_\alpha(\mathbf{k},t)^2 + \frac{1}{2}\omega^2 x_\alpha(\mathbf{k},t)^2\right] d^3k \end{aligned} \quad (11.39)$$

in agreement with (11.34). In terms of the coordinates and momenta, the equations of motion are exactly Hamilton's equations:

$$\frac{dx_\alpha(\mathbf{k},t)}{dt} = p_\alpha(\mathbf{k},t) \quad \text{and} \quad \frac{dp_\alpha(\mathbf{k},t)}{dt} = -\omega^2 x_\alpha(\mathbf{k},t) \quad (11.40)$$

This is the correct canonical description of the free electromagnetic field.

11.2 THE QUANTIZED RADIATION FIELD

11.2.1 The Heisenberg Picture

Requiring the canonical coordinates and momenta to satisfy the canonical commutation relations will ensure that the Heisenberg picture operators $\mathbf{A}(\mathbf{r},t)$ obey the wave equation (11.10). (In the Schrödinger picture this is true of their expectation values.) At the end we will have to transform back to the Schrödinger picture in order to apply the transition rate formulas from Chapter 9.

11.2.2 Canonical Quantization

$x_\alpha(\mathbf{k},t)$ and $p_\alpha(\mathbf{k},t)$ defined in equation (11.33) are both real numbers in classical mechanics; in the quantum theory they are Hermitean operators. The standard prescription is

$$[x_\alpha(\mathbf{k},t), x_{\alpha'}(\mathbf{k}',t)] = [p_\alpha(\mathbf{k},t), p_{\alpha'}(\mathbf{k}',t)] = 0 \quad (11.41a)$$

$$[x_\alpha(\mathbf{k},t), p_{\alpha'}(\mathbf{k}',t)] = i\delta_{\alpha'\alpha}\delta_3(\mathbf{k}' - \mathbf{k}) \quad (11.41b)$$

The coefficient of the Dirac delta function is chosen so that the commutators with the Hamiltonian are the usual Hamilton's equations. For example, the commutator of x_β with the Hamiltonian is[4]

$$[H, x_\beta(\mathbf{k}', t)] = -ip_\beta(\mathbf{k}') = -i\frac{\partial x_\beta(\mathbf{k}', t)}{\partial t} \tag{11.42}$$

The commutation relations are equivalent to the correct equations of motion. That is all that is required for a consistent set of rules. Since the time-translation operator e^{-iHt} is unitary, these rules hold at all times.

It is useful to define the analogs of the raising and lowering operators for the harmonic oscillator. Let

$$a_\alpha(\mathbf{k}, t) = \sqrt{\frac{\omega}{2\pi}} c_\alpha(\mathbf{k}, t) = \sqrt{\frac{1}{2\omega}} [\omega x_\alpha(\mathbf{k}, t) + ip_\alpha(\mathbf{k}, t)] \tag{11.43}$$

The time dependence of these operators is

$$\frac{da_\alpha(\mathbf{k}, t)}{dt} = -i\omega a_\alpha(\mathbf{k}, t) \tag{11.44}$$

The expansion (11.18) becomes

$$\mathbf{A}(\mathbf{r}, t) = \frac{(4\pi)^{1/2}}{(2\pi)^{3/2}} \int \frac{1}{\sqrt{2\omega}} \left[\mathbf{a}(\mathbf{k}, t) e^{i\mathbf{k}\cdot\mathbf{r}} + \mathbf{a}^\dagger(\mathbf{k}, t) e^{-i\mathbf{k}\cdot\mathbf{r}} \right] d^3k \tag{11.45}$$

where

$$\mathbf{a}(\mathbf{k}, t) = \sum_\alpha \hat{\varepsilon}_\alpha(\mathbf{k}) a_\alpha(\mathbf{k}, t) \tag{11.46}$$

The $a_\alpha(\mathbf{k}, t)$ defined in equation (11.43) are related to $x_\alpha(\mathbf{k}, t)$ and $p_\alpha(\mathbf{k}, t)$ just like the lowering operator in the one-dimensional harmonic oscillator problem. From equations (11.41a) and (11.41b) they satisfy the rules

$$[a_\alpha(\mathbf{k}, t), a_{\alpha'}(\mathbf{k}', t)] = [a_\alpha^\dagger(\mathbf{k}, t), a_{\alpha'}^\dagger(\mathbf{k}', t)] = 0$$
$$[a_\alpha(\mathbf{k}, t), a_{\alpha'}^\dagger(\mathbf{k}', t)] = \delta_{\alpha'\alpha}\delta_3(\mathbf{k}' - \mathbf{k}) \tag{11.47}$$

In terms of the $a_\alpha(\mathbf{k}, t)$ and $a_\alpha^\dagger(\mathbf{k}, t)$ the Hamiltonian is a sum of harmonic oscillators, one for each mode labeled by $\mathbf{k}$ and α.

$$H = \int d^3k \sum_\alpha \mathcal{H}_\alpha(\mathbf{k}) \tag{11.48}$$

where

$$\mathcal{H}_\alpha(\mathbf{k}) = \frac{1}{2}\omega \left[a_\alpha(\mathbf{k}, t) a_\alpha^\dagger(\mathbf{k}, t) + a_\alpha^\dagger(\mathbf{k}, t) a_\alpha(\mathbf{k}, t) \right] \tag{11.49}$$

This Hamiltonian (11.48) is the same in the Schrödinger picture as in the Heisenberg picture, since the time dependence cancels. We know its spectrum. Set $N_\alpha(\mathbf{k}) = a_\alpha^\dagger(\mathbf{k}) a_\alpha(\mathbf{k})$. Then

$$\mathcal{H}_\alpha(\mathbf{k}) = \omega \left(N_\alpha(\mathbf{k}) + \frac{1}{2}\delta_3(0) \right) \tag{11.50}$$

---

[4] See Problem 11.2.

The eigenstates are labeled by a quantum number $n_\alpha(\mathbf{k})$ for *each* α and $\mathbf{k}$, and their energy is

$$\sum_\alpha \int d^3k \omega \left(n_\alpha(\mathbf{k}) + \frac{1}{2}\delta_3(0) \right) \tag{11.51}$$

11.2.3 Photons

In the present context $n_\alpha(\mathbf{k})$ is called the **occupation number** of the mode α, $\mathbf{k}$. Increasing $n_\alpha(\mathbf{k})$ by one unit increases the energy by ω, where $\omega = ck$. The eigenstates are states with each possible value for each occupation number, or equivalently $n_\alpha(\mathbf{k})$ **photons** with momentum $\mathbf{k}$ and polarization vector $\hat{\varepsilon}(\mathbf{k})$.

$$N_{\alpha i}(\mathbf{k})\big| n_{\alpha_1}(\mathbf{k}_1), n_{\alpha_2}(\mathbf{k}_2), \cdots \rangle = n_{\alpha i}(\mathbf{k})\big| n_{\alpha_1}(\mathbf{k}_1), n_{\alpha_2}(\mathbf{k}_2), \cdots \rangle \tag{11.52}$$

Each "photon" of a given $\mathbf{k}$ has the same energy *because the harmonic oscillator levels are equally spaced.*

The operators $a_\alpha(\mathbf{k})$ and their Hermitean conjugates are now **annihilation** and **creation** operators; they destroy or create photons with polarization $\hat{\varepsilon}$ and momentum $\mathbf{k}$ (I still have to show some of these properties). And finally, continuing the analogy with the multidimensional harmonic oscillator algebra

$$a^\dagger_{\alpha_i}(\mathbf{k}_i)\big| n_{\alpha_1}(\mathbf{k}_1), n_{\alpha_2}(\mathbf{k}_2), \cdots, n_{\alpha_i}(\mathbf{k}_i), \cdots \rangle$$
$$= \sqrt{n_{\alpha_i}(\mathbf{k}_i) + 1}\,\big| n_{\alpha_1}(\mathbf{k}_1), n_{\alpha_2}(\mathbf{k}_2), \cdots, n_{\alpha_i}(\mathbf{k}_i) + 1, \cdots \rangle \tag{11.53}$$

There appears to be a problem. The state labeled $|0, 0, 0, \cdots\rangle$ has the energy

$$E_o = \sum_\alpha \int d^3k \frac{\omega}{2}\delta_3(0) \tag{11.54}$$

This is the energy of the ground state, the state with no photons at all, the physical vacuum. Each mode, in this formalism, has a ground-state energy of $\omega/2$.

For the one-dimensional oscillator this energy of the ground state was required by the uncertainty principle. Here, it is the first of the infinities that plague relativistic quantum mechanics. It is the easiest of them to get rid of. Since the zero of the energy has no physical meaning here, just change the zero of each mode by $\omega/2$, or equivalently, subtract off this annoying infinite term.

> Note: The zero-point energy of the electromagnetic field does couple the gravitational field. It is part of the energy of the vacuum that shows up as the cosmological constant in general relativity. But there is no way to sort out the various contributions—including higher order electromagnetic corrections—to this unmeasurable constant.

Be careful to distinguish two sources of infinity in the formula for the ground-state energy. The Dirac delta function appears in the expression (11.51) because of the continuum normalization. It is in order to avoid ugly objects like this that some textbooks use a box normalization, putting the whole world inside a large but finite box, replacing $\delta_3(\mathbf{k}' - \mathbf{k})$ by a Kronecker delta $\delta_{\mathbf{k}',\mathbf{k}}$. But that trick will not get rid of the infinity that comes from summing $\omega/2$, the ground-state energy of each mode, over an infinite number of modes. Eliminating that infinity really requires subtracting off the vacuum energy. So from now on H is just redefined as

$$H = \sum_\alpha \int d^3k \, \omega a_\alpha^\dagger(\mathbf{k}) a_\alpha(\mathbf{k}) \tag{11.55}$$

The energy of each eigenstate is just the sum of the energies ω of the photons in it.

Equation (11.55), with the ground-state energy omitted, is not such a revolutionary step as it may seem. You can think of it as an example of the ordering ambiguity of the canonical quantization prescription. It differs from the Hamiltonian we started with only by terms that, like $x_\alpha(\mathbf{k})p_\beta(\mathbf{k}') - p_\beta(\mathbf{k}')x_\alpha(\mathbf{k})$, are zero in the classical theory where everything commutes. "Subtracting off" the ground-state energy is the same as choosing (11.31) instead of (11.34) as the starting point for quantizing the classical theory. This is a good time to read again the discussion of the ground-state energy of the harmonic oscillator in Section 3.2.3.

Photons are Bosons

The creation operators commute; they are like raising operators for independent harmonic oscillators. There is no distinction between the first and second photons. *Photons are bosons.*

The states $a_\alpha^\dagger(\mathbf{k})|0,0,0,\cdots\rangle$ have energy ω. Multiparticle states like

$$a_{\alpha_1}^\dagger(\mathbf{k}_1) a_{\alpha_2}^\dagger(\mathbf{k}_2) \cdots a_{\alpha_N}^\dagger(\mathbf{k}_N)|0,0,0,\ldots\rangle \tag{11.56}$$

form the complete set of energy eigenstates. States like those in (11.56) are the prototype of a basis in the vector space for any collection of bosons.

11.3 PROPERTIES OF THE QUANTUM ELECTROMAGNETIC FIELD

The states in equation (11.56) are a complete set of eigenstates of the Hamiltonian H_o. There still remains a large degeneracy, many states with the same energy. In this section I will isolate a complete set of commuting observables to use in labeling the photon states. Since the free electromagnetic field is invariant under translations and rotations, the momentum and the angular momentum will be among these.

11.3.1 The Momentum of the Field

From (11.45) the fields are

$$\mathbf{E}(\mathbf{r},t) = -\frac{\partial \mathbf{A}}{\partial t} = \frac{(4pi)^{1/2}}{(2\pi)^{3/2}} i \int \frac{1}{\sqrt{2\omega}} \omega \left[\mathbf{a}(\mathbf{k},t) - \mathbf{a}^\dagger(-\mathbf{k},t)\right] e^{i\mathbf{k}\cdot\mathbf{r}} d^3k \tag{11.57}$$

and

$$\mathbf{B}(\mathbf{r},t) = \mathbf{\nabla} \times \mathbf{A} = -\frac{(4pi)^{1/2}}{(2\pi)^{3/2}} i \int \frac{1}{\sqrt{2\omega}} \omega \hat{n} \times \left[\mathbf{a}(-\mathbf{k},t) + \mathbf{a}^\dagger(\mathbf{k},t)\right] e^{-i\mathbf{k}\cdot\mathbf{r}} d^3k \tag{11.58}$$

The energy flux (the energy crossing a square centimeter per second) is the Poynting vector $\mathbf{N} = (1/4\pi)\mathbf{E} \times \mathbf{B}$. Since photons move at speed $c = 1$ and have momentum

366 Chapter 11 The Quantized Electromagnetic Field

$k = \omega$ equal to their energy, the momentum density is $\mathbf{N}/c^2$. Classical electromagnetic theory gives the same result. The momentum of the electromagnetic field is[5]

$$\mathbf{P} = \frac{1}{4\pi}\int \mathbf{E} \times \mathbf{B}\, d^3r = \frac{1}{2(2\pi)^3}\int \sqrt{\omega'\omega}$$

$$\times \left[\mathbf{a}(\mathbf{k},t) - \mathbf{a}^\dagger(-\mathbf{k},t)\right] \times \left[\hat{\mathbf{n}}' \times (\mathbf{a}(-\mathbf{k}',t) + \mathbf{a}^\dagger(\mathbf{k}',t))\right] e^{i(\mathbf{k}-\mathbf{k}')\cdot\mathbf{r}}\, d^3k'\, d^3k\, d^3r$$

$$= \frac{1}{2}\int \omega\,\hat{\mathbf{n}}\Big[\mathbf{a}(\mathbf{k},t)\cdot\mathbf{a}^\dagger(\mathbf{k},t) - \mathbf{a}^\dagger(-\mathbf{k},t)\cdot\mathbf{a}(-\mathbf{k},t)$$

$$+ \mathbf{a}(\mathbf{k},t)\cdot\mathbf{a}(-\mathbf{k},t) - \mathbf{a}^\dagger(\mathbf{k},t)\cdot\mathbf{a}^\dagger(-\mathbf{k},t)\Big]\, d^3k \quad (11.59)$$

In the second term change $\mathbf{k}$ to $-\mathbf{k}$ (and therefore $\hat{\mathbf{n}}$ to $-\hat{\mathbf{n}}$). The last two terms are odd under this substitution, so the integral over them vanishes Thus

$$\mathbf{P} = \frac{1}{2}\int \omega\,\hat{\mathbf{n}}\sum_\alpha \hat{\boldsymbol{\varepsilon}}_\alpha^*(\mathbf{k})\cdot\hat{\boldsymbol{\varepsilon}}_\alpha(\mathbf{k})\left(a_\alpha^\dagger(\mathbf{k},t)a_\alpha(\mathbf{k},t) + a_\alpha(\mathbf{k},t)a_\alpha^\dagger(\mathbf{k},t)\right)d^3k$$

$$= \int \omega\,\hat{\mathbf{n}}\sum_\alpha \hat{\boldsymbol{\varepsilon}}_\alpha^*(\mathbf{k})\cdot\hat{\boldsymbol{\varepsilon}}_\alpha(\mathbf{k})\left(a_\alpha^\dagger(\mathbf{k},t)a_\alpha(\mathbf{k},t) + \frac{1}{2}[a_\alpha(\mathbf{k},t), a_\alpha^\dagger(\mathbf{k},t)]\right)d^3k$$

$$(11.60)$$

The last term has the same origin as the infinite self-energy term in the Hamiltonian (11.50). Formally it is proportional to $\delta_3(0)$, but because of the factor $\hat{\mathbf{n}}$, it vanishes under symmetric integration over the direction of $\mathbf{k}$. You can think of this term also as a consequence of the way we ordered the classical observables $a_\alpha(\mathbf{k})$ and $a_\alpha^\dagger(\mathbf{k})$. It is another example of the ambiguity of the canonical prescription. Thus (in the Schrödinger picture from now on),

$$\boxed{\mathbf{P} = \int d^3k \sum_\alpha \mathbf{k}\, N_\alpha(\mathbf{k})} \quad (11.61)$$

where N_α is the number operator. The photon states are eigenstates of the total momentum $\mathbf{P}$, with each photon contributing $\mathbf{k}$ as expected. $\mathbf{P}$ is independent of the time; and the total momentum of the electromagnetic field is conserved.

11.3.2 The Angular Momentum of the Field

The state $a_\alpha^\dagger(\mathbf{k})|0,0,0,\ldots\rangle$ is a state of energy ω and momentum $\mathbf{k}$ for both values of α. The remaining quantum number is the angular momentum.

For the angular momentum we can take over the expression

$$\boxed{\mathbf{J} = \frac{1}{4\pi}\int \mathbf{r}\times(\mathbf{E}\times\mathbf{B})\, d^3r} \quad (11.62)$$

---
[5] I don't prove this here; but just take it over from classical electromagnetism.

from classical mechanics also.[6]

What is the i-th component of $\mathbf{J}$?

$$\begin{aligned} J_i &= \frac{1}{4\pi}\int d^3r \sum_{jklm} \epsilon_{ijk} r_j \epsilon_{klm} E_l B_m = \frac{1}{4\pi}\int d^3r \sum_{jklmnr} \epsilon_{ijk}\epsilon_{klm}\epsilon_{mnr} r_j E_l \partial_n A_r \\ &= \frac{1}{4\pi}\int d^3r \sum_{jkl} \epsilon_{ijk}\left(r_j E_l \partial_k A_l - r_j E_l \partial_l A_k\right) \\ &= \frac{1}{4\pi}\int \Bigg[\sum_l E_l(\mathbf{r}\times\boldsymbol{\nabla})_i A_l \\ &\quad - \sum_{jkl}\epsilon_{ijk}\left(\partial_l(r_j E_l A_k) - (\partial_l r_j)E_l A_k - r_j(\partial_l E_l)A_k\right)\Bigg] d^3r \end{aligned} \tag{11.63}$$

The second term is the integral of a gradient (it is permissible to ignore surface terms since the normalized physical states all vanish at infinitely large distances, and therefore so do all physical matrix elements; a wave-packet treatment would justify this operation.) The last term vanishes because $\boldsymbol{\nabla}\cdot\mathbf{E}=0$ for a free field. In the third term, use $\partial_l r_j = \delta_{lj}$. Then $\mathbf{J}=\mathbf{L}+\mathbf{S}$, where

$$\boxed{\mathbf{L} = \frac{1}{4\pi}\int \sum_l E_l(\mathbf{r}\times\boldsymbol{\nabla})A_l\, d^3r} \tag{11.64}$$

and

$$\boxed{\mathbf{S} = \frac{1}{4\pi}\int \mathbf{E}\times\mathbf{A}\, d^3r} \tag{11.65}$$

In contrast to the nonrelativistic case, $\mathbf{L}$ and $\mathbf{S}$ do not commute. Furthermore breaking $\mathbf{J}$ up into these two pieces is not Lorentz invariant. But the component of $\mathbf{L}$ along the direction $\hat{n}=\mathbf{k}/k$ of a photon vanishes, so calling these pieces as "orbital" and "spin" angular momentum makes some sense. The component of $\mathbf{J}$ along $\mathbf{k}$ (called the **helicity**) is $\mathbf{J}\cdot\mathbf{k}/k = \mathbf{S}\cdot\mathbf{k}/k$. The demonstration of this fact is left as an exercise (see Problem 11.11).

11.3.3 The Photon Spin

The spin operator is

$$\mathbf{S} = \frac{1}{4\pi}\int \mathbf{E}\times\mathbf{A}\, d^3r \tag{11.65}$$

The $\mathbf{r}$ and $\mathbf{k}'$ integrations can be done as in the $\mathbf{P}$ computation, but they are simpler here because there are no derivatives. From equations (11.57), (11.58), and (11.65),

$$\mathbf{S} = \frac{i}{2}\int \left(\mathbf{a}(\mathbf{k})-\mathbf{a}^\dagger(-\mathbf{k})\right)\times\left(\mathbf{a}(-\mathbf{k})+\mathbf{a}^\dagger(\mathbf{k})\right) d^3k \tag{11.66}$$

---

[6] It is not strictly necessary to borrow these results from the classical theory. It can also be shown that $\mathbf{P}$ and $\mathbf{J}$ as defined here are the generators of translations and rotations in the usual sense. See for example Problem 11.3.

368 Chapter 11 The Quantized Electromagnetic Field

The components of the spin are

$$S_i = \frac{i}{2} \sum_{jk} \epsilon_{ijk} \int \left[a_j(\mathbf{k})a_k(-\mathbf{k}) + a_j(\mathbf{k})a_k^\dagger(\mathbf{k}) - a_j^\dagger(-\mathbf{k})a_k(-\mathbf{k}) - a_j^\dagger(-\mathbf{k})a_k^\dagger(\mathbf{k}) \right] d^3k \tag{11.67}$$

The first and fourth terms vanish because they are odd under $\mathbf{k} \to -\mathbf{k}$. The third term is even under this replacement. Because $\sum_{jk} \epsilon_{ijk} \left[a_j(\mathbf{k}), a_k^\dagger(\mathbf{k}) \right] = 0$ (exercise!) the order of the factors in the second term can be interchanged, and it is odd under $j \leftrightarrow k$. Therefore,

$$S_i = \frac{i}{2} \sum_{jk} \epsilon_{ijk} \int \left[a_k^\dagger(\mathbf{k})a_j(\mathbf{k}) - a_j^\dagger(\mathbf{k})a_k(\mathbf{k}) \right] d^3k = -i \sum_{jk} \epsilon_{ijk} \int a_j^\dagger(\mathbf{k})a_k(\mathbf{k}) d^3k \tag{11.68}$$

or

$$\mathbf{S} = -i \int \mathbf{a}^\dagger(\mathbf{k}) \times \mathbf{a}(\mathbf{k}) d^3k \tag{11.69}$$

In particular the z-component of the spin operator is

$$S_z = -i \sum_{\alpha\beta} \int \left[\epsilon_{\alpha x}^*(\mathbf{k})\epsilon_{\beta y}(\mathbf{k}) a_\alpha^\dagger(\mathbf{k})a_\beta(\mathbf{k}) - \epsilon_{\alpha y}^*(\mathbf{k})\epsilon_{\beta x}(\mathbf{k}) a_\alpha^\dagger(\mathbf{k})a_\beta(\mathbf{k}) \right] d^3k \tag{11.70}$$

Now take $\mathbf{k}$ in the z-direction so that $\mathbf{J} \cdot \mathbf{k} = \omega S_z$. Choose the two polarization vectors as

$$\hat{\boldsymbol{\varepsilon}}_1(k\hat{\mathbf{n}}_z) = \hat{\mathbf{n}}_x \quad \text{and} \quad \hat{\boldsymbol{\varepsilon}}_2(k\hat{\mathbf{n}}_z) = \hat{\mathbf{n}}_y \tag{11.71}$$

There are two states with momentum $k\hat{\mathbf{n}}_z$, each with one of these polarization vectors. Call them

$$|1\rangle = a_1^\dagger(k\hat{\mathbf{n}}_z)|0,0,0,\ldots\rangle \quad \text{and} \quad |2\rangle = a_2^\dagger(k\hat{\mathbf{n}}_z)|0,0,0,\ldots\rangle \tag{11.72}$$

Thus

$$S_z|1\rangle = -i\left[(\epsilon_{1x}^*\epsilon_{1y} - \epsilon_{1y}^*\epsilon_{1x})|1\rangle + (\epsilon_{2x}^*\epsilon_{1y} - \epsilon_{2y}^*\epsilon_{1x})|2\rangle\right] = i|2\rangle \tag{11.73}$$

and

$$S_z|2\rangle = -i\left[(\epsilon_{1x}^*\epsilon_{2y} - \epsilon_{1y}^*\epsilon_{2x})|1\rangle + (\epsilon_{2x}^*\epsilon_{2y} - \epsilon_{2y}^*\epsilon_{2x})|2\rangle\right] = -i|1\rangle \tag{11.74}$$

This is a simple two-state problem. The eigenstates of S_z are

$$\frac{|1\rangle \pm i|2\rangle}{\sqrt{2}} \tag{11.75}$$

with eigenvalues ± 1. These are the usual right and left circularly polarized states. The right circularly polarized states have positive helicity. The component of the angular momentum along the direction of a photon's momentum can be $\pm \hbar$.

There is no state in which the component of $\mathbf{J}$ along the direction of its momentum is zero! For each momentum $\mathbf{k}$, there is a right circularly polarized state and a left circularly polarized state, but no state with longitudinal polarization.

This peculiar situation is impossible for particles with nonzero mass. For a massive particle, one can always make a Lorentz boost to its rest frame; then rotational invariance requires that angular momentum on those states be represented by one of the finite dimensional matrices $D^{(j)}$ and so have $2j+1$ components, where j is the maximum angular momentum along any direction. For a massless particle, this does not happen. A photon or a massless neutrino cannot be boosted to a frame with zero momentum. On the contrary: a massless particle moves at speed c, and after any boost or rotation its speed is still c. It is impossible to catch up with a piece of light. The helicity is a Lorentz invariant for massless particles. That is why neutrinos if they are massless, but not electrons, can have only one value for the helicity.

Photon states must have the two eigenvalues ±1 because a system with particles of only one helicity violates reflection invariance. But no symmetry requires the state with eigenvalue 0 to exist, and indeed it does not.

11.4 ELECTROMAGNETIC DECAYS OF EXCITED STATES

11.4.1 The Unperturbed Hamiltonian

At last we have all the tools needed to compute spontaneous decay rates of excited states by photon emission. I will work out the rates for photon emission by atomic electrons as an important example, and assume that the nonrelativistic description of the electron is adequate.

The unperturbed Hamiltonian has two parts: first, the Hamiltonian (11.55) of the free electromagnetic field, and second, the Hamiltonian of the electron:

$$H_o = H_{\text{photons}} + \frac{p^2}{2m} + V(r) \tag{11.76}$$

The initial state $|i\rangle$ is the atom at rest. The electron is in some excited state $|\psi_i\rangle$. The final state $|f\rangle$ contains an atom with the electron in a state $|\psi_f\rangle$ with lower energy, and a photon with momentum $\mathbf{k}$ and polarization vector $\hat{\boldsymbol{\varepsilon}}$. Since in atomic (and nuclear) decays the photon energy is much smaller than the rest energy of the atom, it is adequate to take the atom to be infinitely heavy; then the final state atom is also at rest. Recoil corrections (see Section 9.3.5) are very small. While total energy is conserved in these calculations, momentum is not. This recoilless approximation will not generally work for electromagnetic decays of elementary particles.

11.4.2 The Vector Potential Interaction

The interaction Hamiltonian is obtained from the minimal coupling rule, replacing the electron momentum $\mathbf{p}$ in H_o with $\mathbf{p} + e\mathbf{A}(\mathbf{r}, t)$. Thus

$$H = H_{\text{photons}} + \frac{1}{2m}\left[\mathbf{p} + e\mathbf{A}(\mathbf{r}, t)\right]^2 + V(r) = H_o + H' \tag{11.77}$$

so

$$\boxed{H' = \frac{e}{m}\mathbf{A}(\mathbf{r}) \cdot \mathbf{p} + \frac{e^2}{2m}\mathbf{A}(\mathbf{r})^2} \tag{11.78}$$

H' has the same *form* as it did when we treated $\mathbf{A}$ as an external field, but the meaning is different here. $\mathbf{A}$ is the quantized electromagnetic potential, in the Schrödinger picture, evaluated at the point $\mathbf{r}$, the electron position operator.

That this is correct is logically a hypothesis. But by taking H' to have the same form as it is in classical electrodynamics, we ensure that the expectation values of the fields and the electron's coordinates will obey Maxwell's equations and the Lorentz force law, as the correspondence principle requires.

$\mathbf{A}$ is linear in the photon creation and annihilation operators; so in lowest order, only the term

$$H' = e\frac{\mathbf{p}}{m} \cdot \mathbf{A}(\mathbf{r}) \tag{11.79}$$

contributes to single photon emission. (Note that $\mathbf{A}$ and $\mathbf{p}$ still commute because of the gauge condition.) The $\mathbf{A}^2$ term will contribute to two-photon emission, to scattering, and to higher-order effects.

11.4.3 The Spin Interaction

There is another term in H', one that has no classical analog. It is the interaction of the electromagnetic field with the electron spin. This term too has the same *form* as the energy of an electron in a classical field:

$$\boxed{H'_{\text{spin}} = \frac{e}{2m}\boldsymbol{\sigma} \cdot \mathbf{B}(\mathbf{r}, t)} \tag{11.80}$$

but here $\mathbf{B} = \nabla \times \mathbf{A}$ is the quantized magnetic field. H'_{spin} is automatically gauge invariant, though not derived from the minimal coupling principle. In the relativistic treatment of the electron, the spin term too is a consequence of minimal coupling, and is on the same footing as the $\mathbf{A} \cdot \mathbf{p}$ term.[7] For particles other than the electron, (11.80) should be multiplied by $g/2$. For the electron, just take the gyromagnetic ratio to be about 2 from experiment.

11.4.4 The Rate for Photon Emission

The matrix element in Born approximation for an atom in an atomic state $|\psi_i\rangle$ to emit a photon $|\alpha, \mathbf{k}\rangle$ and decay into a final atomic state $|\psi_f\rangle$ is

$$H'_{fi} = \left\langle \psi_f; \alpha, \mathbf{k} \left| \frac{e}{m}\mathbf{A} \cdot \mathbf{p} + \frac{e^2}{2m}\mathbf{A}^2 + \frac{e}{2m}\boldsymbol{\sigma} \cdot \mathbf{B}(\mathbf{r}, t) \right| \psi_i \right\rangle \tag{11.81}$$

Usually the second and third terms are important only when the first term vanishes, so I begin by considering decays for which only the first term in equation (11.81) is important.

The initial and final states are eigenstates of the free field. Thus the matrix element that goes into the decay rate is

$$H'_{fi} = \left\langle \psi_f; \alpha, \mathbf{k} \left| \frac{e}{m}\mathbf{A}(\mathbf{r}) \cdot \mathbf{p} \right| \psi_i \right\rangle \tag{11.82}$$

---

[7] See Chapter 12.

Section 11.4 Electromagnetic Decays of Excited States

where

$$\mathbf{A}(\mathbf{r}) = \mathbf{A}_H(\mathbf{r}, 0) = \frac{\sqrt{4\pi}}{(2\pi)^{\frac{3}{2}}} \int \left[a_S(\mathbf{k})e^{i\mathbf{k}\cdot\mathbf{r}} + a_S^\dagger(\mathbf{k})e^{-i\mathbf{k}\cdot\mathbf{r}} \right] \frac{1}{\sqrt{2\omega}} d^3k \quad (11.83)$$

where $\mathbf{a}(\mathbf{k})$ is the time independent operator $\mathbf{a}(\mathbf{k}) = \mathbf{a}_H(\mathbf{k}, 0)$.

The destruction operator $\mathbf{a}(\mathbf{k})$ acting on the initial state vanishes, since $|\psi_i\rangle$ contains no photons. The term in $\mathbf{a}^\dagger(\mathbf{k})$ survives, and the time dependence cancels, as it must:

$$H'_{fi} = \left\langle \psi_f; \alpha, \mathbf{k} \left| \frac{e}{m} \mathbf{A}(\mathbf{r}) \cdot \mathbf{p} \right| \psi_i \right\rangle$$

$$= \frac{e}{m} \frac{\sqrt{4\pi}}{(2\pi)^{3/2}} \sum_{\alpha'} \int \left\langle \psi_f; \alpha, \mathbf{k} \left| a_{\alpha'}^\dagger(\mathbf{k}') \hat{\varepsilon}_{\alpha'}^*(\mathbf{k}') \cdot \mathbf{p} e^{-i\mathbf{k}'\cdot\mathbf{r}} \right| \psi_i \right\rangle \frac{1}{\sqrt{2\omega'}} d^3k' \quad (11.84)$$

$a_{\alpha'}^\dagger(\mathbf{k}')|\psi_i\rangle$ is a state with one photon:

$$a_{\alpha'}^\dagger(\mathbf{k}')|\psi_i\rangle = |\psi_i; \alpha', \mathbf{k}'\rangle \quad (11.85)$$

The matrix element (11.84) contains no more operators explicitly referring to the photons. H' becomes an ordinary matrix element of single-electron operators between electron eigenstates of H_o.

$$H'_{fi} = \frac{e}{m} \frac{\sqrt{4\pi}}{(2\pi)^{3/2}} \sum_{\alpha'} \int \left\langle \psi_f \left| \hat{\varepsilon}_{\alpha'}^*(\mathbf{k}') \cdot \mathbf{p} e^{-i\mathbf{k}'\cdot\mathbf{r}} \right| \psi_i \right\rangle \frac{1}{\sqrt{2\omega'}} \delta_{\alpha',\alpha} \delta(\mathbf{k}' - \mathbf{k}) d^3k'$$

$$= \frac{e}{m} \frac{\sqrt{4\pi}}{(2\pi)^{3/2}} \frac{1}{\sqrt{2\omega}} \left\langle \psi_f \left| \mathbf{p} \cdot \hat{\varepsilon}_\alpha^*(\mathbf{k}) e^{-i\mathbf{k}\cdot\mathbf{r}} \right| \psi_i \right\rangle \quad (11.86)$$

Since the photon states are normalized to a δ-function in $\mathbf{k}$, use

$$\sum_m \rightarrow \sum_\alpha \int d^3k \quad (11.87)$$

for the sum over final states in Fermi's golden rule. The transition rate, in Born approximation, is

$$\Gamma = 2\pi \sum_f |H'_{fi}|^2 \delta(E_f - E_i)$$

$$= 2\pi \frac{e^2}{4\pi^2 m^2} \int \frac{1}{\omega} \sum_\alpha \left| \langle \psi_f | \mathbf{p} \cdot \hat{\varepsilon}_\alpha^*(\mathbf{k}) e^{-i\mathbf{k}\cdot\mathbf{r}} | \psi_i \rangle \right|^2 \delta\left(E_i^o - E_f^o - \omega\right) d^3k \quad (11.88)$$

where E_m^o are the atomic energy levels. Write

$$\int d^3k = \int k^2 \, dk \, d\Omega = \int \omega^2 \, d\omega \, d\Omega \quad (11.89)$$

and use the δ-function to do the ω integral:

$$\Gamma = \frac{e^2\omega}{2\pi m^2} \sum_\alpha \int \left|\langle\psi_f|e^{-i\mathbf{k}\cdot\mathbf{r}}\mathbf{p}\cdot\hat{\boldsymbol{\varepsilon}}_\alpha^*(\mathbf{k})|\psi_i\rangle\right|^2 d\Omega \qquad (11.90)$$

Equation (11.90) is the formula for single photon emission.

11.4.5 Multipole Matrix Elements

The rate Γ is proportional to the matrix element

$$\langle\psi_f|\hat{\boldsymbol{\varepsilon}}\cdot\mathbf{p}\,e^{i\mathbf{k}\cdot\mathbf{r}}|\psi_i\rangle \qquad (11.91)$$

For atoms, r is of the order of the Bohr radius $a = 1/m\alpha$, whereas $|\mathbf{k}| = \omega$ is no greater than the level spacing, of the order $m\alpha^2$. So $\mathbf{k}\cdot\mathbf{r}$ is of the order of the small number α, and you can expand the exponential:

$$e^{i\mathbf{k}\cdot\mathbf{r}} = 1 + i\mathbf{k}\cdot\mathbf{r} + \frac{1}{2}(i\mathbf{k}\cdot\mathbf{r})^2 + \cdots \qquad (11.92)$$

The bound states are angular momentum eigenstates, and the Wigner-Eckart selection rules will usually require some of these terms to vanish. The first term that does not vanish is the leading approximation. It is often the only useful approximation, since the higher order terms in this series will be the same order in α as the higher order terms in the perturbation series for the transition matrix, and also of the same order as radiative corrections like pair production that require a full relativistic electron theory. The multipole expansion is usually not useful for continuum states (ionization problems).

Electric Dipole Transitions

The most rapid decays occur when $e^{i\mathbf{k}\cdot\mathbf{r}}$ can be replaced by 1 and still lead to a nonvanishing result. These are electric dipole decays, and this is the **electric-dipole approximation**:

$$\Gamma = \frac{e^2\omega}{2\pi m^2} \sum_\alpha \int |\langle\psi_f|\mathbf{p}\cdot\hat{\boldsymbol{\varepsilon}}_\alpha^*(\mathbf{k})|\psi_i\rangle|^2 d\Omega \qquad (11.93)$$

As in Section 9.1.2 the computation of the matrix element in (11.93) can be simplified using $[\mathbf{r}, H_o] = i\mathbf{p}/m$. Then[8]

$$\langle\psi_f|\hat{\boldsymbol{\varepsilon}}_\alpha^*\cdot\mathbf{p}|\psi_i\rangle = -im\hat{\boldsymbol{\varepsilon}}_\alpha^*\cdot\langle\psi_f|[\mathbf{r},H_o]|\psi_i\rangle$$
$$= im(E_f^o - E_i^o)\hat{\boldsymbol{\varepsilon}}_\alpha^*\cdot\langle\psi_f|\mathbf{r}|\psi_i\rangle = im\omega_{fi}\hat{\boldsymbol{\varepsilon}}_\alpha^*\cdot\mathbf{r}_{fi} \qquad (11.94)$$

The dipole approximation for the decay rate is

$$\Gamma = \frac{e^2\omega^3}{2\pi}\sum_\alpha \int|\langle\psi_f|\mathbf{r}\cdot\hat{\boldsymbol{\varepsilon}}_\alpha^*(\mathbf{k})|\psi_i\rangle|^2\,d\Omega \qquad (11.95)$$

---

[8]This trick works only for electric dipole transitions.

To compute the total rate into both polarizations, sum over $\hat{\varepsilon}_\alpha^*(\mathbf{k})$. In the dipole approximation

$$\sum_\alpha |\mathbf{r}_{fi} \cdot \hat{\varepsilon}_\alpha^*(\mathbf{k})|^2 = \sum_{ij}(r_{fi})_i(r_{fi}^*)_j \sum_\alpha (\epsilon_\alpha^*)_i(\epsilon_\alpha)_j$$

$$= \sum_{ij}(r_{fi})_i(r_{fi}^*)_j(\delta_{ij} - n_i n_j) \qquad (11.96)$$

where $\mathbf{k} = k\hat{\mathbf{n}}$. For the whole decay rate, the $d\Omega$ integration becomes

$$\int (\delta_{ij} - n_i n_j)d\Omega = 4\pi\left(\delta_{ij} - \frac{\delta_{ij}}{3}\right) = \frac{8\pi}{3}\delta_{ij} \qquad (11.97)$$

and the total rate is

$$\boxed{\Gamma = \frac{e^2\omega^3}{2\pi}\frac{8\pi}{3}|\mathbf{r}_{fi}|^2 = \frac{4\alpha\omega^3}{3}|\mathbf{r}_{fi}|^2} \qquad (11.98)$$

11.5 EXAMPLES

11.5.1 Decay of the 2P State of Atomic Hydrogen

As a first example, I will compute the decay rate of a hydrogen atom in a $2P$ state, ignoring spin effects. This is the Lyman-α line, and is an electric dipole transition. Since the final state is an S-state the total rate will be the same for any initial m. From equation (11.98) the rate is

$$\Gamma = \frac{4\alpha\omega^3}{3}|\mathbf{r}_{fi}|^2 = \frac{4\alpha\omega^3}{3}\langle 100|\mathbf{r}|21m\rangle \cdot \langle 21m|\mathbf{r}|100\rangle \qquad (11.99)$$

Choose $m = 0$, so that only the z-component of $\mathbf{r}$ contributes to equation (11.99). The matrix element is[9]

$$|\langle 100|z|210\rangle|^2 = \frac{2^{15}a^2}{3^{10}} \qquad (11.100)$$

Here $\omega = E_2 - E_1 = 3\alpha^2 m/8$ and a is the Bohr radius $1/\alpha m$. The rate is

$$\Gamma = \frac{4\alpha}{3}\frac{2^{15}}{3^{10}}\omega^3\frac{1}{(am)^2} = \left(\frac{2}{3}\right)^8 \alpha^5 m = 6.27 \times 10^8 \text{ sec}^{-1} = 4.126 \times 10^{-7}\text{eV} \qquad (11.101)$$

The numerical value is the full width at half maximum of the actual distribution of the energies in the final state, according to the discussion of Section 9.4.3. The lifetime is

$$\frac{1}{\Gamma} = 1.6 \times 10^{-9} \text{ sec} \qquad (11.102)$$

This may seem a short time, but it is long on the scale of our problem. The ratio of the width of the line to the level spacing is

$$\frac{\Gamma}{\omega} = \frac{(2/3)^8 \alpha^5 m}{(3/8)\alpha^2 m} = 4.04 \times 10^{-8} \qquad (11.103)$$

---

[9]See, for instance, Problem 7.9.

The dimensionless number in equation (11.103) is a measure of how good the approximations made in Section 9.4 are for this process.

11.5.2 Hyperfine Emission

The total angular momentum of an electron in the ground state of hydrogen, including the proton spin, is (see Section 7.3.6)

$$\mathbf{J} = \mathbf{S} = \mathbf{s}_e + \mathbf{s}_p \tag{11.104}$$

since in the S state $\mathbf{L} = 0$. The eigenstates of H_o are eigenstates of $\mathbf{S}$, with possible eigenvalues $S = 0$ and $S = 1$. The formula for the level splitting to first order in bound-state perturbation theory, (see equation (7.104)) came out to be

$$\Delta_{HFS} = \frac{2}{3}\alpha^4 m \frac{m}{M} g_p g \tag{11.105}$$

where g is the electron gyromagnetic ratio (≈ 2) and g_p the gyromagnetic ratio of the proton (≈ 5.59). The numerical value of Δ_{HFS} is 5.884×10^{-6}eV $= 9.33 \times 10^{-18}$erg, much smaller than any of the fine structure level splittings. The wave length of the emitted photon is about 21 cm.

Now we can calculate the rate. Both states have $n = 1$, $l = m_l = 0$. Label them just by the total spin, here also the total angular momentum, and the z-component of the total spin. The upper states are the three triplet states $|1M\rangle$ ($M = -1, 0, 1$) and the ground state is $|00\rangle$.

What is the matrix element of the interaction Hamiltonian $\langle \gamma; 00| H' |1M\rangle$? The first-order electric term in the expression (11.79) for H' is proportional to

$$\langle \gamma(\alpha, \mathbf{k}), 00| \mathbf{p} \cdot \mathbf{A}(\mathbf{r}) |1M\rangle \sim \langle \gamma(\alpha, \mathbf{k}), 00| \hat{\varepsilon}_\alpha^*(\mathbf{k}) \cdot \mathbf{p}\, a_\alpha^\dagger(\mathbf{k})^\dagger e^{-i\mathbf{k}\cdot\mathbf{r}} |1M\rangle$$
$$= \langle 00| \hat{\varepsilon}_\alpha^*(\mathbf{k}) \cdot \mathbf{p}\, e^{-i\mathbf{k}\cdot\mathbf{r}} |1M\rangle \tag{11.106}$$

First try the dipole approximation $e^{i\mathbf{k}\cdot\mathbf{r}} = 1$. Since the initial and final states are both S states, reflection invariance (parity conservation) requires the matrix elements $\langle 00| \mathbf{p} |LM\rangle$ to vanish.

Next try the quadrupole approximation:

$$\sum_{ij}(\hat{\varepsilon}_\alpha^*)_i k_j \langle 00| p_i r_j |1M\rangle \tag{11.107}$$

$p_i r_j$ are components of a rank-two Cartesian tensor, linear combinations of spherical tensors of ranks 2, 1 and 0. By the second selection rule (the spin coordinates can be ignored in this argument, since $\mathbf{p}$ and $\mathbf{r}$ and act only on the space coordinates) the rank-two part cannot connect two S-states. The rank-one part vanishes also. The scalar part is proportional to $\mathbf{p} \cdot \mathbf{r}\, \delta_{ij}$, and it vanishes because $\hat{\varepsilon}^*(\mathbf{k}) \cdot \mathbf{k} = 0$. The quadrupole term too is identically zero. And so on. The term with n indices will be a linear combination of spherical tensors with all ranks up to n. Because the initial and final states are both S-states, these vanish (by the Wigner-Eckart theorem) unless $n = 0$. The scalar part is always proportional to a factor $\hat{\varepsilon} \cdot \mathbf{k}$ and is therefore zero also. The conclusion is that $\langle \gamma; 00 | \hat{\varepsilon}_\alpha^* \cdot \mathbf{p} e^{-i\mathbf{k}\cdot\mathbf{r}} | 1M \rangle$ is *identically zero*!

The physical explanation is simple, as it usually is when a complicated expression turns out to be zero. The spin of the electron does not appear in this term in H', so it cannot enter into angular momentum conservation. In that case $\langle n'l'm' | \hat{\varepsilon}_\alpha^* \cdot \mathbf{p} e^{-i\mathbf{k}\cdot\mathbf{r}} | nlm \rangle$ would be the amplitude for an atom in the excited state $|nlm\rangle$ to decay to the lower state $|n'l'm'\rangle$ by photon emission. When both $l' = l = 0$ (i.e., suppose the initial and final states are both S states) the component of the electron angular momentum along the direction of the emitted photon is zero in both the initial and final state. Angular momentum conservation requires that the component of the final photon momentum in that direction vanish also. But that cannot be: There are no zero-helicity photons. Transitions by single photon emission from an initial $l = 0$ state to a final $l = 0$ state are forbidden.

The $e^2 \mathbf{A}^2$ term in H' does not contribute here either. It is quadratic in the raising and lowering operators, and cannot connect a state with one photon to a state with no photons. The $\mathbf{A}^2$ term will contribute to two-photon decays, and to scattering of photons off electrons, but neither the $\mathbf{p} \cdot \mathbf{A}$ term nor the $\mathbf{A}^2$ term contribute to the hyperfine splitting decay in the ground state.

When the spins are included the transition is allowed. From equation (11.80), the effective part if the interaction Hamiltonian is

$$H' = -\frac{g_p e}{2M} \mathbf{s}_p \cdot \mathbf{B} + \frac{ge}{2m} \mathbf{s}_e \cdot \mathbf{B} \qquad (11.108)$$

where now $\mathbf{B}$ is the quantized magnetic field. Only the second term is important. The first is smaller by $m/M = 1/1836$.

The magnetic field is

$$\begin{aligned}\mathbf{B} &= \boldsymbol{\nabla} \times \mathbf{A} \\ &= i\frac{\sqrt{4\pi}}{(2\pi)^{3/2}} \int \sqrt{\frac{\omega'}{2}} \, \hat{\mathbf{n}}' \times \left[\mathbf{a}(\mathbf{k}') e^{i\mathbf{k}'\cdot\mathbf{r}} - \mathbf{a}^\dagger(\mathbf{k}') e^{-i\mathbf{k}'\cdot\mathbf{r}} \right] d^3 k'\end{aligned} \qquad (11.109)$$

Since the photon is in the final state, only the term in $\mathbf{a}^\dagger$ counts; operating backward, the operator $\mathbf{a}(\mathbf{k}')$ annihilates the photon and returns a factor $\hat{\varepsilon}_\alpha^*(\mathbf{k}) \delta_3(\mathbf{k}' - \mathbf{k})$. Therefore (set $g = 2$)

$$H'_{fi} = -i\frac{e}{m} \sqrt{\frac{1}{2\pi^2}} \sqrt{\frac{\omega}{2}} \langle J = 0 | \, \hat{\mathbf{n}} \times \hat{\varepsilon}_\alpha^*(\mathbf{k}) \cdot \mathbf{s}_e \, e^{-i\mathbf{k}\cdot\mathbf{r}} | J = 1, M \rangle \qquad (11.110)$$

The exponential can be set equal to 1. This is the *magnetic* dipole approximation.

376 Chapter 11 The Quantized Electromagnetic Field

The total rate, from any of the three initial m_J states, is

$$\Gamma = 2\pi \frac{e^2}{m^2} \frac{1}{2\pi^2} \int \frac{\omega}{2} \sum_\alpha |\langle J=0| \, \hat{\mathbf{n}} \times \hat{\boldsymbol{\varepsilon}}_\alpha^*(\mathbf{k}) \cdot \mathbf{s}_e \, |J=1, m_J\rangle|^2 \, \delta(\Delta_{HFS} - \omega) d^3k$$

$$= 2\pi \frac{e^2}{m^2} \frac{1}{2\pi^2} \int \frac{\omega}{2} \delta(\Delta_{HFS} - \omega)$$
$$\times \sum_{ij} \sum_\alpha (\hat{\mathbf{n}} \times \hat{\boldsymbol{\varepsilon}}_\alpha^*(\mathbf{k}))_i \, (\hat{\mathbf{n}} \times \hat{\boldsymbol{\varepsilon}}_\alpha(\mathbf{k}))_j \, (s_{fi})_i (s_{fi})_j^* d^3k$$

(11.111)

where

$$(s_{fi})_m = \langle J=0| \, (s_e)_m \, |J=1, M\rangle \tag{11.112}$$

Next, sum over the polarizations α of the emitted photon, using[10]

$$\sum_\alpha (\hat{\mathbf{n}} \times \hat{\boldsymbol{\varepsilon}}_\alpha^*(\mathbf{k}))_i \, (\hat{\mathbf{n}} \times \hat{\boldsymbol{\varepsilon}}_\alpha(\mathbf{k}))_j = \delta_{ij} - n_i n_j \tag{11.113}$$

Therefore,

$$\sum_\alpha |\langle J=0|\hat{\mathbf{n}} \times \hat{\boldsymbol{\varepsilon}}_\alpha^*(\mathbf{k}) \cdot \mathbf{s}_e |J=1, M\rangle|^2 = \sum_{ij} (\delta_{ij} - n_i n_j)(s_{fi})_i (s_{fi})_j^* \tag{11.114}$$

Since for any function $f(k^2)$ that is independent of the direction of $\mathbf{k}$,

$$\int n_i n_j f(k^2) d^3k = \frac{1}{3}\delta_{ij} \int d^3k f(k^2) \tag{11.115}$$

the sum and integral in (11.111) become

$$\int \sum_\alpha \left|\langle J=0| \, \hat{\mathbf{n}} \times \hat{\boldsymbol{\varepsilon}}_\alpha^*(\mathbf{k}) \cdot \mathbf{s}_e \, |J=1, M\rangle \right|^2 d^3k$$
$$= \int \frac{2}{3} \sum_i s_i s_i^* d^3k = \frac{2}{3} \int \sum_\mu s^{*(\mu)} s^{(\mu)} d^3k \tag{11.116}$$

The last form is written in the spherical components of vectors introduced in Section 5.1.3.

Write the $J=1$ states in terms of the electron and proton spin states $|s_{ze} s_{zp}\rangle$, using the Clebsch-Gordan coefficients from Section 4.5.2.

$$|\tfrac{1}{2}, \tfrac{1}{2}\rangle \qquad\qquad M=1 \tag{11.117a}$$

$$\frac{|\tfrac{1}{2}, -\tfrac{1}{2}\rangle + |-\tfrac{1}{2}, \tfrac{1}{2}\rangle}{\sqrt{2}} \qquad\qquad M=0 \tag{11.117b}$$

$$|-\tfrac{1}{2}, -\tfrac{1}{2}\rangle \qquad\qquad M=-1 \tag{11.117c}$$

---
[10] The same matrix projects out the unit vectors $\hat{\mathbf{n}} \times \hat{\boldsymbol{\varepsilon}}_\alpha(\mathbf{k})$ as the $\hat{\boldsymbol{\varepsilon}}_\alpha(\mathbf{k})$ themselves.

The $J = 0$ state is

$$\frac{\left|\frac{1}{2}, -\frac{1}{2}\right\rangle - \left|-\frac{1}{2}, \frac{1}{2}\right\rangle}{\sqrt{2}} \qquad M = 0 \qquad (11.117d)$$

$$(11.117e)$$

The rate must be independent of M. Choose $M = 0$. Then only $\mu = 0$ contributes, since s_x and s_y do not connect two states with $m = 0$. Therefore,

$$s^{(0)} = s_z = \langle J = 0 | s_z | J = 1, M = 0 \rangle = +\frac{1}{2}$$

The total decay rate from equation (11.111) now becomes

$$\Gamma = 2\pi \frac{e^2}{m^2} \frac{1}{2\pi^2}$$

$$\times \int \sum_\alpha \frac{\omega}{2} \left| \langle J = 0 | \hat{\boldsymbol{n}} \times \hat{\boldsymbol{\varepsilon}}_\alpha^*(\mathbf{k}) \cdot \boldsymbol{B} s_e | J = 1, M \rangle \right|^2 \delta(\Delta_{HFS} - \omega) d^3k \qquad (11.118)$$

$$= 2\pi \frac{e^2}{m^2} \frac{1}{2\pi^2} \frac{\omega}{2} \frac{2}{3} \frac{1}{4} \int \delta(\Delta_{HFS} - \omega) k^2 \, dk \, d\Omega = \frac{\alpha \omega^3}{3m^2}$$

This width is tiny even compared to the hyperfine splitting

$$\Delta_{HFS} = \frac{2}{3} \alpha^4 m \frac{m}{M} g_p g \qquad (11.105)$$

Indeed, with $\omega = \Delta_{HFS}$ and $g = 2$,

$$\Gamma = \frac{2^5}{3^4} \alpha^{13} g_p^3 (m/M)^3 m = 9.49 \times 10^{-31} \text{eV} = 1.44 \times 10^{-15} \text{sec}^{-1} \qquad (11.119)$$

The lifetime is 6.93×10^{14} seconds, over 200,000 centuries. It should be clear now why this transition is not observed in laboratory experiments. An H atom in the $1S$, $J = 1$ state, will collide with another atom (or, in a near vacuum, with the walls of the container) and get excited to a higher energy level, ionize, or, depending on the temperature, combine to form the more stable H_2 molecule. Only in the near emptiness of outer space does an atom have time between collisions to cascade down to the $1S$ state, and if this is the upper ($J = 1$) level as happens three-quarters of the time, to wait a hundred thousand centuries before emitting the 21 cm radiation. In interstellar space, hydrogen is by far the principal element. Only about a quarter of the primordial protons created in the big bang had time to fuse into deuterons or stable helium nuclei, and a relative handful got cooked into the heavier elements we are made of during the collapse of supernovae. So the 21 cm line is the main signal of the invisible ordinary matter in empty space; it was the signal that was used first to map the spiral arms of our galaxy.[11]

---

[11] See footnote 8 in Chapter 7.

11.6 ABSORPTION AND STIMULATED EMISSION OF RADIATION

Sometimes you want to calculate the rate for absorption of light from incoherent radiation like sunlight, rather than one photon at a time. While it is not impossible to do this using the Dirac delta-function normalization, it is rather awkward, so here I will use the discrete box normalization found in many books. You should take the opportunity to learn this common notational variant.

11.6.1 Periodic Boundary Conditions

A way to avoid continuous energy and momentum eigenvalues, even for free particles, is to divide all space into large cubes with sides of length L, and then impose periodic boundary conditions. Then ordinary wave functions satisfy $\psi(\mathbf{r} + \hat{\mathbf{n}}_i L) = \psi(\mathbf{r})$. The wave functions are normalized inside each cube. The plane wave states are

$$\psi_\mathbf{k}(\mathbf{r}) = \frac{1}{L^{\frac{3}{2}}} e^{i\mathbf{k}\cdot\mathbf{r}} \tag{11.120}$$

The momentum is allowed to have only the components

$$k_i = \frac{2\pi n_i}{L} \tag{11.121}$$

for integers n_i. The states are normalized to[12]

$$\langle \mathbf{k}' | \mathbf{k} \rangle = \delta_{\mathbf{k}',\mathbf{k}} \tag{11.122}$$

Completeness is summarized by

$$\sum_\mathbf{k} |\mathbf{k}\rangle\langle\mathbf{k}| = I \tag{11.123}$$

where the sum is over all different combinations of the integers n_i. In the $L \to \infty$ physically meaningful quantities should be independent of L.

Note: Periodic boundary conditions are not the same as the boundary conditions for a particle confined to a box, like a three-dimensional infinite square well. For a real confined particle, the wave function vanishes at the edge of the box. Using those boundary conditions is more cumbersome.

Application to Photons

With this discrete normalization, the algebra of the quantum field is described by discrete creation and annihilation operators $a_\alpha^\dagger(\mathbf{k})$ and $a_\alpha(\mathbf{k})$ whose commutators are Kronecker delta functions:

$$\left[a_\alpha(\mathbf{k}), a_{\alpha'}^\dagger(\mathbf{k}')\right] = \delta_{\alpha'\alpha}\delta_{\mathbf{k}',\mathbf{k}} \tag{11.124}$$

The single photon states $a_\alpha^\dagger(\mathbf{k})|\rangle$ are normalized to unity instead of to Dirac delta functions.

---

[12] This three-dimensional version of Kronecker's delta symbol means $\delta_{\mathbf{k}',\mathbf{k}} = \delta_{k'_x,k_x}\delta_{k'_y,k_y}\delta_{k'_z,k_z}$.

Section 11.6 Absorption and Stimulated Emission of Radiation 379

What is the energy density of such a state? For ordinary wave functions normalized in a box like this, the states with definite momentum have wave functions

$$\psi(\mathbf{r}) = \frac{1}{L^{3/2}} e^{i\mathbf{k}\cdot\mathbf{r}} \tag{11.125}$$

so the particle density is $\psi^*(\mathbf{r})\psi(\mathbf{r}) = 1/L^3$. This is the right answer for photons too. The free Hamiltonian is

$$H = \int \mathcal{H}(\mathbf{r}) d^3r = \sum_\alpha \sum_\mathbf{k} \omega a_\alpha^\dagger(\mathbf{k}) a_\alpha(\mathbf{k}) \tag{11.126}$$

Since the energy density is constant

$$\mathcal{H}(\mathbf{r}) = \frac{1}{L^3} \sum_\alpha \sum_\mathbf{k} \omega a_\alpha^\dagger(\mathbf{k}) a_\alpha(\mathbf{k}) \tag{11.127}$$

whence the photon number density is $1/L^3$.

11.6.2 Absorption

Photon States in a Cavity

Instead of absorption of photons from a single state, I want to consider a "gas" of photons in a cavity of size L^3, with a known energy density that does not vary much with frequency on the scale of the width of the absorption line. Let there be $N(\omega)d\omega$ states in the frequency range $d\omega$. In statistical mechanics, you sometimes want to compute $N(\omega)$, but that result is not needed here. Let $\bar{n}(\omega)$ be the average occupation number of the occupied states in $d\omega$. Then the energy density due to all the states with frequencies between ω and $\omega + d\omega$ is $I(\omega)d\omega$, where

$$I(\omega) = \frac{1}{L^3} \omega \bar{n}(\omega) N(\omega) \tag{11.128}$$

Suppose the cavity contains an atom or nucleus in its ground state $|\psi_i\rangle$, and compute the rate for excitation to some upper state $|\psi_f\rangle$. The initial and final states are

$$|i\rangle = |\psi_i; n_\alpha(\mathbf{k}), \alpha, \mathbf{k}\rangle \quad \text{and} \quad |f\rangle = |\psi_f; n_\alpha(\mathbf{k}) - 1, \alpha, \mathbf{k}\rangle \tag{11.129}$$

The symbol ψ_a stands for an atomic or nuclear state, α and $\mathbf{k}$ describe the photon, and $n_\alpha(\mathbf{k})$ is the initial occupation number of the state.

The transition rate in Born approximation is

$$\Gamma = 2\pi |H'_{fi}|^2 \delta(E_f - E_i) \tag{11.130}$$

The matrix element of $a_\alpha(\mathbf{k})$ between the initial and final states is $\sqrt{n_\alpha(\mathbf{k})}$, so the matrix element of the interaction Hamiltonian $e\mathbf{p}\cdot\mathbf{A}/m$ is

$$H'_{fi} = \langle f| H' |i\rangle = \sqrt{n_\alpha(\mathbf{k})} \frac{e}{m} \frac{1}{\sqrt{2\omega}} \frac{\sqrt{4\pi}}{L^{3/2}} \langle \psi_f| \mathbf{p}\cdot\hat{\varepsilon}_\alpha(\mathbf{k}) e^{i\mathbf{k}\cdot\mathbf{r}} |\psi_i\rangle \tag{11.131}$$

380 Chapter 11 The Quantized Electromagnetic Field

The absorption rate into this one state is

$$\tilde{\Gamma} = \frac{8\pi^2}{L^3} n_\alpha(\mathbf{k}) \frac{\alpha}{2m^2\omega} \left|\langle\psi_f| \mathbf{p} \cdot \hat{\boldsymbol{\varepsilon}}_\alpha(\mathbf{k})e^{i\mathbf{k}\cdot\mathbf{r}} |\psi_i\rangle\right|^2 \delta(E_f - E_i - \omega) \quad (11.132)$$

Here E_n are the atomic energy levels. Equation (11.132) is Fermi's golden rule, with only one final state. The absorption rate from all photons with energies in $d\omega$ is

$$d\Gamma = \frac{4\pi^2}{L^3} \frac{\alpha}{m^2\omega} \sum_{\omega \text{ in } d\omega} n_\alpha(\mathbf{k}) \left|\langle\psi_f| \mathbf{p} \cdot \hat{\boldsymbol{\varepsilon}}_\alpha(\mathbf{k})e^{i\mathbf{k}\cdot\mathbf{r}} |\psi_i\rangle\right|^2 \delta(E_f - E_i - \omega) \quad (11.133)$$

Multiply and divide by $N(\omega)d\omega$, the number of states in $d\omega$, and write

$$\frac{1}{N(\omega)d\omega} \sum_{\omega \text{ in } d\omega} n_\alpha(\mathbf{k}) \left|\langle\psi_f| \mathbf{p} \cdot \hat{\boldsymbol{\varepsilon}}_\alpha(\mathbf{k})e^{i\mathbf{k}\cdot\mathbf{r}} |\psi_i\rangle\right|^2 = \overline{n_\alpha(\mathbf{k})\left|\langle\psi_f| \mathbf{p} \cdot \hat{\boldsymbol{\varepsilon}}_\alpha(\mathbf{k})e^{i\mathbf{k}\cdot\mathbf{r}} |\psi_i\rangle\right|^2} \quad (11.134)$$

where the bars mean the average over all states in the energy range $d\omega$. The total absorption rate for all photons in the cavity is

$$\boxed{\begin{aligned}\Gamma = \int \frac{d\Gamma}{d\omega} d\omega &= N(\omega_{fi}) \frac{4\pi^2}{L^3} \bar{n}(\omega_{fi}) \frac{\alpha}{m^2\omega_{fi}} \overline{\left|\langle\psi_f| \mathbf{p} \cdot \hat{\boldsymbol{\varepsilon}}_\alpha(\mathbf{k})e^{i\mathbf{k}\cdot\mathbf{r}} |\psi_i\rangle\right|^2} \\ &= I(\omega_{fi}) 4\pi^2 \frac{\alpha}{m^2\omega_{fi}^2} \overline{\left|\langle\psi_f| \mathbf{p} \cdot \hat{\boldsymbol{\varepsilon}}_\alpha(\mathbf{k})e^{i\mathbf{k}\cdot\mathbf{r}} |\psi_i\rangle\right|^2}\end{aligned}} \quad (11.135)$$

The last factor is the average over all directions and polarizations. The rate depends only on the energy density, not the size of the cavity, and as promised, the dimension L has canceled.

Absorption Cross Section

$I(\omega)$ is the energy density, and also (because $c = 1$) the incident energy flux of photons in the range $d\omega$. The incident photon number flux is $I(\omega)/\omega$, so the absorption cross section is

$$\sigma(\omega) = \frac{4\pi^2\alpha}{m^2\omega_{fi}} \overline{\left|\langle\psi_f| \mathbf{p} \cdot \hat{\boldsymbol{\varepsilon}}_\alpha(\mathbf{k})e^{i\mathbf{k}\cdot\mathbf{r}} |\psi_i\rangle\right|^2} \delta(\omega - \omega_{fi}) \quad (11.136)$$

The cross section is independent of the incident intensity.

Absorption by the Lyman-α Line

As an example I will compute the rate for absorption of electromagnetic radiation by atomic hydrogen from the ground state to the states in the $n = 2$ level. This is the Lyman-α line. In the lowest approximation it is legitimate to ignore spin.

All three of the $2P$ states will be excited. Since the initial $1S$ state is spherically symmetric, the total absorption rate is independent of the direction and polarization of the incident beam, and we can take the radiation to be in the z-direction. The polarization vector $\hat{\boldsymbol{\varepsilon}}$ can be any normalized combination of $\hat{\mathbf{n}}_x$ and $\hat{\mathbf{n}}_y$. We choose it to describe circular polarization.

$$\hat{\boldsymbol{\varepsilon}} = -\frac{\hat{\mathbf{n}}_x + i\hat{\mathbf{n}}_y}{\sqrt{2}} \quad (11.137)$$

You can check that another choice gives the same result.

The dipole approximation is valid here. Replace $e^{i\mathbf{k}\cdot\mathbf{r}}$ by 1, and $\mathbf{p}_{fi}$ by $i(E_f - E_i)m\mathbf{r}_{fi}$. Then in spherical vector notation

$$\hat{\varepsilon}\cdot\mathbf{r} = -\frac{x+iy}{\sqrt{2}} = r^{(+1)} \tag{11.138}$$

The matrix element

$$\left\langle 2lm \left| r^{(+1)} \right| 100 \right\rangle \tag{11.139}$$

vanishes unless $l = 1$ and $m = +1$. Parity conservation forbids all components of $\mathbf{r}$ to connect the ground state and the $2S$ states. With the definition (11.137) for $\hat{\varepsilon}$ we need not sum over any final states. Now

$$\left[J_+, r^{(o)}\right] = \sqrt{2}r^{(+1)} \tag{11.140}$$

and

$$\left\langle 211 \left| \left[J_+, r^{(o)}\right] \right| 100 \right\rangle = \sqrt{2}\langle 210| z |100\rangle = \sqrt{2}\sqrt{\frac{2^{15}}{3^{10}}}a \tag{11.141}$$

The absorption rate is

$$\Gamma = 4\pi^2 \alpha I(\omega_{21})\frac{2^{15}}{3^{10}}a^2 \tag{11.142}$$

11.6.3 Stimulated Emission

If there are any atoms in the upper state, there is also a rate for emitting photons with frequency ω_{fi}. Suppose a photon with polarization α and momentum $\mathbf{k}$ is emitted. This time we need the matrix element of the creation operator $a_\beta^\dagger(\mathbf{k}')$ between the initial and final state. If the new final state photon $(\alpha, \mathbf{k})$ is one of the those already present in the incident beam, the matrix element of $a_\beta^\dagger(\mathbf{k}')$ is proportional to $\sqrt{n_\alpha(\mathbf{k}) + 1}$. Except for this factor, which replaces the factor $\sqrt{n_\alpha(\mathbf{k})}$ that occurs in equation (11.131), the calculation imitates the absorption computation.

The rate of absorption, per atom in the lower state, is given in equation (11.135). The emission rate for photons of any of the modes in the original beam, per atom in the upper state, is

$$\boxed{\Gamma = I(\omega_{fi})4\pi^2\frac{\bar{n}(\omega_{fi})+1}{\bar{n}(\omega_{fi})}\frac{\alpha}{m^2\omega_{fi}^2}\left|\langle\psi_f|\mathbf{p}\cdot\hat{\varepsilon}_\alpha(\mathbf{k})e^{i\mathbf{k}\cdot\mathbf{r}}|\psi_i\rangle\right|^2} \tag{11.143}$$

If the occupation numbers are large, so that $\bar{n}(\omega_{fi}) + 1 \approx \bar{n}(\omega_{fi})$, then the rate is about the same answer for the total emission rate as for absorption, in agreement with the semiclassical result.

For absorption, the semiclassical theory gave the correct answer for arbitrarily small $\bar{n}$. That was a bit of good fortune that does not occur in the emission calculation. For small occupation numbers there are real quantum effects.

382 Chapter 11 The Quantized Electromagnetic Field

The emission rate formula gives a nonzero answer even when $\bar{n} \approx 1$. Of course this must be true: Set all the $n_\alpha(\mathbf{k}) = 0$ and sum the rate for emitting a photon of any polarization and momentum. The result is

$$\Gamma = \frac{\alpha}{2\pi m^2 \omega} \int d^3k \sum_\alpha |\langle \psi_f | \hat{\boldsymbol{\varepsilon}}_\alpha(\mathbf{k}) \cdot \mathbf{p}\, e^{i\mathbf{k}\cdot\mathbf{r}} |\psi_i\rangle|^2 \qquad (11.144)$$

This is the rate for "spontaneous" single photon emission that we derived in Section 11.4.

11.6.4 The Blackbody Formula

Consider a cavity containing isotropic radiation in thermal equilibrium. The flux of electromagnetic radiation is the same in any direction. Imagine a box filled with atoms, some in the upper state $|\psi_f\rangle$ and some in the lower state $|\psi_o\rangle$ together with the electromagnetic radiation.

To keep the computation simple, let us assume the dipole approximation is valid. You can check that this assumption is actually unnecessary. The matrix element (take the polarization real) for either emission or absorption is

$$\langle \psi_1 | \hat{\boldsymbol{\varepsilon}}_\alpha(\mathbf{k}) \cdot \mathbf{p}\, e^{i\mathbf{k}\cdot\mathbf{r}} |\psi_o\rangle = -im\omega_{fi}\hat{\boldsymbol{\varepsilon}}_\alpha(\mathbf{k}) \cdot \mathbf{r}_{fi} \qquad (11.145)$$

where $\mathbf{r}_{fi} = \langle \psi_1 | \mathbf{r} | \psi_o \rangle$.

If the energy density in the beam with frequency between ω and $\omega + d\omega$ is $I(\omega)d\omega$, the rate per atom for both absorption and stimulated emission is

$$\Gamma = 4\pi^2 \alpha I(\omega_{fi}) |\mathbf{r}_{fi} \cdot \hat{\boldsymbol{\varepsilon}}|^2 \qquad (11.146)$$

Since the radiation is isotropic, you can average over both $\hat{\boldsymbol{\varepsilon}}$ and the direction of the emitted photon:

$$\epsilon_i \epsilon_j \to \frac{1}{2} \int \sum_\alpha \hat{\varepsilon}_{\alpha i}(\mathbf{k})\hat{\varepsilon}_{\alpha j}(\mathbf{k})\, d\Omega \,\bigg/ \int d\Omega = \frac{\delta_{ij}}{3} \qquad (11.147)$$

so that

$$|\mathbf{r}_{fi} \cdot \hat{\boldsymbol{\varepsilon}}|^2 \to \frac{1}{3}|\mathbf{r}_{fi}|^2 \qquad (11.148)$$

and

$$\Gamma = \frac{4\pi^2 \alpha}{3} I(\omega_{fi}) \mathbf{r}_{fi}^2 \qquad (11.149)$$

Now suppose there are N_1 atoms in the upper state and N_o atoms in the lower state. The number of absorptions and induced emissions per second respectively are

$$\frac{4\pi^2 \alpha}{3} N_o I(\omega_{fi}) \mathbf{r}_{fi}^2 \quad \text{and} \quad \frac{4\pi^2 \alpha}{3} N_1 I(\omega_{fi}) \mathbf{r}_{fi}^2 \qquad (11.150)$$

The rate for spontaneous emissions is

$$N_1 \frac{4\alpha \omega_{fi}^3}{3} \mathbf{r}_{fi}^2 \qquad (11.151)$$

In equilibrium, the total rates of absorption and emission must be equal. Therefore,

$$N_o \frac{4\pi^2 \alpha}{3} I(\omega_{fi}) \mathbf{r}_{fi}^2 = N_1 \left[\frac{4\pi^2 \alpha}{3} I(\omega_{fi}) \mathbf{r}_{fi}^2 + \frac{4\alpha \omega_{fi}^3}{3} \mathbf{r}_{fi}^2 \right] \tag{11.152}$$

and this must be true for any pair of states. Solve for $I(\omega)$:

$$\pi^2 I(\omega) \left(1 - \frac{N_1}{N_o}\right) = \frac{N_1}{N_o} \omega^3 \tag{11.153}$$

The temperature is defined by the ratio of the populations at equilibrium:

$$\frac{N_1}{N_o} = \frac{e^{-E_1/kT}}{e^{-E_o/kT}} = e^{-\omega/kT} \tag{11.154}$$

Therefore,

$$I(\omega) = \frac{\omega^3}{\pi^2} \frac{e^{-\omega/kT}}{1 - e^{-\omega/kT}} = \frac{\omega^3}{\pi^2} \frac{1}{e^{\omega/kT} - 1} \tag{11.155}$$

With the constants $\hbar$ and c put back in, the energy density is

$$\boxed{I(\omega) = \frac{\hbar}{\pi^2} \left(\frac{\omega}{c}\right)^3 \frac{1}{e^{\hbar\omega/kT} - 1}} \tag{11.156}$$

This is the famous formula that was discovered by Max Planck in 1900! Later, Albert Einstein turned the argument around to guess the formula for spontaneous emission even before the invention of the full methods of quantum mechanics.

11.7 SCATTERING OF PHOTONS BY ATOMS

This section contains some relatively complicated calculations, some examples of what can be done with the theory of photons and electrons, and some useful computational techniques. It is not important to memorize all the details.

11.7.1 The Photoelectric Effect

The photoelectric effect, along with Planck's formula for the blackbody spectrum, had no conceivable classical explanation, and led to the understanding that light came in discrete units that came to be dubbed *quanta* in the early days of the twentieth century, giving our subject its esoteric name. I want to illustrate, with some drastic approximations, how the rate for photoionization can be computed. I will restrict the discussion to atomic hydrogen as an illustration. The method is easily generalized to other atoms.

The photoelectric effect is an example of inelastic scattering. The initial state contains a photon with polarization vector $\hat{\varepsilon}$ and momentum $\hat{n}\omega$, and an electron in the ground state.[13] The final states are $|\mathbf{k}_f\rangle$, where $\mathbf{k}_f$ is the momentum of the ejected electron. These are normalized as usual to

$$\langle \mathbf{k}_f | \mathbf{k}'_f \rangle = \delta_3 \left(\mathbf{k}'_f - \mathbf{k}_f\right) \tag{11.157}$$

---
[13] Revert to the Dirac delta function normalization.

384 Chapter 11 The Quantized Electromagnetic Field

The relevant matrix element of the interaction is

$$H'_{fi} = \frac{e}{m} \langle \mathbf{k}_f | \mathbf{A}(\mathbf{r}) \cdot \mathbf{p} | \hat{\mathbf{n}}\omega; \psi_{100} \rangle \tag{11.158}$$

where ψ_{100} is the electron ground state. Then the cross section into a group of final states $d^3 k_f$ follows from equation (9.70), with $v = 1$:

$$d\sigma = \frac{4\pi^2 \alpha}{m^2 \omega} |\langle \mathbf{k}_f | e^{i\omega \hat{\mathbf{n}} \cdot \mathbf{r}} \hat{\boldsymbol{\varepsilon}} \cdot \mathbf{p} | \psi_{100} \rangle|^2 \delta(\omega - \omega_{fi}) d^3 k_f \tag{11.159}$$

Restrict the photon frequency so that the final-state electron can be treated nonrelativistically. Then the final-state energy is $\omega_f = k_f^2/2m$, and

$$\omega_{fi} = \omega_f - E_o = \omega_f + \frac{1}{2}\alpha^2 m \tag{11.160}$$

Write

$$d^3 k_f = k_f^2 \, dk_f \, d\Omega_f = k_f m \, d\omega_f d\Omega_f \tag{11.161}$$

and integrate over ω_f to get the rate into $d\Omega_f$. Then

$$\frac{d\sigma(\omega)}{d\Omega_f} = \frac{4\pi^2 \alpha k_f}{m \omega_{fi}} |\langle \psi_f | e^{i\omega_{fi} \hat{\mathbf{n}} \cdot \mathbf{r}} \hat{\boldsymbol{\varepsilon}} \cdot \mathbf{p} | \psi_i \rangle|^2 \tag{11.162}$$

The initial and final state wave functions are

$$\psi_f(\mathbf{r}) = \frac{1}{(2\pi)^{\frac{3}{2}}} e^{i\mathbf{k}_f \cdot \mathbf{r}} \quad \text{and} \quad \psi_i(\mathbf{r}) = \frac{1}{\sqrt{\pi a^3}} e^{-r/a} \tag{11.163}$$

Actually, a continuum Coulomb wave function would be correct instead of a plane wave for $|\psi_f\rangle$, but it is approximately a plane wave as long as $\omega_f \gg |E_o|$, that is, as long as the final electron energy is not too close to the threshold for ionization. We have also assumed that $\omega_f \ll mc^2$, so that the final electron is nonrelativistic. It is possible to relax both of these conditions, but I want to keep the problem simple.

Since $|\psi_f\rangle$ is an eigenstate of the electron momentum operator, the differential absorption cross section is

$$\frac{d\sigma(\omega)}{d\Omega_f} = \frac{4\pi^2 \alpha k_f}{m\omega} \frac{1}{\pi a^3} \frac{1}{(2\pi)^3} |\hat{\boldsymbol{\varepsilon}} \cdot \mathbf{k}_f|^2 \left| \int e^{-i\mathbf{q} \cdot \mathbf{r}} e^{i\omega \hat{\mathbf{n}} \cdot \mathbf{r}} e^{-r/a} d^3 r \right|^2 \tag{11.164}$$

where $\mathbf{q} = \mathbf{k}_f - \omega \hat{\mathbf{n}}$ is the momentum transferred to the electron. (Momentum is not conserved in this problem since the recoil of the whole atom is being ignored.) The integral in equation (11.164) is computed in equation (A.37a) of the appendix.

$$\frac{d\sigma(\omega)}{d\Omega_f} = \frac{32\alpha k_f a^3}{m\omega} |\hat{\boldsymbol{\varepsilon}} \cdot \mathbf{k}_f|^2 \frac{1}{(1 + q^2 a^2)^4} \tag{11.165}$$

Unpolarized Beam
Usually the incident light is unpolarized. In that case you should average over the two polarizations. Replace $|\hat{\boldsymbol{\varepsilon}} \cdot \mathbf{k}_f|^2$ by

$$\frac{1}{2} \sum_\alpha |\hat{\boldsymbol{\varepsilon}} \cdot \mathbf{k}_f|^2 = \frac{1}{2} \sum_{ij\alpha} \epsilon^*_{\alpha i} \epsilon_{\alpha j} k_{f_i} k_{f_j} = \frac{1}{2} k_f^2 \sin^2 \theta \tag{11.166}$$

where θ is the angle between the direction of the incident radiation $\hat{n}$ and that of the final electron. The momentum transfer squared is

$$q^2 = k_f^2 - 2\omega \hat{n} \cdot \mathbf{k}_f + \omega^2 \tag{11.167}$$

Here ω is the energy transferred to the electron, namely $k_f^2/2m - E_o$, where E_o is the ground-state energy $-\alpha^2 m/2$. Since the electron is nonrelativistic, the last two terms in equation (11.167) are negligible and we can just set $q^2 = k^2$. Because $k_f^2/2m \gg |E_o|$. Therefore $k_f^2 \gg 1/a^2$, so the factor $1/a^2 + q^2$ can be replaced by k_f^2. With these approximations the differential absorption cross section becomes

$$\frac{d\sigma}{d\Omega_f} = \frac{16 m^4 \alpha^6}{\omega} \sin^2\theta \frac{1}{(2m|E_o|)^{\frac{5}{2}}} \left(\frac{E_o}{E_f}\right)^{\frac{5}{2}} \tag{11.168}$$

If in the spirit of all the approximations above we further replace $E_f = \omega - E_o$ by ω, then

$$\frac{d\sigma}{d\Omega_f} = 32\alpha \sin^2\theta\, a^2 \left(\frac{E_o}{\omega}\right)^{\frac{7}{2}} \tag{11.169}$$

The total cross section is

$$\sigma = \frac{256\pi\alpha}{3} a^2 \left(\frac{\omega}{E_i}\right)^{\frac{7}{2}} \tag{11.170}$$

This is how the absorption of light by photoemission of electrons depends on the energy. The dependence on the seven-half power of the radiation frequency is characteristic of the photoelectric effect for a wide range of substances.

11.7.2 Elastic Scattering of Photons

Next I want to work out the scattering of a photon by an atom or nucleus in its ground state. While I will discuss only elastic scattering, inelastic scattering—where the atom is left in an excited state—works the same way.

In order to make a stab at keeping the notation under control, use the letter ψ to label the atomic state, and the letter γ to stand for the photon polarization and momentum α and $\mathbf{k}$. Then $\sum_\gamma$ is shorthand for $\sum_\alpha \int d^3 k$, and $\delta_{\gamma_1 \gamma_2}$ for $\delta_{\alpha_1 \alpha_2} \delta(\mathbf{k}_1 - \mathbf{k}_2)$.

The asymptotic states are the eigenstates of H_o, which contains the free photon term, the electron kinetic energy and the Coulomb interaction that binds the electron in the atom.

$$H_o |\gamma, \psi_o\rangle = (\omega + E_o)|\gamma, \psi_o\rangle \tag{11.171}$$

ω is the energy of the photon, and E_o is the ground-state energy of the atom. The interaction Hamiltonian is the rest:

$$H' = \frac{e}{m} \mathbf{A}(\mathbf{r}) \cdot \mathbf{p} + \frac{e^2}{2m} \mathbf{A}(\mathbf{r}) \cdot \mathbf{A}(\mathbf{r}) + \cdots \tag{11.172}$$

For scattering calculations, the first Born approximation with the term linear in $\mathbf{A}(\mathbf{r})$ vanishes, since $\mathbf{A}(\mathbf{r})$ is linear in $a_\alpha(\mathbf{k})$ and $a_\alpha^\dagger(\mathbf{k})$:

$$\left\langle \gamma', \psi_o \left| \frac{e}{m} \mathbf{A}(\mathbf{r}) \cdot \mathbf{p} \right| \gamma, \psi_o \right\rangle = 0 \tag{11.173}$$

386 Chapter 11 The Quantized Electromagnetic Field

The $\mathbf{A} \cdot \mathbf{A}$ term is quadratic in the creation and annihilation operators, so here it contributes in the first Born approximation to photon scattering, and the amplitude is of order e^2. But that is not the whole story. The $\mathbf{p} \cdot \mathbf{A}$ term, taken to second order, has the same number of powers of e as does the $\mathbf{A}^2$ term taken to first order. Both terms contribute to the same order in α. Fermi's golden rule is only the first Born approximation; here the detailed form of the transition matrix $T(\omega)$ is required.

Second-Order Term in the Scattering Amplitude

Begin with the most complicated part, computing the term in T_{fi} that is second order in H'. According to equation (9.42) this term is

$$T_{fi}^{(II)} = \langle \gamma', \psi_o | H' G^o(\omega_i) H' | \gamma, \psi_o \rangle \qquad (11.174)$$

where $\omega_i = \omega + E_o$. Put into equation (11.174) a complete set of eigenstates of H_o, remembering that $G^o(\omega)$ is diagonal in that basis:

$$T_{fi}^{(II)} = \sum_n \langle \gamma', \psi_o | H' | n \rangle \langle n | G^o(\omega_i) | n \rangle \langle n | H' | \gamma, \psi_o \rangle \qquad (11.175)$$

Each term in $\mathbf{A}(\mathbf{r})$ contains one creation or annihilation operator. Therefore the intermediate states contain either zero or two photons.

Intermediate States with No Photons: The terms with no photons in the intermediate states are

$$T_{fi}^{(II)(0)} = \frac{e^2}{m^2} \sum_n \langle \gamma', \psi_o | \mathbf{p} \cdot \mathbf{A}(\mathbf{r}) | \psi_n \rangle \langle \psi_n | G^o(\omega_i) | \psi_n \rangle \langle \psi_n | \mathbf{p} \cdot \mathbf{A}(\mathbf{r}) | \gamma, \psi_o \rangle \qquad (11.176)$$

Now the sum goes over all atomic states. The individual matrix elements in equation (11.176) are

$$\langle \psi_n | \mathbf{p} \cdot \mathbf{A}(\mathbf{r}) | \gamma, \psi_o \rangle = \frac{\sqrt{4\pi}}{(2\pi)^{3/2}} \frac{1}{\sqrt{2\omega}} \langle \psi_n | \hat{\varepsilon}_{\alpha'} \cdot \mathbf{p} e^{i\mathbf{k}\cdot\mathbf{r}} | \psi_o \rangle \qquad (11.177\text{a})$$

$$\langle \gamma', \psi_o | \mathbf{p} \cdot \mathbf{A}(\mathbf{r}) | \psi_n \rangle = \frac{\sqrt{4\pi}}{(2\pi)^{3/2}} \frac{1}{\sqrt{2\omega'}} \langle \psi_o | \hat{\varepsilon}_\alpha^* \cdot \mathbf{p} e^{-i\mathbf{k}'\cdot\mathbf{r}} | \psi_n \rangle \qquad (11.177\text{b})$$

$$\langle \psi_n | G^o(\omega_i) | \psi_n \rangle = \frac{1}{\omega_i - E_n} = \frac{1}{E_{on} + \omega} \qquad (11.177\text{c})$$

where $E_{on} = E_o - E_n$. The matrix elements of $\mathbf{p} \cdot \mathbf{A}(\mathbf{r})$ are evaluated in the same way as similar matrix elements in Section 11.4.4, except that in equations (11.177) it is the destruction operator term in the Fourier expansion of $\mathbf{A}(\mathbf{r})$ that contributes, so the exponential has the other sign. The contribution to $T_{fi}^{(II)}$ of the no-photon intermediate states is

$$T_{fi}^{(II)(0)} = \frac{4\pi e^2}{m^2} \frac{1}{(2\pi)^3} \frac{1}{\sqrt{4\omega^2}}$$
$$\times \sum_n \langle \psi_o | \hat{\varepsilon}_{\alpha'}^* \cdot \mathbf{p} e^{-i\mathbf{k}'\cdot\mathbf{r}} | \psi_n \rangle \frac{1}{\omega + E_{on}} \langle \psi_n | \hat{\varepsilon}_\alpha \cdot \mathbf{p} e^{i\mathbf{k}\cdot\mathbf{r}} | \psi_o \rangle \qquad (11.178)$$

In these terms the photon is first absorbed, exciting the electron into the state $|\psi_n\rangle$. The process violates energy conservation, so the excited state has to decay quickly into the ground state and a photon. The sequence is illustrated if Figure 11.1a.

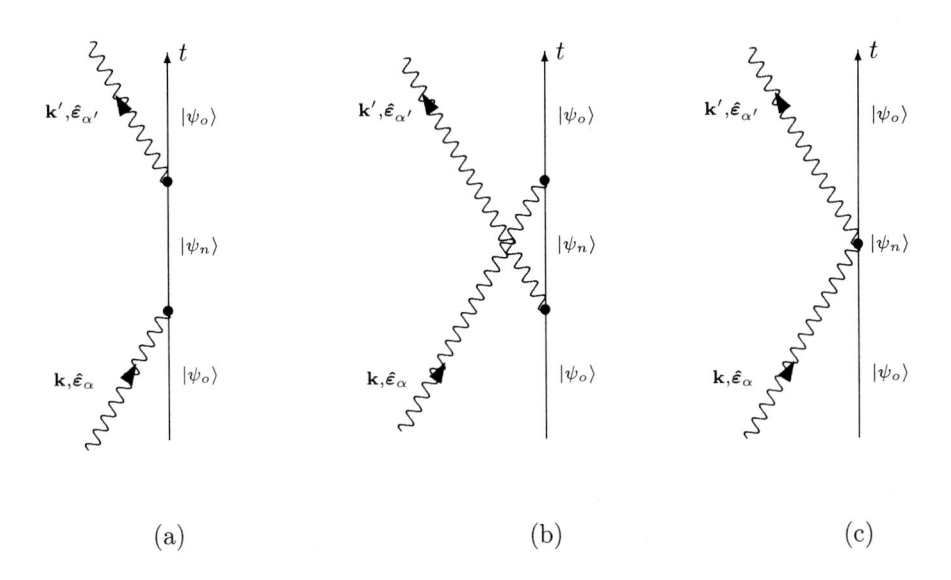

FIGURE 11.1: The amplitude for elastic scattering of a photon off an atomic electron has three terms of order e^2: (a) A term without photons in the intermediate state (11.178), (b) a terms with two photons in the intermediate state (11.185), and (c) a "contact" term (11.186).

Intermediate States with Two Photons: In these states the final photon is created before the initial photon is annihilated. The formalism will force the intermediate-state photons to have the quantum numbers of the initial and final state photons. These intermediate two-photon states are[14]

$$|\gamma_1\gamma_2\rangle = a^\dagger_{\gamma_1} a^\dagger_{\gamma_2} |\rangle \qquad \text{and} \qquad |\gamma_3\gamma_4\rangle = a^\dagger_{\gamma_3} a^\dagger_{\gamma_4} |\rangle \qquad (11.179)$$

so they are normalized to

$$\langle \gamma_1\gamma_2 | \gamma_3\gamma_4 \rangle = \delta_{\gamma_1\gamma_3}\delta_{\gamma_2\gamma_4} + \delta_{\gamma_1\gamma_4}\delta_{\gamma_2\gamma_3} \qquad (11.180)$$

---

[14] The symbol $|\rangle$ means the no-photon state that was previously denoted $|0,0,0,\ldots\rangle$.

388 Chapter 11 The Quantized Electromagnetic Field

The matrix element is[15]

$$T_{fi}^{(II)(2)} = \frac{e^2}{m^2} \sum_n \frac{1}{4} \sum_{\gamma_1 \gamma_2} \sum_{\gamma_3 \gamma_4} \langle \gamma', \psi_o | \mathbf{p} \cdot \mathbf{A}(\mathbf{r}) | \gamma_1 \gamma_2 \psi_n \rangle$$

$$\times \langle \gamma_1 \gamma_2 \psi_n | G^o(\omega_i) | \gamma_3 \gamma_4 \psi_n \rangle \langle \gamma_3 \gamma_4 \psi_n | \mathbf{p} \cdot \mathbf{A}(\mathbf{r}) | \gamma, \psi_o \rangle \quad (11.181)$$

The individual factors are

$$\langle \gamma_3 \gamma_4 \psi_n | \mathbf{p} \cdot \mathbf{A}(\mathbf{r}) | \gamma, \psi_o \rangle$$
$$= \frac{\sqrt{4\pi}}{(2\pi)^{3/2}} \frac{1}{\sqrt{2\omega}} \left\langle \psi_n \left| \mathbf{p} \cdot \left(\hat{\varepsilon}_{\gamma_4}^*(\mathbf{k}_4) e^{-i\mathbf{k}_4 \cdot \mathbf{r}} \delta_{\gamma_3 \gamma} + \hat{\varepsilon}_{\gamma_3}^*(\mathbf{k}_3) e^{-i\mathbf{k}_3 \cdot \mathbf{r}} \delta_{\gamma_4 \gamma} \right) \right| \psi_o \right\rangle$$
(11.182a)

$$\langle \gamma', \psi_o | \mathbf{p} \cdot \mathbf{A}(\mathbf{r}) | \gamma_1 \gamma_2 \psi_n \rangle$$
$$= \frac{\sqrt{4\pi}}{(2\pi)^{3/2}} \frac{1}{\sqrt{2\omega}} \left\langle \psi_o \left| \mathbf{p} \cdot \left(\hat{\varepsilon}_{\gamma_2}(\mathbf{k}_2) e^{i\mathbf{k}_2 \cdot \mathbf{r}} \delta_{\gamma_1 \gamma'} + \hat{\varepsilon}_{\gamma_1}(\mathbf{k}_1) e^{i\mathbf{k}_1 \cdot \mathbf{r}} \delta_{\gamma_2 \gamma'} \right) \right| \psi_n \right\rangle$$
(11.182b)

$$\langle \gamma_1 \gamma_2 \psi_n | G^o(\omega_i) | \gamma_3 \gamma_4 \psi_n \rangle$$
$$= \frac{1}{\omega + E_o - \omega_1 - \omega_2 - E_n} \left(\delta_{\gamma_1 \gamma_3} \delta_{\gamma_2 \gamma_4} + \delta_{\gamma_1 \gamma_4} \delta_{\gamma_2 \gamma_3} \right) \quad (11.182c)$$

Each factor in equations (11.182) has two terms. $T^{(II)(2)}$ has eight terms in all. Here is the contribution of the first of the two terms in each of the three factors displayed in (11.182):

$$\frac{e^2}{4m^2} \sum_n \sum_{\gamma_1 \gamma_2 \gamma_3 \gamma_4} \frac{4\pi}{(2\pi)^3} \frac{1}{2\omega} \delta_{\gamma_3 \gamma} \delta_{\gamma_1 \gamma_3} \delta_{\gamma_2 \gamma_4} \delta_{\gamma_1 \gamma'}$$

$$\times \left[\langle \psi_n | \mathbf{p} \cdot \hat{\varepsilon}_{\gamma_4}^*(\mathbf{k}_4) e^{-i\mathbf{k}_4 \cdot \mathbf{r}} | \psi_o \rangle \frac{1}{\omega + E_o - \omega_1 - \omega_2 - E_n} \langle \psi_o | \mathbf{p} \cdot \hat{\varepsilon}_{\gamma_2}(\mathbf{k}_2) e^{i\mathbf{k}_2 \cdot \mathbf{r}} | \psi_n \rangle \right]$$

$$= \frac{e^2}{4m^2} \sum_n \sum_{\gamma_2} \frac{4\pi}{(2\pi)^3} \frac{1}{2\omega} \delta_{\gamma' \gamma}$$

$$\times \left[\langle \psi_n | \mathbf{p} \cdot \hat{\varepsilon}_{\gamma_2}^*(\mathbf{k}_2) e^{-i\mathbf{k}_2 \cdot \mathbf{r}} | \psi_o \rangle \frac{1}{E_o - \omega - E_n} \langle \psi_o | \mathbf{p} \cdot \hat{\varepsilon}_{\gamma_2}(\mathbf{k}_2) e^{i\mathbf{k}_2 \cdot \mathbf{r}} | \psi_n \rangle \right]$$
(11.183)

For scattering through a nonzero angle, the initial and final photon states are different: $\delta_{\gamma' \gamma} = 0$, and all four of these terms vanish.

---

[15]The piece of the identity operator containing two-photon states is $I = 1/2 \sum_\gamma$. Because the two photons are identical Bosons, omitting the 1/2 would count the states twice. That's why there is an extra factor of 1/4 in the expression for $T_{fi}^{(II)(2)}$.

Section 11.7 Scattering of Photons by Atoms

A different type of combination is obtained by combining the first terms in (11.182a) and (11.182b) with the second term in (11.182c):

$$\frac{e^2}{4m^2} \sum_n \sum_{\gamma_1\gamma_2\gamma_3\gamma_4} \frac{4\pi}{(2\pi)^3} \frac{1}{2\omega} \delta_{\gamma_3\gamma} \delta_{\gamma_1\gamma_4} \delta_{\gamma_2\gamma_3} \delta_{\gamma_1\gamma'}$$

$$\times \left[\langle \psi_n | \mathbf{p} \cdot \hat{\boldsymbol{\varepsilon}}^*_{\gamma_4}(\mathbf{k}_4) e^{-i\mathbf{k}_4 \cdot \mathbf{r}} | \psi_o \rangle \frac{1}{\omega + E_o - \omega_1 - \omega_2 - E_n} \langle \psi_o | \mathbf{p} \cdot \hat{\boldsymbol{\varepsilon}}_{\gamma_2}(\mathbf{k}_2) e^{i\mathbf{k}_2 \cdot \mathbf{r}} | \psi_n \rangle \right]$$

$$= \frac{e^2}{4m^2} \sum_n \frac{4\pi}{(2\pi)^3} \frac{1}{2\omega}$$

$$\times \left[\langle \psi_n | \mathbf{p} \cdot \hat{\boldsymbol{\varepsilon}}^*_{\gamma'}(\mathbf{k}') e^{-i\mathbf{k}' \cdot \mathbf{r}} | \psi_o \rangle \frac{1}{\omega + E_o - \omega - \omega - E_n} \langle \psi_o | \mathbf{p} \cdot \hat{\boldsymbol{\varepsilon}}_{\gamma}(\mathbf{k}) e^{i\mathbf{k} \cdot \mathbf{r}} | \psi_n \rangle \right]$$

(11.184)

There are three more terms exactly like this one too. Therefore the total contribution of these two-photon intermediate states is

$$T^{(II)(2)}_{fi} = \frac{4\pi e^2}{m^2} \frac{1}{(2\pi)^3} \frac{1}{\sqrt{4\omega'\omega}}$$

$$\times \sum_n \langle \psi_o | \hat{\boldsymbol{\varepsilon}}_\alpha \cdot \mathbf{p} e^{i\mathbf{k} \cdot \mathbf{r}} | \psi_n \rangle \frac{1}{E_{on} - \omega} \langle \psi_n | \hat{\boldsymbol{\varepsilon}}^*_\alpha \cdot \mathbf{p} e^{-i\mathbf{k}' \cdot \mathbf{r}} | \psi_o \rangle \quad (11.185)$$

This expression is the amplitude for the electron to emit a photon first. The intermediate state has both photons, as well as the electron in some other state $|\psi_n\rangle$. The intermediate state violates energy conservation, but it is soon reestablished by absorbing the initial photon. See Figure 11.1b.

First-Order Term in the Scattering Amplitude: Finally, there is the contribution of the $\mathbf{A}^2$ term in H', explicitly proportional to e^2.

$$T^{(I)}_{fi} = \frac{e^2}{2m} \langle \gamma', \psi_o | \mathbf{A}(\mathbf{r}) \cdot \mathbf{A}(\mathbf{r}) | \gamma, \psi_o \rangle$$

$$= \frac{e^2}{2m} \frac{4\pi}{(2\pi)^3} \frac{1}{\sqrt{4\omega'\omega}} \sum_{\gamma''\gamma'''} \hat{\boldsymbol{\varepsilon}}^*_{\alpha'''} \cdot \hat{\boldsymbol{\varepsilon}}_{\alpha''} \langle \gamma' | (a_{\gamma''} a^\dagger_{\gamma'''} + a_{\gamma''} a^\dagger_{\gamma'''}) | \gamma \rangle \quad (11.186)$$

$$= \frac{e^2}{m} \frac{4\pi}{(2\pi)^3} \frac{1}{\sqrt{4\omega'\omega}} \hat{\boldsymbol{\varepsilon}}^*_\alpha \cdot \hat{\boldsymbol{\varepsilon}}_\alpha$$

This term is a point interaction between the initial particles and the final ones, as illustrated in Figure 11.1c.

Elastic Scattering of Photons by Atoms at Low Frequencies

I will only work out part of the leading term in the scattering cross section at low photon energies. There one can make the long wavelength dipole approximation and replace all the exponential factors by 1. In this approximation the transition matrix is the sum of the expressions (11.186), (11.185), and (11.178).

$$T_{fi} = \frac{\alpha}{2\pi^2 m} \frac{1}{\sqrt{4\omega'\omega}} \left[\hat{\varepsilon}_\alpha^* \cdot \hat{\varepsilon}_\alpha + \frac{1}{m} \sum_n \left(\langle \psi_o | \hat{\varepsilon}_\alpha \cdot \mathbf{p} | \psi_n \rangle \frac{1}{E_{on} - \omega'} \langle \psi_n | \hat{\varepsilon}_\alpha^* \cdot \mathbf{p} | \psi_o \rangle \right. \right.$$
$$\left. \left. + \langle \psi_o | \hat{\varepsilon}_\alpha^* \cdot \mathbf{p} | \psi_n \rangle \frac{1}{E_{on} + \omega} \langle \psi_n | \hat{\varepsilon}_\alpha \cdot \mathbf{p} | \psi_o \rangle \right) \right] \quad (11.187)$$

The initial photon flux density is $(2\pi)^{-3}$ (since $v = 1$), so according to equation (9.70) the cross section into final momenta in d^3k' for fixed photon polarizations is

$$d\sigma = (2\pi)^4 \int |T_{fi}|^2 \delta(\omega' - \omega) d^3k' = (2\pi)^4 \int |T_{fi}|^2 \omega^2 d\Omega \quad (11.188)$$

The differential cross section is

$$\frac{d\sigma}{d\Omega} = r_o^2 \left| \hat{\varepsilon}_\alpha^* \cdot \hat{\varepsilon}_\alpha + \frac{1}{m} \sum_n \left(\langle \psi_o | \hat{\varepsilon}_\alpha \cdot \mathbf{p} | \psi_n \rangle \frac{1}{E_{on} - \omega} \langle \psi_n | \hat{\varepsilon}_\alpha^* \cdot \mathbf{p} | \psi_o \rangle \right. \right.$$
$$\left. \left. + \langle \psi_o | \hat{\varepsilon}_\alpha^* \cdot \mathbf{p} | \psi_n \rangle \frac{1}{E_{on} + \omega} \langle \psi_n | \hat{\varepsilon}_\alpha \cdot \mathbf{p} | \psi_o \rangle \right) \right|^2 \quad (11.189)$$

$r_o = \alpha/m$ is the **classical radius of the electron**, the radius of a uniformly charged sphere whose electrostatic energy is its mass, about 2.8×10^{-13}cm. Use

$$\frac{1}{E_{on} \pm \omega} = \frac{1}{E_{on}} \left(1 \mp \frac{\omega}{E_{on}} + \frac{\omega^2}{E_{on}^2} + \cdots \right) \quad (11.190)$$

to expand equation (11.189) in powers of the photon frequency.

$$\frac{d\sigma}{d\Omega} = r_o^2 \left| \hat{\varepsilon}_\alpha^* \cdot \hat{\varepsilon}_\alpha \right.$$
$$+ \frac{1}{m} \sum_n \frac{1}{E_{on}} \left(\langle \psi_o | \hat{\varepsilon}_\alpha \cdot \mathbf{p} | \psi_n \rangle \langle \psi_n | \hat{\varepsilon}_\alpha^* \cdot \mathbf{p} | \psi_o \rangle + \langle \psi_o | \hat{\varepsilon}_\alpha^* \cdot \mathbf{p} | \psi_n \rangle \langle \psi_n | \hat{\varepsilon}_\alpha \cdot \mathbf{p} | \psi_o \rangle \right)$$
$$\left. + \frac{\omega}{m} \sum_n \frac{1}{E_{on}^2} \left(\langle \psi_o | \hat{\varepsilon}_\alpha \cdot \mathbf{p} | \psi_n \rangle \langle \psi_n | \hat{\varepsilon}_\alpha^* \cdot \mathbf{p} | \psi_o \rangle - \langle \psi_o | \hat{\varepsilon}_\alpha^* \cdot \mathbf{p} | \psi_n \rangle \langle \psi_n | \hat{\varepsilon}_\alpha \cdot \mathbf{p} | \psi_o \rangle \right) + \cdots \right|^2$$

$$(11.191)$$

and use

$$\langle \psi_n | \mathbf{p} | \psi_o \rangle = im E_{no} \langle \psi_n | \mathbf{r} | \psi_o \rangle \quad (11.192)$$

to compute the first two terms inside the absolute value symbol in equation (11.191) as

$$\sum_{ij} (\hat{\varepsilon}_\alpha^*)_j (\hat{\varepsilon}_\alpha)_i \left[\delta_{ij} - i \sum_n \left(\langle \psi_o | p_i | \psi_n \rangle \langle \psi_n | x_j | \psi_o \rangle - \langle \psi_o | x_j | \psi_n \rangle \langle \psi_n | p_i | \psi_o \rangle \right) \right]$$

$$= \sum_{ij} (\hat{\varepsilon}_\alpha^*)_j (\hat{\varepsilon}_\alpha)_i \left[\delta_{ij} + i \left(\langle \psi_o | x_j p_i - p_i x_j | \psi_o \rangle \right) \right] = 0 \quad (11.193)$$

The leading term in the cross section is

$$\frac{d\sigma}{d\Omega} = r_o^2 \left| \sum_{ij} \left((\hat{\varepsilon}_\alpha^*)_j (\hat{\varepsilon}_\alpha)_i \right) \frac{\omega}{m} \right.$$

$$\left. \times \sum_n \frac{1}{E_{on}^2} \left(\langle \psi_o | p_i | \psi_n \rangle \langle \psi_n | p_j | \psi_o \rangle - \langle \psi_o | p_j | \psi_n \rangle \langle \psi_n | p_i | \psi_o \rangle \right) + \cdots \right|^2 \quad (11.194)$$

Finally, use the electric dipole rule (11.192) in each factor again. The term appearing explicitly in equation (11.194) vanishes also, and the low-frequency expansion of the cross section is

$$\frac{d\sigma}{d\Omega} = \alpha^2 \omega^4$$

$$\times \left| \left(\sum_{ij} (\hat{\varepsilon}_\alpha^*)_j (\hat{\varepsilon}_\alpha)_i \right) \sum_n \frac{1}{E_{on}} \left(\langle \psi_o | x_i | \psi_n \rangle \langle \psi_n | x_j | \psi_o \rangle + \langle \psi_o | x_j | \psi_n \rangle \langle \psi_n | x_i | \psi_o \rangle \right) \right|^2$$
$$(11.195)$$

For long wavelengths, the scattering of light off atoms is proportional to the fourth power of the frequency! Equation (11.195) is called Rayleigh's law. The fourth-power behavior also holds in classical electromagnetism. Rayleigh's law explains why scattered sunlight is polarized, and why the sky is blue.

11.7.3 Scattering by a Free Electron

Finally, consider photon scattering by an atom when the photon energy is much larger than the atomic level spacings, but still much less than mc^2 for the electron. Then the scattering is essentially Thomson scattering, scattering of light by a nonrelativistic free electron. The structure of the computation closely mimics the computation in Section 11.7.2, but there are some interesting differences. (We still ignore the electron spin.)

The initial state contains a photon γ and an electron with momentum $\mathbf{q}$. The final state is $|\gamma', \mathbf{q}'\rangle$. The main technical difference between the present calculation and the one in Section 11.7.2 comes from the way free electron states are normalized:

$$\langle \gamma', \mathbf{q}' | \gamma, \mathbf{q} \rangle = \delta_{\alpha'\alpha} \delta_3(\mathbf{k}' - \mathbf{k}) \delta_3(\mathbf{q}' - \mathbf{q}) \quad (11.196)$$

The second-order term with no intermediate photons is

$$T^{II(o)} = \frac{e^2}{m^2} \int \langle \gamma', \mathbf{q}' | \mathbf{A}(\mathbf{r}) \cdot \mathbf{p} | \mathbf{q}'' \rangle \langle \mathbf{q}'' | G^o(\omega_i) | \mathbf{q}'' \rangle \langle \mathbf{q}'' | \mathbf{A}(\mathbf{r}) \cdot \mathbf{p} | \gamma, \mathbf{q} \rangle \, d^3 q''$$
$$(11.197)$$

392 Chapter 11 The Quantized Electromagnetic Field

The middle factor is

$$\langle \mathbf{q}''| G^o(\omega_i) |\mathbf{q}''\rangle = \frac{1}{\omega + E - E''} \tag{11.198}$$

where $E = q^2/2m$, etc. This time, the electron states are eigenstates of the electron momentum operator $\mathbf{p}$, so for example,

$$\langle \mathbf{q}''| \mathbf{A}(\mathbf{r}) \cdot \mathbf{p} |\gamma, \mathbf{q}\rangle = \frac{\sqrt{4\pi}}{(2\pi)^{3/2}} \frac{1}{\sqrt{2\omega}} \hat{\varepsilon}_\alpha \cdot \mathbf{q} \langle \mathbf{q}''| e^{i\mathbf{k}\cdot\mathbf{r}} |\mathbf{q}\rangle \tag{11.199}$$

The last factor in equation (11.199) can be evaluated using the electron wave functions:

$$\langle \mathbf{q}'| e^{i\mathbf{k}\cdot\mathbf{r}} |\mathbf{q}\rangle = \frac{1}{(2\pi)^3} \int e^{-i\mathbf{q}'\cdot\mathbf{r}} e^{i\mathbf{k}\cdot\mathbf{r}} e^{i\mathbf{q}\cdot\mathbf{r}} d^3r = \delta_3(\mathbf{q} + \mathbf{k} - \mathbf{q}'') \tag{11.200}$$

The matrix element (11.199) is proportional to the momenta of the initial and the final electrons, and vanishes in the low-energy limit. The term with two photons in the intermediate state behaves similarly. The only contribution that survives is from the $\mathbf{A}^2$ term:

$$\begin{aligned} T_{fi}^{(I)} &= 2 \frac{e^2}{2m} \frac{4\pi}{(2\pi)^3} \frac{1}{\sqrt{4\omega'\omega}} \hat{\varepsilon}_\alpha^* \cdot \hat{\varepsilon}_\alpha \langle \mathbf{q}'| e^{-i\mathbf{k}'\cdot\mathbf{r}} e^{i\mathbf{k}\cdot\mathbf{r}} |\mathbf{q}\rangle \\ &= \frac{\alpha}{4\pi^2 m} \frac{1}{\sqrt{\omega'\omega}} \hat{\varepsilon}_\alpha^* \cdot \hat{\varepsilon}_\alpha \delta_3(\mathbf{q}' + \mathbf{k}' - \mathbf{q} - \mathbf{k}) \end{aligned} \tag{11.201}$$

The formal expression for the transition rate into a group of final states is

$$\tilde{\Gamma} = 2\pi \int |T_{fi}|^2 \, \delta\left(E' + \omega' - E - \omega\right) d^3k' \, d^3q' \tag{11.202}$$

The infinity from squaring a Dirac delta function always occurs for scattering with recoil when momentum is conserved, and was discussed in some detail in Section 9.3.5. The rate given by the expression (11.202) is the total rate for the infinite number of particles in the target. The integral for the delta function must be replaced by an integral over a large but finite volume V: The true transition rate, the rate per particle, is

$$\Gamma = \frac{\alpha^2}{(2\pi)^3 m^2} \int \frac{1}{\omega\omega'} |\hat{\varepsilon}_\alpha^* \cdot \hat{\varepsilon}_\alpha|^2 \, \delta\left(E' + \omega' - E - \omega\right) d^3k' \tag{11.203}$$

and the cross section into final photon momenta in d^3k' and final electron momenta in d^3q' is[16]

$$d\sigma = \frac{\alpha^2}{m^2} \int \frac{1}{\omega\omega'} |\hat{\varepsilon}_\alpha^* \cdot \hat{\varepsilon}_\alpha|^2 \, \delta\left(E' + \omega' - E - \omega\right) d^3k' \tag{11.204}$$

$\mathbf{q}'$ is determined by $\mathbf{k}'$ by momentum conservation.

---
[16]The relative speed is 1.

Specialize to the laboratory frame, where $\mathbf{q} = 0$. Then the total final energy is

$$E_f = E' + \omega' = \frac{1}{2m}(\omega^2 + \omega'^2 - 2\omega'\omega\cos\theta) + \omega' \qquad (11.205)$$

so

$$dE_f = \left[\frac{1}{m}(\omega' - \omega\cos\theta) + 1\right]d\omega' \qquad (11.206)$$

The first two terms are small compared to one in the nonrelativistic limit. The differential cross section is the Thomson cross section

$$\frac{d\sigma}{d\Omega} = \frac{\alpha^2}{m^2}\frac{\omega'}{\omega}|\hat{\varepsilon}_\alpha^* \cdot \hat{\varepsilon}_\alpha|^2 \qquad (11.207)$$

To get the unpolarized cross section, average over the initial polarizations and sum over the final ones:

$$\frac{1}{2}\sum_{\alpha\alpha'}|\hat{\varepsilon}_\alpha^* \cdot \hat{\varepsilon}_\alpha|^2 = \frac{1}{2}\sum_{ij}(\delta_{ij} - (\hat{\mathbf{n}}')_i(\hat{\mathbf{n}}')_j)(\delta_{ij} - (\hat{\mathbf{n}})_i(\hat{\mathbf{n}})_j) = \frac{1}{2}(1 + \cos^2\theta) \qquad (11.208)$$

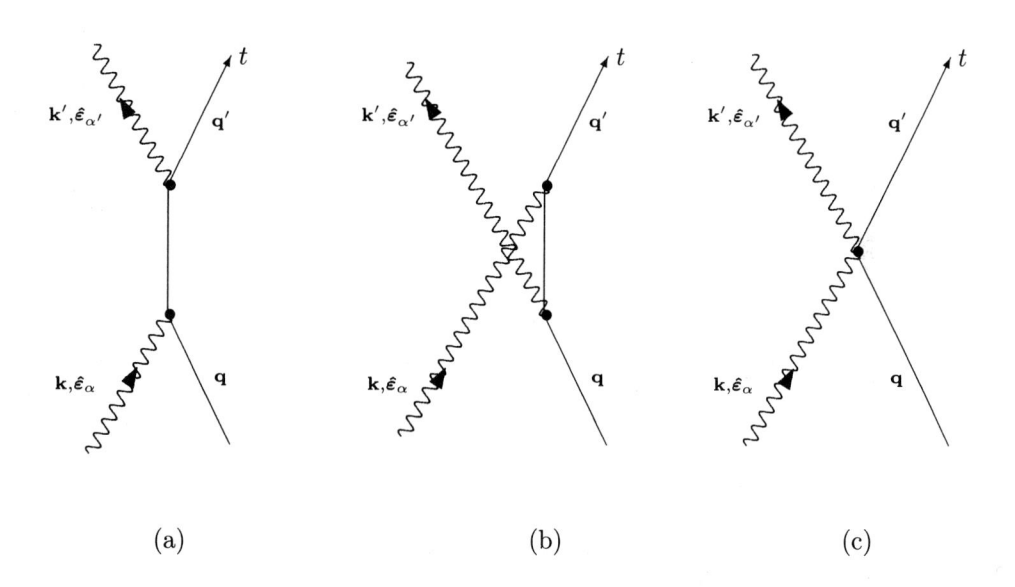

FIGURE 11.2: The amplitude for elastic scattering of a photon off a free electron also has three terms of order e^2: The structure is similar to Rayleigh scattering, Figure 11.1, except that the electron wave functions are the ones appropriate for a free electron instead of a bound state.

The unpolarized differential cross section is

$$\boxed{\frac{d\sigma}{d\Omega} = \frac{\alpha^2}{m^2}\frac{\omega'}{\omega}\frac{1}{2}(1 + \cos^2\theta)} \qquad (11.209)$$

The total cross section, independent of the choice of reference frame, is

$$\sigma = \frac{8\pi}{3} \frac{\alpha^2}{m^2} \frac{\omega'}{\omega} \qquad (11.210)$$

Feynman Diagrams

The three kinds of terms in elastic photon-electron scattering can also be pictured by the diagrams in Figure 11.2. In each term, the lines and vertices stand for a set of simple factors. For electron-photon scattering, there are (in our formalism) two kinds of fundamental interactions, represented by the three- and four-point vertices respectively. These are the "Feynman rules" for this problem. They can be developed into a complete diagrammatic calculus, a set of rules for reading off the algebraic structure of the terms in any order from all possible diagrams you can draw.

You should not be surprised that it is possible to deconstruct the complicated algebra of the perturbation expansion into products of elementary factors and then add them up. The amplitude for a quantum mechanical system to get from one time to a later time is the amplitude for it to go through all possible intermediate states, summed over these possible intermediate states. This was the point of view emphasized by Feynman, building on a suggestion of Dirac, and ultimately on Huygens' superposition principle for light waves. The modern version of this idea, the path integral, was discussed in Chapter 10.

11.8 THE CASIMIR EFFECT

11.8.1 The Ground-State Energy of the Electromagnetic Field

When you apply the methods of this chapter to the electromagnetic field with macroscopic boundary conditions, there are interesting new effects. The simplest configuration is one introduced by Casimir [4, 6, 7] in 1948. Imagine two large parallel square conducting plates, with sides of length L, separated by a distance $a \ll L$. The plates are located in the planes $z = 0$ and $z = a$, and L is large enough so that edge effects are negligible. Such boundary conditions are a way to summarize the detailed interactions of the field with the material inside the conducting plates. The vacuum energy, or at least its dependence on the distance between the plates, turns out to be observable. That is the Casimir effect.

If the very high frequency contributions to the infinite vacuum energy (11.54) are not really observable, it should be possible to suppress them without changing the predictions for any low-energy phenomena. This is the idea of a "cutoff," or regulator, again. In the crudest version we can just modify equation (11.48) to read

$$H = \int_{\omega < \omega_c} \sum_\alpha \mathcal{H}_\alpha(\mathbf{k}) d^3 k \qquad (11.211)$$

With a Hamiltonian like (11.211) the energy of the vacuum state is finite, but it depends strongly on the cutoff ω_c. But no measurable phenomena should depend on the value of ω_c.

I will calculate the difference between the vacuum energy with and without the plates. Both are infinite, but an infinite constant will cancel in the difference and leave a finite remainder that depends on the details of the boundary conditions. This is a remarkable quantum effect that has no classical analog.

Energy between Two Conducting Plates

Quantize the field in the volume L^2a between the plates, using a variant on the box normalization introduced in Section 11.6.1. But here I will use the correct boundary conditions for an electromagnetic field at a conductor, rather than periodic boundary conditions. Then the states are discrete, and you do not have to worry about the Dirac delta function that appears in equation (11.54). The frequency of these modes depends on three integers n_i, which I collect into a vector $\mathbf{n}$:

$$\omega_\mathbf{n}^2 = k_x^2 + k_y^2 + k_z^2 \tag{11.212}$$

where

$$k_x = n_x \pi/L \tag{11.213a}$$
$$k_y = n_y \pi/L \tag{11.213b}$$
$$k_z = n_z \pi/a \tag{11.213c}$$

There is a standing wave solution to the wave equation of the form

$$E_x \sim \cos\frac{n_x \pi x}{L} \sin\frac{n_y \pi y}{L} \sin\frac{n_z \pi z}{a} \sin(\omega_\mathbf{n} t) \tag{11.214a}$$

$$E_y \sim \sin\frac{n_x \pi x}{L} \cos\frac{n_y \pi y}{L} \sin\frac{n_z \pi z}{a} \sin(\omega_\mathbf{n} t) \tag{11.214b}$$

$$E_z \sim \sin\frac{n_x \pi x}{L} \sin\frac{n_y \pi y}{L} \cos\frac{n_z \pi z}{a} \sin(\omega_\mathbf{n} t) \tag{11.214c}$$

for each triplet of positive integers $\mathbf{n}$. Two of the components of $\mathbf{E}$ are independent, but the third one and the components of the magnetic field $\mathbf{B}$ are then fixed by Maxwell's equations. There are also modes where one (but not more) of the n_i are zero. For each n_x and n_y there is only one $n_z = 0$ mode, not two. When $n_z = 0$, equation (11.214) implies that only $E_z \neq 0$.

The ground-state energy is $\omega_\mathbf{n}/2$ for each mode.[17] Thus the total energy in the ground state is

$$E = \frac{1}{2} \sum_{n_x, n_y, n_z = 0}^{\infty} \epsilon(\mathbf{n}) \omega_\mathbf{n} \tag{11.215}$$

where $\epsilon(\mathbf{n}) = 2$ unless one of the n_i vanishes, $\epsilon(\mathbf{n}) = 1$ when one of the n_i is zero, and $\epsilon(\mathbf{n}) = 0$ when two or three components of $\mathbf{n}$ vanish.

Because L is very large the energy $\omega_\mathbf{n}$ does not change much from one n_x or

---

[17] With the discrete normalization the Dirac delta function $\delta(0)$ is replaced by 1.

396 Chapter 11 The Quantized Electromagnetic Field

n_y to the next. The sums over n_x and n_y can be approximated by integrals:

$$E \approx \int_{n_x=0}^{\infty} \int_{n_y=0}^{\infty} \left[\frac{1}{2}\sqrt{k_x^2 + k_y^2} + \sum_{n_z=1}^{\infty} \omega_\mathbf{n} \right] dn_x\, dn_y$$

$$= \frac{L^2}{(2\pi)^2} \int_{-\infty}^{\infty} \int_{-\infty}^{\infty} \left[\frac{1}{2}\sqrt{k_x^2 + k_y^2} + \sum_{n_z=1}^{\infty} \omega_\mathbf{n} \right] dk_x\, dk_y \qquad (11.216)$$

$$= \frac{L^2}{2\pi} \int_0^{\infty} \left[\frac{1}{2}k_\| + \sum_{n_z=1}^{\infty} \sqrt{k_\|^2 + \frac{n^2\pi^2}{a^2}} \right] k_\| dk_\|$$

where $k_\| = \sqrt{k_x^2 + k_y^2}$. The modes with $n_x = 0$ or $n_y = 0$ contribute negligibly and have been omitted. I have kept the $n_z = 0$ mode because there is a finite gap between its energy and the next one.

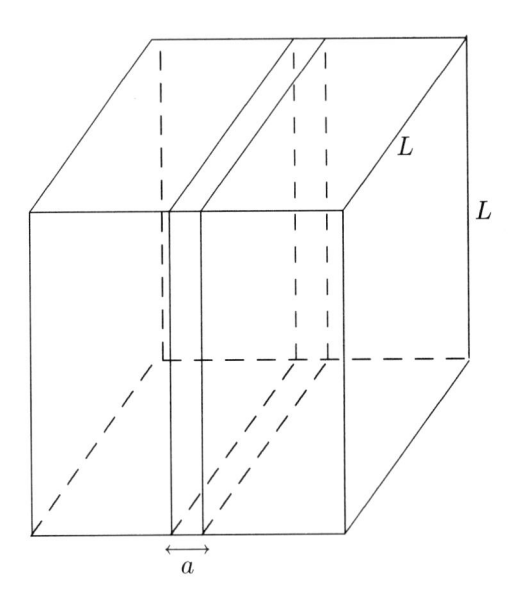

FIGURE 11.3: The energy is calculated between two square parallel plates separated by a, embedded in a very large cube with volume L^3

The idea is to calculate the ground-state energy from (11.216) and subtract from it the ground-state energy in the volume $L^2 a$ in the absence of the conducting plates, using some kind of cutoff to render each of them finite.

Note: The expression (11.216) is not realistic for wavelengths much shorter than the size of the atoms in the metal. For those very short wavelengths, the electromagnetic field should not be affected by the conducting plates at all, and the conducting plate boundary conditions will not hold. The high frequency contributions should be the same here as in empty space.

Energy of the Vacuum

The vacuum energy in the large cube with side L can be computed by the same argument that led to (11.216), except that now the z-direction is treated on the same footing as the other two. It is

$$\sum_{n_x,n_y,n_z=1}^{\infty} \omega_\mathbf{n} \approx \int_{n_x=0}^{\infty}\int_{n_y=0}^{\infty}\int_{n_z=0}^{\infty} \omega_\mathbf{n}\, d^3n = \frac{L^3}{(2\pi)^3}\int_{-\infty}^{\infty}\int_{-\infty}^{\infty}\int_{-\infty}^{\infty} k\, d^3k \quad (11.217)$$

where

$$\omega_\mathbf{n} = \frac{\pi}{L}\mathbf{n} \quad (11.218)$$

The energy that was in the region between the two conducting plates before they were put in place was

$$E_o = \frac{L^2 a}{(2\pi)^3}\int_{-\infty}^{\infty}\int_{-\infty}^{\infty}\int_{-\infty}^{\infty} k\, d^3k \quad (11.219)$$

Note: I will compute the difference between this energy, and the energy in the presence of the plates, in the region between the plates whose volume is $L^2 a$. It might make more sense to calculate the two energies in the whole volume L^3 in the two cases and then subtract. I leave it as an exercise to show that you get the same answer both ways.

11.8.2 The Casimir Force with an Elementary Cutoff

Expressions (11.216) and (11.219) are both divergent. The plan is to cut them both off in some arbitrary but identical way that is convenient for computations. The answer for the difference should not depend on this cutoff as long as it modifies the sums at high enough frequencies.

This energy difference per unit surface area L^2 of the plates should be independent of that area as $L \to \infty$. In that limit a is the only dimensionful parameter in the problem, assuming the cutoff frequency cancels as promised, and dimensional analysis alone is sufficient to show that

$$\frac{E - E_o}{L^2} \sim \frac{1}{a^3} \quad (11.220)$$

The force exerted on the plates (by the vacuum!) is therefore proportional to $1/a^4$. Let us now compute the coefficient.

Energy Between the Plates

Multiply each term in equation (11.216) by a cutoff function $F(\omega)$ with the properties that $F(\omega) = 1$ for $\omega \ll \omega_c$ and $F(\omega) = 0$ when $\omega \gg \omega_c$. The first condition means that $F(\omega)$ is essentially flat for frequencies such that equation (11.216) is valid. $F(\omega)$ cuts the sum off at some very large ω_c. That means replace (11.216) by

$$\frac{E}{L^2} = \frac{1}{2\pi}\int_0^{\infty}\left[\frac{1}{2}k_{\|}F(k_{\|}) + \sum_{n=1}^{\infty}\omega(k_{\|},n)F\left(\omega(k_{\|},n)\right)\right] k_{\|}\, dk_{\|} \quad (11.221)$$

398 Chapter 11 The Quantized Electromagnetic Field

where
$$\omega(k_\parallel, n) = \sqrt{k_\parallel^2 + \frac{n^2\pi^2}{a^2}} \tag{11.222}$$

With $F(\omega)$ chosen this way, the sum in equation (11.221) converges, and it is permissible to interchange sums and integrals. In each term we change the integration variable from $k_\parallel$ to $\omega(k_\parallel, n)$ also :

$$\frac{E}{L^2} = \frac{1}{2\pi}\left[\frac{1}{2}\int_0^\infty \omega^2 F(\omega)d\omega + \sum_{n=1}^\infty \int_{n\pi/a}^\infty \omega^2 F(\omega)d\omega\right] \tag{11.223}$$

A simple choice for the cutoff function is

$$F(\omega) = e^{-\omega/\omega_c} \tag{11.224}$$

Then

$$\frac{E}{L^2} = \frac{1}{2\pi}\left[\frac{1}{2}\int_0^\infty \omega^2 e^{-\omega/\omega_c}d\omega + \sum_{n=1}^\infty \int_{n\pi/a}^\infty \omega^2 e^{-\omega/\omega_c}d\omega\right]$$

$$= \frac{1}{2\pi}\frac{d^2}{dx^2}\left[\frac{1}{2}\int_0^\infty e^{-x\omega}d\omega + \sum_{n=1}^\infty \int_{n\pi/a}^\infty e^{-x\omega}d\omega\right]_{x=1/\omega_c} \tag{11.225}$$

$$= \frac{1}{2\pi}\frac{d^2}{dx^2}\left[\frac{1}{2x} + \frac{1}{x}\sum_{n=1}^\infty e^{-n\pi x/a}\right]_{x=1/\omega_c}$$

The sum in the second term is

$$\sum_{n=1}^\infty e^{-n\pi x/a} = \sum_{n=0}^\infty \left(e^{-\pi x/a}\right)^n - 1 = \frac{1}{1 - e^{-\pi x/a}} - 1$$
$$= \frac{e^{-\pi x/a}}{1 - e^{-\pi x/a}} = \frac{1}{e^{\pi x/a} - 1} \tag{11.226}$$

so

$$\frac{E}{L^2} = \frac{1}{2\pi}\frac{d^2}{dx^2}\left[\frac{1}{2x} + \frac{1}{x}\frac{1}{e^{\pi x/a} - 1}\right]_{x=1/\omega_c} \tag{11.227}$$

The value of E/L^2 is needed for very large ω_c only. An expansion of the expression in (11.225) is available in terms of the ubiquitous Bernoulli numbers defined by

$$\frac{t}{e^t - 1} = \frac{t}{2}\left(\coth\frac{t}{2} - 1\right) = \sum_{k=0}^\infty B_k \frac{t^k}{k!} \tag{11.228}$$

The first few Bernoulli numbers are

$$B_o = 1 \quad B_1 = -\frac{1}{2} \quad B_2 = \frac{1}{6} \quad B_4 = -\frac{1}{30} \quad \cdots \tag{11.229}$$

Section 11.8 The Casimir Effect

All the odd order Bernoulli numbers but B_1 vanish. The energy per unit area is

$$\frac{E}{L^2} = \frac{1}{2\pi}\frac{d^2}{dx^2}\left[\frac{1}{2x} + \left(\frac{1}{x}\sum_{k=0}^{\infty} B_k \frac{1}{k!}\left(\frac{\pi x}{a}\right)^{k-1}\right)\right]_{x=1/\omega_c}$$

$$= \frac{1}{2\pi}\frac{d^2}{dx^2}\left[B_o\frac{a}{\pi x^2} + \frac{1}{x}\left(B_1 + \frac{1}{2}\right) + \frac{1}{2}B_2\frac{\pi}{a} + \frac{1}{24}B_4\frac{\pi^3 x^2}{a^3} + \cdots\right]_{x=1/\omega_c}$$

$$= \frac{1}{2\pi}\left[\frac{6a}{\pi x^4}B_o + \frac{1}{12}B_4\frac{\pi^3}{a^3} + \cdots\right]_{x=1/\omega_c} = \frac{3a\omega_c^4}{\pi^2} - \frac{1}{720}\frac{\pi^3}{a^3} + \cdots$$

(11.230)

The next term is proportional to $1/\omega_c^2$, and the remaining ones to even higher powers of $1/\omega_c^2$. All will vanish in the limit $\omega_c \to \infty$, but the first term remains, and E/L^2 is still infinite in the limit $\omega_c \to \infty$.

Energy of the Vacuum

Next evaluate the divergent expression (11.219) for the vacuum energy in the same region using the cutoff function (11.224). The answer is the analog of equation (11.221) when the modes are allowed to fill the whole volume L^3 of the quantizing box:

$$\frac{E_o}{L^2} = \frac{a}{(2\pi)^3}\int_{-\infty}^{\infty}\int_{-\infty}^{\infty}\int_{-\infty}^{\infty} kF(\omega)d^3k \qquad (11.231)$$

where $F(\omega)$ is the cutoff function introduced in (11.221). The vacuum energy in this region per unit area is

$$\frac{E_o}{L^2} = \frac{a}{(2\pi)^3}\int_{-\infty}^{\infty}\int_{-\infty}^{\infty}\int_{-\infty}^{\infty} ke^{-\omega/\omega_c}d^3k = \frac{a}{2\pi^2}\int_0^{\infty} k^3 e^{-\omega/\omega_c}dk$$

$$= \frac{a}{2\pi^2}\int_0^{\infty} \omega^3 e^{-\omega/\omega_c}d\omega = 3\omega_c^4\frac{a}{\pi^2}$$

(11.232)

The Regularized Energy

Only the difference between E and E_o as a function of the separation a is physically meaningful. The energy per unit area between the plates, with the vacuum energy subtracted off, is

$$\boxed{\frac{E - E_o}{L^2} = -\frac{1}{720}\frac{\pi^2}{a^3} + \cdots} \qquad (11.233)$$

All reference to the large cutoff ω_c has vanished! A finite energy like this, with an infinite but constant term removed, is sometimes called a **regularized** energy.

If the plates are moved slightly, the energy changes, so there is an attractive macroscopic force on the plates:

$$\boxed{F = -\frac{\pi^2}{240a^4}} \qquad (11.234)$$

The vacuum fluctuations of the electromagnetic field cause two parallel conducting plates to move toward each other! The force F is unambiguous and does not depend on the cutoff. This is the Casimir effect, and it has been verified experimentally [5].

400 Chapter 11 The Quantized Electromagnetic Field

11.8.3 The General Calculation

You might wonder whether the marvelous result (11.233) is special to our choice of (11.224) for the cutoff function. The answer is no. Combine equations (11.221) and (11.231) without specifying the exact form of $F(\omega)$ except to require that it vanish for large ω and that it be arbitrarily flat in the low energy region.

$$\frac{E - E_o}{L^2} = \frac{1}{2\pi} \int_0^\infty k_\| dk_\| \left[\frac{1}{2} k_\| F(k_\|) + \sum_{n_z=1}^\infty \sqrt{k_\|^2 + \frac{n^2 \pi^2}{a^2}} F\left(\sqrt{k_\|^2 + \frac{n^2 \pi^2}{a^2}}\right) \right.$$

$$\left. - \frac{a}{\pi} \int_0^\infty \sqrt{k_\|^2 + k_z^2} F\left(\left(\sqrt{k_\|^2 + k_z^2}\right)\right) dk_z \right] k_\| dk_\| \tag{11.235}$$

Change the integration variable in the last term (coming from the vacuum energy density) from k_z to $n = ak_z/\pi$.

$$\frac{E - E_o}{L^2} = \frac{1}{2\pi} \left[\frac{1}{2} f(0) + \sum_{n=1}^\infty f(n) - \int_0^\infty f(n) dn \right] \tag{11.236}$$

where

$$f(n) = \int_0^\infty k_\| dk_\| \sqrt{k_\|^2 + \frac{n^2 \pi^2}{a^2}} F\left(\sqrt{k_\|^2 + \frac{n^2 \pi^2}{a^2}}\right) k_\| dk_\|$$

$$= \frac{1}{2} \int_{n^2 \pi^2/a^2}^\infty \sqrt{w} F(w) dw \tag{11.237}$$

With the cutoff in place, interchanging the orders of integration and summation like this are justified.

The quantity in brackets in equation (11.236) is the difference between the integral and the trapezoidal approximation to it. This difference has an expansion in derivatives of $f(n)$ similar to the ordinary Taylor-Maclaurin power series expansion. It is called the Euler-Maclaurin series, sometimes used in deriving the asymptotic series for the gamma function. Since it is not a standard part of mathematical physics education, I have put a derivation in Section A.1.4 of the appendix.

For a function $f(n)$ continuous and differentiable for $0 \leq n \leq N$, the Euler-Maclaurin rule for the difference between an integral and the trapezoidal approximation to the integral is

$$\int_0^N f(n) dn - \frac{1}{2} f(0) + f(1) + f(2) + \cdots f(N-1) + \frac{1}{2} f(N)$$

$$= \frac{B_2}{2!} (f'(0) - f'(N)) + \frac{B_4}{4!} \left(f^{(3)}(0) - f^{(3)}(N) \right) \tag{11.238}$$

$$+ \frac{B_6}{6!} \left(f^{(5)}(0) - f^{(5)}(N) \right) + \cdots$$

$f(n)$

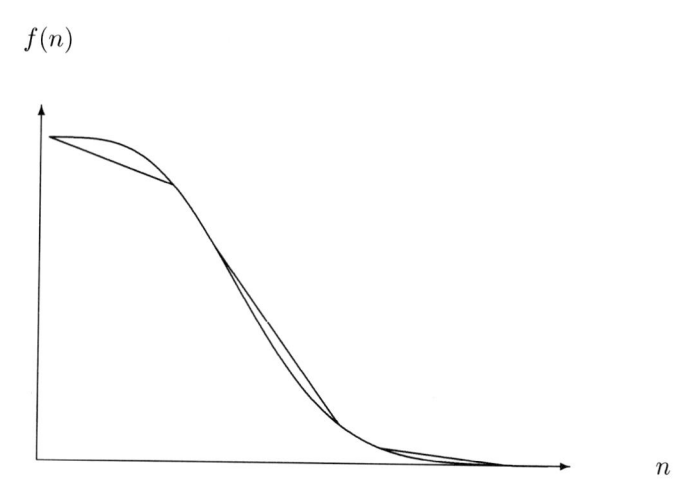

FIGURE 11.4: The Euler-MacLaurin series evaluates the difference between the areas under the smooth curve $f(n)$ and the trapezoidal approximation to it.

where B_k are the Bernoulli numbers. Because in the present case $f(n)$ together with all its derivatives is supposed to vanish as $n \to \infty$, the limit of (11.238) as $N \to \infty$ is

$$\int_0^\infty f(n)dn - \frac{1}{2}f(0) - \sum_{n=1}^\infty f(n) = \frac{B_2}{2!}f'(0) + \frac{B_4}{4!}f^{(3)}(0) + \frac{B_6}{6!}f^{(5)}(0) + \cdots \quad (11.239)$$

Since the cutoff is not supposed to modify $f(n)$ for small n, assume that $F(0) = 1$ and that all its derivatives vanish at $n = 0$. In the simple example of Section 11.8.2 above, this was true only in the limit $\omega_c \to \infty$. The cutoff is flat for low frequencies. Then all the derivatives of $f(n)$ vanish at $n = 0$ except the third. From equation (11.237)

$$f'(n) = -\frac{n^2\pi^3}{a^3}F\left(\frac{n^2\pi^2}{a^2}\right) \quad (11.240)$$

and so forth, so that

$$f^{(3)}(0) = -\frac{2\pi^3}{a^3} \quad (11.241)$$

I am cheerfully interchanging all sorts of limits here, but the method can be developed systematically. Therefore, it follows from equation (11.239) that the energy per unit area in the volume between the plates, equation (11.236), is

$$\frac{E - E_o}{L^2} = -\frac{1}{2\pi}\frac{B_4}{4!}f^{(3)}(0) = -\frac{1}{2\pi}\frac{1}{30}\frac{1}{24}\frac{2\pi^3}{a^3} = -\frac{\pi^2}{720a^3} \quad (11.242)$$

The result (11.233) is quite general.

The derivation is still crude and open to some criticism, but more careful treatment shows that it is correct. It might be objected that it is improper to use macroscopic boundary conditions in the quantum theory of the electromagnetic field, which is essentially a microscopic description. It has been shown that you get the same result by studying higher order corrections to intermolecular forces and the van der Waals effect. There is a large literature on this subject, but the essential physics is illustrated by the calculation presented above.

I have discussed only the simplest configuration. The Casimir effect has been calculated for a variety of geometries, and the predictions have been confirmed by experiments.

PROBLEMS

11.1. Commutators of Field Components

(a) For the free electromagnetic field, compute the commutator

$$[E_i(\mathbf{r},t), B_j(\mathbf{r}',t)]$$

(b) For the free electromagnetic field, show that the commutator

$$[E_i(\mathbf{r},t), A_j(\mathbf{r}',t)]$$

is proportional to the "transverse delta function"

$$\frac{1}{(2\pi)^3}\int (\delta_{ij} - n_i n_j) e^{i\mathbf{k}\cdot(\mathbf{r}-\mathbf{r}')}\, d^3k$$

11.2. Maxwell's Equations in Quantum Mechanics

(a) Use Part (a) of Problem 11.1 to compute the commutators of $E_i(\mathbf{r},t)$ and $B_i(\mathbf{r},t)$ with the Hamiltonian

$$H = \frac{1}{8\pi}\int \left(\mathbf{E}^2 + \mathbf{B}^2\right) d^3r$$

and thus find the time dependence of the fields from the Heisenberg picture rule $i\partial A/\partial t = [A, H]$. Check that your answers agree with Maxwell's equations.

(b) Complete the proof of equation (11.42).

11.3. Momentum as Generator of Translations

From equations (11.59) and (11.61), the momentum operator for the free electromagnetic field is

$$\mathbf{P} = \frac{1}{8\pi}\int \mathbf{E}\times\mathbf{B}\, d^3r = \sum_\alpha \int \mathbf{k}\, a_\alpha^\dagger(\mathbf{k}) a_\alpha(\mathbf{k})\, d^3k$$

Show that $\mathbf{P}$ is indeed the generator of translations, as follows: Compute the commutator $[P_i, A_j(\mathbf{r},t)]$ of a component of the momentum operator $\mathbf{P}$ with a component of the field $\mathbf{A}(\mathbf{r},t)$. Then define the translation operator in the usual way by

$$T(\mathbf{a}) = e^{-i\mathbf{P}\cdot\mathbf{a}}$$

and compute $T(\mathbf{a})\mathbf{A}(\mathbf{r},t)T(\mathbf{a})^{-1}$.
   Hint: Use the Baker-Hausdorff lemma [see Problem 2.10]

$$e^{i\theta B} A e^{-i\theta B} = A + i\theta[B,A] - \frac{\theta^2}{2}[B,[B,A]] + \cdots + \frac{i^n \theta^n}{n!}[B,[B,[B\ldots[B,A]]]\ldots]+\cdots$$

11.4. Angular Dependence of 2P Decay
Suppose a hydrogen atom is somehow prepared in the $2P$ state, and it is known that the value of L_z is definitely $+1$, that is, the atom is in the state $m = 1$.
(a) Calculate the differential decay rate into any solid angle $d\Omega$.
(b) Integrate over all angles and check that the total rate is the same as the one computed in Section 11.5.1.

11.5. Lyman-β Decay Rate
The transitions from the $n = 3$ states to the $1S$ state of atomic hydrogen are components of the Lyman-β line. Compute the rate for the spontaneous decay of a $3P$ state to the $1S$ state. What is the width of these components of the Lyman-β line? (Here you can ignore spin—these are electric transitions.) Compare this rate to the rate for the Lyman-α decay transition.

11.6. D Component of the Lyman-β Line
Compute the rate for the spontaneous decay of the $3D$ state to the $1S$ state. What is the width of this component of the Lyman-β line? (This is an electric quadrupole transition; the width should be smaller than that of the Lyman-α line, an electric dipole transition.)

   Hint: Choose the $m = 2$ initial state, and use the combinations of the operators suggested by the Wigner-Eckart theorem. Use the rule

$$\sum_\alpha \hat{\varepsilon}_\alpha^*(\mathbf{k})_i \hat{\varepsilon}_\alpha(\mathbf{k})_j = \delta_{ij} - n_i n_j$$

11.7. Rare 2P Decay
Review the fine structure splitting (Section 7.3) to first order in bound-state perturbation theory between the $j = 3/2$ and the $j = 1/2$ states in the $n = 2$ level of atomic hydrogen. Calculate the lifetime for the $2P_{\frac{3}{2}}$ state to decay into the $2S_{\frac{1}{2}}$ state. What is the branching ratio for this decay mode of the $2P_{\frac{3}{2}}$ state compared to its principal decay mode into the $1S$ state?
   Hint: Start from the $m_j = 3/2$ state.

   Note: If you calculate the matrix element

$$\left\langle \phi^{2l}_{\frac{3}{2}m'} \middle| \mathbf{p} \middle| \phi^{2l}_{\frac{1}{2}m} \right\rangle$$

using the unperturbed states, you will get zero, since the matrix element of $\mathbf{p}$ between states of the same energy vanishes. It might seem that the correct way to proceed would be to use the wave functions computed in bound-state perturbation theory to first order; these states are eigenstates of H_o where now H_o includes the Coulomb term *and* the spin-orbit coupling term to first order, with eigenvalues differing by the fine structure splitting Δ_{FS}. But that is complicated and unnecessary. This is an electric dipole

11.8. Single Photon 2S Decay

Without spin, the $2S$ state of atomic hydrogen is completely stable against single photon emission. This is an example of the "no zero-zero transitions" rule. But when spin is included, there is a finite rate for the $2S_{\frac{1}{2}}$ state to decay into a photon and the $1S_{\frac{1}{2}}$ state. Calculate the rate for this decay as follows:

(a) First convince yourself that the matrix element of $\mathbf{p} \cdot \mathbf{A}(\mathbf{r}) e^{i\mathbf{k}\cdot\mathbf{r}}$ indeed vanishes to all orders in the expansion of $e^{i\mathbf{k}\cdot\mathbf{r}}$. Write down the magnetic part of the interaction (take $g = 2$):

$$H' = \frac{e}{m}\mathbf{s}\cdot\mathbf{B}(\mathbf{r})$$

The magnetic dipole and quadrupole terms vanish, so this is a magnetic *octupole* transition. Expect the rate to be very small.

(b) Show that the transition rate Γ is proportional to

$$\sum_{m\alpha}\sum_{ij}\int \left(\hat{\mathbf{n}}\times\hat{\boldsymbol{\varepsilon}}_\alpha^*(\mathbf{k})\right)_i \left(\hat{\mathbf{n}}\times\hat{\boldsymbol{\varepsilon}}_\alpha(\mathbf{k})\right)_j$$

$$\times \left\langle 1S_{\frac{1}{2}}m\left|s_i(\mathbf{k}\cdot\mathbf{r})^2\right|2S_{\frac{1}{2}}m'\right\rangle\left\langle 1S_{\frac{1}{2}}m\left|s_j(\mathbf{k}\cdot\mathbf{r})^2\right|2S_{\frac{1}{2}}m'\right\rangle^* \delta(\Delta-\omega)\,d^3k$$

where m is the eigenvalue of s_z in the final state and Δ is the energy difference $E_2 - E_1$. The matrix element factors into a spin part and a space part. Simplify the space part using the fact that $\langle 100|(\mathbf{k}\cdot\mathbf{r})^2|200\rangle$ is proportional to $k^2\langle 100|r^2|200\rangle$.

(c) Use the rule

$$\sum_\alpha \left(\hat{\mathbf{n}}\times\hat{\boldsymbol{\varepsilon}}_\alpha^*(\mathbf{k})\right)_i \left(\hat{\mathbf{n}}\times\hat{\boldsymbol{\varepsilon}}_\alpha(\mathbf{k})\right)_j = \delta_{ij} - n_i n_j$$

and the rule that for any function $f(k^2)$ of the magnitude of $\mathbf{k}$ only

$$\int n_i n_j f(k^2)\,d^3k = \frac{1}{3}\delta_{ij}\int f(k^2)\,d^3k$$

to do the sum over α and m.

(d) Evaluate the remaining factors and express the result as a form like

$$\Gamma = \alpha^A m 2^B 3^C 5^D$$

What is the numerical value of the rate and the line width?

11.9. Absorption of Sunlight—I

The intensity of the sun's radiation at the earth's surface is measured by the "solar constant," about 1.94 calories per square centimeter per minute.[18]

(a) About how many calories of the sun's heat fall on the earth per second?

---

[18] One calorie is the heat equivalent of 4.184 joules.

(b) The oceans cover about 60% of the earth's surface, and have an average depth of about 3.6 km. If all the sun's radiation that falls on the earth could be used to warm the seas, what would be the daily rise in the water's temperature?

(c) If instead all this energy could be converted into electricity, how many 100-watt lightbulbs could be kept lit?

(d) If instead all this energy could be converted into chemical rocket fuel that could be used with 100% efficiency, how many 100 metric ton space shuttles could be launched per day into low earth orbit? (Ignore the energy necessary to lift some of the fuel off the earth.)

(e) The spectrum of the sun's radiation follows Planck's blackbody law very closely. The energy spectrum has the form ($\hbar = c = 1$)

$$I(\omega)d\omega = \beta \frac{\omega^3\, d\omega}{e^{\omega/kT} - 1}$$

with a characteristic temperature T of about 6000 Kelvin. Here k is Boltzmann's constant and β is a dimensionless number.[19] What is the numerical value of β?

Hint: Remember that

$$\int_0^\infty \frac{x^3}{e^x - 1} dx = \frac{\pi^4}{15}$$

(f) Let $I(\lambda)d\lambda$ be the energy flux in the wavelength interval $d\lambda$. What value of λ (in Ångstrøms) maximizes $I(\lambda)$? [This is not the same as the value of ω that maximizes $I(\omega)$.]

(g) Suppose sunlight as described above falls on atomic hydrogen in its ground state. Calculate the rate of absorption, per atom, into the $2P$ state (ignore spin). Write an algebraic expression, then the numerical value.

Hint: Since the target is an S state, the absorption should be independent of the initial direction and polarization. So take the beam to have circular polarization and to be moving in the z-direction.

11.10. Absorption of Sunlight—II

In all the following, derive an algebraic expression, then the numerical value if you can.

(a) Under the same circumstances as in Problem 11.9, compute the rate of absorption per atom into the $3P$ state. What is the ratio of this rate to the $2P$ absorption rate you computed above?

Hint: Since the target is in an S state, the answer must be independent of the directions of the absorbed photon's momentum or polarization. Take the initial momentum in the z-direction, and either take the photon to be circularly polarized, or average over polarizations.

(b) What is the ratio of the energy absorbed into the $3D$ state to the energy absorbed into the $3P$ state? (Some of the steps in Problem 11.6 can be taken over directly here.)

---

[19] Boltzmann's constant is about 8.62×10^{-5} electron volts per degree.

(c) What is the ratio of the energy absorbed into the $3S$ state to the energy absorbed into the $3P$ state? Here it is probably easier to average over the initial polarizations.

11.11. Orbital Angular Momentum of the Electromagnetic Field
Show that the orbital angular momentum of the free electromagnetic field, defined in equation (11.64), indeed satisfies $\mathbf{L} \cdot \mathbf{P} = 0$ as claimed.

11.12. Casimir Force
Two identical large square copper plates, each 1 mm thick, are placed in empty space, parallel, exactly at rest, and electrically neutral, with a 1 mm gap between them. The density of copper is about 9 grams per cubic centimeter.

(a) Derive a formula for the initial acceleration of the plates due to the Casimir force. If the acceleration were constant, and equal to its initial value, how much time would elapse before the plates collide?

(b) Since the force on the plates increases as they get closer, the actual time will be less than the time computed as in part (a). What is the correct time before the plates collide? (You probably have to estimate an integral numerically.)

REFERENCES

Planck's constant first appeared in his formula for the black body spectrum

[1] M. PLANCK, Verh. der Deutch. Phys. Ges. **2** (1900), 237.

"Quanta" of radiation energy, now photons, were introduced by

[2] A. EINSTEIN, Annalen der Physik **17** (1905), 132.

Both these papers are translated into English in

[3] D. TER HAAR, *The Old Quantum Theory*, Oxford, 1967.

The Casimir effect was proposed by

[4] H. B. K. CASIMIR, Proc. Kon. Ned. Akad. Wet. **51** (1948), 793.

and verified experimentally by

[5] M. J. SPAARNAY, Physica **24** (1958), 751.

There are more references in

[6] V. M. MOSTEPANENKO AND N. N. TRUNOV, *The Casimir effect and its applications*, Clarendon Press, 1997.

[7] E. ELIZALDE AND A. ROMEO, Am. Jour. of Phys. **59** (1991), 711.

[8] S. K. LAMOREAUX, *Comments on Modern Physics*, **1** (2000) 247–261. Also as quant-ph/9907076 at http://arxiv.org/.

CHAPTER 12

Relativistic Wave Equations

12.1 LORENTZ TRANSFORMATIONS

This chapter is about relativistic generalizations of the Schrödinger wave equation. One of these, the Dirac equation, is enough to explain the arbitrary parameters in the nonrelativistic theory of the electron. It is also a first step toward a complete theory.

12.1.1 Four-Vectors and Tensors

This first section is a quick tour of the mathematics of Lorentz transformations. I will draw heavily on the analysis of rotations in Chapters 4 and 5. The Minkowski or four-vector notation is essential for any but the most elementary manipulations in special relativity.[1] The components of a four-vector are V^μ, $\mu = 0, 1, 2, 3$. The statement $V^\mu = (X, \mathbf{Y})$ means $V^o = X$ and $V^i = Y^i$. Latin superscripts and subscripts by convention go from 1 to 3 (or over x, y, and z); Greek ones go from 0 to 3. The "true" components of a vector are indicated by superscripts. The prototypes of a four-vector are $x^\mu = (t, \mathbf{r})$ and $p^\mu = (E, \mathbf{p})$.

Vector components with subscripts are defined in terms of the components with superscripts by $V_o = V^o$ but $V_i = -V^i$. This rule becomes more elegant upon introducing the **summation convention**—repeated indices, one a subscript and one a superscript, are to be summed over—and the **metric tensor** $g_{\mu\nu}$ ("metric" for short):[2]

$$g_{oo} = 1 \qquad g_{ii} = -1 \qquad g_{\mu\nu} = 0 \quad (\mu \neq \nu) \tag{12.1}$$

Then

$$V_\mu = g_{\mu\nu} V^\nu \tag{12.2}$$

Define $g^{\mu\nu} = g_{\mu\nu}$ so that

$$V^\mu = g^{\mu\nu} V_\nu \tag{12.3}$$

also. The notation comes from general relativity, but here these are just algebraic rules.

The same rules hold for tensors of any rank:

$$T^{\mu\nu} = g^{\mu'\mu} g^{\nu'\nu} T_{\mu'\nu'} \tag{12.4}$$

Quantities that obey these rules are called Minkowski tensors, to distinguish them from Cartesian tensors. In particular

$$g^{\mu\nu} = g^{\mu'\mu} g^{\nu'\nu} g_{\mu'\nu'} \tag{12.5}$$

---
[1] There are several variants of the notational conventions, so you should not be alarmed to find discussions elsewhere with different signs and factors of 4π here and there.
[2] $\eta_{\mu\nu}$ is a common alternative notation.

so that $g^{\mu\nu}$ itself turns out to be a second-rank tensor.

Indices may be raised and lowered freely provided it is done the same way in every term. For example, if $T^{\mu\nu\lambda} = U^{\mu\nu\lambda}$, then $T^{\mu\nu}{}_\lambda = U^{\mu\nu}{}_\lambda$, where

$$T^{\mu\nu}{}_\lambda = g_{\lambda\lambda'} T^{\mu\nu\lambda'} \tag{12.6}$$

etc. Similarly

$$T^\mu{}_\mu{}^\lambda = T_\mu{}^{\mu\lambda} \tag{12.7}$$

12.1.2 Lorentz Transformations

For a collection of numbers to qualify as a Minkowski tensor, it is not enough that the subscript and superscript versions obey all these algebraic conventions. The components of a tensor in two different frames of reference must be related by Lorentz transformations. The rule for a four-vector under a Lorentz transformation is

$$x^\mu \to x'^\mu = \bar{L}^\mu{}_\nu x^\nu \tag{12.8}$$

The 4×4 matrices $\bar{L}^\mu{}_\nu$ are analogous to the 3×3 rotation matrices $\bar{R}_{ij}$ first introduced in Section 1.2, and they can be found from the way $\mathbf{r}$ and t transform under Lorentz transformations. The summation implied on the right-hand side of equation (12.8) is ordinary matrix multiplication; the conventions for raising and lowering indices require that a matrix be written with the first index as a superscript and the second as a subscript. Expressions with the same number of free indices of the same altitude transform the same way under Lorentz transformations. Expressions with no free indices are invariants.

For any four-vector, $V^\mu V_\mu = (V^o)^2 - \mathbf{V}^2$ is an invariant. In particular

$$x^\mu x_\mu = t^2 - \mathbf{r}^2 \quad \text{and} \quad p^\mu p_\mu = E^2 - \mathbf{p}^2 = m^2 \tag{12.9}$$

are invariant. One of the consequences of special relativity is that for two vectors x^μ and y^μ, the combination

$$x_\mu y^\mu = g_{\mu\nu} x^\mu y^\nu \tag{12.10}$$

is the same in all reference frames. This fact defines the matrices $\bar{L}^\mu{}_\nu$. It says that the $\bar{L}$ are the real 4×4 matrices that preserve the $g_{\mu\nu}$ scalar product. With x' and y' defined as in equation (12.8), $x'^\mu y'_\mu = x^\mu y_\mu$, or

$$x'^\mu g_{\mu\nu} y'^\nu = \bar{L}^\mu{}_{\mu'} \bar{L}^\nu{}_{\nu'} x^{\mu'} y^{\nu'} g_{\mu\nu} = x^{\mu'} g_{\mu'\nu'} y^{\nu'} \tag{12.11}$$

Since x and y are arbitrary,[3]

$$\boxed{g_{\mu\nu} = \bar{L}^{\mu'}{}_\mu \bar{L}^{\nu'}{}_\nu g_{\mu'\nu'} = \bar{L}^{\mu'}{}_\mu \bar{L}_{\mu'\nu}} \tag{12.12}$$

Any $\bar{L}$ that satisfies the rule (12.12) is a Lorentz transformation matrix.

The $\bar{L}$'s leave the metric $g_{\mu\nu}$ invariant. If $g_{\mu\nu}$ were $\delta_{\mu\nu}$, (12.12) would be the definition of orthogonal matrices in four dimensions.[4] The structure defined by the matrices $\bar{L}$ is the **Lorentz Group**.

---

[3] The components of the matrix $\bar{L}$ are $\bar{L}^\mu{}_\nu$. The array $\bar{L}_{\mu\nu}$ is *defined* as $g_{\mu'\mu} \bar{L}^{\mu'}{}_\nu$.
[4] Problem 4.8.

Under ordinary three-dimensional rotations there are two constant Cartesian tensors, the rank-two tensor δ_{ij} and the rank-three tensor ϵ_{ijk}. There are two constant tensors here also. One is $g_{\mu\nu}$. The other is the four-dimensional totally antisymmetric symbol $\epsilon^{\alpha\beta\gamma\delta}$, defined so that it changes sign if two indices are interchanged, and so that $\epsilon^{0123} = 1$. It is a simple exercise to show that $\epsilon^{\beta\gamma\delta\alpha} = -\epsilon^{\alpha\beta\gamma\delta}$ and that $\epsilon_{0123} = -1$.

Explicit Form of the Lorentz Transformations

What are these transformations? The rotations rotate x, y, and z in the usual way, but the time t is invariant. The pure rotations can be written as 4×4 matrices

$$\bar{R} = \begin{pmatrix} 1 & 0 & 0 & 0 \\ 0 & & & \\ 0 & & \bar{R} & \\ 0 & & & \end{pmatrix} \tag{12.13}$$

where inside the matrix $\bar{R}$ means the 3×3 matrix $\bar{R} = e^{-i\theta \hat{n} \cdot \mathbf{J}}$.

A Lorentz boost in the x-direction by velocity v is

$$x' = \frac{x + vt}{\sqrt{1 - v^2}} \tag{12.14a}$$

$$y' = y \tag{12.14b}$$

$$z' = x \tag{12.14c}$$

$$t' = \frac{t + vx}{\sqrt{1 - v^2}} \tag{12.14d}$$

The two others look the same, with y or z replacing x. They all leave the form $t^2 - \mathbf{r}^2$ unchanged. The Lorentz group is a six-parameter group, with three parameters for rotations about each axis, and three more for the boost along each direction.

Next, define the generators for the boosts as well as the rotations and calculate their commutators. But be careful. Although it is sufficient just to expand the boosts to first order in v/c, you cannot write the boosts in direction $\hat{n}$ as $\bar{B}(v\hat{n})$ and hope to exponentiate $\bar{B}(v\hat{n}) = \exp(-iv\hat{n} \cdot \mathbf{A})$ for some matrices A_i. If you could, then the product of two boosts in the x-direction would be

$$\bar{B}(v_1 \hat{n}_x)\bar{B}(v_2 \hat{n}_x) = \exp(-iv_1 A_x)\exp(-iv_2 A_x)$$
$$= \exp(-i(v_1 + v_2)A_x) = \bar{B}((v_1 + v_2)\hat{n}_x) \tag{12.15}$$

which is not right. The correct rule for the relativistic addition of velocities is not $v' = v_1 + v_2$ but rather

$$v' = \frac{v_1 + v_2}{1 + v_1 v_2} \tag{12.16}$$

The solution is to parametrize the boosts by ω instead of v, where

$$v = \tanh \omega = \frac{e^\omega - e^{-\omega}}{e^\omega + e^{-\omega}} \tag{12.17}$$

since then

$$\tanh(\omega_1 + \omega_2) = \frac{v_1 + v_2}{1 + v_1 v_1} \tag{12.18}$$

and these have the correct multiplication law for two boosts in the same direction. A pure boost in the x-direction is the matrix

$$\bar{B} = \begin{pmatrix} \cosh\omega & \sinh\omega & 0 & 0 \\ \sinh\omega & \cosh\omega & 0 & 0 \\ 0 & 0 & 1 & 0 \\ 0 & 0 & 0 & 1 \end{pmatrix} \quad (12.19)$$

The Lorentz group has six generators $\bar{J}_i$ and $\bar{K}_i$, and all elements of the Lorentz group depend on six parameters. The rotations and boosts have the forms

$$\bar{R} = e^{-i\theta\hat{n}\cdot\bar{J}} \quad (12.20a)$$

$$\bar{B} = e^{-i\omega\hat{n}\cdot\bar{K}} \quad (12.20b)$$

Explicitly:

$$(\mathbf{J}_i)^0{}_i = (\mathbf{J}_i)^i{}_0 = 0 \quad (12.21a)$$

$$(\mathbf{J}_i)^j{}_k = -i\epsilon_{ijk} \quad (12.21b)$$

$$(\mathbf{K}_i)^i{}_j = 0 \quad (12.21c)$$

$$(\mathbf{K}_i)^i{}_0 = (\mathbf{K}_i)^0{}_i = i \quad (12.21d)$$

An arbitrary Lorentz transformation is obtained by compounding rotations and boosts. A product of two rotations is still a rotation, but a rotation followed by a boost is not generally either a pure rotation or a pure boost. Nevertheless, an arbitrary Lorentz transformation can be obtained by exponentiating *some* linear combinations of the J_i and the K_i.

The generators satisfy these commutation relations (Problem 12.1)

$$[\bar{J}_i, \bar{J}_j] = i\sum_k \epsilon_{ijk}\bar{J}_k \quad (12.22a)$$

$$[\bar{J}_i, \bar{K}_j] = i\sum_k \epsilon_{ijk}\bar{K}_k \quad (12.22b)$$

$$[\bar{K}_i, \bar{K}_j] = -i\sum_k \epsilon_{ijk}\bar{J}_k \quad (12.22c)$$

Mathematical note: A form for the generators that looks more covariant is sometimes useful. Write a general Lorentz transformation matrix as

$$\bar{L} = \exp\left(-\frac{i}{2}\omega_{\alpha\beta}\bar{J}^{\alpha\beta}\right) \quad (12.23)$$

where $\omega_{\alpha\beta}$ is an antisymmetric array of real numbers (so only six of them are independent). Identify

$$\omega_{ij} = \theta\sum_k \epsilon_{ijk}n_k \quad (12.24)$$

for rotations, and
$$\omega_{0i} = -\omega_{i0} = \hat{n}\omega \tag{12.25}$$
for a boost in the direction $\hat{n}$ by a velocity $\tanh \omega$. Then
$$\bar{J}^{ij} = \sum_k \epsilon_{ijk} \bar{J}_k \quad \text{and} \quad \bar{J}^{0i} = -\bar{J}^{i0} = \bar{K}_i \tag{12.26}$$

With these definitions,
$$\left(\bar{J}^{\alpha\beta}\right)^\mu{}_\nu = i\left(g^{\alpha\mu}g^\beta{}_\nu - g^{\beta\mu}g^\alpha{}_\nu\right) \tag{12.27}$$
and to first order
$$\bar{L}^\mu{}_\nu = g^\mu{}_\nu + \omega^\mu{}_\nu + \cdots \tag{12.28}$$

12.1.3 Spin

Any collection of matrices $D(\bar{J}_i)$ and $D(\bar{K}_i)$ that satisfy the commutation relations (12.22) generate a representation of the Lorentz transformation algebra. The matrices
$$D(\bar{R}) = e^{-i\theta \hat{n} \cdot D(\bar{J})} \quad \text{and} \quad D(\bar{B}) = e^{-i\omega \hat{n} \cdot D(\bar{K})} \tag{12.29}$$
will satisfy the group multiplication rules.

All this works just as it did for rotations. A wave function $\psi(x)$ is a column vector. Under $\bar{L}$, $\psi(x) \to \psi'(x)$, where $x'^\mu = \bar{L}^\mu{}_\nu x^\nu$, and
$$\psi'(x') = D(\bar{L})\psi(x) \tag{12.30}$$

Irreducible Representations

What are the irreducible representations? This is a problem we solved long ago. Review the algebra of the Runge-Lenz vector and the angular momentum vector, equations (3.150). These equations are the rules of the $O(4)$ algebra; they differ from the Lorentz algebra rules just by the sign in the third commutator. The representations are obtained by replacing K_i with iK_i. That is, set
$$\mathbf{X} = \frac{\mathbf{J} + i\mathbf{K}}{2} \quad \text{and} \quad \mathbf{Y} = \frac{\mathbf{J} - i\mathbf{K}}{2} \tag{12.31}$$
Then the algebra becomes
$$[X_i, X_j] = \sum_k \epsilon_{ijk} X_k \quad [Y_i, Y_j] = \sum_k \epsilon_{ijk} Y_k \quad [X_i, Y_j] = 0 \tag{12.32}$$

$\mathbf{X}$ and $\mathbf{Y}$ satisfy two independent angular-momentum algebras. The most general representation is specified by two nonnegative integers or half-integers (x, y), in which, in analogy to equation (3.153)

$$\begin{aligned}
\mathbf{X}^2 |x, m_x, y, m_y\rangle &= x(x+1)|x, m_x, y, m_y\rangle \\
\mathbf{Y}^2 |x, m_x, y, m_y\rangle &= y(y+1)|x, m_x, y, m_y\rangle \\
X_z |x, m_x, y, m_y\rangle &= m_j |x, m_x, y, m_y\rangle \\
Y_z |x, m_x, y, m_y\rangle &= m_k |x, m_x, y, m_y\rangle \\
X_+ |x, m_x, y, m_y\rangle &= \sqrt{x(x+1) - m_x(m_x+1)}|x, m_x+1, y, m_y\rangle \\
Y_+ |x, m_x, y, m_y\rangle &= \sqrt{y(y+1) - m_y(m_y+1)}|x, m_x, y, m_y+1\rangle
\end{aligned} \tag{12.33}$$

In this representation the angular momentum operators are

$$D^{(x,y)}(\bar{\mathbf{J}}) = \mathbf{X} + \mathbf{Y} \tag{12.34}$$

and the generators of the boosts are

$$D^{(x,y)}(\bar{\mathbf{K}}) = -i(\mathbf{X} - \mathbf{Y}) \tag{12.35}$$

Note the following:

- The Runge-Lenz algebra is the algebra of rotations in four dimensions [see part (b) of Problem 4.8]. The Lorentz transformations would be the same if $g_{\mu\nu}$ were $\delta_{\mu\nu}$ instead of equation (12.1).

- The representations of $\mathbf{K}$ are antihermitean, not Hermitean.

- These irreducible representations of the Lorentz transformations are not irreducible representations of the rotations. Rather, they contain all angular momenta between $x+y$ and $|x-y|$. Constructing relativistic wave equations for particles with spin is not quite trivial, since the wave functions have to describe a single spin and the Lorentz transformations on the states ought to be unitary.

Wave Functions

A spin-zero wave function has a single component, and transforms under the $(0,0)$ representation. Spin-half wave functions can transform as either $(0,\frac{1}{2})$ or $(\frac{1}{2},0)$. Wave functions can be constructed using either of these, but they will turn out to describe only massless spin-half particles. Real electrons will need a reducible representation.

The four-component representation $(\frac{1}{2},\frac{1}{2})$ has spin one and spin zero. The 4×4 matrices $\bar{L}^{\mu}{}_{\nu}$ are the matrices in this representation, though you have to change the basis to see this explicitly, as in Problem 4.2. This is the representation in which a four-vector like V^{μ} transforms. The combination $\partial_{\mu}V^{\mu}$ transforms like a scalar under rotations, and the three other components transform like a vector. So you can describe a massive spin-one particle by a four component wave function $V^{\mu}(x)$ subject to the condition that $\partial_{\mu}V^{\mu}(x) = 0$. Higher spin particles require more of these "subsidiary conditions."

For massless spin-one particles, the usual idea of spin—the angular momentum in the rest frame—has to be modified, and the analysis is more complicated. Photons transform like a sum of the two three-dimensional representations $(1,0)$ and $(0,1)$. The standard "spin" basis vectors are linear combinations of the six components of the antisymmetric tensor $F_{\mu\nu}$. Maxwell's equations are the subsidiary conditions.

Fields

There are also **fields**, vector, scalar, or tensor functions of x^{μ}. In analogy with the way wave functions transform, scalar fields $\phi(x)$ and vector fields $V^{\mu}(x)$ should transform under Lorentz transformations like

$$\phi'(x') = \phi(x) \quad \text{and} \quad V'^{\mu}(x') = \bar{L}^{\mu}{}_{\nu}V^{\nu}(x) \tag{12.36}$$

with $x'^\mu = \bar{L}^\mu{}_\nu x^\nu$. Two important vector fields are the electromagnetic potential strength $A^\mu = (\phi, \mathbf{A})$ and the current density $j^\mu = (\rho, \mathbf{j})$.

Fields can be differentiated: The generalization of the gradient is

$$\partial_\mu = \frac{\partial}{\partial x^\mu} = \left(\frac{\partial}{\partial t}, \boldsymbol{\nabla}\right) \tag{12.37}$$

∂_μ is defined with a lower index, since if the charge and current density satisfy a continuity equation (charge conservation), then

$$\partial_\mu j^\mu(x) = \frac{\partial \rho}{\partial t} + \boldsymbol{\nabla} \cdot \mathbf{j} = 0 \tag{12.38}$$

Equation (12.38) will hold in all frames if it is true in one frame.

12.2 VECTOR AND SCALAR FIELDS

12.2.1 The Electromagnetic Field

The potential A^μ is easier to use than the electromagnetic field strengths $\mathbf{E}$ and $\mathbf{B}$, since $\mathbf{E}$ and $\mathbf{B}$ are components of a rank-two Lorentz tensor $F^{\mu\nu}$, called the field strength tensor, defined by[5]

$$F_{\mu\nu}(x) = \partial_\mu A_\nu(x) - \partial_\nu A_\mu(x) \tag{12.39}$$

$F_{\mu\nu}$ is antisymmetric: $F_{\mu\nu} = -F_{\nu\mu}$, so it has only six independent nonvanishing components. Define

$$E_i = F_{oi} = -F_{io} = F^{io} = -F^{oi} \tag{12.40}$$

and

$$F_{ij} = F^{ij} = -\sum_k \epsilon_{ijk} B_k \tag{12.41}$$

With these definitions

$$\mathbf{E} = -\boldsymbol{\nabla}\phi - \frac{\partial \mathbf{A}}{\partial t} \tag{11.3a}$$

and

$$\mathbf{B} = \boldsymbol{\nabla} \times \mathbf{A} \tag{11.3b}$$

as required.

In the covariant notation there are only two Maxwell's equations. The homogeneous equation is a consequence of the definition of $F_{\mu\nu}$, and it is equivalent to the statement that A^μ exists. From the antisymmetry of $F_{\mu\nu}$ it follows that

$$\partial_\lambda F_{\mu\nu} + \partial_\mu F_{\nu\lambda} + \partial_\nu F_{\lambda\mu} = 0 \tag{12.42}$$

provided the λ, μ, and ν are all different. More elegantly,

$$\epsilon^{\alpha\beta\gamma\delta} \partial_\beta F_{\gamma\delta} = 0 \tag{12.43}$$

---

[5] A definition with μ and ν interchanged is not uncommon.

There are two cases of equation (12.43). If $\alpha = i$ (1, 2, or 3) then one of β, γ, or δ must be 0. Explicitly:

$$\epsilon^{i0\mu\nu}\partial_0 F_{\mu\nu} + \epsilon^{i\lambda 0\nu}\partial_\lambda F_{0\nu} + \epsilon^{i\lambda\mu 0}\partial_\lambda F_{\mu 0} = 0 \tag{12.44}$$

which is equivalent to equation (11.1a). The remaining possibility is that $\alpha = 0$ in (12.43). Then

$$\epsilon^{0ijk}\partial_i F_{jk} = 0 \tag{12.45}$$

which is equation (11.1b).

The inhomogeneous equation relates $F^{\mu\nu}$ to its source, the current density j^μ.

$$\partial_\mu F^{\mu\nu} = 4\pi j^\nu \tag{12.46}$$

In terms of A^μ this equation is

$$\partial_\mu \left(\partial^\mu A^\nu - \partial^\nu A^\mu\right) = 4\pi j^\nu \tag{12.47}$$

Set $\nu = i$:

$$\partial_o F^{oi} + \partial_j F^{ji} = 4\pi j^i \quad \text{or} \quad -\frac{\partial E_i}{\partial t} + \sum_{jk}\partial_j \epsilon_{ijk} B_k = 4\pi j^i \tag{12.48}$$

which is the same as equation (11.2a). Set $\nu = 0$ to obtain the remaining Maxwell's equation (11.2b)

Differentiate (12.46) and use the antisymmetry of $F_{\mu\nu}$ to get

$$\partial_\nu \partial_\mu F^{\mu\nu} = 4\pi \partial_\nu j^\nu = 0 \quad \text{or} \quad \partial_\nu j^\nu = 0 \tag{12.49}$$

which is the continuity equation for the charge-current density

$$\frac{\partial \rho}{\partial t} + \boldsymbol{\nabla} \cdot \mathbf{j} = 0 \tag{12.50}$$

Electric charge conservation follows from Maxwell's equations.

Finally, note that $F_{\mu\nu}$ and therefore Maxwell's equations are invariant under a gauge transformation

$$A^\mu \to A^\mu - \partial^\mu \lambda \tag{12.51}$$

where $\lambda(x)$ is any scalar field. This is the covariant form of the gauge transformation (11.4).

12.2.2 The Klein-Gordon Equation

The simplest representation of the Lorentz transformations is $D(\bar{L}) = 1$. This is the $(0,0)$ representation in the notation of Section 12.1.3. The wave functions, ordinary numbers $\phi(x)$, describe spinless particles. We still impose the basic principles of quantum mechanics in the form

$$i\frac{\partial \phi(x)}{\partial t} = H\phi(x) \tag{12.52a}$$

$$-i\boldsymbol{\nabla}\phi(x) = \mathbf{p}\phi(x) \tag{12.52b}$$

which fortunately can be written together as one covariant equation:

$$i\partial^\mu \phi(x) = p^\mu \phi(x) \tag{12.53}$$

If (12.52) are true in one frame of reference, they will hold in another.

In the absence of interactions, ω and $\mathbf{k}$, the eigenvalues of H and $\mathbf{p}$, satisfy $\omega^2 = k^2 + m^2$, so a sensible wave equation for a free particle is

$$\left(-\frac{\partial^2}{\partial t^2} + \boldsymbol{\nabla}^2 - m^2\right)\phi(x) = 0 \tag{12.54}$$

If $\mathbf{p} = 0$ these wave functions are invariant under rotations, so they indeed describe spinless particles. Rotations here are just the coordinate transformation. There is no spin matrix.

Equation (12.54) is the **Klein-Gordon equation**. It was discovered by Schrödinger [1, 2, 3]. In covariant notation the Klein-Gordon equation becomes

$$\boxed{\left(-\partial^\mu \partial_\mu - m^2\right)\phi(x) = 0} \tag{12.55}$$

Equation (12.55) describes free particles. When interactions are included, the right-hand side of the Klein-Gordon equation (12.55) will no longer be zero.

Equation (12.54) has plane wave solutions

$$\phi(x) = e^{i(\mathbf{k}\cdot\mathbf{r}-\omega t)} = e^{-ik^\mu x_\mu} \tag{12.56}$$

that are eigenfunctions of $p^\mu = i\partial^\mu$. These solutions require

$$\left(\omega^2 - k^2 - m^2\right)\phi(x) = 0 \tag{12.57}$$

so they exist provided $E = \pm\sqrt{p^2 + m^2}$. The Klein-Gordon equation allows negative energy solutions, and a complete set of solutions must include these negative energy solutions.

As in the nonrelativistic case, there is a conserved probability current, conserved by virtue of the equation of motion. This current is normalized to agree with the nonrelativistic form in the appropriate limit, j^μ is

$$j^\mu = i\left(\phi^* \partial^\mu \phi - \phi \partial^\mu \phi^*\right) \tag{12.58}$$

Then $\partial_\mu j^\mu = 0$, and

$$\mathbf{j} = -i\left(\phi^* \boldsymbol{\nabla}\phi - \phi \boldsymbol{\nabla}\phi^*\right) \tag{12.59}$$

and

$$\rho = j^0 = i\left(\phi^* \frac{\partial \phi}{\partial t} - \phi \frac{\partial \phi^*}{\partial t}\right) \tag{12.60}$$

are the current and density. While the current (12.59) agrees with the nonrelativistic form, the fact that $\rho < 0$ for negative energy plane-wave solutions is a problem.

Electromagnetic Interactions

In general it looks complicated to introduce interactions in a covariant way, but for electromagnetic interactions that is easy. The minimal coupling rule we have been using for the vector potential $\mathbf{A}(\mathbf{r})$, together with the standard electrostatic term in the energy, combine naturally into a four-dimensional minimal coupling prescription: For a particle with electric charge q, make the replacement

$$p^\mu \to p^\mu - qA^\mu \tag{12.61}$$

The substitution (12.61) then insures that physics is gauge invariant.[6] Then in the presence of an electromagnetic potential $A^\mu(x)$, the interacting Klein-Gordon equation is

$$\left[(\partial_\mu + iqA_\mu)(\partial^\mu + iqA^\mu) + m^2\right]\phi(x) = 0 \tag{12.62}$$

Equation (12.62) is manifestly covariant—if it is true in one frame it is true in another.

The stationary wave functions are $\phi(\mathbf{r})$, where with the notation $\mathbf{p} = -i\boldsymbol{\nabla}$,

$$\left[(\omega - qA^o)^2 - (\mathbf{p} - q\mathbf{A})^2 - m^2\right]\phi(\mathbf{r}) = 0 \tag{12.63}$$

Set $\omega = m + E'$. Then E' is the usual nonrelativistic energy. Equation (12.63) becomes

$$\left[\frac{(\mathbf{p} - q\mathbf{A})^2}{2m} - qA^o\right]\phi(\mathbf{r}) = E'\phi(\mathbf{r}) + \frac{(E' - qA^o)^2}{2m}\phi(\mathbf{r}) \tag{12.64}$$

Equation (12.64) is the nonrelativistic Schrödinger equation for a spinless particle in an external electromagnetic field as long as the last term can be ignored, that is, as long as A^o is small and $E' \ll m$.

The Hydrogen Atom

What are the stationary states of the Klein-Gordon equation for a particle with charge $-e$ moving in the electrostatic field of a proton? In the general equation (12.63), set $\mathbf{A} = 0$ and $qA^o = -\alpha/r$:

$$\left(\boldsymbol{\nabla}^2 + \frac{2\omega\alpha}{r} + \frac{\alpha^2}{r^2} + \omega^2 - m^2\right)\psi(\mathbf{r}) = 0 \tag{12.65}$$

Write

$$\boldsymbol{\nabla}^2 = \frac{1}{r^2}\frac{\partial}{\partial r}r^2\frac{\partial}{\partial r} - \mathbf{L}^2 \tag{12.66}$$

where $\mathbf{L} = \mathbf{r} \times \mathbf{p}$ as usual, and choose the wave functions to be eigenstates of $\mathbf{L}^2$ and L_z. The radial wave equation for $u(r) = rR(r)$ is

$$\left(\frac{1}{2m}\frac{d^2}{dr^2} - \frac{l(l+1) - \alpha^2}{2mr^2} + \frac{\omega\alpha}{mr} - \frac{\omega^2 - m^2}{2m}\right)u(r) = 0 \tag{12.67}$$

---

[6] This is proved in Section 10.2.2 for the nonrelativistic Schrödinger equation, but the statement is more general.

Equation (12.67) has exactly the same form as the corresponding nonrelativistic Schrödinger equation (3.161) for the hydrogen atom, with the constants there replaced as follows:

$$l(l+1) \longrightarrow l(l+1) - \alpha^2$$
$$Z\alpha \longrightarrow \frac{\omega\alpha}{m} \qquad (12.68)$$
$$E \longrightarrow \frac{\omega^2 - m^2}{2m}$$

The nonrelativistic equation has bound-state solutions with eigenvalues $E = -\alpha^2 m/2n^2$, whenever $n - l - 1$ is a nonnegative integer. So equation (12.67) has bound-state solutions with

$$\frac{\omega^2 - m^2}{2m} = -\frac{\alpha^2 \omega^2}{2mn^2} \qquad (12.69)$$

whenever $n - \nu - 1$ is a nonnegative integer n', where

$$\nu(\nu+1) = l(l+1) - \alpha^2 \quad \text{or} \quad \nu = -\frac{1}{2} + \sqrt{\left(l + \frac{1}{2}\right)^2 - \alpha^2} \qquad (12.70)$$

The predicted energies for the positive energy bound states are

$$\omega = m \left[1 + \frac{\alpha^2}{\left(n' + 1/2 + \sqrt{(l+1/2)^2 - \alpha^2}\right)^2} \right]^{-\frac{1}{2}} \qquad (12.71)$$

This solution, first found by Schrödinger, gives correctly the gross structure of the levels, but not the observed corrections higher order in α^2, and of course not the fine structure, since there is no spin quantum number.

12.3 RELATIVISTIC SPIN-HALF EQUATIONS

The Klein-Gordon equation has solutions with negative probability densities because it is second order in the time. The Schrödinger equation did not have a negative probability density because it is first order in the time. Dirac originally argued that a relativistic wave equation without the negative-probability problem would have to be first order in the time, and then that Lorentz invariance would require it to be first order in **r** also. He discovered that the program was impossible for spinless particles, but works for spin half.

12.3.1 Two Component Spin-Half Equations

Let us start from the two-dimensional representation $(0, \frac{1}{2})$ in the notation of equations (12.33). That is,

$$X_i = 0 \quad \text{and} \quad Y_i = D^{(\frac{1}{2})}\left(\bar{J}_i\right) \qquad (12.72)$$

or

$$D^{(0,\frac{1}{2})}(\bar{J}_i) = D^{(\frac{1}{2})}(\bar{J}_i) = \frac{1}{2}\sigma_i \quad \text{and} \quad D^{(0,\frac{1}{2})}(\bar{K}_i) = iD^{(\frac{1}{2})}(\bar{J}_i) = i\frac{1}{2}\sigma_i \qquad (12.73)$$

It is trivial to check that these matrices satisfy the commutators (12.22). The spin is one-half. The transformation law for the wave functions is [see equation (12.30)]

$$\psi'(x') = D^{(0,\frac{1}{2})}(\bar{L})\psi(x) \qquad (12.74)$$

But this is not the whole story. There are new constraints from translation invariance and time-translation invariance.

The Poincaré Group

It must be possible to represent translations and time-translations as well as rotations and boosts on these wave functions. In the nonrelativistic case this was never a complication, but here it is not trivial. This larger group has for generators not only the J_i and the K_i, but also P_i and H. It is a ten-parameter group, called the Poincaré group. The property needed here is the obvious one that $H^2 - \mathbf{P}^2$ is an invariant, that is, that the particles have a definite mass m.

Helicity, Parity, and Mass

Recall from Section 11.3 that helicity is the projection of the angular momentum along the direction of a particle's momentum, and it is invariant under any Lorentz transformation. Consider a particle with some definite momentum $\mathbf{k}$, and a definite helicity λ:

$$\mathbf{P}|\mathbf{k},\lambda\rangle = \mathbf{k}|\mathbf{k},\lambda\rangle \qquad \text{and} \qquad \mathbf{P}\cdot\mathbf{J}|\mathbf{k},\lambda\rangle = \lambda k|\mathbf{k},\lambda\rangle \qquad (12.75)$$

In the present case $\lambda = \pm\frac{1}{2}$. If $m > 0$, the particle has some velocity

$$\mathbf{v} = \frac{\mathbf{k}}{E} = \frac{\mathbf{k}}{\sqrt{\mathbf{k}^2 + m^2}} < 1 \qquad (12.76)$$

Boosting this particle by $-2\mathbf{v}$ turns it into a state with the opposite momentum, but the same spin (boosts commute with angular momentum along the same direction), and therefore *opposite* helicity. The collection of states for any (free) massive spin-half particle contains both helicities.

But a reflection through the plane perpendicular to $\mathbf{v}$ does the same thing—changing the sign of $\mathbf{v}$ but not the spin. So it is always possible to define the reflection operator $\mathcal{R}$ on states of a massive particle, and there is a matrix $D^{(0,\frac{1}{2})}(\mathcal{R})$ such that

$$\psi(x) \to \psi'(x) \qquad \text{where} \qquad \psi'(t,-\mathbf{r}) = D^{(0,\frac{1}{2})}(\mathcal{R})\psi(t,\mathbf{r}) \qquad (12.77)$$

analogous to (12.74) for proper Lorentz transformation.

$\mathbf{K}$ is a polar vector, not an axial vector like $\mathbf{J}$, and reflections commute with rotations, but not with boosts:

$$\mathcal{R}e^{-i\omega\cdot\mathbf{K}}\mathcal{R} = e^{i\omega\cdot\mathbf{K}} \qquad \text{or} \qquad \mathcal{R}K_i\mathcal{R} = -K_i \qquad (12.78)$$

$D^{(0,\frac{1}{2})}(\mathcal{R})$ must be a matrix such that

$$D^{(0,\frac{1}{2})}(\mathcal{R})D^{(0,\frac{1}{2})}(\mathbf{K}_i)D^{(0,\frac{1}{2})}(\mathcal{R}) = -D^{(0,\frac{1}{2})}(\mathbf{K}_i) \qquad (12.79)$$

$D^{(0,\frac{1}{2})}(\mathbf{K}_i)$ is given in equation (12.73) above, so $D^{(0,\frac{1}{2})}(\mathcal{R})$ is a 2×2 matrix that anticommutes with all three Pauli matrices:

$$D^{(0,\frac{1}{2})}(\mathcal{R})\sigma_i = -\sigma_i D^{(0,\frac{1}{2})}(\mathcal{R}) \qquad (12.80)$$

but there is no such matrix except zero. The upshot of this tortuous reasoning is that the representation $(0, \frac{1}{2})$ cannot be extended to include reflections, and therefore cannot describe particles with mass!

Massless Spin-Half Wave Functions

So these two component wave functions describe massless particles with one helicity only. Neutrinos, if they were massless, could be described by such wave functions, since as far as is known only one helicity state interacts with anything.

A wave function with definite momentum looks like

$$\psi(x) = \chi e^{-ik \cdot x} \tag{12.81}$$

χ is some constant two-component spinor (a **Weyl spinor**), and $k \cdot x = \omega t - \mathbf{k} \cdot \mathbf{r}$, where $\omega = |\mathbf{k}|$ since $m = 0$. In order that the state have definite helicity,

$$\mathbf{k} \cdot D^{(0,\frac{1}{2})}(\mathbf{J})\chi = \pm \frac{1}{2}\omega\chi \quad \text{or} \quad (E \mp \mathbf{k} \cdot \boldsymbol{\sigma})\chi = 0 \tag{12.82}$$

The two signs correspond to positive and negative (or right-handed and left-handed) helicity respectively. Write the condition (12.82) in the covariant-looking form

$$k_\mu \beta^\mu \chi = 0 \tag{12.83}$$

with

$$\beta^0 = I \quad \text{and} \quad \beta^i = \pm \sigma_i \tag{12.84}$$

In terms of the wave function,

$$i\beta^\mu \partial_\mu \psi(x) = 0 \tag{12.85}$$

Equations (12.85) are the two-component wave equations for spin-half particles.

Which is the correct sign? The helicity of a state is unchanged by a boost. This is true for the top sign in equation (12.82), but not the bottom one. The $(0, \frac{1}{2})$ representation describes right-handed particles. I leave the detailed demonstration as an exercise.[7]

A similar computation shows that the $(\frac{1}{2}, 0)$ representation describes left-handed particles. For this representation, equations (12.73) are replaced by

$$D^{(\frac{1}{2},0)}(\bar{J}_i) = D^{(\frac{1}{2})}(\bar{J}_i) = \frac{1}{2}\sigma_i \quad \text{and} \quad D^{(\frac{1}{2},0)}(\bar{K}_i) = -iD^{(\frac{1}{2})}(\bar{J}_i) = -i\frac{1}{2}\sigma_i \tag{12.86}$$

In order not to forget which is which, I hereby rename these representations with a more transparent notation:

$$D^{(\frac{1}{2},0)} \longrightarrow D_L \quad \text{and} \quad D^{(0,\frac{1}{2})} \longrightarrow D_R \tag{12.87}$$

---

[7] Occasionally you may find the roles of the two representations interchanged. There are several arbitrary choices of signs, most obviously the sign of the velocity on the definitions (12.14a). It is also easy to make an odd number of mistakes here.

12.3.2 The Dirac Equation

A representation of the Lorentz group with a nonzero mass can be constructed from these irreducible representations. Combine the left-handed and right-handed two-component spinors into a four-component object.

$$\psi = \begin{pmatrix} \psi_L \\ \psi_R \end{pmatrix} \tag{12.88}$$

The upper and lower components transform differently under Lorentz transformations, so the reducible representation matrices are

$$D(\bar{L}) = \begin{pmatrix} D_L(\bar{L}) & 0 \\ 0 & D_R(\bar{L}) \end{pmatrix} \tag{12.89a}$$

or, in terms of the generators,

$$D(\bar{J}_i) = \frac{1}{2}\begin{pmatrix} \sigma_i & 0 \\ 0 & \sigma_i \end{pmatrix} \quad D(\bar{K}_i) = \frac{i}{2}\begin{pmatrix} -\sigma_i & 0 \\ 0 & \sigma_i \end{pmatrix} \tag{12.89b}$$

Equations (12.89a) and (12.89b) are the Dirac representations of the Lorentz group and Lorentz algebra. The four-dimensional wave equation is

$$i\begin{pmatrix} \partial_o - \boldsymbol{\sigma}\cdot\boldsymbol{\nabla} & 0 \\ 0 & \partial_o + \boldsymbol{\sigma}\cdot\boldsymbol{\nabla} \end{pmatrix}\psi = 0 \tag{12.90}$$

So far the wave function defined by equation (12.90) describes two independent massless spin-half particles, one right-handed and one left-handed, but now it is possible to introduce a reflection matrix that exchanges the two. Define reflections by

$$\mathcal{R}\psi(\mathbf{r},t) = D(\mathcal{R})\psi(-\mathbf{r},t) \tag{12.91}$$

with

$$D(\mathcal{R}) = \begin{pmatrix} 0 & I \\ I & 0 \end{pmatrix} \tag{12.92}$$

The matrix in (12.92) has the required property: It commutes with $D(\mathbf{J})$ and anticommutes with $D(\mathbf{K})$.

Next add a term that mixes up the two pieces:

$$i\begin{pmatrix} \partial_o - \boldsymbol{\sigma}\cdot\boldsymbol{\nabla} & 0 \\ 0 & \partial_o + \boldsymbol{\sigma}\cdot\boldsymbol{\nabla} \end{pmatrix}\psi - m\begin{pmatrix} 0 & I \\ I & 0 \end{pmatrix}\psi = 0 \tag{12.93}$$

Here I is the 2×2 identity. Equation (12.93) is the Dirac equation, and it transforms—by definition—under the Dirac representation of the Lorentz algebra. Dirac defined the four Hermitean matrices[8]

$$\alpha_i = \begin{pmatrix} -\sigma_i & 0 \\ 0 & \sigma_i \end{pmatrix} \quad \beta = \begin{pmatrix} 0 & I \\ I & 0 \end{pmatrix} \tag{12.94}$$

---

[8]Reference [9] of Chapter 2, page 259.

so that (12.93) becomes

$$i\left(\frac{\partial}{\partial t} + \boldsymbol{\alpha}\cdot\boldsymbol{\nabla}\right)\psi - m\beta\psi = 0 \tag{12.95}$$

The square of each of these four matrices is the identity, and they anticommute:

$$\alpha_i\alpha_j + \alpha_j\alpha_i = 2\delta_{ij} \qquad \beta\alpha_i + \alpha_i\beta = 0 \tag{12.96}$$

It is convenient to introduce a more covariant-looking notation. Following Dirac down the Greek alphabet, define

$$\gamma^o = \beta \quad \text{and} \quad \gamma^i = \gamma^o\alpha_i = \begin{pmatrix} 0 & \sigma_i \\ -\sigma_i & 0 \end{pmatrix} \tag{12.97}$$

and multiply equation (12.95) on the left by γ^o:

$$i\left(\gamma^o\frac{\partial}{\partial t} + \boldsymbol{\gamma}\cdot\boldsymbol{\nabla}\right)\psi - m\psi = 0 \tag{12.98}$$

or

$$i\gamma^\mu\partial_\mu\psi - m\psi = 0 \tag{12.99}$$

Equation (12.99) is the covariant form of the Dirac equation.

The matrices γ^μ have the following properties:

$$(\gamma^o)^2 = I \qquad (\gamma^i)^2 = -I \qquad \gamma^o\gamma^i + \gamma^i\gamma^o = 0 \tag{12.100}$$

and for $i \neq j$

$$\gamma^i\gamma^j + \gamma^j\gamma^i = 0 \tag{12.101}$$

or more succinctly

$$\boxed{\gamma^\mu\gamma^\nu + \gamma^\nu\gamma^\mu = 2g^{\mu\nu}} \tag{12.102}$$

12.3.3 Free Particle Solutions

Conventional Basis

The basis for the Dirac spinors used in Section 12.3.2 is convenient in the ultrarelativistic limit, where the states of fixed energy have fixed helicity. For the nonrelativistic limit, it is more convenient to use a basis in which the lower components vanish and the upper two components become the Pauli spinors. Change the basis using the unitary matrix

$$U = \frac{1}{\sqrt{2}}\begin{pmatrix} I & -I \\ I & I \end{pmatrix} \tag{12.103}$$

that is,

$$\alpha_i \to U^{-1}\alpha_i U \quad \text{and} \quad \beta \to U^{-1}\beta U \tag{12.104}$$

The commutation and anticommutation relations are unchanged. In this new basis

$$\gamma^o = \begin{pmatrix} I & 0 \\ 0 & -I \end{pmatrix} \quad \text{and} \quad \gamma^i = \begin{pmatrix} 0 & \sigma_i \\ -\sigma_i & 0 \end{pmatrix} \tag{12.105}$$

and the representation matrices for the generators are

$$D(\bar{J}_i) = \frac{1}{2}\begin{pmatrix} \sigma_i & 0 \\ 0 & \sigma_i \end{pmatrix} = \frac{1}{2}\Sigma_i \qquad D(\bar{K}_i) = \frac{i}{2}\begin{pmatrix} 0 & \sigma_i \\ \sigma_i & 0 \end{pmatrix} = \frac{i}{2}\alpha_i \qquad (12.106)$$

The Σ_i are the four-dimensional generalizations of the Pauli matrices. Reflections are represented by γ^o in (12.105). The form of the Dirac equation is unchanged:

$$(i\gamma^\mu \partial_\mu - m)\psi(x) = 0 \qquad (12.99)$$

Plane-Wave Solutions

Plane-wave solutions of the free-particle Dirac equation are

$$\psi(x) = w(\mathbf{k})e^{-ik^\mu x_\mu} = w(\mathbf{k})e^{-i(\omega t - \mathbf{k}\cdot\mathbf{r})} \qquad (12.107)$$

$\psi(x)$ is an eigenfunction of H and $\mathbf{p}$. The four-component constant Dirac spinors $w(\mathbf{k})$ satisfy

$$(\gamma^\mu k_\mu - m)w(\mathbf{k}) = 0 \qquad (12.108)$$

Write

$$w(\mathbf{k}) = \begin{pmatrix} \phi_1(\mathbf{k}) \\ \phi_2(\mathbf{k}) \end{pmatrix} \qquad (12.109)$$

where the ϕ_i have two components. Equation (12.108) is equivalent to two equations connecting ϕ_1 and ϕ_2.

$$(E - m)\phi_1 - \boldsymbol{\sigma}\cdot\mathbf{k}\phi_2 = 0 \qquad (12.110a)$$

$$-(E + m)\phi_2 + \boldsymbol{\sigma}\cdot\mathbf{k}\phi_1 = 0 \qquad (12.110b)$$

There is no solution unless the determinant of the coefficients vanishes:

$$E^2 - k^2 - m^2 = 0 \qquad (12.111)$$

Solve equation (12.110b) for ϕ_1 and insert the result into (12.110a):

$$(E - m)\phi_1 - \frac{k^2}{E + m}\phi_1 = 0 \qquad (12.112)$$

Choose

$$\phi_1 = \begin{pmatrix} 1 \\ 0 \end{pmatrix} \quad \text{or} \quad \phi_1 = \begin{pmatrix} 0 \\ 1 \end{pmatrix} \qquad (12.113)$$

Then

$$\phi_2 = \frac{1}{E + m}\boldsymbol{\sigma}\cdot\mathbf{k}\begin{pmatrix} 1 \\ 0 \end{pmatrix} \quad \text{or} \quad \phi_2 = \frac{1}{E + m}\boldsymbol{\sigma}\cdot\mathbf{k}\begin{pmatrix} 0 \\ 1 \end{pmatrix} \qquad (12.114)$$

These are actually four solutions for each real $\mathbf{k}$, two for each sign of the energy. When $E > 0$, the factor $E + m$ in the denominator assures that bottom components are small compared to the top ones when $p/m \ll 1$. In that limit ϕ_1 becomes the nonrelativistic Pauli spinor.

In the nonrelativistic formulation the components of the spin are constants, but not here. The eigenstates of H and $\mathbf{p}$ cannot be chosen eigenstates of s_z, but will be approximately eigenstates of s_z in the low momentum limit. For a free Dirac particle, only the total angular momentum is conserved.

12.3.4 Probability Current and Hole Theory

The obvious choice for the probability density is

$$\rho(x) = \psi^\dagger(x)\psi(x) \qquad (12.115)$$

since it reduces to the usual form in the nonrelativistic limit. Because the boost matrices are not unitary, $\psi^\dagger(x)\psi(x)$ is not invariant under a boost.

Since $(\gamma^o)^2 = I$, the generalization of equation (12.115) to four functions $j^\mu(x)$ is

$$j^\mu(x) = \psi^\dagger(x)\gamma^o\gamma^\mu\psi(x) \qquad (12.116)$$

The space components are

$$\mathbf{j} = \psi^\dagger \gamma^o \boldsymbol{\gamma} \psi = \psi^\dagger \boldsymbol{\alpha} \psi \qquad (12.117)$$

By virtue of the equation of motion (12.99) the current density is conserved.

$$\partial_\mu j^\mu = 0 \quad \text{or} \quad \frac{\partial \rho}{\partial t} + \boldsymbol{\nabla} \cdot \mathbf{j} = 0 \qquad (12.118)$$

This is the continuity equation. $\mathbf{j}(x)$ reduces to the nonrelativistic form if ψ is a plane wave with $E > 0$ and $(E-m)/m$ small. The probability density is always positive definite.

> Historical note: Dirac suggested thinking about the negative-energy solutions as follows: Suppose the true zero of the energy is not the one we have been using, but infinitely far below; and all the states below our zero of the energy are filled with electrons. Because of the Pauli exclusion principle, there is only one electron in each state. These states are sometimes called the Dirac sea. All the negative-energy states are filled with electrons, and all the positive-energy states are empty. In our ignorance we call this configuration the vacuum state. A positive-energy electron, described by the Dirac equation with $E > 0$, is stable; it cannot emit radiation and fall into the sea, because all those states are filled. On the other hand, you can excite a negative-energy electron state. Then there is a positive-energy electron and a hole in the sea, corresponding to the absence of an electron. The hole has all the properties of a positron; its charge is opposite the electron's charge, and any other additive quantum numbers have opposite signs too. A negative-energy wave function in an eigenstate of s_z with eigenvalue $\pm 1/2$ would be a positron (hole!) with eigenvalue $\mp 1/2$. This description, while somewhat informal, turns out to be essentially correct; it predicts the existence of the positron with all the observed properties, as well as the mechanism for pair production and the electromagnetic interactions of electrons and positrons.

12.4 DIRAC ELECTRON IN AN ELECTROMAGNETIC FIELD

12.4.1 Second-Order Form of the Dirac Equation

Electromagnetic interactions can be introduced into the Dirac equation with the minimal coupling prescription (12.61). With $p^\mu = i\partial^\mu$ the wave equation is

$$\boxed{\left[\gamma^\mu \left(p_\mu + eA_\mu(x)\right) - m\right]\psi(x) = 0} \qquad (12.119)$$

Chapter 12 Relativistic Wave Equations

Some properties of solutions to the Dirac equation can be found efficiently by converting it into a second-order equation, as follows: Multiply (12.119) on the left by $\gamma^\nu [p_\nu + eA_\nu(x)] + m$:

$$\left[\gamma^\mu \gamma^\nu (p_\mu + eA_\mu)(p_\nu + eA_\nu) - m^2\right] \psi(x) = 0 \qquad (12.120)$$

Define

$$\Sigma^{\mu\nu} = \frac{i}{2}[\gamma^\mu, \gamma^\nu] \qquad (12.121)$$

so that

$$\Sigma^{0i} = i\alpha_i = D(\bar{K}_i) \quad \text{and} \quad \Sigma^{ij} = \sum_k \epsilon_{ijk} \Sigma_k \qquad (12.122)$$

with

$$\Sigma_i = \begin{pmatrix} \sigma_i & 0 \\ 0 & \sigma_i \end{pmatrix} = 2s_i = 2D(\bar{J}_i) \qquad (12.123)$$

as in equation (12.106).

Write the product of two γ matrices as half the sum of an anticommutator and a commutator,

$$\gamma^\mu \gamma^\nu = \frac{1}{2}(\gamma^\mu \gamma^\nu + \gamma^\nu \gamma^\mu) + \frac{1}{2}(\gamma^\mu \gamma^\nu - \gamma^\nu \gamma^\mu) = g^{\mu\nu} - i\Sigma^{\mu\nu} \qquad (12.124)$$

so that equation (12.120) becomes

$$\left[(p^\mu + eA^\mu)(p_\mu + eA_\mu) - i\Sigma^{\mu\nu}(p_\mu + eA_\mu)(p_\nu + eA_\nu) - m^2\right]\psi(x) = 0 \qquad (12.125)$$

Using the antisymmetry of $\Sigma^{\mu\nu}$, the second term in (12.125) can be written as

$$-i\Sigma^{\mu\nu}(i\partial_\mu + eA_\mu)(i\partial_\nu + eA_\nu)\psi(x) = e\Sigma^{\mu\nu}(\partial_\mu A_\nu)\psi$$

$$= \frac{e}{2}\Sigma^{\mu\nu} F_{\mu\nu} \psi(x) = e(-\boldsymbol{\Sigma}\cdot \mathbf{B} + i\boldsymbol{\alpha}\cdot \mathbf{E})\psi(x) \qquad (12.126)$$

Therefore, the wave equation (12.120) is

$$\left[(E + eA^o)^2 - (\mathbf{p} + e\mathbf{A})^2 - e\boldsymbol{\Sigma}\cdot\mathbf{B} + ie\boldsymbol{\alpha}\cdot\mathbf{E} - m^2\right]\psi(x) = 0 \qquad (12.127)$$

Equation (12.127) is almost the Klein-Gordon equation (12.63), but with the addition of extra terms containing the interaction of the external fields with the electron spin.

12.4.2 The Gyromagnetic Ratio

Let $\mathbf{E} = 0$ and let $\mathbf{B}$ be a constant field. Let $E' = E - m$ be the nonrelativistic energy and assume $|E'| \ll m$. Then $A^o = 0$, $\mathbf{B} = \boldsymbol{\nabla}\times\mathbf{A}$, and

$$E^2 - m^2 \approx 2mE' \qquad (12.128)$$

Divide equation (12.127) by $2m$ and set $\mathbf{E} = 0$:

$$\left[E' - \frac{1}{2m}(\mathbf{p} + e\mathbf{A})^2 - \frac{e}{2m}\boldsymbol{\Sigma}\cdot\mathbf{B}\right]\psi(x) = 0 \qquad (12.129)$$

The second term in (12.129) contains the usual orbital magnetic moment coupling $(e/2m)(\mathbf{r} \times \mathbf{p}) \cdot \mathbf{B}$. The Dirac electron has the same orbital magnetic moment as the Schrödinger electron. The remaining term proportional to the magnetic field is

$$\frac{e}{2m} \mathbf{\Sigma} \cdot \mathbf{B} = \frac{ge}{2m} \mathbf{s} \cdot \mathbf{B} \tag{12.130}$$

where $\mathbf{s} = \mathbf{\Sigma}/2$ is the spin, and g the gyromagnetic ratio, automatically 2 for a Dirac electron. Here is the explanation of the value of the coupling of the electron spin to a magnetic field introduced in Section 6.1.4.

12.4.3 The Nonrelativistic Limit and the Fine Structure

The corrections to atomic energy levels due to the electron's motion in the electrostatic field of the nucleus itself, in the absence of external fields, are the atomic fine structure. To keep the computation simple, I will restrict it to the case of a spherically symmetric electrostatic field, like that in a hydrogen atom. Here the second-order form (12.127) is not the most useful starting point, since it is quadratic in both the energy and the potential. Instead, use the Dirac equation (12.119) itself, but eliminate two of the four components. Set $\mathbf{A}(\mathbf{r}) = 0$ and multiply (12.119) on the left by γ^o, so that it looks more like a Hamiltonian problem:

$$i\frac{\partial}{\partial t}\psi(\mathbf{r}, t) = \left[-i\boldsymbol{\alpha} \cdot \boldsymbol{\nabla} - eA^o + \gamma^o m \right] \psi(\mathbf{r}, t) \tag{12.131}$$

Since there are no explicitly time-dependent terms, there are energy eigenfunctions $\psi(\mathbf{r})$:

$$\psi(\mathbf{r}, t) = e^{-iEt}\psi(\mathbf{r}) \tag{12.132}$$

where

$$E\psi(\mathbf{r}) = \left[-i\boldsymbol{\alpha} \cdot \boldsymbol{\nabla} - eA^o(r) + \gamma^o m \right] \psi(\mathbf{r}) \tag{12.133}$$

Write the Dirac wave function in terms of two two-component spinors:

$$\psi(\mathbf{r}) = \begin{pmatrix} \phi_1(\mathbf{r}) \\ \phi_2(\mathbf{r}) \end{pmatrix} \tag{12.134}$$

Set $V(\mathbf{r}) = -eA^o$, the electrostatic potential energy of the electron. Then equation (12.133) becomes

$$(E - V(r) - m)\phi_1(\mathbf{r}) + i\boldsymbol{\sigma} \cdot \boldsymbol{\nabla}\phi_2(\mathbf{r}) = 0 \tag{12.135a}$$

$$(E - V(r) + m)\phi_2(\mathbf{r}) + i\boldsymbol{\sigma} \cdot \boldsymbol{\nabla}\phi_1(\mathbf{r}) = 0 \tag{12.135b}$$

Replace $\phi_2(\mathbf{r})$ in equation (12.135a) with its value obtained from (12.135b):

$$(E - V(r) - m)\phi_1(\mathbf{r}) + \boldsymbol{\sigma} \cdot \boldsymbol{\nabla}\frac{1}{E - V(r) + m}\boldsymbol{\sigma} \cdot \boldsymbol{\nabla}\phi_1(\mathbf{r}) = 0 \tag{12.136}$$

In terms of the nonrelativistic energy $E' = E - m$,

$$E'\phi_1(\mathbf{r}) = V(r)\phi_1(\mathbf{r}) - \frac{1}{2m}\boldsymbol{\sigma} \cdot \boldsymbol{\nabla}F(r)\boldsymbol{\sigma} \cdot \boldsymbol{\nabla}\phi_1(\mathbf{r}) \tag{12.137}$$

426 Chapter 12 Relativistic Wave Equations

where
$$F(r) = \frac{1}{1 + (E' - V(r))/2m} \qquad (12.138)$$

Use some of the rules (4.77) for the Pauli matrices to write

$$\boldsymbol{\sigma} \cdot \boldsymbol{\nabla} F(r) \boldsymbol{\sigma} \cdot \boldsymbol{\nabla} \phi_1(\mathbf{r}) = \sum_{ij} \sigma_i \sigma_j \partial_i F(r) \partial_j \phi_1(\mathbf{r})$$

$$= \sum_{ij} \left(\delta_{ij} + i \sum_k \epsilon_{ijk} \sigma_k \right) \left([\partial_i F(r)] \partial_j \phi_1(\mathbf{r}) + F(r) \partial_i \partial_j \phi_1(\mathbf{r}) \right) \qquad (12.139)$$

Then

$$E' \phi_1(\mathbf{r}) = V(r) \phi_1(\mathbf{r})$$
$$- \frac{1}{2m} \left([\boldsymbol{\nabla} F(r)] \cdot \boldsymbol{\nabla} \phi_1(\mathbf{r}) + i\boldsymbol{\sigma} \cdot [\boldsymbol{\nabla} F(r)] \times \boldsymbol{\nabla} \phi_1(\mathbf{r}) + F(r) \boldsymbol{\nabla}^2 \phi_1(\mathbf{r}) \right) \qquad (12.140)$$

We will find the solution for the special case $V(r) = -\alpha/r$ in the next section. Here I want to assume $\phi_1(\mathbf{r})$ is a solution to the nonrelativistic eigenvalue problem, and estimate the energy shifts using perturbation theory through order $\alpha^4 m$.

Kinetic Energy Correction

In an atom like hydrogen, $E' - V$ is the kinetic energy, whose expectation value is of order $\alpha^2 m$, so we can write

$$F(r) \approx 1 - \frac{E' - V(r)}{2m} + \cdots \qquad (12.141)$$

plus terms of order $\alpha^4 m$. Then through order $\alpha^4 m$ the correction due to the last term inside the big bracket in equation (12.140) is

$$-\frac{1}{2m} \int \phi_1^\dagger(\mathbf{r}) \left[1 - \frac{E' - V(r)}{2m} \right] \boldsymbol{\nabla}^2 \phi_1(\mathbf{r}) d^3 r$$
$$= \frac{1}{2m} \int \phi_1^\dagger(\mathbf{r}) \left[1 - \frac{\mathbf{p}^2}{8m^2} \right] \mathbf{p}^2 \phi_1(\mathbf{r}) d^3 r = \int \phi_1^\dagger(\mathbf{r}) \left[\frac{\mathbf{p}^2}{2m} - \frac{\mathbf{p}^4}{8m^2} \right] \phi_1(\mathbf{r}) d^3 r \qquad (12.142)$$

Therefore,

$$E' = \int \phi_1^\dagger \left[\frac{\mathbf{p}^2}{2m} + V(r) \right] \phi_1(\mathbf{r}) d^3 r + \Delta_{kin} + \Delta_1 + \Delta_2 \qquad (12.143)$$

The first term is the lowest order energy. The second is the expectation value of the first relativistic correction to the kinetic energy found in Section 7.3.3.

Spin-Orbit Correction

The remaining terms from equation (12.140) in the energy shift are

$$\Delta_1 = -\frac{1}{2m} \int \phi_1^\dagger(\mathbf{r}) \left[\boldsymbol{\nabla} F(r) \right] \cdot \boldsymbol{\nabla} \phi_1(\mathbf{r}) d^3 r \qquad (12.144)$$

and
$$\Delta_2 = -i\frac{1}{2m}\int \phi_1^\dagger(\mathbf{r})\boldsymbol{\sigma}\cdot\left[\boldsymbol{\nabla}F(r)\right]\times\boldsymbol{\nabla}\phi_1(\mathbf{r})\,d^3r \tag{12.145}$$

The second term is the spin-orbit energy:
$$\Delta_2 = \frac{1}{2m}\int \phi_1^\dagger(\mathbf{r})\frac{1}{r}F'(r)\,\boldsymbol{\sigma}\cdot\mathbf{L}\,\phi_1(\mathbf{r})\,d^3r \tag{12.146}$$

From the approximation (12.141), $F'(r) \approx V'(r)/2m$, and
$$\Delta_2 = \frac{1}{4m^2}\int \phi_1^\dagger(\mathbf{r})\frac{1}{r}V'(r)\,\boldsymbol{\sigma}\cdot\mathbf{L}\,\phi_1(\mathbf{r})\,d^3r \tag{12.147}$$

For a hydrogen atom, $V'(r) = \alpha/r^2$, and $\boldsymbol{\sigma} = 2\mathbf{s}$:
$$\Delta_2 = \frac{\alpha}{2m^2}\int \phi_1^\dagger(\mathbf{r})\frac{1}{r^2}\mathbf{s}\cdot\mathbf{L}\,\phi_1(\mathbf{r})\,d^3r \tag{12.148}$$

Here is the anticipated explanation of the numerical coefficient in the nonrelativistic spin-orbit energy shift (7.64).

For S states, $\phi_1(\mathbf{r})$ does not vanish at the origin, and the integral is divergent. Since $\mathbf{s}\cdot\mathbf{L} = 0$, the value of Δ_2 is undetermined. The nonrelativistic theory provided no guidance to evaluating this term when $l = 0$.

The integral is infinite when $l = 0$ because the potential contributes all the way down to $r = 0$. But then the approximation (12.141) was not justified, and the approximate form (12.147) should be replaced by the original form (12.146). The expansion was invalid because near the nucleus, $E' - V(r)$ is no longer small. As $r \to 0$, the function $F'(r)$ approaches a finite value, the integral in the (12.146) converges, and when $l = 0$ the energy shift Δ_2 vanishes.

For $l \neq 0$, the spin orbit shift for the hydrogen atom is indeed given by the formula (7.68):

$$\Delta_2 = \frac{\alpha^4 m}{4n^3(l+1)(l+\frac{1}{2})} \qquad \left(j = l + \frac{1}{2}\right) \tag{12.149a}$$

$$\Delta_2 = -\frac{\alpha^4 m}{4n^3 l(l+1)^2(l+\frac{1}{2})} \qquad \left(j = l - \frac{1}{2}\right) \tag{12.149b}$$

Darwin Term

There still remains the term Δ_1 from equation (12.144). It is

$$\Delta_1 = -\frac{1}{2m}\int \phi_1^\dagger(\mathbf{r})\frac{dF(r)}{dr}\frac{\partial\phi_1(\mathbf{r})}{\partial r}\,d^3r \approx -\frac{1}{4m^2}\int \phi_1^\dagger(\mathbf{r})\frac{dV(r)}{dr}\frac{\partial\phi_1(\mathbf{r})}{\partial r}\,d^3r \tag{12.150}$$

For the hydrogen atom, this is

$$\Delta_1 = -\frac{\alpha}{4m^2}\int_0^\infty R_{nl}(r)\frac{d}{dr}R_{nl}(r)\,dr = -\frac{\alpha}{8m^2}\int_0^\infty \frac{d}{dr}R_{nl}(r)^2\,dr = \frac{\alpha}{8m^2}R_{nl}(0)^2 \tag{12.151}$$

428 Chapter 12 Relativistic Wave Equations

Δ_1 is proportional to the square of the wave function at the origin; it vanishes *except* when $l = 0$. It can be shown, for example, from the properties of the associated Laguerre polynomials, that $R_{n0}^2 = 4/(na)^3 = 4m^3\alpha^3/n^3$. (You can verify this assertion for the radial functions listed explicitly in Table 3.1.) Thus

$$\Delta_1 = \frac{\alpha^4 m}{2n^3}\delta_{l0} \tag{12.152}$$

in agreement with (12.149a), when $l = 0$. The level shift given in equation (12.149a) turns out to be valid for all l.

In the nonrelativistic Pauli theory, Δ_1 can be obtained from a term in the Hamiltonian like

$$H_D = -\frac{1}{8m^2}\boldsymbol{\nabla}^2 V(\mathbf{r}) \tag{12.153}$$

which for hydrogen is

$$H_D = -\frac{\alpha}{8m^2}\boldsymbol{\nabla}^2\left(\frac{1}{r}\right) = \frac{\pi\alpha}{2m^2}\delta_3(\mathbf{r}) \tag{12.154}$$

Equation (12.154) is the **Darwin term** [4].

While the derivation here is not particularly elegant, it can be found in several classic textbooks [5, 6]. That it gives the right answer is perhaps fortuitous, since the term in equation (12.140) that eventually became the Darwin term is not Hermitean. But I hope this treatment makes clear the physical origin of the various terms. A correct procedure for approximating the Dirac Hamiltonian to the desired accuracy is the **Foldy-Wouthuysen transformation** [7].

12.5 THE DIRAC HYDROGEN ATOM

12.5.1 Second-Order Equation

The quickest way to get the Dirac hydrogen atom spectrum is to use the second-order differential equation derived in Section 12.4.1 and then follow closely the techniques used for a Klein-Gordon particle at the end of Section 12.2.2. First specialize equation (12.127) to the case $\mathbf{A} = 0$ and $eA^o = \alpha/r$:

$$\left(\boldsymbol{\nabla}^2 + \frac{2E\alpha}{r} + \frac{\alpha^2}{r^2} + E^2 + i\alpha\frac{\boldsymbol{\alpha}\cdot\hat{\mathbf{r}}}{r^2} - m^2\right)\psi(\mathbf{r}) = 0 \tag{12.155}$$

or[9]

$$\left(\frac{1}{2mr^2}\frac{\partial}{\partial r}r^2\frac{\partial}{\partial r} - \frac{\mathbf{L}^2 - \alpha^2}{2mr^2} + \frac{E\alpha}{mr} + i\alpha\frac{\boldsymbol{\alpha}\cdot\hat{\mathbf{r}}}{2mr^2} + \frac{E^2 - m^2}{2m}\right)\psi(\mathbf{r}) = 0 \tag{12.156}$$

Next restrict the calculation to the positive energy states and choose a basis of eigenstates of $\mathbf{J}^2$, J_z, and $\mathbf{L}^2$. These are convenient even though they are not eigenstates of the Hamiltonian. For each value of j and m, there are two eigenstates of $\mathbf{L}^2$, with $l = j \pm 1/2$ respectively. The $\boldsymbol{\alpha}\cdot\hat{\mathbf{r}}$ term is completely off diagonal. It changes an odd spherical harmonic into an even one, and vice versa, so it changes

---
[9]Sorry about the notation. α is the fine structure constant. $\boldsymbol{\alpha}$ is the matrix defined in (12.106).

the value of l from $j \pm 1/2$ to $j \mp 1/2$. $\mathbf{L}^2$ and $\boldsymbol{\alpha} \cdot \hat{\mathbf{r}}$ do not commute, and so $[\mathbf{L}^2, H] \neq 0$. The energy eigenstates are linear combinations of these states with different l.

Since $(\boldsymbol{\alpha} \cdot \hat{\mathbf{r}})^2 = 1$, its off-diagonal value is ± 1, and for now we do not care which.[10] The radial equation is

$$\left(\frac{1}{2mr^2} \frac{\partial}{\partial r} r^2 \frac{\partial}{\partial r} - \frac{1}{2mr^2} M + \frac{E\alpha}{mr} - \frac{E^2 - m^2}{2m} \right) \psi(\mathbf{r}) = 0 \qquad (12.157)$$

where

$$M = \begin{pmatrix} \mathbf{L}^2 - \alpha^2 & \pm i\alpha \\ \pm i\alpha & \mathbf{L}^2 - \alpha^2 \end{pmatrix} = \begin{pmatrix} (j+1/2)(j+3/2) - \alpha^2 & \pm i\alpha \\ \pm i\alpha & j^2 - 1/4 - \alpha^2 \end{pmatrix} \qquad (12.158)$$

The two rows and columns correspond to the two possible values of l for fixed $\mathbf{J}^2$ and J_z. The states that diagonalize the Hamiltonian are a linear combination of the two values of l. To get the radial equations, replace M by its eigenvalues[11]

$$\lambda_1 = \mu(\mu + 1) \quad \text{and} \quad \lambda_2 = \mu(\mu - 1) \qquad (12.159)$$

where

$$\mu = \sqrt{(j + 1/2)^2 - \alpha^2} \qquad (12.160)$$

In each case, the equation is identical to the nonrelativistic spinless hydrogen atom equation, with the substitutions

$$\begin{aligned} l &\longrightarrow l' \qquad (\text{with} \quad l' = \mu \quad \text{or} \quad l' = \mu - 1) \\ \alpha &\longrightarrow \frac{E\alpha}{m} \\ E &\longrightarrow \frac{E^2 - m^2}{2m} \end{aligned} \qquad (12.161)$$

The other two algebraic solutions for l are negative and do not provide solutions with normalizable behavior near the origin.

With these substitutions

$$\frac{E^2 - m^2}{2m} = -\frac{E^2 \alpha^2 m}{(n' + l' + 1)^2} \qquad (12.162)$$

or

$$E = m \left[1 + \frac{\alpha^2}{(n' + l' + 1)^2} \right]^{-1/2} \qquad (12.163)$$

Here n' is a nonnegative integer. For the solutions with $l' = \mu$, set $n = n' + j + 3/2$. For the solutions with $l' = \mu - 1$, set $n = n' + j + 1/2$. Then in either case the

---

[10]The sign was calculated in part (c) of Problem 4.5.
[11]Just check that $\lambda_1 + \lambda_2 = 2\mu^2$ is the trace of the matrix, and $\lambda_1 \lambda_2 = \mu^2(\mu^2 - 1)$ is its determinant.

energies are given by **Dirac's fine-structure formula**:

$$E = m \left[1 + \frac{\alpha^2}{\left(n - j - \frac{1}{2} + \sqrt{(j+1/2)^2 - \alpha^2}\right)^2} \right]^{-\frac{1}{2}} \qquad (12.164)$$

for some positive integer n, now identified with the principal quantum number.

There are two combinations of n' and l' for each pair of n and j except for the highest allowed value of j for each n. The energies are a function of j only to *all* orders in α. The degeneracy discovered in Section 7.3.4 between states of different l but the same j to order α^2 in perturbation theory is an exact property of the Dirac equation.

The magnitude of the spin-orbit coupling, introduced earlier without derivation, is predicted automatically by the Dirac equation. Expand equation (12.164) in powers of α:

$$E = m - \frac{\alpha^2 m}{2n^2} + \frac{\alpha^4 m}{2n^4}\left(\frac{3}{4} - \frac{2n}{2j+1}\right)\cdots \qquad (12.165)$$

This was the final result of Section 7.3.4.

Relation to the First-Order Equation

The energies in (12.165) are the eigenvalues of the second-order version (12.127) of the Dirac equation, but that does not quite prove that they are eigenvalues of the Dirac equation itself. Solutions to the Dirac equation must be solutions to (12.127), but not necessarily the other way around. In fact the wave function that is a solution to (12.157) is not a solution to the original equation. The eigenvalues of equation (12.156) are nevertheless eigenvalues of the Dirac equation. Here is a proof:

Proof. A solution ψ to the Dirac equation satisfies $(H - E)\psi = 0$, where

$$H = \boldsymbol{\alpha} \cdot \mathbf{p} - eA^o(r) + \gamma^o m \qquad (12.166)$$

which we wrote as

$$\gamma^o(E - H)\psi = 0 \qquad (12.167)$$

The eigenfunctions and eigenvalues of the second-order equation (12.120) satisfy

$$\left[\gamma^o(E-H) + 2m\right]\gamma^o(E-H)\phi = 0 \qquad (12.168)$$

If ϕ solves (12.120) you can interchange the order of the two factors (they commute) to get

$$\gamma^o(E-H)\left[\gamma^o(E-H) + 2m\right]\phi = 0 \qquad (12.169)$$

The function $\psi = \left[\gamma^o(E-H) + 2m\right]\phi$ satisfies the Dirac equation, with the same energy! So the energies (12.164) are indeed the spectrum of the Dirac equation. ■

Section 12.5 The Dirac Hydrogen Atom   431

This solution also provides a method for computing the wave functions: Make the required substitutions in the nonrelativistic wave function, and then operate on it with $\gamma^o(E - H) + 2m$ to get the four-component Dirac wave function. But this method is cumbersome, and does not display the quantum numbers of the wave functions very transparently. In the next section I will find the wave functions for the Dirac hydrogen atom exactly, using a more direct approach.

12.5.2 Spherically Symmetric Potentials

For some purposes one wants the Dirac wave functions as well as the energies in a convenient form.[12] Since this is an important problem, I will work it out in some detail. First I will find the radial equations for the Dirac equation in any central electrostatic potential, and later specialize to the Coulomb potential for hydrogenlike atoms.

For these potentials the magnetic field vanishes, so $\mathbf{A} = 0$, but $A^o \neq 0$. As in the previous section set $V(r) = -eA^o(r)$. and write equations (12.135) as the two coupled two-component equations:

$$(E - V(r) - m)\, \phi_1(\mathbf{r}) + i\boldsymbol{\sigma} \cdot \boldsymbol{\nabla}\phi_2(\mathbf{r}) = 0 \quad (12.135a)$$

$$(E - V(r) + m)\, \phi_2(\mathbf{r}) + i\boldsymbol{\sigma} \cdot \boldsymbol{\nabla}\phi_1(\mathbf{r}) = 0 \quad (12.135b)$$

Equations (12.135) are the central potential problem for the Dirac equation.

In the nonrelativistic version of the hydrogen atom, $\mathbf{L}$ and $\mathbf{S}$ commuted with H, and there were simultaneous eigenstates of H, $\mathbf{J}^2$, J_z, and $\mathbf{L}^2$. Here, while $[J_i, H] = 0$, $\mathbf{L}^2$ and $\mathbf{s}^2$ do not separately commute with H. We need a quantum number to replace $\mathbf{L}^2$.

What quantum number can replace $\mathbf{L}^2$? The nonrelativistic solution has two states for each j, with $l = j \pm 1/2$ respectively. The states were distinguished not only by their eigenvalues of $\mathbf{L}^2$ but also by the parity $(-1)^l$. $\mathbf{L}^2$ is not a good relativistic quantum number, but the parity still is. States of definite E, $\mathbf{J}^2$, J_z and parity do not have fixed l. The large components, the ones that survive in the nonrelativistic limit, should have the appropriate l value.

First exploit rotational invariance to turn these partial differential equations into a collection of one-dimensional radial wave equations. Both ϕ_1 and ϕ_2 satisfy $[(E - V)^2 - m^2 + \boldsymbol{\nabla}^2]\phi(\mathbf{r}) = 0$. Since $V(r)$ is spherically symmetric, ϕ_1 and ϕ_2 can be chosen separately as eigenfunctions of $\mathbf{L}^2$ and L_z, though not with the same l. Thus

$$\left(L_z + \frac{1}{2}\Sigma_z\right)\begin{pmatrix}\phi_1 \\ \phi_2\end{pmatrix} = m\begin{pmatrix}\phi_1 \\ \phi_2\end{pmatrix} \quad (12.171)$$

and

$$J^2\begin{pmatrix}\phi_1 \\ \phi_2\end{pmatrix} = \left(\mathbf{L} + \frac{1}{2}\Sigma\right)^2\begin{pmatrix}\phi_1 \\ \phi_2\end{pmatrix} = j(j+1)\begin{pmatrix}\phi_1 \\ \phi_2\end{pmatrix} \quad (12.172)$$

None of these operators connects components of ϕ_1 with components of ϕ_2. Each is a function of r alone multiplied by a two-component spinor that is an eigenfunction

---

[12] Try computing the scattering of a photon off a hydrogen atom with correct Dirac wave functions for the electron.

432 Chapter 12 Relativistic Wave Equations

of $\mathbf{L}^2$, $\mathbf{J}^2$, and J_z, but ϕ_1 and ϕ_2 have different l. With $l = j \mp 1/2$, the ϕ_i are proportional to the two-component spinors

$$\mathcal{Y}_l^{l\pm\frac{1}{2},m} = \frac{1}{\sqrt{2l+1}} \begin{pmatrix} \sqrt{l\pm m+1/2}\, Y_l^{m-\frac{1}{2}} \\ \pm\sqrt{l\mp m+1/2}\, Y_l^{m+\frac{1}{2}} \end{pmatrix} \qquad (12.173)$$

that transform as indicated under $\mathbf{J}$, $\mathbf{L}$, and $\mathbf{S}$:[13]

$$J_z \mathcal{Y}_l^{l\pm\frac{1}{2},m} = m\mathcal{Y}_l^{l\pm\frac{1}{2},m} \qquad (12.174a)$$

$$\boldsymbol{\sigma}\cdot\mathbf{L}\,\mathcal{Y}_l^{l+\frac{1}{2},m} = l\mathcal{Y}_l^{l+\frac{1}{2},m} \qquad (12.174b)$$

$$\boldsymbol{\sigma}\cdot\mathbf{L}\,\mathcal{Y}_l^{l-\frac{1}{2},m} = -(l+1)\mathcal{Y}_l^{l-\frac{1}{2},m} \qquad (12.174c)$$

Equations (12.135) are invariant under the correct reflection transformation [see equation (12.92)]

$$\mathcal{R}\begin{pmatrix}\psi_1(\mathbf{r})\\ \psi_2(\mathbf{r})\end{pmatrix} = \gamma^o \begin{pmatrix}\psi_1(-\mathbf{r})\\ \psi_2(-\mathbf{r})\end{pmatrix} = \begin{pmatrix}\psi_1(-\mathbf{r})\\ -\psi_2(-\mathbf{r})\end{pmatrix} \qquad (12.175)$$

Under $\mathbf{r} \to -\mathbf{r}$, $\mathcal{Y}_l^{j,m} \to (-1)^l \mathcal{Y}_l^{j,m}$, so the upper and lower components of these functions transform as eigenstates of $\mathbf{L}$ with opposite parity. Since they have the same j, the l values differ by unity.

So the solutions to the spherically-symmetric potential problem are of the form

$$\psi_a(\mathbf{r}) = \begin{pmatrix} A(r)\mathcal{Y}_{j-\frac{1}{2}}^{j,m} \\ -ia(r)\mathcal{Y}_{j+\frac{1}{2}}^{j,m} \end{pmatrix} \quad \text{or} \quad \psi_b(\mathbf{r}) = \begin{pmatrix} B(r)\mathcal{Y}_{j+\frac{1}{2}}^{j,m} \\ -ib(r)\mathcal{Y}_{j-\frac{1}{2}}^{j,m} \end{pmatrix} \qquad (12.176)$$

where $A(r)$, $a(r)$, $B(r)$, and $b(r)$ are functions of r alone. $\psi_a(\mathbf{r})$ is a solution with parity $(-1)^{j-1/2}$, and $\psi_b(\mathbf{r})$ is a solution with parity $(-1)^{j+1/2}$. The large components have the nonrelativistic association of orbital angular momentum with parity; the nonrelativistic limit of these wave functions has $l = j - 1/2$ for ψ_a and $l = j + 1/2$ for ψ_b. The lower components have the "wrong" parity.

12.5.3 Radial Equations

There is a trick to make the states manifestly eigenstates of $\mathbf{J}^2$ and J_z. From equations (12.135), a solution of the type ψ_a in (12.176) satisfies

$$(E - V - m)A(r)\mathcal{Y}_{j-\frac{1}{2}}^{j,m} = -\boldsymbol{\sigma}\cdot\boldsymbol{\nabla} a(r)\mathcal{Y}_{j+\frac{1}{2}}^{j,m} \qquad (12.177a)$$

$$(E - V + m)\,a(r)\mathcal{Y}_{j+\frac{1}{2}}^{j,m} = +\boldsymbol{\sigma}\cdot\boldsymbol{\nabla} A(r)\mathcal{Y}_{j-\frac{1}{2}}^{j,m} \qquad (12.177b)$$

Use one of the rules (4.77) to write

$$\boldsymbol{\sigma}\cdot\boldsymbol{\nabla} = \frac{1}{r^2}\boldsymbol{\sigma}\cdot\mathbf{r}\boldsymbol{\sigma}\cdot\mathbf{r}\boldsymbol{\sigma}\cdot\boldsymbol{\nabla} = \boldsymbol{\sigma}\cdot\hat{\mathbf{r}}\left(\hat{\mathbf{r}}\cdot\boldsymbol{\nabla} - \frac{\boldsymbol{\sigma}\cdot\mathbf{L}}{r}\right) \qquad (12.178)$$

---
[13]See Problems 4.4 and 4.5.

Section 12.5 The Dirac Hydrogen Atom

where $\hat{r} = \mathbf{r}/r$. Now $\hat{r} \cdot \nabla = \partial/\partial r$, and the $\boldsymbol{\sigma} \cdot \mathbf{L}$ operating on the $\mathcal{Y}_l^{j,m}$ are (see Problem 4.4)

$$\boldsymbol{\sigma} \cdot \mathbf{L}\, \mathcal{Y}_{j+\frac{1}{2}}^{j,m} = -(j+\tfrac{3}{2})\mathcal{Y}_{j+\frac{1}{2}}^{j,m} \quad \text{and} \quad \boldsymbol{\sigma} \cdot \mathbf{L}\, \mathcal{Y}_{j-\frac{1}{2}}^{j,m} = (j-\tfrac{1}{2})\mathcal{Y}_{j-\frac{1}{2}}^{j,m} \quad (12.179)$$

Therefore equations (12.177) are

$$(E - V - m)A(r)\mathcal{Y}_{j-\frac{1}{2}}^{j,m} = -\boldsymbol{\sigma}\cdot\hat{r}\left(\frac{\partial}{\partial r} + \frac{j+3/2}{r}\right)a(r)\mathcal{Y}_{j+\frac{1}{2}}^{j,m} \quad (12.180a)$$

$$(E - V + m)a(r)\mathcal{Y}_{j+\frac{1}{2}}^{j,m} = \boldsymbol{\sigma}\cdot\hat{r}\left(\frac{\partial}{\partial r} - \frac{j-1/2}{r}\right)A(r)\mathcal{Y}_{j-\frac{1}{2}}^{j,m} \quad (12.180b)$$

Now[14]

$$\boldsymbol{\sigma}\cdot\hat{r}\,\mathcal{Y}_{j-\frac{1}{2}}^{j,m} = -\mathcal{Y}_{j+\frac{1}{2}}^{j,m} \quad (12.181)$$

so we finally obtain simple first-order coupled radial equations:

$$(E - V - m)A(r) = \left(\frac{d}{dr} + \frac{j+3/2}{r}\right)a(r) \quad (12.182a)$$

$$(E - V + m)a(r) = -\left(\frac{d}{dr} - \frac{j-1/2}{r}\right)A(r) \quad (12.182b)$$

Apply the same argument to $\psi_b(\mathbf{r})$:

$$(E - V - m)B(r) = \left(\frac{d}{dr} - \frac{j-1/2}{r}\right)b(r) \quad (12.183a)$$

$$(E - V + m)b(r) = -\left(\frac{d}{dr} + \frac{j+3/2}{r}\right)B(r) \quad (12.183b)$$

The two sets of equations (12.182) and (12.183) correspond to parities $(-1)^{j \mp \frac{1}{2}}$ respectively. It is not necessary to solve both. If $A(r)$ and $a(r)$ are a solution for some j, then $B(r) = A(r)$ and $b(r) = a(r)$ are a solution for $j \to -j - 1$.

12.5.4 The Hydrogen Atom

For hydrogenlike atoms, $V(r) = -Z\alpha/r$. There are many ways to solve this problem. Perhaps the simplest is to imitate the procedure for the nonrelativistic hydrogen wave functions in Section 3.5.3 as closely as possible. Write

$$A(r) = r^\nu P(r)e^{-\beta r} \qquad a(r) = r^\nu p(r)e^{-\beta r} \quad (12.184)$$

The constants β and ν dominate the asymptotic behavior of $A(r)$ as $r \to \infty$ and $r \to 0$ respectively, provided $P(r)$ and $p(r)$ are polynomials.

$$\left(E + \frac{Z\alpha}{r} - m\right)P(r) = h'(r) - \beta p(r) + \nu\frac{1}{r}p(r) + \frac{j+3/2}{r}p(r) \quad (12.185a)$$

$$\left(E + \frac{Z\alpha}{r} + m\right)p(r) = -H'(r) + \beta P(r) - \nu\frac{1}{r}P(r) + \frac{j-1/2}{r}P(r) \quad (12.185b)$$

$$\quad (12.185c)$$

---

[14] See part (c) of Problem 4.4.

434   Chapter 12   Relativistic Wave Equations

If $P(r)$ and $p(r)$ are indeed polynomials, they will grow faster than their derivatives for large r. The nonvanishing terms in equations (12.185) in that limit are

$$(E-m)P(r) = -\beta p(r) \quad \text{and} \quad (E+m)p(r) = \beta P(r) \quad (12.186)$$

Therefore,
$$\beta^2 = m^2 - E^2 \quad (12.187)$$

For normalizable bound-state solutions, $|E|$ must be less than m. Since β must be positive

$$\beta = \sqrt{m^2 - E^2} \quad (12.188)$$

In the nonrelativistic limit $\beta \to \sqrt{-2mE'}$.

Next examine equations (12.185) in the small r limit:

$$Z\alpha P(r) = (\nu + j + \tfrac{3}{2})p(r) \quad (12.189a)$$

$$Z\alpha p(r) = (j - \tfrac{1}{2} - \nu)P(r) \quad (12.189b)$$

Therefore,
$$\nu = -1 \pm \sqrt{(j+1/2)^2 - (Z\alpha)^2} \quad (12.190)$$

As long as $Z\alpha$ is less than $j+1/2$, the negative sign above must be excluded to avoid an unacceptable singularity at the origin. In the nonrelativistic limit $\nu \to l$ as predicted by (3.134). But the exact index ν in equation (12.190) is not an integer.

With these values of ν and β, the differential equations (12.185) can be solved by the power series method. Here there are two coupled series:

$$P(r) = \sum_n C_n r^n \quad \text{and} \quad p(r) = \sum_n c_n r^n \quad (12.191)$$

Equate coefficients of powers of r:

$$(E-m)C_n = (n+\nu+j+5/2)c_{n+1} - Z\alpha C_{n+1} - \beta c_n \quad (12.192a)$$

$$(E+m)c_n = -(n+\nu-j+3/2)C_{n+1} - Z\alpha c_{n+1} + \beta C_n \quad (12.192b)$$

Equations (12.192) are coupled recursion relations for C_n and c_n. Unless the series terminates, the behavior of $P(r)$ and $p(r)$ at $r \to \infty$ will be dominated by $e^{+2\beta r}$ and the functions would not be normalizable. The eigenvalue condition is that both functions are polynomials, and it is not hard to see that both series terminate at the same value n'. Suppose that for some integer n', both $C_n = 0$ and $c_n = 0$ for all $n > n'$. Set $n = n'$ to get

$$(E-m)C_{n'} = -\beta c_{n'} \qquad (E+m)c_{n'} = \beta C_{n'} \quad (12.193)$$

Since the two rules are the same, it was indeed correct to assume that the two polynomials are of the same degree. In either case

$$C_{n'} = \sqrt{\frac{m+E}{m-E}}\, c_{n'} \quad (12.194)$$

Write the recursion relations (12.192) for $n = n' - 1$; multiply the first by β and the second by $E - m$ and add them. The left-hand side of the sum vanishes identically. Use the result to eliminate C_n, leaving

$$(n' - 1 + \nu + j + \frac{5}{2})c_{n'} - Z\alpha\sqrt{\frac{m+E}{m-E}}c_{n'}$$

$$= -(n' - 1 + \nu - j + \frac{3}{2})c_{n'} - Z\alpha\sqrt{\frac{m-E}{m+E}}c_{n'} \quad (12.195)$$

By hypothesis, $c_{n'} \neq 0$, so (12.195) is the eigenvalue condition, and after a little algebra it reduces to

$$Z\alpha E = \sqrt{m^2 - E^2}(n' + \nu + 1) \quad (12.196)$$

Finally, solve for E:

$$E = m\frac{1}{\sqrt{1 + \frac{(Z\alpha)^2}{(n'+\nu+1)^2}}} = m\left[1 + \frac{(Z\alpha)^2}{\left(n' + \sqrt{(j+1/2)^2 - (Z\alpha)^2}\right)^2}\right]^{-\frac{1}{2}} \quad (12.197)$$

in agreement with (12.164) when $Z = 1$.

From the remarks just below equation (12.183), the eigenvalue condition for the solution with parity $(-1)^{l+1}$ is obtained by repeating the computation with $j \to -j - 1$. The formula (12.197) for the energies depends on j only in the combination $(j+1/2)^2$, and $(j+1/2)^2$ doesn't change under this replacement. *The energy levels for the same total angular momentum but opposite parity are given by the same formula.* The energies are a function only of j to *all* orders in α.

PROBLEMS

12.1. Lorentz transformations

(a) By comparing the first-order terms of the expansions of the finite boosts, find the three 4×4 matrices $\bar{K}_i$, $(i = 1, 2, 3)$ explicitly.

(b) Compute the commutators:

$$[\bar{J}_i, \bar{J}_j] = i\sum_k \epsilon_{ijk}\bar{J}_k$$

$$[\bar{J}_i, \bar{K}_j] = i\sum_k \epsilon_{ijk}\bar{K}_k$$

$$[\bar{K}_i, \bar{K}_j] = \quad ?$$

The first two lines just confirm that the $\bar{J}_i$ and the $\bar{K}_i$ transform under rotations as components of vectors.

(c) Write the 4×4 matrix for a pure boost in the z direction by a finite (not just first-order) velocity v.

12.2. Algebra of Generators in the Dirac Representation

Define the 4×4 matrices

$$\Sigma_i = \begin{pmatrix} \sigma_i & 0 \\ 0 & \sigma_i \end{pmatrix} \quad \text{and} \quad \alpha_i = \begin{pmatrix} 0 & \sigma_i \\ \sigma_i & 0 \end{pmatrix}$$

436 Chapter 12 Relativistic Wave Equations

where each entry stands for a 2×2 matrix: 0 is the two-dimensional zero matrix, and σ_i are the Pauli matrices. Define

$$D(\bar{J}_i) = \frac{1}{2}\Sigma_i \quad \text{and} \quad D(\bar{K}_i) = \frac{i}{2}\alpha_i$$

(a) Show that the six matrices $D(\bar{J}_i)$ and $D(\bar{K}_i)$ are indeed a representation of the generators of the Lorentz group [see equations (12.22)].

(b) For the three-dimensional rotation group you learned that there is only one representation for each finite dimension, up to a unitary transformation. This is not true for the Lorentz group. In particular, there are two different 4-dimensional representations:

Show that no change of basis can transform the six four-dimensional matrices $D(\bar{J}_i)$ and $D(\bar{K}_i)$ in the Dirac representation into the six four-dimensional matrices $\bar{J}$ and $\bar{K}$.

12.3. Two Component Spin-Half Equations

(a) Let $\psi(x)$ be a two-component spinor transforming according to the $D^{(0,\frac{1}{2})}$ representation of the Lorentz transformations. Show that if $\psi(x)$ satisfies equation (12.85) for any four matrices β^μ, the condition that it satisfy the same equation in any other reference frame is

$$D^{(0,\frac{1}{2})}(\bar{L})^\dagger \beta^\mu D^{(0,\frac{1}{2})}(\bar{L}) = \bar{L}^\mu{}_\nu \beta^\nu$$

for any Lorentz transformation $\bar{L}$.

Hint: Require that $\psi(x)^\dagger \beta^\mu \partial_\mu \psi(x)$ transform like a scalar field.

(b) Now write this equation out to first order in the generators of the Lorentz transformations. Verify that either sign in equation (12.84) is consistent as long as you just consider rotations.

(c) Finally, show that for the condition in part (a) to be true for boosts, you must choose the upper sign, corresponding to right-handed (positive helicity) particles, as claimed at the end of Section 12.3.1.

12.4. Schrödinger's Relativistic Hydrogen Atom

(a) Expand the eigenvalues (12.71) of the Klein-Gordon equation in a Coulomb potential powers of α through terms in α^4. What is the physical meaning of each term?

(b) If you did the calculation correctly, the extra degeneracy of the nonrelativistic Coulomb problem has disappeared—in this model states of different angular momentum no longer have exactly the same energy. When Schrödinger first solved this problem, he hoped that the splitting between the $2S$ and the $2P$ levels might explain the observed spectrum. What is the difference in energy between these two levels (to order α^4)? How does it compare with the observed splitting?

(c) Are there any negative energy solutions to this problem? If so, what are their energies?

12.5. The Dirac Relativistic Hydrogen Atom

(a) In the Dirac hydrogen atom, write down the explicit four-component normalized wave function for the spin-up state in the lowest energy level.

(b) How does the probability density in the lowest level differ qualitatively from the nonrelativistic hydrogen atom? At what radius is the relativistic probability density, averaged over angle, twice as great as the nonrelativistic probability density?

REFERENCES

The Klein-Gordon equation is due to

[1] E. SCHRÖDINGER, Annalen der Physik **81** (1926), 109.

[2] W. GORDON, Zeitschrift für Physik **40** (1926), 117.

[3] O. KLEIN, Zeitschrift für Physik **41** (1927), 407.

The Dirac matrices are defined on page 259 of reference [9] of Chapter 2.

The Darwin term was suggested in

[4] C. G. DARWIN, Proc. Roy. Soc. **A118** (1928), 654.

The reduction of the Dirac equation to get the Darwin term and the other nonrelativistic effects is done carefully in

[5] E. CONDON AND G. SHORTLEY, *The Theory of Atomic Spectra*, Cambridge, 1935, 1951-1970, page 130. See also Appendix D.

[6] L. SCHIFF, *Quantum Mechanics*, McGraw Hill, 1968, page 482. See also Appendix D.

The correct method for extacting the nonrelativistic limit of the Dirac equation is due to

[7] L. L. FOLDY AND S. A. WOUTHUYSEN, Phys. Rev. **78** (1950), 29.

CHAPTER 13

Identical Particles

13.1 NONRELATIVISTIC IDENTICAL-PARTICLE SYSTEMS

The free electromagnetic field is a collection of oscillators, one for each mode.[1] The level of each mode is the **occupation number**, and the raising and lowering operators become creation and annihilation operators. All the states can be obtained by repeated application of the creation operators to the ground state, or vacuum state. They are states with a definite number of photons, and the photons automatically come out to satisfy Bose-Einstein statistics.

The construction could have been done the other way around, defining the electromagnetic field in terms of the states with a definite number of photons. That is a way to describe any kind of physical system containing many identical particles, and it is the subject of this chapter. I will begin with ordinary nonrelativistic particles interacting through a local potential, then move on to the construction of relativistic identical-particle systems. For the nonrelativistic systems, the occupation number description is an efficient way to write down the physics we already know. For relativistic systems like the Dirac electron, it is the foundation for a consistent relativistic theory.

13.1.1 Creation and Annihilation Operators for Bosons

Start with a collection of free, spinless particles. The single particle states are momentum eigenstates $|\mathbf{p}\rangle$. Let $|\rangle$ be the ground state, the vacuum state with no particles, and define the creation operators by

$$a^\dagger(\mathbf{p})|\rangle = |\mathbf{p}\rangle \tag{13.1}$$

The creation and annihilation operators satisfy rules like equations (11.47) for photons, but without the spin index:

$$[a(\mathbf{p}), a^\dagger(\mathbf{p}')] = \delta_3(\mathbf{p}' - \mathbf{p}) \tag{13.2a}$$

$$[a(\mathbf{p}), a(\mathbf{p}')] = 0 \tag{13.2b}$$

$$[a^\dagger(\mathbf{p}), a^\dagger(\mathbf{p}')] = 0 \tag{13.2c}$$

The coefficient of the Dirac delta function comes from the normalization

$$\langle \mathbf{p}' | \mathbf{p} \rangle = \delta_3(\mathbf{p}' - \mathbf{p}) \tag{13.3}$$

A basis for the two-particle states is

$$|\mathbf{p}_1, \mathbf{p}_2\rangle = a^\dagger(\mathbf{p}_1) a^\dagger(\mathbf{p}_2)|\rangle \tag{13.4}$$

---

[1] See Section 11.2.

Section 13.1 Nonrelativistic Identical-Particle Systems 439

which is normalized to (exercise!)

$$\langle \mathbf{p}_1, \mathbf{p}_2 | \mathbf{p}_3, \mathbf{p}_4 \rangle = \langle | a(\mathbf{p}_1)a(\mathbf{p}_2)a^\dagger(\mathbf{p}_3)a^\dagger(\mathbf{p}_4) | \rangle$$
$$= \delta_3(\mathbf{p}_1 - \mathbf{p}_3)\delta_3(\mathbf{p}_2 - \mathbf{p}_4) + \delta_3(\mathbf{p}_1 - \mathbf{p}_4)\delta_3(\mathbf{p}_2 - \mathbf{p}_3) \quad (13.5)$$

These particles are bosons because of equation (13.2c). The operator

$$N = \int a^\dagger(\mathbf{p})a(\mathbf{p})d^3p \quad (13.6)$$

is the total number of particles. The Hamiltonian is

$$H = \int \omega(\mathbf{p})a^\dagger(\mathbf{p})a(\mathbf{p})d^3p \quad (13.7)$$

with

$$\omega(\mathbf{p}) = \frac{\mathbf{p}^2}{2m} \quad (13.8)$$

A quantum mechanical space like this, with states containing arbitrary numbers of particles connected by creation and annihilation operators, is called a **Fock space**.

Quantum Fields

In analogy to the electromagnetic field define a *field* operator, an operator function of the coordinates, by

$$\Psi(\mathbf{r}) = \frac{1}{(2pi)^{3/2}} \int e^{i\mathbf{p}\cdot\mathbf{r}}a(\mathbf{p})d^3p \quad (13.9)$$

Then

$$\Psi^\dagger(\mathbf{r})|\rangle = \frac{1}{(2pi)^{3/2}} \int e^{-i\mathbf{p}\cdot\mathbf{r}}a^\dagger(\mathbf{p})|\rangle d^3p = \frac{1}{(2pi)^{3/2}} \int e^{-i\mathbf{p}\cdot\mathbf{r}}|\mathbf{p}\rangle d^3p \quad (13.10)$$

This is a one-particle state whose wave function is

$$\psi(\mathbf{r}') = \langle \mathbf{r}' | \Psi^\dagger(\mathbf{r}) | \rangle = \frac{1}{(2pi)^{3/2}} \int e^{-i\mathbf{p}\cdot\mathbf{r}}\langle \mathbf{r}' | \mathbf{p} \rangle d^3p = \delta_3(\mathbf{r} - \mathbf{r}') \quad (13.11)$$

The field operator $\Psi^\dagger(\mathbf{r})$ creates a state "at" $\mathbf{r}$, just as $a^\dagger(\mathbf{p})$ creates a momentum eigenstate "at" $\mathbf{p}$. It satisfies commutation relations analogous to (13.2)

$$[\Psi(\mathbf{r}), \Psi^\dagger(\mathbf{r}')] = \delta_3(\mathbf{r}' - \mathbf{r}) \quad (13.12a)$$

$$[\Psi(\mathbf{r}), \Psi(\mathbf{r}')] = 0 \quad (13.12b)$$

$$[\Psi^\dagger(\mathbf{r}), \Psi^\dagger(\mathbf{r}')] = 0 \quad (13.12c)$$

Then if $|\psi\rangle$ is any normalizable state

$$\langle | \Psi(\mathbf{r}) | \psi \rangle = \langle \mathbf{r} | \psi \rangle = \psi(\mathbf{r}) \quad (13.13)$$

Chapter 13 Identical Particles

The matrix element of the field operator between a one-particle state and the vacuum is the ordinary wave function. That is, for any normalizable one-particle state

$$|\psi\rangle = \int \psi(\mathbf{r})\Psi^\dagger(\mathbf{r})\,|\rangle\, d^3r = \int \phi(\mathbf{p})a^\dagger(\mathbf{p})\,|\rangle\, d^3p \tag{13.14}$$

Here $\psi(\mathbf{r})$ is the wave function, and

$$\phi(\mathbf{p}) = \frac{1}{(2pi)^{3/2}} \int \psi(\mathbf{r}) e^{-i\mathbf{p}\cdot\mathbf{r}}\, d^3r \tag{13.15}$$

is its Fourier transform, the momentum space wave function.

Normalized two-particle states can be constructed as follows:

$$|\psi\rangle = \frac{1}{\sqrt{2}} \int \psi(\mathbf{r}_1, \mathbf{r}_2) \Psi^\dagger(\mathbf{r}_1) \Psi^\dagger(\mathbf{r}_2)\, |\rangle\, d^3r_1\, d^3r_2 \tag{13.16}$$

The function $\psi(\mathbf{r}_1, \mathbf{r}_2)$, symmetric in its two arguments, is the conventional wave function, like the one we used for the helium atom in Sections 7.4 and 7.5. Because of the identity (exercise!)

$$\langle|\,\Psi(\mathbf{r}'_1)\Psi(\mathbf{r}'_2)\Psi^\dagger(\mathbf{r}_1)\Psi^\dagger(\mathbf{r}_2)\,|\rangle = \delta_3(\mathbf{r}'_1 - \mathbf{r}_1)\delta_3(\mathbf{r}'_2 - \mathbf{r}_2) + \delta_3(\mathbf{r}'_1 - \mathbf{r}_2)\delta_3(\mathbf{r}'_2 - \mathbf{r}_1) \tag{13.17}$$

these two-particle states are properly normalized:

$$\langle\psi|\psi\rangle = \frac{1}{2}\int \langle|\,\psi(\mathbf{r}'_1,\mathbf{r}'_2)^*\psi(\mathbf{r}_1,\mathbf{r}_2)\Psi(\mathbf{r}'_1)\Psi(\mathbf{r}'_2)\Psi^\dagger(\mathbf{r}_1)\Psi^\dagger(\mathbf{r}_2)\,|\rangle\, d^3r'_1\, d^3r'_2\, d^3r_1\, d^3r_2$$

$$= \frac{1}{2}\int [\psi^*(\mathbf{r}_1,\mathbf{r}_2) + \psi^*(\mathbf{r}_2,\mathbf{r}_1)]\,\psi(\mathbf{r}_1,\mathbf{r}_2)\, d^3r_1\, d^3r_2 = 1 \tag{13.18}$$

provided the wave function is symmetric in $\mathbf{r}_1$ and $\mathbf{r}_2$.

You can also construct a basis starting from discrete one-particle states. Let $|\psi_i\rangle$ be an orthonormal basis for the single-particle states. Define

$$a_i = \int \psi_i^*(\mathbf{r})\Psi(\mathbf{r})d^3r \tag{13.19}$$

Then

$$\left[a_i, a_j^\dagger\right] = \iint \psi_i^*(\mathbf{r}')\psi_j(\mathbf{r})\left[\Psi(\mathbf{r}'), \Psi^\dagger(\mathbf{r})\right] d^3r'\, d^3r = \int \psi_i^*(\mathbf{r}')\psi_j(\mathbf{r})d^3r = \delta_{ij} \tag{13.20}$$

This is the algebra of a collection of discrete oscillators. Since

$$\left\langle\left|\,a_i a_j a_i^\dagger a_j^\dagger\,\right|\right\rangle = 1 + \delta_{ij} \tag{13.21}$$

the normalized two-particle states are

$$|i,i\rangle = \frac{1}{\sqrt{2}} a_i^\dagger a_i^\dagger |\rangle$$

$$|i,j\rangle = a_i^\dagger a_j^\dagger |\rangle \qquad (i \neq j) \tag{13.22}$$

Section 13.1 Nonrelativistic Identical-Particle Systems   441

If $i \neq j$, the wave function defined as in equation (13.16) is

$$\psi(\mathbf{r}_1, \mathbf{r}_2) = \frac{[\psi_i(\mathbf{r}_1)\psi_j(\mathbf{r}_2) + \psi_i(\mathbf{r}_2)\psi_j(\mathbf{r}_1)]}{\sqrt{2}} \tag{13.23}$$

while if $i = j$

$$\psi(\mathbf{r}_1, \mathbf{r}_2) = \psi_i(\mathbf{r}_1)\psi_i(\mathbf{r}_2) \tag{13.24}$$

Hamiltonian

In terms of the field the free Hamiltonian (13.7) is

$$H = \frac{1}{2m}\int \mathbf{p}' \cdot \mathbf{p}\, a^\dagger(\mathbf{p})\, a(\mathbf{p})\, \delta_3(\mathbf{p}' - \mathbf{p})\, d^3p\, d^3p'$$

$$= \frac{1}{(2\pi)^3}\frac{1}{2m}\int \left[\int \nabla a^\dagger(\mathbf{p})e^{-i\mathbf{p}\cdot\mathbf{r}}d^3p\right]\cdot\left[\int \nabla a(\mathbf{p}')e^{i\mathbf{p}'\cdot\mathbf{r}}d^3p'\right]d^3r \tag{13.25}$$

$$= \frac{1}{2m}\int \nabla \Psi^\dagger(\mathbf{r}) \cdot \nabla \Psi(\mathbf{r})d^3r$$

If the particles are subject to an external potential $v_1(r)$, the Hamiltonian has another term

$$V_1 = \int \Psi^\dagger(\mathbf{r})v_1(\mathbf{r})\Psi(\mathbf{r})d^3r \tag{13.26}$$

More interesting is an interaction between pairs of particles, like a repulsive electrostatic potential energy. In the usual wave-function language, the wave function for the Hamiltonian acting on a two-particle state is

$$\langle \mathbf{r}_1, \mathbf{r}_2 | H | \psi \rangle = \left[-\frac{1}{2m}\nabla_1^2 - \frac{1}{2m}\nabla_2^2 + v_2(\mathbf{r}_1, \mathbf{r}_2)\right]\psi(\mathbf{r}_1, \mathbf{r}_2) \tag{13.27}$$

The potential function $v_2(\mathbf{r}_1, \mathbf{r}_2)$ must be real and symmetric in its two arguments. Usually it is a function of $|\mathbf{r}_1 - \mathbf{r}_2|$. The quantum field provides a simple way to write the contribution of this potential to the Hamiltonian in any state. You can check that

$$H = -\frac{1}{2m}\int \nabla \Psi^\dagger(\mathbf{r}) \cdot \nabla \Psi(\mathbf{r})d^3r$$

$$+ \frac{1}{2}\int v(\mathbf{r}_1, \mathbf{r}_2)\Psi^\dagger(\mathbf{r}_1)\Psi^\dagger(\mathbf{r}_2)\Psi(\mathbf{r}_1)\Psi(\mathbf{r}_2)d^3r_1 d^3r_2 \tag{13.28}$$

Heisenberg Picture

Remember that in the Heisenberg picture, the time dependence of the observables, rather than their expectation values, is the same as in classical mechanics. The electromagnetic field was quantized first in the Heisenberg picture for this reason. Here, for free particles the time dependence in the Heisenberg picture is elementary:

$$a(\mathbf{p}, t) = a(\mathbf{p})e^{-i\omega t} \tag{13.29}$$

$$\Psi(\mathbf{r}, t) = \frac{1}{(2pi)^{3/2}}\int e^{i\mathbf{p}\cdot\mathbf{r}}a(\mathbf{p}, t)d^3p = \frac{1}{(2pi)^{3/2}}\int e^{i\mathbf{p}\cdot\mathbf{r}}a(\mathbf{p})e^{-i\omega t}d^3p \tag{13.30}$$

442 Chapter 13 Identical Particles

The field satisfies an equation of motion:

$$i\frac{\partial \Psi(\mathbf{r},t)}{\partial t} = \frac{1}{(2pi)^{3/2}} \int \frac{\mathbf{p}^2}{2m} e^{i\mathbf{p}\cdot\mathbf{r}} a(\mathbf{p}) e^{-i\omega t} d^3p = -\frac{\nabla^2}{2m}\Psi(\mathbf{r},t) \qquad (13.31)$$

When there is a potential, the equation of motion for the field can be found using the general formula (2.147):

$$i\frac{\partial \Psi(\mathbf{r},t)}{\partial t} = [\Psi(\mathbf{r},t), H] \qquad (13.32)$$

I leave it as an exercise to show that

$$i\frac{\partial \Psi(\mathbf{r},t)}{\partial t} = -\frac{\nabla^2}{2m}\Psi(\mathbf{r},t) + v_1(\mathbf{r})\Psi(\mathbf{r},t) + \int v_2(\mathbf{r}',\mathbf{r})\Psi^\dagger(\mathbf{r}',t)\Psi(\mathbf{r}',t)\Psi(\mathbf{r},t)d^3r' \qquad (13.33)$$

Because the operator $\Psi(\mathbf{r},t)$ satisfies an equation of motion similar to the one for the one-particle wave function, introducing the field in this way is often (but confusingly) called **second quantization**.

Spin

The generalization to particles with spin is straightforward. The one-particle states are $|\mathbf{p},s\rangle = a_s^\dagger(\mathbf{p})|\rangle$, and they commute according to

$$\left[a_s(\mathbf{p}), a_{s'}^\dagger(\mathbf{p}')\right] = \delta_3(\mathbf{p}'-\mathbf{p})\delta_{s's} \qquad (13.34a)$$

$$[a_s(\mathbf{p}), a_{s'}(\mathbf{p}')] = 0 \qquad (13.34b)$$

$$\left[a_s^\dagger(\mathbf{p}), a_{s'}^\dagger(\mathbf{p}')\right] = 0 \qquad (13.34c)$$

It is possible to invent a field $\Psi_s(\mathbf{r})$ for each value of the spin index. These commute like equations (13.12), but with the index added as in (13.34):

$$\left[\Psi_s(\mathbf{r}), \Psi_{s'}^\dagger(\mathbf{r}')\right] = \delta_3(\mathbf{r}'-\mathbf{r})\delta_{s's} \qquad (13.35a)$$

$$[\Psi_s(\mathbf{r}), \Psi_{s'}(\mathbf{r}')] = 0 \qquad (13.35b)$$

$$\left[\Psi_s^\dagger(\mathbf{r}), \Psi_{s'}^\dagger(\mathbf{r}')\right] = 0 \qquad (13.35c)$$

But usually we combine the wave functions into a vector or spinor function with several rows. A basis with definite spin and momentum are wave functions like

$$\psi_s(\mathbf{r}) = V_s(\mathbf{p})e^{i\mathbf{p}\cdot\mathbf{r}} \qquad (13.36)$$

where $V(\mathbf{p})$ is some constant vector, independent of $\mathbf{r}$. Then $\psi_s(\mathbf{r})$ is not a single function, but a vector:

$$\Psi_s(\mathbf{r}) = \frac{1}{(2pi)^{3/2}} \int V_s(\mathbf{p}) e^{i\mathbf{p}\cdot\mathbf{r}} a_s(\mathbf{p}) d^3p \qquad (13.37)$$

In simple cases equations that look like (13.35) will hold also where s is a row or column index instead of a spin index, but you have to be careful.[2]

---
[2] Problem 11.1 part (b) is a counterexample.

Section 13.1 Nonrelativistic Identical-Particle Systems   443

13.1.2 Creation and Annihilation Operators for Fermions

The many-particle states constructed in Section 13.1.1 describe a collection of bosons. You cannot use them for fermions, like electrons in a atom, or neutrons and protons in a nucleus. For fermions, we can still invent a creation operator to construct a one-particle state from the vacuum state:

$$|\mathbf{p}\rangle = b^\dagger(\mathbf{p})|\rangle \qquad (13.38)$$

but the operators have to satisfy *anticommutation* relations. Define the anticommutator symbol

$$\{A, B\} = AB + BA \qquad (13.39)$$

and then, instead of equations (13.2) set

$$\{b(\mathbf{p}), b^\dagger(\mathbf{p}')\} = \delta_3(\mathbf{p}' - \mathbf{p}) \qquad (13.40a)$$

$$\{b(\mathbf{p}), b(\mathbf{p}')\} = 0 \qquad (13.40b)$$

$$\{b^\dagger(\mathbf{p}), b^\dagger(\mathbf{p}')\} = 0 \qquad (13.40c)$$

The number operator and the free Hamiltonian have the same form as for bosons:

$$N = \int b^\dagger(\mathbf{p}) b(\mathbf{p}) d^3 p \quad \text{and} \quad H = \int \omega(\mathbf{p}) b^\dagger(\mathbf{p}) b(\mathbf{p}) d^3 p \qquad (13.41)$$

but the two-body states are normalized to

$$\langle \mathbf{p}_1, \mathbf{p}_2 | \mathbf{p}_3, \mathbf{p}_4 \rangle = \delta_3(\mathbf{p}_1 - \mathbf{p}_3)\delta_3(\mathbf{p}_2 - \mathbf{p}_4) - \delta_3(\mathbf{p}_1 - \mathbf{p}_4)\delta_3(\mathbf{p}_2 - \mathbf{p}_3) \qquad (13.42)$$

The fields are defined by equation (13.9), with $a(\mathbf{p})$ replaced by $b(\mathbf{p})$. They satisfy

$$\{\Psi(\mathbf{r}), \Psi^\dagger(\mathbf{r}')\} = \delta_3(\mathbf{r}' - \mathbf{r}) \qquad (13.43a)$$

$$\{\Psi(\mathbf{r}), \Psi(\mathbf{r}')\} = 0 \qquad (13.43b)$$

$$\{\Psi^\dagger(\mathbf{r}), \Psi^\dagger(\mathbf{r}')\} = 0 \qquad (13.43c)$$

Here too normalized two-particle states can be constructed:

$$|\psi\rangle = \frac{1}{\sqrt{2}} \int \psi(\mathbf{r}_1, \mathbf{r}_2) \Psi^\dagger(\mathbf{r}_1) \Psi^\dagger(\mathbf{r}_2) |\rangle \, d^3 r_1 \, d^3 r_2 \qquad (13.44)$$

but now the wave function has to be antisymmetric in its argument. If $\psi(\mathbf{r}_1, \mathbf{r}_2)$ is symmetric, it is easy to see that the state is zero.

Construct a basis starting from discrete one-particle states of these fermions. Let $|\psi_i\rangle$ be an orthonormal basis for the single-particle states, and

$$b_i = \int \psi_i^*(\mathbf{r}) \Psi(\mathbf{r}) d^3 r \qquad (13.45)$$

Then

$$\{b_i, b_j^\dagger\} = \iint \psi_i^*(\mathbf{r}') \psi_j(\mathbf{r}) \{\Psi(\mathbf{r}'), \Psi^\dagger(\mathbf{r})\} d^3 r' \, d^3 r = \int \psi_i^*(\mathbf{r}') \psi_j(\mathbf{r}) d^3 r = \delta_{ij}$$
$$(13.46)$$

444 Chapter 13 Identical Particles

and
$$\left\{b_i^\dagger, b_j^\dagger\right\} = \left\{b_i, b_j\right\} = 0 \tag{13.47}$$

Instead of equation (13.21)
$$\left\langle \left| b_i b_j b_i^\dagger b_j^\dagger \right| \right\rangle = 1 - \delta_{ij} \tag{13.48}$$

If $i \neq j$, the normalized two-particle states are
$$|i,j\rangle = b_i^\dagger b_j^\dagger \qquad (i \neq j) \tag{13.49}$$

and there is an antisymmetric two-particle wave function:
$$\psi(\mathbf{r}_1, \mathbf{r}_2) = \frac{[\psi_i(\mathbf{r}_1)\psi_j(\mathbf{r}_2) - \psi_i(\mathbf{r}_2)\psi_j(\mathbf{r}_1)]}{\sqrt{2}} \tag{13.50}$$

The two-particle states with $i = j$ do not exist, since from equation (13.47)
$$b_i^\dagger b_i^\dagger = 0 \tag{13.51}$$

The occupation number of a discrete quantum number is restricted to be zero or one—this is the Pauli exclusion principle.

Finally, a spin index can be added to the quantum fields for fermions just as for bosons.

13.2 ELEMENTARY APPLICATIONS

13.2.1 Ideal Gas Distributions

Bosons

Consider a collection of bosons in a large volume, in thermal equilibrium with its surroundings. It is convenient to put them in a cubic box of length L, with periodic boundary conditions. Each one-particle state has some energy ϵ_i. The states of the whole system are characterized by a collection of occupation numbers n_i, and have energies
$$E_{(n)} = \sum_i n_i \epsilon_i \tag{13.52}$$

The subscript (n) means any collection of nonnegative integers n_i.

According to statistical mechanics, for a fixed number of particles in thermal equilibrium, the relative population of any two states is $\exp[\beta(E_{(n_1)} - E_{(n_2)})]$, where $\beta = kT$. If our volume is part of a much larger volume, the exact total number of particles will be allowed to vary, and the probability to be in a state (n) will be
$$P_{(n)} = \frac{1}{Z} e^{-\alpha N_{(n)} - \beta E_{(n)}} = \frac{1}{Z} e^{-\alpha \sum_i n_i - \beta \sum_i n_i \epsilon_i} \tag{13.53}$$

Since the sum of all the probabilities is
$$\sum_{(n)} P_{(n)} = 1 \tag{13.54}$$

Section 13.2 Elementary Applications 445

Z is the grand canonical partition function

$$Z = \sum_{(n)} e^{-\alpha \Sigma_i n_i - \beta \Sigma_i n_i \epsilon_i} \tag{13.55}$$

If $|\psi_{(n)}\rangle$ is the state with n_i particles in the one-particle state $|\psi_i\rangle$, then

$$\rho_{(n)} = \langle \psi_{(n)} | \rho | \psi_{(n)} \rangle \tag{13.56}$$

where

$$\rho = \frac{1}{Z} e^{-\alpha N - \beta H} \tag{13.57}$$

is the density matrix introduced in Section 6.2. The average total number of particles is determined by α. The operators N and H can be written in terms of the occupation numbers of the levels:

$$N = \sum_i a_i^\dagger a_i \quad \text{and} \quad H = \sum_i \epsilon_i a_i^\dagger a_i \tag{13.58}$$

For any observable, the average value for a measurement of A over the ensemble is

$$\langle A \rangle = \sum_{(n)} \langle \psi_{(n)} | A | \psi_{(n)} \rangle \rho_{(n)} = \text{Tr } A\rho \tag{13.59}$$

In particular, the average number of particles in level i is

$$\langle n_i \rangle = \text{Tr}\left(\rho a_i^\dagger a_i\right) = \frac{1}{Z}\text{Tr}\left(e^{-\alpha N - \beta H} a_i^\dagger a_i\right) \tag{13.60}$$

Now use the Baker-Hausdorff lemma from Problem 2.10:

$$e^{\theta B} A e^{-\theta B} = A + \theta[B, A] + \frac{\theta^2}{2}[B, [B, A]] + \cdots \tag{13.61}$$

with

$$A = a_i^\dagger \quad \text{and} \quad B = -\alpha N - \beta H \tag{13.62}$$

The commutator is[3]

$$[B, A] = \left[-\alpha N - \beta H, a_i^\dagger\right] = \sum_j \left[-\alpha n_j - \beta \epsilon_j n_j, a_i^\dagger\right] = -(\alpha + \epsilon_i \beta) a_i^\dagger \tag{13.63}$$

and

$$e^{-\alpha N - \beta H} a_i^\dagger e^{\alpha N + \beta H} = e^{-\alpha - \beta \epsilon_i} a_i^\dagger \tag{13.64}$$

Since Tr AB = Tr BA, the trace (13.60) can be written

$$\langle n_i \rangle = \frac{1}{Z} e^{-\alpha - \beta \epsilon_i} \text{Tr}\left(a_i^\dagger e^{-\alpha N - \beta H} a_i\right) = \frac{1}{Z} e^{-\alpha - \beta \epsilon_i} \text{Tr}\left(e^{-\alpha N - \beta H} a_i a_i^\dagger\right) \tag{13.65}$$

---

[3] $\left[n_j, a_i^\dagger\right] = a_i^\dagger \delta_{ij}$

Finally, use $\left[a_i, a_i^\dagger\right] = 1$ to get

$$\langle n_i \rangle = \frac{1}{Z} e^{-\alpha-\beta\epsilon_i} \text{Tr}\left(e^{-\alpha N-\beta H} a_i^\dagger a_i\right) + \frac{1}{Z} e^{-\alpha-\beta\epsilon_i} \text{Tr}\left(e^{-\alpha N-\beta H}\right) \qquad (13.66)$$
$$= e^{-\alpha-\beta\epsilon_i}\left[\langle n_i\rangle + 1\right]$$

Therefore

$$\boxed{\langle n_i \rangle = \frac{1}{e^{\alpha+\beta\epsilon_i} - 1}} \qquad (13.67)$$

This is the Bose-Einstein distribution. For a given temperature, the average total number of particles in the volume is determined indirectly by α through

$$\langle N \rangle = \sum_i \langle n_i \rangle = \sum_i \frac{1}{e^{\alpha+\beta\epsilon_i} - 1} \qquad (13.68)$$

Fermions

Most of the previous argument goes through for fermions just as for bosons. $b_i^\dagger b_i$ is the number operator for each mode, and we arrive at the analog of equation (13.65) for the average number of particles in each one-particle mode:

$$\langle n_i \rangle = \frac{1}{Z} e^{-\alpha-\beta\epsilon_i} \text{Tr}\left(b_i^\dagger e^{-\alpha N-\beta H} b_i\right) = \frac{1}{Z} e^{-\alpha-\beta\epsilon_i} \text{Tr}\left(e^{-\alpha N-\beta H} b_i b_i^\dagger\right) \qquad (13.69)$$

But now b_i and $b_i^\dagger$ anticommute, so

$$\langle n_i \rangle = -\frac{1}{Z} e^{-\alpha-\beta\epsilon_i} \text{Tr}\left(e^{-\alpha N-\beta H} b_i^\dagger b_i\right) + \frac{1}{Z} e^{-\alpha-\beta\epsilon_i} \text{Tr}\left(e^{-\alpha N-\beta H}\right) \qquad (13.70)$$
$$= e^{-\alpha-\beta\epsilon_i}\left(-\langle n_i\rangle + 1\right)$$

Therefore

$$\boxed{\langle n_i \rangle = \frac{1}{e^{\alpha+\beta\epsilon_i} + 1}} \qquad (13.71)$$

This is the Fermi-Dirac distribution.

13.2.2 Ideal Electron Gas

It is customary to rewrite equations (13.67) and (13.71) with $\alpha = -\mu\beta$:

$$\langle n_i \rangle = \frac{1}{e^{\beta(\epsilon_i-\mu)} \mp 1} \qquad (13.72)$$

so that μ is what is usually called the **chemical potential**. Like α, μ is a function $\mu(kT)$ of the temperature or β. In the zero temperature limit, $\beta \to \infty$, the Fermi-Dirac distribution approaches zero for $\epsilon_i > \mu(0)$ and approaches 1 for $\epsilon_i < \mu(0)$. All the states with energies below $\mu(0)$ are filled, and all those above are empty. $\mu(0)$ is the **Fermi energy** E_F, and the magnitude of the momentum of fermions with this energy is called the **Fermi momentum** p_F. The ground state for a gas of N electrons is:

$$|\psi_o\rangle = \prod_i b_i^\dagger b_i \Theta(E_F - \epsilon_i) |\rangle = \prod_i b_i^\dagger b_i \Theta(p_F - |\mathbf{p}_i|) |\rangle \tag{13.73}$$

What is the Fermi momentum for a gas of nonrelativistic electrons in a volume $V = L^3$? Each possible state can be labeled by the eigenvalue of s_z (or s_x or s_y) and a vector $\mathbf{n}$ whose components are integers. The momentum and energy of these states depend on $n = |\mathbf{n}|$ but not on the spin index:

$$\mathbf{p_n} = \frac{2\pi}{L}\mathbf{n} \quad \text{and} \quad E_\mathbf{n} = \frac{\mathbf{p}^2}{2m} = \frac{4\pi^2 n^2}{2mL^2} \tag{13.74}$$

Thus

$$N = 2\sum_\mathbf{n} \Theta\left(\frac{Lp_F}{2\pi} - n\right) \tag{13.75}$$

N is supposed to be very large, so replace the sums by integrals. Then N is proportional to the volume of a sphere:

$$N = 2\int_{-\infty}^\infty dn_x \int_{-\infty}^\infty dn_y \int_{-\infty}^\infty dn_z \Theta\left(\frac{Lp_F}{2\pi} - n\right) = \frac{1}{3\pi^2} p_F^3 V \tag{13.76}$$

The intrinsic properties of this electron gas depend only on the density $\rho = N/V$, or equivalently on the Fermi momentum:

$$\boxed{p_F = \left(3\pi^2 \frac{N}{V}\right)^{\frac{1}{3}} = \left(3\pi^2 \rho\right)^{\frac{1}{3}}} \tag{13.77}$$

If the gas is nonrelativistic the Fermi energy is

$$E_F = \frac{1}{2m}\left(3\pi^2 \rho\right)^{\frac{2}{3}} \tag{13.78}$$

Degeneracy Pressure

Now you can understand in a qualitative way why ordinary solids do not collapse. The total energy of a cold gas of free electrons is

$$E = 8\pi \int \frac{p^2}{2m} \Theta\left(p_F - \frac{2\pi n}{L}\right) n^2 dn = \frac{V}{10m\pi^2} p_F^5 \tag{13.79}$$

and the energy density is

$$e = \frac{E}{V} = \frac{1}{10m\pi^2} p_F^5 = \frac{(3\pi^2)^{5/3}}{10m\pi^2} \rho^{5/3} \tag{13.80}$$

For fixed N the total energy $E = eV$ is proportional to $V^{-2/3}$. It increases as the volume goes down. You have to add energy to compress a Fermi gas.

448 Chapter 13 Identical Particles

The pressure is

$$P = -\frac{\partial E(V,N)}{\partial V} = -\frac{\partial (eV)}{\partial V} = -e - V\frac{de}{d\rho}\frac{\partial \rho}{\partial V} = \rho\frac{de}{d\rho} - e \qquad (13.81)$$

which for our nonrelativistic electron gas is

$$P = \frac{5}{3}e - e = \frac{2}{3}e = \frac{p_F^5}{15\pi^2 m} \qquad (13.82)$$

The exclusion principle prevents solids from collapsing. But this "degeneracy pressure" is not a consequence of any repulsive force between the electrons or atoms.

This pressure can be very large—it requires a huge amount of energy to compress most solid metals even a little bit. How big is it in a typical metal? Metallic copper, for example, has atomic weight 63.5 and a density, at ordinary temperatures, about 8.92 g/cm. Only one valence electron per copper atom belongs to the degenerate Fermi gas—the others are more tightly bound to individual atoms. For copper, the electron density is (N_A is Avogadro's number)

$$\rho = 8.92\, N_A/63.5\,\text{cm}^{-3} \approx 6.50 \times 10^8 \text{eV}^3 \qquad (13.83)$$

From equation (13.78) the Fermi energy is $E_F \approx 7.03\,\text{eV}$ and the energy density is $e \approx 2.740 \times 10^9\,\text{eV}^4$. The pressure is

$$P = 1.827 \times 10^9 \text{eV}^4 = 3.81 \times 10^{10} \text{pascal} = 3.78 \times 10^5 \text{atmospheres} \qquad (13.84)$$

You can check that the zero temperature approximation was consistent: The Fermi energy in degrees is the **Fermi temperature** T_F. The electron distribution is almost flat to the extent $T \ll T_F$.[4] For copper, $T_F = 81,500$ deg, so at room temperature, or indeed any temperature below the metal's melting point, the degenerate gas is a good approximation.

You can also check that is was consistent to treat all the valence electrons nonrelativistically. The speed of the most energetic electron is $v = p_F/m \approx .0052\,c$.

13.2.3 Collapsed Stars

One of the most spectacular consequences of the exclusion principle for many-fermion systems can be found in the history of a star that has spent most of its hydrogen fuel.

White Dwarf Stars

By the time a star like our sun has used up all its hydrogen, most of its protons have been fused to the stable isotope ^{4}He. Then the thermonuclear reactions cease, the star cools down, the thermal pressure is no longer enough to counter the gravitational pressure, and the star begins to collapse. The collapse itself heats up the star again, fusing the lighter elements into heavier and heavier ones, until it consists only of the most stable nuclei, like iron. At that point no more nuclear reactions are

---
[4]The distribution is significantly different from the zero temperature distribution only in the region $|E - E_F| \ll kT$. The fraction of particles with energies greater than the Fermi energy is therefore of the order kT/E_F.

possible, and the gravitational collapse takes over once more. Only the degeneracy pressure of the electrons stops the collapse. The star becomes one big solid.

Let R be the radius of a star with mass M at some instant. The degeneracy pressure is given by equation (13.82). The number of nucleons (not nuclei) is M/m_N, where m_N is the nucleon mass. The number of electrons is $M/\mu m_N$, where μ is the number of nucleons per electron. For iron, this number is 56/26, and for most nuclei heavier than hydrogen, $\mu \approx 2$. The degeneracy pressure is

$$P = \frac{1}{5mR^5}(3\pi^2)^{2/3}\left(\frac{3M}{4\pi\mu m_N}\right)^{5/3} \tag{13.85}$$

At ordinary densities and radii P is small compared to the inward gravitational pressure of the star. To keep the calculations simple, model the star as a sphere of uniform density (correct only qualitatively, of course). The gravitational potential energy is

$$W = -\frac{3}{5}GM^2\left(\frac{4\pi}{3V}\right)^{1/3} \tag{13.86}$$

and the gravitational pressure is

$$P_G = \frac{dW}{dV} = \frac{3}{4\pi}\frac{GM^2}{5R^4} \tag{13.87}$$

While a real star does not have a strictly uniform density, the form of the gravitational pressure should be given by (13.87) up to a dimensionless constant of order unity.

As the star collapses, the inward gravitational pressure increases as the inverse fourth power of the radius, but the degeneracy pressure increases even faster, as the inverse fifth power. So at some radius the two effects balance, and the collapse ceases. This radius, which is also the radius that minimizes the total energy, is

$$R = \frac{1}{GmM_N^2}\left(\frac{9\pi}{4}\right)^{\frac{2}{3}}\left(\frac{1}{\mu}\right)^{\frac{5}{3}}\left(\frac{M_N}{M}\right)^{\frac{1}{3}} \tag{13.88}$$

Note: Let us evaluate R for a star whose mass is XM_S, where M_S is the mass of the sun, about 1.99×10^{33} g; and take $\mu \approx 2$. Then

$$R = 7.57 \times 10^{27}\left(\frac{M_N}{M}\right)^{\frac{1}{3}} \text{ cm} = \frac{7.15 \times 10^8 \text{ cm}}{X^{\frac{1}{3}}} \tag{13.89}$$

R comes out comparable to the radius of the earth for stars whose mass is of the order of sun's. The mass density such a collapsed star, called a **white dwarf star**, is

$$\rho = \frac{3M}{4\pi R^3} \approx 1.3 \times 10^7 X^2 \frac{\text{g}}{\text{cm}^3} \tag{13.90}$$

vastly greater than the density of ordinary matter in the earth or the sun. For comparison, the density of the earth and the present sun are

$$\rho_S = \frac{3M_S}{4\pi R_S^3} \approx 1.4 \frac{\text{g}}{\text{cm}^3} \qquad \rho_E = \frac{3M_E}{4\pi R_E^3} \approx 5.5 \frac{\text{g}}{\text{cm}^3} \tag{13.91}$$

450 Chapter 13 Identical Particles

Was it really consistent to use the zero-temperature limit for the electron gas in a white dwarf star? The Fermi energy is

$$E_F = \frac{1}{2m}\left(3\pi^2\frac{N}{V}\right)^{\frac{2}{3}} = 2 \times 10^5 X^{\frac{4}{3}} \text{eV} \qquad (13.92)$$

corresponding to a Fermi temperature about 2×10^9 degrees. Since the typical temperature of a white dwarf star is 10^7 degrees or less, the degenerate Fermi gas is a good approximation.

This analysis contains many approximations, so it can be only qualitative. It omits, for example, the effects on the calculation of the Coulomb repulsion between electrons, the density dependence on the radius, the star's rotation, its temperature, and its chemical composition.

The treatment above neglects relativity also. For stars of moderate mass, the electrons at the Fermi surface have nonrelativistic kinetic energies, speeds small compared to c. For more massive stars, the energies become relativistic and we need to redo the whole calculation replacing $p^2/2m$ with $\sqrt{p^2+m^2}$. In fact, the Fermi energy in equation (13.92) for a collapsed star with even the mass of the sun is a good fraction of the electron rest mass, and the result (13.89) can be only qualitative.

Relativity becomes important when the most energetic electrons, those near the Fermi surface, have momenta comparable to their mass, that is [from equation (13.77)], when

$$p_F^2 = \left(3\pi^2\frac{N}{V}\right)^{\frac{2}{3}} \approx m^2 \qquad (13.93)$$

or

$$\rho \approx \rho_c = \frac{M_N \mu m^3}{3\pi^2} = 1.96 \times 10^6 \frac{\text{g}}{\text{cm}^3} \qquad (13.94)$$

where ρ here is the mass density of the star. When the density of a star is of the order of ρ_c a relativistic computation is required.

According to equation (13.89), when the sun collapses its equilibrium radius will be about 7.15×10^8 cm. This is somewhat higher than the critical density, and you should suspect that there are relativistic corrections. For stars of larger mass, it is quite important to take relativity into account.

More Massive Stars

A relativistic expression for the Fermi energy is

$$E_F = \sqrt{p_F^2 + m^2} \qquad (13.95)$$

From now on I will include the rest mass in the energy, so each level has the energy

$$E_\mathbf{n} = \sqrt{\frac{n^2\pi^2}{L^2} + m^2} \qquad (13.96)$$

Instead of equation (13.79), the total energy of the electrons is

$$E = \frac{Vm^4}{\pi^2}F(x_F) \qquad (13.97)$$

where
$$F(x) = \int_0^x t^2\sqrt{1+t^2}\,dt \tag{13.98}$$

and $x_F = p_F/m$. Using

$$\frac{dp_F}{dV} = -\frac{1}{3V}\left(\frac{3\pi^2 M}{m_N\mu V}\right)^{\frac{1}{3}} = -\frac{1}{3V}p_F \tag{13.99}$$

compute

$$\frac{dF(x_F)}{dV} = \frac{dF(x_F)}{dx_F}\frac{dx_F}{dV} = -\frac{1}{3V}x_F^3\sqrt{1+x_F^2} \tag{13.100}$$

and

$$\frac{dE}{dV} = \frac{m^4}{\pi^2}\left[F(x_F) - \frac{1}{3}x_F^3\sqrt{1+x_F^2}\right] \tag{13.101}$$

For small p_F,

$$\begin{aligned}\frac{dE}{dV} &= \frac{m^4}{\pi^2}\left[\frac{x_F^3}{3} + \frac{x_F^5}{10} + \cdots - \frac{1}{3}x_F^3 - \frac{1}{6}x_F^5 - \cdots\right] = -\frac{1}{15}\frac{m^4}{\pi^2}\frac{p_F^5}{m^5} \\ &= -\frac{1}{5mR^5}(3\pi^2)^{\frac{2}{3}}\left(\frac{3M}{4\pi m_N\mu}\right)^{\frac{5}{3}}\end{aligned} \tag{13.102}$$

in agreement with equation (13.85).

At the other extreme, for large p_F,

$$\begin{aligned}\frac{dE}{dV} &= \frac{m^4}{\pi^2}\left[\frac{x_F^4}{4} + \frac{x_F^2}{4} + \cdots - \frac{x_F^4}{3} - \frac{x_F^2}{3} - \cdots\right] \\ &= -\frac{m^4}{12\pi^2}[x_F^4 + x_F^2] = -\frac{1}{12\pi^2}\left[\frac{1}{R^4}\left(\frac{9\pi M}{4m_N\mu}\right)^{\frac{4}{3}} + m^2\frac{1}{R^2}\left(\frac{9\pi M}{4m_N\mu}\right)^{\frac{2}{3}}\right]\end{aligned} \tag{13.103}$$

which should equal the gravitational pressure

$$P_G = \frac{dW}{dV} = \frac{3}{4\pi}\frac{GM^2}{5R^4} \tag{13.87}$$

But now the leading term in the degeneracy pressure goes only as the inverse fourth power of the radius also. There is no solution unless

$$\frac{3GM^2}{20\pi} \leq \frac{1}{12\pi^2}\left(\frac{9\pi M}{4m_N\mu}\right)^{\frac{4}{3}} \tag{13.104}$$

or[5]

$$M \le \left(\frac{5}{9\pi G}\right)^{\frac{3}{2}} \left(\frac{9\pi}{4m_N\mu}\right)^2 = 5^{\frac{3}{2}} \frac{3\sqrt{\pi}}{16} \frac{1}{G^{\frac{3}{2}}} \left(\frac{1}{m_N\mu}\right)^2$$

$$= \sqrt{\frac{1125\pi}{256}} \left(\frac{1}{G}\right)^{\frac{3}{2}} \left(\frac{1}{\mu m_N}\right)^2$$

$$\approx .929\ldots m_N \left(\frac{M_{Pl}}{M_N}\right)^3 = 3.4 \times 10^{33} \text{g} \quad (13.105)$$

This mass is a universal natural constant that depends only on the nucleon mass, and the very large—and very mysterious—ratio of the Planck mass to the nucleon mass. In the extreme relativistic limit, the gravitational pressure of a star with a mass greater than this critical value will always exceed the degeneracy pressure. It will not become a white dwarf star because the fermion degeneracy pressure is not enough to resist the pressure of gravity. A more careful computation, taking into account the star's structure, and so forth, gives the critical mass, called the **Chandrasekhar limit** [1, 2], as about 2.4×10^{33} g. or 1.4 times the mass of the sun.

Neutron Stars

A star whose mass exceeds the limit (13.105) may not collapse forever. As it begins to get smaller than the typical white dwarf star radius, the temperature again increases, and nuclear reactions are possible. These are the stars that become supernovae, cosmic explosions so bright that they can be seen here even if they happen in another galaxy. After burning the original hydrogen at a more or less steady rate for billions of years, these stars collapse in a few seconds. It is in supernovae that the heaviest nuclei, gold, lead, uranium, are born and thrust out into interstellar space, perhaps to be reused in a distant second-generation solar system like our own. These implosions also produce neutrinos in vast numbers, and presumably are responsible for some of the very high energy cosmic rays that reach the earth. Such neutrinos were actually detected in the supernova explosion in 1987.

A supernova explosion can leave a substantial fraction of the original star in a highly compressed remnant. The star is so compressed that the energy is actually lowered by inverse beta decay, $p + e^- \to n + \nu$. The neutrinos escape, leaving a star made up mostly of cold, degenerate neutrons.

If the mass is small enough, the star's collapse is halted by the degeneracy pressure of the Fermi gas of neutrons. You can compute the radius R_N of such a neutron star directly from the radius R of a white dwarf star given equation (13.88), replacing the electron mass m with the neutron mass M_N, and μ by 1. The radius of a neutron star comes out to be

$$R_N = \frac{1}{GM_N^3} \left(\frac{9\pi}{4}\right)^{\frac{2}{3}} \left(\frac{M_N}{M}\right)^{\frac{1}{3}} = \frac{m}{M_N} \mu^{\frac{5}{3}} R \approx \frac{12\,\text{km}}{X^{\frac{1}{3}}} \quad (13.106)$$

---
[5]The **Planck mass** is $M_{Pl} = 1/\sqrt{G} = 1.221 \times 10^{28}$ eV.

The Fermi momentum for a neutron star at equilibrium, calculated nonrelativistically, is

$$p_{F,n} = \frac{M_N}{m\mu^{\frac{4}{3}}} p_F \qquad (13.107)$$

where p_F is the Fermi momentum of a white dwarf star at equilibrium, which according to equation (13.93) is of the order m. Therefore the Fermi momentum for a stable neutron star is of the order of the neutron mass, and the nonrelativistic calculation cannot be quantitatively correct. Furthermore, the gravitational potential at the surface is not small, and general relativity is needed to model the star correctly. But this calculation gives the order of magnitude of a neutron star's size correctly.

A neutron star with mass greater than about 0.7 times the mass of our sun is not stable [3]. Neutron stars, presumably observed as pulsars, are only a few kilometers in diameter, and they can exist only in a narrow mass range near m_{Pl}^3/M_N^2. Bigger neutron stars will continue to collapse forever, since the gravitational pressure always exceeds the neutron degeneracy pressure, and becomes a singularity in the space-time continuum, a black hole.

13.3 RELATIVISTIC SPINLESS PARTICLES

13.3.1 The Neutral Scalar Field

The single-particle wave functions in Chapter 12 are an improvement on the nonrelativistic equations at low energies, but they are fundamentally flawed because they have negative energy states. The Klein-Gordon equation and the Dirac equation were one-particle theories, and relativistic systems cannot really conserve the number of particles. The method of creation and annihilation operators should really have been used from the outset.

Relativity requires that interactions between particles be *local*, taking place at one space-time point. Otherwise there will be violation of the relativistic principle of causality in some reference frame. The only way anyone knows how to do this is to use relativistic quantum fields, variations on the idea developed in Section 13.1. We have already studied a fairly complicated example, photons, but I have not yet presented a complete relativistic theory of other particles like electrons.

The treatment of spin-zero particles will be discussed first. This section and the next few are of course only an introduction to the vast subject of relativistic quantum mechanics, or quantum field theory.[6]

13.3.2 The Classical Theory

In quantum mechanics we will construct generalizations of the fields introduced in Section 13.1.1 to fields that satisfy a relativistic equation of motion. The electromagnetic potential $A^\mu(x)$ is the model. Historically the Maxwell field is special: There actually exist stable, massless spin-one particles.

There is also a classical theory of these relativistic fields. Think of $\phi(x)$ as a classical field that satisfies the Klein-Gordon equation, and then turn it into a quantum observable. The quickest way to turn this classical theory into a sensible

---
[6]See, for example, references [6,7,8] in Chapter 10.

quantum theory is to use the canonical method, starting from the classical action introduced in Section 1.1.2. Start from an action that is relativistically invariant, so the theory will be guaranteed to be invariant no matter how complicated we manage to make it look.[7]

What can the action for a neutral relativistic scalar field look like? It should be a quadratic functional of $\phi(x)$ and its first derivatives, just as nonrelativistic Lagrangians depend on q and $\dot{q}$. To be a local theory, the action should depend on products of fields only at the same space-time point:

$$S = \int L\,dt = \int \mathcal{L}\,d^4x \qquad (13.108)$$

where $\mathcal{L}$ is a Lorentz scalar field, a function of the field and its first derivatives. The only possibility is

$$\mathcal{L} = \frac{1}{2}\left[\partial^\mu\phi(x)\partial_\mu\phi(x) - m^2\phi(x)^2\right] = \frac{1}{2}\left[\dot{\phi}(x)^2 - [\boldsymbol{\nabla}\phi(x)]^2 - m^2\phi(x)^2\right] \qquad (13.109)$$

The coefficients $1/2$ and m^2 are arbitrary, chosen here so that the kinetic energy and mass have the conventional forms below. The **Lagrangian density** $\mathcal{L}$ is a function of an infinite number of "coordinates," the field $\phi(\mathbf{r})$ at each point. There is a boundary condition that these classical fields vanish for large $|\mathbf{r}|$. If ϕ and its derivatives vary arbitrarily at each point, their values at the initial and final times held fixed, the change in the action is

$$\delta S = \frac{1}{2}\int\left[\frac{\partial\mathcal{L}}{\partial\phi}\delta\phi + \frac{\partial\mathcal{L}}{\partial(\partial_\mu\phi)}\delta\partial_\mu\phi\right]d^4x = \frac{1}{2}\int\left[\frac{\partial\mathcal{L}}{\partial\phi}\delta\phi - \partial_\mu\left(\frac{\partial\mathcal{L}}{\partial(\partial_\mu\phi)}\right)\delta\phi\right]d^4x \qquad (13.110)$$

Since the variations are arbitrary, the Euler-Lagrange equations follow from Hamilton's principle

$$\frac{\partial\mathcal{L}}{\partial\phi} - \partial_\mu\frac{\partial\mathcal{L}}{\partial(\partial_\mu\phi)} = 0 \qquad (13.111)$$

which here is the Klein-Gordon equation

$$\partial_\mu\partial^\mu\phi + m^2\phi = 0 \qquad (13.112)$$

This is the correct Lagrangian.

The canonical momenta too will be fields:

$$\Pi(x) = \frac{\partial\mathcal{L}}{\partial\dot{\phi}(x)} = \dot{\phi}(x) \qquad (13.113)$$

so the Hamiltonian is

$$H = \int \mathcal{H}\,d^3x \qquad (13.114)$$

where

$$\mathcal{H} = \Pi\dot{\phi} - \mathcal{L} = \frac{1}{2}\left[\dot{\phi}(x)^2 + [\boldsymbol{\nabla}\phi(x)]^2 + m^2\phi(x)^2\right] \qquad (13.115)$$

This is the classical theory of a Klein-Gordon field.

---

[7] I didn't use this method for the photon field because gauge invariance introduces some subtleties that would have taken us too far afield. You can find a Lagrangian formalism for classical electromagnetism in many books on that subject.

13.3.3 The Quantum Theory

The quantum theory is obtained from the canonical commutation rules:

$$[\phi(\mathbf{r}'), \Pi(\mathbf{r})] = \left[\phi(\mathbf{r}'), \dot\phi(\mathbf{r})\right] = i\delta_3(\mathbf{r} - \mathbf{r}') \tag{13.116}$$

Note: You might worry whether the coefficients in all these equations, obtained as the continuum limit of a theory with a finite number of degrees of freedom, are correct. They are, and the proof is that in the Heisenberg picture the time dependence of the field is correctly given using equation (13.116) in[8]

$$i\dot\phi(x) = [\phi(x), H] \tag{13.117}$$

A similar issue was discussed in Section 11.2.2.

Notice that for relativistic scalar fields the commutation relations (13.116) look quite different from the nonrelativistic ones (13.12).

The Heisenberg picture fields are obtained by making a unitary transformation on the fields at time zero. At any t,

$$[\phi(\mathbf{r}', t), \Pi(\mathbf{r}, t)] = \left[\phi(\mathbf{r}', t), \dot\phi(\mathbf{r}, t)\right] = i\delta_3(\mathbf{r} - \mathbf{r}') \tag{13.118}$$

Next make a Fourier transform of the field, in the style of equation (11.45):

$$\begin{aligned}\phi(\mathbf{r}, t) &= \frac{1}{(2\pi)^{3/2}} \int \frac{1}{\sqrt{2\omega}} \left[a(\mathbf{k}, t)e^{i\mathbf{k}\cdot\mathbf{r}} + a^\dagger(\mathbf{k}, t)e^{-i\mathbf{k}\cdot\mathbf{r}}\right] d^3k \\ &= \frac{1}{(2\pi)^{3/2}} \int \frac{1}{\sqrt{2\omega}} \left[a(\mathbf{k})e^{-ik\cdot x} + a^\dagger(\mathbf{k})e^{ik\cdot x}\right] d^3k\end{aligned} \tag{13.119}$$

where $\omega = \sqrt{k^2 + m^2}$ and $k \cdot x = \omega t - \mathbf{k} \cdot \mathbf{r}$. Then

$$\Pi(\mathbf{r}, t) = \dot\phi(\mathbf{r}, t) = -i\frac{1}{(2\pi)^{3/2}} \int \sqrt{\frac{\omega}{2}} \left[a(\mathbf{k})e^{-ik\cdot x} - a^\dagger(\mathbf{k})e^{ik\cdot x}\right] d^3k \tag{13.120}$$

Therefore,

$$\left[a(\mathbf{r}'), a^\dagger(\mathbf{r})\right] = \delta_3(\mathbf{r}' - \mathbf{r}) \tag{13.121}$$

This commutator is equivalent to (13.116). The factor $1/\sqrt{2\omega}$ in (13.119) makes the canonical commutator come out right. It also ensures that $\phi(x)$ transforms like a scalar field under Lorentz transformations. A straightforward calculation gives the Hamiltonian in terms of the occupation number operators:

$$\begin{aligned}H &= \frac{1}{2} \int \left[\dot\phi(x)^2 + (\boldsymbol{\nabla}\phi(x))^2 + m^2\phi(x)^2\right] d^3x \\ &= \frac{1}{2} \int \omega \left[a(\mathbf{k})a^\dagger(\mathbf{k}) + a^\dagger(\mathbf{k})a(\mathbf{k})\right] d^3k\end{aligned} \tag{13.122}$$

---
[8] See equation (2.147).

Again, just drop the infinite "zero-point" energy, and argue that it is not observable. The ground-state energy is another example of the ordering ambiguity in the canonical prescription, and we could just as well have quantized the classical theory in a way that makes all the creation operators come to the left of all the annihilation operators.[9] Either way, the Hamiltonian of the quantum theory is

$$H = \int \omega a^\dagger(\mathbf{k}) a(\mathbf{k}) d^3 k \tag{13.123}$$

Note on units: A factor $\sqrt{4\pi}$ that occurs in the analogous equation (11.45) and elsewhere in the expansions of the electromagnetic field is missing in the definition (13.119). When you start from the Lagrangian, it is conventional to use such "rationalized" units.

The electromagnetic field could have been defined without the 4π factor in (11.45) also, but then Coulomb's law would have an extra 4π in the denominator, as indeed it does in MKS units. By the time you get to the Hamiltonian in terms of the creation and annihilation operators, the discrepancy disappears. Compare equations (13.123) and (11.48).

13.3.4 Charged Particles

In the discussion (Section 12.2.2) of a Klein-Gordon wave function coupled to the electromagnetic field, it was important to use a *complex* wave function. A real wave function does not have a conserved current like equation (12.58). The model described by the Lagrangian density (13.109) cannot describe charged particles.

A complex scalar field is easily built out of two real fields:

$$\begin{aligned}\phi(x) &= \frac{\phi_1(x) + i\phi_2(x)}{\sqrt{2}} \\ \phi^\dagger(x) &= \frac{\phi_1(x) - i\phi_2(x)}{\sqrt{2}}\end{aligned} \tag{13.124}$$

The Lagrangian density is

$$\boxed{\begin{aligned}\mathcal{L} &= \frac{1}{2}\left[\partial_\mu \phi_1(x)\partial^\mu \phi_1(x) - m^2 \phi_1(x)^2\right] + \frac{1}{2}\left[\partial_\mu \phi_2(x)\partial^\mu \phi_2(x) - m^2 \phi_2(x)^2\right] \\ &= \partial_\mu \phi(x)^\dagger \partial^\mu \phi(x) - m^2 \phi^\dagger(x)\phi(x)\end{aligned}}$$
(13.125)

Since ϕ and $\phi^\dagger$ are independent functions of ϕ_1 and ϕ_2, the equations of motion can be obtained by varying either pair. One of the nice features of the Lagrange method is that one can choose any convenient set of independent coordinates, even combinations that are not real. So instead of

$$\delta S = \frac{\delta S}{\delta \phi_1}\delta\phi_1 + \frac{\delta S}{\delta \phi_2}\delta\phi_2 \tag{13.126}$$

---

[9] The analogous discussion for the photon operators is in Section 11.2.3.

write
$$\delta S = \frac{\delta S}{\delta \phi^\dagger} \delta \phi^\dagger + \frac{\delta S}{\delta \phi} \delta \phi \tag{13.127}$$

The Euler-Lagrange equations are
$$\partial_\mu \partial^\mu \phi + m^2 \phi = 0 \tag{13.128}$$
and
$$\partial_\mu \partial^\mu \phi^\dagger + m^2 \phi^\dagger = 0 \tag{13.129}$$

Now there is a conserved current, so electromagnetic interactions can be added with the minimal coupling prescription. The conserved current is
$$j^\mu = i \left[\phi^\dagger \partial^\mu \phi - \left(\partial^\mu \phi^\dagger\right) \phi \right] \tag{13.130}$$
and $\partial_\mu j^\mu = 0$ is a consequence of the equation of motion.

The momentum conjugate to ϕ is
$$\Pi(x) = \frac{\partial \mathcal{L}}{\partial \dot\phi(x)} = \dot\phi^\dagger \tag{13.131}$$

so the fundamental commutator is
$$\left[\phi(\mathbf{r}', t), \dot\phi^\dagger(\mathbf{r}, t) \right] = i \delta_3 (\mathbf{r} - \mathbf{r}') \tag{13.132}$$

The Fourier expansion is
$$\phi(\mathbf{r}, t) = \frac{1}{(2\pi)^{3/2}} \int \frac{1}{\sqrt{2\omega}} \left[b(\mathbf{k}) e^{-ik\cdot x} + d^\dagger(\mathbf{k}) e^{ik\cdot x} \right] d^3 k \tag{13.133}$$

Since ϕ is not Hermitean, b is not the conjugate of d. Rather
$$\phi^\dagger(\mathbf{r}, t) = \frac{1}{(2\pi)^{3/2}} \int \frac{1}{\sqrt{2\omega}} \left[d(\mathbf{k}) e^{-ik\cdot x} + b^\dagger(\mathbf{k}) e^{ik\cdot x} \right] d^3 k \tag{13.134}$$

You can check that the canonical commutator (13.132) is equivalent to
$$[b(\mathbf{k}), b^\dagger(\mathbf{k}')] = [d(\mathbf{k}), d^\dagger(\mathbf{k}')] = \delta_3(\mathbf{k} - \mathbf{k}') \tag{13.135a}$$
$$[b(\mathbf{k}), d(\mathbf{k}')] = [b(\mathbf{k}), d^\dagger(\mathbf{k}')] = 0 \tag{13.135b}$$

The model automatically contains two types of particle, those created by $d^\dagger$ and those created by $b^\dagger$. The conserved charge is
$$Q = \int j^0 d^3 x = i \int \left[\phi^\dagger \dot\phi - \dot\phi^\dagger \phi \right] d^3 x = \int \left[b^\dagger(\mathbf{k}) b(\mathbf{k}) - d^\dagger(\mathbf{k}) d(\mathbf{k}) \right] d^3 k \tag{13.136}$$

The two types of particles have opposite charges.

The Hamiltonian density is
$$\mathcal{H} = \Pi \dot\phi + \dot\phi^\dagger \Pi^\dagger - \mathcal{L} = \dot\phi^\dagger(x) \dot\phi(x) + \boldsymbol{\nabla} \phi^\dagger(x) \cdot \boldsymbol{\nabla} \phi(x) + m^2 \phi^\dagger(x) \phi(x) \tag{13.137}$$

458 Chapter 13 Identical Particles

and again a short calculation gives the Hamiltonian in terms of the occupation number operators:

$$H = \int \mathcal{H} d^3x = \int \omega \left[b^\dagger(\mathbf{k}) b(\mathbf{k}) + d^\dagger(\mathbf{k}) d(\mathbf{k}) \right] d^3k \qquad (13.138)$$

Magically, the negative energy states that occurred in the one-particle theory have disappeared. More precisely they turned into negative "charge" states. Both kinds of particle have ordinary, positive energy. Except for their charge the particles are identical: They have the same mass and spin.

The particles created by $d^\dagger(\mathbf{k})$ are the **antiparticles** of the ones created by $d^\dagger(\mathbf{k})$. There is a profound theorem that says it is impossible to create a local relativistic quantum theory without antiparticles.

Interchanging the two kinds of particles is a discrete symmetry of this system. Define $\mathcal{C}$ by

$$\mathcal{C} b^\dagger |\rangle = d^\dagger |\rangle \quad \text{and} \quad \mathcal{C} d^\dagger |\rangle = b^\dagger |\rangle \qquad (13.139)$$

and so forth. $\mathcal{C}$ is the **charge conjugation** operator, and has the same simple algebra as reflections: $\mathcal{C}^2 = I$. The eigenvalues of $\mathcal{C}$ are called the "charge parity," and can have the values ± 1. In the neutral model at the beginning of this section the particles are their own antiparticles. This can never happen for a particle with electric charge.

13.4 THE QUANTIZED DIRAC FIELD

13.4.1 The Dirac Action

The four-component Dirac wave function in Section 12.3.2 transforms under a rotation about the axis $\hat{\mathbf{n}}_i$ like

$$\psi(x) \to e^{-i\theta \Sigma_i / 2} \psi(x) \qquad (13.140)$$

while under a boost along that direction

$$\psi(x) \to e^{\omega \alpha_i / 2} \psi(x) \qquad (13.141)$$

the α_i are defined in equation (12.94). So the form $\psi^\dagger(x)\psi(x)$ is not a scalar field, but rather

$$\psi^\dagger(x)\psi(x) \to \psi^\dagger(x) e^{\omega \alpha_i} \psi(x) \qquad (13.142)$$

Notice that there is no i in the exponent. Since the matrices α_i are Hermitean, the boost matrices in the Dirac representation are not unitary. But the α_i anticommutes with the matrix β or γ_o also defined in (12.94). Therefore,

$$\psi^\dagger(x) \gamma^o \psi(x) \to \psi^\dagger(x) e^{\omega \alpha_i / 2} \gamma^o e^{\omega \alpha_i / 2} \psi(x)$$
$$= \psi^\dagger(x) \gamma^o e^{-\omega \alpha_i / 2} e^{\omega \alpha_i / 2} \psi(x) = \psi^\dagger(x) \gamma^o \psi(x) \qquad (13.143)$$

is a scalar field. The correct invariant form for the Lagrangian density is

$$\mathcal{L} = \psi^\dagger(x) \left(i \partial_o + i \boldsymbol{\alpha} \cdot \boldsymbol{\nabla} - m \gamma^o \right) \psi(x) = \psi^\dagger(x) \gamma^o \left(i \gamma^\mu \partial_\mu - m \right) \psi(x) \qquad (13.144)$$

Define
$$\bar{\psi}(x) = \psi^\dagger(x)\gamma^o \tag{13.145}$$
so that
$$\mathcal{L} = \bar{\psi}(x)\left(i\gamma^\mu \partial_\mu - m\right)\psi(x) \tag{13.146}$$

The four ψ_i and $\psi_i^\dagger$ can be varied independently. Then one of the Euler-Lagrange equations is
$$\frac{\partial \mathcal{L}}{\partial\left(\partial_\mu \bar{\psi}_i\right)} = 0 \tag{13.147}$$

and the equation of motion is
$$0 = \frac{\partial \mathcal{L}}{\partial \bar{\psi}_i} = \left[(i\gamma^\mu \partial_\mu - m)\,\psi(x)\right]_i \tag{13.148}$$

as required. The remaining Euler-Lagrange equations are equivalent. The canonical momenta are
$$\bar{\Pi}_i = \frac{\partial \mathcal{L}}{\partial\left(\partial_o \bar{\psi}_i\right)} = 0 \quad \text{and} \quad \Pi_i = \frac{\partial \mathcal{L}}{\partial\left(\partial_o \psi_i\right)} = i\sum_j \bar{\psi}_j \gamma^o_{ji} = i\psi_i^\dagger \tag{13.149}$$

since $(\gamma^o)^2 = 1$. So the Hamiltonian density is
$$\mathcal{H} = \sum_i \Pi_i \dot{\psi}_i - \mathcal{L} = \psi^\dagger\left(-i\boldsymbol{\alpha}\cdot\boldsymbol{\nabla} + m\gamma^o\right)\psi \tag{13.150}$$

There is also a conserved current
$$j^\mu = \bar{\psi}\gamma^\mu \psi \tag{13.151}$$

and therefore a conserved charge
$$Q = \int j^o(x) d^3x = \int \psi^\dagger(x)\psi(x) d^3x \tag{13.152}$$

If the quanta of the Dirac field were bosons, the fields would satisfy a relation like[10]
$$\left[\psi_i(\mathbf{r}), \psi_j^\dagger(\mathbf{r}')\right] = \delta_{ij}\delta_3\left(\mathbf{r} - \mathbf{r}'\right) \tag{13.153}$$

13.4.2 The Plane Wave Expansion

From equation (13.148) the field satisfies the free Dirac equation
$$(i\gamma^\mu \partial_\mu - m)\psi(x) = 0 \tag{12.99}$$

There are four solutions for every $\mathbf{k}$ of the form
$$\psi(x) = w_i(\mathbf{k}, t)e^{i\mathbf{k}\cdot\mathbf{r}} = w_i(\mathbf{k})e^{i(\mathbf{k}\cdot\mathbf{r} - Et)} \tag{13.154}$$

---
[10] The subscripts are the actual rows of the Dirac spinors, not a spin index like the ones in equations (13.35).

as in Section 12.3.3. Two of the solutions have $E = \omega$, and the other two have $E = -\omega$, where $\omega = +\sqrt{k^2 + m^2}$. So the Dirac field could be expanded in plane waves as

$$\psi(x) = \frac{1}{(2pi)^{3/2}} \int \frac{1}{\sqrt{2\omega}} \left[\sum_{s=1}^{2} b_s(\mathbf{k}) w_s(\mathbf{k}) e^{-i\omega t} + \sum_{s=3}^{4} d_s^\dagger(\mathbf{k}) w_s(\mathbf{k}) e^{i\omega t} \right] e^{i\mathbf{k}\cdot\mathbf{r}} d^3k \tag{13.155}$$

where $w_1(\mathbf{k})$ and $w_2(\mathbf{k})$ are the positive energy solutions, satisfying

$$[\gamma^o \omega - \boldsymbol{\gamma} \cdot \mathbf{k} - m] \, w_i(\mathbf{k}) = 0 \tag{13.156}$$

while $w_3(\mathbf{k})$ and $w_4(\mathbf{k})$ are the negative energy solutions, satisfying

$$[-\gamma^o \omega - \boldsymbol{\gamma} \cdot \mathbf{k} - m] \, w_i(\mathbf{k}) = 0 \tag{13.157}$$

Like the Klein-Gordon field, there are components of both positive and negative frequencies, requiring two kinds of particles—though for a different reason.

Here these negative solutions are awkward, since they satisfy an equation that does not look covariant. Instead, for the two positive energy solutions we write

$$u_i(\mathbf{k}) = w_i(\mathbf{k}) \tag{13.158}$$

but for the negative energy solutions

$$v_i(\mathbf{k}) = w_{i+2}(-\mathbf{k}) \tag{13.159}$$

so that

$$(\gamma \cdot k - m) \, u(\mathbf{k}) = 0 \tag{13.160a}$$

$$(\gamma \cdot k + m) \, v(\mathbf{k}) = 0 \tag{13.160b}$$

Now the plane wave expansion has the form[11]

$$\psi(x) = \frac{1}{(2pi)^{3/2}} \sum_{s=1}^{2} \int \frac{1}{\sqrt{2\omega}} \left[b_s(\mathbf{k}) u_s(\mathbf{k}) e^{-ik\cdot x} + d_s^\dagger(\mathbf{k}) v_s(\mathbf{k}) e^{ik\cdot x} \right] d^3k \tag{13.161}$$

where the normalization is chosen in analogy to equation (13.133) for a charged scalar field. I have not yet shown that $b_s(\mathbf{k})$ and $d_s^\dagger(\mathbf{k})$ are really annihilation operators for particles and creation operators for antiparticles, but it will work out correctly provided the plane wave spinors are normalized to

$$u_s^\dagger(\mathbf{k}) u_{s'}(\mathbf{k}) = v_s^\dagger(\mathbf{k}) v_{s'}(\mathbf{k}) = 2\omega \delta_{s's} \tag{13.162}$$

Equations (13.160) can be written

$$[\boldsymbol{\alpha} \cdot \mathbf{k} + \gamma^o m] \, u(\mathbf{k}) = \omega u_s(\mathbf{k}) \tag{13.163a}$$

$$[\boldsymbol{\alpha} \cdot \mathbf{k} + \gamma^o m] \, v(-\mathbf{k}) = -\omega v_s(-\mathbf{k}) \tag{13.163b}$$

---

[11] The index s is now a spin label, not a row label. For given s, u_s and v_s (but not the b_s or the $d_s^\dagger$) are four-component spinors.

Section 13.4 The Quantized Dirac Field

The spinors $u_s(\mathbf{k})$ and $v_s(-\mathbf{k})$ are eigenvectors of the same 4×4 Hermitean matrix with different eigenvalues, and so are orthogonal. For each k, there are two independent $u(\mathbf{k})$ and two independent $v(\mathbf{k})$. When $\mathbf{k} = 0$ take them to be

$$u_1(0) = \begin{pmatrix} 1 \\ 0 \\ 0 \\ 0 \end{pmatrix} \quad u_2(0) = \begin{pmatrix} 0 \\ 1 \\ 0 \\ 0 \end{pmatrix} \quad v_1(0) = \begin{pmatrix} 0 \\ 0 \\ 1 \\ 0 \end{pmatrix} \quad v_2(0) = \begin{pmatrix} 0 \\ 0 \\ 0 \\ 1 \end{pmatrix} \qquad (13.164)$$

and for other $\mathbf{k}$ apply a Lorentz boost to these four.

The Hamiltonian

Now

$$(-i\boldsymbol{\alpha}\cdot\boldsymbol{\nabla} + m\gamma^o)\,u_s(\mathbf{k})e^{i\mathbf{k}\cdot\mathbf{r}} = \omega u_s(\mathbf{k})e^{i\mathbf{k}\cdot\mathbf{r}} \qquad (13.165)$$

while

$$(-i\boldsymbol{\alpha}\cdot\boldsymbol{\nabla} + m\gamma^o)\,v_s(\mathbf{k})e^{-i\mathbf{k}\cdot\mathbf{r}} = -\omega v_s(\mathbf{k})e^{-i\mathbf{k}\cdot\mathbf{r}} \qquad (13.166)$$

so the Hamiltonian is

$$\begin{aligned}
H &= \int \mathcal{H}(x)d^3x = \int \psi^\dagger(x)\,(-i\boldsymbol{\alpha}\cdot\boldsymbol{\nabla} + m\gamma^o)\,\psi(x) \\
&= \frac{1}{(2\pi)^3}\sum_{s,s'}\int d^3k\,d^3k'\frac{1}{2}\sqrt{\frac{\omega'}{\omega}}\left[b_s^\dagger(\mathbf{k})u_s^\dagger(\mathbf{k})e^{i\mathbf{k}\cdot x} + d_s(\mathbf{k})v_s^\dagger(\mathbf{k})e^{-i\mathbf{k}\cdot x}\right] \\
&\quad \times \left[b_{s'}(\mathbf{k'})u_{s'}(\mathbf{k'})e^{-i\mathbf{k'}\cdot x} - d_{s'}^\dagger(\mathbf{k'})v_{s'}(\mathbf{k'})e^{i\mathbf{k'}\cdot x}\right]d^3x \\
&= \frac{1}{2}\sum_{s,s'}\int \left[b_s^\dagger(\mathbf{k})b_{s'}(\mathbf{k})u_s^\dagger(\mathbf{k})u_{s'}(\mathbf{k}) - b_s^\dagger(\mathbf{k})d_{s'}^\dagger(-\mathbf{k})u_s^\dagger(\mathbf{k})v_{s'}(-\mathbf{k})e^{-2i\omega t}\right. \\
&\quad \left. + d_s(\mathbf{k})b_{s'}(-\mathbf{k})v_s^\dagger(\mathbf{k})u_{s'}(-\mathbf{k})e^{2i\omega t} - d_s(\mathbf{k})d_{s'}^\dagger(\mathbf{k})v_s^\dagger(\mathbf{k})v_{s'}(\mathbf{k})\right]d^3k
\end{aligned}$$
$$(13.167)$$

The cross terms vanish because $u_s(\mathbf{k})^\dagger v_{s'}(-\mathbf{k}) = 0$.

$$\begin{aligned}
H &= \frac{1}{2}\sum_{s,s'}\int \left[b_s^\dagger(\mathbf{k})b_{s'}(\mathbf{k})u_s^\dagger(\mathbf{k})u_{s'}(\mathbf{k}) - d_s(\mathbf{k})d_{s'}^\dagger(\mathbf{k})v_s^\dagger(\mathbf{k})v_{s'}(\mathbf{k})\right]d^3k \\
&= \sum_s \int \omega\left[b_s^\dagger(\mathbf{k})b_s(\mathbf{k}) - d_s(\mathbf{k})d_s^\dagger(\mathbf{k})\right]d^3k
\end{aligned} \qquad (13.168)$$

Here is the specter of the negative energy solutions to the Dirac equation. If

$$\left[d_s(\mathbf{k}), d_{s'}^\dagger(\mathbf{k'})\right] = \delta_{s's}\delta_3\left(\mathbf{k'} - \mathbf{k}\right)$$

were correct, then (except for the spin indices) the Hamiltonian would look like free charged boson field Hamiltonian (13.138), but with a negative energy for the antiparticles! Since the theory contains states with any number of antiparticles, the energy has no lower bound.

This is clearly not right. The canonical commutation rule (13.153) must be wrong for Dirac particles.

462 Chapter 13 Identical Particles

The Connection Between Spin and Statistics

In order that all the excited levels have energies higher than the energy of the vacuum, the Dirac field must be quantized with anticommutation relations, and the particles associated with the field obey Fermi-Dirac statistics [4]. Instead of equation (13.153), the Dirac fields satisfy

$$\left\{\psi_i(\mathbf{r}), \psi_j^\dagger(\mathbf{r}')\right\} = \delta_{ij}\delta_3\left(\mathbf{r}-\mathbf{r}'\right) \tag{13.169}$$

Then it is a straightforward computation to invert the plane wave expansion (13.161) and show that

$$\left\{b_s(\mathbf{k}), b_{s'}^\dagger(\mathbf{k}')\right\} = \left\{d_s(\mathbf{k}), d_{s'}^\dagger(\mathbf{k}')\right\} = \delta_{s's}\delta_3\left(\mathbf{k}-\mathbf{k}'\right) \tag{13.170a}$$

$$\left\{b_s(\mathbf{k}), d_{s'}^\dagger(\mathbf{k}')\right\} = \{b_s(\mathbf{k}), d_s(\mathbf{k}')\} = 0 \tag{13.170b}$$

The free Dirac Hamiltonian becomes

$$H = \sum_s \int \omega \left[b_s^\dagger(\mathbf{k})b_s(\mathbf{k}) + d_s^\dagger(\mathbf{k})d_s(\mathbf{k})\right] d^3k \tag{13.171}$$

up to the constant ground-state energy.

This is the spin-statistics theorem: In order that the Hamiltonian have a lower bound, particles with half-integer spin must be fermions. If you go back and try to use anticommutators for boson fields, you will find a similar catastrophe.

The Conserved Charge

A similar computation shows that

$$Q = \int \psi^\dagger(x)\psi(x)d^3x = \sum_s \int \left[b_s^\dagger(\mathbf{k})b_s(\mathbf{k}) - d_s^\dagger(\mathbf{k})d_s(\mathbf{k})\right] d^3k \tag{13.172}$$

The conserved charge is the number of particles minus the number of antiparticles. Q doesn't have to have a lower bound.

13.5 INTERACTING RELATIVISTIC FIELDS

13.5.1 Normal Ordering

I want to conclude with a simple example, just to show you the power of relativistic quantum field methods.

The simplest model consists of a single field, a Hermitean scalar field $\phi(x)$, as in Section 13.3.2. The free Lagrangian density is given in equation (13.109):

$$\mathcal{L}_o = \frac{1}{2}\partial_\mu\phi\partial^\mu\phi - \frac{1}{2}m^2\phi^2 \tag{13.173}$$

The Hamiltonian (13.122) that follows literally from this Lagrangian density has an infinite ground-state energy, $\omega/2$ coming from each mode. It is better to order the operators so the vacuum energy is zero, as in equation (13.123). To guarantee that the destruction operators always appear to the right of the creation operators introduce the **normal ordering** symbol and replace (13.173) by

$$\mathcal{L}_o = :\frac{1}{2}\partial_\mu \phi \partial^\mu \phi: - \frac{1}{2}m^2:\phi^2: \tag{13.174}$$

The fields between colons are "normal ordered." That means, write the fields out as if everything commuted, and then move all the creation operators to the left. Explicitly, break of the field into two parts:

$$\phi(x) = \phi(x)^+ + \phi(x)^- \tag{13.175}$$

where

$$\phi(x)^+ = \int f(\mathbf{k})^* a^\dagger(\mathbf{k},t) d^3k \tag{13.176}$$

and

$$\phi(x)^- = \int f(\mathbf{k}) a(\mathbf{k},t) d^3k \tag{13.177}$$

with the single particle "wave function"

$$f(\mathbf{k}) = \frac{1}{(2pi)^{3/2}} \frac{1}{\sqrt{2\omega}} e^{-i\mathbf{k}\cdot\mathbf{r}} \tag{13.178}$$

Then

$$:\phi^2: = :\left(\phi^+ + \phi^-\right)^2: = \phi^+\phi^+ + 2\phi^+\phi^- + \phi^-\phi^- \tag{13.179}$$

The free Hamiltonian automatically comes out to be

$$H_o = \int \omega a^\dagger(\mathbf{k}) a(\mathbf{k}) d^3k \tag{13.180}$$

The energy of the vacuum state is zero.

13.5.2 Example: The ϕ^4 interaction

Forces that act instantaneously between two different points, so called "action at a distance," like Newton's law or Coulomb's law, are unacceptable in a relativistic theory. These interactions are instantaneous between two different points, information travels faster than light, and the order of "cause" and "effect" depends on the frame of reference. The interaction from products of quantum fields has to ensure that the interaction be local.

A local interaction has the form $\mathcal{L}(x)$, not something like $\int \mathcal{L}(x,y)dy$. Furthermore, $\mathcal{L}(x)$ must be a scalar field under Lorentz transformations. As a first example start with a quartic term, writing

$$\mathcal{L}(x) = \mathcal{L}_o(x) + \mathcal{L}'(x) \tag{13.181}$$

with

$$\mathcal{L}'(x) = -\frac{\lambda}{24}:\phi(x)^4: \tag{13.182}$$

464   Chapter 13   Identical Particles

Note on dimensions: The action is dimensionless ($\hbar = c = 1$), so the units of $\mathcal{L}$ are cm^{-4} or eV^4, the same as the dimension of an energy density. Then a glance at equation (13.173) tells you that the dimension of the field $\phi(x)$ is the same as mass or energy. The constant λ in equation (13.182) is dimensionless, just like the electric charge e.

The Hamiltonian is

$$H = \int \mathcal{H}(x) d^3x = H + H' = \int \mathcal{H}_o(x) d^3x + \int \mathcal{H}'(x) d^3x \qquad (13.183)$$

where according to the discussion in Section 13.3.2

$$\mathcal{H} = \Pi \dot{\phi} - \mathcal{L} = \frac{1}{2}\left[\dot{\phi}(x)^2 + (\boldsymbol{\nabla}\phi(x))^2 + m^2\phi(x)\right] - \mathcal{L}'(x) \qquad (13.184)$$

The interaction Hamiltonian density is

$$\mathcal{H}'(x) = -\mathcal{L}'(x) = \frac{\lambda}{24}{:}\phi(x)^4{:} \qquad (13.185)$$

Scattering in Born Approximation

Let us compute the elastic scattering of two of these scalar particles. The initial and final states are described by the momenta of the two particles. The cross section into a group of final states is given by the Born approximation to equation (9.93):

$$\sigma = \frac{(2\pi)^4}{v_1 + v_2} \int |\bar{H}'_{ba}|^2 \delta(\omega_b - \omega_a) d^3p_3 \qquad (13.186)$$

where

$$\bar{H}_{ba} = \delta_3(\mathbf{p}_1 + \mathbf{p}_2 - \mathbf{p}_3 - \mathbf{p}_4) \bar{H}'_{ba} \qquad (13.187)$$

Work in the center of mass frame, $\mathbf{p}_1 + \mathbf{p}_2 = 0$, and

$$v_1 + v_2 = 2\frac{p}{E} \qquad (13.188)$$

where $p = |\mathbf{p}_1|$ is the magnitude of the momentum, and $E = E_1 = E_2$ is the energy, of either particle.

The Hamiltonian operator is a four-dimensional Fourier transform in terms of the creation and annihilation operators, and the terms that have nonzero matrix elements are those with two creation operators and two annihilation operators. In the binomial expansion of $\phi^4 = (\phi^+ + \phi^-)^4$ there are six such terms, each of the form

$$\frac{\lambda}{24} \int d^3k_1 d^3k_2 d^3k_3 d^3k_4 f(\mathbf{k}_3)^* f(\mathbf{k}_4)^* f(\mathbf{k}_1) f(\mathbf{k}_2)$$
$$\times \langle \mathbf{p}_3, \mathbf{p}_4 | a^\dagger(\mathbf{k}_3) a^\dagger(\mathbf{k}_4) a(\mathbf{k}_1) a(\mathbf{k}_2) | \mathbf{p}_1, \mathbf{p}_2 \rangle d^3r \qquad (13.189)$$

$a(\mathbf{k}_1)$ can annihilate the particle with momentum $\mathbf{p}_1$, and $a(\mathbf{k}_2)$ can annihilate the particle with momentum $\mathbf{p}_2$, or the other way around:

$$a(\mathbf{k}_1) a(\mathbf{k}_2) |\mathbf{p}_1, \mathbf{p}_2\rangle = \left[\delta_3(\mathbf{k}_1 - \mathbf{p}_1)\delta_3(\mathbf{k}_2 - \mathbf{p}_2) + \delta_3(\mathbf{k}_1 - \mathbf{p}_2)\delta_3(\mathbf{k}_2 - \mathbf{p}_3)\right]|\rangle$$
$$(13.190)$$

Each term contributes equally to H_{ba}. Similarly, there are two ways the creation operators can create the final two particle state. Altogether, there are twenty-four identical terms, which is why I put this number in the denominator in the definition of the interaction. The momentum delta functions allow us to do the integrals over the momenta:

$$\begin{aligned} H_{ba} &= \lambda \int f(\mathbf{p}_3)^* f(\mathbf{p}_4)^* f(\mathbf{p}_1) f(\mathbf{p}_2) d^3 r \\ &= \lambda \frac{1}{(2\pi)^6} \frac{1}{\sqrt{16 E_1 E_2 E_3 E_4}} \int e^{i((\mathbf{p}_3 + \mathbf{p}_4 - \mathbf{p}_1 - \mathbf{p}_2) \cdot \mathbf{r})} d^3 r \\ &= \lambda \frac{1}{(2\pi)^3} \frac{1}{\sqrt{16 E_1 E_2 E_3 E_4}} \delta_3 (\mathbf{p}_3 + \mathbf{p}_4 - \mathbf{p}_1 - \mathbf{p}_2) \end{aligned} \quad (13.191)$$

Now insert the expression (13.191) into equation (13.186).

$$d\sigma = \frac{(2\pi)^4}{v_1 + v_2} \frac{\lambda^2}{(2\pi)^6} \frac{1}{16 E_1 E_2 E_3 E_4} \int \delta(E_3 + E_4 - E_1 - E_2) d^3 p_3 \quad (13.192)$$

The energy-conserving delta function allows us to set $E_3 = E_4 = E$ also and take all these factors outside the integral. The integral itself is

$$\int p_3^2 dp_3 d\Omega_3 \delta\left(2\sqrt{p_3^2 + m^2} - 2E\right) = \frac{1}{2} \int p_3 E_3 dE_f \delta(E_f - 2E) d\Omega = \frac{1}{4} pW d\Omega \quad (13.193)$$

where $W = E_f = 2E$ is the total energy in the center of mass frame, and $p^2 = E^2 - m^2$. The differential cross section, here independent of the scattering angle, is

$$\frac{d\sigma}{d\Omega} = \frac{\lambda^2}{64 W^2 \pi^2} \quad (13.194)$$

The total cross section is

$$\sigma_{tot} = \frac{1}{2} \int \frac{d\sigma}{d\Omega} d\Omega = \frac{\lambda^2}{32 W^2 \pi} \quad (13.195)$$

The states $|\mathbf{p}'_1, \mathbf{p}'_2\rangle$ and $|\mathbf{p}'_2, \mathbf{p}'_1\rangle$ are the same, so integrating over all $d^3 p_3 d^3 p_4$ counts each state twice. So divide by two in the total cross section. Because this is a point interaction, there is only an S-wave phase shift. The scattering amplitude is independent of the scattering angle.

PROBLEMS

13.1. Nonrelativistic Quantum Fields
Fill in the missing steps in equations (13.5), (13.17), and (13.33).

13.2. The Fermion Number Operator
Show that the

$$N = \int b^\dagger(\mathbf{p}) b(\mathbf{p}) d^3 p$$

is in fact the number operator for the fermion Hamiltonian given in equation (13.41), as claimed.

13.3. A Simple Supersymmetry Model
Let
$$H = \omega\left(a^\dagger a + b^\dagger b\right)$$
where a are b are have the algebra of a single boson and fermion mode, respectively:
$$\left[a, a^\dagger\right] = 1 \quad \text{and} \quad \left\{b, b^\dagger\right\} = 1$$
a and $a^\dagger$ are independent of b and $b^\dagger$, so they commute:
$$\left[a, b\right] = \left[a, b^\dagger\right] = 0$$

The eigenstates of this Hamiltonian are
$$|n_B, n_F\rangle = \frac{\left(a^\dagger\right)^{n_B}}{\sqrt{n_B!}} \left(b^\dagger\right)^{n_F} |0, 0\rangle$$

n_B can be any nonnegative integer, while n_F is 0 or 1. The state has n_F "fermions" and n_B "bosons." These are not really bosons and fermions of course. To do that, a and b would have to be functions of the momentum **p**. Rather, they have the properties of fermions and bosons with a single momentum.

The state with one fermion has the same energy as the state with one boson. When there is extra degeneracy, you know there must be extra symmetry.

(a) Define the operator
$$Q = b^\dagger a$$
Show that Q and $Q^\dagger$ commute with the Hamiltonian, and that
$$\left\{Q, Q^\dagger\right\} = \beta H$$
What is the constant β? Set
$$Q = Q_1 + iQ_2 \quad \text{and} \quad Q^\dagger = Q_1 - iQ_2$$
Show that
$$\left\{Q_i, Q_j\right\}$$
is proportional to $\delta_{ij} H$.

(b) $|1, 0\rangle$ is the one-boson state. Show that $Q|1, 0\rangle$ is proportional to $|0, 1\rangle$.

> Note: These two states have the same energy, and Q changes one into the another. A symmetry like this that changes a boson state into a fermion state is called a **supersymmetry**.

(c) On the usual vector space of one-dimensional wave functions, or equivalently the space for which the states $|x\rangle$ are a basis, you can define a and $a^\dagger$ in the usual way, but there is no way to define operators with the properties of b and $b^\dagger$. Instead, you have to add a discrete quantum number, so that a basis is $|x, m\rangle$, where $m = \pm\frac{1}{2}$. Think of the states as a column vector, or spinor, with two rows:
$$\langle x, m | \psi\rangle = \psi_m(x)$$

and
$$\psi(x) = \begin{pmatrix} \psi_{+\frac{1}{2}}(x) \\ \psi_{-\frac{1}{2}}(x) \end{pmatrix} = \psi_{+\frac{1}{2}}(x)\chi_+ + \psi_{-\frac{1}{2}}(x)\chi_-$$

with
$$\chi_+ = \begin{pmatrix} 1 \\ 0 \end{pmatrix} \quad \text{and} \quad \chi_- = \begin{pmatrix} 0 \\ 1 \end{pmatrix}$$

Set
$$a = \frac{1}{\sqrt{2m\omega}}(m\omega x + ip) \quad \text{and} \quad b = \frac{1}{2}(\sigma_x - i\sigma_y)$$

Show that b and $b^\dagger$ satisfy the correct anticommutation rule, and then write the Hamiltonian in terms of x, p, and the Pauli matrices.

(d) What is the two-component wave function for the lowest energy state, and what is its energy?

13.4. Supersymmetry Model with Interaction

Problem 13.3 describes a boson excitation and a fermion excitation with the same energy (level spacing). But those were just independent free particles. Here is a more interesting model (but still only a single mode). Again consider a space of wave functions that are two component spinors. Let[12]

$$Q_1 = \frac{1}{\sqrt{8m\omega}}\left(p\sigma_y - m\omega^3 x^3 \sigma_x\right) \qquad Q_2 = \frac{1}{\sqrt{8m\omega}}\left(p\sigma_x + m\omega^3 x^3 \sigma_y\right)$$

(a) Show that
$$\{Q_i, Q_j\} = \beta \delta_{ij} H$$

where
$$H = \frac{p^2}{2m} + V_1(x) + V_2(x)\sigma_z$$

What is H? What is the constant β?

(b) It follows that Q_1 and Q_2 are symmetries of this problem: $[Q_i, H] = 0$. Why? Show that there is an eigenstate of H with energy zero. What is its wave function?

Hint: If $H|\psi\rangle = 0$, then $Q_i|\psi\rangle = 0$ also. Why?

(c) The state with zero energy is the ground state. Why?

13.5. Hamiltonian for Boson Fields
Derive the form (13.122) of the Hamiltonian for a free Bose-Einstein field in terms of the creation and annihilation operators.

13.6. Fermion Pair Correlations
Consider a collection of noninteracting nonrelativistic spin-half fermions, confined to a cubic box with side L, and satisfying periodic boundary conditions at the walls, as in Section 13.2.2. The particle annihilation operators are $b_s(\mathbf{p_n})$, where s is the spin index and the components of the momentum are $p_i = 2\pi n_i/L$

---
[12] Suggested in a slightly different form by J. M. Cornwall.

468 Chapter 13 Identical Particles

where n_i are any integers. These operators have a discrete normalization

$$\{b_r(\mathbf{p}), b_s^\dagger(\mathbf{p}')\} = \delta_{rs}\delta_{\mathbf{p},\mathbf{p}'} \tag{13.196a}$$

$$\{b_r(\mathbf{p}), b_s(\mathbf{p}')\} = 0 \tag{13.196b}$$

$$\{b_r^\dagger(\mathbf{p}), b_s^\dagger(\mathbf{p}')\} = 0 \tag{13.196c}$$

where

$$\delta_{\mathbf{p},\mathbf{p}'} = \delta_{n_x,n_x'}\delta_{n_y,n_y'}\delta_{n_z,n_z'}$$

And instead of continuous normalization as in equation (13.9), the fields should be

$$\Psi_s(\mathbf{r}) = \frac{1}{L^{3/2}}\sum_{\mathbf{n}} e^{i\mathbf{p}\cdot\mathbf{r}} b_s(\mathbf{p}) \tag{13.197}$$

Let $|N_o\rangle$ be the ground state of a large number of fermions in the box, with number density ρ, and such that all the single-particle states are filled up to the Fermi momentum p_F, as in the discussion of the ideal Fermi gas in Section 13.2.2.

(a) Compute the single particle correlation function

$$C_r(|\mathbf{r} - \mathbf{r}'|) = \left\langle N_o \middle| \Psi_r^\dagger(\mathbf{r})\Psi_r(\mathbf{r}') \middle| N_o \right\rangle \tag{13.198}$$

as a function of ρ, p_F, and the distance $|\mathbf{r} - \mathbf{r}'|$. Work out the cases $r = s$ and $r \neq s$ separately. Show that for $r = s$ this function approaches the density of particles of a single spin (why?).

(b) For the same state, work out the pair correlation function

$$C_{rs}(|\mathbf{r} - \mathbf{r}'|) = \left\langle N_o \middle| \Psi_r^\dagger(\mathbf{r})\psi_s^\dagger(\mathbf{r}')\Psi_s(\mathbf{r}')\Psi_r(\mathbf{r}) \middle| N_o \right\rangle \tag{13.199}$$

Again, separate the two cases of equal spins and unequal spins. Show that for equal spins, this correlation function vanishes as $\mathbf{r} \to \mathbf{r}'$. At what separation does the correlation function first reach a maximum?

13.7. Boson Pair Correlations

Now consider a collection of spinless bosons in a cubic box with side L, with periodic boundary conditions. The creation and annihilation operators $a^\dagger(\mathbf{p})$ and $a(\mathbf{p})$ have a discrete normalization as in Problem 13.6, but with commutators instead of anticommutators. The fields are normalized in the box:

$$\Psi(\mathbf{r}) = \frac{1}{L^{3/2}}\sum_{\mathbf{n}} e^{i\mathbf{p}\cdot\mathbf{r}} a(\mathbf{p}) \tag{13.200}$$

Let $|\psi_o\rangle$ be some state, with $n_\mathbf{p}$ particles in the single-particle state with momentum $\mathbf{p}$:

$$a^\dagger(\mathbf{p})a(\mathbf{p})|\psi_o\rangle = n_\mathbf{p}|\psi_o\rangle \tag{13.201}$$

(a) Show that

$$\left\langle \psi_o \middle| \Psi^\dagger(\mathbf{r})\Psi(\mathbf{r}) \middle| \psi_o \right\rangle = \rho \tag{13.202}$$

where ρ is the particle number density.

(b) Calculate the pair correlation function

$$C(\mathbf{r},\mathbf{r}') = \left\langle \psi_o \middle| \Psi^\dagger(\mathbf{r})\Psi^\dagger(\mathbf{r}')\Psi(\mathbf{r}')\Psi(\mathbf{r}) \middle| \psi_o \right\rangle \tag{13.203}$$

for arbitrary $n_\mathbf{p}$.

(c) Suppose the bosons in the gas are in a beam with a gaussian distribution about some momentum $\mathbf{p}_o$:

$$n_\mathbf{k} \sim e^{-\beta(\mathbf{k}-\mathbf{k}_o)^2/2} \tag{13.204}$$

What is the form of the pair correlation function now?

Hint: The volume appears in some terms in a combination other than $\rho = N/V$. Let $V \to \infty$ while holding ρ fixed.

What is the ratio of the correlation function when $\mathbf{r} \to \mathbf{r}'$ compared to its value as $|\mathbf{r}-\mathbf{r}'| \to \infty$?

REFERENCES

The place to learn about the Chandrasekhar limit is

[1] S. CHANDRASEKHAR, *Stellar structure*, Dover, 1957, Chapter 9.

The original reference is

[2] S. CHANDRASEKHAR, Mon. Not. Royal Astron. Soc. **95** (1935), 207.

Neutron stars were first suggested by

[3] J. R. OPPENHEIMER AND G. M. VOLKOFF, Phys. Rev. **55** (1939), 374.

The first proof of connnection between spin and statistics was by

[4] W. PAULI, Phys. Rev. **58** (1920), 716, reprinted in reference [4] of Chapter 10.

There is a vast literature on relativistic quantum fields. References [6,7,8] in Chapter 10 are some recent textbooks.

APPENDIX A

Mathematical Tools

A.1 MISCELLANEOUS TOOLS

From time to time I have used some mathematical formulas or techniques without derivation. Some of them are collected in this appendix for easy reference, even though with some effort they can all be found in standard mathematical physics references.

A.1.1 The Dirac Delta Function

The Dirac delta function $\delta(x)$ was defined in Section 2.3.1 as a generalization of the Kronecker delta function δ_{ij}. Its basic property is that

$$\int_a^b f(x)\delta(x-c)dx = f(c) \qquad (a \leq c \leq b) \tag{A.1}$$

and it is zero otherwise. It follows that

$$\int f(x)\delta(x-c)dx = f(c)\int \delta(x-c)dx \tag{A.2}$$

To integrate over a delta function of a *function* of x, it is usually enough to change variables:

$$\int \delta(f(x))dx = \int \frac{1}{df(x)/dx}\delta[f(x)]\,df(x) = \left|\frac{df(x)}{dx}\right|_{f(x)=0}^{-1}\int \delta(y)dy \tag{A.3}$$

but care must be taken when $f(x)$ has more than one root in the integration region.

Derivatives are defined by partial integration.

$$\int f(x)\delta'(x)dx = -\int f'(x)\delta(x)dx \tag{A.4}$$

Since $\delta(x)$ vanishes for $x \neq 0$, there is no end-point term.

There are many representations of the delta function, the most important being

$$\frac{1}{2\pi}\int_{-\infty}^{\infty} e^{i(x-z)y}\,dy = \delta(x-z) \tag{2.104}$$

which was demonstrated in Section 2.3.3.

Several representations of the $\delta(x)$ are especially useful in the theory of scattering and decays. If a and b are both positive, or both negative,

$$\lim_{t \to \infty} \int_a^b \frac{e^{-ixt}}{x}\,dx = 0 \tag{A.5}$$

because the increasingly rapid oscillation causes the positive and negative parts of the integral to cancel. Furthermore, by contour integration

$$\int_{-\infty}^{\infty} \frac{e^{-ixt}}{x + i\epsilon} dx = -2\pi i e^{-\epsilon t} \xrightarrow[\epsilon \to 0]{} -2\pi i \quad (A.6)$$

So for large t and small ϵ,

$$\frac{e^{-ixt}}{x + i\epsilon} \to -2\pi i \delta(x) \quad (A.7)$$

Another useful representation is obtained by rewriting equation (2.104) as

$$\delta(x) = \frac{1}{2\pi} \lim_{t \to \infty} \int_{-t}^{t} e^{ixy} \, dy = \frac{1}{\pi} \lim_{t \to \infty} \frac{\sin xt}{x} \quad (A.8)$$

so that

$$\lim_{t \to \infty} \frac{\sin xt}{x} = \pi \delta(x) \quad (A.9)$$

Furthermore, if $x \neq 0$, then

$$\lim_{\epsilon \to 0} \frac{\epsilon}{x^2 + \epsilon^2} = 0 \quad (A.10)$$

Since

$$\int_{-\infty}^{\infty} \frac{\epsilon}{x^2 + \epsilon^2} dx = \int_{-\infty}^{\infty} \frac{dt}{1 + t^2} = \pi \quad (A.11)$$

yet another representation is

$$\lim_{\epsilon \to 0} \frac{\epsilon}{x^2 + \epsilon^2} = \pi \delta(x) \quad (A.12)$$

The left-hand side of equation (2.104) above is not really well defined, while for finite ϵ expressions like those on the left-hand sides of equations (A.7) and (A.12) are unambiguous. You can multiply the integrand on the left-hand side of equation (2.104) by $e^{-\epsilon|y|}$, where $\epsilon > 0$, and take the limit later. Then the integral makes sense for all $x - z$, and is easily seen to be

$$\frac{1}{2\pi} \int_{-\infty}^{\infty} e^{i(x-z)y} e^{-\epsilon|y|} \, dy = \frac{1}{\pi} \frac{\epsilon}{(x-z)^2 + \epsilon^2} \quad (A.13)$$

in agreement with equation (A.12).

Three Dimensions

The three-dimensional Dirac delta function is

$$\delta_3(\mathbf{r}) = \delta(x)\delta(y)\delta(z) = \frac{1}{(2\pi)^3} \int e^{i\mathbf{k} \cdot \mathbf{r}} d^3k \quad (A.14)$$

In spherical coordinates the function $1/r$ is the electrostatic potential of a unit point charge at the origin. The gradient of $1/r$ in spherical coordinates is[1]

$$\nabla\left(\frac{1}{r}\right) = -\frac{1}{r^2}\hat{\mathbf{n}}_r \quad (A.15)$$

---

[1] See equation (A.158).

Apply Gauss' law to this function inside a sphere of radius ϵ:

$$\int_{r<\epsilon} \nabla^2 \left(\frac{1}{r}\right) d^3 r = \int \nabla \left(\frac{1}{r}\right) \cdot \hat{\mathbf{n}}_r \bigg|_{r=\epsilon} \epsilon^2 d\Omega = -4\pi \tag{A.16}$$

Since $\nabla^2(1/r) = 0$ except at the origin, one can conclude that

$$\nabla^2 \left(\frac{1}{r}\right) = -4\pi \delta_3(\mathbf{r}) \tag{A.17}$$

A.1.2 The Levi-Civita Symbol

The symbol ϵ_{ijk} is defined to be antisymmetric in the exchange of any pair of indices, with the rule that $\epsilon_{123} = 1$. In particular, ϵ_{ijk} is zero if any pair of its indices are the same. The nonzero values are

$$\epsilon_{123} = \epsilon_{231} = \epsilon_{312} = 1 \qquad \epsilon_{213} = \epsilon_{321} = \epsilon_{132} = -1 \tag{A.18}$$

The components of a curl or cross product are

$$(\mathbf{A} \times \mathbf{B})_i = \sum_{jk} \epsilon_{ijk} A_j B_k \tag{A.19}$$

I have frequently used the identities

$$\sum_i \epsilon_{ijk} \epsilon_{imn} = \delta_{jm}\delta_{kn} - \delta_{jn}\delta_{km} \tag{A.20a}$$

$$\sum_{ij} \epsilon_{ijk} \epsilon_{ijm} = 2\delta_{km} \tag{A.20b}$$

The second follows from the first. The proof of the first is left as an exercise.

A.1.3 Some Integrals

Gaussian Integrals
The normalization of Gaussian integrals follows from

$$\int_{-\infty}^{\infty} e^{-t^2} dt = \sqrt{\pi} \tag{A.21}$$

To demonstrate equation (A.21), square both sides and then use plane polar coordinates on the left. A common generalization is

$$\int_{-\infty}^{\infty} e^{-(at^2+bt+c)} dt = \sqrt{\frac{\pi}{a}} \exp\left(\frac{b^2 - 4ac}{4a}\right) \tag{A.22}$$

To derive equation (A.22), just expand the exponent about its minimum and use (A.21).

Consider multiple integrals of the form

$$I = \int \cdots \int \prod_n dq_n \exp\left(-\sum_{ij} A_{ij} q_i q_j\right) \quad (A.23)$$

If A is a real symmetric matrix it can be diagonalized. The new variables are y_i and

$$\sum_j A_{ij} y_j = \lambda_i y_i \quad (A.24)$$

The transformation is effected by an orthogonal matrix, so its Jacobian is unity, and

$$I = \int \cdots \int \prod_n dy_n \exp\left(-\sum_i \lambda_i y_i^2\right) = \prod_n \left(\frac{\pi}{\lambda_n}\right)^{1/2} = \frac{\pi^{n/2}}{\sqrt{\det A}} \quad (A.25)$$

Miscellaneous Integrals

Here are some other integrals that were used in the body of the text without derivation: Integrals of the form

$$\int_0^\infty x^n e^{-a^2 x^2} \, dx \quad (A.26)$$

occur frequently. Their values follow directly from equation (A.55).

$$\int_0^\infty x^{2n} e^{-a^2 x^2} \, dx = \frac{1}{2 a^{2n+1}} \Gamma\left(\frac{2n+1}{2}\right) = \frac{(2n-1)!!}{2^{n+1} a^{2n+1}} \sqrt{\pi} \quad (A.27)$$

An elementary calculation gives the integral

$$\int_0^\infty e^{-\mu x} \sin(kx) dx = \frac{1}{2i} \int_0^\infty \left[e^{-(\mu-ik)x} - e^{-(\mu+ik)x}\right] dx = \frac{k}{\mu^2 + k^2} \quad (A.28)$$

Differentiate (A.28) to obtain

$$\int_0^\infty x e^{-\mu x} \sin(kx) dx = \frac{2k\mu}{(\mu^2 + k^2)^2} \quad (A.29a)$$

$$\int_0^\infty x^2 e^{-\mu x} \sin(kx) dx = \frac{2k(3\mu^2 - k^2)}{(\mu^2 + k^2)^3} \quad (A.29b)$$

$$\int_0^\infty x^3 e^{-\mu x} \sin(kx) dx = \frac{24 k \mu (\mu^2 - k^2)}{(\mu^2 + k^2)^4} \quad (A.29c)$$

and so forth. By differentiating

$$\int_0^\infty \frac{dx}{(a^2 + x^2)} = \frac{\pi}{2a} \quad (A.30)$$

repeatedly with respect to a, you get the rules for the integrals

$$\int_0^\infty \frac{dx}{(1+x^2)^n} = \frac{(2n-3)!!}{2^n (n-1)!} \pi \quad (A.31)$$

474 Appendix A Mathematical Tools

The integral (from Problem 11.9)

$$\int_0^\infty \frac{x^3 \, dx}{e^x - 1} \tag{A.32}$$

is easily evaluated as follows:

$$\int_0^\infty \frac{x^3 \, dx}{e^x - 1} = \int_0^\infty x^3 e^{-x} \frac{1}{1 - e^{-x}} = \int_0^\infty x^3 \sum_{n=1}^\infty e^{-nx} dx$$
$$= 6 \sum_{n=1}^\infty \frac{1}{n^4} = \frac{\pi^4}{15} \tag{A.33}$$

Some Fourier Transforms
From equations (A.14) and (A.17)

$$\nabla^2 \left(\frac{1}{r}\right) = -\frac{4\pi}{(2\pi)^3} \int e^{i\mathbf{k}\cdot\mathbf{r}} d^3 k \tag{A.34}$$

whence

$$\frac{1}{r} = \frac{4\pi}{(2\pi)^3} \int \frac{1}{k^2} e^{i\mathbf{k}\cdot\mathbf{r}} d^3 k \tag{A.35}$$

and the inverse transform is[2]

$$\frac{1}{k^2} = \frac{1}{4\pi} \int \frac{1}{r} e^{-i\mathbf{k}\cdot\mathbf{r}} d^3 r \tag{A.36}$$

Three-dimensional Fourier transforms of functions like $e^{-\mu r}$ follow from equations (A.29) and the expansion (8.75):

$$\int e^{-\mu r} e^{i\mathbf{k}\cdot\mathbf{r}} d^3 r = \frac{4\pi}{k} \int_0^\infty r e^{-\mu r} \sin(kr) dr = \frac{8\pi \mu}{(\mu^2 + k^2)^2} \tag{A.37a}$$

$$\int r e^{-\mu r} e^{i\mathbf{k}\cdot\mathbf{r}} d^3 r = \frac{4\pi}{k} \int_0^\infty r^2 e^{-\mu r} \sin(kr) dr = \frac{8\pi(3\mu^2 - k^2)}{(\mu^2 + k^2)^3} \tag{A.37b}$$

A.1.4 The Trapezoidal Approximation Series

The integral of any differentiable function $f(x)$ between any two successive integers can be approximated by the area of a trapezoid:

$$\int_{n-1}^n f(x) dx \approx \frac{1}{2} f(N-1) + \frac{1}{2} f(N) \tag{A.38}$$

The **Euler-Maclaurin** series is a power series for the error in the trapezoidal approximation (A.38). I made use of it in detail in Section 11.8. Define the Bernoulli polynomials $B_n(t)$ by

$$\frac{x e^{xt}}{e^x - 1} = \sum_{n=0}^\infty B_n(t) \frac{x^n}{n!} \tag{A.39}$$

---
[2] These forms are slightly illegal. When in doubt, multiple $1/r$ by $e^{-\epsilon r}$ and let $\epsilon \to 0$ later when it is safe.

The Bernoulli *numbers* were also defined in (11.228). They appear frequently in Taylor expansions of the trigonometric functions. Set $t = 0$ in (A.39) to identify $B_n = B_n(0)$. Setting $t = 1$, we obtain

$$\sum_{n=0}^{\infty} B_n(1) \frac{x^n}{n!} = \frac{xe^x}{e^x - 1} = \frac{-x}{e^{-x} - 1} = \sum_{n=0}^{\infty} B_n(0) \frac{(-x)^n}{n!} \tag{A.40}$$

whence $B_n(1) = (-1)^n B_n$. Differentiate equation (A.39) with respect to t and compare the coefficients of each power of x to get a recursion relation for the Bernoulli functions:

$$B'_n(s) = nB_{n-1}(s) \tag{A.41}$$

Use $B_o(x) = 1$ and (A.41) to write

$$\int_n^{n+1} f(x)dx = \int_0^1 f(x+n)B'_1(x)dx \tag{A.42}$$

Integrate by parts:

$$\int_n^{n+1} f(x)dx = f(n+1)B_1(1) - f(n)B_1(0) - \int_0^1 f'(x+n)B_1(x)dx$$
$$= \frac{1}{2}[f(n+1) + f(n)] - \int_0^1 f'(x+n)B_1(x)dx \tag{A.43}$$

Do it once more:

$$\int_n^{n+1} f(x)dx = \frac{1}{2}[f(n+1) + f(n)] - \frac{1}{2}\int_0^1 f'(x+n)B'_2(x)dx$$
$$= \frac{1}{2}[f(n+1) + f(n)] - \frac{1}{2}B_2[f'(n+1) - f'(n)] \tag{A.44}$$
$$+ \frac{1}{2}\int_0^1 f''(x+n)B_2(x)dx$$

Except for B_1, all the odd order Bernoulli numbers vanish. Therefore,

$$\int_n^{n+1} f(x)dx = \frac{1}{2}[f(n+1) + f(n)] - \frac{1}{2}B_2[f'(n+1) - f'(n)]$$
$$+ \frac{1}{3!}\int_0^1 f''(x+n)B'_3(x)dx$$
$$= \frac{1}{2}[f(n+1) + f(n)] - \frac{1}{2}B_2[f'(n+1) - f'(n)]$$
$$- \frac{1}{4!}\int_0^1 f^{(3)}(x+n)B'_4(x)dx \tag{A.45}$$
$$= \frac{1}{2}[f(n+1) + f(n)] - \frac{1}{2}B_2[f'(n+1) - f'(n)]$$
$$- \frac{1}{4!}B_4\left[f^{(3)}(n+1) - f^{(3)}(n)\right] + \frac{1}{4!}\int_0^1 f^{(4)}(x+n)B_4(x)dx$$

476 Appendix A Mathematical Tools

Continuing over and over again, you can develop the whole series

$$\int_n^{n+1} f(x)dx = \frac{1}{2}[f(n+1) + f(n)]$$
$$- \sum_{m=1}^N \frac{1}{(2m)!} B_{2m} \left[f^{(2m-1)}(n+1) - f^{(2m-1)}(n) \right] + R_N(n) \quad \text{(A.46)}$$

The remainder term is

$$R_N(n) = \frac{1}{(2N)!} \int_n^{n+1} f^{(2N)}(x) B_{2N}(x) dx \quad \text{(A.47)}$$

The integral from zero to $n+1$ can be obtained by adding up the pieces to get the Euler-Maclaurin series:

$$\int_0^{n+1} f(x)dx = \sum_{n'=1}^n \int_{n'}^{n'+1} f(x)dx$$
$$= \frac{1}{2}f(0) + f(1) + f(2) + \cdots + f(n) + \frac{1}{2}f(n+1)$$
$$- \sum_{m=1}^N \frac{1}{(2m)!} B_{2m} \left[f^{(2m-1)}(n+1) - f^{(2m-1)}(0) \right] + R_N \quad \text{(A.48)}$$

where

$$R_N = \frac{1}{(2N)!} \int_0^{n+1} f^{(2N)}(x) B_{2N}(x) dx \quad \text{(A.49)}$$

If the remainder term is well behaved the limit $N \to \infty$ is allowed. If, furthermore, the integral converges as $n \to \infty$, then

$$\int_0^\infty f(x)dx = \frac{1}{2}f(0) + \sum_{n=1}^\infty f(n) + \sum_{m=1}^\infty \frac{1}{(2m)!} B_{2m} f^{(2m-1)}(0) \quad \text{(A.50)}$$

This is the same as equation (11.238).

A.2 SPECIAL FUNCTIONS

In this section I have collected properties of some of the special functions that were used in the body of the text. It is quite possible to learn quantum mechanics without knowing these details—I include brief derivations here for those readers who cannot just accept them on faith. Read this section when you want to know where recursion relations, asymptotic behaviors, normalizations, and so forth, of the functions come from. Most of these results date back at least to the nineteenth century.

A.2.1 Gamma Function

The Integral Representation
The integral

$$\Gamma(z) = \int_0^\infty e^{-t} t^{z-1} dt \quad \text{(A.51)}$$

is the gamma function. It converges whenever the real part of z is positive, but not at $z = 0$. For real positive z, you can integrate by parts and drop the endpoint term, obtaining the recursion relation

$$z\Gamma(z) = \Gamma(z+1) \tag{A.52}$$

Since $\Gamma(1) = 1$, the gamma function is an integral representation for the factorial:

$$\Gamma(n+1) = n! \qquad (n = 0, 1, 2, \ldots) \tag{A.53}$$

Equation (A.53) defines the factorial for $n = 0$.

The value of $\Gamma(z)$ for $z = \frac{1}{2}$ is obtained by substituting $t = x^2$ in equation (A.51) and using equation (A.21):

$$\Gamma\left(\frac{1}{2}\right) = \sqrt{\pi} \tag{A.54}$$

Then from (A.52)

$$\Gamma\left(\frac{2n+1}{2}\right) = \frac{(2n-1)!!}{2^n}\sqrt{\pi} = \frac{(2n)!}{(2^n)^2 n!}\sqrt{\pi} \tag{A.55}$$

The rule (A.52) can be used to extend the definition of $\Gamma(z)$ to values of z for which the integral in (A.52) does not converge. Near $z = 0$

$$\epsilon\Gamma(\epsilon) = \Gamma(1+\epsilon) \xrightarrow[\epsilon \to 0]{} 1 \tag{A.56}$$

so $\Gamma(z)$ has a simple pole at $z = 0$ with unit residue. There are simple poles at the negative integers also. Near $z = -n$

$$(-n+\epsilon)\Gamma(-n+\epsilon) = \Gamma(-n+1+\epsilon) \tag{A.57}$$

so near $z = -n$

$$\Gamma(z) \approx \frac{(-1)^n}{n!}\frac{1}{z+n} \tag{A.58}$$

$\Gamma(z)$ is analytic everywhere except for these simple poles at the nonpositive integers.

Contour Integral Representation

Define a branch of the function $e^u u^{z-1}$ with a cut along the negative real axis, so that the phase of u is everywhere between $-\pi$ and π. Let C_o be a contour (Figure A.1) that starts at $u = -\infty - i\epsilon$, loops around the origin in the clockwise sense, and ends up at $u = -\infty + i\epsilon$. Then the integral

$$\int_{C_o} e^u u^{z-1} du \tag{A.59}$$

defines a function of z wherever it converges. When $\operatorname{Re} z > 0$ there is no contribution from the loop around the origin and

$$\int_{C_o} e^u u^{z-1} du = 2i \sin \pi z \int_0^\infty e^{-u} u^{-z-1} du = 2i \sin \pi z \Gamma(z) \tag{A.60}$$

478 Appendix A Mathematical Tools

Since both sides of (A.60) are consistent with the relation (A.52) the integral can be continued to $\mathrm{Re}\, z < 0$ and is an integral representation for $\Gamma(z)$ anywhere it is not singular:

$$\Gamma(z) = \frac{1}{2i \sin \pi z} \int_{C_o} e^u u^{z-1} du \qquad (A.61)$$

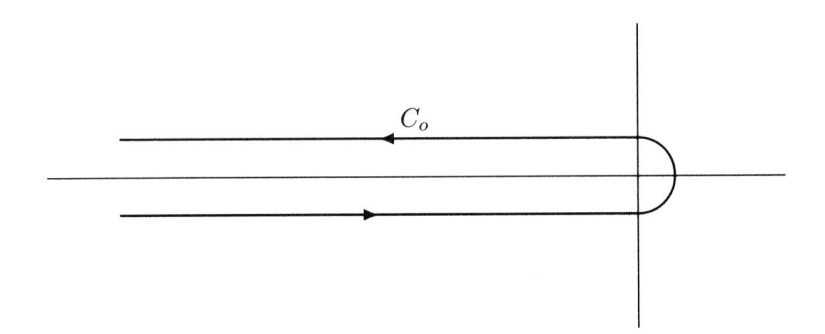

FIGURE A.1: Contour for the integral in equation (A.60)

Infinite Product Representation
Integrate the identity

$$(z+n) \int_0^1 t^{z+n-1} dt = 1 \qquad (A.62)$$

repeatedly by parts to get

$$\int_0^1 (1-t)^n t^{z-1} dt = n! \prod_{k=0}^n \frac{1}{z+k} \qquad (A.63)$$

which in the limit becomes[3]

$$\Gamma(z) = \lim_{n \to \infty} n^z n! \prod_{k=0}^n \frac{1}{z+k} \qquad (A.64)$$

The representation (A.64) explicitly exhibits the poles at the nonnegative integers. Then

$$\Gamma(z) = \frac{1}{z} \lim_{n \to \infty} \left[n^z \prod_{k=1}^n \frac{k}{z+k} \right] = \frac{1}{z} \lim_{n \to \infty} \left[e^{z \log n} \prod_{k=1}^n \left(e^{z/k} \frac{1}{1+z/k} e^{-z/k} \right) \right]$$

$$= \frac{1}{z} \lim_{n \to \infty} \exp\left[z \left(\log n - \sum_{k=1}^n \frac{1}{k} \right) \right] \prod_{k=1}^n \left(e^{z/k} \frac{1}{1+z/k} \right)$$

$$= \frac{1}{z} e^{-\gamma z} \prod_{k=1}^\infty e^{z/k} \frac{1}{1+z/k}$$

$$(A.65)$$

---
[3]Change the integration variable to $y = nt$.

where $\gamma \approx 0.577215664901533\ldots$ is the Euler or Mascheroni constant. The infinite product representation (A.65) converges everywhere except at the negative integers. To prove this, write the logarithmic derivative of the product and use standard methods for proving the convergence of a sum.

There is also an infinite product representation for the sine function:

$$\sin \pi z = \pi z \prod_{k \neq 0} e^{z/k} \left(1 - \frac{z}{k}\right) = \pi z \prod_{k=1}^{\infty} \left(1 - \frac{z^2}{k^2}\right) \tag{A.66}$$

The product in (A.66) has the same zeros as the sine function, and like the sine it has no poles. Even so, it might differ by a function $\exp f(z)$ for some entire $f(z)$. By computing the logarithmic derivative of (A.66) and looking at enough limits, it is not difficult to see that $f(z)$ must vanish.

From this representation for the sine functions and the representation (A.65) for $\Gamma(z)$, it follows immediately that

$$\Gamma(z)\Gamma(-z) = -\frac{1}{z^2} \prod_{k=1}^{\infty} \left(1 - \frac{z^2}{k^2}\right)^{-1} = -\frac{\pi}{z \sin \pi z} \tag{A.67}$$

or, using (A.52),

$$\Gamma(z)\Gamma(1-z) = \frac{\pi}{\sin \pi z} \tag{A.68}$$

Notice that (A.54) follows from (A.68) also.

A.2.2 Legendre Polynomials

Definition and Normalization

The Legendre polynomials[4] are polynomials of order l, solutions to

$$\frac{d}{dz}(z^2 - 1)\frac{d}{dz} P_l(z) = l(l+1) P_l(z) \tag{A.69}$$

and are normalized so that $P(1) = 1$.

Let $V(\mathbf{r})$ be some function that satisfies Laplace's equation and is independent of the azimuthal angle ϕ. In spherical coordinates[5]

$$\nabla^2 V(r, \cos\theta) = \frac{1}{r^2}\frac{\partial}{\partial r} r^2 \frac{\partial}{\partial r} V(r, \cos\theta) - \frac{1}{r^2} L(\cos\theta) V(r, \cos\theta) = 0 \tag{A.70}$$

where

$$L(z) = \frac{\partial}{\partial z}(z^2 - 1)\frac{\partial}{\partial z} \tag{A.71}$$

V can be expanded in a Legendre series (with $z = \cos\theta$)

$$V(r, z) = \sum_{l=0}^{\infty} a_l(r) P_l(z) \tag{A.72}$$

---

[4]See also Problem 2.5.
[5]Cf. equations (3.81) and (3.82).

480 Appendix A Mathematical Tools

with coefficients that are functions of r. Then

$$\nabla^2 V(r,z) = \frac{1}{r^2}\sum_l \left[\frac{\partial}{\partial r}r^2\frac{da_l(r)}{dr} - a_l(r)l(l+1)\right]P_l(z)V(r,z) \quad (A.73)$$

Therefore the coefficients $a_l(r)$ all satisfy

$$\frac{d}{dr}r^2\frac{da_l(r)}{dr} = l(l+1)a_l(r) \quad (A.74)$$

The general solution is

$$a_l(r) = A_l r^l + B_l/r^{l+1} \quad (A.75)$$

for any constants A_l and B_l.

Let V be the potential of a unit point charge located on the z-axis $z=a$:

$$V(r,z) = \frac{1}{|\mathbf{r}-\mathbf{a}|} \quad (A.76)$$

where $\mathbf{a} = a\hat{\mathbf{n}}_z$. The function (A.76) is a solution to Laplace's equation everywhere except at $\mathbf{r} = \mathbf{a}$. Since $V(r,z) \to 0$ as $r \to \infty$,

$$\frac{1}{|\mathbf{r}-\mathbf{a}|} = \sum_l \frac{B_l}{r^{l+1}}P_l(z) \quad (A.77)$$

for $r > a$. When $\mathbf{r}$ is along the z-axis, $z = 1$ and $P_l(z) = 1$ also. Then

$$\frac{1}{r-a} = \frac{1}{r}\sum_l \left(\frac{a}{r}\right)^l = \sum_l \frac{B_l}{r^{l+1}} \quad (A.78)$$

it follows that $B_l = a^l$, and

$$\frac{1}{|\mathbf{r}-\mathbf{a}|} = \frac{1}{r}\sum_l \left(\frac{a}{r}\right)^l P_l(z) \quad (A.79)$$

When $r < a$, a similar argument leads to

$$\frac{1}{|\mathbf{r}-\mathbf{a}|} = \frac{1}{a}\sum_l \left(\frac{r}{a}\right)^l P_l(z) \quad (A.80)$$

Then from equation (A.79) for $r > 1$,

$$\frac{1}{\sqrt{1+r^2-2rz}} = \sum_l \frac{1}{r^{l+1}}P_l(z) \quad (A.81)$$

The different P_l must be orthogonal, To find the normalization, square both sides of equation (A.81) and integrate:

$$\int_{-1}^{1}\frac{dz}{1+r^2-2rz} = \frac{1}{r^2}\sum_l \left(\frac{1}{r}\right)^{2l}\int_{-1}^{1}[P_l(z)]^2\,dz \quad (A.82)$$

The integral on the left is

$$\int_{-1}^{1} \frac{dz}{1+r^2-2rz} = \frac{1}{r}\log\left[\frac{1+1/r}{1-1/r}\right] = 2\left[\frac{1}{r^2} + \frac{1}{3}\frac{1}{r^4} + \frac{1}{5}\frac{1}{r^6} + \cdots\right] \quad \text{(A.83)}$$

so

$$\int_{-1}^{1} P_l(z)^2 dz = \frac{2}{2l+1} \quad \text{(A.84)}$$

Recursion Relations

Differentiate equation (A.81) with respect to r:

$$\frac{z-r}{\sqrt{1+r^2-2rz}} = -\frac{1+r^2-2rz}{r}\sum_l \frac{l+1}{r^{l+1}}P_l(z) \quad \text{(A.85)}$$

Use equation (A.81) again on the left-hand side and collect powers of r:

$$\sum_l \left[\frac{(2l+1)z}{r^l} - \frac{l}{r^{l-1}} - \frac{l+1}{r^{l+1}}\right]P_l(z) = 0 \quad \text{(A.86)}$$

The coefficient of each power of r must vanish independently:

$$(2l+1)zP_l(z) = (l+1)P_{l+1}(z) + lP_{l-1}(z) \quad \text{(A.87)}$$

This is a recursion relation among the Legendre polynomials.

Next differentiate (A.81) with respect to z instead of r:

$$\frac{r}{\sqrt{1+r^2-2rz}} = \frac{1+r^2-2rz}{r}\sum_l \left(\frac{1}{r}\right)^l P'_l(z) \quad \text{(A.88)}$$

Use (A.81) again on the left-hand side and collect the coefficients of powers of r. The result is

$$P'_{l+1} + P'_{l-1} = 2zP'_l(z) + P_l(z) \quad \text{(A.89)}$$

Differentiate the first recursion relation (A.87):

$$(2l+1)P_l(z) + z(2l+1)P'_l(z) = (l+1)P'_{l+1}(z) + lP'_{l-1}(z) \quad \text{(A.90)}$$

Use equation (A.89) to eliminate the term in $P'_l(z)$ in equation (A.90) and obtain the second recursion relation in a concise form:

$$(2l+1)P_l(z) = P'_{l+1}(z) - P'_{l-1}(z) \quad \text{(A.91)}$$

Series Expansion and Rodrigues Formula

Inserting the series expansion

$$P_l(z) = \sum_{n=0}^{l} a_n z^n \quad \text{(A.92)}$$

into equation (A.69) and comparing coefficients of powers of z gives a recursion relation for the coefficients

$$a_{n+2} = \frac{n(n+1) - l(l+1)}{(n+1)(n+2)} a_n \tag{A.93}$$

The first coefficient a_o will be determined by the normalization.

Define the functions

$$f_l(z) = \frac{d^l}{dz^l} (z^2 - 1)^l \tag{A.94}$$

These are polynomials of order l. From the binomial expansion

$$(z^2 - 1)^l = \sum_{m=0}^{l} (-1)^{l-m} \frac{l!}{m!(l-m)!} z^{2m} \tag{A.95}$$

and the rule

$$\frac{d^l}{dz^l} z^N = \frac{N!}{(N-l)!} z^{N-l} \tag{A.96}$$

write $f_l(z)$ as

$$f_l(z) = \sum_m (-1)^{l-m} \frac{l!}{m!(l-m)!} \frac{(2m)!}{(2m-l)!} z^{2m-l} \tag{A.97}$$

The sum over m goes from the lowest integer such that $2m \geq l$ up to l. With $n = 2m - l$

$$f_l(z) = \sum_{n=0 \text{ or } 1}^{l} a_n z^n \tag{A.98}$$

Only even or odd powers occur according as l is even or odd. The coefficients are

$$a_n = (-1)^{(l-n)/2} \frac{l!}{\left(\frac{l+n}{2}\right)! \left(\frac{l-n}{2}\right)!} \frac{(n+l)!}{n!} \tag{A.99}$$

These a_n satisfy the recursion relation (A.93), and so the $f_l(z)$ are indeed proportional to the Legendre polynomials.

Rewrite equation (A.94) in terms of $w = z - 1$ and take the limit $w \to 0$ to get

$$\lim_{z \to 1} f_l(z) = 2^l l! \tag{A.100}$$

Since $P_l(1) = 1$, Rodrigues' formula for the Legendre polynomials follows directly:

$$P_l(z) = \frac{1}{2^l l!} \frac{d^l}{dz^l} (z^2 - 1)^l \tag{A.101}$$

Spherical Harmonics

The spherical harmonics $Y_l^m(\theta, \phi)$ are solutions to[6]

$$-i \frac{\partial}{\partial \phi} Y_l^m(\theta, \phi) = m Y_l^m(\theta, \phi) \tag{A.102}$$

---

[6]See equations (3.118) and (3.123).

and

$$-\left(\frac{1}{\sin\theta}\frac{\partial}{\partial\theta}\sin\theta\frac{\partial}{\partial\theta}+\frac{1}{\sin^2\theta}\frac{\partial^2}{\partial\phi^2}\right)Y_l^m(\theta,\phi)=l(l+1)Y_l^m(\theta,\phi) \quad \text{(A.103)}$$

They are conventionally normalized on the unit sphere so that

$$\int Y_{l'}^{m'}(\theta,\phi)^* Y_l^m(\theta,\phi)d\Omega = 1 \quad \text{(A.104)}$$

$Y_l^0(\theta,\phi)$ is proportional to $P_l(\cos\theta)$. The normalizations (A.84) and (A.104) together imply

$$Y_l^0(\theta,\phi) = \sqrt{\frac{2l+1}{4\pi}} P_l(\cos\theta) \quad \text{(A.105)}$$

The choice (A.105) specifies the one phase that is still arbitrary. The phases of the remaining Y_l^m are then determined by the requirement that $|lm\rangle$ can be obtained from $|l0\rangle$ by repeated application of the raising and lowering operators.[7]

$$Y_0^0(\theta,\phi) = \frac{1}{\sqrt{4\pi}}$$

$$Y_1^0(\theta,\phi) = \sqrt{\frac{3}{4\pi}}\cos\theta \qquad Y_1^{\pm 1}(\theta,\phi) = \mp\sqrt{\frac{3}{8\pi}}\sin\theta e^{\pm i\phi}$$

$$Y_2^0(\theta,\phi) = \sqrt{\frac{5}{16\pi}}(3\cos^2\theta - 1) \qquad Y_2^{\pm 1}(\theta,\phi) = \mp\sqrt{\frac{15}{8\pi}}e^{\pm i\phi}\sin\theta\cos\theta$$

$$Y_2^{\pm 2} = \sqrt{\frac{15}{32\pi}}e^{\pm 2i\phi}\sin^2\theta$$

TABLE A.1: Some low-order spherical harmonics

$$P_0(z) = 1$$

$$P_1(z) = z \qquad P_2(z) = \frac{3}{2}z^2 - \frac{1}{2}$$

$$P_3(z) = \frac{5}{2}z^3 - \frac{3}{2}z \qquad P_4 = \frac{35}{8}z^4 - \frac{15}{4}z^2 + \frac{3}{8}$$

TABLE A.2: Some low-order Legendre polynomials

---

[7]See also equation (B.6).

484 Appendix A   Mathematical Tools

A.2.3 Solutions to the Free Radial Equation

Spherical Hankel Functions

The spherical Bessel, Neumann, and Hankel functions satisfy the free radial wave equation (see Section 8.3)

$$\left[\frac{1}{\rho^2} \frac{d}{d\rho} \rho^2 \frac{d}{d\rho} - \frac{l(l+1)}{\rho^2} + 1 \right] R_l(\rho) = 0 \tag{A.106}$$

each with different boundary conditions. The functions $u_l(\rho) = \rho R_l(\rho)$ satisfy the simpler differential equation

$$\left[\frac{d^2}{d\rho^2} - \frac{l(l+1)}{\rho^2} + 1 \right] u_l(\rho) = 0 \tag{A.107}$$

The most efficient way to derive the properties of this collection of functions is to define them as contour integrals. In particular, the spherical Hankel function is[8]

$$h_l^{(1)}(\rho) = -\frac{\rho^l}{2^l l!} \int_1^{1+i\infty} e^{i\rho z} (1-z^2)^l dz \tag{A.108}$$

The integral is along a vertical straight line from $z = 1$ to $z = 1 + i\infty$.

First we have to show that the function defined in equation (A.108) satisfies the differential equation (A.106). To that end, differentiate $\rho h_l(\rho)$ twice:

$$\frac{d^2}{d\rho^2} \left[\rho h_l^{(1)}(\rho) \right] = \left[\frac{l(l+1)}{\rho^2} - 1 \right] \rho h_l^{(1)}(\rho) + 2i(l+1)\rho^l \int_1^{1+i\infty} z e^{i\rho z} (1-z^2)^l dz$$

$$+ \rho^{l+1} \int_1^{1+i\infty} e^{i\rho z} (1-z^2)^{l+1} dz \tag{A.109}$$

Integrate the last term by parts. The surface term vanishes, and then the last two terms cancel. $\rho h_l(\rho)$ satisfies equation (A.107), so $h_l(\rho)$ itself is a solution to equation (A.106).

The integral in (A.108) diverges in the limit $\rho \to 0$. To find the leading term for small ρ set $x = -i\rho z$ and rewrite equation (A.108) as

$$h_l^{(1)}(\rho) = -\frac{\rho^l}{2^l l!} \frac{i}{\rho^{2l+1}} \int_{-i\rho}^\infty e^{-x} (x^2 + \rho^2)^l dx \xrightarrow[\rho \to 0]{} \frac{-i(2l)!}{2^l l! \rho^{l+1}} = -\frac{i(2l-1)!!}{\rho^{l+1}} \tag{A.110}$$

The large ρ limit is obtained by setting $z = 1 + it/\rho$ in equation (A.108), so that $h_l^{(1)}(\rho)$ is represented by an integral along the positive real axis:

$$h_l^{(1)}(\rho) = -\frac{\rho^l}{2^l l!} \int_0^\infty e^{i\rho} e^{-t} \left[\frac{t^2}{\rho^2} - \frac{2it}{\rho} \right]^l \frac{i}{\rho} dt$$

$$\xrightarrow[\rho \to \infty]{} -i \frac{\rho^l}{2^l l!} \frac{(-2i)^l}{\rho^{l+1}} e^{i\rho} \int_0^\infty e^{-t} t^l dt = \frac{1}{\rho} e^{i[\rho - (l+1)\pi/2]} \tag{A.111}$$

---

[8] In the main part of the text I often use the common notation $h_l(\rho)$ instead of $h_l^{(1)}(\rho)$.

For real ρ the spherical Hankel functions of the second kind are the complex conjugates of those of the first kind. Make a change of variable so that the complex conjugate can be computed in a straightforward way: Set $z = 1 + it$ in equation (A.108). The definition of the spherical Hankel functions becomes

$$h_l^{(1)}(\rho) = -\frac{i\rho^l}{2^l l!} e^{i\rho} \int_0^\infty \left[t^2 - 2it\right]^l e^{-\rho t} dt \tag{A.112}$$

The spherical Hankel functions of the second kind are

$$h_l^{(2)}(\rho) = \frac{i\rho^l}{2^l l!} e^{-i\rho} \int_0^\infty \left[t^2 + 2it\right]^l e^{-\rho t} dt = \frac{\rho^l}{2^l l!} \int_{-1}^{-1+i\infty} e^{i\rho z}(1-z^2)^l dz \tag{A.113}$$

The second form in equation (A.113) is obtained substituting $z = it - 1$. Since the argument below equation (A.109) works as long as the integrand vanishes at the endpoints of the integral, $h_l^{(2)}(\rho)$ is also a solution to equation (A.106).

For large ρ, the limiting behaviors of $h_l^{(2)}(\rho)$ are

$$h_l^{(2)}(\rho) \xrightarrow[\rho \to 0]{} \frac{i(2l-1)!!}{\rho^{l+1}} \quad \text{and} \quad h_l^{(2)}(\rho) \xrightarrow[\rho \to \infty]{} \frac{1}{\rho} e^{-i[\rho - (l+1)\pi/2]} \tag{A.114}$$

Equation (A.113) defines $h_l^{(2)}(\rho)$ for any complex ρ. The two kinds of spherical Hankel functions are two linearly independent solutions of the second-order differential equation (A.106).

Spherical Bessel and Neumann Functions

When ρ is real and positive, the spherical Bessel functions are defined as the real part of $h_l^{(1)}(\rho)$:

$$j_l(\rho) = \frac{1}{2}\left[h_l^{(1)}(\rho) + h_l^{(2)}(\rho)\right] = \frac{\rho^l}{2^{l+1} l!} \left[\int_{1+i\infty}^1 + \int_{-1}^{-1+i\infty}\right] e^{i\rho z}(1-z^2)^l dz \tag{A.115}$$

The two integrals in equation (A.115) go along two sides of an infinitely long rectangle. Since the integrand is an entire function, Cauchy's theorem implies that

$$\lim_{R \to \infty} \left[\int_{1+iR}^1 - \int_{-1}^1 + \int_{-1}^{-1+iR} + \int_{-1+iR}^{1+iR}\right] e^{i\rho z}(1-z^2)^l dz = 0 \tag{A.116}$$

For real positive ρ the last integral vanishes as $R \to \infty$. Therefore

$$j_l(\rho) = \frac{\rho^l}{2^{l+1} l!} \int_{-1}^1 e^{i\rho z}(1-z^2)^l dz \tag{A.117}$$

The spherical Hankel functions and the spherical Bessel functions have representations as integrals with the same integrand over different contours, as shown in Figure A.2. Equation (A.117) defines the spherical Bessel functions for any ρ. $j_l(\rho)$ is an entire function, real when ρ is real. For real positive ρ, the first few spherical Bessel functions are shown in Figure 8.2.

486 Appendix A Mathematical Tools

$$f(z) = \frac{\rho^l}{2^{l+1}l!} \int e^{i\rho z}(1-z^2)^l dz$$

$h_l^{(2)}(\rho)$ $\quad h_l^{(1)}(\rho)$

$j_l(\rho)$

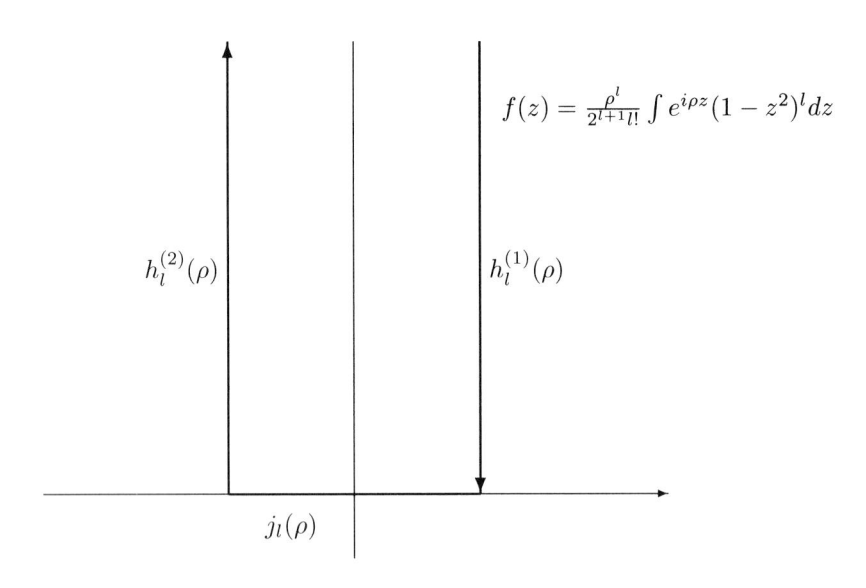

FIGURE A.2: Contours for the spherical Hankel and Bessel Functions

What are the limits of the spherical Bessel function for large and small ρ? When ρ is large, the asymptotic behavior is the real part of equation (A.111):

$$j_l(\rho) \xrightarrow[\rho\to\infty]{} \frac{1}{\rho}\cos\left[\rho - \frac{\pi}{2}(l+1)\right] \tag{A.118}$$

Since $h_l^{(1)}(\rho)$ is imaginary as $\rho \to 0$, equation (A.110) tells us that $j_l(\rho)$ vanishes in the limit. It is possible to do better, since in that limit the integral in equation (A.117) is a constant.

$$\lim_{\rho\to 0} j_l(\rho) = \frac{\rho^l}{2^{l+1}l!}\int_{-\frac{\pi}{2}}^{\frac{\pi}{2}} \cos^{l+1}\theta\, d\theta = \rho^l\frac{2^l l!}{(2l+1)!} = \frac{\rho^l}{(2l+1)!!} \tag{A.119}$$

Finally, the spherical Neumann functions are

$$n_l(\rho) = \frac{1}{2i}\left[h_l^{(1)}(\rho) - h_l^{(2)}(\rho)\right] \tag{A.120}$$

The behavior of these functions in the limits is

$$n_l(\rho) \xrightarrow[\rho\to 0]{} -\frac{(2l-1)!!}{\rho^{l+1}} \quad \text{and} \quad n_l(\rho) \xrightarrow[\rho\to\infty]{} \frac{1}{\rho}\sin\left[\rho - \frac{\pi}{2}(l+1)\right] \tag{A.121}$$

Relation to the Legendre Polynomials

Integrate equation (A.117) l times by parts so that it becomes

$$j_l(\rho) = \frac{\rho^l}{2^l l!}\left(\frac{-i}{\rho}\right)^l \int_{-1}^{1} e^{i\rho z}\frac{d^l}{dz^l}(1-z^2)^l dz \tag{A.122}$$

Then compare equation (A.122) with Rodrigues' formula (A.101) for the Legendre polynomials to get the important relation

$$j_l(\rho) = \frac{1}{2i^l} \int_{-1}^{1} e^{i\rho z} P_l(z) dz \qquad (A.123)$$

It is equation (A.123) that lets us find the expansion of plane waves in a Legendre series of the form

$$e^{ikz} = e^{ikr\cos\theta} = \sum_l a_l(r) P_l(\cos\theta) \qquad (A.124)$$

Since the plane wave is a solution to Laplace's equation, the coefficients $a_l(r)$ satisfy the free radial equation (A.106) with $\rho = kr$, and they are finite as $r \to 0$. For some constants a_l.

$$e^{ikr\cos\theta} = \sum_l a_l j_l(kr) P_l(\cos\theta) \qquad (A.125)$$

Multiply equation (A.125) by $P_l(\cos\theta)$ and integrate from $\cos\theta = -1$ to $\cos\theta = 1$, using (A.123) on the left and (A.84) on the right. The result is the solution for the coefficients:

$$e^{ikr\cos\theta} = \sum_l i^l (2l+1) j_l(kr) P_l(\cos\theta) \qquad (A.126)$$

Another way to obtain equation (A.126) is described in Section 8.3.1.

$$j_0(\rho) = \frac{\sin\rho}{\rho} \qquad j_1(\rho) = \frac{\sin\rho}{\rho^2} - \frac{\cos\rho}{\rho}$$

$$j_2(\rho) = \left(\frac{3}{\rho^3} - \frac{1}{\rho}\right) \sin\rho - \frac{3}{\rho^2} \cos\rho$$

TABLE A.3: Some low-order spherical Bessel functions

$$n_0(\rho) = -\frac{\cos\rho}{\rho} \qquad n_1(\rho) = -\frac{\cos\rho}{\rho^2} - \frac{\sin\rho}{\rho}$$

$$n_2(\rho) = -\left(\frac{3}{\rho^2} - \frac{1}{\rho}\right) \cos\rho - \frac{3}{\rho^3} \sin\rho$$

TABLE A.4: Some low-order spherical Neumann functions

A.2.4 Hermite Polynomials

The equation

$$H_n(z) = (-1)^n e^{z^2} \frac{d^n}{dz^n} e^{-z^2} \qquad (3.76)$$

$$h_0^{(1)}(\rho) = -i\frac{e^{i\rho}}{\rho} \qquad h_1^{(1)}(\rho) = -\left(\frac{1}{\rho} + \frac{i}{\rho^2}\right)e^{i\rho}$$

$$h_2^{(1)}(\rho) = -\left(-\frac{3}{\rho^2} + \frac{i}{\rho} - \frac{3i}{\rho^3}\right)e^{i\rho}$$

TABLE A.5: Some low-order spherical Hankel functions

$$j_l(\rho) \rightarrow \frac{\rho^l}{(2l+1)!!} \qquad n_l(\rho) \rightarrow -\frac{(2l-1)!!}{\rho^{l+1}}$$

$$h_l^{(1)}(\rho) \rightarrow -i\frac{(2l-1)!!}{\rho^{l+1}}$$

TABLE A.6: Limits of $j_l(\rho)$, $n_l(\rho)$, and $h_l^{(1)}(\rho)$ as $\rho \to 0$

$$j_l(\rho) \rightarrow \frac{1}{\rho}\cos\left[\rho - \frac{\pi}{2}(l+1)\right] \qquad n_l(\rho) \rightarrow \frac{1}{\rho}\sin\left[\rho - \frac{\pi}{2}(l+1)\right]$$

$$h_l^{(1)}(\rho) \rightarrow \frac{1}{\rho}\exp i\left[\rho - \frac{\pi}{2}(l+1)\right]$$

TABLE A.7: Limits of $j_l(\rho)$, $n_l(\rho)$, and $h_l^{(1)}(\rho)$ as $\rho \to \infty$

is a Rodrigues formula for the Hermite polynomials. It provides a representation analogous to equation (A.82) for the Legendre polynomials. Start from

$$e^{-(z-w)^2} = \sum_{n=0}^{\infty} \frac{w^n}{n!} \frac{\partial^n}{\partial w^n} e^{-(z-w)^2}\bigg|_{w=0} = \sum_{n=0}^{\infty} (-1)^n \frac{w^n}{n!} \frac{\partial^n}{\partial z^n} e^{-(z-w)^2}\bigg|_{w=0}$$

$$= \sum_{n=0}^{\infty} (-1)^n \frac{w^n}{n!} \frac{d^n}{dz^n} e^{-z^2} = e^{-z^2} \sum_{n=0}^{\infty} \frac{w^n}{n!} H_n(z) \tag{A.127}$$

Differentiate equation (A.127) with respect to z and compare the coefficients of powers of w in each term to get the rule

$$H_n'(z) = 2n H_{m-1}(z) \tag{A.128}$$

Alternatively, you can differentiate (A.127) with respect to w and compare the coefficients of powers of w in each term to get the recursion relation

$$H_{n+1}(z) = 2z H_n(z) - 2n H_{n-1}(z) \tag{A.129}$$

From the discussion in Section 3.2.4, it follows that

$$\int_{-\infty}^{\infty} H_n(z)^2 e^{-z^2} dz = 2^n n! \sqrt{\pi} \tag{A.130}$$

$$H_o = 1 \qquad\qquad H_1 = 2z$$
$$H_2 = 4z^2 - 2 \qquad\qquad H_3 = 8z^3 - 12z$$
$$H_4 = 16z^4 - 48z^2 + 12 \quad H_5 = 32z^5 - 160z^3 + 120z$$

TABLE A.8: Some low-order Hermite polynomials

A.2.5 Bessel Functions

Several properties of solutions to Bessel's equation were used in discussing the WKB approximation in Section 7.7. They can be derived efficiently following a procedure like the one in Section A.2.3 for the spherical functions.

Integral Representation

There are a number of equivalent integral representations for these functions. Start by defining

$$J_\nu(x) = \frac{1}{2\pi i} \int_{C_o} e^{\frac{x}{2}(t-\frac{1}{t})} t^{-\nu-1} dt \qquad (A.131)$$

The integrand is analytic in the t-plane except for a cut along the negative t-axis. The contour C_o starts at $t = -\infty - i\epsilon$, loops around the origin in the clockwise sense, and ends up at $t = -\infty + i\epsilon$. It is the same contour used for the gamma function shown in Figure A.1.

The integral in the definition (A.131) converges for any ν as long as $\operatorname{Re} z > 0$. If ν is an integer, the discontinuity across the negative real axis vanishes, and the contour can be deformed into a circle about the origin. Then equation (A.131) is the definition of the coefficients in a Laurent series, so

$$e^{\frac{x}{2}(t-\frac{1}{t})} = \sum_{n=-\infty}^{\infty} J_n(x) t^n \qquad (A.132)$$

providing a generating function for the Bessel functions of integer order.

Bessel's Equation

Set $t = 2u/x$. Then equation (A.131) becomes

$$J_\nu(x) = \frac{1}{2\pi i} \left(\frac{x}{2}\right)^\nu \int_{C_o} e^{u - x^2/4u} u^{-\nu-1} du \qquad (A.133)$$

The first and second derivatives are

$$J'_\nu(x) = \frac{\nu}{2} \left(\frac{x}{2}\right)^{\nu-1} \frac{1}{2\pi i} \int_{C_o} e^{u-x^2/4u} u^{-\nu-1} du$$

$$- \left(\frac{x}{2}\right)^{\nu+1} \frac{1}{2\pi i} \int_{C_o} \frac{1}{u} e^{u-x^2/4u} u^{-\nu-1} du \qquad (A.134)$$

490 Appendix A Mathematical Tools

$$J''_\nu(x) = \frac{\nu(\nu-1)}{4}\left(\frac{x}{2}\right)^{\nu-2}\frac{1}{2\pi i}\int_{C_o}e^{u-x^2/4u}u^{-\nu-1}du$$

$$-\frac{2\nu+1}{2}\left(\frac{x}{2}\right)^\nu\frac{1}{2\pi i}\int_{C_o}\frac{1}{u}e^{u-x^2/4u}u^{-\nu-1}du$$

$$+\left(\frac{x}{2}\right)^{\nu+2}\frac{1}{2\pi i}\int_{C_o}\frac{1}{u^2}e^{u-x^2/4u}u^{-\nu-1}du \quad \text{(A.135)}$$

Therefore,

$$x^2 J'_\nu(x) + x J_\nu(x) + \left(x^2-\nu^2\right)J_\nu(x)$$
$$= 4\left(\frac{x}{2}\right)^{\nu+2}\frac{1}{2\pi i}\int_{C_o}e^{u-x^2/4u}\left[u^{-\nu-1}+\frac{d}{du}u^{-\nu-1}+\frac{x^2}{4}u^{-\nu-3}\right]du \quad \text{(A.136)}$$

Integrate by parts the second term in the square brackets to get Bessel's equation

$$x^2 J''_\nu(x) + x J'_\nu(x) + \left(x^2-\nu^2\right)J_\nu(x) = 0 \quad \text{(A.137)}$$

Small x

Since the contour in the integral representation (A.133) avoids the origin, one can let $x \to 0$ inside the integral:

$$J_\nu(x) \to \frac{1}{2\pi i}\left(\frac{x}{2}\right)^\nu\int_{C_o}e^u u^{-\nu-1}du \quad \text{(A.138)}$$

Then use equations (A.60) and (A.68) to obtain

$$J_\nu(x) \to 2i\frac{1}{2\pi i}\left(\frac{x}{2}\right)^\nu\sin(-\pi\nu)\Gamma(-\nu) = \frac{1}{\Gamma(\nu+1)}\left(\frac{x}{2}\right)^\nu \quad \text{(A.139)}$$

The small x behavior of $J_\nu(x)$ is given by equation (A.139) for any ν except the negative integers. There, one should use (A.132), obtaining immediately $J_{-n}(x) = (-1)^n J_n(x)$.

You can get the entire expansion about $x = 0$ by expanding the factor $e^{-x^2/4u}$ in equation (A.133):

$$J_\nu(x) = \frac{1}{2\pi i}\sum_{n=0}^\infty (-1)^n\left(\frac{x}{2}\right)^{\nu+2n}\int_{C_o}e^u u^{-\nu-1-n}du$$

$$= \frac{1}{\pi}\sum_{n=0}^\infty (-1)^n\left(\frac{x}{2}\right)^{\nu+2n}\sin[\pi(\nu+n)]\Gamma(-\nu-n) \quad \text{(A.140)}$$

$$= \sum_{n=0}^\infty (-1)^n\left(\frac{x}{2}\right)^{\nu+2n}\frac{1}{\Gamma(\nu+n+1)}$$

Large x

The points $t = \pm i$ are saddle points of the function in the exponent of equation (A.131); that is,

$$\frac{d}{dt}\left[t-\frac{1}{t}\right]_{t=\pm i} = 0 \quad \text{(A.141)}$$

Deform the contour so that it passes through these two points. Near $t = -i$, let the tangent to the curve make an angle ϕ with the x-axis. That is, as t passes through the point $-i$, parametrize the contour by the real parameter ϵ, defined by $t = -i + \epsilon \exp i\phi$. ϵ increases from negative to positive values as the contour passes through $-i$. Then near $t = -i$

$$t = -2i + i\epsilon^2 e^{2i\phi} + \cdots \tag{A.142}$$

and there is a contribution amounting to

$$\frac{1}{2\pi i} e^{-ix} \int e^{\frac{x}{2} i \epsilon^2 e^{2i\phi}} \left(-i + \epsilon e^{i\phi}\right)^{-\nu-1} e^{i\phi} d\epsilon \tag{A.143}$$

to the integral (A.131). Deform the contour so that its slope at $t = -i$ is $\phi = \pi/4$. The expression (A.143) becomes

$$\frac{1}{2\pi i} e^{-ix} e^{i\frac{\pi}{4}} \int e^{-\frac{x}{2}\epsilon^2} \left(-i + \epsilon e^{i\frac{\pi}{4}}\right)^{-\nu-1} d\epsilon \tag{A.144}$$

As x gets larger and larger the range in ϵ that contributes significantly to the integral gets smaller and smaller, and asymptotically (A.144) is

$$\frac{1}{2\pi i} e^{-ix} e^{i\frac{\pi}{4}} e^{i\frac{\pi}{2}(\nu+1)} \int_{-\infty}^{\infty} e^{-\frac{x}{2}\epsilon^2} d\epsilon = \frac{1}{\sqrt{2\pi x}} e^{-i(x+\pi/4-\pi(\nu+1)/2)} \tag{A.145}$$

Similarly, deform the contour so that it also passes through the point $t = i$. The upshot of a similar computation is that the contour should pass through that point at an angle $\phi = 3\pi/4$, and there is a second large contribution to the integral (A.131) that is

$$\frac{1}{\sqrt{2\pi x}} e^{i(x+\pi/4-\pi(\nu+1)/2)} \tag{A.146}$$

for large real x. In this way we have determined the asymptotic behavior or these Bessel functions by the method of steepest descent:

$$J_\nu(x) \xrightarrow[x \to \infty]{} \sqrt{\frac{2}{\pi x}} \cos\left(x + \frac{\pi}{4} - (\nu+1)\frac{\pi}{2}\right) \tag{A.147}$$

A.3 ORTHOGONAL CURVILINEAR COORDINATES

A.3.1 Vector Calculus in Orthogonal Curvilinear Coordinates

Formulas from vector calculus in general orthogonal coordinates have been used throughout the main part of the text. Here is their derivation for cylindrical and spherical coordinates, as well as the less familiar parabolic and elliptic coordinate systems, with an application to the hydrogen atom spectrum and to the evaluation of some integrals.

In a coordinate system with coordinates q_1, q_2, and q_3, the path length between nearby points is ds, where

$$ds^2 = dx^2 + dy^2 + dz^2 = \sum_{ij} g_{ij} dq_i dq_j \tag{A.148}$$

in the limit of small s for some real symmetric matrix g. Here the discussion is restricted to the case where g is diagonal everywhere, in which case the coordinates are said to be **orthogonal**, and the path length has the form [9]

$$ds^2 = h_1^2\, dq_1^2 + h_2^2\, dq_2^2 + h_3^2\, dq_3^2 \tag{A.149}$$

The length of a short line between $q_i + \Delta q_i$ and q_i, the other two coordinates being held fixed, is $h_i \Delta q_i$. The three lines from (q_1, q_2, q_3) to $(q_1 + \Delta q_1, q_2, q_3)$, $(q_1, q_2 + \Delta q_2, q_3)$, and $(q_1, q_2, q_3 + \Delta q_3)$ are three legs of a rectangular parallelepiped, as in Figure A.3.

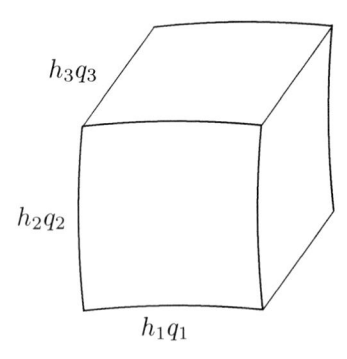

FIGURE A.3: The surfaces of constant q_i almost form a parallelepiped in the limit of small Δq_i.

- In Cartesian coordinates, $h_x = h_y = h_z = 1$.
- Cylindrical or axial coordinates are defined by

$$x = \rho \cos \phi \quad \text{and} \quad y = \rho \sin \phi \tag{A.150}$$

from which

$$ds^2 = d\rho^2 + \rho^2 d\phi^2 + dz^2 \tag{A.151}$$

or $h_\rho = h_z = 1$ and $h_\phi = \rho$.

- Spherical coordinates are defined by

$$x = r \sin \theta \cos \phi \quad y = r \sin \theta \sin \phi \quad z = r \cos \theta \tag{A.152}$$

the path length satisfies

$$ds^2 = dr^2 + r^2\, d\theta^2 + r^2 \sin^2 \theta\, d\phi^2 \tag{A.153}$$

so $h_r = 1$, $h_\theta = r$, and $h_\phi = r \sin \theta$.

---

[9] The functions $h_i(q) = +\sqrt{g_{ii}(q)}$ are **scale factors**.

Gradient

Let $f(\mathbf{r})$ be a function defined in three-dimensional space (a scalar field). Along any path from, $\mathbf{r}_1$ to $\mathbf{r}_2$

$$\int_{\mathbf{r}_1}^{\mathbf{r}_2} \boldsymbol{\nabla} f \cdot d\mathbf{s} = f(\mathbf{r}_2) - f(\mathbf{r}_1) \tag{A.154}$$

At every point define unit vectors $\hat{\boldsymbol{n}}_{q_i}$ that point in the direction of increasing q_i, the other two coordinates being held fixed, and resolve the gradient along these directions:

$$\boldsymbol{\nabla} f = \hat{\boldsymbol{n}}_{q_1} \boldsymbol{\nabla}_{q_1} f + \hat{\boldsymbol{n}}_{q_2} \boldsymbol{\nabla}_{q_2} f + \hat{\boldsymbol{n}}_{q_3} \boldsymbol{\nabla}_{q_3} f \tag{A.155}$$

The components of a small displacement are

$$\Delta \mathbf{r} = h_1 \Delta q_1 \hat{\boldsymbol{n}}_{q_1} + h_2 \Delta q_2 \hat{\boldsymbol{n}}_{q_2} + h_3 \Delta q_3 \hat{\boldsymbol{n}}_{q_3} \tag{A.156}$$

Along a short path from $\mathbf{r} = (q_1, q_2, q_3)$ to $\mathbf{r} + \Delta \mathbf{r} = (q_1 + \Delta q_1, q_2, q_3)$ the line integral of the gradient is

$$\int_{\mathbf{r}}^{\mathbf{r}+\Delta\mathbf{r}} \boldsymbol{\nabla} f \cdot \mathbf{s} = \boldsymbol{\nabla}_{q_1} f h_1 dq_1 = f(\mathbf{r} + \Delta \mathbf{r}) - f(\mathbf{r}) \tag{A.157}$$

and the same for q_2 and q_3. Therefore

$$\boldsymbol{\nabla}_{q_i} = \frac{1}{h_1} \frac{\partial}{\partial q_i} \tag{A.158}$$

Divergence

Let a surface S enclose a volume V, and $\hat{\boldsymbol{n}}$ at any point on the surface be an outward-pointing normal unit vector. Then Gauss' law says that for a vector field $\mathbf{A}(\mathbf{r})$

$$\int \boldsymbol{\nabla} \cdot \mathbf{A}(\mathbf{r}) \, dV = \int \mathbf{A}(\mathbf{r}) \cdot \hat{\boldsymbol{n}} \, dS \tag{A.159}$$

The volume of the little parallelepiped in Figure A.3 is $h_1 h_2 h_3 \Delta q_1 \Delta q_2 \Delta q_3$. The surface integral has six pieces. The two sides with fixed q_1 contribute

$$\Delta q_1 \frac{\partial}{\partial q_1} \left[A_{q_1}(q_1, q_2, q_3) h_2(q_1, q_2, q_3) h_3(q_1, q_2, q_3) \right] \Delta q_2 \Delta q_3 \tag{A.160}$$

to the right-hand side. The contributions from the other two pairs of surfaces can be obtained by cyclic permutation of the indices. The left hand side is

$$\boldsymbol{\nabla} \cdot \mathbf{A}(\mathbf{r}) h_1 h_2 h_3 \Delta q_1 \Delta q_2 \Delta q_3 \tag{A.161}$$

Thus

$$\boldsymbol{\nabla} \cdot \mathbf{A}(\mathbf{r}) = \frac{1}{h_1 h_2 h_3} \left[\frac{\partial}{\partial q_1}(A_{q_1} h_2 h_3) + \frac{\partial}{\partial q_2}(A_{q_2} h_1 h_3) + \frac{\partial}{\partial q_3}(A_{q_3} h_1 h_2) \right] \tag{A.162}$$

Laplacian

The Laplacian is the divergence of a gradient. From equations (A.158) and (A.162)

$$\nabla^2 f(\mathbf{r}) = \frac{1}{h_1 h_2 h_3} \left[\frac{\partial}{\partial q_1}\left(\frac{h_2 h_3}{h_1} \frac{\partial f}{\partial q_1}\right) + \frac{\partial}{\partial q_2}\left(\frac{h_1 h_3}{h_2} \frac{\partial f}{\partial q_2}\right) + \frac{\partial}{\partial q_3}\left(\frac{h_1 h_2}{h_3} \frac{\partial f}{\partial q_3}\right) \right] \tag{A.163}$$

- In Cartesian coordinates,

$$\nabla^2 f(\mathbf{r}) = \frac{\partial^2 f}{\partial x^2} + \frac{\partial^2 f}{\partial y^2} + \frac{\partial^2 f}{\partial z^2} \tag{A.164}$$

- In cylindrical coordinates,

$$\nabla^2 f(\mathbf{r}) = \frac{1}{\rho}\frac{\partial}{\partial \rho}\rho\frac{\partial f}{\partial \rho} + \frac{1}{\rho^2}\frac{\partial^2}{\partial \phi^2} + \frac{\partial^2}{\partial z^2} \tag{A.165}$$

- and in spherical coordinates, the rule is

$$\nabla^2 f(\mathbf{r}) = \frac{1}{r^2}\frac{\partial}{\partial r}r^2\frac{\partial f}{\partial r} + \frac{1}{\sin\theta}\frac{\partial}{\partial \theta}\sin\theta\frac{\partial f}{\partial \theta} + \frac{1}{\sin^2\theta}\frac{\partial^2 f}{\partial \phi^2} \tag{A.166}$$

Curl

The same method can be used to find the curl of a vector field. Consider a small rectangle whose opposite points are $\mathbf{r} = (q_1, q_2, q_3)$ and $\mathbf{r} + \Delta\mathbf{r} = (q_1 + \Delta q_1, q_2 + \Delta q_2, q_3)$. Its area is $h_1 h_2 \Delta q_1 \Delta q_2$. If $\mathbf{V}(\mathbf{r})$ is any continuous vector field, Stokes' theorem says that

$$\int \mathbf{V} \cdot d\mathbf{r} = \int \hat{\mathbf{n}} \cdot \nabla \times \mathbf{V} \, dS \tag{A.167}$$

The line integral on the left goes around the rectangle, the integral on the right is over the rectangle's surface, and the unit vector $\hat{\mathbf{n}}$ is orthogonal to the rectangle. Because the coordinate system is locally orthogonal, the right-hand side is

$$\int \hat{\mathbf{n}} \cdot \nabla \times \mathbf{V} \, dS = \int (\nabla \times \mathbf{V})_{q_3} \, dS \tag{A.168}$$

The line integral has four pieces. The two lines with fixed q_2 contribute

$$-\int_{q_1}^{q_1+\Delta q_1}\int_{q_2}^{q_1+\Delta q_2} \frac{\partial}{\partial q_2'}\left(V_{q_1}(q_1', q_2', q_3)h_1(q_1', q_2', q_3)\right) dq_1' dq_2' \tag{A.169}$$

and the remaining two sides contribute

$$\int_{q_2}^{q_2+\Delta q_2}\int_{q_1}^{q_1+\Delta q_1} \frac{\partial}{\partial q_1'}\left(V_{q_2}(q_1', q_2', q_3)h_2(q_1', q_2', q_3)\right) dq_1' dq_2' \tag{A.170}$$

to the line integral. So for small Δq_i,

$$\int \mathbf{V} \cdot d\mathbf{r} = \Delta q_1 \Delta q_2 \left[\frac{\partial}{\partial q_1}\left(V_{q_2}(q_1, q_2, q_3)h_2(q_1, q_2, q_3)\right) \right.$$
$$\left. - \frac{\partial}{\partial q_2}\left(V_{q_1}(q_1, q_2, q_3)h_1(q_1, q_2, q_3)\right)\right] \tag{A.171}$$

The surface integral in the same limit is

$$\int \hat{\mathbf{n}} \cdot \nabla \times \mathbf{V} \, dS = \Delta q_1 \Delta q_2 \, (\nabla \times \mathbf{V})_{q_3} h_1 h_2 \tag{A.172}$$

Section A.3 Orthogonal Curvilinear Coordinates 495

so the components of the curl are

$$(\nabla \times \mathbf{V})_{q_3} = \frac{1}{h_1 h_2} \left[\frac{\partial}{\partial q_1}(h_2 V_{q_2}) - \frac{\partial}{\partial q_2}(h_1 V_{q_1}) \right] \tag{A.173}$$

and two more by cyclic permutation of the three coordinates.

- In Cartesian coordinates, this is the defining rule.
- In cylindrical coordinates,

$$(\nabla \times V)_\rho = \frac{1}{\rho}\frac{\partial V_z}{\partial \phi} - \frac{\partial V_\phi}{\partial z} \tag{A.174a}$$

$$(\nabla \times V)_\phi = \frac{\partial V_\rho}{\partial z} - \frac{\partial V_z}{\partial \rho} \tag{A.174b}$$

$$(\nabla \times V)_z = \frac{1}{\rho}\frac{\partial(\rho V_\phi)}{\partial \rho} - \frac{1}{\rho}\frac{\partial V_\rho}{\partial \phi} \tag{A.174c}$$

- and in spherical coordinates,

$$(\nabla \times V)_r = \frac{1}{r\sin\theta}\left[\frac{\partial(\sin\theta V_\phi)}{\partial \theta} - \frac{\partial V_\theta}{\partial \phi}\right] \tag{A.175a}$$

$$(\nabla \times V)_\phi = \frac{1}{r}\frac{\partial}{\partial r}(rV_\theta) - \frac{1}{r}\frac{\partial V_r}{\partial \theta} \tag{A.175b}$$

$$(\nabla \times V)_\theta = \frac{1}{r\sin\theta}\frac{\partial V_r}{\partial \phi} - \frac{1}{r}\frac{\partial}{\partial r}(rV_\phi) \tag{A.175c}$$

A.3.2 Hydrogen Atom in Parabolic Coordinates

The hydrogen atom bound-state problem can be solved in closed form in parabolic coordinates as well as in spherical coordinates, providing another insight into the special degeneracy this problem possesses. The exact solution of the Coulomb scattering problem, which requires the hydrogen atom continuum (positive energy) eigenstates, is somewhat simpler in parabolic coordinates.

Define the variables u and v in terms of the usual spherical coordinates by[10]

$$u = r(1 - \cos\theta) = r - z \quad \text{and} \quad v = r(1 + \cos\theta) = r + z \tag{A.176}$$

The surfaces of constant ϕ are planes through the z-axis. In terms of the cylindrical coordinate ρ, defined as $\rho^2 = r^2 - z^2$,

$$u = \sqrt{\rho^2 + z^2} - z \tag{A.177}$$

or

$$z = \frac{\rho^2 - u^2}{2u} \tag{A.178}$$

For constant u, this is a paraboloid of revolution about the z-axis with focus at the origin, opening in the $+z$ direction. The surfaces of constant v are paraboloids

---
[10] Some books use the squares of these coordinates.

496 Appendix A Mathematical Tools

opening in the opposite direction. Parabolic coordinates are an orthogonal coordinate system in the sense defined in Section A.3.1.

Differential Path Length and Laplacian

Spherical coordinates are related to these parabolic coordinates by

$$r = \frac{v+u}{2} \quad \text{and} \quad z = r\cos\theta = \frac{v-u}{2} \tag{A.179}$$

So

$$r^2 \sin^2\theta = r^2 - z^2 = vu \tag{A.180}$$

The Cartesian coordinates are

$$x = \sqrt{vu}\cos\phi \qquad y = \sqrt{vu}\sin\phi \qquad z = \frac{v-u}{2} \tag{A.181}$$

Then

$$dx = \frac{1}{2}\sqrt{\frac{u}{v}}\cos\phi\,dv + \frac{1}{2}\sqrt{\frac{v}{u}}\cos\phi\,du - \sqrt{uv}\sin\phi\,d\phi \tag{A.182}$$

$$dy = \frac{1}{2}\sqrt{\frac{u}{v}}\sin\phi\,dv + \frac{1}{2}\sqrt{\frac{v}{u}}\sin\phi\,du + \sqrt{uv}\cos\phi\,d\phi \tag{A.183}$$

$$dz = \frac{1}{2}(dv - du) \tag{A.184}$$

and the square of the path length is

$$ds^2 = dx^2 + dy^2 + dz^2 = \frac{1}{4}\left(\frac{u}{v}+1\right)dv^2 + \frac{1}{4}\left(\frac{v}{u}+1\right)du^2 + uv\,d\phi^2 \tag{A.185}$$

The scale factors as defined in equation (A.149) are

$$h_u = \sqrt{\frac{u+v}{4u}} \qquad h_v = \sqrt{\frac{u+v}{4v}} \qquad h_\phi = \sqrt{uv} \tag{A.186}$$

so from the general rule (A.163), the Laplacian is

$$\nabla^2 = \frac{4}{u+v}\left(\frac{\partial}{\partial v}v\frac{\partial}{\partial v} + \frac{\partial}{\partial u}u\frac{\partial}{\partial u}\right) + \frac{1}{uv}\frac{\partial^2}{\partial\phi^2} \tag{A.187}$$

Schrödinger Equation

Set $E = -\alpha^2 mc^2/2n^2$, and $a = \hbar/mc\alpha$. Schrödinger's equation is

$$\left(\nabla^2 + \frac{4}{a(u+v)} - \frac{1}{a^2 n^2}\right)\psi(u,v,\phi) = 0 \tag{A.188}$$

The solutions will be eigenfunctions of L_z:

$$\psi(u,v,\phi) = \Psi(u,v)e^{im\phi} \tag{A.189}$$

After a bit of rearrangement the equation for Ψ can be written

$$\left(\frac{\partial}{\partial v}v\frac{\partial}{\partial v} - \frac{m^2}{4v} - \frac{v}{4a^2 n^2}\right)\Psi + \left(\frac{\partial}{\partial u}u\frac{\partial}{\partial u} - \frac{m^2}{4u} - \frac{u}{4a^2 n^2}\right)\Psi = -\frac{1}{a}\Psi \tag{A.190}$$

The first term on the left depends on v alone and the second on u alone. The differential operator in each bracket commutes with the other, and therefore with the sum. These two operators are the remaining operators that commute with the Hamiltonian and with each other. That is why parabolic coordinates are useful for the hydrogen atom problem.

Write $\Psi(u,v) = U(u)V(v)$, and find solutions to the two eigenvalue equations

$$\left(\frac{\partial}{\partial v}v\frac{\partial}{\partial v} - \frac{m^2}{4v} - \frac{v}{4a^2n^2}\right)V(v) = \lambda_v V(v) \tag{A.191}$$

and

$$\left(\frac{\partial}{\partial u}u\frac{\partial}{\partial u} - \frac{m^2}{4u} - \frac{u}{4a^2n^2}\right)U(u) = \lambda_u U(u) \tag{A.192}$$

The original equation is satisfied provided

$$\lambda_u + \lambda_v = -\frac{1}{a} \tag{A.193}$$

Solution

To solve equation (A.192) change variables to $x = u/2na$:

$$xU'' + U' - \frac{m^2}{4x}U - xU = 2na\lambda_u U \tag{A.194}$$

Find the asymptotic forms by arguments analogous to those used for spherical coordinates in Section 3.4.2. For large u or x,

$$xU'' = xU \tag{A.195}$$

The solutions go like $e^{\pm x}$. The boundary condition requires the minus sign. So with $U(x) = e^{-x}G(x)$

$$xG'' + (1-2x)G' - G - \frac{m^2}{4x}G = 2an\lambda_u G \tag{A.196}$$

If $G(x)$ goes like x^s for small x, then $s^2 = m^2/4$. Since s cannot be negative,

$$s = \frac{|m|}{2} \tag{A.197}$$

Next write $G(x) = x^s F(x)$:

$$xF'' + (2s - 2x + 1)F' - (2s + 1 + 2an\lambda_u)F = 0 \tag{A.198}$$

or with $y = 2x$:

$$yF''(y) + (|m|+1-y)F' - \left(na\lambda_u + \frac{|m|+1}{2}\right)F(y) = 0 \tag{A.199}$$

This is the same—with different coefficients—as equation (3.168) for the polynomial factor in the radial wave equation for the hydrogen atom in ordinary spherical coordinates. It is solved by series expansion. Set

$$F(y) = \sum_{k=0} a_k y^k \tag{A.200}$$

The condition that the coefficient of each y^k must vanish is a recursion relation among the coefficients:

$$\frac{a_{k+1}}{a_k} = \frac{k + na\lambda_u + \frac{|m|+1}{2}}{(k+1)(k+|m|+1)} \tag{A.201}$$

The eigenvalue condition—the condition that the series is really a polynomial—is

$$na\lambda_u + \frac{|m|+1}{2} = -N_u \tag{A.202}$$

for some nonnegative integer N_u.

The analysis of the equation for $V(v)$ is identical:

$$na\lambda_v + \frac{|m|+1}{2} = -N_v \tag{A.203}$$

for some nonnegative integer N_v. The original eigenvalue condition (A.193) becomes

$$n = |m| + 1 - N_u - N_v \tag{A.204}$$

Thus n is a positive integer; the energies satisfy the Bohr energy level formula.

Degeneracy

The levels are degenerate, and the degeneracy depends on the number of ways m, N_u, and N_v can be chosen to add up to the same nonnegative integer $n-1$. If $m = 0$, there are n ways to choose N_u, and N_v is then fixed. Otherwise, $N_u + N_v$ must add up to $n+1-|m|$; N_u can be any integer from 0 to $n-|m|$, leaving $n-|m|$ choices; and there are two possible values of m, $|m|$ and $-|m|$. The total is

$$n + \sum_{|m|=1}^{n-1} 2(n-|m|) = n^2 \tag{A.205}$$

A.3.3 Elliptic Coordinates

In Section 7.6 I referred to an orthogonal system called **elliptic coordinates**. Let $\mathbf{R}$ be a vector in the z-direction, and let $r_\pm$ be the distance from the coordinate point $\mathbf{r}$ to $\pm \mathbf{R}/2$. Define new coordinates u and v by

$$r_\pm = \left|\mathbf{r} \mp \frac{\mathbf{R}}{2}\right| = \frac{R}{2}(u \mp v) \tag{A.206}$$

or

$$u = \frac{r_- + r_+}{R} \tag{A.207}$$

$$v = \frac{r_- - r_+}{R} \tag{A.208}$$

The third coordinate is the azimuthal angle ϕ. The ranges of the new coordinates are $1 \leq u \leq \infty$ and $-1 \leq v \leq 1$. The surfaces of constant u are ellipsoids with foci at **R**/2. The surfaces of constant v are hyperboloids. In terms of the orthogonal cylindrical coordinates defined in equation (A.150) above,

$$r_\pm = \sqrt{\rho^2 + \left(z \mp \frac{R}{2}\right)^2} = \frac{R}{2}(u \mp v) \tag{A.209}$$

or

$$z = \frac{R}{2}uv \tag{A.210}$$

$$\rho = \frac{R}{2}\sqrt{u^2 - 1}\sqrt{1 - v^2} \tag{A.211}$$

$r_\pm$ are not orthogonal coordinates, but u and v are. To see this, define

$$u = \cosh \xi_1 \quad \text{and} \quad v = \cos \xi_2 \tag{A.212}$$

with $0 \le \xi_1 \le \infty$ and $0 \le \xi_2 \le \pi$. Then

$$z = \frac{R}{2} \cosh \xi_1 \cos \xi_2 \tag{A.213}$$

$$\rho = \frac{R}{2} \sinh \xi_1 \sin \xi_2 \tag{A.214}$$

and

$$dz = \frac{R}{2}(\sinh \xi_1 \cos \xi_2 \, d\xi_1 - \cosh \xi_1 \sin \xi_2 \, d\xi_2) \tag{A.215}$$

$$d\rho = \frac{R}{2}(\cosh \xi_1 \sin \xi_2 \, d\xi_1 + \sinh \xi_1 \cos \xi_2 \, d\xi_2) \tag{A.216}$$

The path length is

$$ds^2 = dz^2 + d\rho^2 + \rho^2 d\phi^2$$

$$= \frac{R^2}{4}(\cosh^2 \xi_1 - \cos^2 \xi_2)(d\xi_1^2 + d\xi_2^2) + \frac{R^2}{4}\sinh^2 \xi_1 \sin^2 \xi_2 d\phi^2 \tag{A.217}$$

$$= \frac{R^2}{4}\left[\frac{u^2 - v^2}{u^2 - 1}du^2 + \frac{u^2 - v^2}{1 - v^2}dv^2 + (u^2 - 1)(1 - v^2)\,d\phi^2\right]$$

In either form, the scale factors can be read off equation (A.217). In particular, the three-dimensional integration volume has the simple form

$$dx\,dy\,dz = \frac{R^3}{8}(u^2 - v^2)\,du\,dv\,d\phi \tag{A.218}$$

This is the identity you need to compute integrals like (7.180).

APPENDIX B

Rotation Matrices

B.1 ROTATION MATRICES—I

This appendix is a collection of some of the rotation matrix lore alluded to in the body of the text. It begins with a variety of results about spherical harmonics and rotation matrices. Then comes the proof of the general form of these matrices promised in Section 4.3.7. The projection theorem is useful for some calculations of properties of atoms and molecules. The last section, on averages over rotations, is an introduction to an more elegant treatment of the rotation group than I have discussed in the main text.

B.1.1 Rotation Matrices and Spherical Harmonics

Introduction

There are some relations between the spherical harmonics $Y_m^l(\theta, \phi)$ and the Wigner matrices $D^{(l)}(\alpha, \beta, \gamma)$ when l is an integer. Consider a spinless particle described by the coordinates **r**. If the Hamiltonian is spherically symmetric, the energy eigenfunctions are

$$\langle \mathbf{r} | E, l, m \rangle = R_l(r) Y_l^m(\theta, \phi) \tag{B.1}$$

Define states $|\theta, \phi\rangle$, or just $|\hat{\mathbf{n}}\rangle$ for short, that correspond to points on the unit sphere. Then

$$\langle \hat{\mathbf{n}}' | \hat{\mathbf{n}} \rangle = \delta(\cos\theta' - \cos\theta)\delta(\phi' - \phi) \tag{B.2}$$

The identity operator on these states is

$$I = \int |\hat{\mathbf{n}}\rangle\langle \hat{\mathbf{n}}| \, d\Omega = \int |\hat{\mathbf{n}}\rangle\langle \hat{\mathbf{n}}| \, d\cos\theta d\phi \tag{B.3}$$

Since the radial coordinate does not change under rotations, the rotations can be represented on the states $|\hat{\mathbf{n}}\rangle$:

$$R|\hat{\mathbf{n}}\rangle = |\hat{\mathbf{n}}'\rangle \tag{B.4}$$

where

$$n_i' = \sum_j \bar{R}_{ij} n_j \tag{B.5}$$

is also a unit vector.

Some Matrix Elements of the $D^{(l)}$ in Terms of the Spherical Harmonics

The spherical harmonics are the angular part of the wave functions associated with $|l, m\rangle$:

$$\langle \hat{\mathbf{n}} | lm \rangle = Y_l^m(\hat{\mathbf{n}}) \tag{B.6}$$

The wave function for the rotated state $R|l, m\rangle$ is

$$\langle \hat{n}'| R |l, m\rangle = \langle \hat{n}| l, m\rangle \tag{B.7}$$

The states $|l, m\rangle$ are also the basis for the $2l + 1$ dimensional irreducible representation of the rotations:[1]

$$R(\hat{n}, \theta)|l, m\rangle = \sum_{m'} |l, m'\rangle D^{(l)}(\hat{n}, \theta)_{m'm} \tag{B.8}$$

Therefore,

$$Y_l^m(\theta, \phi) = \sum_{m'} Y_l^{m'}(\hat{n}') D^{(l)}(\bar{R})_{m'm} \tag{B.9}$$

where $\hat{n}'$ and $\hat{n}$ are related by equation (B.5). Interchange $\hat{n}'$ and $\hat{n}$ and use the unitary property of the $D^{(l)}$ matrices:

$$Y_l^m(\theta', \phi') = \sum_{m'} Y_l^{m'}(\hat{n}) D^{(l)}(\bar{R}^{-1})_{m'm} = \sum_{m'} D^{(l)}(\bar{R})^*_{mm'} Y_l^{m'}(\hat{n}) \tag{B.10}$$

Now choose $\hat{n} = \hat{n}_z$. All the spherical harmonics vanish in the z-direction except those with $m = 0$, so

$$Y_l^m(\theta', \phi') = D^{(l)}(\bar{R})^*_{m0} Y_l^0(\hat{n}_z) = D^{(l)}(\bar{R})^*_{m0} \sqrt{\frac{2l+1}{4\pi}} P_l(1) \tag{B.11}$$

and since $P_l(1) = 1$,

$$D^{(l)}(\bar{R})_{m0} = \sqrt{\frac{4\pi}{2l+1}} Y_l^m(\theta', \phi')^* \tag{B.12}$$

Choose $\bar{R}$ to be a rotation around the y-axis followed by a rotation about the z-axis:

$$\bar{R} = \bar{R}(\hat{n}_z, \alpha) \bar{R}(\hat{n}_y, \beta) \tag{B.13}$$

so that $\hat{n}'$ points in the direction $\theta = \beta$, $\phi = \alpha$. In terms of the matrices $d^{(l)}(\beta)$ introduced in Section 4.3.5,

$$e^{-im\alpha} d^{(l)}(\beta)_{m0} = \sqrt{\frac{4\pi}{2l+1}} Y_l^m(\beta, \alpha)^* \tag{B.14}$$

Set $m = 0$ to obtain the important special case

$$d^{(l)}(\theta)_{00} = P_l(\cos \theta) \tag{B.15}$$

The element at the center of $d^{(l)}$ matrix is the Legendre polynomial of order l.

Addition Theorem for Spherical Harmonics

One more result illustrates the power of the algebraic language. Describe the rotations by the Euler angles. If $\hat{n}$ points in the direction of the spherical coordinates θ and ϕ, then

$$|\hat{n}\rangle = R(\phi, \theta, 0)|\hat{n}_z\rangle \tag{B.16}$$

---

[1] The relative phases of the spherical harmonics have to be chosen to satisfy (B.9).

Let
$$|\hat{n}_1\rangle = R(\phi_1, \theta_1, 0)|\hat{n}_z\rangle \quad \text{and} \quad |\hat{n}_2\rangle = R(\phi_2, \theta_2, 0)|\hat{n}_z\rangle \tag{B.17}$$

and set
$$\hat{n}' = \bar{R}(\phi_1, \theta_1, 0)^{-1}\hat{n}_2 \tag{B.18}$$

$\hat{n}'$ points in some direction Θ, Φ, where Θ is the angle $\hat{n}'$ makes with $\hat{n}_z$. But this is also the angle between $\hat{n}_1$ and $\hat{n}_2$:

$$\hat{n}' \cdot \hat{n}_z = \hat{n}_1 \cdot \hat{n}_2 = \cos\Theta \tag{B.19}$$

Furthermore,
$$Y_l^m(\Theta, \Phi) = \langle \hat{n}' | l, m \rangle = \langle \hat{n}_n | \bar{R}(\Phi_1, \theta_1, 0) | l, m \rangle$$
$$= \sum_{m'} Y_l^{m'}(\theta_2, \phi_2) D^{(l)}{}_{m'm}(\phi_1, \theta_1, 0) \tag{B.20}$$

Equation (B.20) relates spherical harmonics in three different directions. The most useful case is $m = 0$:

$$Y_l^0(\Theta, \Phi) = \sum_{m'} Y_l^{m'}(\theta_2, \phi_2) e^{-im'\phi_1} d^{(l)}{}_{m'0}(\theta_1) \tag{B.21}$$

From equation (B.14),
$$d^{(l)}{}_{m'0}(\theta_1) = e^{im'\theta_1} \sqrt{\frac{4\pi}{2l+1}} Y_l^{m'}(\theta_1, \phi_1)^* \tag{B.22}$$

Therefore,
$$Y_l^0(\Theta, \Phi) = \sum_{m'} Y_l^{m'}(\theta_2, \phi_2) e^{-im'\phi_1} e^{im'\phi_1} \sqrt{\frac{4\pi}{2l+1}} Y_l^{m'}(\theta_1, \phi_1)^* \tag{B.23}$$

and in particular,
$$P_l(\cos\Theta) = \sum_m \frac{4\pi}{2l+1} Y_l^m(\theta_1, \phi_1)^* Y_l^m(\theta_2, \phi_2) \tag{B.24}$$

The rule (B.24) is the **addition theorem** for spherical harmonics.

B.1.2 The Explicit Form of the Rotation Matrices

There are several ways to obtain the explicit formula (4.91). One of the more efficient is to construct tensors as products of operators that transform as spin half.

In Chapter 5 I considered tensors T_κ^q of integer rank only. It is possible to define tensors of half-integral rank also. Simply let u_i, where $i = \pm\frac{1}{2}$, be a pair of operators that transform like the spin-half representation of the rotations:

$$u'_i = R(\hat{n}, \psi) u_i R^\dagger(\hat{n}, \psi) = \sum_j u_j D^{(\frac{1}{2})}(\hat{n}, \psi)_{ji} \tag{B.25}$$

These two-dimensional rotation matrices can be parametrized by any two complex numbers a and b such that $|a|^2 + |b|^2 = 1$. They are the elements of an arbitrary 2×2 unitary unimodular matrix:

$$D^{(\frac{1}{2})} = \begin{pmatrix} a & -b \\ b^* & a^* \end{pmatrix} \tag{B.26}$$

In terms of ψ and the angles θ and ϕ that specify $\hat{n}$,

$$a = \cos\frac{\psi}{2} - i \sin\frac{\psi}{2} \cos\theta \quad \text{and} \quad b = i \sin\frac{\psi}{2} \sin\theta e^{-i\phi} \tag{B.27}$$

Example: The Three-Dimensional Representation

I will illustrate the method for $j = 1$ and then proceed to the general case. Consider the two-index tensor

$$T_{ij} = u_i u_j \tag{B.28}$$

Like the Cartesian tensors discussed in Section 5.2.1, the four quantities T_{ij} transform as a *reducible* representation. Separate T_{ij} into its symmetric and antisymmetric parts:[2]

$$T_{ij} = \frac{1}{2}(T_{ij} + T_{ji}) + \frac{1}{2}(T_{ij} - T_{ji}) \tag{B.29}$$

Under a rotation, the three symmetric combinations mix up only with each other. The antisymmetric combination is invariant. The symmetric combination transforms as $D^{(1)}$, and the antisymmetric combination as $D^{(0)}$.

Suppose you did not know the explicit forms of the $D^{(1)}$ matrices. You could find them from the symmetric part of T_{ij} above as follows: First find the three combinations T^q that transform as a rank-one tensor. The method is analogous to the construction in Section 5.2.2. Start with

$$T^1 = u_+ u_+ \tag{B.30}$$

and commute both sides with J_-:

$$[J_-, T^1] = \sqrt{2} T^o = u_- u_+ + u_+ u_- \tag{B.31}$$

and so forth, so that

$$T^o = \frac{u_- u_+ + u_+ u_-}{\sqrt{2}} \quad \text{and} \quad T^{-1} = u_- u_- \tag{B.32}$$

These rules, together with $[J_z, T^q] = qT^q$, guarantee that the three operators T^q are components of a rank-one tensor:

$$T'^q = R(a,b) T^q R(a,b)^\dagger = \sum_{q'=-1}^{1} T^{q'} D^{(1)}(a,b)_{q'q} \tag{B.33}$$

But also (for $q = 1$)

$$T'^1 = R(a,b) T^1 R(a,b)^\dagger = u'_+ u'_+ \tag{B.34}$$

---

[2] But here it is less complicated to do that than in equation (5.40).

Similarly

$$T'^0 = \frac{u'_- u'_+ + u'_+ u'_-}{\sqrt{2}} \quad \text{and} \quad T'^{-1} = R(a,b) T^{-1} R(a,b)^\dagger = u'_- u'_- \quad \text{(B.35)}$$

Now make a further simplification, that the two components u_+ and u_- commute with each other, so that $T^0 = \sqrt{2} u_+ u_-$. Some information appears to be lost, since the antisymmetric combination is then zero, but for the present calculation it is not needed.

$$T'^1 = u'^2_+ = a^2 T^1 + \sqrt{2} ab^* T^0 + b^{*2} T^{-1} \quad \text{(B.36a)}$$

$$T'^0 = \sqrt{2} u'_+ u'_- = -\sqrt{2} ab T^1 + (|a|^2 - |b|^2) T^0 + \sqrt{2} a^* b^* T^{-1} \quad \text{(B.36b)}$$

$$T'^{-1} = u'^2_- = b^2 T^1 - \sqrt{2} a^* b T^0 + a^{*2} T^{-1} \quad \text{(B.36c)}$$

The coefficients form the matrix elements of $D^{(1)}$:

$$D^{(1)} = \begin{pmatrix} a^2 & -\sqrt{2} ab & b^2 \\ \sqrt{2} ab^* & (|a|^2 - |b|^2) & -\sqrt{2} a^* b \\ b^{*2} & \sqrt{2} a^* b^* & a^{*2} \end{pmatrix} \quad \text{(B.37)}$$

For a rotation by ψ around the z-axis, $a = e^{-i\psi/2}$ and $b = 0$.

$$D^{(1)} = \begin{pmatrix} e^{-i\psi} & 0 & 0 \\ 0 & 1 & 0 \\ 0 & 0 & e^{i\psi} \end{pmatrix} \quad \text{(B.38)}$$

while for a rotation about the y-axis, $a = \cos \psi/2$ and $b = \sin \psi/2$.

$$D^{(1)} = \begin{pmatrix} \cos^2 \frac{\psi}{2} & -\sqrt{2} \sin \frac{\psi}{2} \cos \frac{\psi}{2} & \sin^2 \frac{\psi}{2} \\ \sqrt{2} \sin \frac{\psi}{2} \cos \frac{\psi}{2} & \cos \psi & -\sqrt{2} \sin \frac{\psi}{2} \cos \frac{\psi}{2} \\ \sin^2 \frac{\psi}{2} & \sqrt{2} \sin \frac{\psi}{2} \cos \frac{\psi}{2} & \cos^2 \frac{\psi}{2} \end{pmatrix} \quad \text{(B.39)}$$

in agreement with equation (4.90).

The General Case: Tensors of any Rank:

Construct an N-index tensor out of the $u_\pm$:

$$T_{i_1 i_2 \cdots i_N} = u_{i_1} u_{i_2} \cdots u_{i_N} \quad \text{(B.40)}$$

There are 2^N combinations, but only one that can be the top component of an N-index tensor, namely the combination that has N spin-up factors u_+. Commuting that product with J_z is equivalent to multiplying it by $N/2$, so this one is a component of an irreducible tensor[3] with $q = N/2$.

$$T_{\frac{N}{2}}^{\frac{N}{2}} = u_+ u_+ \cdots u_+ \quad \text{(B.41)}$$

---

[3] The superscript is the value of q, where $[J_z, T^q] = q T^q$.

This object is one of $2j+1$ combinations of products of the u_i that mix up among themselves as an irreducible representation of rotations. The generalization of (B.31) is

$$[J_-, u_+u_+ \cdots u_+] = u_-u_+u_+ \cdots u_+ + u_+u_-u_+ \cdots u_+ \cdots u_+u_+u_+ \cdots u_+u_- \quad \text{(B.42)}$$

There are precisely N terms, each with a u_- in one place, and u_+ in the rest. Again the computation is simpler if $[u_+, u_-] = 0$. Then equation (B.42) becomes

$$[J_-, u_+u_+ \cdots u_+] = Nu_-u_+u_+ \cdots u_+ \quad \text{(B.43)}$$

Now

$$[J_-, u_+u_+ \cdots u_+] = \left[J_-, T_{\frac{N}{2}}^{\frac{N}{2}}\right] = \sqrt{N} T_{\frac{N}{2}}^{\frac{N}{2}-1} \quad \text{(B.44)}$$

so the next component of this rank $N/2$ tensor is

$$T_{\frac{N}{2}}^{\frac{N}{2}-1} = \sqrt{N} u_-u_+u_+ \cdots u_+ = \sqrt{N}(u_+)^{N-1} u_- \quad \text{(B.45)}$$

Commute both sides again with J_-, using

$$\left[J_-, T_{\frac{N}{2}}^{\frac{N}{2}-1}\right] = \sqrt{2}\sqrt{N-1}\, T_{\frac{N}{2}}^{\frac{N}{2}-2} \quad \text{(B.46)}$$

and

$$\left[J_-, (u_+)^{N-1}\right] = (N-1)(u_+)^{N-2} u_- \quad \text{(B.47)}$$

so that

$$T_{\frac{N}{2}}^{\frac{N}{2}-2} = \sqrt{\frac{N(N-1)}{2}} (u_+)^{N-2} (u_-)^2 \quad \text{(B.48)}$$

The general rule should be evident. It is

$$T_{\frac{N}{2}}^{\frac{N}{2}-k} = \sqrt{\binom{N}{k}} (u_+)^{N-k} (u_-)^k \quad \text{(B.49)}$$

Proof of Equation (B.49). The proof is by a simple induction. Assume the result for some k: Then

$$\left[J_-, T_{\frac{N}{2}}^{\frac{N}{2}-k}\right] = \sqrt{(N-k)(k+1)}\, T_{\frac{N}{2}}^{\frac{N}{2}-k-1} \quad \text{(B.50)}$$

while

$$\left[J_-, \sqrt{\binom{N}{k}}(u_+)^{N-k}(u_-)^k\right] = \sqrt{\binom{N}{k}}(u_-)^k \left[J_-, (u_+)^{N-k}\right]$$

$$= \sqrt{\binom{N}{k}}(u_-)^{k+1}(N-k)(u_+)^{N-k-1} \quad \text{(B.51)}$$

So

$$T_{\frac{N}{2}}^{\frac{N}{2}-k-1} = \sqrt{\binom{N}{k+1}}(u_+)^{N-k-1}(u_-)^{k+1} \quad \text{(B.52)}$$

506 Appendix B Rotation Matrices

or, with $N = 2j$ and $q = \frac{N}{2} - k$,

$$T_j^q = \sqrt{\binom{2j}{j-q}}(u_+)^{j+q}(u_-)^{j-q} = \sqrt{\frac{(2j)!}{(j+q)!(j-q)!}}(u_+)^{j+q}(u_-)^{j-q} \tag{B.53}$$

∎

Now calculate the rotation matrices: Under a rotation (set $N = 2j$)

$$T'^q_j = R(a,b)T_j^q R(a,b)^\dagger = \sum_{q'} T_j^{q'} D^{(j)}(a,b)_{q'q} \tag{B.54}$$

This is also

$$T'^q_j = \sqrt{\frac{(2j)!}{(j+q)!(j-q)!}}(au_+ + b^*u_-)^{j+q}(-bu_+ + a^*u_-)^{j-q}$$

$$= \sqrt{\frac{(2j)!}{(j+q)!(j-q)!}}$$

$$\times \sum_{\mu=0}^{j+q} \binom{j+q}{\mu}(au_+)^{j+q-\mu}(b^*u_-)^\mu \sum_{\nu=0}^{j-q}\binom{j-q}{\nu}(-bu_+)^{j-q-\nu}(a^*u_-)^\nu$$

$$= \frac{\sqrt{(2j)!(j+q)!(j-q)!}}{(j+q-\mu)!(j-q-\nu)!\mu!\nu!}a^{j+q-\mu}(-b)^{j-q-\nu}a^{*\nu}b^{*\mu}(u_+)^{2j-\mu-\nu}(u_-)^{\mu+\nu} \tag{B.55}$$

The sums go over all integers μ and ν such that none of the four factors in the denominator is the factorial of a negative number.

Next set $\nu = j - \mu - q'$:

$$T'^q_j = \sum_{\mu,q'} \sqrt{\frac{(j+q')!(j-q')!}{(2j)!}} \frac{\sqrt{(2j)!(j+q)!(j-q)!}}{(j+q-\mu)!(q'+\mu-q)!\mu!(j-\mu-q')!}$$

$$\times a^{j+q-\mu}(-b)^{\mu+q'-q}a^{*j-\mu-q'}b^{*\mu} T_j^{q'} \tag{B.56}$$

This is the final form of the matrix.

$$D^{(j)}(a,b)_{q'q} = \sum_\mu \frac{\sqrt{(j+q')!(j-q')!(j+q)!(j-q)!}}{(j+q-\mu)!(q'+\mu-q)!\mu!(j-\mu-q')!}$$

$$\times a^{j+q-\mu}(-b)^{\mu+q'-q}a^{*j-\mu-q'}b^{*\mu} \tag{B.57}$$

It is enough to know the rule for rotations about the y-axis. Set $a = \cos\beta/2$ and

Section B.1 Rotation Matrices—I 507

$b = \sin \beta/2$ to get

$$d^{(j)}_{q'q}(\beta) = (-1)^{q'-q} \sum_{\mu} (-1)^{\mu} \frac{\sqrt{(j+q')!(j-q')!(j+q)!(j-q)!}}{(j+q-\mu)!(q'+\mu-q)!\mu!(j-\mu-q')!}$$

$$\times \left(\cos\frac{\beta}{2}\right)^{2j-2\mu+q-q'} \left(\sin\frac{\beta}{2}\right)^{2\mu+q'-q} \quad (B.58)$$

These are the elements of the Wigner matrices.

Example

Since

$$e^{-i\pi J_y} J_z e^{i\pi J_y} = -J_z \quad (B.59)$$

The state $e^{-i\pi J_y}|j,q\rangle$ is proportional to $|j,-q\rangle$. That is, $d^{(j)}(\pi)_{q',q}$ vanishes unless $q' = -q$. In the general formula above, when $\beta \to \pi$ the factor $\sin \beta/2$ is always unity, while the factor $(\cos \beta/2)^{2j-2\mu+q-q'}$ vanishes unless the exponent is zero. So the only term in the sum is $\mu = j+q$, and

$$d^{(j)}_{-q,q} = (-1)^{j-q} \quad (B.60)$$

B.1.3 The Projection Theorem

This handy theorem is useful for computing matrix elements of magnetic moments in atomic spectroscopy. The content of the Wigner-Eckart theorem is that the ratios of the matrix elements of T^q_κ to the Clebsch-Gordan coefficients are independent of m', m, and q. Write the theorem in the form

$$\langle \alpha', j', m'| T^q_\kappa |\alpha, j, m\rangle = \frac{1}{\sqrt{2j'+1}} \langle j, \kappa; j', m'| j, \kappa; m, q\rangle \langle \alpha', j'||T_\kappa||\alpha, j\rangle \quad (5.59)$$

The projection theorem is a theorem about diagonal matrix elements of vector operators: $\alpha = \alpha'$, and $j' = j$. For any vector operator $\mathbf{V}$ equation (5.59) becomes

$$\langle \alpha, j, m'| V^q |\alpha, j, m\rangle = \frac{1}{\sqrt{2j+1}} \langle j, 1; j, m'| j, 1; m, q\rangle \langle \alpha, j||V||\alpha, j\rangle \quad (B.61)$$

and in particular this is true for $\mathbf{J}$ itself. The ratios of matrix elements of the same vector components of $\mathbf{V}$ and $\mathbf{J}$ are

$$\frac{\langle \alpha, j, m'| V^q |\alpha, j, m\rangle}{\langle \alpha, j, m'| J^q |\alpha, j, m\rangle} = \frac{\langle \alpha, j||V||\alpha, j\rangle}{\langle \alpha, j||J||\alpha, j\rangle} \quad (B.62)$$

In spherical components (see Section 5.1.3.)

$$\mathbf{V} \cdot \mathbf{J} = J^\circ V^\circ - J^+ V^- - J^- V^+ = J_z V^\circ + \sqrt{2} J_+ V^- - \sqrt{2} J_- V^+ \quad (B.63)$$

The matrix elements of (B.63) are

$$\langle \alpha, j, m'| \mathbf{V} \cdot \mathbf{J} |\alpha, j, m\rangle = m\langle \alpha, j, m'| V^\circ |\alpha, j, m\rangle$$

$$+ \sqrt{j(j+1) - m(m+1)} \langle \alpha, j, m'| V^- |\alpha, j, m+1\rangle$$

$$- \sqrt{j(j+1) - m(m-1)} \langle \alpha, j, m'| V^+ |\alpha, j, m-1\rangle \quad (B.64)$$

Each term on the right is proportional to the reduced matrix element $\langle \alpha, j || V || \alpha, j \rangle$, with a proportionality constant independent of the vector operator $\mathbf{V}$:

$$\langle \alpha, j, m' | \mathbf{V} \cdot \mathbf{J} | \alpha, j, m \rangle = c_{j,m',m} \langle \alpha, j || V || \alpha, j \rangle \tag{B.65}$$

The left-hand side is the matrix element of a scalar operator, so it vanishes unless $m' = m$, and it is independent of m. Therefore,

$$\langle \alpha, j, m' | \mathbf{V} \cdot \mathbf{J} | \alpha, j, m \rangle = c_j \delta_{m'm} \langle \alpha, j || V || \alpha, j \rangle \tag{B.66}$$

The ratio of the two matrix elements is independent of c_j:

$$\frac{\langle \alpha, j, m' | V^q | \alpha, j, m \rangle}{\langle \alpha, j, m' | J^q | \alpha, j, m \rangle} = \frac{\langle \alpha, j, m | \mathbf{V} \cdot \mathbf{J} | \alpha, j, m \rangle}{j(j+1)} \tag{B.67}$$

or

$$\langle \alpha, j, m' | V_i | \alpha, j, m \rangle = \frac{\langle \alpha, j, m | \mathbf{V} \cdot \mathbf{J} | \alpha, j, m \rangle}{j(j+1)} \langle \alpha, j, m' | J_i | \alpha, j, m \rangle \tag{B.68}$$

Equation (B.68) is the projection theorem.

B.2 ROTATION MATRICES—II

There are some powerful theorems about *averages* of products of matrix elements of the $D^{(j)}(\alpha, \beta, \gamma)$ over the rotation group.

Let $F(\bar{R})$ be some function of the rotations $\bar{R}$. $\bar{R}$ itself is a function of the Euler angles α, β, and γ, where[4] $0 \leq \beta \leq \pi$ while $0 \leq \alpha, \gamma \leq 4\pi$. Define

$$\langle F(\bar{R}) \rangle = \frac{\iiint F(\bar{R}) d\bar{R}}{\iiint d\bar{R}} \tag{B.69}$$

where

$$\iiint F(\bar{R}) d\bar{R} = \int_0^{4\pi} d\alpha \int_0^{\pi} \sin\beta \, d\beta \int_0^{4\pi} d\gamma \, F[\bar{R}(\alpha, \beta, \gamma)] \tag{B.70}$$

The denominator is the "volume" of the rotation group:

$$\iiint d\bar{R} = 32\pi^2 \tag{B.71}$$

B.2.1 Averages over Products of Rotation Matrices

Average of One $D^{(j)}$ Matrix Element

$$\begin{aligned}\left\langle D^{(j)}(\alpha, \beta, \gamma)_{mn} \right\rangle &= \frac{1}{32\pi^2} \int_0^{4\pi} d\alpha \int_0^{\pi} \sin\beta \, d\beta \int_0^{4\pi} d\gamma \, D^{(j)}(\alpha, \beta, \gamma)_{mn} \\ &= \frac{1}{32\pi^2} \int_0^{4\pi} d\alpha \int_0^{\pi} \sin\beta \, d\beta \int_0^{4\pi} d\gamma \, e^{-i(\alpha m + \gamma n)} d^{(j)}(\beta)_{mn}\end{aligned} \tag{B.72}$$

---

[4] Even though I use the notation $\bar{R}$, I really mean the 2×2 unitary unimodular matrices.

Since $4m$ is always an integer, the α and γ integrals vanish unless $m = n = 0$:

$$\int_0^{4\pi} e^{-i\alpha m} d\alpha = 4\pi \delta_{m,0} \quad \text{and} \quad \int_0^{4\pi} e^{-i\gamma n} d\gamma = 4\pi \delta_{n,0} \tag{B.73}$$

In particular the integral vanishes for all m and m' when j is a half-integer. Then from equation (B.15)

$$\left\langle D^{(j)}(\alpha,\beta,\gamma)_{m,n} \right\rangle = \frac{1}{2}\delta_{n,0}\delta_{m,0} \int_0^\pi \sin\beta\, d^{(j)}(\beta)_{00} d\beta = \frac{1}{2}\delta_{n,0}\delta_{m,0} \int_{-1}^1 P_j(z) dz$$

$$= \delta_{n,0}\delta_{m,0}\delta_{j,0} \tag{B.74}$$

Average of a Product of Two $D^{(j)}$ Matrix Elements

To compute this average, one needs the coefficient of the singlet $j = 0$ state in the Clebsch-Gordan expansion of states with angular momenta j_1 and j_2. This is[5]

$$\left\langle \psi_{m_1,m_2}^{j_1,j_2} \middle| \phi_{00}^{j_1,j_2} \right\rangle \tag{B.75}$$

The coefficient vanishes unless $m_1 + m_2 = 0$, and $j = 0$ occurs in the product only if $j_1 = j_2$. The expansion of $\left|\phi_{00}^{j,j}\right\rangle$ in the $\left|\psi_{m_1,m_2}^{j,j}\right\rangle$ has the form

$$\left|\phi_{00}^{j,j}\right\rangle = \sum_\mu a_\mu \left|\psi_{\mu,-\mu}^{j,j}\right\rangle \tag{B.76}$$

Applying $J_+ = J_{1+} + J_{2+}$ to both sides

$$a_{m-1}\sqrt{j(j+1) - m(m-1)} + a_m\sqrt{j(j+1) - m(m-1)} = 0 \tag{B.77}$$

from which $a_{m-1} = -a_m$. Each term appears once, and the signs alternate. The normalization requires that up to a phase

$$\left\langle \psi_{m,-m}^{j,j} \middle| \phi_{00}^{j,j} \right\rangle = (-1)^{j-m}\sqrt{\frac{1}{2j+1}} \tag{B.78}$$

Use equation (B.74) and the Clebsch-Gordon series (4.147) to integrate over the product of two D-matrix elements:

$$\left\langle D^{(j_1)}(\alpha,\beta,\gamma)_{m_1',m_1} D^{(j_2)}(\alpha,\beta,\gamma)_{m_2',m_2} \right\rangle$$

$$= \sum_{j,m,m'} \left\langle \psi_{m_1',m_2'}^{j_1,j_2} \middle| \phi_{j,m'}^{j_1,j_2} \right\rangle \left\langle \psi_{m_1,m_2}^{j_1,j_2} \middle| \phi_{j,m}^{j_1,j_2} \right\rangle \left\langle D^{(j)}(\alpha,\beta,\gamma)_{m'm} \right\rangle \delta_{j,0}\delta_{m',0}\delta_{m,0}$$

$$= \left\langle \psi_{m_1',m_2'}^{j_1,j_2} \middle| \phi_{0,0}^{j_1,j_2} \right\rangle \left\langle \psi_{m_1,m_2}^{j_1,j_2} \middle| \phi_{0,0}^{j_1,j_2} \right\rangle \tag{B.79}$$

---

[5] The notation is defined in Section 4.5.1.

510 Appendix B Rotation Matrices

Finally, put in the values (B.78) for these Clebsch-Gordan coefficients:

$$\left\langle D^{(j_1)}(\alpha,\beta,\gamma)_{m'_1,m_1} D^{(j_2)}(\alpha,\beta,\gamma)_{m'_2,m_2} \right\rangle$$
$$= \frac{1}{2j_1+1}(-1)^{j_1+j_2-m_1-m'_1}\delta_{j_1,j_2}\delta_{m_1,-m_2}\delta_{m'_1,-m'_2} \quad (B.80)$$

Average of $D^{(j)*}D^{(j)}$

More useful than (B.80) is the average of a product of a matrix element of one $D^{(j)}$ with the complex conjugate of another. First use equation (B.60) to get the following identity:

$$d^{(j)}(\beta)_{m,m'} = \langle j,m|e^{i\pi J_y}e^{-i\beta J_y}e^{-i\pi J_y}|j,m'\rangle = (-1)^{m-m'}d^{(j)}_{-m,-m'}(\beta) \quad (B.81)$$

Therefore,

$$D^{(j)}(\alpha,\beta,\gamma)^*_{m,m'} = (-1)^{m-m'}D^{(j)}(\alpha,\beta,\gamma)_{-m,-m'} \quad (B.82)$$

Then from equation (B.80)

$$\left\langle D^{(j_1)}(\alpha,\beta,\gamma)^*_{m_1,m'_1} D^{(j_2)}(\alpha,\beta,\gamma)_{m_2,m'_2}\right\rangle$$
$$= (-1)^{m_1-m'_1}\left\langle D^{(j_1)}(\alpha,\beta,\gamma)_{-m_1,-m'_1} D^{(j_2)}(\alpha,\beta,\gamma)_{m_2,m'_2}\right\rangle \quad (B.83)$$
$$= \frac{1}{2j_1+1}\delta_{j_1,j_2}\delta_{m_1,m_2}\delta_{m'_1,m'_2}$$

The algebraically initiated reader will recognize (B.83) as an orthogonality theorem for irreducible representations of finite groups, generalized to the continuous case.

Average of a Product of Three $D^{(j)}$ Matrix Elements

Use the Clebsch-Gordan series (4.147) again to compute the average of three rotation matrix elements:

$$\left\langle D^{(j')}(\bar{R})^*_{\mu',m'} D^{(j)}(\bar{R})_{\mu,m} D^{(\kappa)}(\bar{R})_{q'q}\right\rangle$$
$$= \sum_{JMM'}\left\langle \psi^{j,\kappa}_{\mu,q'}\bigg|\phi^{j,\kappa}_{J,M'}\right\rangle\left\langle\psi^{j,\kappa}_{m,q}\bigg|\phi^{j,\kappa}_{J,M}\right\rangle\left\langle D^{(j')}(\bar{R})^*_{\mu',m'}D^{(J)}(\bar{R})_{M'M}\right\rangle \quad (B.84)$$
$$= \frac{1}{2J+1}\left\langle\psi^{j,\kappa}_{\mu,q'}\bigg|\phi^{j,\kappa}_{j',\mu'}\right\rangle\left\langle\psi^{j,\kappa}_{m,q}\bigg|\phi^{j,\kappa}_{j',m'}\right\rangle$$

B.2.2 The Wigner-Eckart Theorem Again

Here is a constructive, and instructive, proof of the theorem. It exhibits an explicit form for the reduced matrix element:

Write the rule that defines a spherical tensor as

$$T^q_\kappa = \sum_{q'} R^\dagger T^{q'}_\kappa R D^{(\kappa)}(\bar{R})_{q'q} \quad (B.85)$$

Section B.2 Rotation Matrices—II

Then

$$\langle \alpha', j', m' | T^q_\kappa | \alpha, j, m \rangle$$

$$= \sum_{q',\mu',\mu} \langle \alpha', j', m' | R^\dagger | \alpha', j', \mu' \rangle \langle \alpha', j', \mu' | T^{q'}_\kappa | \alpha, j, \mu \rangle \langle \alpha, j, \mu | R | \alpha, j, m \rangle D^{(\kappa)}(\bar{R})_{q'q}$$

$$= \sum_{q',\mu',\mu} D^{(j')}(\bar{R})^*_{\mu',m'} \langle \alpha', j', \mu' | T^{q'}_\kappa | \alpha, j, \mu \rangle D^{(j)}(\bar{R})_{\mu,m} D^{(\kappa)}(\bar{R})_{q'q}$$

(B.86)

The left-hand side is independent of $\bar{R}$, so the right-hand side must be also.

$$\langle \alpha', j', m' | T^q_\kappa | \alpha, j, m \rangle = \frac{1}{\sqrt{2j'+1}} \langle \psi^{j,\kappa}_{m,q} | \phi^{j,\kappa}_{j',m'} \rangle \langle \alpha', j' || T_\kappa || \alpha, j \rangle \quad \text{(B.87)}$$

with

$$\langle \alpha', j' || T_\kappa || \alpha, j \rangle = \sqrt{2j'+1} \sum_{q'\mu'\mu} \langle \alpha', j', \mu' | T^{q'}_\kappa | \alpha, j, \mu \rangle \langle \psi^{j,\kappa}_{\mu,q'} | \phi^{j,\kappa}_{j',\mu'} \rangle \quad \text{(B.88)}$$

Equation (B.88) is the Wigner-Eckart theorem, with an explicit expression for the reduced matrix element.

APPENDIX C

SU(3)

C.1 THE GROUP AND ALGEBRA

$SU(n)$ is the collection of $n \times n$ unitary matrices with unit determinant. Several of the problems at the end of Chapter 5 refer to the group $SU(3)$, its generators, and their representations. The group is the generalization of $SU(2)$ to three dimensions. It has applications in nuclear, atomic, molecular, and particle physics. $SU(3)$ is also the symmetry group of the three-dimensional harmonic oscillator, and that was the subject of these problems.

All the useful information about these $SU(3)$ matrices is contained in the generators. Any matrix in $SU(3)$ can be written in the form

$$U = e^{-i\Lambda} \tag{C.1}$$

where, as in $SU(2)$, Λ is Hermitean and Tr $\Lambda = 0$. Persuade yourself that there are eight independent matrices with these properties. The dimension of $SU(3)$ is 8. What is the dimension of $SU(n)$?

It is customary to develop the notation in as close analogy as possible to $SU(2)$.

$$U = \exp\left(-\frac{i}{2} \sum_{i=1}^{8} \theta_i \lambda_i\right) \tag{C.2}$$

The standard choice for eight traceless, Hermitean matrices λ_i are the Gell-Mann matrices

$$\lambda_1 = \begin{pmatrix} 0 & 1 & 0 \\ 1 & 0 & 0 \\ 0 & 0 & 0 \end{pmatrix} \quad \lambda_2 = \begin{pmatrix} 0 & -i & 0 \\ i & 0 & 0 \\ 0 & 0 & 0 \end{pmatrix} \quad \lambda_3 = \begin{pmatrix} 1 & 0 & 0 \\ 0 & -1 & 0 \\ 0 & 0 & 0 \end{pmatrix}$$

$$\lambda_4 = \begin{pmatrix} 0 & 0 & 1 \\ 0 & 0 & 0 \\ 1 & 0 & 0 \end{pmatrix} \quad \lambda_5 = \begin{pmatrix} 0 & 0 & -i \\ 0 & 0 & 0 \\ i & 0 & 0 \end{pmatrix} \quad \lambda_6 = \begin{pmatrix} 0 & 0 & 0 \\ 0 & 0 & 1 \\ 0 & 1 & 0 \end{pmatrix} \tag{C.3}$$

$$\lambda_7 = \begin{pmatrix} 0 & 0 & 0 \\ 0 & 0 & -i \\ 0 & i & 0 \end{pmatrix} \quad \lambda_8 = \frac{1}{\sqrt{3}} \begin{pmatrix} 1 & 0 & 0 \\ 0 & 1 & 0 \\ 0 & 0 & -2 \end{pmatrix}$$

Like the two-dimensional Pauli matrices, these are normalized so that

$$\text{Tr }(\lambda_i \lambda_j) = 2\delta_{ij} \tag{C.4}$$

The standard generators of the group $SU(3)$ can be taken to be

$$\bar{F}_i = \frac{1}{2}\lambda_i \tag{C.5}$$

and the "structure constants" f_{ijk} are defined by

$$[\bar{F}_i, \bar{F}_j] = i \sum_k f_{ijk} \bar{F}_k \tag{C.6}$$

In $SU(2)$, the f_{ijk} were the antisymmetric symbols ϵ_{ijk}, but that is not the case here.

C.2 SOME REPRESENTATIONS

A representation of $SU(3)$ is a collection of linear operators that have the same multiplication rule as the three-dimensional matrices that define the group. Here again, it is enough to find a collection of eight linear operators that have the same commutators as the eight $\bar{F}_a$ defined above. The demonstration of this idea in Section 4.2.4 is not special to rotations.

The key to understanding representations of algebras like this one lies in the following observation: When you combine identical representations to make a larger one—for example, in $SU(2)$ when you add spin half and spin half, using the Clebsch-Gordan coefficients—the operation of *exchanging* any pair of the constituent representations commutes with all the elements of the group, and with the generators. So eigenstates of the generators must be eigenstates of the permutation operation. They are even or odd under exchange of any pair of labels. There is a simple example in Section 4.5.2, on adding two spin half representations, and another in Problem 4.6.

$SU(2)$: First look at $SU(2)$ from this point of view. Higher dimensional representations of $SU(2)$ can be obtained by combining N of the fundamental two-dimensional representations. The states are labeled $|a_1, a_2, \cdots a_N\rangle$ where the a_i take on the values $\pm\frac{1}{2}$. These are eigenstates of $J_z = J_{1z} + J_{2z} + \cdots J_{Nz}$. The individual entries are eigenvalues of J_{iz}. In the top state, all the a_i are $+\frac{1}{2}$, and the eigenvalue of J_z is $j = N/2$:

$$J_z |++\cdots+\rangle = (J_{1z} + J_{2z} + \cdots J_{Nz})|++\cdots+\rangle = \frac{N}{2}|++\cdots+\rangle \tag{C.7}$$

By repeatedly applying J_- you can generate the irreducible representation with this maximum eigenvalue. All the states are symmetric in the a_i, and the dimension of the representation is the number of different states with N indices that are symmetric in all the a_i.

For example, if you start with two "spin up" states, and repeatedly apply J_-, you get the representation with three states

$$|++\rangle \qquad \frac{|+-\rangle + |-+\rangle}{\sqrt{2}} \qquad |--\rangle \tag{C.8}$$

514   Appendix C   SU(3)

and starting with three "spin up" states, and repeatedly apply J_-, you get the representation with four states

$$|+++\rangle \qquad \frac{|++-\rangle+|+-+\rangle+|-++\rangle}{\sqrt{3}}$$

$$\frac{|+--\rangle+|--+\rangle+|-+-\rangle}{\sqrt{3}} \qquad |---\rangle \tag{C.9}$$

Each state is completely symmetric under any permutation of the labels.

What is the dimension of this irreducible representation in general? If you did not already know the answer you could argue as follows: All the states can be obtained by repeated application of the lowering operator J_- to this top state, and they will all be completely symmetric. Furthermore, every completely symmetric state must occur in this representation. How many states are there? How many different ways can you choose N pluses and minuses? The number N_+ of pluses can be any integer from zero through N, so the answer is $N+1$. The dimension of this representation is $N+1 = 2j+1$.

$SU(3)$: In any representation of $SU(3)$ both F_3 and F_8 can be taken diagonal. The states must be labeled by the eigenvalues of both F_3 and F_8.

In the fundamental representation of $SU(2)$, the eigenstates are labeled by $|m\rangle$, which can have the values $\pm\frac{1}{2}$. The fundamental representation of $SU(3)$ has *three* states, $|a\rangle$, where $a = 1, 2, 3$. The labels a are the analogs of the labels $\pm$, and stand for the eigenvalues as follows:

$$\bar{F}_3|a\rangle = \mu_a|a\rangle \qquad \text{and} \qquad \bar{F}_8|a\rangle = \nu_a|a\rangle \tag{C.10}$$

where from the two diagonal matrices λ_3 and λ_8 above

$$(\mu_1, \nu_1) = \left(\frac{1}{2}, \frac{1}{\sqrt{12}}\right) \qquad (\mu_2, \nu_2) = \left(-\frac{1}{2}, \frac{1}{\sqrt{12}}\right) \qquad (\mu_3, \nu_3) = \left(0, -\frac{1}{\sqrt{3}}\right) \tag{C.11}$$

Other representations of $SU(3)$ can be obtained by combining N three-dimensional representations. The states are labeled $|a_1, a_2, \ldots a_N\rangle$ where the a_i take on the values $1, 2, 3$. These are eigenstates of $F_3 = F_{31} + F_{32} + \cdots F_{3N}$ and $F_8 = F_{81} + F_{82} + \cdots F_{8N}$. The individual entries are eigenvalues of F_{3i} and F_{8i}. In the top state, the eigenvalue of F_3 is $N/2$, and the eigenvalue of F_8 is $N/\sqrt{12}$. By repeatedly applying the various other operators, one generates an irreducible representation. This representation has no state with higher eigenvalues of F_3 or of F_8. Furthermore, this top state is symmetric in the a's (since they are all the same) and therefore so are all the other states, because the result of applying one of the F_a to a symmetric state is a symmetric state. All the states are symmetric in the a, and all completely symmetric states occur in a representation.

For example, label the states in the fundamental three-dimensional representation $|a\rangle$ as above, with the indicated eigenvalues of $\bar{F}_3$ and $\bar{F}_8$. If you put two of these together, the top state is $|1, 1\rangle$, and the others in that representation are the totally symmetric combinations

$$\frac{|1,2\rangle + |2,1\rangle}{\sqrt{2}} \quad \frac{|1,3\rangle + |3,1\rangle}{\sqrt{2}} \quad \frac{|3,2\rangle + |2,3\rangle}{\sqrt{2}} \quad |2,2\rangle \quad |3,3\rangle \quad \text{(C.12)}$$

As predicted, there are $3 \cdot 4/2 = 6$ states. You can work out the eigenvalues of F_3 and F_8 on these states, and the action of the other F_i, but all we need here is to count the dimension. That is the subject of Problem 5.10.

APPENDIX D
References

These are some quantum mechanics textbooks I am familiar with. I have tried to indicate the level of the treatment. The list is not meant to be exhaustive.

Introductory

B. H. Bransden and C. J. Joachain, *Quantum Mechanics* (Prentice Hall, 2000). A long and thorough textbook, with many interesting examples. Recently updated.

R. Dicke and J. Wittke, *Introduction to Quantum Mechanics* (Addison-Wesley, 1960). A bit dated in places, but still a good reference.

S. Gasiorowicz, *Quantum Physics* (Wiley, 2003) A new edition of this excellent textbook looks even better than the earlier ones. Very well written with lots of good problems.

D. Griffiths, *Introduction to Quantum Mechanics* (Prentice-Hall, 1995). An elegant and popular textbook.

Robert L. Liboff, *Introductory Quantum Mechanics* (Addison-Wesley, 2003). A popular and somewhat more formal textbook, recently updated.

David Park, *Introduction to the Quantum Theory* (McGraw-Hill, 1990). An excellent textbook covering some unusual and interesting applications. Unfortunately hard to come by.

Richard W. Robinett, *Quantum Mechanics: Classical Results, Modern Systems, and Visualized Examples* (Oxford, 1997). An good textbook with an unusual number of drawings and diagrams.

D. Saxon, *Elementary Quantum Mechanics* (Holden-Day, 1968). A classic undergraduate textbook.

Intermediate

H. Bethe and R. Jackiw, *Intermediate Quantum Mechanics* (Benjamin, 1968). An intermediate level book with much detailed material on atomic structure, collisions, etc.

Claude Cohen-Tannoudji, Bernard Diu, and Franck Laloë, *Quantum Mechanics* (Wiley, 1977). A thorough book with a good review of undergraduate level material; comprehensive, but less formal than Messiah. Many interesting applications. Two volumes, paperback.

R. P. Feynman and A. R. Hibbs, *Quantum Mechanics and Path Integrals* (McGraw-Hill, 1965). An interesting book, presenting quantum mechanics from the Lagrangian, or path integral, viewpoint.

R. J. Finkelstein, *Nonrelativistic Mechanics* (Benjamin, 1973). This short book emphasizes the formal connections between classical and quantum mechanics, and has a clear and original treatment of the nonrelativistic hydrogen atom.

J. Schwinger, *Quantum Mechanics: Symbolism of Atomic Measurements*, ed. B.-G. Englert (Springer, 2001). Notes from an undergraduate course by a master, with an emphasis on the theory of measurements and on the quantum action principle.

Graduate

Gordon Baym, *Lectures on Quantum Mechanics* (Benjamin, 1973). A good textbook with many interesting and unusual examples. Especially lucid on the transformation properties of vector and tensor operators.

Alexander Davydov, *Quantum Mechanics* (Moscow, 1963). Expositions of many subjects, especially from nuclear and solid-state physics, which cannot be found in most textbooks.

H. Georgi, *Lie Algebras and Particle Physics* (Perseus, 1999). An excellent introduction to continuous groups, algebras, and representations in quantum mechanics. You don't have to know a lot about particle physics.

Kurt Gottfried, *Quantum Mechanics, Volume I* (Benjamin,1966). Deeper than most, especially on the theory of measurement and scattering theory. A sophisticated treatment of the Coulomb problem is especially good. There is no Volume II.

Walter Greiner and Berndt Müller, *Quantum Mechanics (Symmetries)* (Springer, 1994). One of the Greiner series of German textbooks, this volume explains in detail continuous groups and their applications in physics.

Barry Holstein, *Topics in Advanced Quantum Mechanics* (Addison-Wesley, 1992). This book covers some advanced material on transitions and scattering.

Rubin H. Landau, *Quantum Mechanics II* (Wiley, 1996). A text with more applications to many body theory and quantum field theory than most.

Eugen Merzbacher, *Quantum Mechanics* (Wiley, 1998). A new edition of an excellent and thorough textbook first published in 1961. The first part covers the subject in the way usually done in undergaduate quantum mechanics; the rest is the more powerful formal approach suitable for a second course. Examples in part one of the more formal methods in part two are especially instructive.

Albert Messiah, *Quantum Mechanics* (Dover, 2000). Volume I translated by G. M. Tenner. Volume II translated by J. Potter. A good, standard textbook in the French style, long on formalism, skimpy on applications. Clear, coherent, and careful, Messiah is still a useful reference. There is a new paperback edition.

Jun J. Sakurai, *Modern Quantum Mechanics* (Addison-Wesley, 1994). A clear and deep exposition of the formalism accompanied by unusual examples from recent experiments. Intended for a graduate course and written just before the author's untimely death in 1982, it was never quite finished.

Jun J. Sakurai, *Advanced Quantum Mechanics* (Addison-Wesley, 1967). A superb little book on a handful of selected topics, mostly suitable as a supplement to a graduate course. Both elegant and intuitive, but a little out of date by now.

Leonard Schiff, *Quantum Mechanics* (McGraw-Hill, 1968). This thorough book was the standard graduate textbook for two decades. A bit old-fashioned in places, but useful nonetheless.

Ramamurti Shankar, *Principles of Quantum Mechanics* (second edition, Plenum, 1994). A good graduate level textbook. Long and comprehensive, the new edition has many up-to-date applications.

Michael Tinkham, *Group Theory and Quantum Mechanics* (McGraw Hill, 1964). A classic text on applications of group theory to physics, with an emphasis on atoms, molecules, and solids.

Classics

D. Bohm, *Quantum Theory* (Prentice Hall, 1951). An early undergraduate level book before there were many undergraduate courses in quantum mechanics, Bohm is very careful about measurement theory.

E. U. Condon and G. Shortley, *The Theory of Atomic Spectra* (Cambridge, 1935, 1951–1970). An ancient monument by now, it still has much useful material.

Paul A. M. Dirac, *The Principles of Quantum Mechanics* (Oxford, 1968). This is the first and in some ways still the best textbook. Everything is clearly explained, since the author couldn't expect any previous knowledge on the reader's part. From the basic fundamentals to the introduction of second quantization in the relativistic theory. Here you find the bra-ket notation, the Dirac delta function, and the Dirac equation for the electron.

R. P. Feynman. R. B. Leighton, and M. Sands, *The Feynman Lectures in Physics, Volume III* (Addison-Wesley, 1965). The third volume of the Big Red Books, written for sophomores at the California Institute of Technology, is peerless in its physical insight into quantum mechanics. Formally an undergraduate text; but some sophistication is needed to appreciate what Feynman is saying.

L. Landau and E. Lifshitz, *Quantum Mechanics: Nonrelativistic Theory* (Pergamon Press, Oxford, 1965–1977), translated by J. B. Sykes and J. S. Bell. One of the famous Russian series of textbooks. Rather advanced, but an excellent addition to anyone's library.

Eugene Wigner, *Group Theory and its Application to the Quantum Mechanics of Atomic Spectra*, translated by J. J. Griffin (Academic Press, 1959). This is the classic work that explained to us why continuous groups and their representations are important in quantum mechanics.

Index

Absorption of radiation
    cross section, 380
    from sunlight, 404, 405
    in a black body, 378–383
    quantum theory, 378
    semiclassical calculation, 289
Action, 2, 12, 334, 350, 454
Action at a distance, 463
Addition of angular momenta, see Angular momentum
Adiabatic theorem, 230, 343
Aharonov, Y., 338, 354
Aharonov-Bohm effect, 338, 351
    bound states, 352
Airy functions, 241
Ammonia molecule, 63, 245, 249
Angular momentum
    addition of, 120–130
        spin half + spin l, 126
        spin half + spin half, 123
        spin half + spin one, 125
        three spins, 133
    and helicity, see Helicity
    and impact parameter, 280
    and reflections, 152
    and rotations, 102, 114
    and spherical symmetry, 75
    commutation with momentum, 74
    commutator algebra, 12, 76, 80
    conservation, 120, 140, 375
    electromagnetic field, 366
    in beta decay, 185
    ladders, 78
    orbital, 74–81, 113, 120, 432
    selection rules, 148
    spin, 22, 55, 114–120
Annihilation operators, 364, 438
Antilinear operator, 154
Antiparticles, 162, 169, 185, 458, 460, 462
Antiunitary operator, 155
Appendices, xiv, 241
Arago, F., 272
Aspect, A., 200
Atomic orbitals, 233
Axial vector, see Pseudovector

Baker-Hausdorff lemma, 59, 100, 141, 403, 445
Balmer, J., 88, 101
Balmer formula, 88, 90
Balmer lines, 98, 216
Baryon number, 139, 162
Bates, D. R., 236, 262

Baym, G., 517
Bell, J., 192, 200
Bell's inequality, 192–195
Bernoulli
    numbers, 398, 401, 474
    polynomials, 474
Berry, M., 232, 342, 355
Berry's phase, 341–350
Bessel functions, 485–491
Bessel's equation, 490
Bethe, H., 262, 516
Bitter, T., 355
Blackbody formula, 382
Bohm, D., 192, 200, 338, 354, 518
Bohr, N., xiii, 39, 88
Bohr energy level formula, 88, 498
Bohr magneton, 176, 199
Bohr radius, 88, 89, 100, 372
Boltzmann's constant, 98, 405
Boosts
    Galilean, 7, 17, 102, 151
    Lorentz, 7, 102, 368
Born, M., xiii, 24, 60, 229, 233, 262
Born approximation, 270, 281, 285, 301, 317, 370, 379, 385, 386, 464
Born-Oppenheimer method, 229–237, 257, 342–350, 353
Bose-Einstein statistics, 158, 159
    photons, 365
    second quantization, 442
Bosons, 158, 365, 438, 444
Bragg scattering, 22
Bransden, B.H., 516
Brillouin, L., 237, 262
Brink, D.M., 170

Cahn, R. N., 137
Canonical quantization, 1, 42, 73, 362, 454
    Dirac field, 459
    electromagnetic field, 356, 362
    scalar fields, 454, 457
Casimir, H.B.K., 394, 406
Casimir effect, 394–402, 406
Causality, 192, 453
Cayley-Hamilton theorem, 261
Central potential, 14, 15, 80, 120
    Dirac equation, 431
    Scattering, 270
Chadwick, J., 185
Chandrasekhar, S., 469
Chandrasekhar limit, 452
Charge conjugation, 7, 102, 458
Chemical potential, 447

519

Chiao, R., 355
Chingas, G. C., 355
Classical radius of electron, 390
Clebsch-Gordan coefficients, 122–130, 134, 149–151, 214, 217, 221, 250, 376, 507, 510
    recursion relation for, 129
Clebsch-Gordan series, 129, 150
Cohen-Tannoudji, C., 516
Coherent states, 100
Collapse of the wave function, 191, 313
Commutation relations
    and Poisson brackets, 16
    angular momentum, 111, 170
    canonical, 42, 50, 362, 457
    coordinates and momenta, 42
    creation and annihilation operators, 439
    Dirac field, 461
    electromagnetic field, 363, 402
    generators of rotations, 12
    group generators, 111
    harmonic oscillator algebra, 69
    Klein-Gordon field, 455
    Lorentz transformations, 410
    orbital angular momentum, 76
    Runge-Lenz vector, 85
    spherical components of vectors, 142
    spherical tensor components, 146–148
    spin matrices, 114
    vector observable, 141
Compton wavelength, 96, 250, 271, 286
Condon, E. U., 263, 518
Connection formula, see WKB method
Continuity equation, 53, 414
Continuous eigenvalues, 36
Cornwall, J. M., 467
Correspondence principle, 39, 289, 370
Cosmic rays, 161, 185, 189, 423, 452
Cosmological constant, 364
Coulomb gauge, 358
Coupled oscillators, 93
Creation operators, 364, 438
Cross section, 162, 265, 300, 385, 391, 393
    absorption, 380
    electron-photon, 392
    neutron-proton, 285
    photoelectric effect, 384
    Rayleigh, 390
    relativistic scalar particles, 464
    Rutherford, 271
    solar neutrino, 189
Curvilinear coordinates, 491

Dalgarno, A., 210, 261
Dalibard, J., 195, 200
Darwin, C. G., 437
Darwin term, 214, 428
Davisson, C., 60
Davisson-Germer experiment, 22, 60

Davydov, A., 517
De Broglie, L., 21, 193, 324
Decay
    final state distribution, 314
    in hydrogen atom
        $2P$ state, 373, 403
        $2P_{1/2}$ to $2S_{1/2}$, 403
        $2S$ state, 404
        $3D$ state, 403
        hyperfine transition, 374
    Λ decay, 322
    neutron decay, 322
    π^+ decay, 320
    rate calculation, 308–316
    time dependence, 310
Degeneracy
    and symmetries, 159
    Dirac hydrogen atom, 430
    free particle, 66
    half-integral spin states, see Kramers degeneracy
    hydrogen atom, 85–88, 97, 134, 495
    Landau levels, 174
    three-dimensional oscillator, 135, 167, 169
    two-dimensional oscillator, 136, 167
Degeneracy pressure, 447
Degenerate perturbation theory, 207
Delta shell, 286
Delta-function potential, 92
Delta-shell potential, 96
    and nuclear forces, 96, 286
Density matrix, 178–184, 445
    and spin-half systems, 181–184, 199
    statistical mechanics, 180–181
Deuteron, 96, 161, 286
Diamagnetism, 252
Dicke, R., 516
Dirac, P.A.M., xiii, 24, 60, 69, 158, 196, 354, 394, 417, 423, 437, 518
Dirac brackets, 24, 26
Dirac delta function, 34, 35, 49, 92, 96, 268, 365, 392, 395, 470
    definition, 35
    representations, 37, 38, 60, 298, 300, 470–472
Dirac equation, 420–435
    and gyromagnetic ratio, 177, 213, 424
    and spin-orbit interaction, 213, 426
    central potentials, 431
    free particle, 421
    hydrogen atom, 433
    plane waves, 459
    spherically symmetric potentials, 431
Dirac fine-structure formula, 430
Dirac γ matrices, 421
Dirac matrices, 420
Dirac sea, 423
Dirac spinor, 421

Dirac string, 348
Discrete normalization, 378
Dispersion, 47
Distributions, 35
Diu, B., 516
Double delta-function potential, 92
Double-slit experiment, 19–22, 324
Dubbers, D., 355

Ehrenberg, W., 355
Ehrenfest's principle, 39
Einstein, A., xiii, 158, 192, 200, 383, 406
Einstein-Podolsky-Rosen paradox, 192–193
Electric-dipole approximation, 291, 372, 381, 390, 404
Electric-dipole moment, 253
Electric-dipole transitions, 372
Electromagnetic field
    canonical coordinates, 361
    canonical quantization, 362–365
    classical coordinates and momenta, 356
    classical Hamiltonian, 358–361
    gauge invariance, 172, 357
    ground-state energy, 365
    independent modes, 358
    minimal coupling, *see* Minimal coupling
    momentum, 366
    quantum theory, 362–402
    semiclassical treatment, 288–291
    transverse gauge, 357
    transverse modes, 359
    vacuum energy, 394–402
Electron gas, *see* Ideal electron gas
Electron magnetic resonance, *see* Magnetic resonance
Electron volt, 56, 171
Elementary particles, 22
    scattering, 292
    table, 139
Elizade, E., 406
Elliptic coordinates, 234, 498
Euler angles, 118, 131, 165, 501, 508
Euler-Lagrange equations, 3, 362, 454, 459
Euler-Maclaurin series, 400, 474–476
Euler-Mascheroni constant, 479
Euler's theorem, 9
Ewen, H. I., 220, 262
Exchange term, 307
Exclusion principle, *see* Pauli exclusion principle
Expectation value, 31

Fermi, E., 158, 185, 300, 323
Fermi constant, 186, 189, 317, 320
Fermi-Dirac distribution, 446
Fermi-Dirac statistics, 158, 221
    second quantization, 443
Fermi energy, 447
Fermi gas, 446–452

Fermi momentum, 447
Fermi's golden rule, 300, 309, 317, 323, 371, 386
Fermi temperature, 448
Fermions, 158
    and anticommutators, 462
    and time inversion, 156
    helium atom, 221
    many body theory, 443
Feynman, R., 264, 324, 336, 354, 394, 516
Feynman diagrams, 394
Feynman-Hellmann theorem, 99
Fields
    and Lorentz transformations, 412
    definition, 10
    quantum, 439
    relativistic, 412
Fine structure constant, 84, 202, 428
Fine structure of atomic levels, 212–218, 252, 425
    Dirac's formula, 430
Finite nuclear size effect, 250
Finkelstein, R., 517
Fock space, 439
Foldy, L. L, 437
Foldy-Wouthuysen transformation, 428
Four-vectors, 407
Fourier transform, 37, 222, 234, 292, 342, 358, 455, 464, 474
Fresnel, A., 272
Full width at half maximum, 316, 373
Function spaces, 33, 57
Functional integral, 326

Galilean boosts, *see* Boosts, Galilean
Gamma function, 476–479
Gamow, G., 247, 263
Gasiorowicz, S., 516
Gauge
    boson, 317
    Coulomb, 358
    invariance, 7, 172, 337, 370, 413, 414, 454
    and path integrals, 338
    radiation, 358
    transformation, 172, 198, 351, 357, 414
    transverse, 176, 197, 356, 357, 359, 363
Gaussian integrals, 52, 326–334, 472
Gaussian units, 3
Gaussian wave function, 51, 58, 256
Gell-Mann matrices, 512
Generators
    and conservation laws, 140
    of classical symmetries, 105
    of continuous transformations, 103
    of Lorentz transformations, 410–413
        Dirac representation, 435
    of one-dimensional subgroups, 104
    of Poincaré group, 418

522 Index

Generators, *continued*
    of rotations, 11, 106–113
        electromagnetic field, 367
        spin half, 116
    of transformations, 140
    of translations, 17, 105, 107
        electromagnetic field, 365, 402
Geometric phase, 345–350
Georgi, H., 137, 517
Gerlach, W., 23
Germer, H., 60
Girvin, S. M., 200
Goldberger, M., 268, 287, 298
Golden rule, *see* Fermi's golden rule
Goldstein, H., 18
Gordon, W., 437
Gottfried, K., 268, 287, 517
Goudsmit, S., 23, 60
Grangier, P., 195, 200
Green's function, 269, 281, 311
Green's operator, 294
Greiner, W., 517
Griffiths, D., 230, 263, 516
Group, 8
    Abelian, 159
    continuous, 103
    definition, 103
    generators, 104, 106
    Lie, 103
    representation, 106
        irreducible, 106
    subgroup, 103
Group velocity, 53
Gurney, R. W., 263
Gyromagnetic ratio, 177
    electron, 197, 200, 370, 424
    μ meson, 319
    proton, 218, 374

Hadron, 161, 317
Hamilton's equations, 5
Hamilton's principle, 2–4, 454, 459
    for electromagnetic field, 362
Harmonic oscillator, 68–74
    energy levels, 70
    in three dimensions, 97
    in two dimensions, 135–137
    level-number operator, 70
    path integral, 328, 332
    series method for wave functions, 94
    wave functions, 73
    WKB approximation, 245
Harris, R. A., 355
Hecht, E., 287
Heisenberg picture, 45, 140, 362, 402, 441, 455
    equations of motion, 45
Heisenberg, W., xiii, 24, 47, 60, 160, 170

Helicity, 367, 369, 418–421
    of neutrinos, 186
    of photons, 369
Helium atom, 221–223
    excited states, 227
    variational method for, 225–227
Hermite polynomials, 74, 488
Hermitean operators
    definition, 29
    properties, 29
Hibbs, A. R., 336, 354, 516
Hidden variable theories, 193
Hilbert space, 24–33, 38
Hole theory, 423
Holstein, B., 517
Hulbert, H., 236, 262
Hund, F., 262
Hund's rule, 227
Huygens, C., 324, 394
Hydrogen atom, 84–90
    decay rates for excited states, *see* Decay, in H atom
    degeneracy of the spectrum, 86
    Dirac equation, 428–435
    energy levels
        Dirac theory, 425
        Klein-Gordon theory, 436
        nonrelativistic, 84–90
    expectation values of powers of r, 99–100
    fine structure, 212–217, 425–428, 430, 435
    hyperfine splitting, *see* Hyperfine structure
    in parabolic coordinates, 495–498
    Klein-Gordon equation, 416, 436
    radial eigenfunctions, 88
    Stark effect, 206
    wave functions, *see* Radial wave functions
    Zeeman effect, *see* Zeeman effect
H^- ion, 221
H_2^+ molecular ion, 229–237
Hylleraas, E. A., 227, 262
Hyperfine structure, 218, 255, 374

Ideal electron gas, 446–450
Identical particles, 157, 221, 438–462
Identity operator, 28, 36, 103, 230, 325, 388
Impact parameter, 280
Incoherent electromagnetic radiation, 290, 378
Index of refraction, 273
Induced emission, *see* Stimulated emission, 289
Infinite square well, 68, 249
    uncertainty principle for, 68
Infinitesimal transformations, *see* Generators of transformations

Internal symmetries, 159–164
Invariance, 102
    and conservation laws, 6, 102, 138
    and degeneracy, 159
    and relativity principle, 6
    and selection rules, 138
    and symmetries, 7
    approximate, 140
    classical and quantum symmetry, 105
    discrete, 151
    Galilean boosts, 7
    gauge, 172, 413
    identical particles, 221, 458
    in classical mechanics, 6–14
    in quantum mechanics, 7, 102–130, 138
    internal symmetries, 7, 102
    isospin, 159
    multiplication by constant phase, 105
    of Hamiltonian, 7
    reflections, 7, 64, 67, 151, 369, 374
    rotations, 74, 120, 251, 369
    Runge-Lenz vector, 85
    symmetries and transformations, 6, 102
    three-dimensional oscillator, 137
    time inversion, 153
    translations, 59
Irreducible representations, 106, 145, 169, 501, 505, 510, 514
    of Lorentz transformations, 411
Isospin, 7, 159, 170
Isotope effect, 250

$\bar{J}$, 11
Jackiw, R., 262, 516
Jacobi identity, 16, 110
Joachain, C. J., 516
Jordan, P., 24, 60

$\bar{K}$, 410
Kamiokande, 188
Kinoshita, T., 177, 200
Klein, O., 437
Klein-Gordon equation, 414, 436, 454
    negative energy solutions, 415
Kramers, H., 157, 237, 262
Kramers degeneracy, 157

$\bar{L}$, 408
Ladsham, K., 262
Lagrangian, 1
    and path integrals, 327–334
    charged scalar field, 454–456
    Dirac field, 458
    electromagnetic field, 361
    for charged particle in electromagnetic field, 3, 337
    molecules, 349
Laguerre polynomials, 90, 97, 428
Laloë, F., 516

Lamb shift, 217
Λ decay, 319
Lamoreaux, S. K., 406
Landau, L. D., 18, 176, 200, 518
Landau, R. H., 517
Landau levels, 173–175, 197, 198
Landé, A., 261
Landé interval rule, 213
Larmor frequency, 174
Lawrence, E. O., 264
LCAO molecular wave functions, 233
Legendre polynomials, 57, 479–483
    and spherical harmonics, 82
    Rodrigues formula, 276
Legendre transformation, 5
Leighton, R.B., 518
Lenz, W., 101
Lepton, 161, 185, 318
Levi-Civita symbol, 472
Lewis, J. T., 261
Liboff, R. L., 516
Lie algebra, 16, 110
Lie group, 103
Lifetimes of excited states, see Decay
Lifshitz, E. M., 18, 200, 518
Lindquist, W. B., 177, 200
Linear operators, 27
    and matrices, 28
Lippman, B., 323
Lippmann-Schwinger equation, 295–298, 302
Li^+ ion, 221
Lorentz algebra, 410, 420
Lorentz boosts, see Boosts, Lorentz
Lorentz force, 4, 6
Lorentz group, 408
    Dirac representation, 420, 436
    irreducible representations, 411
    massless spin-half representations, 419
    two-component representation, 436
Lorentz transformations, 45, 105, 134, 158, 357, 408–413
Lowering operator, 70
Lüders, G., 158
Lyman lines, 216
    Lyman-α line, 98, 373, 380
    Lyman-β line, 98, 403

Magnetic moment
    anomalous, 177
    of electron, 22, 218, 425
    of neutron, 177
    of proton, 177, 218
    orbital, 176
    spin, 176
Magnetic monopole, 196
Magnetic resonance, 183, 199, 317, 343
Magnetic-dipole approximation, 375
Maximal set of commuting observables, 31

Index

Maxwell's equations, 356, 412–414
    for quantized fields, 402
Measurement, 19–24, 39, 55
    and Stern-Gerlach experiment, 55
    and uncertainty principle, 47
    and wave-function collapse, 313
    in quantum mechanics, 191–195
    in statistical ensembles, 178, 445
Mermin, N. D., 195, 200
Merzbacher, E., 210, 261, 517
Messiah, A., 230, 263, 517
Metric tensor, 407
Mikheyev, S. P., 201
Minimal coupling, 172, 218, 357, 369, 416, 423, 457
Minkowski notation, 407
Mohr, P. J., 177, 200
Molecules, 229–237
    and geometrical phase, 349–350
Momentum
    and canonical commutation relations, 42
    and translations, 49, 402
    canonical, 4, 350, 454, 459
    conservation, 392
    in spherical coordinates, 98
    of photons, 365
    radial, 14, 98–99
    space, 51
Moody, J., 229, 262
Mostepanenko, V. M., 406
μ meson, 161, 185, 319
Müller, B., 517
Muller, C. A., 220, 262
Multipole expansion, 372

Natural units, 171
Negative energy states, 415, 423, 436, 453, 458, 460
Neutrinos, 319, 322, 369, 419, 452
    masses, 187–189
    mixing, 186–189
    solar, 189, 353
Neutron
    and isotopes, 251
    as fermion, 158
    beta decay, 185, 322
    inverse beta decay, 185
    isospin, 160
    magnetic moment, 177
    scattering with proton, 285–287, 317
    stars, 452
Newton, R., 268, 287
Noether, E., 18
Noether's theorem, 14, 17, 18
Normal ordering, 463
Nuclear magneton, 177
Nuclear rotational levels, 236
Nuclear vibrational levels, 236

Nucleons
    forces, 96, 271
    isospin, 159
    magnetic moment, 177
Number operator, 70, 366, 443, 458, 465

$O(4)$, 134, 159, 167
$O(n)$, 106, 134
Occupation number, 364, 438
One electron atoms, *see* Hydrogen atom
Oort, J. H., 220, 262
Operator ordering ambiguity, 42, 72, 85, 326, 365, 456
Oppenheimer, J. R., 229, 262, 469
Optical theorem, 272–275, 301, 302, 323
    and partial waves, 279
    and refractive index, 275
Orbital angular momentum, *see* Angular momentum
Orthogonal group, *see* $O(n)$
Orthohelium, 227
Oscillator strength sum rule, 92

Parabolic coordinates, 495–498
Parahelium, 227
Parity, 1, 67, 91, 92, 151, 152
    and spin-half relativistic wave equation, 418
    conservation, 186, 317, 374, 381
    Dirac electrons, 431
    selection rules, 152, 153
    violation, 322
Park, D., 516
Partial waves, 275–280
Partition function, 180
Paschen lines, 98
Path integral, 324–341
    Aharonov-Bohm effect, 338
    Born series, 334
    Euclidean formalism, 331–334
    Euclidean propagator, 332
    free particle, 328
    gauge invariance, 338
    harmonic oscillator, 328–331
    propagator, 326
Pauli, W., 85, 101, 158, 185, 469
Pauli exclusion principle, 158, 159, 423, 444
Pauli matrices, 116, 132, 136, 156, 160, 169, 187, 418, 436
Pauli spinor, 115, 132, 133, 160, 320, 421, 422
Periodic boundary conditions, 378, 395, 444, 468
Perturbation theory
    bound state, 202–206
    degenerate, 207
    first-order energy shift, 203
    first-order wave function shift, 204
    second-order energy shift, 204

second-order wave function shift, 248
    time dependent, 288, 316
Peskin, M., 354
Phase shift, 278–280
Phase velocity, 54
Photoelectric effect, 383–385
Photon, 364–369
    absorption, 378–383
    emission, 371, 378–383
    momentum, 365
    scattering, 383–394
    spin, 367
π meson, 139, 140, 161, 164, 185, 191, 318–322
Pines, A., 355
Planck, M., 383, 406
Planck mass, 452
Planck's constant, 43
Plane wave expansion, 276, 277
Podolsky, B., 192, 200
Poincaré group, 418
Poisson, S., 272
Poisson brackets, 16–17, 105, 110
    and commutators, 43, 47
Poisson-Arago spot, 272
Polar vector, 152
Polarizability of H atom
    exact calculation, 210
    perturbation calculation, 208
Polarizabilty tensor, 209
Polarization
    circular, 380
    electromagnetic field, 288
    of spin-half particles, 181–184, 199
    photons, 195, 373, 383, 390, 405
    spin-half particles, 323
    vector, 364, 368, 369
Poole, C. P., 18
Positron, 185, 192
    in β-decay, 319, 322
    in Dirac's hole theory, 423
Poynting vector, 366
Prange, R. E., 200
Probability
    amplitude, 27
    conservation, 59, 272, 302
    current, 53, 239, 264, 415, 423, 457
    density, 49
    in quantum mechanics, 19–24
Projection operators, 132
Projection theorem, 151, 218, 255, 507
Propagator, 324–341
Proton, 177
    as fermion, 158
    bound in deuteron, 96–97
    charge, 56
    finite size, 202, 250
    gyromagnetic ratio, 218, 374
    in cosmic rays, 185

inverse beta decay, 185
isospin, 159
lifetime, 140
magnetic field, 219
magnetic moment, 177
    scattering by neutron, 285–287, 317
    spin, 114
Pseudoscalar, 152
Pseudovector, 11, 152
Purcell, E. M., 220, 262

Quantum fields, 439
Quantum Hall effect, 200
Quantum virial theorem, 94
Quantum Zeno effect, 313
Quartic potential
    eigenvalue-variational estimate, 259
    variational method for energies, 256
    WKB method for energies, 257

$\bar{R}$, 7
$\mathcal{R}$, 7, 57, 64, 91, 151
Rabi, I. I., 184
Rabi's formula, 184, 199
Radial wave equation
    Dirac equation, 432
    Schrödinger equation, 83
Radial wave functions, 275, 281
    delta-shell potential, 96
    harmonic oscillator, 97
    hydrogen atom, 88, 97
        table, 88, 483
    spherical square well, 95
    three-dimensional oscillator, 97
Radiation gauge, 358
Raising operator, 70
Rayleigh, Lord, 202, 223, 262, 391
Rayleigh-Ritz method, 223
Rayleigh-Schrödinger perturbation method, 202
Rayleigh's law, 391
Reduced matrix element, 149, 151, 507, 511
Reducible representations, 106
Reflections, 57, 64, 66, 91, 102, 151
    and massless particles, 418
    electromagnetic field, 369
    in classical mechanics, 7
    in Dirac theory, 420, 432
Refractive index, 273
Regulator, 60, 271, 274, 293, 394, 399, 471
Relativistic correction to energy levels, 215, 426
Relativity
    and stellar collapse, 450
    invariant interval, 408
    Lorentz transformations, 407–413
    Minkowski notation, 26, 407

Representations
    Dirac algebra, 420
    irreducible, 106
    of $SU(3)$, 515
    of Dirac delta function, see Dirac delta function
    of groups, 105
    of isospin, 160
    of Lorentz transformations, 411–413
    of rotations, 105–112, 131, 500–510
    spin, 113–119
Resonance curve, 316
Ritz, W., 262
Robinett, R. W., 516
Rodrigues formula
    Hermite polynomials, 488
    Legendre polynomials, 276, 482
Roger, G., 195, 200
Romeo, A., 406
Rosen, N., 192, 200
Rotation group, 106–112
Rotation matrices, 9, 11, 58, 500–508
    averages over products, 508
    infinitesimal, 12
Rotations, 102
    and angular momentum, 114
    as Lie group, 103
    in n dimensions, 134
    in classical mechanics, 7, 9–12
    representations, 502
Runge, C., 101
Runge-Lenz vector, 17, 84–88, 94, 135, 170, 207, 411
    defined, 15
Rutherford, E., xiii, 271
Rutherford scattering formula, 271, 285, 308
Ryder, L. H., 354

Safko, J. L., 18
Sakurai, J. J., 517
Sands, M., 518
Saxon, D., 516
Scattering
    amplitude, 269–275
    Born approximation, 270
    Born series, 270, 295, 334
    central potentials, 270
    Coulomb potential, 271
    delta shell, 286
    elastic, 265
    electron by atom, 302
    Green's function, 269
    identical particles, 305
    impenetrable sphere, 283
    inelastic, 303
    length, 284
    matrix, 299, 336
    neutron by proton, 317
    partial waves, 276–280

    photons by atoms, 385–391
    photons by electrons, 391
    radial wave functions, 281–284
    square well, 285
    Thomson, 391
    wave packets, 268, 292
    with recoil, 303
Schiff, L., 437, 518
Schroeder, D., 354
Schrödinger, E., xiii, 21, 24, 60, 202, 261, 415, 437
Schrödinger picture, 45, 312, 362, 366
Schrödinger's cat, 191
Schrödinger's equation, 42
    in one dimension, 52
    in three dimensions, 52
Schwartz, C., 210, 261
Schwinger, J., 158, 177, 200, 323, 354, 517
Second quantization, 442
Selection rules, 138, 148, 372
    tensor operators, 148
    vector operators, 142
Semiclassical approximation, 288–291, 381
Separation of variables, 82
Shankar, R., 518
Shapere, A., 229, 262, 355
Shortley, G., 518
Siday, R., 355
Simple harmonic oscillator, see Harmonic oscillator
Sky, why it is blue, 391
Smirnov, A. Y., 201
$SO(3)$, 117, 159
$SO(n)$, 106
Solar constant, 404
Solar neutrinos, 189
Sommerfeld, A., 287
Spaarnay, M. J., 406
Spectroscopic nomenclature, 121
Spherical Bessel functions, 84, 95, 276–279, 485
    table, 487
Spherical Hankel functions, 282, 484
    table, 488
Spherical harmonics, 81, 482
    addition theorem, 277, 280, 501
    and Legendre polynomials, 82
    table, 483
Spherical Neumann functions, 486
    table, 487
Spherical symmetry, 74–80
    and angular momentum, 75
    and rotations, 102
Spherical tensors, 146–148
Spherical vector components, 142
Spin, 22, 55, 114–119
    and statistics theorem, 462
    magnetic resonance, 183, 199, 343

precession in constant magnetic field, 181
spin half, 116, 121, 126, 156, 169, 184, 418
spin one, 119, 131, 221
spin zero, 116, 155, 156, 221
Spin-orbit coupling, 212, 214, 258, 426, 430
Spinor
- Dirac, see Dirac spinor
- Pauli, see Pauli spinor
- Weyl, see Weyl spinor

Spontaneous emission, 369, 371
Square well
- and nuclear forces, 96, 285
- in one dimension, 91
- in three dimensions, 95

Stark effect, 206–212
- quadratic, 252
- variational estimate for ground state shift, 257

Stark, J., 261
Stellar collapse, 448–453
Step function, 35
Stern, O., 23
Stern-Gerlach experiment, 22, 60, 165
Stewart, A. L., 262
Stimulated emission, 289, 378, 381
$SU(2)$, 159
$SU(3)$, 117, 137, 167
- representations, 168, 512–515

Summation convention, 407
Sunlight
- incoherent, 378
- polarization, 391

Superposition principle, 1, 19, 22, 27, 39, 394
Superselection rule, 157
Supersymmetry, 466
Suter, D., 355
Symmetries, see Invariance

$\mathcal{T}$, 7, 153
Taylor, B. N., 177, 200
Tensor, 144–151
- Cartesian, 144–146, 407
- field strength, 413
- Lorentz, 407
- rank-two, 144–148, 165, 166, 219
- spherical, 147

ter Haar, D., 406
Thomas precession, 213
Thomson scattering, 391, 393
Time inversion, 102, 153
- as antiunitary operator, 154
- for spin half, 156
- for spin zero, 156
- in classical mechanics, 7

Time reversal, see Time inversion

Time translation, 102
- in classical mechanics, 7
- in quantum mechanics, 39
- operator, 40, 44

Tinkham, M., 518
Tomita, A., 355
Transition
- amplitude, 292, 299
- external fields, 288
- matrix, 292, 295
- rate, 300

Transition matrix, 302
Translations, 102, 151
- and momentum, 49, 104, 402, 403
- as Lie group, 103
- in classical mechanics, 7, 8
- in time, see Time translations

Transmission coefficient, 248
Transverse gauge, see Gauge
Trunov, N. N., 406
21 cm line, 220, 377
Two-state systems
- ammonia molecule, 63
- H atom $n = 2$ level, 208
- neutrinos, 187
- spin-half, 181

Tycho, R., 355

Uhlenbeck, G., 23, 60
Uncertainty principle, 47–48
- energy-time, 316
- exponential wave function, 93
- Gaussian wave function, 52
- hydrogen atom, 98
- infinite square well, 68

Unitary operators, 30
Units, 3, 43, 57, 171, 318, 356, 464

Vacuum energy, 365
- electromagnetic field, 394–402

Van der Waals effect, 253, 254, 402
Van der Waerden, B. L., 24, 61
Variational method, 223–237, 256
- and eigenvalues of truncated Hamiltonian, 228
- ground state, 224
- helium atom, 225–227
- molecules, 230
- quartic potential, 256
- Stark effect, 257

Vector
- observables, 10, 17, 74, 140
- operator
  - selection rules, 142–144
  - spherical components, 141
  - under rotations, 140–141
- space, 24–34

Virial theorem, 94
Volkoff, G. M., 469

Wallis, R. F., 236, 262
Watson, K. M., 268, 287, 298
Wave function, 49
    collapse, *see* Collapse of the wave function
Wave packet, 53, 292
    scattering, 268
    spread, 55
Weak interactions, 7, 140, 151, 159, 186, 317
Weinberg, S., 354
Wenzel, G., 237, 262
Weyl spinor, 419
White dwarf stars, 449
Wigner, E., 112, 119, 130, 137, 150, 154, 170, 507, 518
Wigner-Eckart theorem, 149, 152, 166, 207, 218, 250, 254, 372, 375, 403
    and selection rules, 149
    proof, 130, 150, 510
Wigner matrices, 112, 500, 502–507
Wilczek, F., 229, 262, 355

Wittke, J., 516
WKB method, 237–247
    and bound states, 243
    connection formula, 238–243
    double well, 245
    quartic potential, 257
Wolfenstein, L., 201
Wouthuysen, S. A., 437

Young, T., 21, 272
Yukawa, H., 270
Yukawa potential, 271

Zajac, A., 287
Zeeman, P., 200
Zeeman effect, 177–178, 217
    anomalous, 23, 177
    intermediate magnetic fields, 258
    Paschen-Back limit, 252
    quadratic, 252
Zumino, B., 158